John R Freeman
408 Twin Towers

L - C.
V, H,

Kevin — Units Book

Modern Concepts in Biochemistry

This book is part of the

ALLYN AND BACON CHEMISTRY SERIES

Consulting Editors: Daryle H. Busch
Harrison Shull

Allyn and Bacon, Inc.

Boston
London
Sydney
Toronto

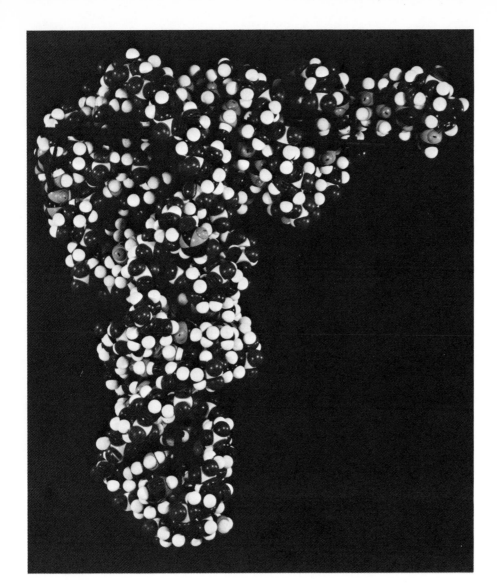

Modern Concepts in

BIOCHEMISTRY

Iris Brown — 424-2380

Third Edition

Robert C. Bohinski
John Carroll University

ART CREDITS

Chapter openers
Introduction: From J. C. Kendrew, "Myoglobin and the Structure of Proteins," *Science,* **139,** 1259–1266 (1963); photograph supplied by J. C. Kendrew. *Chapter 1:* Adapted with permission from R. D. Dyson, *Cell Biology,* 2nd edition, Boston: Allyn and Bacon, 1978. *Chapter 4:* Reproduced with permission from R. Walter, C. W. Smith, and J. Roy, *Proc. Nat. Acad. Sci. (USA),* **73,** 3054 (1976); drawing supplied by Dr. Roderich Walter. *Chapter 5:* Adapted with permission from R. D. Dyson, *Cell Biology,* 2nd Edition, Boston: Allyn and Bacon, 1978. *Chapter 6:* Reproduced with permission from "Crystal Structure of Human Erythrocyte Carbonic Anhydrase C," *Cold Spring Harbor Symposium on Quantitative Biology,* **36,** 221–231 (1971); drawing supplied by Dr. K. K. Kannan. *Chapter 7:* Taken with permission from J. D. Watson and F. H. C. Crick, "Genetic Implications of the Structure of Deoxyribonucleic Acid," *Nature,* **171,** No. 4361, 964–967 (1953). *Chapter 8:* Reproduced with permission from *Cold Spring Harbor Symposium on Quantitative Biology,* **28,** 43 (1963); photograph supplied by John Cairns. *Chapter 10:* Reproduced with permission from Daniel Branton, Department of Biology, Harvard University. *Chapter 15:* Reproduced with permission from L. K. Shumway, Department of Botany and Program in Genetics, Washington State University, Pullman, Washington.

Title page: Photograph provided through the courtesy of Dr. Sung Hou Kim.

Cover: Micrograph reproduced with permission from D. W. Fawcett, *The Cell: An Atlas of Fine Structure,* Philadelphia: W. B. Saunders, 1966, and generously supplied by D. W. Fawcett.

Manufacturing buyer: Karen Mason

Production editor: Judith Fiske

Printing number and year (last digits):
10 9 8 7 6 5 4 3 2   85 84 83 82 81 80

Library of Congress Cataloging in Publication Data

Bohinski, Robert C
  Modern concepts in biochemistry.

  Includes bibliographies and index.
  1. Biological chemistry. I. Title.
[DNLM: 1. Biochemistry. QU4.3 B676m]
QP514.2.B63 1979   574.1'92   78-21639
ISBN 0-205-06541-X
ISBN (international) 0-205-06610-0

Printed in the United States of America.

To the students enrolled in
biochemistry courses at
John Carroll University

CONTENTS IN BRIEF

CONTENTS

CHAPTER 1

Enzymes

Nucleotides and Nucleic Acids

PREFACE

The preparation of this third edition three years after the second reflects the need to offer a biochemistry textbook that keeps abreast of current developments and improves on the previous editions while retaining their successful features. One of the most important of these features is the emphasis on two themes: (1) the relation of structure and function and (2) the regulation and integration of metabolism. These themes not only unify the study of a complex subject, but do it in an understandable way.

This textbook can be used effectively in either the one-semester or two-quarter introductory course under the guidance of the professor aware of the backgrounds of the students. It is also still appropriate for the longer two-semester course. By streamlining the coverage of many topics, new material has been added to this edition without an overall increase in the amount of material covered. Among these additions are discussions of

- radioimmunoassay
- enkephalins and somatostatin
- protein binding (Scatchard and Hill plots)
- negative versus positive cooperativity
- carboxypeptidase structure and function
- the induced-fit model of allosterism
- restriction enzymes
- recombinant DNA technology
- DNA sequencing
- new knowledge about eukaryotic mRNA
- post-transcriptional and post-translational modifications

- lectins
- prostaglandin biosynthesis
- thromboxane and prostacyclin
- the peroxidation of unsaturated fatty acids
- phosphonolipids
- Vitamins D and E
- the adenylate charge hypothesis
- the anatomy and function of pyruvate dehydrogenase complex
- oxygen toxicity and superoxide dismutase
- the structure and mechanism of ribulose diphosphate carboxylase
- CO_2 fixation in C4 plants
- photorespiration
- the anatomy and function of fatty acid synthetase
- sulfur metabolism—activated sulfate and biosynthesis of methionine, cysteine, and Coenzyme A
- the use of creatine phosphokinase in clinical diagnosis

There are two distinct organizational changes in this edition. One is the early (Chapter 1) description of cell structure and the other is the earlier (Chapter 8) discussion of RNA, DNA, and protein biosynthesis. The latter change in particular—the treatment of biochemical genetics immediately after coverage of protein structure, enzymes, and nucleic acid structure—has met with a favorable response from students in my own classes. However, the material of Chapter 8 is not a prerequisite for Chapters 9 through 17. Those who prefer to finish the course with the study of biochemical genetics can do so without concern.

The number of exercises has been maintained, and an instructor's manual with complete solutions to most of these exercises is available. Students who wish to pursue certain topics can consult the literature listings at the end of each chapter; *Scientific American* offprints provide another excellent source.

I would like to acknowledge again the contributions of S. Aust, C. Dunne, P. Heyman, P. Khairallah, M. Konrad, J. Kúc, A. Rule, G. Toraballa, and C. Vestling in reviewing previous editions. Only authors can appreciate the value of a good typist; it has been my good fortune through three editions to have an excellent one in the person of Elaine DiCillo. The work done by Vantage Art on the line drawings and Progressive Typographers on the chemical structures and diagrams for this revision is most appreciated. I would again like to extend my thanks to the production staff at Allyn and Bacon, especially Judith Fiske. My gratitude is also extended to Harvey Pantzis as editor for his coordination of the revision and for his patience and understanding.

Modern Concepts in Biochemistry

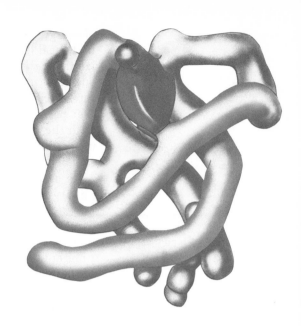

INTRODUCTION

The proper beginning of a textbook is to define what the book is about. In this case the task is not easy since the nature of modern biochemistry defies reduction to a short, meaningful description. The usual approach is to state simply that biochemistry is the *study of the chemistry of life.* Although true, to many the statement is not very informative. Consider, then, an alternative statement: *biochemistry is a laboratory science with a physico-biochemical focus on all types of cellular activities.* Although this may appear to say little more than the previous statement, it does convey considerably more. There are three elements in this definition. After completing this course, hopefully you will appreciate and understand each.

The *first element* of the definition is the designation of *biochemistry as a laboratory science.* As such it requires an inquisitive, imaginative, and (frequently) patient investigator. Other criteria for successful and productive research are the availability of proper tools (lab equipment) and supplies and the skill to use them. The point is that the practice of biochemistry, originating with keen thoughts and ideas, ultimately depends on the performance of analytical procedures in the laboratory and the interpretation of collected data.

The *second element* of the definition is represented by the word *physico-biochemical,* implying that biochemical studies are *multidisciplinary in nature.* This means that a complete analysis and subsequent understanding of living systems is predicated on a multipronged approach employing physical, biological, and chemical principles. A good illustration of this point is provided by the phenomena of energy production and energy utilization in a living cell. On the physical level, we must first recognize the basic concept of energy, its various forms, and the reality of their interconversions. On the biological level, the cellular processes are studied to identify and characterize those that require energy, those that yield energy, and the intricate balance and relationship between both types. On the chemical level, we are interested in defining the flow of energy through the chemical molecules that participate in processes. Notice the trend here. The analysis began with the physical concept of energy, a general application was made to living systems, and finally it was *reduced to the molecular level*. More will be said about this molecular focus a little later.

As a student of biochemistry, you will clearly discern this interdisciplinary character as you proceed into future chapters. This approach does not mean that this is a sophisticated and difficult textbook. It is an approach demanded by the reality of modern biological science, which is not merely a systematic collection of superficial and unexplainable observations of the living world. This was the extent of biology until the beginning of the twentieth century, but it is no longer. Before proceeding with a further analysis of the definition of biochemistry, let us digress a little to expand on this aspect in a historical context.

Until the 1860s *vitalism* was the popular biological theory. The proponents of this theory claimed that the generation of life, and to some extent the maintenance of the essential life processes, were controlled by an unexplainable and unmeasurable vital force. In the latter part of the nineteenth century the vitalistic attitude was finally refuted, and an alternative explanation for the living process was demanded. The demise of vitalism created a climate in the scientific community that resulted in changing attitudes between the biologists on one side and the physicists and chemists on the other. They started to talk and listen to each other. The biologists finally began to realize that life processes were not necessarily so mysterious that they required explanations equally mysterious and even mystical. Those in the then more exact sciences of physics and chemistry had been saying this for years and were somewhat miffed by the reluctance of the vitalists to listen.

The basic premise of the physical scientists was simple. They argued that natural (living) phenomena could be explained physically. Difficult as it must have been—human nature being what it is—the limits of professional polarization began to shrink, and ultimately the marriage of the exact physical sciences of physics and chemistry and the then nonexact natural sciences occurred. The early years were not completely harmonious, and scientific history reveals that heated debate was frequent. The important part, however, is that these sciences began to share a common scientific philosophy and common methods. Through it all biology matured as an exact science, and only then did it advance. This interdisciplinary exchange is manifested today, more so than ever, by the emergence of hybrid specializations such as biochemistry, biophysics, biotechnology, bioengineering, theoretical biology, biosystematics, and so on. This hybridization has caused a problem in communication (see margin), but it does not represent a fragmentation in modern biology, nor is it based on a true divergence in philosophy. Each represents a specialized division with a specialized strategy and specialized tactics. An excellent summary of these thoughts is provided by the words of Samuel Devons (see suggested readings at the end of the chapter):

Perhaps the biggest problem is the massive amount of research literature that is published every year. Because of it, keeping abreast of recent developments (even in one's special field) is a time-consuming chore. Furthermore, because of high subscription costs it is also economically difficult for both individuals and libraries.

. . . The victorious advances of physics and chemistry in the present century, their penetration into much of what was erstwhile biological territory, and their culmination in the spectacular discoveries of the past decade or two, have completely changed the whole character of the debate. [The debate between natural and physical scientists.] The power of physical science, its concepts, methods and techniques, are firmly established. There is no longer any feature of living matter which presents a clearly demarked frontier beyond which physics or chemistry cannot pass. Science may not be fully unified, but the boundaries are not the fixed, impenetrable ones that some imagined and even hoped for a century ago.

The *third element* of the definition we are working with refers to the subject of this scientific analysis, the *living cell*. Note that there is no particular reference to any specific type of cellular activity or biological phenomenon as the sole domain of biochemistry. On the contrary, biochemical investigations encompass a wide spectrum of topics, ranging from multicellular organisms to unicellular organisms; from whole cellular systems to the compartmentalized subcellular parts; and from subcellular molecular aggregates to molecules themselves. In short, it can be said that modern biochemistry, which serves as the basis for all modern biology, is concerned with any phenomenon associated with any organism, be it animal, plant, bacteria, or virus.

The massive scope of biochemical research is evidenced by the large number of regular journal and review publications that provide the major means of communication among researchers in different areas. (A partial listing is given in Table 1.) The literature listings of Table 1 include many publications that have been started within the past two decades, a fact reflecting the fantastic growth of biochemistry as a scientific discipline and also the increase in biochemical knowledge in this period. During the years of 1961–76 inclusive, the published biochemical research in the 13 major international journals has shown an annual growth of about 10%, representing a doubling time of about 7.5 years! Further evidence of this growth is provided by comparing the size of the three editions of the *Handbook of Biochemistry and Molecular Biology*. The first edition, published in 1968, contained 950 pages. The second edition (1970) grew to 1,600 pages. The third edition (1976) exploded into eight volumes having a total of 5,000 pages!

A FOCUS ON MOLECULES

All life forms are material. This means simply that organisms are composed of chemical substances of both the inorganic and organic classes. The thrust of biochemistry is to understand processes of the living state in terms of these substances.

The *biochemical diversity* of living things is obvious to all of us. Our biosphere consists of animal, plant, and bacterial organisms and the viruses. In addition there is an extensive variety of organisms within each of these classes. Further, the multicellular organism is typified by various types of specialized cells (brain, kidney, heart, liver, spleen, leaf, stem, and so on), each with specialized functions. Not so obvious, however, is the extensive amount of *biochemical unity* that exists. In what sense? Well, for one

TABLE 1. Literature sources in biochemistry and related sciences. (Asterisks identify those journals used most frequently)

Research Journals

Acta Chemica Scandinavica (Acta Chem. Scand.)
*Analytical Biochemistry (Anal. Biochem.)
*Archives of Biochemistry and Biophysics (Arch. Biochem. Biophys.)
*Biochemical and Biophysical Research Communications (Biochem. Biophys. Res. Commun.)
Biochemical Genetics (Biochem. Genet.)
*Biochemical Journal (Biochem. J.)
*Biochemistry (Biochemistry)
*Biochimica et Biophysica Acta (Biochim. Biophys. Acta)
Biokhimiya (USSR) (Biokhimiya)
Bioorganic Chemistry (Bioorg. Chem.)
Biopolymers (Biopolymers)
BioScience (BioScience)
Canadian Journal of Biochemistry (Can. J. Biochem.)
Canadian Journal of Microbiology (Can. J. Microbiol.)
Cancer Research (Cancer Res.)
Chemistry and Physics of Lipids (Chem. Phys. Lipids)
Clinical Chemistry (Clin. Chem.)
Comparative Biochemistry and Physiology (Comp. Biochem. Physiol.)
Comptes Rendus (Compt. Rend.)
*European Journal of Biochemistry (Eur. J. Biochem.; formerly Biochemische Zeitschrift)
Experimental Cell Research (Exp. Cell Res.)
Federation Proceedings (Fed. Proc.)
Immunology (Immunology)
Indian Journal of Biochemistry (Indian J. Biochem.)
*Journal of the American Chemical Society (J. Am. Chem. Soc.)
*Journal of Bacteriology (J. Bacteriol.)
Journal of Biochemistry (Japan) (J. Biochem. (Tokyo))
Journal of Bioenergetics (J. Bioenerg.)
*Journal of Biological Chemistry (J. Biol. Chem.)
*Journal of Cell Biology (J. Cell Biol.)
*Journal of Cellular Physiology (J. Cell. Physiol.)
Journal of Chromatography (J. Chromatogr.)
Journal of Electron Microscopy (J. Electron Microsc.)
Journal of Experimental Biology (J. Exp. Biol.)
Journal of General Microbiology (J. Gen. Microbiol.)
*Journal of General Physiology (J. Gen. Physiol.)
Journal of Histochemistry and Cytochemistry (J. Histochem. Cytochem.)
Journal of Immunology (J. Immunol.)
Journal of Lipid Research (J. Lipid Res.)
Journal of Medicinal Chemistry (J. Med. Chem.)
Journal of Membrane Biology (J. Membrane Biol.)
*Journal of Molecular Biology (J. Mol. Biol.)
Journal of Neurochemistry (J. Neurochem.)
Journal of Nutrition (J. Nutr.)
Journal of Pharmacology (J. Pharmacol.)
Journal of Physiology (J. Physiol.)
Journal of Theoretical Biology (J. Theor. Biol.)
Journal of Ultrastructure Research (J. Ultrastruct. Res.)
Journal of Virology (J. Virol.)
Lipids (Lipids)
Molecular Pharmacology (Mol. Pharmacol.)
Mycologia (Mycologia)
*Nature (Nature)
*Nature New Biology (Nat. New Biol.)
*New England Journal of Medicine (New. Engl. J. Med.)
Plant Physiology (Plant Physiol.)
*Proceedings of the National Academy of Sciences of the United States of America (Proc. Nat. Acad. Sci. U.S.)
Proceedings of the Royal Society (Proc. Roy. Soc.)
Proceedings of the Society for Experimental Biology and Medicine (Proc. Soc. Exp. Biol. Med.)
Prostaglandins
*Science (Science)
Steroids (Steroids)
Trends in Biochemical Sciences (T.I.B.S.)

Review Sources (Most are annual publications)

Accounts of Chemical Research (Accounts Chem. Res.)
Advances in Carbohydrate Chemistry (Advan. Carbohydrate Chem.)
Advances in Cell and Molecular Biology (Advan. Cell Mol. Biol.)
Advances in Comparative Physiology and Biochemistry (Advan. Comp. Physiol. Biochem.)
Advances in Enzyme Regulation (Advan. Enzyme Regul.)
Advances in Enzymology (Advan. Enzymol.)
Advances in Experimental Medicine and Biology (Advan. Exp. Med. Biol.)
Advances in Genetics (Advan. Genet.)
Advances in Immunology (Advan. Immunol.)
Advances in Lipid Research (Advan. Lipid Res.)
Advances in Microbial Physiology (Advan. Microbiol. Physiol.)
Advances in Protein Chemistry (Advan. Protein Chem.)
Annual Review of Biochemistry (Ann. Rev. Biochem.)
Annual Review of Entomology (Ann. Rev. Entomol.)
Annual Review of Genetics (Ann. Rev. Genet.)
Annual Review of Medicine (Ann. Rev. Med.)
Annual Review of Microbiology (Ann. Rev. Microbiol.)
Annual Review of Pharmacology (Ann. Rev. Pharmacol.)
Annual Review of Physiology (Ann. Rev. Physiol.)
Annual Review of Plant Physiology (Ann. Rev. Plant Physiol.)

Bacteriological Reviews (Bacteriol. Rev.)	*Physiological Reviews (Physiol. Rev.)*
Biochemical Society Symposia (Biochem. Soc. Symp.)	*Progress in Biophysics and Molecular Biology (Progr. Biophys. Mol. Biol.)*
Biological Reviews (Biol. Rev.)	*Progress in the Chemistry of Fats and Other Lipids (Progr. Chem. Fats Lipids)*
Chemical Reviews (Chem. Rev.)	
Cold Spring Harbor Symposia on Quantitative Biology (Cold Spring Harbor Symp. Quant. Biol.)	*Progress in Nucleic Acid Research and Molecular Biology (Progr. Nucleic Acid Res. Mol. Biol.)*
Current Topics in Cellular Regulation (Current Topics Cell. Reg.)	*Subcellular Biochemistry (Subcell. Biochem.)*
Essays in Biochemistry (Essays Biochem.)	*Vitamins and Hormones (Vitam. Horm.)*

thing, all of these organisms are composed of the same type of substances that perform the same type of general function. In fact, there is one specific inorganic substance that occurs in all organisms, namely, *water*, which in many ways is the most important substance for life. The major classes of organic substances are *proteins, nucleic acids, carbohydrates,* and *lipids.*

The study of these essential organic *biomolecules* is what biochemistry is all about. How are they synthesized in a living cell? How are they degraded in a cell? How is their degradation linked to the production of useful chemical energy within a cell? How is this energy produced? How are these substances interconverted? How do they enter and leave a cell? How can they be isolated in the laboratory? How can they be synthesized in the laboratory? What is their molecular structure? What specific functions do they perform? What is the explanation for how they function in terms of their molecular structure? How do genes control the biochemical individuality of an organism? How can the genetic information of an organism be modified by laboratory manipulation? How do cells (and organisms) communicate with each other on a chemical basis? What are the explanations of the abnormal state, that is, the disease state (including mental illness)? How do antibiotics work? What biochemical distinctions exist between normal and cancer cells? How can medical diagnosis be improved by using precise biochemical criteria? How do hormones regulate the activities of an organism and in what other ways are organisms capable of self-regulation?

These are some of the questions that are being dealt with in the various aspects of biochemical research. Hopefully, the combination of your professor and this book will help you understand some of what we know about these and other questions. A lot is known and in many cases our understanding is quite detailed. However, there is still much to be done.

PROTEINS AND NUCLEIC ACIDS—SPECIAL BIOMOLECULES

Throughout this course you will become familiar with the names and structures of a few hundred substances (such as the nonpolymeric, low-molecular weight types shown on p. 6) and with the participation of each substance in life processes. Although all types of biomolecules are important, the *proteins and nucleic acids* (ribonucleic acid RNA and deoxyribonucleic acid DNA) are especially important, because they *carry the information* that determines what happens in a cell.

Some examples of low molecular weight, nonpolymeric compounds found in nature

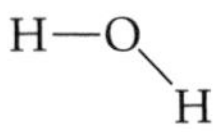

water
H_2O

molecular weight
is 18

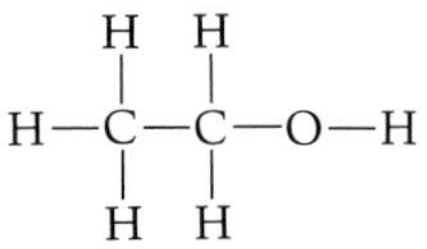

ethyl alcohol
C_2H_6O

mol. wt. = 46

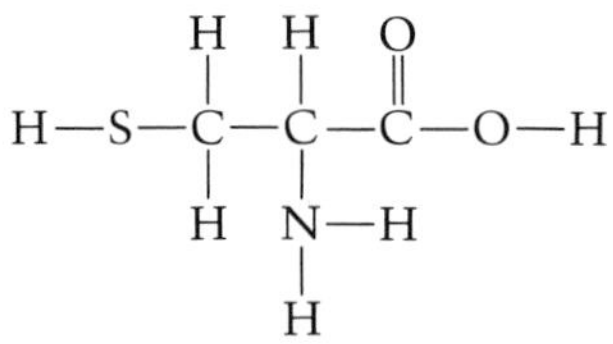

cysteine
(an amino acid)
$C_3H_7NO_2S$

mol. wt. = 121

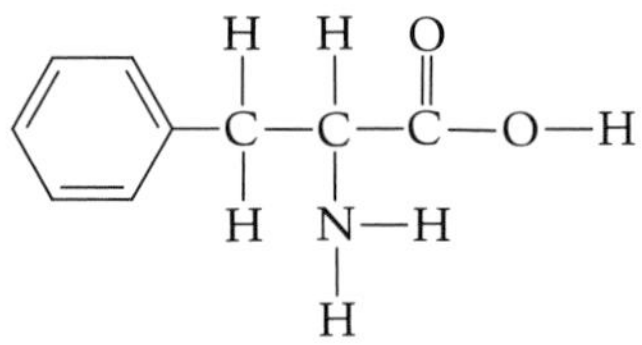

phenylalanine
(another amino acid)
$C_9H_{11}NO_2$

mol. wt. = 165

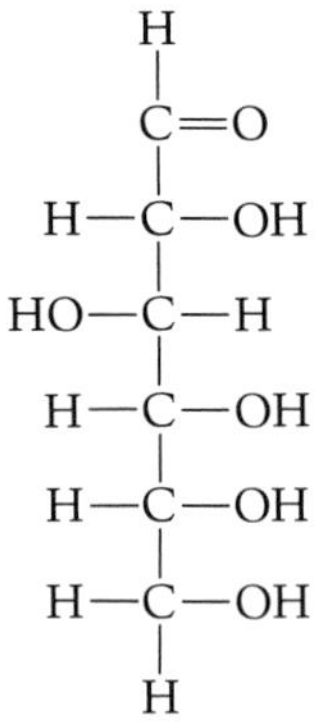

glucose
(a sugar)
$C_6H_{12}O_6$

mol. wt. = 180

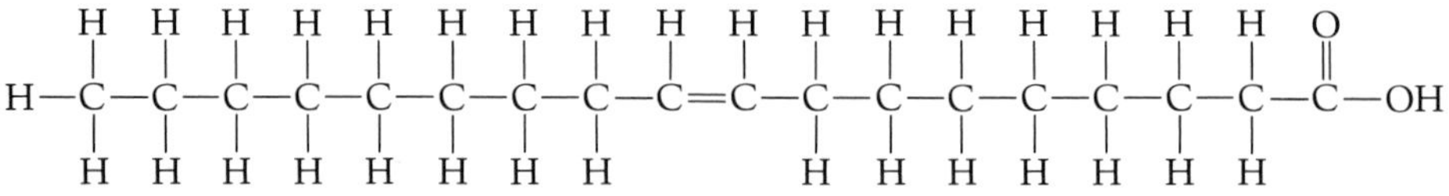

oleic acid
(a fatty acid)
$C_{18}H_{34}O_2$

mol. wt. = 282

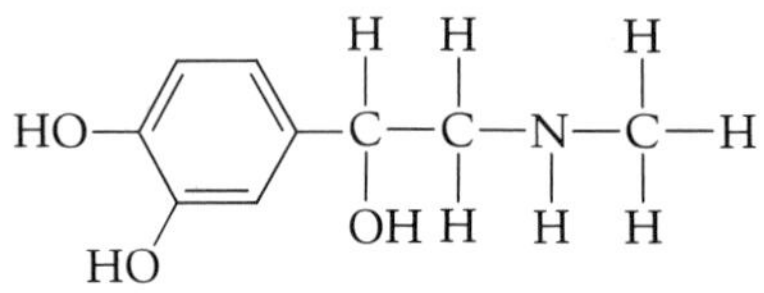

adrenalin
(a mammalian hormone
produced by the adrenal gland)
$C_9H_{13}NO_3$

mol. wt. = 183

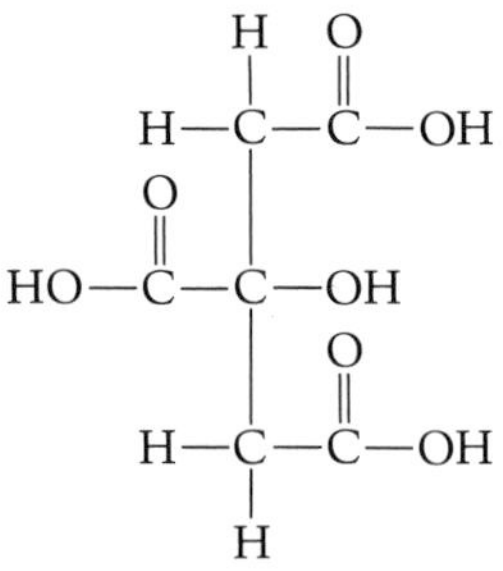

citric acid
(an important intermediate
of metabolism)
$C_6H_8O_7$

mol. wt. = 192

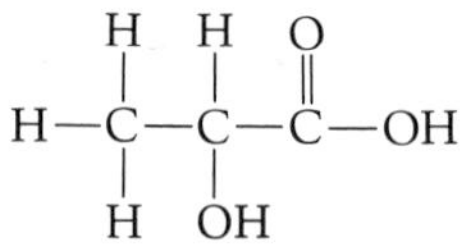

lactic acid
(an end product
of carbohydrate
degradation in
muscle)
$C_3H_6O_2$

mol. wt. = 74

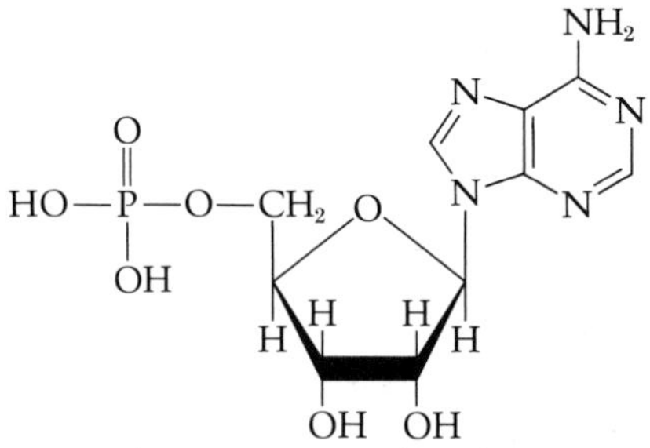

adenosine monophosphate
(a nucleotide)
$C_{10}H_{12}N_5O_5P$

mol. wt. = 313

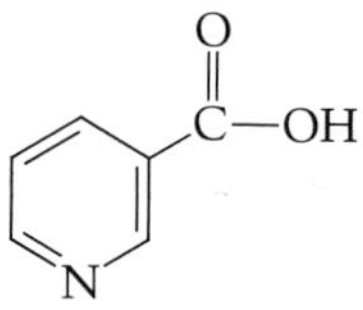

nicotinic acid
(an essential
vitamin)
$C_6H_5NO_2$

mol. wt. = 123

Because of their very large size, it is not feasible to represent structural formulas of high-molecular-weight polymers in this way. For example, the protein hemoglobin has a formula of $C_{2952}H_{4664}N_{812}O_{832}S_8Fe_4$ corresponding to a molecular weight of 65,248. Many proteins are even larger.

As you know, the individuality of a cell is due to the presence of *chromosomes*, which contain the *genes*. A chromosome is a nucleic acid molecule, and more specifically, it is a *deoxyribonucleic acid (DNA) molecule*. The individual genes are segments of the intact DNA chromosome. The number of genes in the 46 paired chromosomes (46 DNA molecules) of each human somatic cell is several thousand. The production of identical copies in the replication of cells is based on the production of an identical set of gene-carrying chromosomes. That is, the DNA molecules in the parent cell are duplicated during their biosynthesis to give exact copies. The phenomenon of mutation, which results in the formation of a modified gene or contingent of genes, involves alterations in this process.

What is genetic information? Well, a specific gene provides the specific instructions for the assembly of a specific protein molecule. In other words, *genetic information, packaged in the molecular structure of the chromosomes containing the genes, is ultimately expressed in the biosynthesis of all the proteins that are present in the cell*. Each protein, unique in its own structure and hence unique in its function, then participates in the processes that characterize the individuality of the cell and the entire organism. The functions performed by proteins are many (see p. 118), but the most important proteins are those that serve as *enzymes*. An enzyme is a protein that serves as a *catalyst, that is, it speeds up a reaction*. Virtually every reaction in a living cell requires an enzyme and most reactions require a specific enzyme. It stands to reason then that a substantial amount of biochemical research is directed at the isolation and study of enzymes.

Protein and nucleic acid molecules share one common characteristic, namely, *both are polymers*. A polymer (from the Greek *poly*, meaning several; *mer*, meaning a unit) is a large compound of high molecular weight. It is composed of several smaller compounds of low molecular weight, called monomers, which are successively linked to each other by covalent bonding. Think of a polymer as a chain and the links of the chain as the monomers. The monomeric units of a protein are the *amino acids* and they are linked to each other by *peptide bonds*. The monomeric units of a nucleic acid are *nucleotides* and they are linked to each other by *phosphodiester bonds*. In due course we will examine these biopolymers in greater detail. For now it is important that you develop only a general understanding of the nature of a polymer. The following diagram should help.

Thirty years ago knowledge of DNA was confined primarily to scientists and students in higher education. Today sixth grade school children learn about it. In fact, DNA even made the network TV news in February, 1977, when the city council in Cambridge, Mass. debated whether or not to allow certain types of DNA recombinant research to be performed in the laboratories of Harvard University. This sort of national attention to and public interest in DNA research is most justified. Indeed, DNA may very well become the hottest news item of this century.

1. The basic structural feature of a polymer is the successive linkage of monomers.

the line represents that successive monomers are linked to each other by covalent bonding

the value of *n* is characteristic of the size of a polymer;

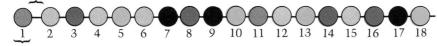

1 2 3 4 5 6 7 8 9 10 11 12 13 14 15 16 17 18   *n*

each sphere represents a monomer

heterogeneous biopolymer
(the sequence determines everything)

among proteins it can range from 50 to several hundred; among nucleic acids it can range from 90 to several thousand

an amino acid in proteins
a nucleotide in nucleic acids

7

2. The representation on p. 7 is that of a *heterogeneous* polymer, which typifies the proteins and nucleic acids. Heterogeneous means that each monomer is not identical (conveyed here by different colors and shading). In proteins, for example, there may be as many as 18 different amino acids. Furthermore, of those present, some may appear in only one or two positions while others may appear in several positions. The monomer composition of a polymer is, of course, a characteristic of the polymer, but it is the specific *sequence of monomers that determines the unique individuality of the polymer.*

3. The actual shape of a protein or nucleic acid molecule does not correspond to a completely straight chain. Rather, the actual shape is due to the way the chain folds, twists, and turns. Chains may also associate with and intertwine with other chains. We will examine all of this later. The point to be made at this time, however, is that the *individual functions performed by these biopolymers are a result of this overall three-dimensional structure, which in turn is governed by the sequence of the monomeric units.* It is in this context that the sequence of monomers is **informational.**

4. Another type of naturally occurring biopolymer is diagramed below. It is a *homogeneous* polymer, composed of identical monomeric units (same color and shading). The only naturally occurring substances of this type are some of the polymeric carbohydrates such as cellulose, starch, and glycogen. If you grasp the significance of the principles stated in 2 and 3 above, you should understand why, in contrast to proteins and nucleic acids, homogeneous polymers are noninformational. A further distinction of homogeneous polymers is that some are branched. The branching can occur at various positions along the main chain. Branches can also have branches. Glycogen and one type of starch typify branched polymers.

1 2 3 4 5 6 7 8 9 10 11 12 13 14 15 16 17 18   *n*

homogeneous biopolymer

(the same monomer is in every position;
in cellulose it is β-D-glucose; in starch
and glycogen it is α-D-glucose)

This description of the structure of biopolymers merely scratches the surface and is intended only as an introduction to some important basic concepts. Additional details will be presented later.

THE CHEMICAL COMPOSITION OF CELLS

A brief examination of the materials that make up living organisms is a fitting conclusion to this introduction. First, let us consider the most fundamental aspect of this subject, namely, the chemical *elements* of which organisms are composed. This does not require an extensive description, since only about 20 elements are involved. Furthermore, most of the molecules in an organism are composed of only six elements, all nonmetals: *oxygen, carbon, hydrogen, nitrogen, phosphorus,* and *sulfur.* Most of the other elements are metals that exist as ions (refer to the material below on minerals).

The subject does become more complicated when one considers the overwhelming number of chemical substances in which these elements occur, particularly those of an organic nature. Our biosphere contains an incredible variety of different cellular organisms and each has a unique set of molecules. However, there is a significant generalization that can be made: *all organisms are composed of the same types of chemical substances and most contain them in similar proportions.* Listed in decreasing order of occurrence in terms of the approximate percent of total organism weight they are as follows: *water* (70%), *proteins* (15%), *nucleic acids* (7%), *carbohydrates* and their metabolites (3%), *lipids* and their metabolites (2%), *inorganic ions* (1%), free *amino acids* and their metabolites (0.8%), free *nucleotides* and their metabolites (0.8%), and a variety of other organic molecules such as the *vitamins* and *small peptides* (0.4%). Although the percentages correspond to the estimated composition of the unicellular bacterium, *Escherichia coli* (see Table 2), the values are not greatly different for

TABLE 2. **Approximate chemical composition of a rapidly dividing cell of *E. coli*[a]**

Material	% of total wet wt.[b]	Average mol. wt.	Approx. no. of molecules/cell[c]	Different kinds of molecules/cell[c]
Water	70	18	40 billion	1
Proteins	15	40,000 (several much larger)	1 million	2,000–3,000
Nucleic acids (DNA and RNA total 7%)				
DNA	1	2.5 billion[d]	2 or 4	1
RNA	6			
5S ribosomal RNA		40,000[d]	30,000	1[e]
16S ribosomal RNA		500,000[d]	30,000	1[e]
23S ribosomal RNA		1 million[d]	30,000	1[e]
transfer RNA		25,000	400,000	40
messenger RNA		1 million	1 million	1,000
Carbohydrates and metabolites	3	150 (excluding polymers)	200 million	200
Lipids and metabolites	2	750	25 million	50
Inorganic ions	1	40	250 million	20
Amino acids and metabolites	0.8	120	30 million	100
Nucleotides and metabolites	0.8	300	12 million	200
Others	0.4	150	15 million	200

[a] Adapted with permission from James D. Watson, *Molecular Biology of the Gene*, 2nd edition, Philadelphia: W. B. Saunders Co., 1972.
[b] Approximate values apply to the majority of living organisms existing in a normal state.
[c] Numbers for some materials vary considerably from bacterial cells to cells of higher organisms.
[d] Corresponding substances in cells of higher organism are larger in size.
[e] Not yet certain whether all molecules are identical.

other organisms including the normal human body. However, on a cellular level the composition of an *E. coli* cell is quite different than that of a liver cell, which in turn is different than that of a fat cell of adipose tissue, and so on. In other words, we must make the distinction between intact organism and individual cells.

While recognizing this distinction, other interesting generalizations can be made from the data of Table 1.

(a) Organisms contain a greater variety of proteins than any other type of material.
(b) Approximately 50% of the solid matter of an organism is protein.
(c) Organisms contain many, many more protein molecules than DNA molecules.
(d) The largest biomolecule is DNA.
(e) *E. coli* and similar bacteria contain only a single kind of chromosome, that is, one particular type of DNA (applying the implied relationship of one chromosome–one DNA to the cells of organisms means that each human diploid cell contains 46 different DNA molecules in the nucleus).
(f) About 99% of the molecules in a cell are H_2O molecules.

In view of the last item it can be said that the science of biochemistry is concerned with only 1% of the molecules in a living cell. Keep in mind, however, that the specific molecules comprising this 1% and the totality of chemical events in which they participate are unique for each cell.

Although the study of biochemistry emphasizes organic substances, the inorganic substances (frequently termed *minerals*) are also important. Indeed, several are essential nutrients for all organisms and must be supplied in the diet (or natural surroundings) because they cannot be produced within the organism. The inorganic elements are present as ionic forms (see Table 3) and exist as free ions or complexed to an organic grouping. At this time we will not itemize the many and varied processes in which they function. It will become quite clear that the vital roles of many inorganic ions are associated with their effect on the activity of proteins in general and of enzymes in particular.

By now you should have the impression that biochemistry is a vast subject and this is precisely the case. Indeed, it boggles the mind even for those of us who carry the label of professional biochemist and particularly for authors of biochemistry textbooks. Don't let yourself be intimidated by it, however. You may find it difficult, but you will also discover that it is a most fascinating and exciting scientific discipline. I am confident you will be satisfied that you decided to study biochemistry.

I would like to offer some study tips and other thoughts. A lot of your study will require memorization. Indeed, you must become as familiar with certain things—such as the names, structures, and characteristics of the amino acids—as you are with the alphabet; without hesitation you should be able to rattle off the events of the major metabolic pathways. Do not, however, equate your understanding of the subject by how much and how well you have memorized.

To truly appreciate and understand things it is necessary to *think* about them. There is *a logic that permeates the principles of biochemistry*. Try to reason, to gain the sense of things. For example, you will discover that (a) establishing the relationship of function to structure (that is, relating what a molecular or multimolecular aggregate does to its composition and arrangement) and (b) gaining an understanding of the operation of various regula-

TABLE 3. A partial listing of inorganic substances of biological importance.

Mineral (ionic form)
Calcium (Ca^{2+})
Chlorine (Cl^-)
Cobalt (Co^{2+})
Copper (Cu^+, Cu^{2+})
Iodine (I^-)
Iron (Fe^{2+}, Fe^{3+})
Magnesium (Mg^{2+})
Manganese (Mn^{2+})
Molybdenum (Mo^{6+})
Phosphorus ($H_2PO_4^-$, HPO_4^{2-})
Potassium (K^+)
Sodium (Na^+)
Sulfur (SO_4^{2-}, S^{2-})
Zinc (Zn^{2+})

All of the above substances are definitely essential nutrients for humans and, with the exception of iodine, for all other organisms as well. Trace amounts of about a dozen additional minerals have been detected in various organisms. However, the importance and functions of many trace elements have yet to be clarified, so continued research in nutrition is still very active. For example, it was recently established that nickel (Ni) and silicon (Si) are essential minerals in trace quantities for several organisms, including humans.

tory mechanisms by which a cell or organism controls its own activities are two central themes of modern biochemistry. Neither of these topics can be truly learned by mere rote memorization. Rather, based on other principles, you should strive to develop an understanding of the rationale for such phenomena in much the same sense as you can appreciate and understand why a basketball is round instead of square, or why each lock has its own key, or why you can't put a size 6 glove on a size 12 hand, or why a right hand will fit snugly only in a right-handed glove. What I am saying is that there are reasons for things and in many instances the reasons are fundamentally simple ones. Your professor and I will try to assist you in this endeavor but we can't do the thinking for you.

Do a lot of work with pencil and paper. Solve problems. Design your own problems and exchange them with other students. Talk biochemistry with your associates. Finally, your approach to this subject should anticipate that topics in later chapters are very much dependent on earlier chapters. In other words, strive to carry over your knowledge from one chapter to the next and always refer back for review and reinforcement of your understanding. The many topics of an introductory biochemistry course comprise an integrated package, perhaps more so than any other course you have taken or will take.

The modern history of the human race is full of remarkable and admirable (and some not so admirable) achievements in the areas of science, engineering, medicine, agriculture, communications, space exploration, social progress, economics, and the arts. The future will certainly record many more. However, of all the progress that is anticipated, the greatest impact may result from the quest to learn how a living organism operates in terms of the molecules that compose it and to then manipulate the molecules and thus manipulate the organism. A profound statement to say the least, but yet this is a real prospect for the future. The core of this quest is biochemical knowledge.

I am pleased to have obtained the necessary permission to be able to close this introduction with an essay written by Dr. Harold J. Morowitz, who is Professor of Molecular Biophysics and Biochemistry at Yale University. It is a fitting conclusion to this introduction. Moreover, I am certain you will find it entertaining. The essay concludes on an appropriately profound philosophical note.

Another annual cycle inevitably passed and the pain was eased by a humorous birthday card from my daughter and son-in-law. The front bore the caption "According to BIO-CHEMISTS the materials that make up the HUMAN BODY are only worth 97¢" (Hallmark 25B 121-8, 1975). Before I could get to the birthday greeting I began to think that if the materials are only worth ninety-seven cents, my colleagues and I are really being taken by the biochemical supply companies. Lest the granting agencies were to find out first, I decided to make a thorough study of the entire matter.

I started by sitting down with my catalogue from the (name deleted) Biochemical Co. and began to list the ingredients. Hemoglobin was $2.95 a gram, purified trypsin was $36 a gram, and crystalline insulin was $47.50 a gram. I began to look at slightly less common constituents such as acetate kinase at $8,860 a gram, alkaline phosphatase at $225 a gram, and NADP at $245 a gram. Hyaluronic acid was $175 a gram, while bilirubin was a bargain at $12 a gram. Human DNA was $768 a gram, while collagen was as little as $15 a gram. Human albumin was down at $3 a gram, whereas bradykinin was $12,000 for a gram. The real shocker came when I got to follicle-stimulating hormone at $4,800,000 a gram—clearly outside the reach of anything that Tiffany's could offer. I'm going to suggest it as a gift for people who have everything. For the really wealthy, there is prolactin at $17,500,000 a gram, street price.

Not content with a brief glance at the catalogue, I averaged all the constituents over the best estimate of the percent composition of the human body and arrived at $245.54 as the average value of a gram dry weight of human being. With that fact burning in my head I rushed over to the gymnasium and jumped on the scale. There it was, 168 pounds, or, after a quick go-round with my pocket calculator, 76,364 grams. Remembering that I was 68% water, I calculated my dry weight to be 24,436 grams. The next computation was done with a great sense of excitement. I had to multiply $245.54 per gram dry weight by 24,436 grams. The number literally jumped out at me—$6,000,015.44. I was a Six Million Dollar Man—no doubt about it—and really an enormous upgrade to my ego after the ninety-seven cent evaluation!

Assuming that the profits of the biochemical companies are considerably less than the 618,558,239% indicated above, we must still strike a balance between the ninety-seven cent figure and the six million dollar figure. The answer is at the same time very simple and very profound: information is much more expensive than matter. In the six million dollar figure I was paying for my atoms in the highest informational state in which they are commercially available, while in the ninety-seven cent figure I was paying for the informationally poorest form of coal, air, water, lime, bulk iron, etc.

This argument can be developed in terms of protein as an example. The macromolecules of amino acid subunits cost somewhere between $3 and $20,000 a gram in purified form, yet the simpler, information-poorer amino acids sell for about twenty-five cents a gram. The proteins are linear arrays of the amino acids that must be assembled and folded. Thus we see the reason for the expense. The components such as coal, air, water, limestone and iron nails are, of course, simple and correspondingly cheap. The small molecular weight monomers are much more complex and correspondingly more expensive, and so on for larger molecules.

This means that my six million dollar estimate is much too low. The biochemical companies can sell me their wares for a mere six million because they isolate them from natural products. Doubtless, if they had to synthesize them from ninety-seven cents worth of material they would have to charge me six hundred million or perhaps six billion dollars. We have, to date, synthesized only insulin and ribonuclease. Larger proteins would be even more difficult.

A moment's reflection shows that even if I bought all the macromolecular components, I would not have purchased a human being. A freezer full of unstable molecules at −70 °C does not qualify to vote or for certain other unalienable rights. At six billion it would certainly qualify for concern over my −70 °C deep freezer, which is always breaking down.

The next step is to assemble the molecules into organelles. Here the success of modern science is limited as we are in a totally new area of research. A functionally active subunit of ribosomes has been assembled from the protein and RNA constituents. Doubtlessly other cellular structures will similarly yield to intensive efforts. The ribosome is perhaps the simplest organelle so that considerably more experimental sophistication will be required to get at the larger cell components. One imagines that if I wanted to price the human body in terms of synthesized cellular substructures, I would have to think in terms of six hundred billion or perhaps six trillion dollars. Lest my university begin to salivate about all the overhead they would get on these purchases, let me point out that this is only a thought exercise, and I have no plans to submit a grant request in this area.

Continuing the argument to its penultimate conclusion, we must face the fact that my dry-ice chest full of organelles (I have given up the freezer, at six trillon it simply can't be trusted) cannot make love, complain, and do all those other things that constitute our humanity. Dr. Frankenstein was a fraud. The task is far more difficult than he ever realized. Next, the organelles must be assembled into cells. Here we are out on a limb estimating the cost, but I cannot imagine that it can be done for less than six thousand trillon dollars. Do you hear me, Treasury Secretary Simon, Federal Reserve Chairman Burns? Are these thoughts taking a radical turn?

A final step is necessary in our biochemical view of man. An incubator of 76,364 grams of cell culture at 37 °C still does not measure up, even in the crassest material terms, to what we consider a human being. How would we assemble the cells into tissues, into organs, and organs into a person? The very task staggers the imagination. Our ability to ask the question in dollars and cents has immediately disappeared. We suddenly and sharply face the realization that each human being is priceless. We are led cent by dollar from a lowly pile of common materials to a grand pilosophical conclusion— the infinite preciousness of every person. The scientific reasons are clear. We are, at the molecular level, the most information-dense structures around, surpassing by many orders of magnitude the best that computer engineers can design or even contemplate by miniaturization. The result must, however, go beyond science and color our view of the world. It might even lead us to Alfred North Whitehead's conclusion that "the human body is an instrument for the production of art in the life of the human soul."

This essay, entitled "The Six Million Dollar Man," appeared in the March, 1976 issue of *Hospital Practice*. It is reproduced here with the permission of Professor Morowitz and the publisher of *Hospital Practice*.

LITERATURE

DEVONS, S., ed. *Biology and the Physical Sciences*. New York, London: Columbia University Press, 1969. A collection of technical essays covering diverse subjects, delivered by authorities in each field at a symposium on the relationship between biology and physical science. The brief Foreword and Postscript written by S. Devons focus attention on the interdisciplinary nature of modern biology.

FLORKIN, M., and E. H. STOTZ, eds. *A History of Biochemistry*. Volumes 30 through 33 of Comprehensive Biochemistry. New York: Elsevier, 1973. A most thorough treatment.

FRUTON, J. S. *Molecules and Life*. New York: John Wiley, 1972. An authoritative and well-written account of scientific developments in biochemistry covering the period from 1800–1950. Many excerpts from the original literature are included. A uniquely invaluable source to gain a historical perspective and appreciation of the foundations of modern biochemistry.

———. "The Emergence of Biochemistry." *Science*, **192**, 327–334 (1976). An interesting and authoritative discussion of the growth of biochemistry focusing on several aspects of the interplay and conflicts of chemistry and biology since 1800.

GREEN, D. E., and R. F. GOLDBERGER. *Molecular Insights Into the Living Process*. New York, London: Academic Press, 1967. This book presents a unified picture of biochemistry from the standpoint of universal principles that apply to all living organisms. Best read after you know some biochemistry.

KARLSON, P. "From Vitalism to Intermediary Metabolism." *Trends Biochem. Sci.*, **1**, N184 (1976). A brief historical article.

1. Many brief items concerning the history of biochemistry and the persons involved are routinely included in the new journal, *Trends in Biochemical Sciences*, published by Elsevier/North-Holland Biomedical Press. Available at reasonable subscription rates.

2. Autobiographical reflections of prominent biochemists are included in each issue of *Annual Review of Biochemistry*. They provide interesting reading.

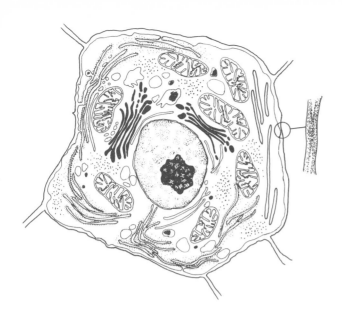

CHAPTER 1

CELLULAR ORGANIZATION

It is common knowledge that our biosphere contains a tremendous variety of cells. For example, in the advanced multicellular (many-celled) organisms of the animal and plant kingdoms, we can identify liver cells, fat cells, pancreas cells, kidney cells, heart cells, muscle cells, retina cells, nerve cells, brain cells, red blood cells, white blood cells, leaf cells, stem cells, root cells, and several more. Considering just the microbial world, which is composed of both unicellular (single-cell) and multicellular organisms, we see that there are a countless number of individual cell types.

Despite this diversity and individuality, there are among living cells many gross similarities at the level of both cell structure and cell function. Thus it is possible to give a coherent and meaningful discussion of cells according to a general format that highlights elements of cellular unity.

At the functional level the most fundamental similarity is that any cell, sustained by the ingestion and utilization of nutrients and energy from the external environment, is capable of growth and cell division. Because an intact cell is the smallest entity capable of these processes, the cell is termed the *unit of life*. In other words, *all living organisms have a cellular structure, and the activity of the whole organism is the result of the individual and collective activities of cells.* It is interesting to note that the first meaningful statements of this *cellular theory of life* were not made until

1838 by M. J. Schleiden and T. Schwann, and then again in the late 1850s by R. Virchow. Now in the twentieth century, particularly during the past 25 years, the cellular basis of life has been extrapolated to the level of molecules. Thus, in a physiocochemical sense, a cell is what it is and does what it does only because it is composed of a certain set of molecules that function according to the principles of chemistry and physics. Nevertheless, the cell is regarded as the *basic unit* of life, since it is only at the cellular level that all of the basic characteristics of life are expressed.

Starting at the level of structure we must first realize that a living cell is not an amorphous, indivisible, continuous molecular blob. Quite the contrary; the cell is a highly organized entity consisting of separate and distinguishable parts, each of which performs an important function in the overall living process. In other words, the cell is like any other machine, with the operation of the complete unit a result of the combined and integrated operation of its individual component parts. In this sense, the cell is a complex unit of life, and, depending on the type and source of the cell, the degree of complexity is quite variable.

However, most cells do share a fundamental level of organization in that they all possess a nonrigid but continuous molecular barrier (*cell membrane*) that separates the interior of the cell from the external surroundings. Many cells, mostly plants and bacteria, contain a second barrier in the form of a *cell wall*, which is exterior to and more rigid than the cell membrane. Although both the structure and the function of the cell membrane and cell wall vary from cell to cell, the major basis for the differentiation among cells is the nature of the intracellular environment. Here again, however, there is still a basic level of subcellular organization represented by the fact that all cells can be classified as either of only two types—the *prokaryotes* and the *eukaryotes*.

Prokaryotic cells are distinguished by the *absence of any membrane-bound subcellular compartments*. In this type of cell, most of the functional biomolecules and particles are found in the aqueous intracellular fluid, called the *cytoplasm*, while the others are embedded in the main cell membrane. Most representative of this group of cells are the unicellular bacteria, believed to be the first type of living organisms, having evolved some 3–3.5 billion years ago. (*Note:* At present the most reliable estimate of the age of the earth is 4.8 billion years.)

Eukaryotic cells, on the other hand, *do contain distinct subcellular particulate bodies*, called *organelles*. These either are entirely membranous in nature or are organized units surrounded by a membrane. The most common membrane-bound organelle is the *cell nucleus*. Others include the *mitochondrion*, the *chloroplast*, the *lysosome*, and the *peroxisome*. The principal organelles that are entirely membranous bodies are the *endoplasmic reticulum* and the *Golgi apparatus*. Obviously, eukaryotic cells represent a more advanced state of the living cell. Indeed, the eukaryotic cells are believed to have evolved from the smaller and less specialized prokaryotes. The group includes nearly all animal and plant cells, and advanced microbes such as the fungi, many algae, and some forms of bacteria.

Our objective in this chapter will be to study the ultrastructural details of both groups of cells. Such an analysis is possible primarily because of the great strides that have been made in the past 10 to 20 years both in the technological development of the *electron microscope* and in the laboratory preparation of biological samples for viewing. The specific coverage of this

Recent studies suggest that the evolution of living organisms may have involved a third line of descent. In addition to prokaryotes and eukaryotes, it has been proposed that *methanogens* may represent a third class of life. These primitive microorganisms possess some unusual anatomical, metabolic, and molecular features indicating that the methanogens are no more related to the prokaryotic bacteria than bacteria are related to higher organisms. There will, however, be much debate as to whether the differences warrant a reclassification of the methanogens as a third form of life rather than simply an early form of bacteria. For further information refer to the article by Maugh cited at the end of the chapter.

Structural organization[a] of living cells
Prokaryotic cells
Cytoplasm
Ribosomes (free)
Cell (plasma) membrane
Cell wall
Outer (coat) layer
Eukaryotic cells
Cytoplasm
Ribosomes (free and bound)
Nucleus
Mitochondria
Chloroplasts[b]
Lysosomes
Peroxisomes
Endoplasmic reticulum
Golgi body
Cell (plasma) membrane
Cell wall[c]
Outer (coat) layer

[a] This list includes only the major subcellular entities; each type, but particularly eukaryotic cells, contain a few other specialized structures; for these and additional detail over this chapter see the text by Dyson cited at the chapter end.
[b] In photosynthetic plant cells only.
[c] In plants and bacterial cells; not in animal cells.

Our current understanding of cell ultrastructure and the functions associated with the various organelles is the result of an extensive amount of research over the past 30–35 years. In recognition of the impact this knowledge has had in our quest to conquer the secrets of life and particularly for their pioneering achievements in the area, Nobel prizes in physiology and medicine were awarded in 1974 to Albert Claude, George Palade, and Christian de Duve.

$1 \text{ Å} = 10^{-8} \text{ cm} = 10^{-10} \text{ m}$

(Å is the symbol for angstrom, a unit of length)

$1 \ \mu\text{m} = 10^{-6} \text{ m}$

therefore,

$1 \text{ Å} = 10^{-4} \ \mu\text{m}$

subject here is made possible only through the courtesy and generosity of several individuals who provided prints of electron micrographs from their personal files. As you will discover, the structure of a living cell—particularly that of a eukaryotic cell—certainly is not that of continuous blob.

ELECTRON MICROSCOPY

The unaided human eye (normal) is capable of observing two or more objects as separate objects as long as the distance between them is about 0.1 mm, or 100 micrometers (i.e., 100 μm). Below this limit of *resolving power* the eye sees only one object. Furthermore, the eye is incapable of any magnification.

However, through various types of microscopy we have been able to enormously extend our observations of small objects. A microscope provides both increased resolving power and magnification of the specimen. The effective resolving power varies from one type to another. For example, the best conventional light microscope cannot distinguish between objects that are closer than about one-half the wavelength of illuminating light. Thus, with the most perfectly ground lenses and with ordinary white-light illumination (having an average wavelength of 5000 Å), the resolution of the light microscope is about 2500 Å or 0.25 μm. By using ultraviolet illumination (of shorter wavelength) and special quartz lenses, a resolution of about 0.17 μm can be achieved. Obviously both are better than the unaided eye. Indeed, intact living cells with a diameter ranging from about 1 to 20 μm can be seen, and the larger ones in considerable detail. However, many still remain invisible and many details are fuzzy. For example, a cell membrane is only about 0.01 μm thick, a ribosome is only about 0.02 μm in diameter, and the massive DNA molecule is only about 0.002 μm thick.

Increased resolving power is provided by the electron microscope. The basic principle of the electron microscope (see Fig. 1-1) is the same as that of any microscope—an object is illuminated by radiation and an enlarged image of the interaction is produced. In the electron microscope, however, the radiation does not consist of typical light waves, but rather is composed of *rays of high speed electrons that are focused on a specimen under the influence of a magnetic field.* As these electrons pass through the specimen being viewed, a differential absorption of electrons will occur due to structural variations in the specimen, thus forming an image of the object that can be detected on an electron-sensitive photographic film. The photograph of the image is called an *electron micrograph.* In the study of biological specimens, the quality of the image is very much related to how the specimen was prepared for viewing. Frequently cellular specimens are prepared as thin-sections with an ultramicrotome. This involves fixing the specimen (dipping it in a solution of glutaraldehyde, for example, to render protein components insoluble), then dehydrating the specimen, and finally embedding the specimen in a plastic block, which is then sliced in the ultramicrotome. To improve contrast the sample may also be stained or shadowed with heavy electron-dense substances such as tungsten salts, osmium salts, or platinum. Consult one of the references for details concerning the various modes of sample preparation. (A special technique, called *freeze etching,* is described in Chap. 10; see p. 332.) What-

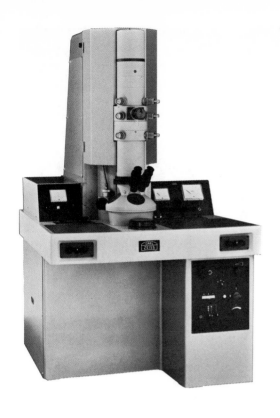

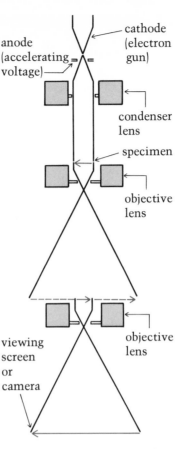

anode (accelerating voltage)

cathode (electron gun)

condenser lens

specimen

objective lens

viewing screen or camera

objective lens

FIGURE 1–1 *Right:* a diagramatic outline of the major components of an electron microscope. The accelerated beam of electrons is focused on the specimen (arrow) by a magnetic field (condenser lens). A series of two objective lenses (each a magnetic coil also) magnifies and focuses the emerging beam onto an electron sensitive screen or photographic plate, which depicts the image after development. The entire column assembly is maintained under a very high vacuum. *Left:* photograph of electron microscope—model EM-9S-2 manufactured by Carl Zeiss, Inc. (Photograph provided by Carl Zeiss, Inc., New York.)

ever the procedures, by the time the sample is prepared for viewing it has undergone some rather harsh treatments and there is always the question as to how "real" the image is or whether artifacts are being observed.

As with the ordinary light microscope, the resolving power of the electron microscope is also dependent on the wavelength of incident radiation, which in this case is determined by the voltage used to accelerate the electrons. For example, if the electron beam was produced under a potential (V) of 50,000 volts, the approximate wavelength (λ) of a ray of high speed electrons would be

$$\lambda \text{ (in Å)} = \frac{12.3}{\sqrt{V}} = 0.05 \text{ Å}$$

and the resolving power (estimated by $\frac{1}{2}\lambda$) would be about 0.025 Å or 0.0000025 μm. If the instrument were 100% efficient, this would mean that we would be able to see individual hydrogen atoms. Because of technical difficulties in instrument design, however, this degree of resolution cannot be achieved. The best resolution that can be delivered is about 5 Å or 0.0005 μm. This is still fantastic and is sufficient to see objects in tremendous detail, including tiny viruses and even some molecules (see Fig. 1-2).

The photographs in Fig. 1-2 clearly depict the usefulness of electron microscopy. Several others are displayed in this and other chapters.

The electron microscope is, or course, not a tool used only by the biochemist. Nevertheless, it does have very valuable application to biochemistry in establishing relationships between structure and function.

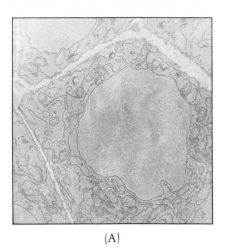

(A)

(B)

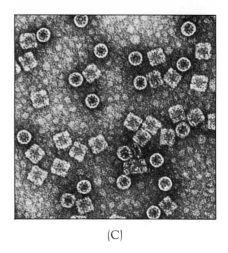

(C)

FIGURE 1-2 Electron micrographs of (A) the interior of a rat kidney cell, 21,000×; (B) several rod-shaped, tobacco mosaic virus particles; 222,000×; (C) molecules of hemocyanin, a copper-containing protein found in invertebrates and responsible for the transport of oxygen; obtained from snails; 140,000×. (Photograph A was provided by Hilton H. Mollenhauer, Charles F. Kettering Research Laboratory, Yellow Springs, Ohio. B and C were provided by Carl Zeiss, Inc., New York.)

Other advantages in working with *E. coli* are that it is noninfectious, has a short generation time, and is easy to culture reproducibly in small or large quantities. For these reasons, it has proven and continues to prove to be an extremely useful model system for basic biochemical research.

PROKARYOTIC CELLS

The organization of a typical prokaryotic cell is depicted in the electron micrographs of the bacterium *Escherichia coli*, shown in Figs. 1–3 and 1–4. *E. coli*, one of the simplest of the bacteria, is perhaps the organism that we understand best. We are interested in *E. coli* not merely because it is a normal and vital constituent of the intestinal tract of humans, but primarily because it is an aerobic organism whose basic biochemistry is, for the most part, representative of that of all other aerobic organisms including humans. Obviously, *E. coli* and a human are different organisms, but remember that they are nevertheless composed of the same major classes of biomolecules. Moreover, if we neglect for the moment all of the specialized activities of each organism, we recall further that the basic function of these biomolecules is the same in both. In view of the fact that *E. coli* evolved approximately 1.5 billion years before humans, this is a rather fantastic generalization, to say the least.

The electron micrograph of Fig. 1–3, showing the surface of two intact *E. coli* cells from different views, indicates that the organism has a cylindrical or rodlike shape. The cells are quite small, having a length of about 1–2 μm and a diameter of about 0.5–1 μm. The cell surface actually corresponds to the *cell wall*, a rigid sheath with a polysaccharide-peptide composition (see p. 310) that is coated with an outer layer of carbohydrate and lipid substances. Although the primary function of the cell wall is to provide physical protection, recent studies suggest that the wall (including its outer layer) may have important functions in the general physiology of bacteria, such as participation in a molecular communication system between the exterior and the interior of the cell. The cell wall and its outer covering layer also confer the antigenic specificities of the cell, determine the reaction with dyes that are used to stain bacteria for microscopic examination, and provide the sites of binding for bacterial viruses (see Fig. 1–4).

Figure 1–3 also reveals a second ultrastructural feature of the cell surface common to *E. coli* and several other bacteria, namely, the presence of long (up to several micrometers in length) and thin (approximately 100 Å) filamentous strands, called *fimbriae* or *pili*. Known to consist primarily of protein and lipid, these appendages are proposed to be channels whereby DNA

18

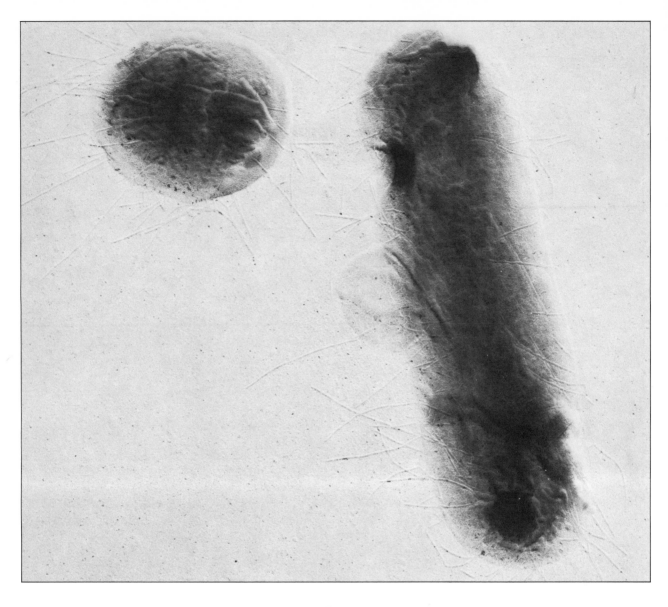

FIGURE 1–3 An electron micrograph of cells of *Escherichia coli* B. These are surface views depicting the rodlike shape of this organism. The image at the upper left represents a view from one end of the cell. The image at the right represents a view of the long axis of the cell. Faintly visible projections from each cell are pili (see text). Magnification: 45,000×.

passes between two mating cells during bacterial sexual conjugation. There is considerable debate as to whether the pili are specialized elaborations of the cell wall, of the interior cell membrane, or of the cytoplasm. Finally, the pili are similar in appearance to, but structurally distinct from, another group of specialized protein appendages, called *flagella*, which are responsible for the mobility of certain bacteria.

Although not clearly, the electron micrograph of a sectioned *E. coli* cell (Fig. 1–4) reveals the *cell membrane* (CM) adjacent to the interior of the cell wall. The dual structure is most visible at left center. At both the upper left and top of the cell, the wall appears irregularly detached from the membrane. (The cell membrane is also referred to as the *plasma membrane* or the *cytoplasmic membrane.*) In addition to contributing to the protection of the cell, the plasma membrane contains several enzymes and is also

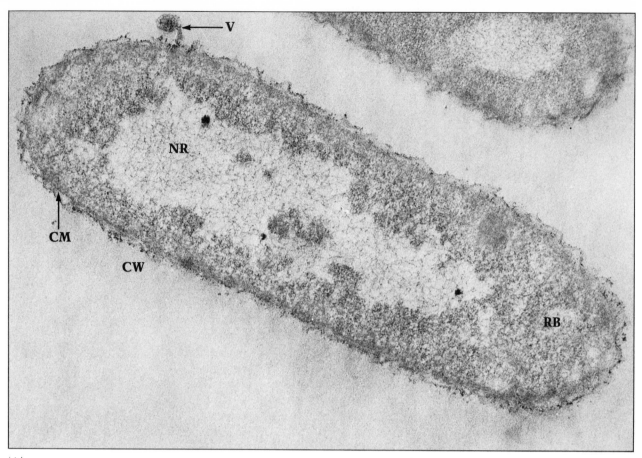

(A)

(B)

FIGURE 1–4 (A) An electron micrograph of an *Escherichia coli* cell in cross-sectional view, showing the cell wall (CW); the cytoplasmic cell membrane (CM); the DNA-containing nuclear region (NR); and the ribosomes (RB) distributed throughout the inner cytoplsm (CYP). The object attached to the cell at the upper right is a small bacterial virus (V). Magnification: 88,000×. Horizontal line at the bottom represents a distance of 1 micrometer. (Photograph generously supplied by Carl Zeiss, Inc., New York.) (B) An electron micrograph depicting the DNA that has seeped out of a disrupted *E. coli* cell. Most of the DNA is in the form of a single large molecule of double-strand DNA, which constitutes the chromosome of this prokaryotic organism. (From C. Grobstein, ''The Recombinant-DNA Debate,'' *Scientific American*, **237**, No. 1, 22–23 (July 1977). Micrograph prepared and supplied by Dr. Jack Griffith, University of North Carolina.)

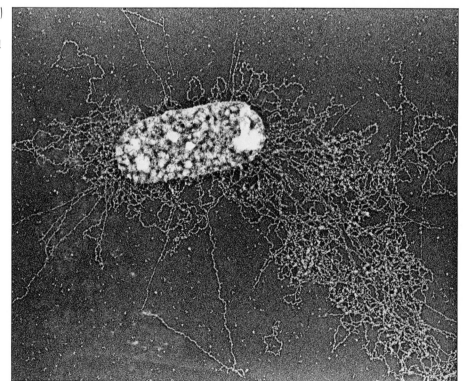

responsible for determining what molecules move into and out of the cell, i.e., it determines the *permeability* characteristics of the cell.

The conspicuous body at the upper right of the cell represents an infecting bacterial virus (V) that has become attached to the cell wall. Following attachment, the virus injects its own chromosome into the cell, whereupon the viral genes begin to express themselves, resulting in the production of new viral particles within the host cell. Since no viral particles appear in this view of the cell, we would conclude that infection has not yet occurred or that, if it has, viral assembly within the cell has not yet taken place to any appreciable degree.

Figure 1–4(A) also illustrates the nonspecialized interior of a prokaryotic cell. That is to say, the inside of the cell contains no membrane-bound compartments. The appearance of distinct light and dark regions within the cell is created by selective fixing and staining methods in the preparation of the sample for viewing. The less dense region seen along the longitudinal axis of the cell corresponds to the *nuclear region* (NR) of the cell, which contains the DNA chromosomal material dissolved in the *cytoplasm* (CYP). The massive and faintly visible filamentous network appearing against the light background is interpreted as representing DNA itself. (The three intense black spots are granules of unknown composition and function.) Depending on the growth conditions, an *E. coli* cell may contain one, two, or four DNA molecules, each of which is a single circular chromosome (see p. 9). The massive size of the *E. coli* DNA chromosome is vividly displayed in the micrograph of Fig. 1–4(B). The term *nuclear region* is used here because of the absence of a limiting membrane that would literally separate it from the rest of the cell interior. When the nuclear region is so bounded, the resulting entity is called a *nucleus* (see eukaryotic cells, p. 22). The darker, more electron-dense area surrounding the nuclear region is interpreted as representing the presence of *ribosomes* (RB). The ribosomes are known to be the cellular site for the assembly of amino acids into protein molecules (see p. 230, Chap. 7, and Chap. 8). It has been estimated that an actively metabolizing *E. coli* cell may contain anywhere from 10,000–15,000 ribosomes distributed throughout the cytoplasm.

In addition to the dissolved DNA and dispersed ribosomal particles, the *cytoplasm* (cellular fluid) contains thousands of other dissolved materials, many of which are enzyme proteins. Dissolved materials are invisible to the electron microscope and so in electron micrographs this colloidal aqueous solution appears only as a background of low electron density against the remainder of the cell structure.

EUKARYOTIC CELLS

Whereas a prokaryotic cell is conspicuous by the absence of distinct membrane-bound particulates within the cytoplasm, just the opposite is true of a eukaryotic cell. This is shown by the drawing in Fig. 1–5, which illustrates an idealized cross-sectional view of a typical eukaryotic cell. Although there is really no such thing as a typical cell of any type, the subcellular organization depicted in Fig. 1–5 is common to most eukaryotic cells. In other words, neglecting any specialized characteristics that many cells display, it is a valid generalization that most specimens of eukaryotic cells contain those features identified in the sketch, namely, the *nucleus,*

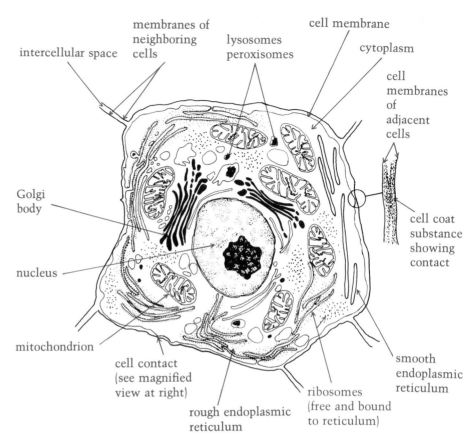

intercellular space

membranes of neighboring cells

lysosomes peroxisomes

cell membrane

cytoplasm

cell membranes of adjacent cells

Golgi body

nucleus

cell coat substance showing contact

mitochondrion

cell contact (see magnified view at right)

rough endoplasmic reticulum

ribosomes (free and bound to reticulum)

smooth endoplasmic reticulum

FIGURE 1–5 A simplified sketch of a generalized eukaryotic cell. Photosynthetic cells would also contain chloroplasts (not shown here). See text for descriptive information and representative electron micrographs. (Adapted with permission from R. D. Dyson, *Cell Biology*, 2nd edition, Boston: Allyn and Bacon, Inc., 1978.)

the *mitochondrion,* the *endoplasmic reticulum,* the *Golgi body,* the *lysosome,* the *peroxisome,* the *ribosomes,* and, of course, the *cell membrane* and the *cytoplasm.* With the aid of high-magnification electron micrographs, let us now examine each of these organelles in greater detail. In addition, we will also consider the *chloroplast,* a specialized organelle of photosynthetic cells.

Nucleus

The nucleus of a cell (generally there is only one per cell) is the largest and hence the most conspicuous subcellular organelle. It is easily seen with a good light microscope and suitable stains. Yet it is only with the electron microscope that the ultrastructure of the nucleus is revealed with good resolution (see Fig. 1–6). The first noteworthy feature, one that typifies every compartmentalized region of a eukaryotic cell, is the presence of a membrane segregating the nucleus from the cytoplasm. The *nuclear membrane* (sometimes called the *nuclear envelope*) is distinct from most other membranes in that it is clearly a *double membrane* separated by an *intermembranous space.* Another feature is the presence of membrane *pores,* which are interruptions of the double membrane systems. These pores apparently result from the fusion of the individual membranes of the double membrane network. The existence of nuclear pores accounts for the exit into the cytoplasm of the large RNA molecules (messenger, transfer, and ribosomal) that are synthesized in the nucleus.

The inside of the nucleus, called the *nucleoplasm,* also contains a degree

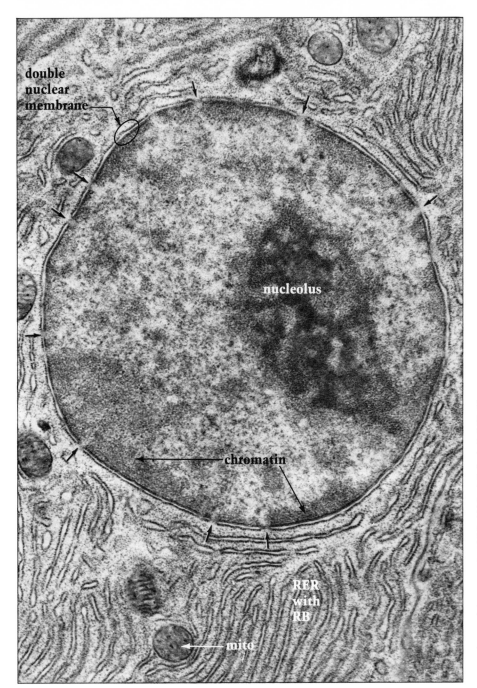

double
nuclear
membrane

nucleolus

chromatin

RER
with
RB

mito

FIGURE 1–6 An electron micrograph of a cross-sectional view of an intact cell nucleus. Clearly visible are the two membranes of the nucleus, with several pores indicated by the arrows around the circumference of the nucleus. Visible within the nucleus and dispersed in the nucleoplasm are the nucleolus and chromatin regions. A large population of rough endoplasmic reticulum (RER) with attached ribosomes (RB) is visible in the cytoplasm surrounding the nucleus. A small number of mitochondria (MITO) are also present. Specimen was obtained from the pancreas of a bat. Magnification: 22,000×. (Taken with permission from D. W. Fawcett, *An Atlas of Fine Structure: The Cell*, Philadelphia: W. B. Saunders Company, 1966. Photograph generously supplied by D. W. Fawcett.)

of internal organization. Particularly evident is the very electron-dense *nucleolus*, a region that is quite rich in RNA. Current belief is that the nucleolus is the site within the nucleus where RNA biosynthesis occurs. Also note that, throughout the nucleoplasm, but particularly near the nuclear membrane, there appear regions of less electron density than that found in the nucleolus. These are termed *chromatin* and are known to contain the major portion (95% or more) of the total DNA found in the cell associated with *histone proteins* (see p. 232, Chap. 7). The more or less random distribution of the chromatin is characteristic of the nucleus when the cell is not dividing. At the onset of cell division and during the staged mitotic process, the chromatin regions become highly organized into the hereditary units commonly known to all as the *chromosomes*. Ultimately, the chromosomes replicate to produce two identical sets. The enzyme required for the replication of DNA is also found in the nucleoplasm.

In describing the structural organization of the nucleus, note that we also established the primary functions of this specialized organelle. *To summarize:* the nucleus is the cellular site where *genetic informatin* is (a) *stored* as DNA, (b) *transmitted* to the rest of the cell (DNA → RNA → proteins; with protein biosynthesis occurring in the cytoplasm), and (c) *replicated* to insure perpetuation of the cell line (DNA → DNA).

Mitochondrion

It has been suggested that even in aerobic cells the number of mitochondria that are actually present *in vivo* is quite small and indeed, there may be only one per cell. Proponents of this idea, termed the *unit-mitochondrion hypothesis*, argue that the preparation of samples for observation by electron microscopy destroys these massive mitochondria and that the hundreds to thousands that appear are only artifact fragments of the original whole. While it appears that certain yeast cells are so characterized when grown under particular conditions and prepared and examined by special electron microscopic techniques, there is controversy as to the general occurrence of this unit mitochondrial morphology in other eukaryotic cells.

Although mitochondria are found in virtually all eukaryotic cells, their size, shape, and number are quite variable from one cell to another. In fact, all three characteristics appear to change in response to shifts in metabolism and as a result of cell aging. In addition various pathologies are associated with alterations in these characteristics. The significance of such changes to any of these situations is not yet clear.

In animal cells, the mitochondrion is frequently a rod-shaped particle with a length of $1.5-2$ μm and a diameter of $0.5-1$ μm. In other words, it is approximately one-twentieth the size of the cell nucleus and about equal in size to a prokaryotic bacterial cell such as *E. coli*. Certain generalizations can also be made regarding the number of mitochondria per cell. In cells characterized by a high degree of aerobic metabolism, the number per cell may be quite large. For example, each cell of liver tissue is estimated to contain close to a thousand mitochondria. On the other hand, cells participating primarily in anaerobic metabolism, such as the cells of skeletal muscle tissue, contain only a few mitochondria.

The origin of the word mitochondrion (from the Greek *mitos*, threadlike; *chondros*, grain) is found in the the gross structural features of this organelle recorded when it was first observed as a stained body under a light microscope some 65 years ago. In view of our current understanding of mitochondrial structure, the name is pitifully inappropriate. The electron micrograph of Fig. 1–7, showing a cross section of a typical mitochondrion magnified approximately 95,000 times, illustrates why. Immediately apparent is the presence of a defining membrane system and an extensive amount of *intra*mitochondrial structure. Close inspection reveals that the membrane system is also a *double membrane*—two unit membranes separated by an intermembranous space. Whereas the outer membrane appears to be smooth and continuous around the mitochondrion, the inner membrane undergoes an extensive and irregular folding within the mitochondrion. In cross section, the numerous foldings of the inner membrane, called *cristae*, appears as two electron-dense invaginations. The region having a homogeneous granular appearance surrounding the cristae represents the inner fluid of the mitochondrion, and is called the *mitochondrial matrix*. The very dense spots that appear within the matrix are small granules of unknown composition and function. These important features of mitochondrial ultrastructure are diagramatically depicted in the sketch on p. 26.

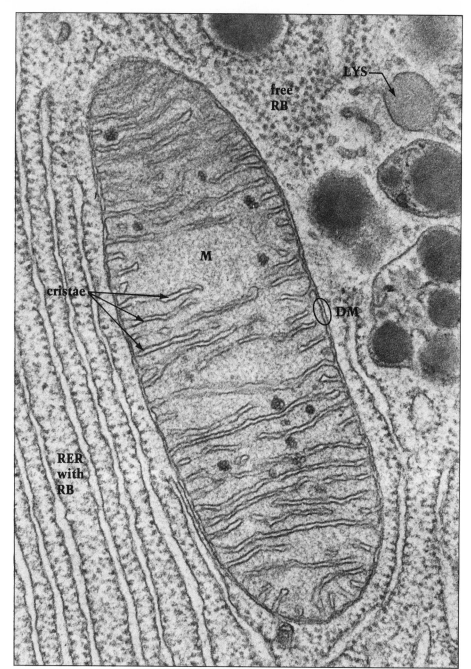

Labels on figure: LYS, free RB, cristae, M, DM, RER with RB

FIGURE 1–7 Electron micrograph of a longitudinal section of a mitochondrion and surrounding cytoplasm from the pancreas of a bat. Note the distinct double membrane (DM) of the mitochondrion and the numerous foldings (cristae) of the inner membrane that project into the matrix (M) of the mitochondrion. The heavily stained small granules in the matrix are of unknown composition and function. They are not found in all mitochondria. Visible at the left from top to bottom is a region of rough endoplasmic reticulum (RER) with attached ribosomes (RB). Free ribosomes are also present in the upper portion of the micrograph. A lysosome (LYS) can be seen in the upper right corner. Magnification: 95,000×. (Taken with permission from D. W. Fawcett, *An Atlas of Fine Structure: The Cell*, Philadelphia: W. B. Saunders Company, 1966. Photograph generously supplied by D. W. Fawcett. Original micrograph prepared by Dr. K. R. Porter.)

Mitochondria are responsible for the bulk of the *aerobic* (oxygen-dependent) *metabolism* of the cell, which includes the crucial biochemical processes of the *citric acid cycle* and *oxidative phosphorylation*. Together these activities produce nearly all of the energy required to sustain the growth and viability of the entire cell. Among other important activities known to be associated with mitochondria is the degradation of fatty acids by the process of β *oxidation*. The details of all three of these processes will be explored in subsequent chapters. Although most of the enzymes of the citric acid cycle and β oxidation are contained in the mitochondrial ma-

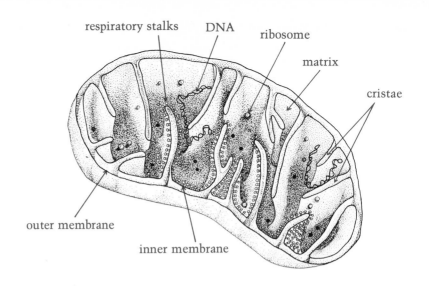

respiratory stalks   DNA   ribosome   matrix   cristae

outer membrane

inner membrane

The mitochondrion. The drawing shows features usually associated with mitochondria. Note that the cristae are formed from folds in the inner membrane. (Adapted with permission from R. D. Dyson, *Cell Biology*, 2nd edition, Boston: Allyn and Bacon, Inc., 1978.)

trix, the enzymes and other factors participating in the process of oxidative phosphorylation are known to be compartmentalized within the inner membrane, particularly within the cristae. In the mitochondrial sketch the mushroom-shaped subunits (respiratory stalks) attached to the cristae (on the side facing the matrix) are associated with this all-important process, for which additional details are presented in Chap. 14 (see p. 467).

As indicated in the drawing, the matrix of the mitochondrion also contains DNA. The amount is small, about 2–4% of the total DNA in the cell. It has been proposed that the mitochondrial DNA contains much of the necessary genetic information required for the development of this organelle. The presence also in mitochondria of ribosomes and some enzymes involved in nucleic acid biosynthesis supports this possibility. In other words, within the main cell the mitochondrion may be a secondary minicell capable of producing most of its own proteins from its own genetic program. Given this, it has been argued further that mitochondria may have evolved from primitive bacteria. Their presence in eukaryotic cells may have occurred via an infection of a developing eukaryotic organism early in its evolution by a parasitic, mitochondrionlike bacterial cell. At this stage, the primitive eukaryotic cell may have acquired a natural selective advantage in that the activities of the host cell and the invading bacterial cell proved to be complementary, and the two may then have continued to evolve as a unit cell.

Endoplasmic Reticulum

Whereas the nucleus and the mitochondrion were first observed with light microscopy, the discovery of the endoplasmic reticulum had to await the development of electron microscopy. First observed in 1953, it is now recognized as an organelle that occurs in nearly all types of higher plant and animal eukaryotic cells. What is it? Well, the simplest description is that it is a *netlike system* (reticulum) of flattened, *membrane-bound regions* that are localized within the cytoplasm (endoplasmic) of the cell. Thus, the endoplasmic reticulum is not a singular, highly ordered entity as the nucleus and mitochondrion are, but an irregular and interconnected array of membranous vesicles. In many cells it is quite profuse, occupying much of the available intracellular space. Two types of endoplasmic reticulum are known. One is called *rough endoplasmic reticulum* (RER) due to the presence of small dense granules that appear to be attached to the outer surface of the membrane vesicle. The granules are *ribosome* particles. The second type is termed *smooth endoplasmic reticulum* (SER) and is characterized by the absence of attached ribosomes. In schematic cross section, each type would appear as already shown in Fig. 1–5, with solid lines representing the membranous system and the black dots on the RER representing the ribosomes. The appearance of the reticulum, particularly the rough type, under the electron microscope is represented in Fig. 1–6 (surrounding the nucleus), in Fig. 1–7 (to the left and lower right of the mitochondrion), and at greater magnification in the micrograph shown here in the margin.

In addition to a variety of enzymes that are localized in the reticular membranes, the reticulum has functions in the biosynthesis of proteins (this is confined to the rough type with its attached ribosomes) and in the storage and transport of proteins that are destined for secretion from the cell. After their synthesis proteins enter the inner cavity of the reticulum (called the *cisternal space*) and then move through the catacomb array of cisternae to the surface of the cell. Secretion occurs in the form of small vesicles of packaged proteins that are pinched off the SER. Alternatively protein transport proceeds through the cisternae of the SER to the Golgi body (see below), which packages the protein into the secretory vesicles. A fact consistent with these roles is that the reticulum, especially the rough variety, is most abundant in those cells known to be specialized sites for the synthesis of various proteins. For example, the partial views in Figs. 1–6 and 1–7 (both of which show a high RER content) are of a pancreas cell, which daily produces and secretes rather large amounts of several enzymes that participate in the digestion of ingested foodstuffs by mammals. It does not appear, however, that the RER is the cellular site for the biosynthesis of all protein. The bulk of the proteins produced by the cell for use within the cell itself are believed to be assembled at ribosomal clusters found free in the cytoplasm. Supporting this notion is the observation that significantly smaller levels of RER are found in those cells where little protein is produced for secretion. A region of free ribosomes is seen at the top right of the micrograph shown in Fig. 1–7.

When the whole cells are disrupted by most methods, the reticulum network is randomly cleaved and smaller fragments are eventually isolated rather than an intact reticulum system. The reticulum pieces are termed *microsomes*.

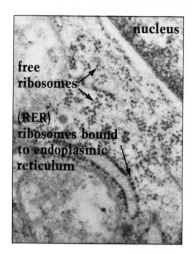

High magnification (130,000×) view of section of rough endoplasmic reticulum with bound ribosomes. Many free ribosomes are also visible. (Micrograph supplied by Donald Mullaly.)

Golgi Body

The *Golgi body is* also a network of flattened, membrane-bound vesicles found in the cytoplasm. The membrane surface is of the smooth type. Unlike the endoplasmic reticulum, however, the Golgi body is not an extensive system permeating large regions of the cytoplasm. Rather, it is more restricted in size. Moreover, the vesicles of the apparatus are frequently stacked in a small cluster and are usually localized near the nucleus or near the apex of specialized secreting cells. Its electron microscopic appearance in cross section is shown in Fig. 1–8. (Refer to Fig. 1–5 for an idealized representation.)

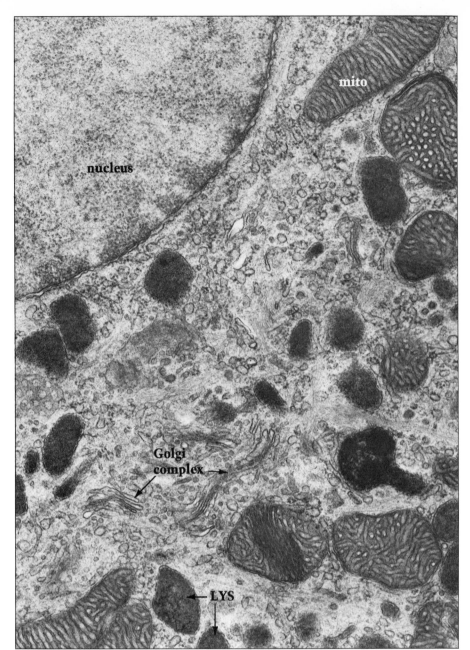

FIGURE 1–8 An electron micrograph of a cell interior showing several organelles. Clearly visible are lysosomes (LYS), three distinct Golgi bodies, several mitochondria (MITO), and a portion of the nucleus. The specimen was obtained from the suprarenal cortex of a hamster. Magnification: 25,000×. (Taken with permission from D. W. Fawcett, *An Atlas of Fine Structure: The Cell*, Philadelphia: W. B. Saunders Company, 1966. Photograph generously supplied by D. W. Fawcett.)

As stated earlier, one of the known functions of the Golgi apparatus is that it accepts proteins from the RER-SER system for concentrating and packaging into dense granules, which are then secreted from the cell. Another function is that it concentrates certain proteins and packages them in membrane-bound bodies called *lysosomes* (see below), which remain in the cell. The Golgi body has also been implicated as the cellular site for the biosynthesis of complex carbohydrate materials, which are ultimately secreted from the cell and then deposited in the exterior coating of the cell. Considerable research is in progress to clarify these roles and to investigate others. Since the Golgi apparatus is present in cells that do not specialize in secreting protein as well as in those that do, it may have a general and essential function in the biochemistry and physiology of the cell.

Lysosomes and Peroxisomes

Lysosomes and *peroxisomes* are small vesicles characterized by a single confining membrane and a granular matrix. Both particles are membrane-bound packages of protein, but different types of proteins are found in each. Discovered in 1952, lysosomes (see Fig. 1–8) contain various types of enzymes, several of which degrade proteins and nucleic acids by catalyzing the hydrolysis (bond breaking by water) of bonds in these polymers. As long as such enzymes are confined within the lysosome they are in a latent state because they are isolated from the substances they degrade. Obviously, if these types of enzymes were not segregated after their synthesis, the cell would be faced with a built-in, self-destruct capability, since the cell's own nucleic acids and proteins would be the target of the hydrolytic action of the enzymes. However, under certain conditions (which we will not go into here) the lysosome enzymes are released to catalyze the intracellular degradation of materials that enter the cell from the outside in unhydrolyzed form. The degradation products are then used in the general metabolism of the cell. The known occurrence of some 30 mammalian diseases caused by deficiencies of single lysosomal enzymes is strong evidence that the function of lysosomes is rather important to a normal cell.

Discovered in 1966, peroxisomes (see Fig. 1–9) contain a variety of enzymes that participate in reactions that either produce or degrade peroxides (such as hydrogen peroxide, H_2O_2); hence their name. The most prevalent enzyme that occurs is *catalase*, which catalyzes the conversion of toxic H_2O_2 to H_2O and O_2. Although peroxisomes are found in almost all types of eukaryotic cells, aside from detoxification their exact biological function is yet unknown. There is some evidence that links certain peroxisome enzymes to the metabolic conversion of lipids to carbohydrates in plants and bacteria. In animals they may also be associated with lipid metabolism.

Cell Membrane

Every cell is surrounded by a single membrane that is referred to as the *plasma membrane* or simply the *cell membrane*. It is obviously a distinct part of the cell, but there is evidence (see Fig. 1–10) that the cell membane is not completely separated from other membranes in the cell. It is quite clear in the micrograph of Fig. 1–10 that the main plasma membrane is *continuous* with the outer membrane of the nucleus. It may very well be that this continuity applies to all intracellular membranes. That is to say, all of the individual membranes throughout a eukaryotic cell are specialized segments of a single massive membrane system, part of which is the plasma membrane, another part the nuclear membrane, another part the endoplasmic reticulum, and so on. This continuity does not mean, however, that there is a continuity of chemical composition, molecular ultrastructure, and biochemical function. In each of these characteristics, individual segments of this massive membrane (i.e., the cell membrane, the nuclear membrane, the Golgi body, and so forth) are unique.

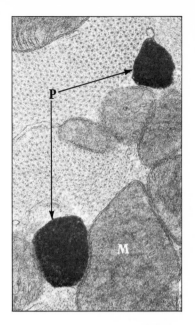

FIGURE 1–9 Peroxisomes (P) as observed in an electron micrograph of a transverse section of mouse myocardial fiber. The peroxisomes appear here as very electron-opaque bodies because the tissue was incubated with a catalase-specific staining mixture prior to preparation for electron microscopy. Obviously then these peroxisomes are rich in catalase. The significance of the close association of peroxisomes with mitochondria (M) is not clear. Magnification is 42,000×. (Reproduced with permission from *Science*, **185,** 271–273 (1974). Copyright 1974 by the American Association for the Advancement of Science. Photograph generously supplied by Dr. H. Dariush Fahimi.)

29

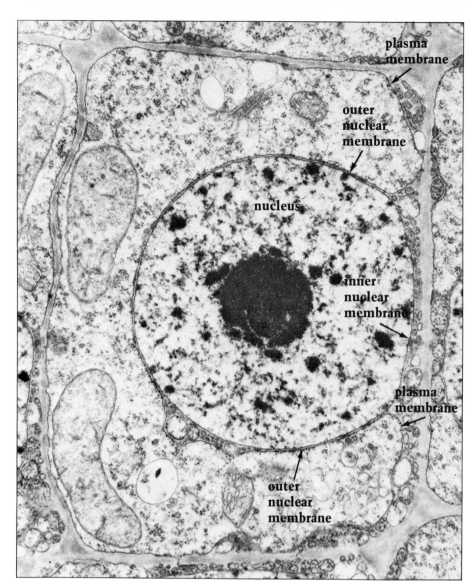

FIGURE 1–10 The continuity of the plasma membrane and the outer nuclear membrane can be seen at two and four o'clock. (Reproduced with permission from Carothers, Z. B., "Membrane Continuity between Plasmalemma and Nuclear Envelope in Spermatogenic Cells of Blasia," *Science,* **175,** 652–654 (1972). Photograph supplied by Dr. Carothers.)

In addition to partitioning one cell from another and providing it with some protection, the cell membrane also plays an important role in regulating the passage of materials into and out of the cell. This is true, of course, for all of the membranes of both eukaryotic and prokaryotic cells and also the membranes of subcellular organelles. In recent years, the list of other functions attributed to the membrane has been growing. These functions vary from species to species, from cell to cell, and from one physiological state to another. Most membrane functions involve the participation of highly specific proteins in the membrane called *receptor-proteins,* which operate by binding to a specific substance on one side of the membrane. This event then triggers other events in the membrane and then in the cell, thus signalling the cell to do (or not do) something and controlling the extent of what happens. This and other aspects of membrane biochemistry—including the molecular ultrastructure of membranes—are described more fully in later chapters.

Cell Wall and Cell Surface Coat

In addition to the cell membrane, many plant and bacterial eukaryotic cells also contain additional protection in the form of a cell wall external to the surface membrane. Similar to that found in prokaryotic bacteria, the wall of the higher eukaryotic cells is a rigid, covalent network composed largely of polysaccharide (polymeric carbohydrate) material. In plants the major components are closely packed cellulose fibers, cemented together by other substances. Animal cells generally do not have a cell wall.

In most cells (eukaryotes and prokaryotes alike), the surface cell membrane and cell wall are frequently covered with a viscous (thick) substance called a *cell coat*. The composition of this substance is quite variable from one cell to another, but is usually known to contain complex polysaccharides in conjugation with other constituents such as proteins and lipids. In addition, some of the membrane proteins protrude into this surface layer. In a multicellular tissue, adjacent cells make an irregular contact through their respective coats across the intracellular space (see Fig. 1–5). It has been proposed that this cellular contact serves as a form of communication among neighboring cells, resulting in a control of their growth and division. The phenomenon has been termed *contact inhibition*. Normal cells exhibit this ability but tumor cells do not. There is clear evidence that an alteration in the composition of the cell coat can result in the transformation of normal cells to tumor cells (see Burger reference at the end of the chapter).

Chloroplast

Although eukaryotic cells of higher animals and plants share most of the structural organization described above, many plant cells, particularly photosynthesizing cells, do contain a unique organelle, the *chloroplast* (from the Greek *chloros*, green; *plast*, formed mass). Chloroplasts contain the green *chlorophyll* molecules and are the specialized membrane-bound bodies that function in the crucial processes of harnessing and converting the energy of the sun into metabolic energy, which is then used in the fixation and conversion of atmospheric CO_2 into carbohydrate material (Chap. 15).

The shape and size of the chloroplast are quite variable. In some cells, the chloroplast is similar in both respects to the nucleus—somewhat spherical and very large. In other cells, it is more cylindrical in shape but significantly (2–5 times) larger than a typical mitochondrion. As indicated in Fig. 1–11, the chloroplast also displays a considerable amount of fine inner structure. The most prominent feature is the presence of several electron-dense stackings. These are called *grana* and represent ordered pilings of flattened membranous systems that presumably originate from the main membrane surrounding the periphery of the chloroplast. The white bodies in the interior of the chloroplast represent large storage vacuoles of the carbohydrate end products of photosynthesis. They are not common to all chloroplasts. The structural and functional characteristics of the chloroplast will be discussed in greater detail in Chap. 15.

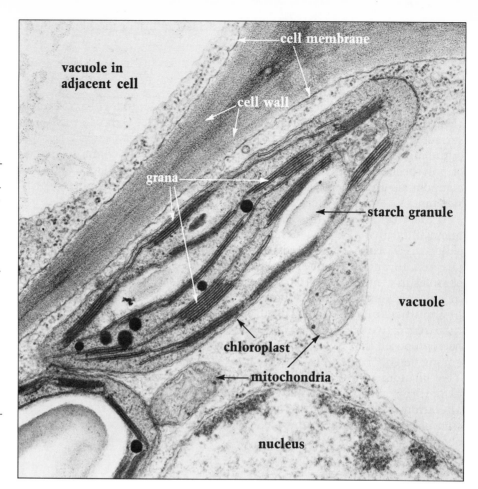

FIGURE 1–11 An electron micrograph showing, in cross-sectional view, pieces of adjacent cells in tomato stem tissue. An intact chloroplast containing several grana can be seen in the lower cell, situated very close to the cell membrane. The thick, cellulose-containing cell wall of each cell is also visible. Other organelles visible in the lower cell are two mitochondria and a portion of the cell nucleus. The large clear region at the right is portion of a vacuole surrounded by a membrane. Vacuoles are present in most plant cells, particularly in older cells, functioning as storage compartments for dissolved sugars, proteins, oxygen, carbon dioxide, and other substances. See Figs. 15–2 and 15–3 for more detail of chloroplast ultrastructure. (Photograph generously supplied by Hilton H. Mollenhauer, Charles F. Kettering Research Laboratory, Yellow Springs, Ohio.)

CELLULAR FRACTIONATION

Biochemical research often requires the isolation of a particular subcellular organelle either (a) to study the organelle intact or (b), as is more commonly the case, to isolate and then study a specific substance from that organelle. The isolation of any cell part in a pure and undamaged state can at present be achieved by well-established, and for the most part, relatively routine fractionation procedures. The general approach consists of two phases: (1) whole cells are disrupted (broken, lysed) to yield a *cell-free system*, frequently termed a *homogenate*; and (2) *differential centrifugation* is used to separate the subcellular parts from each other. If a protein is to be isolated from an organelle, then the isolated organelle is disrupted. The isolation of a protein that is originally localized in a membrane can also be accomplished, but often the procedures must be more tedious to avoid damaging the protein during disruption of the membane.

Cell Disruption

A variety of methods are available to lyse cells. The most common techniques are (a) blending, (b) grinding, (c) exposure to ultrasonic frequencies, (d) osmotic shock, (e) high-pressure extrusion, and (f) treatment with lysozyme. The first three (a,b,c) are generally used in the processing of animal and plant tissue, while the latter four (c,d,e,f) are employed in the lysis of the smaller bacterial cells. Whatever the case, the decision to choose one procedure over another is based on the particular type of cell, the objective of the experimenter, and the mass of the material to be processed. In the absence of published guidelines in the literature, the choice is essentially based on trial and error. What procedure will give the best results? The desired objectives are *maximum disruption of whole cells* and *minimum damage to subcellular components*, particularly the organelles to be studied.

Blending is generally accomplished with electrical devices of various construction, each offering different advantages and disadvantages. The basis of rupture is quite simple; it is a shearing of cellular tissue by rotating blades.

In grinding methods, the cells are merely rubbed against an abrasive and hence against each other. The simplest tools are a mortar and pestle. Typical abrasives are ground-glass beads, sand, or alumina. For the routine processing of small samples, specially constructed glass homogenizers are often used. These consist of a ground-glass barrel with a fitted ground-glass piston providing a clearance of 0.005 inch. On moving the piston up and down the barrel containing the sample, the tissue is forced through this small clearance. A glass and teflon unit is shown in the margin.

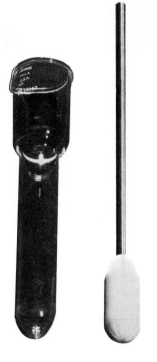

The use of the ultrasonic vibrations is a relatively recent development and is widely used. Two reasons for its popularity are that (a) in most cases the subcellular organelles can be recovered in a reasonably intact, undamaged state, and (b) the severity of the treatment can be finely controlled.

One of the mildest techniques is that of osmotic shock. Here the cells are first suspended in a solution of high solute concentration, causing the migration of water out of the cell. Then they are transferred to pure water, whereupon the water rushes into the cell and it bursts open. With the high-pressure extrusion method, small cells such as bacteria are efficiently and gently broken by forcing a concentrated suspension through a small opening under several thousand pounds of pressure.

The most delicate procedure is to use lysozyme, but unfortunately its use is confined to bacterial cells. Lysozyme is an enzyme that breaks up the rigid cell wall structure of bacteria. The destruction of the cell wall leaves a cell protected only by its membrane, called a protoplast, which is highly susceptible to being ruptured by osmotic shock.

Regardless of the method utilized, all operations are conducted under carefully controlled conditions carried out at reduced temperature (2–4 °C) to minimize damage to the particulate organelles and to proteins.

Separation of Organelles

The separation of the soluble cell fluid from the particulate matter, as well as the further fractionation of the latter by differential centrifugation, is based on a simple principle. Since virtually all of the components in the cell-free system have a different mass-to-volume ratio, that is, a different density, heavier bodies will sediment under low speeds and low gravitational forces, while lighter substances will require higher speeds and higher gravitational forces. Particles of intermediate density will obviously require intermediate conditions. Consequently, an efficient fractionation can be achieved by starting on the low side and performing a series of separate and successive centrifugations toward the high side. Such a scheme is diagramed below. The specified steps and their conditions approximate only a general and simplified situation. The precise conditions are generally more numerous and complex.

Outline of a typical fractionation procedure

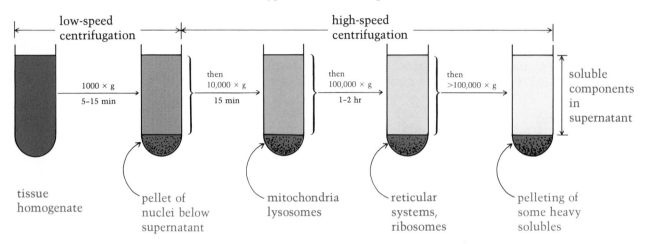

LITERATURE

ALLISON, A. "Lysosomes and Disease." *Scientific American*, **217,** 62–72 (1967). Description of the structure of lysosomes and their function in normal and pathological cells.

BECK, F., and J. B. LLOYD, eds. *The Cell in Medical Science*, Vol. 3. New York: Academic Press, 1976. Considers cells specialized for a variety of organs.

BLOOM, W., and D. W. FAWCETT. *Textbook of Histology*. 10th edition. Philadelphia: W. B. Saunders, 1975. Includes some 1200 illustrations, mostly electron micrographs.

BRACHET, J., and A. E. MIRSKY, eds. *The Cell*. New York: Academic Press. A collection of six volumes on the biochemistry, physiology, and morphology of cells. Volume 2 is devoted to the component parts of cells. Although published in 1960, this is still a valuable reference work.

BURGER, M. M. "Surface Properties of Neoplastic Cells." *Hospital Practice*, **8,** 55–62 (July, 1973). Excellent article (at the introductory level) discussing the transformation of normal cells to tumor cells by manipulation of the composition of the cell surface coat.

DYSON, R. D. *Cell Biology*. 2nd edition. Boston: Allyn and Bacon, 1978. An excellent textbook. Chapter 1 is devoted to cellular and subcellular ultrastructure and has many electron micrographs. Appendix 1 covers the theory and practice of electron microscopy.

FAWCETT, D. W. *An Atlas of Fine Structure: The Cell— Its Organelles and Inclusions*. Philadelphia: W. B. Saunders, 1966. A collection of illustrations, obtained from electron microscopy, of various types of cells. Emphasis given to the major subcellular components of mammalian cells. Text descriptions of the illustrations are also included.

The Living Cell and From Cell to Organism. San Francisco: W. H. Freeman, 1965 and 1967. Two books containing a collection of articles from *Scientific American* on cell biology. Excellent introductory material on a variety of topics dealing with cellular and subcellular ultrastructure and biological function.

MARGULIS, M. "Symbiosis and Evolution." *Scientific American*, **225,** 48–57 (1971). A discussion of the origin and symbiotic evolution of specialized organelles (chloroplasts and mitochondria) of the cells of higher plants and animals.

MAUGH, T. H. "Phylogeny: Are Methanogens a Third Class of Life?" *Science*, **198,** 812 (1977). A brief review of some of the biochemical distinctions of methanogens.

NEUTRA, M., and C. P. LEBLOND. "The Golgi Apparatus." *Scientific American*, **220,** 100–107 (1969). A well-illustrated article describing the ultrastructure and cellular function of this organelle.

SCHOPF, J. W., and D. Z. OEKLER. "How Old are the Eukaryotes?" *Science*, **193,** 47–49 (1976). Evidence that the eukaryotic organisms may have existed as early as 1.5 billion years ago.

SCHWARTZ, R. M., and M. O. DAYHOFF. "Origins of Prokaryotes, Eukaryotes, Mitochondria, and Chloroplasts." *Science*, **199,** 395–402 (1978). A review article describing the tracing of evolutionary history of organisms and organelles using the amino acid sequence of proteins and the nucleotide sequence of nucleic acids.

SJOSTRAND, F. S. *Electron Microscopy of Cells and Tissues.* New York: Academic Press, 1967. First volume of a two-volume work describes the instrumentation and techniques of electron microscopy. Excellent.

UMBREIT, W. W., R. H. BURRIS, and J. F. STAUFFER. *Manometric and Biochemical Techniques.* 5th edition. Minneapolis: Burgess, 1972. Contains an excellent coverage of the many techniques available for the preparation of animal, plant, and bacterial cell-free extracts.

WHALEY, W. G., M. DAUWALDER, and J. E. KEPHART. "Golgi Apparatus: Influence on Cell Surfaces." *Science,* **175,** 596–599 (1972). A review article devoted to the involvement of the Golgi body in the assembly of proteins destined for use outside the cell.

EXERCISES

1–1. Assuming that the overall shape of a single bacterial ribosome is approximately spherical, with a diameter of 200Å, calculate the percent of the total cell volume occupied by ribosomes in a single *E. coli* cell. Assume further that the shape of an *E. coli* cell is approximated by a cylinder about 2 μm long and about 0.8 μm in diameter. (1 μm = 10^{-6} meter.)

1–2. What fraction of the total cell volume of an *E. coli* cell is ocupied by its DNA? Assume that one DNA molecule is present and that the length of the molecule is 1 millimeter. An additional piece of information can be found on p. 222 in Chap. 7.

1–3. Take a piece of paper and cut a 3½ inch circle out of the center. Position the paper over the sketch in Fig. 1–5 and identify the parts of the cell. How many different types of membranes and membranous bodies are depicted?

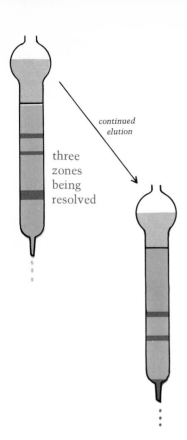

three
zones
being
resolved

*continued
elution*

CHAPTER 2

METHODS OF BIOCHEMISTRY

In any exact science where ideas are tested and discoveries are made in the laboratory, there is a definite correlation between both how much is done and how well it is done and the techniques that are available to perform the necessary work. Thus it is that the rapid growth of biochemistry is—in great part—closely linked to the development of better and new laboratory methods and instruments used in various types of qualitative and quantitative procedures, and also for the task of isolating pure substances. Moreover, many procedures have been automated, permitting laboratory studies to be performed more efficiently and quickly.

In this chapter then, we will examine the distinguishing principles of some major techniques vital to biochemical research. This can be done without extensive knowledge of the nature of biologically occurring compounds or systems. In fact, in the study of this material you will learn a few preliminary things about the chemical nature of proteins that will be amplified in future chapters. It is anticipated, however, that you will refer to these pages again and again, and you should. Only by repeated crossreference will you achieve maximum value from this chapter.

The methods we will examine—*ultracentrifugation, electrophoresis, chromatography, spectroscopy, radioisotope labeling, radioimmunoassay,*

TABLE 2–1 **Applications of methods.**   Methods of Biochemistry

Method	Applications	For the study of
Ultracentrifugation (there are various procedures)	Separation and isolation Determination of purity Measurement of molecular size and shape Nucleotide composition of nucleic acids	Primarily polymers, that is, high molecular weight substances
Electrophoresis[a] (there are various procedures)	Separation and isolation Determination of purity Measurement of molecular size To learn something about the ionic nature of a substance	Both polymeric and smaller substances that are ionic in nature
Chromatography[a] (there are various procedures)	Separation and isolation Determination of purity Measurement of molecular size	Both polymeric and small substances
Spectroscopy[a] (there are various procedures)	Identification of the structure of substances Quantitative measurements on the amount of a substance	Both polymeric and smaller substances
Radioisotopes	Tracing the paths of synthesis and degradation of substances Quantitative measurements at very low levels	Potentially any substance
Radioimmunoassay[a]	Quantitative analysis of specific substances in very small amounts	Potentially any substance for which an antibody can be prepared
X-Ray diffraction	Determining the complete three-dimensional structure of a molecule	Any substance that can be crystallized
Electron microscopy (described in previous chapter)	To see images of whole cells, subcellular parts, viruses, membranes, multimolecular aggregates, and even individual molecules of very high molecular weight	

[a] Many analytical procedures performed in clinical laboratories use these methods.

and *X-ray diffraction*—are used extensively in biochemical research. However, they are not the only ones; nor should they be considered as methods used exclusively by the biochemist.

Table 2–1 summarizes what these methods are used for. You will note that more than one method is used for separation, isolation, purity determination, and the measurement of structural characteristics (such as size and shape) of substances. The point is that a lot of biochemical research involves these objectives. For example, before you proceed with a study on the activity or structure of any protein you must *first isolate it* from its natural source. Note also that some of these methods are in routine use in clinical laboratories for both qualitative and quantitative analyses of various constituents of body fluids (blood and urine).

ULTRACENTRIFUGATION

The separation of substances from a solution by subjecting them to a centrifugal force under high rotational speeds is called *centrifugation*. It is one of the oldest and one of the most common techniques used in all types of laboratories. In biochemical applications the substances may be cells, subcellular organelles, or large polymeric molecules—all of which we will call "particles" to simplify the terminology.

The instrument, of which there are many types and sizes, is called a centrifuge. Tubes containing the liquid sample are placed in a rotor; the rotor is mounted on a shaft that spins very fast; under the influence of the centrifugal force that results, materials move toward the bottom of the tube (they sediment). Factors that determine the rate of sedimentation of a substance are revealed in the Stokes' law, which states that the velocity v for the settling of a spherical particle in a gravitational field is given by the equation

$$v = \frac{d^2(\rho_p - \rho_l)}{\mu} \times g$$

where
v = sedimentation rate of the particle
d = diameter of the particle
ρ_p = density of the particle
ρ_l = density of the liquid in which movement occurs
μ = viscosity of the liquid medium
g = gravitational force applied

From this equation, it can be seen that

1. The sedimentation rate of a given particle is directly proportional to the size of the particle.
2. The sedimentation rate is directly proportional to the difference between the density of the particle and the density of the liquid medium.
3. The sedimentation rate is zero when the density of the particle is equal to the density of the liquid medium.
4. The sedimentation rate decreases as the viscosity of the liquid medium increases. This is due to greater frictional forces that would retard sedimentation.
5. The sedimentation rate increases as the gravitational force field increases.

An *ultracentrifuge* is distinguished by the production of intense forces (in the range of 50,000–500,000 times the force of gravity—$\times g$) at extremely high speeds (anywhere from 20,000–70,000 revolutions per minute—rpm). Two instrument types are available, a *preparative* ultracentrifuge and an *analytical* ultracentrifuge. The preparative instrument, which permits the processing of large volumes of material, is used primarily for isolation purposes. The analytical instrument, which uses a very small volume of material (1–3 ml or less), is used to determine molecular weight (size), shape, density, and purity. Both types have greatest application in the study of large biomolecules (biopolymers), particularly the proteins and nucleic acids.

Depending on the density of the liquid medium in which sedimentation occurs, all techniques of ultracentrifugation can be classified as *differential* or *density-gradient* methods. In the differential methods, the sedimentation of particles occurs in a medium of unchanging uniform density. In density-gradient methods, sedimentation occurs in a medium with a gradually changing density, increasing with distance from the center axis of rota-

The description in the previous chapter (p. 33) of the use of centrifugation to fractionate a cell-free homogenate is an example of differential centrifugation with a preparative instrument.

tion. Either mode of operation is possible with a preparative instrument (see note in margin).

The differential mode of operation applies to the use of an analytical ultracentrifuge. One particular method is called the *sedimentation velocity* technique for which the essential events are diagrammed on p. 40. The drawing depicts what is happening in the open chamber of the centrifuge cell. (The term cell is used to refer to the small chamber in which a sample of the solution is placed.) The margin shows a photo of an actual cell and the rotor in which the cell (in its housing) would be placed. The instrument itself is shown in Fig. 2–1.

During centrifugation different regions of composition with distinct *boundaries* are produced. The boundary represents the point where one region ends and the other begins. If the original solution consisted of only one dissolved solute (that is to say, it was a pure solution of a single protein, for example), only one boundary would be produced. If the original solution consisted of two dissolved solutes (it was not a pure solution of either one; that is, it was impure), two boundaries would result.

An analytical ultracentrifuge is capable of detecting these boundaries and following the rate of their displacement prior to complete pelleting at the bottom. Boundaries can be detected (diagram, p. 41) by passing light through the entire length of the cell and measuring the variation in light absorption or light refraction along the length of the compartment. At the end and

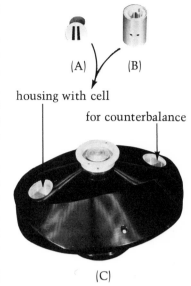

(A)   (B)

housing with cell

for counterbalance

(C)

Analytical ultracentrifuge cell and rotor. (A) Top view of the cell, containing two compartments, one for the sample and one for the reference solution; (B) the housing in which the cell is placed; (C) the rotor.

eye-viewer

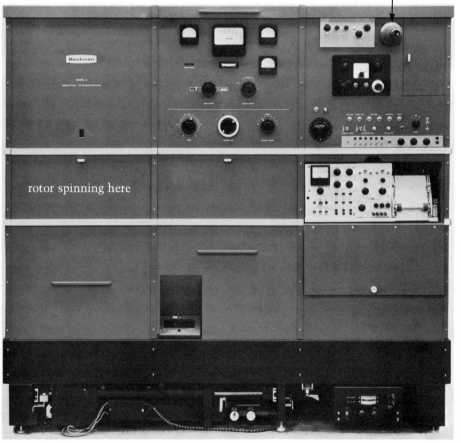

rotor spinning here

FIGURE 2–1 Model E analytical ultracentrifuge. The eye-viewer (upper right) permits the visual observation of the progress of sedimentation in the centrifuge cell as the rotor (see margin) spins at high speeds under an intense gravitational force in an evacuated and refrigerated chamber. The price of the instrument ranges from $35,000–65,000, depending on accessories. See Fig. 2–2 for photographs taken with this instrument. Dimensions of instrument: 74 × 27 × 60 inches. (Photograph supplied through the courtesy of Beckman Instruments, Inc.)

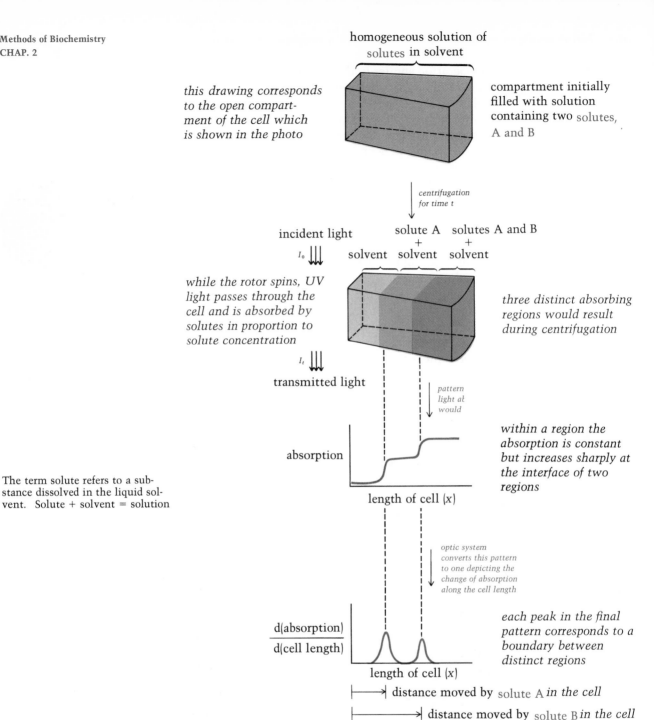

homogeneous solution of
solutes in solvent

*this drawing corresponds
to the open compart-
ment of the cell which
is shown in the photo*

compartment initially
filled with solution
containing two solutes,
A and B

*centrifugation
for time t*

incident light

I_0

solute A   solutes A and B
   +   +
solvent   solvent   solvent

*while the rotor spins, UV
light passes through the
cell and is absorbed by
solutes in proportion to
solute concentration*

*three distinct absorbing
regions would result
during centrifugation*

I_t

transmitted light

*pattern
light at
would*

absorption

*within a region the
absorption is constant
but increases sharply at
the interface of two
regions*

length of cell (x)

The term solute refers to a substance dissolved in the liquid solvent. Solute + solvent = solution

*optic system
converts this pattern
to one depicting the
change of absorption
along the cell length*

$$\frac{d(absorption)}{d(cell\ length)}$$

*each peak in the final
pattern corresponds to a
boundary between
distinct regions*

length of cell (x)

⊢——→⊣ distance moved by solute A *in the cell*

⊢————→⊣ distance moved by solute B *in the cell*

beginning of two regions a distinct transition in absorption or refraction
would occur. A system of lenses then analyzes the rate of change in absorption or refraction with respect to the length of the cell, producing an image
of a *peak*. This image can be viewed and photographed at any time during
the run. If only one sharp, symmetrical peak is observed at all times, the
conclusion is the sample is pure (see Fig. 2–2). Were it impure, two distinct
peaks or a skewed peak would be observed.

sedimentation

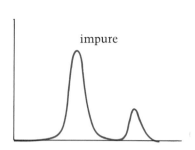

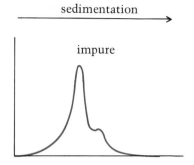

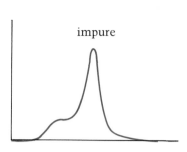

From the theory of ultracentrifugation an equation can be derived that relates the size (that is, molecular weight) of a substance to its rate of sedimentation. That equation is

$$\text{MW} = \frac{RTs}{D(1 - \overline{V}\rho)}$$

where

MW = molecular weight of the solute (units are g/mole)

R = molar gas constant (value is 8.3×10^7 g-cm²/sec²/deg/mole)

T = absolute temperature; that is degrees Kelvin, which equals °C + 273

D = diffusion constant of the solute (units are cm²/sec)

$\overline{V}$ = partial specific volume of the solute; this is the volume resulting from the addition of 1 gram of the solute to a large volume of the solvent (units are ml/g)

ρ = density of the liquid solvent (units are g/ml)

s = sedimentation coefficient of the solute (unit is sec)

The value of the sedimentation coefficient (s) is what is measured with the analytical ultracentrifuge. The molecular weight is then calculated by plugging in the value of s along with values for R, T, D, $\overline{V}$, and ρ. The values of D, $\overline{V}$, and ρ must be experimentally measured by other methods. Of these, D is the most difficult to measure. Familiarize yourself with this computation by solving Exercise 2–1 at the end of the chapter.

The sedimentation coefficient of most biopolymers is on the order of 10^{-13} sec. To eliminate the cumbersome exponent, the s value is normally multiplied by 10^{+13}, and the resulting number is designated as S, called the *Svedberg constant* in tribute to T. Svedberg, who pioneered in the development of the analytical ultracentrifuge. For example, a solute with a sedimentation coefficient of 2.5×10^{-13} sec would have a 2.5S value. As the particle

$$s = \frac{dx/dt}{\omega^2 x}$$

where
ω is the angular velocity of the rotor in units of radians per second and x is the distance in centimeters from the solute boundary to the center of rotation.

The determination of s is made by measuring (a) the rate of sedimentation of the solutes (that is, how fast the peak boundary is moving) at different times during the run and (b) the actual distance it has moved in those times.

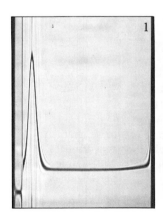

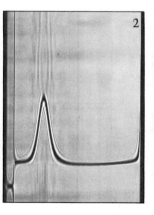

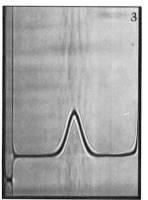

FIGURE 2–2 Analytical photographs of a solution consisting of a pure protein; note the single symmetrical peak. The displacement of the peak from left to right represents the displacement of the single boundary with increased time of centrifugation. (Photographs supplied through the courtesy of Beckman Instruments, Inc.)

becomes larger, the S value increases. (While working on the ultracentrifuge, Svedberg was awarded the Nobel Prize (1926) for his early studies in solution chemistry.)

Let us now turn our attention to density gradient methods, wherein the sedimentation of substances occurs in a solvent medium that has a gradually changing concentration. Remember, in the previous discussion the concentration of the solvent medium was the same throughout the length of the cell and the duration of the run. There are two types of density gradient methods, depending on when the gradient is formed in the centrifuge tube. (Cylindrical centrifuge tubes of various sizes rather than the small analytical cells are used in these methods.) Either method is commonly used in the operation of a preparative ultracentrifuge. The *rate zonal* technique utilizes a *density gradient that is formed prior to centrifugation*. The gradient itself can be formed manually or by automatic devices. Usually sucrose solutions are used to obtain the different densities. A small volume of the sample under study is then layered on top of the sucrose gradient. As illustrated below, the result of centrifugation is that solutes having different sedimentation characteristics move through the medium as discrete zones or bands. The operational significance of centrifuging according to the density gradient principle is obvious—separation of solutes can be achieved. By stopping the centrifuge before the zones merge together and pellet at the bottom of the tube, individual zones can be isolated by (1) merely puncturing the tube bottom and collecting small fractions in a dropwise fashion or by (2) freezing the tube and then slicing it in small sections.

An elegant variation of this theme is the *isopycnic* technique in which the *density gradient is formed during centrifugation*. Initially the entire centrifuge tube is uniformly filled with a homogeneous mixture containing the solute(s) under study, which are dissolved in a solution of a very dense inorganic salt such as cesium chloride. As illustrated below, the events during centrifugation are that (1) a density gradient of the cesium chloride is

Density gradient techniques

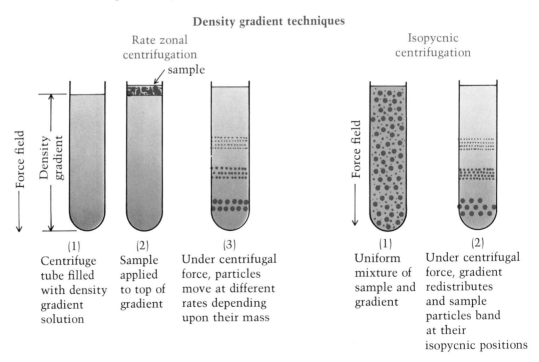

Rate zonal centrifugation

sample

Isopycnic centrifugation

Force field

Density gradient

Force field

(1) Centrifuge tube filled with density gradient solution

(2) Sample applied to top of gradient

(3) Under centrifugal force, particles move at different rates depending upon their mass

(1) Uniform mixture of sample and gradient

(2) Under centrifugal force, gradient redistributes and sample particles band at their isopycnic positions

self-generating and (2) as long as the density of the solutes is within the range of this gradient, individual solutes will seek out a position in the tube where the density of the gradient is equal to their own density. A condition of equilibrium is attained. The bands will remain as long as the gradient remains. The equilibrium method has been extremely valuable in the study of the nucleic acids RNA and DNA.

There are other methods of ultracentrifugation and also many examples of applications besides those of determining size and purity and achieving separations. Indeed, reports of new applications appear regularly in the research literature. This is of course true of most laboratory methods. Limitations of space and time prevent any further discussion. Those interested can get more information from some of the sources listed at the end of this chapter.

ELECTROPHORESIS

Electrophoresis involves the movement of a charged particle (ion) in an electric field. It is one of the most effective methods for separation, isolation, and determination of size and purity. There are also other applications. The basic requirement for electrophoresis is that the substance(s) under study be capable of existing in a charged (negative or positive) state. There are various types of electrophoresis, each having its own advantages and disadvantages for specific applications. One of the most popular procedures is called *polyacrylamide gel electrophoresis.* We will examine it after a brief consideration of basic theory for electrophoresis in general.

In electrophoresis the movement of substances occurs in a liquid medium that is supported by an inert solid substance such as paper or a semisolid gel substance. The liquid serves as a conducting medium for electric current when an external voltage V is applied. The extent of movement of a charged substance (molecule) in an electric field is termed its *electrophoretic mobility* (symbolized μ), and for a spherical molecule not experiencing any strong electrostatic interaction from surrounding ions, it can be shown that

$$\mu = \frac{Q}{6\pi\eta r} V$$

where Q = the net charge on the molecule
   r = the radius of the molecule in cm
   η = the viscosity of the liquid medium in which movement occurs
   V = the applied voltage

This fundamental equation reveals the essence of an electrophoretic analysis. If we assume (1) that the conducting liquid solution remains unchanged during the analysis and thus the viscosity of the medium is constant throughout the procedure, and (2) that the applied voltage is held constant, then we can say that the migration of a particle is controlled by the ratio of two variables, specifically the *charge-to-size* ratio. That is,

$$\mu \propto \frac{Q}{r}$$

As this ratio increases, the mobility increases. In considering a mixture of several charged molecules, if we further assume that the components of the

The technique of electrophoresis was first developed by Arne Tiselius approximately 45 years ago. Tiselius, who early in his career was a research assistant to Svedberg, was awarded the Nobel Prize in 1948.

$$CH_2 = CH$$
$$|$$
$$C = O$$
$$|$$
$$NH_2$$

acrylamide
(monomer)

+

$$CH_2 = CH$$
$$|$$
$$C = O$$
$$|$$
$$NH$$
$$|$$
$$CH_2$$
$$|$$
$$NH$$
$$|$$
$$C = O$$
$$CH_2 = CH$$

bis-acrylamide
(cross linker)

polymerization

acrylamide
chain

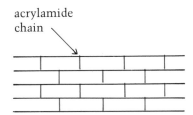

diagram of
the molecular structure of the
polyacrylamide gel; the
cross-linking contributes to
the strength of the gel (it is not a
jelly) and confers a porous
network to the gel medium

mixture do not differ appreciably in molecular size, the equation is reduced to a direct relationship of the form

$$\mu \propto Q \qquad \text{(for different molecules of the same size in the same medium)}$$

This is nothing more than a statement that the migration of charged molecules similar in size is directly dependent on their respective charges. The greater the charge, the greater the migration.

Polyacrylamide gel electrophoresis (PAGE) is typical of what is called a *zonal technique* of electrophoresis. This means that a solid or semisolid system is used to support the liquid medium, and this permits a small amount of the sample to be applied. The result is that molecules of the same charge-to-size ratio will move through the medium as a single *zone.* If the sample consists of two substances of different Q/r values and therefore different μ values, two zones would result, and so on. Polyacrylamide gel electrophoresis is so named simply because the liquid medium is supported on a gel system of polyacrylamide, a copolymer of acrylamide and bis-acrylamide (see margin). When these two materials copolymerize, a highly cross-linked (netlike) polymeric matrix is formed. The physical state of the polymerized material is a transparent semisolid gel that can be handled easily. It is important to recognize that the gel matrix is porous because of the cross-linked structure. This porous network provides channels through which molecules must move. Obviously smaller molecules would move more readily than larger ones.

The stages of a PAGE analysis are diagramed on p. 45. The events depict a procedure that uses a small tube supported in a vertical position. Since individual substances migrate through the tube column as sharp, narrow zones, the method is also called *disc gel electrophoresis.* The resolving power of this method in analyzing complex mixtures can be appreciated by inspecting Fig. 2–3.

The composition of the original sample can be determined merely by measuring the amount of material in each zone. The sensitivity of detection is excellent. For example, as little as 5–10 μg (micrograms) of protein in a zone can be detected. The method can be applied on a large scale (that is, using larger tubes and applying a larger sample) to achieve the isolation of a specific substance. In this case the zone containing the substance of interest is simply cut out. It should be obvious how the method can be used to ascertain the purity of a substance—the stained gel will show only a single band.

In *SDS-polyacrylamide gel electrophoresis* the porous matrix of the gel is exploited to achieve separation primarily on the basis of differences in size of particles; this permits a measurement of molecular weight of a substance. The method is particularly useful in the study of proteins. Sodium dodecyl sulfate (SDS, an ionic detergent) is, for example, used to bind to proteins and convert them to forms that have the same charge/mass ratio. This means that differential electrophoretic mobility is based then on the differential passage of the SDS-protein complex through the channels of the porous gel matrix. In effect the gel functions as a molecular sieve. The result is that the relative mobility of the protein is a log function of the molecular weight of the protein. The molecular weight of a protein is determined by comparing its mobility relative to the mobilities of other proteins of known molecular weight—with all mobilities measured under the same conditions (see Fig. 2–4, p. 46).

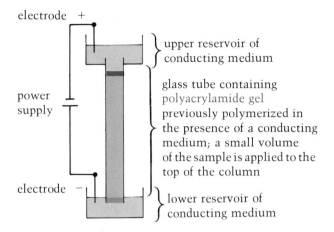

electrode +

power supply

electrode −

upper reservoir of conducting medium

glass tube containing polyacrylamide gel previously polymerized in the presence of a conducting medium; a small volume of the sample is applied to the top of the column

lower reservoir of conducting medium

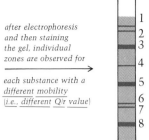

after electrophoresis and then staining the gel, individual zones are observed for

→

each substance with a different mobility (i.e., different Q/r value)

1
2
3
4
5
6
7
8

this shows eight zones, implying that the original sample contained eight different substances—three major ones and five minor ones; obviously, the sample was not pure

the initial setup looks like this, although the actual apparatus can accommodate several tubes simultaneously; as implied, the sample in this case consists of materials that are positively charged and migration occurs toward the reservoir of opposite charge; see Fig. 2-3 for size of tube

substance 8 has the greatest mobility and 1 has the smallest; if all substances were of the same (or nearly the same) size, then 8 has the greatest net positive charge and 1 has the smallest; whatever the case, the point is that substances with different μ values will exhibit <u>differential movement</u>

Another method is *paper electrophoresis,* in which a strip or sheet of filter paper is used as the inert solid support. The paper, uniformly moistened with the conducting medium, is suspended between two electrode compartments also containing the conducting medium. Samples are applied to the paper at the designated origin near either end or at the middle. Under an applied voltage the solutes would move as zones toward the appropriate pole. The zones are not as sharp and narrow as in disc gel electrophoresis. Despite this and other disadvantages, the technique is useful in the study of mixtures containing low molecular weight solutes such as amino acids, peptide fragments of proteins, and nucleotide fragments of nucleic acids. Figure 2–5 illustrates the capability of the method for a mixture of four amino acids electrophoresed at pH 6. At present it is not important to understand why each of these amino acids has a different net charge. We will examine the reasons in Chap. 4.

Two other zonal methods use *starch* or *cellulose acetate* as the solid supporting medium. Cellulose acetate strips are used extensively in clinical laboratories for the quantitative analysis of blood proteins.

CHROMATOGRAPHIC METHODS

At present, chromatographic methods are undoubtedly the most widely used of all separation techniques. This popularity is due to a combination of factors. First, chromatographic techniques do not necessarily require expensive apparatus; second, most of the methods require a minimum of technical skill; third, most of the methods have a good capability of separating similar substances in a homogeneous mixture, which is to say that the resolving power of some methods is very high; and fourth, there is now available a large selection of different chromatographic procedures that permit the resolution of all classes of biological compounds.

FIGURE 2–3 Photo of stained polyacrylamide gel of human serum proteins. Twenty bands (some very faint in this reproduction) can be detected. The standard length of gel in PAGE is about 3 inches. (EG&G ORTEC)

45

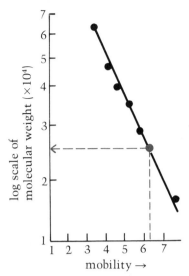

FIGURE 2–4 Estimation of molecular weight by SDS-polyacrylamide gel electrophoresis. The line constructed through black points corresponds to the mobilities of proteins of known molecular weight. Comparing the mobility of protein of unknown size (color point) yields a molecular weight of 25,000.

Before discussing different techniques, it is necessary to first establish the general principle of chromatography. Chromotography is a *process characterized by the uninterrupted flow of a mixture (gas or liquid) through a region of immobilized substance (liquid or solid), which, through various means, allows for the differential migration of the components of the mixture.* This definition refers to two common characteristics of all chromatographic techniques.

1. They are *dynamic processes* involving the continual movement of one phase (the moving phase), which originally contained the mixture, relative to a second phase (stationary phase), which is immobile.
2. Separation is achieved due to differences in the extent to which individual components in the moving phase interact with the stationary phase. This *differential interaction results in a differential movement* of the components in a mixture.

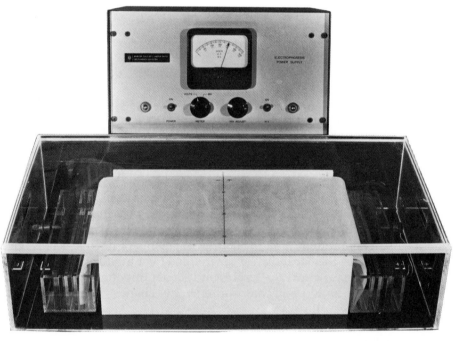

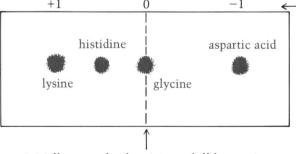

FIGURE 2–5 An actual paper electrophoresis apparatus and a diagramatic representation of the results for a mixture of amino acids. Note that the origin can also be at either the anode or cathode end of the system. Ninhydrin is a substance that produces a blue-violet color when it reacts with amino acids (see p. 101).

46

The feature that distinguishes one chromatographic procedure from another is the basis for the interaction between the stationary and moving phases. This interaction is usually based on one of the following principles: *solubility*, *ion exchange*, *adsorption*, and *sieving*. Some procedures utilize a combination of factors.

A common strategy in chromatographic methodology is to pack the stationary phase in a column and then allow the moving phase to flow through the column (sometimes with the aid of applied pressure). For an obvious reason, such methods are collectively referred to as *column chromatography*. The essential events of a column procedure are diagrammed in the margin. Depending on the application, one can use a column of nearly any size.

Gel-permeation Column Chromatography

Referred to by various names—gel-permeation, molecular-sieve, or molecular-exclusion chromatography—this is perhaps the simplest of all chromatographic techniques to understand, because the basis of differential migration is variation in the size of different molecules. (This assumes that components of a mixture do not differ too greatly in their molecular shape; otherwise, it too is a factor.)

The solid stationary phase consists of tiny gel particles, spherical in shape, that can be regarded as molecular sieves. The particles may be composed of cross-linked polyacrylamide (as just described), or cross-linked dextran, a carbohydrate material. When placed in water, the particles absorb water and swell into larger *granules having a porous network* with the size of the pores determined by the degree of cross-linking. As substances in the applied sample (dissolved in the moving phase) enter the stationary phase they begin to interact with the gel granules by entering the granule through its pores. This interaction with the stationary granules is what will *retard* the movement of the solutes down the column. Small molecules would be retarded more than larger ones because they would enter more easily and permeate more of the interior of granules. In fact, molecules with sizes greater than the molecular dimensions of the pores would not penetrate at all and thus would move, dissolved in the moving phase, right through the column. The essential events are diagramed on p. 48. For the sake of simplicity, only a two-component mixture is shown.

The results of a column chromatographic procedure are usually presented by plotting whatever property the fractions are measured for against the eluted volume as it is collected. The typical profile of such a plot is shown in Fig. 2–6, which shows a peak corresponding to each zone that is resolved. The volume of effluent collected that corresponds to the apex of a peak is termed the *elution volume* of that substance.

In addition to its obvious applications for separation and isolation and the evaluation of purity, this technique can also be used for the measurement of the molecular weight of a solute. In this regard it is a particularly good analytical tool for biopolymers, especially proteins. First a column of swollen gel is calibrated by determining the elution volumes required to displace several proteins of known molecular weight from the column. Then the protein of unknown moleculear weight is passed through the same column in the same eluting solvent, and the elution volume required for its displacement is determined. By comparing this volume to the calibrated elution

reservoir of the eluting solvent, that is, the liquid moving phase

column packed with stationary phase in contact with the eluting solvent throughout its length; a small volume of the sample is applied to the top of the bed of solid phase

liquid emerging (the effluent) from the column is collected

with continued flow of the moving phase, a differential migration of zones of different solutes is observed

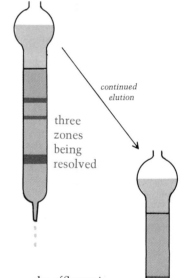

three zones being resolved

continued elution

the effluent is automatically collected and analyzed for the presence of solute; at the right the zone of the fastest moving substance is just emerging

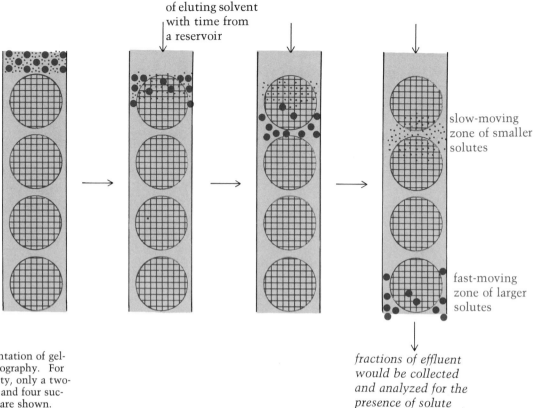

Sample of mixture applied to top of gel column

continuous flow of eluting solvent with time from a reservoir

slow-moving zone of smaller solutes

fast-moving zone of larger solutes

fractions of effluent would be collected and analyzed for the presence of solute

Diagramatic representation of gel-permeation chromatography. For the sake of simiplicity, only a two-component mixture and four successive gel granules are shown. (The exaggerated size and idealized stacking of the granules are not realistic. Actually in any one plane of the column there would be several granules; in the whole column, a countless number.) The grid network represents pores of the granule. After elution begins the smaller solute molecules would enter and penetrate the gel granules to a greater extent than the larger molecules. Because of this greater interaction with the stationary phase, the small molecules have a slower rate of migration. Of course, if any one of the molecules had dimensions larger than the pore size, they would be completely excluded from the gel and move right through the column.

volumes of the control samples, a reasonable estimate of the molecular weight can be made. See Fig. 2–7.

Ion-exchange Column Chromatography

When the solutes to be separated are capable of existing as positively or negatively charged ions, the procedure of ion-exchange chromatography is a very effective technique and one with a high resolving power. The stationary phase consists of a solid substance that actually participates in the process by interacting directly with the components of a mixture. For reasons that will become obvious, these materials are called *ion exchangers*.

The majority of ion exchangers are synthetic materials (called resins) that are fabricated in the form of very small beads (see margin). Chemically each solid bead is composed of large polymeric chains. The ability of the resin material to function as an ion exchanger is due to the presence of several ionizable chemical groupings that are attached along the length of the polymeric chain. Thousands of these groupings would exist in a single bead.

One of the most widely used resins contains sulfonic acid groupings ($-SO_3H$) that are strongly acidic. Usually this resin is used in a salt form (such as $-SO_3^-Na^+$) obtained by passing a solution containing sodium or lithium ions (Na^+ or Li^+) through a column of the resin. When a sample containing positively charged solutes is applied to the top of a packed col-

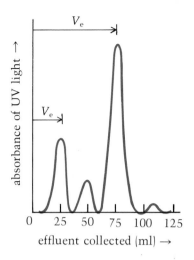

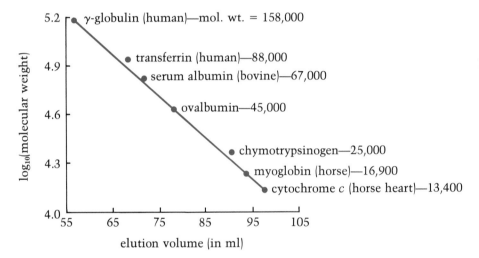

FIGURE 2–6 Elution profile of a column chromatographic separation. The area under each peak is related to the amount of substance eluted in that zone. The elution volume (V_e) of the fastest moving substance is 25 ml. If this were a gel-permeation procedure, this substance would be the one with the largest molecular weight (i.e., largest size). By simply combining the fractions that correspond to a specific peak, you isolate that substance from the others. The measurement of absorption of ultraviolet (UV) light is a common method of detecting the presence of proteins or nucleic acids.

FIGURE 2–7 Gel-permeation column chromatography of proteins. A gel column can be calibrated by chromatographing pure proteins of known molecular weight and plotting the data as shown here (i.e. log scale of molecular weight vs. elution volume). The elution volume determined on the same column for a pure protein of unknown molecular weight can then be used to determine the molecular weight. (Data taken from the *Handbook of Chromatography*, Vol. 1, Cleveland, Ohio: CRC Press, 1972.)

umn and begins to move through the resin bed, the movement of each solute will be retarded as the solute interacts with the resin. The interaction is one of *electrostatic association, in which the positively charged solute exchanges with the Na^+ for the negatively charged $—SO_3^-$ grouping.* As a result of this interaction (an ion exchange, and, specifically, a cation exchange) the solute becomes bound to the resin. Thus, because the resin particles are stationary, the effect of the exchange is to retard the movement of ionic solutes through the column as dissolved materials in the moving phase. Throughout the operation of the column, solutes are continually in a cycle of being bound, released, being bound, released, etc. Obviously, the release into the moving phase is important in order for the solutes to move through the column. The extent of binding will be controlled primarily by the magnitude of the electrostatic force of attraction between X^+ and the resin particle. Consequently, *if a mixture contains a number of solutes, each with a different net positive charge, they will exchange to different degrees with the cation-exchange resin.* Solutes with a greater net positive

A cation is a positively charged ion; an anion is a negatively charged ion.

a photograph of spherical ion-exchange resin particles

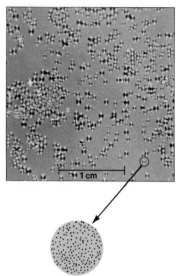

1 cm

each dot represents a functional group (e.g. $-SO_3^-$) attached to the polymeric structure; these groupings are present on the exterior surface and also inside the bead; interior sites would be accessible because the bead is porous; bead particles with a diameter as small as 10 μm (micrometers) are available and they give excellent resolution (see Fig. 2-8)

charge will interact with the resin particles to a greater degree and thus be retarded on the column for a longer time. This difference in exchange constitutes the primary basis of selective and differential migration through the column. (Although migration is affected by other secondary factors, we will not consider them.) As described earlier, the effluent emerging from the bottom of the column can be systematically collected in small-volume fractions, and the presence of a solute can be detected by a suitable procedure.

Various types of ion-exchange resins are available for the study of proteins, nucleic acids, nucleotides, amino acids, and other substances. A specific example of the high resolving power of this technique is shown in Fig. 2–8 (below), which depicts the analysis of amino acids in human plasma.

Anion-exchangers consist of a polymeric matrix containing positively charged functional groupings that are usually substituted aminoethyl groups in chloride salt form, that is

$$\text{polymer} \left(\text{CH}_2\text{CH}_2 - \overset{\displaystyle R}{\underset{\displaystyle H}{\overset{+}{N}}} - R \ \text{Cl}^- \right)_n$$

A popular anion-exchanger is (diethylaminoethyl)-cellulose (DEAE-cellulose) where each R = $-CH_2CH_3$. It is particularly efficient in the ion-exchange chromatography of protein mixtures. With negatively charged solutes(X^-), the basic exchange reaction would be as follows:

$$X^- + \text{polymer} \left(\text{CH}_2\text{CH}_2 - \overset{\displaystyle R}{\underset{\displaystyle H}{\overset{+}{N}}} - R \ \text{Cl}^- \right)_n \rightleftharpoons$$

$$\text{polymer} \left(\text{CH}_2\text{CH}_2 - \overset{\displaystyle R}{\underset{\displaystyle H}{\overset{+}{N}}} - R \ \text{Cl}^- \right)_{n-1} \left(\text{CH}_2\text{CH}_2 - \overset{\displaystyle R}{\underset{\displaystyle H}{\overset{+}{N}}} - R \ X^- \right) + \text{Cl}^-$$

The principle of operation is, of course, the same as for cation-exchangers.

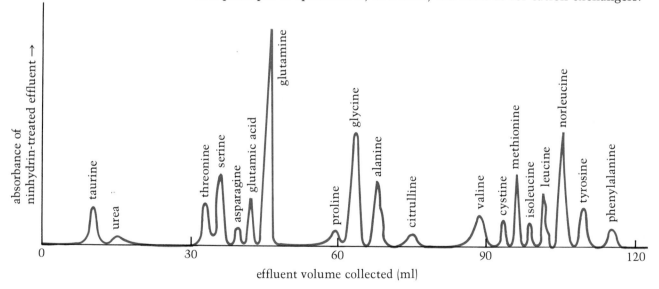

effluent volume collected (ml)

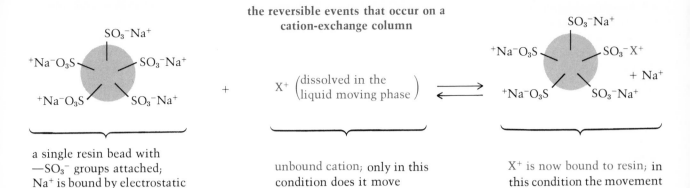

the reversible events that occur on a
cation-exchange column

a single resin bead with
—SO_3^- groups attached;
Na^+ is bound by electrostatic
association

unbound cation; only in this
condition does it move
through the column

X^+ is now bound to resin; in
this condition the movement
of X^+ is retarded

exchange in → direction binds X^+ to the resin
exchange in ← direction releases X^+ back to moving phase

*these events occur many times per bead
and throughout the entire length of
the column with solutes encountering
new beads as they move*

Gas-Liquid Column Chromatography

In gas-liquid chromatography an inert gas such as helium or argon is used as
the moving phase. The gas migrates through a coiled or linear column
packed with an inert solid material. This solid material is coated with a
high-boiling liquid, which serves as the stationary phase. During operation
the complete column is maintained at a high temperature. After injection
of a sample, the solutes in the mixtures are vaporized and dissolved in the
carrier gas. As the carrier gas with the dissolved solutes flows through the
column containing the stationary liquid phase, very sharp zoning (separation) of the solutes occurs on the basis of *differences in solubilities in the
moving gas and the stationary liquid.* The movement through the column
of solutes that are more soluble in the liquid phase is retarded to a greater
degree than the movement of those that are more soluble in the gas phase.
A simplified sketch of the essential components of a gas chromatograph is
shown in Fig. 2–9.

FIGURE 2–8 Cation-exchange
column chromatographic analysis
of free amino acids in deproteinized (protein-removed) human
plasma. The column size was
2.8×300 mm ($\frac{1}{8}'' \times 12''$); the
height of the resin bed was 200
mm; the eluting solvent was a Li^+
salt solution; the rate of flow of
the moving phase was 7 ml/hr;
the analysis was completed in
about $4\frac{1}{2}$ hours.

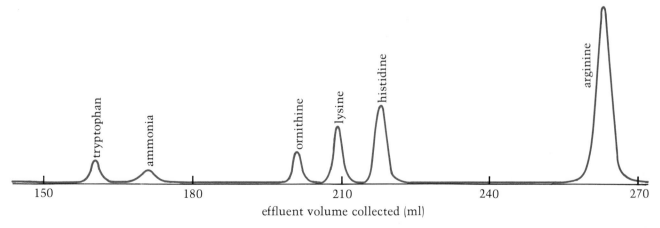

effluent volume collected (ml)

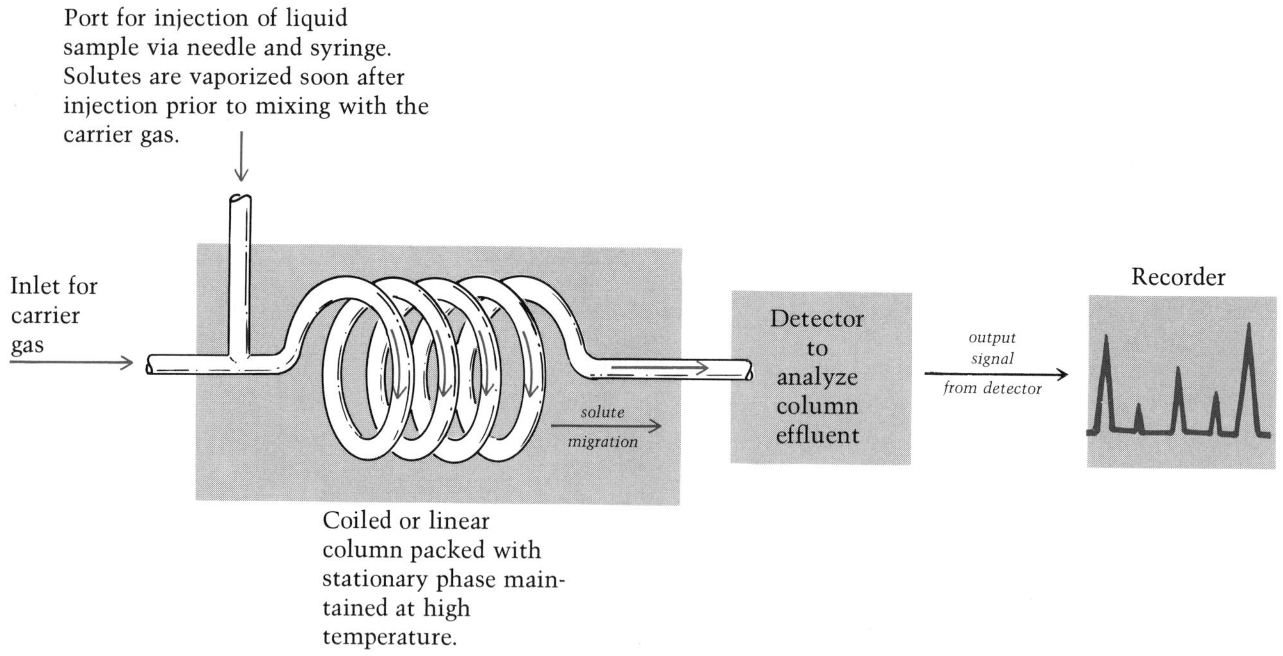

Port for injection of liquid sample via needle and syringe. Solutes are vaporized soon after injection prior to mixing with the carrier gas.

Inlet for carrier gas

Recorder

solute migration

Detector to analyze column effluent

output signal from detector

Coiled or linear column packed with stationary phase maintained at high temperature.

FIGURE 2–9 A schematic diagram of the essential components of a gas chromatograph. Each peak in the recorder printout would represent a different solute.

A variety of methods are available for the detection and quantitative analysis of the compounds in the moving phase as it exists the column. Some devices are extremely sensitive, capable of detecting quantities of material as small as 10^{-12} mole. For a material with a molecular weight of 100 this corresponds to 0.0000000001 gram (i.e., 100 picograms). The resolving power—the clear separation of zones that move close together—is also excellent. Another reason for its popularity is that almost any type of substance can be studied, with the notable exception of high molecular weight substances (such as proteins and nucleic acids), which cannot be vaporized.

Other Column Methods

There is another column technique that uses a solid stationary phase. In this case the solid phase interacts with substances by *adsorption*, which involves the binding of the substances at the surface of the solid particles. Some common adsorbents are silica gel, calcium phosphate gel, and alumina. Other solid materials have been fabricated that operate (i.e., interact with moving solutes) by a combination of principles, such as *sieving and ion exchange*. Depending on the substances under study, one type of column gives better results than another.

Affinity chromatography is a recently developed technique of great importance. It involves the *specific binding of a single substance to the stationary phase*, with all other substances in the mixture moving right through the column. As such, it stands out from all of the other methods, which take advantage of differences in a property that is common to all substances in the mixture. We will defer further discussion of this technique to Chap. 5, where we will discuss proteins, for which this method is ideally suited. The nature of the specific binding will also be better understood at that time.

The technique of chromatogaphy was described for the first time in 1903 by Tswett, a Russian chemist-botanist. Although he called the method chromatography because his studies dealt with the separation of plant pigments (from the Greek *chromato*, meaning color), the procedure applies to all types of chemical compounds, colored or colorless.

In all of the previous methods (except the gas-liquid technique) the stationary phase was a solid confined in a column, and a liquid moving phase flowed through. In paper chromatography the function of the column is assumed by a sheet of paper, which serves as an immobile support for a liquid stationary phase.

The procedure is as follows. A small volume of the sample to be studied is applied to a position (the origin) near one end of the paper. The spotted origin is then dried, and the paper is hung in a sealed chamber that is saturated with vapor of the developing solvent. The vapor (and the developing solvent they originate from) consists of water and an organic substance(s). A flap at the hanging end of the paper is anchored in a trough by a glass rod. During this equilibration period the paper becomes moistened by absorbing some of the vapors, primarily water vapor. After equilibration, the trough is filled with the developing solvent immersing the flap. Thereafter, the developing solvent will begin to move down the paper by capillary action. As it moves, the paper takes up even more water.

When the solvent front reaches the origin, the material spotted there will be dissolved to different extents between the two phases. That is to say, there will be an initial distribution of all solutes at the origin between the paper-supported stationary phase (which is primarily water) and the moving phase (which is primarily organic). This distribution is based on the relative *differences in the solubilities of each solute in the two phases.* This process will continue along the entire length of the paper until the chamber is opened. Substances that are more soluble in water will move less than those that are more soluble in the moving organic phase. The capability of the method is depicted in Fig. 2–10.

The extent of migration of a substance is expressed by the R_f value, defined as follows:

$$R_f = \frac{\text{distance traveled by substance}}{\text{total distance traveled by developing solvent}}$$

(both distances measured from the origin of sample application).

For samples containing many substances the resolution of zones can be increased by drying the paper and then allowing a second solvent to migrate

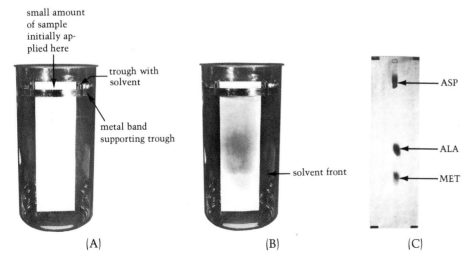

small amount of sample initially applied here

trough with solvent

metal band supporting trough

solvent front

(A)   (B)   (C)

ASP

ALA

MET

FIGURE 2–10 One-dimensional, descending paper chromatography of an amino acid mixture containing aspartic acid (ASP), methionine (MET), and alanine (ALA). *Panel A:* paper strip suspended in trough of developing solvent; no migration has occurred; rectangle designates origin. *Panel B:* solvent front has migrated nearly the length of the paper. *Panel C:* appearance of paper strip after removal and treatment with ninhydrin. (Actual size of the jar is 12″ × 24″.)

FIGURE 2–11 A comparison of the resolving power of one-dimensional (*panel A*) and two-dimensional (*panel B*) paper chromatography. A mixture of 12 amino acids was applied to each origin (rectangle), and the papers were developed under identical conditions and then sprayed with a ninhydrin solution. In each photograph the ninhydrin-positive spots appear as dark areas against the white background. The developing solvents used for each direction are indicated on the borders. In the two-dimensional analysis, the paper was developed first with the phenol solvent. (Amino acid abbreviations are identified in Table 4–1, p. 90.)

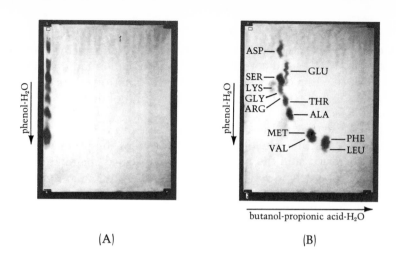

(A)

(B)

in a direction at a right angle to the first solvent. This is referred to as a two-dimensional analysis. The photos in Figure 2–11 clearly depict the advantages of this second step.

Although paper chromatography can be used to examine most types of substances, it cannot be used for the separation of proteins or nucleic acids.

Thin-layer Chromatography

The use of a solid stationary phase spread as a thin coating on the flat surface of a firm support typifies the method called *thin-layer chromatography* (TLC). The solid phase may function as an adsorbent (silica gel is common), an ion exchanger (modified cellulose materials are used), or as a sieving agent (polyacrylamide gel is common). The procedure (see margin) is similar to that of paper chromatography. The wide popularity of thin-layer chromatography is due to several features: solute resolution and reproducibility of results are good to excellent; most separations are achieved quickly, in 0.5–3 hours; it can be used for the analysis of nearly all types of substances; very small amounts (microgram levels) of solutes can be detected; and last but not least, the method is inexpensive and easy to do.

This concludes our analysis of chromatographic methods. Collectively they comprise a potent arsenal of methods for the study of biomolecules,

TABLE 2–2 **A summary of chromatographic methods.**

Procedure	Stationary phase	Moving phase	Property exploited to achieve separation
Gel-permeation chromatography	Solid	Liquid	Differences in molecular size
Ion-exchange chromatography	Solid	Liquid	Differences in ionic charge
Adsorption chromatography	Solid	Liquid	Differences in surface binding
Affinity chromatography	Solid	Liquid	A unique binding to the stationary phase of only one component in a mixture (see p. 149)
Gas-liquid chromatography	Liquid	Gas	Differences in solubilities
Paper chromatography	Liquid	Liquid	Differences in solubilities
Thin layer chromatography	Solid (or liquid on solid)	Liquid	Adsorption, ion exchange, sieving, solubility

much of which, I remind you, involves separation, isolation, purity check, size characterization, and the quantitative analysis of complex mixtures. For more detail the interested student is referred to the reference source by Heftmann (see p. 60). A summary of the distinctive features of chromatographic methods is given in Table 2–2.

SPECTROSCOPY

As light (electromagnetic radiation) passes through a substance, one or more of the following phenomena can occur: reflection, refraction, absorption, interference, diffraction, fluorescence, and ionization. Spectroscopic methods of analysis are based on *absorption*. When light is absorbed, energy is absorbed. Since the pattern of light absorption is unique for each substance, it can be used as a distinguishing feature (a sort of chemical fingerprint) in identifying a substance. This is a qualitative application. Equally important is the fact that the amount of absorption is directly related to the amount of the substance. This is a quantitative application. Instruments called spectrophotometers are used in these measurements, which can be made with a high degree of sensitivity and reproducibility.

The energy (E) content of light is governed by its *wavelength* (λ) in an inverse relationship; that is, the lower the wavelength, the higher the energy, and vice versa.

$$E \propto \frac{1}{\lambda}$$

(We will have more to say about this in Chapter 15; see p. 476.) Thus, when a substance absorbs a specific amount of light energy, it is absorbing at a specific wavelength of light. At any wavelength, however, the absorption of energy is an all or none situation. In other words, absorption will occur at some wavelengths but not at others. In addition, when absorption does occur, it may be weak, moderate, or intense. The measurement of these variations, which are due to the unique chemical structure of the absorbing substance, is what provides an *absorption spectrum*. The measurement is usually made in either the *ultraviolet* (UV), *visible*, or *infrared* (IR) region of the electromagnetic spectrum (see Fig. 2–12).

What may seem to be a slight structural difference can have a significant effect on the absorption spectrum. For example, let us consider a specific set of compounds, namely the oxidized and reduced forms of nicotinamide adenine dinucleotide (symbolized NAD^+ and NADH, respectively). (The + indicates a net molecular charge of +1 in NAD due to an electron deficiency.) These are two forms of an important compound that participates in a large number of biological reactions to be discussed in subsequent chapters. The formula for NAD^+ is $C_{21}H_{26}O_{14}N_7P_2$, and that for NADH is $C_{21}H_{27}O_{14}N_7P_2$. The complete structures can be examined on p. 359. All atoms are arranged in the same way, but note that the reduced form has one more hydrogen atom (and two more electrons) than the oxidized form. Because they have nearly identical structures, your common sense should suggest that they should have similar absorption spectra. On the other hand, because of their different structures, their spectra should be different. Figure 2–13, showing the UV-visible spectrum of each substance, reveals that this reasoning is correct. Both materials absorb maximally at 260 nm,

In thin-layer chromatography several different samples can be applied at individual spots, or a larger amount of the same sample can be applied across the entire width of the plate. The standard plate size is 20×20 cm and the standard thickness of the thin layer is 250 nm (a fourth of a millimeter). The support is glass or plastic.

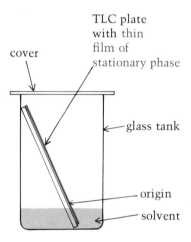

After sample application the thin-layer plate is placed in a chamber containing the chromatography solvent on the floor of the chamber. The solvent migrates up the thin-layer plate by capillary action. As with paper chromatography the analysis can be one- or two-dimensional.

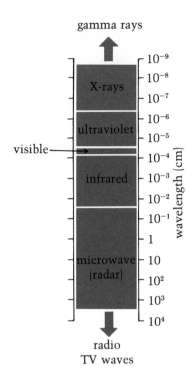

gamma rays

X-rays

ultraviolet

visible

infrared

microwave
(radar)

10^{-9}
10^{-8}
10^{-7}
10^{-6}
10^{-5}
10^{-4}
10^{-3}
10^{-2}
10^{-1}
1
10
10^{2}
10^{3}
10^{4}

wavelength (cm)

radio
TV waves

FIGURE 2–12 The regions of the electromagnetic spectrum. For additional detail and discussion see Chap. 15, p. 476.

$$\epsilon = \frac{A}{cl}$$

Since the absorbance is a unitless number, the units of the absorption extinction coefficient (ϵ) are $c^{-1}\, l^{-1}$; the light path usually used is 1 cm and the concentration is expressed in molarity (M) or milli-molarity (mM).

In vivo studies (from the Latin *vivus,* meaning alive or living) use systems consisting of intact living cells. *In vitro* studies (from the Latin *vitrum,* meaning glass) use a component previously isolated from intact cells, and the system is referred to as being cell-free.

but the extent of absorption with the reduced form is somewhat less than with the oxidized form. More remarkable, however, is the appearance of a new absorption band at a maximum of 340 nm in the reduced form. This is totally absent in the spectrum of the oxidized form. Thus, the reduced absorption at 260 nm and the unique absorption at 340 nm are spectrophotometric properties that, indeed, distinguish NADH from NAD$^+$ and, under the conditions of the analysis, distinguish both from many other compounds.

Quantitative measurements are based on the *Beer-Lambert* law, which states that the absorption of light energy at any wavelength is solely dependent on the concentration of the absorbing material and the length of the light path through the absorbing medium. A statement of this relationship is given below, where the *absorbance A* is defined as the logarithm of the ratio of incident light intensity (I_0) to transmitted light intensity (I_T).

$$\underset{(A)}{\text{absorbance}} = \log_{10}\frac{(I_0)}{(I_T)} = \epsilon cl$$

The concentration is symbolized as c, l is the light path, which usually is 1 centimeter, and ϵ is the absorption extinction coefficient. The latter is a constant corresponding to the light absorption of a known concentration of the compound at a specific wavelength. Given the extinction coefficient of a substance at a wavelength corresponding to maximum absorption, any solution of this substance can have its concentration determined by measuring the absorbance of the unknown solution at the same wavelength. Alternatively, an unknown concentration of a compound in solution can be determined if the absorbance of a solution of the same compound of known concentration is measured. Again the same conditions would have to apply to use this approach.

For two solutions of the same compound, one with a known (K) concentration and the other with an unknown (UK) concentration

$$A_K = \epsilon c_K l \qquad \text{and} \qquad A_{UK} = \epsilon c_{UK} l$$

If the absorbance of each solution of the same compound is measured under the same conditions (that is, at the same wavelength and in the same size tube so that l is the same; all of which means that ϵ is the same), then

$$\frac{A_K}{c_K} = \epsilon l = \frac{A_{UK}}{c_{UK}} \qquad \text{or simply} \qquad \frac{A_K}{c_K} = \frac{A_{UK}}{c_{UK}}$$

You measure A_K and A_{UK} with a spectrophotometer; you know c_K; therefore, you can calculate c_{UK}.

RADIOISOTOPES

The use of radioisotopes has had an enormous impact on the development and growth of biochemistry. This is due to the fact that radioactivity permits the labeling (or tagging) of compounds in order to determine (easily, accurately, and with great sensitivity) what happens to a compound during reactions occurring *in vivo* or *in vitro*. In other words, their use can be likened to that of a homing device. They are also very useful in measuring the activity of proteins, particularly if only a small amount of the protein is present. Since their initial use some thirty-five years ago, radioisotopes have been employed in the design of countless experiments and assays, most of which would be difficult, if not impossible, to perform using other means.

Radioisotopes that have extensive application in biochemical investigations are carbon 14, hydrogen 3 (tritium), sulfur 35, and phosphorus 32, all of which are pure β emitters (see Table 2–3). These are four of the six primary elements of which most biomolecules are composed. Radioisotopic forms of the other two elements, oxygen and nitrogen, are known but their practical use is obviated because they decay very rapidly. Instead, stable isotopic forms of oxygen and nitrogen, ^{18}O and ^{15}N, are used, and detection and measurement are based on their contribution through their larger atomic masses. In the same way, considerable use is also made of deuterium (^{2}H), the stable isotope of hydrogen, and ^{13}C, a stable isotope of carbon.

To illustrate briefly the application of radioisotopes, let us consider the following simple example. (Another example, *radioimmunoassay*, is described in the next section.) A biochemist wishes to verify (a) that an enzyme just isolated is capable of converting pyruvic acid and aspartic acid to alanine and oxaloacetic acid, and (b) that all of the carbon atoms in alanine originate from pyruvic acid. This can be accomplished by mixing the enzyme with pyruvic acid labeled with ^{14}C in every position (shown in color) and with aspartic acid that contains no ^{14}C, and then demonstrating, after the enzyme has acted and the products alanine and oxaloacetic acid have been separated, that the ^{14}C label is found exclusively in alanine and not at all in oxaloacetic acid. Obviously, this is categorical proof of what occurs during the reaction. Tracer experiments and the design of assay systems are made possible by the commercial availability of hundreds of compounds labeled in specific positions; most of these materials are, however, rather expensive. If the compound is commercially unavailable, you can synthesize it yourself.

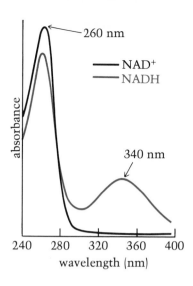

FIGURE 2–13 Ultraviolet absorption spectra of equimolar solutions of oxidized (NAD⁺) and reduced (NADH) nicotinamide adenine dinucleotide. See p. 359 for structures.

C = ^{14}C C = ^{12}C

$$\underset{\text{pyruvic acid}}{CH_3\overset{\overset{\displaystyle O}{\|}}{C}COOH} + \underset{\text{aspartic acid}}{HOOCCH_2\overset{\overset{\displaystyle NH_2}{|}}{C}HCOOH} \xrightarrow{\text{enzyme}} \underset{\text{alanine}}{CH_3\overset{\overset{\displaystyle NH_2}{|}}{C}HCOOH} + \underset{\text{oxaloacetic acid}}{HOOCCH_2\overset{\overset{\displaystyle O}{\|}}{C}COOH}$$

TABLE 2–3 **A comparison of radioisotopic forms of the primary biological elements. Note that isotopes differ not only in relative instability but also in the energy value E of the emitted radiation. The energy content is directly correlated to hazardous biological effects as well as the efficiency with which the radiation can be detected. (1 MeV = 3.83×10^{-14} calories.)**

Isotope	Radiation emitted	Half-life[a]	E (in MeV)	Naturally abundant stable isotope (% occurrence)
^{14}C	β	5770 yr	0.156	^{12}C (98.89%)
^{3}H	β	12.26 yr	0.0186	^{1}H (99.985%)
^{35}S	β	86.7 days	0.167	^{32}S (95.0%)
^{32}P	β	14.3 days	1.71	^{31}P (100%)
^{19}O	β,γ	29 sec	3.25 (β); 0.20 (γ)	^{16}O (99.76%)
^{16}N	β,γ	7.35 sec	4.3 (β); 6.13 (γ)	^{14}N (99.63%)

[a] Time required for 50% of the material to decay; a unique physical property of the isotope.

The quantitative measurement of radioactivity (particularly β radiation) can be performed by *Geiger-Muller counting* or the more sensitive method of *liquid scintillation*. Interested students will find the book by Wang et al. an excellent source (see end of chapter).

RADIOIMMUNOASSAY

Antibodies are proteins produced in the blood of animals for the purpose of defending against infection by foreign substances called *antigens*, which usually are also proteins. In response to the presence of a foreign substance in the bloodstream, the body produces an antibody (Ab) that *reacts specifically* with that antigen (Ag) to produce an Ab-Ag complex, with the antigen bound to antibody. In the lab, antibodies can be obtained by injecting an animal (e.g., rabbits) with the antigen and then bleeding the animal. An antibody against any nonprotein substance (symbolized as X) can be prepared by first attaching X to a protein foreign to the rabbit—for example, bovine serum albumin—and then injecting the rabbit with the X-albumin complex. The serum obtained will contain an antibody that will usually react with X alone.

For the quantitative measurement of small amounts of a substance present in a complex mixture, the sensitivity and specificity of *radioimmunoassay* (RIA) are equal to or better than all other procedures. It is also quick and easy to do. Developed by R. S. Yalow (Nobel Prize in 1977), the RIA method can be applied to detect any substance that can be prepared in a radioactive form (radio) and for which an antibody serum preparation (immuno) can be obtained. The radioactivity confers sensitivity and the use of an antibody confers specificity.

In describing how the method works, let us assume we are dealing with an assay for morphine (M)—an assay that would obviously require (1) a serum antibody preparation against morphine (Ab_M) and (2) radioactively labeled morphine such as ${}^3H$-morphine (**M**). To assay something—urine, for example—for morphine content, a small sample is added to a solution containing a known amount of Ab_M and excess **M**. Since both M and **M** are present, they both can bind to Ab_M. Sometimes M-Ab_M will form; sometimes **M**-Ab_M will form. The amount of each type of complex will be governed by the ratio of M/**M**. With more M there will be less **M**-Ab_M formed. After equilibration, the M-Ab_M and **M**-Ab_M are separated from unreacted M and **M**. Then the radioactivity level in the mixture of M-Ab_M and **M**-Ab_M is measured and compared to control systems where the same events are done with several known concentrations of M. See the diagram below.

In addition to its use in research, the RIA method is used in many clinical laboratories for accurate assays on drug and hormone levels in physiological fluids.

$$\underbrace{\mathbf{M} + Ab_M}\; + M \longrightarrow \underbrace{\mathbf{M} + M}\; + \;\underbrace{\mathbf{M} - Ab_M + M - Ab}$$

a precise amount of these are present and the same amounts are used in preparing the standard curve

unbound species

bound species that are recovered and measured for radioactivity, which is compared to standard curve to determine the concentration of M

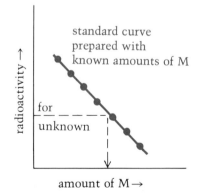

standard curve prepared with known amounts of M

radioactivity →

for unknown

amount of M →

X-RAY DIFFRACTION

We are all familiar with the tremendous diagnostic utility of X-rays in the fields of medicine and dentistry, where images of the human anatomy are produced and analyzed for structural defects. This technique is related to, but not identical with, a special type of spectroscopy called X-ray diffraction, which measures the interaction of materials in the solid state with X-radiation. It is by far the most powerful tool available for the complete determination of fine structural details on the molecular level.

In the solid crystalline state, the atoms of any pure material are arranged in a lattice configuration that consists of repeating units forming an ordered, three-dimensional structure framework. X-Rays are nonparticulate, high-energy radiations of very short wavelength. Because of these properties, X-rays, like gamma rays, have a very high degree of penetrating power. As X-rays pass through a lattice-type structure they strike atoms and are then scattered or reflected at various angles. If the crystal exposed to X-rays has an ordered arrangement of atoms, the X-rays will be reflected (diffracted) in an ordered fashion commensurate with the structure of the material. In addition, the intensity of the scattered radiation is dependent on the number and arrangement of electrons in the scattering atoms.

In X-ray diffraction analysis, a solid sample is exposed to a beam of X-radiation of known wavelength and the rays, which may be scattered at any angle, are detected photographically. The intensities of each scattered ray can then be measured from the film. Depending on the structural complexity of the material, thousands of images will be collected. The data are then fed to a computer programed for analysis. In effect, the output of the computer analysis can be used to construct a two-dimensional map of the variation in electron density within any one of several planes in the crystal. By superimposing several of these two-dimensional contour maps of electron density for successive planes, a three-dimensional representation of the substance can be constructed.

For reasons cited previously, we will not embark on an analysis of the theory or the mechanics of X-ray diffraction. The important point is for you to appreciate that this method can be used effectively to reveal the structural aspects of a material in intimate detail. It can be envisioned as a form of molecular microscopy characterized by a high degree of resolving power. With modern instruments and with highly refined techniques, it is possible to attain a resolving power of 1.5×10^{-8} cm (1.5 Å). At this resolution, most of the atoms (except the small hydrogen atoms) in a molecule may be unambiguously located by the images of X-ray diffraction. Since 1960, the method has proven extremely useful in deciphering the complete structure of proteins and nucleic acids. Such information contributes greatly to an understanding of how these molecules function. A good example of the uses of X-ray diffraction in the study of protein structure is given in Chap. 5 (p. 142).

OTHER METHODS

Although we will not consider their basis or application, there are other lab techniques that are used (less frequently than the ones just described) in biochemical studies. Some of these are *nuclear magnetic resonance spectroscopy* (NMR), *fluorescence spectroscopy, mass spectroscopy, electron spin resonance spectroscopy, optical rotatory dispersion* (ORD), and *neutron diffraction.* The point is that modern biochemistry is an extensive, and frequently a highly specialized and sophisticated, science. Those of you who might choose to pursue a career in biochemistry will discover this first hand.

LITERATURE

ABBOTT, D., and R. S. ANDREWS. *An Introduction to Chromatography.* Boston: Houghton Mifflin, 1965. Theoretical and practical aspects of various chromatographic techniques are expertly covered in this paperback. The use of many diagrams makes the book very informative.

BREWER, J. M., A. J. PESCE, and R. B. ASHWORTH. *Experimental Techniques in Biochemistry.* Englewood, N. J.: Prentice-Hall, Inc., 1974. Coverage of chromatography, electrophoresis, ultracentrifugation, spectroscopy, radioactivity, and immunological procedures. Rigorous physicochemical treatment for the advanced student.

CLARK, J. M., and R. L. SWITZER. *Experimental Biochemistry.* 2nd edition. San Francisco: W. H. Freeman, 1977. One of the best laboratory manuals available. Contains a section on the theory and application of biochemical methods.

DYER, J. R. *Application of Absorption Spectroscopy of Organic Compounds.* Englewood, N.J.: Prentice Hall, 1965. A short book, available in paperback, that provides a good introduction to the theory and use of absorption spectroscopic methods.

FINLAYSON, J. S. *Basic Biochemical Calculations.* Reading, Mass.: Addison-Wesley, 1969. A book dealing with quantitative aspects of general biochemistry and designed to serve students with a minimum background in chemistry. Approximately one-fifth of the book is devoted to the basic theory, techniques, and biological applications of radioisotopic methods.

FREIFELDER, D. *Physical Biochemistry: Applications to Biochemistry and Molecular Biology.* San Francisco: W. H. Freeman, 1976. The theory and application of several laboratory techniques are discussed. A good source and available in paperback.

GAUCHER, G. M. "An Introduction to Chromatography." *J. Chem. Ed.,* **46,** 729–733 (1969). A brief but informative article summarizing the important historical, practical, and theoretical aspects of chromatographic analysis.

KELLER, R. A. "Gas Chromatography." *Scientific American,* **205,** 58–67 (1961). Clearly written article describing the principles of gas-liquid partition chromatography.

MORRIS, C. J., O. R. MORRIS, and P. MORRIS. *Separation Methods in Biochemistry.* New York: Interscience-John Wiley, 1964. A complete coverage of the theories, techniques, and applications of a wide range of methods for the separation of biochemically important substances. Several references to important mono-graphs and over 2000 references to the original literature are given.

SILVERSTEIN, R. M., C. G. BASSLER, and T. C. MORRILL. *Spectrophotometric Identification of Organic Compounds.* 3rd edition. New York: John Wiley, Inc., 1974. A popular introductory text to the theory, application, and interpretation of spectroscopic techniques (excluding X-rays). Relatively noncomplicated presentation of complicated topics.

UMBRIET, W. W., R. H. BURRIS, and J. F. STAUFFER. *Manometric and Biochemical Techniques.* 5th edition. Minneapolis: Burgess, 1972. A revised and updated edition of a long-respected book. Although the major thrust is directed toward the theory and use of manometric techniques in the study of biological systems, this new edition covers most of the topics discussed in this chapter.

WANG, C. H., D. L. WILLIS, and W. D. LOVELAND. *Radiotracer Methodology in the Biological, Environmental, and Physical Sciences.* Englewood, N.J.: Prentice-Hall, 1975. Textbook treatment of the nature, measurement, and application of radioactivity. One of the few good sources in this area.

WHARTON, D. C., and R. E. McCARTY. *Experiments and Methods in Biochemistry.* New York: The Macmillan Co., 1972. An introductory level laboratory manual containing excellent descriptive material on basic techniques of biochemical studies.

WILLIAMS, V. R., W. L. MATTICE, and H. B. WILLIAMS. *Basic Physical Chemistry for the Life Sciences.* 3rd edition. San Francisco-London: W. H. Freeman, 1978. An excellent text of physicochemical principles designed for students in life sciences. Chapter 9 deals with the principles of electrophoresis, ultracentrifugation, and other topics; Chap. 8 with radioisotopes; and Chap. 7 with spectroscopy.

YALOW, R. S. "Radioimmunoassay: A Probe for the Fine Structure of Biologic Systems." *Science,* **200,** 1236–1245 (1978). The Nobel lecture of R. S. Yalow describing the development and applications of radioimmunoassay.

ZWEIG, G., and J. SHEMA, ed. *Handbook of Chromatography.* Volumes I and II. Cleveland: Chemical Rubber Company Press, 1972. Volume I: a compilation of chromatographic data for various types of substances by gas, column, paper, and thin-layer chromatography. Volume II: primary focus is on discussion of principles and techniques of the various chromatographic techniques.

EXERCISES

2–1. The diagram below is a sketch of the ultraviolet absorption pattern from an analytical ultracentrifugal analysis of a protein solution after a 60-minute running time. Given this and the following data calculate the Svedberg constant and the molecular weight of the protein: partial specific volume of protein = 0.74 ml/g; diffusion constant of protein = 1×10^{-7} cm²/sec; density of solvent = 1.0 g/ml; angular rotation = 10,000 radians/sec; distance from center of rotation to meniscus of solution in centrifuge tube = 5 cm; and gas constant = 8.3×10^7

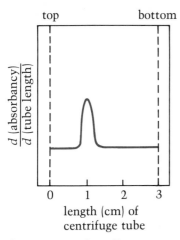

top bottom

$$\frac{d \text{ (absorbancy)}}{d \text{ (tube length)}}$$

0 1 2 3

length (cm) of
centrifuge tube

ergs/deg/mole. Assume that all measurements are expressed for 20 °C. (In dimensional analysis, note that 1 erg = 1 dyne-cm and 1 dyne = 1 g-cm/sec² and radians are to be considered unitless.)

2-2. A protein solution contains two different proteins, P_A and P_B. The concentration of P_A is greater than P_B. The molecular size of P_A is also greater than P_B. Moreover, the difference in size is sufficient to resolve clearly the protein boundaries when P_A has sedimented halfway down the cell. Draw the analytical optical pattern that would correspond to this information. (Assume that both proteins have the same molecular shape.)

2-3. Analytical ultracentrifugation of a protein solution exhibits a single symmetrical peak during sedimentation. However, polyacrylamide gel electrophoresis of the same protein solution reveals two zones after the gel is stained. What can be concluded about the protein solution?

2-4. Four pure proteins were used as standards to construct a standard curve for a molecular weight analysis via SDS-gel electrophoresis. Protein 1, with a molecular weight of 15,000, was the smallest protein. Protein 2 (mol. wt. = 35,000) moved only 39% as far as protein 1. Protein 3 (mol. wt. = 25,000) moved only 63% as far as protein 1. Protein 4 (mol. wt. = 20,000) moved only 81% as far as protein 1. Construct the standard curve and then determine the molecular weight of an unknown protein that had a mobility (under the same conditions) midway between that of proteins 2 and 3.

2-5. Ion-exchange column chromatography was being considered for the separation of material A from B. To determine the suitability of the method, the chromato-

for A		for B	
cpm of ¹⁴C (at volume)		*cpm of ³H (at volume)*	
0	(at 10 ml)	0	(at 45 ml)
100	(at 20 ml)	200	(at 55 ml)
400	(at 30 ml)	600	(at 60 ml)
1150	(at 50 ml)	200	(at 65 ml)
400	(at 70 ml)	100	(at 68 ml)
0	(at 80 ml)	0	(at 75 ml)

graphic behavior of pure A (carbon-14 labeled) and pure B (hydrogen-3 labeled) was separately determined on the same column operated under the same conditions (i.e., temperature, solvent, and flow rate). The run-off (the eluent) of the column was continuously monitored for the presence of radioactivity. Some of the data are presented here. Plot both sets of data on the same graph and evaluate whether A could be cleanly separated from B if a mixture of the two were applied to the column.

2-6. If a solution of the same unknown protein identified in Exercise 2-4 were examined by gel permeation column chromatography under conditions applicable to the standard curve shown in Fig. 2-7 on p. 49, what would be the elution volume for the unknown protein?

2-7. A small sample of a pure material was subjected to thin-layer chromatography. After drying the plate was sprayed with sulfuric acid and heated. The appearance of the plate is represented by the drawing below. What is the R_f value of this material in the solvent system that was used?

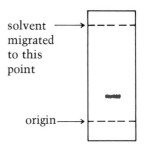

solvent
migrated
to this
point

origin

2-8. Given the information below, determine which combination of developing solvents (Ph + BuAc or Ph + BuP) would provide optimum resolution by two-dimensional paper chromatography of a mixture containing all seven of the amino acids listed. Proceed by making a sketch of the paper sheet as it would appear after ninhydrin treatment. (Represent the paper sheet as a 5-inch square and the ninhydrin zones as spheres with a diameter of ⅜ inch. You will find it convenient to use graph paper with ten squares to the inch.)

R_f values of amino acids on Whatman No. 1 paper

	Developing solvent		
Amino acid	*Ph[a]*	*BuAc[b]*	*BuP[c]*
glutamic acid	0.33	0.28	0.20
lysine	0.42	0.12	0.13
glycine	0.42	0.23	0.29
alanine	0.58	0.30	0.37
valine	0.78	0.51	0.48
serine	0.35	0.22	0.33
methionine	0.80	0.50	0.58

[a] Phenol saturated with water.
[b] *n*-Butanol-glacial acetic acid-water (12:3:5).
[c] *n*-Butanol-pyridine-water (1:1:1).

61

2–9. From the data below, construct the ultraviolet absorption spectrum of adenine—a nitrogen base found in all nucleic acids—and graphically estimate the wavelength at which maximum absorption occurs. Then, given that the extinction coefficient of a 1% solution is 890 at this wavelength, calculate the concentration of adenine in a solution that gives an absorbance of 0.15 at the same wavelength. (Assume a 1-cm light path.)

Spectral data for adenine

Wavelength	Absorbance
235 nm	0.45
240	0.20
250	0.26
260	0.50
270	0.47
280	0.25
290	0.07

2–10. A 0.1-ml sample of a glucose standard solution containing 50 micromoles of glucose/ml was analyzed via a spectrophotometric procedure that produces a colored product with an absorption maximum at 540 nm. The absorbance at this wavelength was 0.46. If a 0.1-ml sample of a glucose-containing solution of unknown concentration was assayed in exactly the same manner and gave an absorbance of 0.18, what is its glucose concentration? Express your answer first in micromoles/ml and then in terms of the millimolarity (mM) of the solution.

2–11. *Escherichia coli,* a bacterium, was allowed to grow in a medium that contained radioactive phosphorus in the form of $^{32}PO_4^{3-}$. These cells will incorporate the label into every phosphorus-containing compound. After a 45-minute incubation period, the cells were harvested and extracted with an organic solvent to remove the phospholipids—a class of P-containing lipids. A small aliquot of the extract was then chromatographed on a silica gel thin-layer plate. After the position of the radioactive regions was detected, they were scraped off, transferred to scintillation vials, and then counted by liquid scintillation. From the data below, determine the relative amounts of the phospholipids in the extract.

Component	R_f	Cpm in extract
phosphatidyl glycerol (PG)	0.3	96,695
phosphatidyl ethanolamine (PE)	0.5	421,402
cardiolipin (CL)	0.7	5,973

2–12. Most manufacturers of radioactively labeled pure materials offer the same compound with different labeling patterns. For example, it is possible to purchase methionine (an amino acid) as either (colored C atoms represent C-14)

$$CH_3SCH_2CH_2CHCOOH \quad \text{or} \quad CH_3SCH_2CH_2CHCOOH$$
$$\qquad\qquad\quad | \qquad\qquad\qquad\qquad\qquad\quad |$$
$$\qquad\qquad NH_2 \qquad\qquad\qquad\qquad\qquad\quad NH_2$$

How might these materials be utilized in tracing the reactions of methionine?

2–13. An antibody is prepared against the drug mescaline. A precise amount of the antibody is mixed with an excess of ^3H-labeled mescaline. When various known amounts of unlabeled mescaline are then added and the systems are allowed to equilibrate, the amounts of ^3H label that are found bound to antibody are as follows: with 1 microgram of mescaline added, 70,000 cpm are measured; with 2 μg added, 55,000 cpm; with 3 μg added, 38,000 cpm; with 4 μg added, 21,000 cpm. If the same conditions of analysis were applied to a specimen of urine and 49,000 cpm of ^3H were detected as bound to antibody, what amount of mescaline was present in the urine sample that was examined?

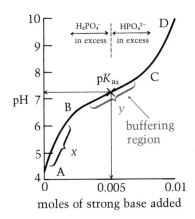

moles of strong base added

CHAPTER 3

pH, BUFFERS, WATER, AND NONCOVALENT BONDING

A great majority of the organic compounds that occur in living organisms exist in a charged state. That is, they are *ions*. Depending on the nature of the charge, they are *anions* (negatively charged), *cations* (positively charged) or *ampholytes* (both negatively and positively charged).

The presence of organic biomolecules in ionic forms is determined by the hydrogen ion concentration $[H^+]$, commonly expressed as the pH of the system. In addition, when two or more ionic forms of the same substance are possible, only one will usually predominate. Because of this control by pH of the population of organic ions, it is only logical that living organisms be capable of preventing excessive changes in the pH of intracellular and extracellular fluids, which are normally maintained at a pH value of approximately 7—*physiological pH*. This is accomplished through the action of *buffer systems*. Without buffers, the pH and the ionic environment would be in a constant state of flux, a condition that could have serious consequences. For example, a prolonged disturbance of the pH of blood (normally maintained at a value of 7.3–7.4) could be fatal.

The use of buffers in biochemical (biological) research is also of immense importance. It is totally impractical to undertake the study of whole cells, of subcellular particulates, or in many cases to even isolate and characterize

The concentration of H^+ in a solution is expressed as pH, which is defined as

$$pH = -\log [H^+] = \log \frac{1}{[H^+]}$$

This mathematical manipulation eliminates the use of expressing concentrations that have exponential powers of 10. For example, a hydrogen ion concentration of 1×10^{-7} mole/liter would correspond to a pH of 7.

purified biomolecules—particularly the biopolymers—without attention to pH control. The molecular focus is particularly important. Indeed, the sensitivity of the proteins and the nucleic acids constitutes the molecular basis for the significance of pH control to living systems. Gradually it will become quite clear that the prevailing pH confers a specific ionic nature to these molecules that is important to both their overall chemical structure and to their biological functions. In other words, the *harmful effects of variations in pH are due to harmful effects on molecules.*

General physicochemical principles that apply to the nature and operation of buffer systems are described in this chapter. The specific topics of the selection and preparation of buffers for biological research and the buffering system in blood are also discussed. In addition, some attention is given to the most important substance of all—water. The discussion of water provides a basis to introduce the very important topic of noncovalent bonding.

In addition to ionic forms of organic molecules, many inorganic ions are also essential for the normal growth and development of organisms. In fact, the buffering of physiological fluids involves two inorganic systems, phosphate and bicarbonate, both of which are described in this chapter. The occurrence and essential roles of the ionic forms of the inorganic minerals were emphasized in the Introduction (see p. 10).

IONIC EQUILIBRIA

A knowledge of the chemistry of acids and bases, of ionization, and of ionization equilibria is basic to an understanding of the dynamics of pH regulation by buffer systems. Consequently, we will begin by reviewing some of these points.

Ionization. Acids and Bases

A process that results in the formation of ions is called *ionization*. The phenomenon of ionization was first described by Svante Arrhenius in the nineteenth century, and served as the basis for his classification of acids and bases. An acid was defined as a substance that yields hydrogen ions (H^+, protons) upon ionization and a base as a substance that yields hydroxide ions (OH^-) upon ionization. He offered a further distinction in terms of the extent to which the ionization occurred. A *strong* acid or base undergoes complete ionization (essentially 100%), and a *weak* acid or base undergoes partial ionization (that is, considerably less than 100%). These statements are summarized by the equations at the top of p. 65.

A single arrow corresponds to complete dissociation and represents the absence of any undissociated acid or base in solution. A set of arrows corresponds to a dynamic equilibrium between the forward and reverse processes and represents the coexistence of an undissociated acid or base in solution with the corresponding ions. The longer arrow in the set indicates which process is favored and thus specifies which components will be in excess at equilibrium. The (aq) notation designates a water (aqueous) solution.

The majority of acids and bases found in nature are organic compounds and are weak acids or bases. Accordingly, emphasis will be given to these types of compounds. Before proceeding further, however, it would be bene-

General examples   Specific examples

$$Acids \begin{cases} HA\ (strong) \xrightarrow{H_2O} H^+(aq) + A^-(aq) \\ \\ HA\ (weak) \underset{H_2O}{\rightleftharpoons} H^+(aq) + A^-(aq) \end{cases}$$

$$HCl \xrightarrow{H_2O} H^+(aq) + Cl^-(aq)$$
(hydrogen   (chloride
chloride)   ion)

$$CH_3COOH \underset{H_2O}{\rightleftharpoons} H^+(aq) + CH_3COO^-(aq)$$
(acetic acid)   (acetate ion)

$$Bases \begin{cases} MOH\ (strong) \xrightarrow{H_2O} M^+(aq) + OH^-(aq) \\ \\ MOH\ (weak) \underset{H_2O}{\rightleftharpoons} M^+(aq) + OH^-(aq) \end{cases}$$

$$NaOH \xrightarrow{H_2O} Na^+(aq) + OH^-(aq)$$
(sodium   (sodium
hydroxide   ion)

$$NH_4OH \underset{H_2O}{\rightleftharpoons} NH_4^+(aq) + OH^-(aq)$$
(ammonium   (ammonium
hydroxide)   ion)

Color identifies what species would exist in solution

ficial to review an alternative and more useful theory of acids and bases, the *Bronsted proposal*, which defines an *acid* as any material that *donates a proton, and a *base* as any material that *accepts a proton*. To illustrate, the ionization of a hypothetical weak acid can be written (and in a more correct way) as

$$HA\ (aq) + H_2O \rightleftharpoons A^-\ (aq) \quad + \quad H_3O^+\ (aq)$$
hydronium ion

It is clear that HA is donating a proton and thus is functioning as a Bronsted acid. H_2O is accepting the proton from HA and is functioning as a Bronsted base. Similarly, for the reverse process, H_3O^+ functions as a Bronsted acid and A^- as a Bronsted base. Because of the relationship between HA and A^-, they are referred to as a *conjugate acid-base pair*. That is to say, A^- is the conjugate base formed from the dissociation of the original Bronsted acid, HA. A similar relationship exists between H_2O and H_3O^+. (*Note:* The hydronium ion, H_3O^+, is the common representation of the hydrated proton, $H^+(H_2O)$). For reasons of simplification, the H_3O^+ notation will not be routinely employed in the discussions to follow and the presence of water will be assumed. Hence we will merely write

conjugate acid $\rightleftharpoons$ conjugate base + H^+

Two very common weak acid groupings that occur in organic compounds are the *carboxyl group* (—COOH) and the *protonated amino group* (—$\overset{+}{N}H_3$). Their ionization is summarized by the following equations. Here, arrows of equal length merely specify that the system has the potential to reach an equilibrium condition when the rate of reaction in each direction is the same. The letter R is merely a symbol for the rest of the molecule.

$$RCOOH \rightleftharpoons RCOO^- + H^+$$
$$RNH_3^+ \rightleftharpoons RNH_2 + H^+$$

Relative Strengths of Acids

The relative strength of a weak acid depends on the percent of its ionization. An acid that is 50% ionized will yield a greater amount of its conjugate base and H^+ than one that is 5% ionized. Under a condition of unchanging temperature, the extent of ionization is a constant value that is characteristic of the acid. Since we are dealing with a system that can attain a state of equilibrium (that is, a state of balance, with reactants yielding products and being reformed from products at the same rate), the *law of mass action* can be applied to express the system. This is done by defining what is called an *equilibrium constant* (K) in terms of the amounts of products and reactants that are present. Since we are dealing in this case with an equilibrium system that results from the ionization of a weak acid, the equilibrium constant is called an *acid ionization constant* and symbolized as K_a. It is expressed as follows, where [ ] represents molar concentration (moles/liter) of each substance.

For the acid ionization written as

conjugate acid $\rightleftarrows$ conjugate base + H^+

$$K = \frac{[\text{conjugate base}][H^+]}{[\text{conjugate acid}]} = K_a \qquad (3-1)$$

This expression is *valid for any weak monoprotic acid.* The term monoprotic refers to an acid (HA) that consists of only one ionizable hydrogen. H_2A and H_3A acids—as the formulas imply—would be classified as diprotic and triprotic acids respectively, or polyprotic acids generally. In solution, *polyprotic acids undergo ionization through two or three separate and sequential steps.* Each step can be considered as the ionization of a monoprotic species, and hence for each step a K_a expression can be written. This is illustrated below for phosphoric acid, H_3PO_3, a triprotic acid.

	Ionization	Expression	Value (at 25 °C)
		Ionization constant	
Step 1:	$H_3PO_4 \rightleftarrows H_2PO_4^- + H^+$	$K_{a_1} = \dfrac{[H_2PO_4^-][H^+]}{[H_3PO_4]}$	$= 7.56 \times 10^{-3}$
Step 2:	$H_2PO_4^- \rightleftarrows HPO_4^{2-} + H^+$	$K_{a_2} = \dfrac{[HPO_4^{2-}][H^+]}{[H_2PO_4^-]}$	$= 6.12 \times 10^{-8}$
Step 3:	$HPO_4^{2-} \rightleftarrows PO_4^{3-} + H^+$	$K_{a_3} = \dfrac{[PO_4^{3-}][H^+]}{[HPO_4^{2-}]}$	$= 5.0 \times 10^{-13}$
Overall:	$H_3PO_4 \rightleftarrows PO_4^{3-} + 3H^+$	$K_{a_{net}} = \dfrac{[PO_4^{3-}][H^+]^3}{[H_3PO_4]}$	$= (K_{a_1})(K_{a_2})(K_{a_3})$

The subscripts attached to the K_a expressions correspond to the successive steps of the ionization, which occur in the order given. The sequence is based on acid strength, with the strongest Bronsted acid dissociating first and the weakest last. Thus, the three distinct monoprotic acids would be arranged as follows according to their relative acid strength:

$$H_3PO_4 > H_2PO_4^- > HPO_4^{2-}$$

This hierarchy of acid strength is reflected in the trend of the K_a values, with

$$K_{a_1} > K_{a_2} > K_{a_3}$$

Thus, we see that relative acid strength can be predicted by merely inspecting K_a values. The rule is *the greater the K_a value, the stronger the acid.*

Having completed this brief digression concerning polyprotic acids, let us return to our generalized equation (3–1) for any weak monoprotic acid. An important conclusion regarding the nature of this system is apparent if both sides of the equation are multiplied by $1/[H^+]$. This gives

$$\frac{[\text{conjugate base}]}{[\text{conjugate acid}]} = \frac{K_a}{[H^+]}$$

and since K_a is a constant, and also since $1/[H^+]$ is expressed as pH,

$$\frac{[\text{conjugate base}]}{[\text{conjugate acid}]} \propto \frac{1}{[H^+]} \propto pH$$

This proportionality states that at any given temperature the *ratio of the equilibrium concentrations of the conjugate acid-base pair is solely dependent on the hydrogen ion concentration (pH) of the solution.* A corollary of this is that the *pH of a solution consisting of a conjugate acid-base pair is solely dependent on the ratio of their equilibrium concentrations.* Since a living cell is an equilibrium mixture consisting of an abundance of weak acids and their conjugate bases, it is clear that any change in the physiological pH can cause a significant change in the ionic composition of the cell. This point will be reiterated later in the chapter and a specific example will be given. Having established these general principles of ionization equilibria, we are now in a position to consider the nature of buffer systems.

pH Buffers

Suppose you added 0.1 mole of solid KH_2PO_4 and 0.1 mole of solid K_2HPO_4 to water and adjusted the final volume to 1000 ml. The salts would dissolve and yield 0.1 mole $H_2PO_4^-$ and 0.1 mole HPO_4^{2-} and a total of 0.3 mole of K^+, all present in 1 liter of solution. The pH of this solution would be about 7.2. If now you added 1 ml of a NaOH solution containing 0.01 mole of OH^-, the pH of the solution would increase to about 7.3—an increase of 0.1 pH unit. In contrast, if the same amount of OH^- (0.01 mole) were added to the same volume of pure water with a pH of 7, the pH would change to 12—an increase of 5 pH units. Similar degrees of change would apply if 1 ml of an HCl solution containing 0.01 mole of H^+ were added to each system, with the phosphate solution changing from pH 7.2 to 7.1 (0.1 pH unit decrease) and pure water changing from pH 7 to 2 (5 pH units decrease). The less drastic pH changes are due to the phenomenon of *buffering* that occurs in the phosphate solution but not in pure water. This leads us to the following definition.

A buffer solution is a mixture of a conjugate acid-base pair that is capable of resisting large changes in pH when small amounts of another acid or base are added.

The reason for buffering is that a solution consisting of a conjugate acid-base pair can scavenge the addition of H^+ or OH^-. In the case of H^+ addition the HPO_4^{2-} present would act as an H^+ acceptor to form $H_2PO_4^-$. Some of the newly formed $H_2PO_4^-$ ionizes to yield back $HPO_4^{2-} + H^+$, but the amount of each that is returned is much less than what was consumed. Most of the added H^+ remains as $H_2PO_4^-$. Of course, this chemistry alters the composition of the conjugate acid-base solution—$H_2PO_4^-$ is increased and HPO_4^{2-} is decreased, i.e., the base/acid ratio of $HPO_4^{2-}/H_2PO_4^-$ decreases. It can be shown that such a solution will still be capable of buffering further additions of H^+ until the ratio gets lower than 1:9.

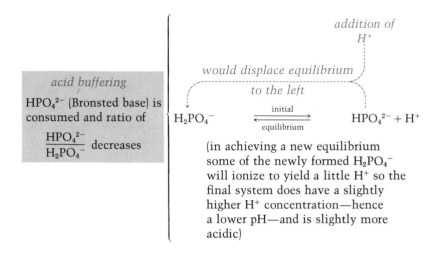

acid buffering

HPO_4^{2-} (Bronsted base) is consumed and ratio of

$\dfrac{HPO_4^{2-}}{H_2PO_4^-}$ decreases

addition of H^+

would displace equilibrium to the left

$H_2PO_4^- \underset{\text{equilibrium}}{\overset{\text{initial}}{\rightleftharpoons}} HPO_4^{2-} + H^+$

(in achieving a new equilibrium some of the newly formed $H_2PO_4^-$ will ionize to yield a little H^+ so the final system does have a slightly higher H^+ concentration—hence a lower pH—and is slightly more acidic)

In the case of OH^- addition the $H_2PO_4^-$ present would act as donor of H^+, which would consume the OH^- base. More HPO_4^{2-} would also be formed, a small amount of which would also react with a small amount of H^+ to reform a small amount of $H_2PO_4^-$. The result is that the amount of free H^+ would be slightly less than what was originally present. The change in composition in this case is that $H_2PO_4^-$ is decreased and HPO_4^{2-} is increased, i.e., the base/acid ratio of $HPO_4^{2-}/H_2PO_4^-$ increases. It can also be shown that such a solution will still be capable of buffering further additions of OH^- until the ratio gets larger than 9:1.

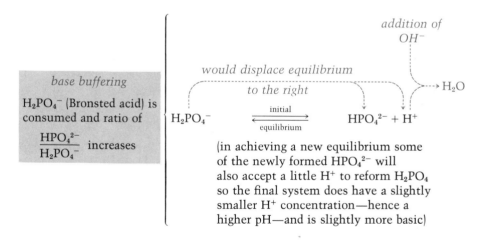

base buffering

$H_2PO_4^-$ (Bronsted acid) is consumed and ratio of

$\dfrac{HPO_4^{2-}}{H_2PO_4^-}$ increases

addition of OH^-

would displace equilibrium to the right

$H_2PO_4^- \underset{\text{equilibrium}}{\overset{\text{initial}}{\rightleftharpoons}} HPO_4^{2-} + H^+$ $\cdots\!\rightarrow H_2O$

(in achieving a new equilibrium some of the newly formed HPO_4^{2-} will also accept a little H^+ to reform H_2PO_4 so the final system does have a slightly smaller H^+ concentration—hence a higher pH—and is slightly more basic)

Within the effective range of 1:9 to 9:1 for the base/acid ratio where buffering occurs, the effectiveness of buffering is not constant. In fact, at each of the stated limits the efficiency is at a minimum. Obviously, this implies that a maximum efficiency exists within these limits. This efficiency does exist, and it corresponds to a base/acid ratio of 1:1. This 50-50 (equimolar) composition has still another meaning that is provided by the following analysis.

For the system

$$H_2PO_4^- \rightleftarrows HPO_4^{2-} + H^+$$

which is a specific case of

$$\text{conjugate acid} \rightleftarrows \text{conjugate base} + H^+$$

We have stated that

$$K_a = \frac{[\text{base}][H^+]}{[\text{acid}]} \quad \text{or} \quad [H^+] = K_a \frac{[\text{acid}]}{[\text{base}]}$$

Taking the log of both sides of the $[H^+]$ equation we have

$$\log [H^+] = \log K_a + \log \frac{[\text{acid}]}{[\text{base}]}$$

Multiplying through by a minus sign yields

$$-\log [H^+] = -\log K_a - \log \frac{[\text{acid}]}{[\text{base}]}$$

which can be rewritten as

$$-\log [H^+] = -\log K_a + \log \frac{[\text{base}]}{[\text{acid}]}$$

Now first we recognize that $-\log [H^+] = \text{pH}$. Then in a similar fashion, the p notation (meaning $-\log$ of) can be applied to the $-\log K_a$ term; hence, $-\log K_a = pK_a$; with these substitutions we have

$$\text{pH} = pK_a + \log \frac{[\text{base}]}{[\text{acid}]} \tag{3-2}$$

Finally, a meaning of a base/acid ratio of 1:1 is evident:

$$\text{pH} = pK_a + \log (1)$$

and since $\log (1) = 0$,

$$\text{pH} = pK_a$$

Thus, the pH of an equimolar (50-50) solution of a conjugate acid-base pair is equal to its pK_a value. Since such a solution has maximum buffering effectiveness, it follows then that this is also the pH at which the solution buffers best. The values of $\log (1:9)$ and $\log (9:1)$ correspond roughly to -1 and $+1$. Hence, *the range of pH where buffering is effective corresponds to $pK_a \pm 1$* and is maximum at the pK_a value. This applies to any buffer system. For a $HPO_4^{2-}/H_2PO_4^-$ solution the range is 6.2–8.2 ($pK_{a_2} = 7.2$).

A graphical summary of our analysis is contained in Fig. 3–1. The titration curve in this graph depicts the pattern of pH change (a) from low pH to high pH as strong base is added to a solution of a weak acid (such as $H_2PO_4^-$), or, looking at it the other way, (b) from high pH to low pH as strong acid is added to a solution of a weak base (such as HPO_4^{2-}). The flatter midsection of the curve corresponds to the buffering region, with the midpoint of that region corresponding to the pK_a value. The shape of this curve applies to *any weak monoprotic acid.* The excess of the base species at pH values on the high side of the pK_a value, and of the acid species at pH values on the low side of the pK_a value, is also a universal interpretation.

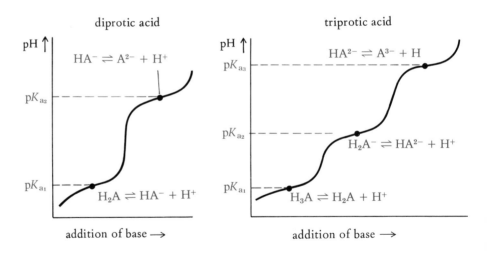

FIGURE 3–1 A titration curve for a weak monoprotic acid. Buffering action is confined to the pH range of $pK_a + 1$ to $pK_a - 1$. This is a graphical representation of the Henderson-Hasselbalch equation for the two dependent variables—pH and composition of HA and A. As pH changes, the composition changes and vice versa.

nonbuffering region

small rate of pH change as acid = buffering or base is added region

nonbuffering region

as base is added to HA

as acid is added to A

$pK_a + 1$

pK_a

$pK_a - 1$

% acid species HA in solution → 100 90   50   10 0
% base species A in solution → 0 10   50   90 100

HA in excess ← | → A in excess
e.g., $H_2PO_4^-$ e.g., HPO_4^{2-}

The curves below depict the process of titrating a weak diprotic and triprotic acid. Note that each ionization would represent a buffering system in different pH regions.

diprotic acid

$HA^- \rightleftharpoons A^{2-} + H^+$

pK_{a_2}

pK_{a_1}

$H_2A \rightleftharpoons HA^- + H^+$

addition of base →

triprotic acid

$HA^{2-} \rightleftharpoons A^{3-} + H$

pK_{a_3}

pK_{a_2}

$H_2A^- \rightleftharpoons HA^{2-} + H^+$

pK_{a_1}

$H_3A \rightleftharpoons H_2A + H^+$

addition of base →

A Recapitulation

The expression

$$pH = pK_a + \log \frac{[\text{base}]}{[\text{acid}]}$$

sometimes called the *Henderson-Hasselbalch equation*, defines pH for a solution of a conjugate acid-base pair in terms of the ionization constant of the acid and the relative amounts of the acid and base. Since K_a is a constant, the pH for a solution of this type must be fixed by the composition of the mixture. Thus, by controlling the composition, you can control the pH. This is exactly what you do in the preparation of a buffer at a specific pH for laboratory use. We will apply this principle in a later section.

There is also an important corollary to this principle, namely, that the composition of the solution is fixed by the pH. Thus, by controlling the pH, you can control the composition of the solution. That is, you can control the type of chemical species that are present in solution. It is for this reason that living organisms utilize buffers to control pH. To put it simply, *the control of physiological pH insures that the amino acids, nucleotides, proteins, nucleic acids, lipids, and most other biomolecules are maintained in certain ionic forms that are most suited for their structure, function, and even their solubility in water*, the liquid in which they are dissolved. As one example, consider the phosphate esters of the simple sugars (see Chap. 9, p. 295) represented below by glucose-6-phosphate. The fully protonated species contains two ionizable hydrogens with $pK_{a_1} = 0.94$ and $pK_{a_2} = 6.11$. In a system (i.e., a living cell) that maintains pH at about 7, which is much greater than 0.94 and about 1 pH unit greater than 6.11, the fully protonated form would not exist at all, about 10% would exist as the monoanion, and about 90% would exist as the dianion.

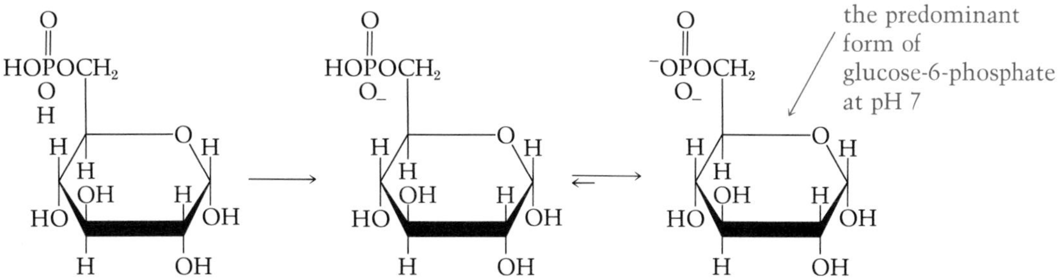

glucose-6-phosphate
(fully protonated,
non-ionic species)

monoanion
species

dianion
species

the predominant
form of
glucose-6-phosphate
at pH 7

none present
at pH 7

10% present
at pH 7

90% present
at pH 7

$pK_{a_1} = 0.94$

$pK_{a_2} = 6.11$

Buffering of Blood

An interesting illustration of buffering in biological systems is provided by what occurs in blood, where the pH is maintained within the relatively narrow limits of 7.36–7.40 for venous blood and 7.38–7.42 for arterial blood. This is accomplished through the action of three distinct conjugate acid-base pair systems: (1) H_2CO_3 and HCO_3^-, (2) the acid and base species of oxygenated hemoglobin, and (3) the acid and base species of deoxygenated hemoglobin. Before examining the intricate manner in which these systems operate, a brief description of hemoglobin is called for.

Hemoglobin is an iron-containing protein found in the red blood cells (erythrocytes). It is the most abundant protein in blood, with normal values

Although the inorganic phosphate buffer system of $H_2PO_4^- \rightleftarrows HPO_4^{2-} + H^+$ is also operative in blood, it is of secondary importance because the phosphate concentration in blood is not large enough to accommodate all of the H^+ produced.

in the range of 14–16 grams/100 ml of whole blood. As you undoubtedly know, its primary function is to transport oxygen from the lungs to respiring tissues. At the lungs, oxygen molecules enter the erythrocyte and become bound to the iron atoms of the hemoglobin molecule. A maximum of four O_2 molecules can bind to a single hemoglobin molecule (see Chap. 5 for additional details). This *oxygenated hemoglobin* (HbO_2) is then carried in the arterial blood to the various tissues of the body, where the oxygen becomes dissociated and enters the cell. The deoxygenated hemoglobin (Hb) then returns to the lungs in the venous blood, ready to participate in the same cycle.

To appreciate what all of this has to do with buffering the pH of blood, consider the following items.

1. Some years ago Bohr determined that the efficiency of O_2 binding was significantly reduced by lowering the pH. Known as the *Bohr effect*, this phenomenon means that the dissociation of oxygenated hemoglobin to deoxygenated hemoglobin would be favored by the presence of H^+.

$$HbO_2 \xrightarrow[\substack{for\ O_2\ and\ thus \\ dissociation\ is \\ favored}]{\substack{H^+\ decreases\ the \\ affinity\ of\ Hb}} Hb + O_2$$

2. When CO_2 enters the erythrocyte, it is efficiently converted to H_2CO_3 by action of the important Zn^{2+}-dependent enzyme, *carbonic anhydrase* (see p. 141). How much of the H_2CO_3 ionizes? Well, since the pH of the cell is about 7.4 and since the pK_{a_1} of H_2CO_3 is 6.35, about 90% or more will be ionized to HCO_3^- plus H^+. Thus the input of CO_2 causes an increase in the H^+ concentration and this could cause a potential increase in the acidity of blood if the free H^+ were not eliminated.

$$CO_2 + H_2O \xrightarrow[\substack{anhydrase \\ (Zn^{2+})}]{carbonic} H_2CO_3 \underset{\substack{at\ pH\ of \\ blood}}{\overset{dissociation}{\rightleftharpoons}} HCO_3^- + H^+$$

3. As does every protein, hemoglobin contains ionizable groupings contributed by some of the amino acids of which it is composed (see Chap. 4). However, a titration curve of a protein is similar in shape to that for a simple compound containing only a single ionizable group, such as that shown in Fig. 3–1 for $H_2PO_4^-$. The curve would show a single broad buffering region and the midpoint would be the pK_a of the protein. Oxygenated hemoglobin ($HHbO_2$) has a pK_a of 6.62 while deoxygenated hemoglobin (HHb) has a pK_a of 8.18. (*Note:* The symbolic abbreviation for each of the two hemoglobins now includes an H referring to an ionizable hydrogen.) The difference in these pK_a values means that the *binding of oxygen has changed a property of the hemoglobin molecule. Specifically, oxygenated hemoglobin is a stronger acid than deoxygenated hemoglobin.* Moreover, at the pH of blood the equilibrium concentration of each respective conjugate acid-base pair will be quite different. For oxygenated hemoglobin the base form (HbO_2^-) would predominate, whereas for deoxygenated hemoglobin the acid form (HHb) would predominate.

Remember, acid strength can be evaluated from K_a and pK_a values. The rule is: the stronger the acid, the larger the K_a and the smaller (because $pK_a = \log 1/K_a$) the pK_a.

for oxygenated Hb with $pK_a = 6.62$
$$HHbO_2 \underset{}{\overset{at\ pH\ \approx\ 7.4}{\rightleftharpoons}} HbO_2^- + H^+$$
$$\text{(predominates)}$$

for deoxygenated Hb with $pK_a = 8.18$
$$HHb \xleftarrow{at\ pH\ \approx\ 7.4} \rightarrow Hb^- + H^+$$
$$\text{(predominates)}$$

All three of these items fit nicely into an explanation of pH control in blood. The H^+ that is generated from H_2CO_3 when CO_2 diffuses in the cell can react with the predominating base form of oxygenated hemoglobin (HbO_2^-) to form $HHbO_2$. $HHbO_2$ has a reduced affinity for oxygen (the Bohr effect) and dissociates to yield the acid form of deoxygenated hemoglobin (HHb) and free oxygen. The oxygen diffuses out of the cell into the tissues. Because of its high pK_a value, most of the HHb would not ionize at the pH of blood but rather remain as HHb. Thus the increased amount of H^+ caused by the diffusion of CO_2 into the cell has been scavenged. The CO_2 is carried in the plasma of the venous blood as HCO_3^-. To prevent an imbalance of the ionic environment, this diffusion of HCO_3^- out of the erythrocyte is counterbalanced by the movement of Cl^- from the plasma into the erythrocyte, a phenomenon called the *chloride shift*.

events in erythrocyte *when arterial blood is delivered to respiring tissue*

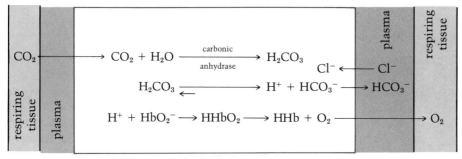

Note: an erythrocyte is actually a disk with a diameter of about 8 microns

When the venous blood reaches the lungs, O_2 and HCO_3^- enter the erythrocyte and Cl^- exits. The O_2 binds to the major hemoglobin species that is present at the pH of blood—namely, HHb—to produce $HHbO_2$. Now, however, the $HHbO_2$ would function as an acid in the presence of HCO_3^- to yield HbO_2^- and H_2CO_3. By the action, again of carbonic anhydrase, the H_2CO_3 is converted to H_2O and CO_2, and the latter diffuses into the plasma and ultimately into the lungs.

events in erythrocyte *when venous blood is delivered to lungs*

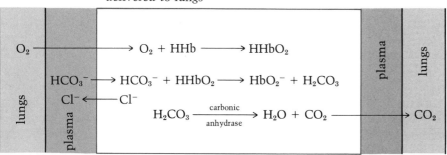

Although the preceding description oversimplifies things a little, it is a reasonable summary of important events associated with the processes by which O_2 is delivered and CO_2 is eliminated by the circulating blood without any serious alteration in blood pH—despite the production of H^+ from H_2CO_3. At the very most, the pH of venous blood is decreased by only a few hundredths of pH unit. Thus, hemoglobin is not only responsible for the transport of oxygen but also participates as an efficient buffer.

Before leaving this subject, I want to focus again on the change of the acid strength of hemoglobin upon oxygenation. This is a specific illustration of how the binding of a small, nonprotein molecule to a much larger protein molecule can cause a significant change in the function of that protein molecule. In view of the structure-function principle, you should appreciate—at least in general terms for now—that the explanation of this is that the event of binding (of oxygen in this case) modifies the structure of the protein in some fashion and that this in turn alters the activity of the protein. More will be said of this immensely important principle in later chapters.

Requirements for Biological Buffers

The primary natural buffer systems in cellular fluids are the phosphate $(H_2PO_4^-/HPO_4^{2-})$ system, dissolved proteins, and many weak organic acids, such as citric acid. However, as was pointed out earlier, the routine use of buffers in laboratory studies is equally important. Indeed, it is fundamental to both *in vivo* and *in vitro* studies to first establish a physiological pH in the system and then to prevent it from changing appreciably. Moreover, in certain instances the intention may be to determine the effect of pH on the system under study. This would require the establishment and maintenance of specific pH values within the range to be studied.

What considerations apply in selecting a buffer? The most basic requirement is that the pK_a value be *close to the pH at which you intend to use the buffer.* Why? Well, remember that a buffer system buffers best at a pH that equals its pK_a value. For physiological buffering the pK_a should be close to 7 or in a range of about 6–8. Other requirements are *nontoxicity, stability,* and *water solubility.* Although the criteria assume a research application, they are obvious requirements of natural buffer systems as well.

The various factors (the ones mentioned and some others) to be considered in selecting a buffer for use in biological research were clearly stated by N. E. Good and coworkers in their report (see legend of Fig. 3–2) on the development of new buffers for laboratory use. The study evaluated several substances according to the above criteria with the hope of finding some that would be superior in one or more respects to the small number of then available buffers—phosphate, TRIS, and citrate being at that time the most widely used. Their study succeeded and 12 new buffers were reported. Table 3–1 illustrates the primary conjugate acid-base pair of three of these, as well as other materials mentioned previously.

All of Good's buffers were zwitterionic buffers. The term *zwitterionic* refers to the presence of both a positive and a negative charge within the same molecular species. A useful symbolic representation would be $H\overset{+}{A}$. Because of this dual electrostatic character, such ions are often referred to as *dipolar ions.* In acid-base chemistry, a zwitterion is considered *amphoteric*

TABLE 3–1 pK_a values for some conjugate acid-base pairs. The abbreviations in capital letters for the last four materials are routinely used shorthand designations.

Acid		Base	pK_a	
Acetic acid CH_3COOH	$\rightleftharpoons$	Acetate ion CH_3COO^-	4.74	
Carbonic acid H_2CO_3	$\rightleftharpoons$	Bicarbonate ion HCO_3^-	6.35	
Bicarbonate ion HCO_3^-	$\rightleftharpoons$	Carbonate ion CO_3^{2-}	10.3	
Dihydrogen phosphate $H_2PO_4^-$	$\rightleftharpoons$	Monohydrogen phosphate HPO_4^{2-}	7.2	an important buffer system of natural occurrence and for years the standard buffer used in research

Citric acid

$$
\begin{array}{l}
H_2C-COOH \\
\quad | \\
HO-C-COOH \\
\quad | \\
H_2C-COOH
\end{array}
\quad \rightleftharpoons \quad
\begin{array}{l}
H_2C-COOH \\
\quad | \\
HO-C-COO^- \\
\quad | \\
H_2C-COOH
\end{array}
\quad 3.09
$$

Citrate monoanion → Citrate monoanion

$$
\begin{array}{l}
H_2C-COOH \\
\quad | \\
HO-C-COO^- \\
\quad | \\
H_2C-COOH
\end{array}
\quad \rightleftharpoons \quad
\begin{array}{l}
H_2C-COO^- \\
\quad | \\
HO-C-COO^- \\
\quad | \\
H_2C-COOH
\end{array}
\quad 4.75
$$

Citrate dianion → Citrate trianion

$$
\begin{array}{l}
H_2C-COO^- \\
\quad | \\
HO-C-COO^- \\
\quad | \\
H_2C-COOH
\end{array}
\quad \rightleftharpoons \quad
\begin{array}{l}
H_2C-COO^- \\
\quad | \\
HO-C-COO^- \\
\quad | \\
H_2C-COO^-
\end{array}
\quad 5.41
$$

N-Tris(hydroxymethyl)aminomethane [TRIS]

TRIS·H$^+$ (*protonated form*) $(HOCH_2)_3CNH_3^+$	$\rightleftharpoons$	TRIS (*free amine*) $(HOCH_2)_3CNH_2$	8.3	also a long-time, standard buffer used in research

N-Tris(hydroxymethyl)methyl-2-aminoethane sulfonate [TES]

| $\overline{T}E\overset{+}{S}·H$ (*Zwitterionic form*) $(HOCH_2)_3C\overset{+}{N}H_2CH_2CH_2SO_3^-$ | $\rightleftharpoons$ | TES (*anionic form*) $(HOCH_2)_3CNHCH_2CH_2SO_3^-$ | 7.55 | |

N-2-Hydroxyethylpiperazine-N'-2-ethane sulfonate [HEPES]

| $HE\overline{P}E\overset{+}{S}·H$ (*Zwitterionic form*) | $\rightleftharpoons$ | HEPES (*anionic form*) | 7.55 | three examples of some new buffers developed for research use by N. E. Good and workers in 1966; for various reasons, these are markedly superior to phosphate and TRIS for several types of studies |

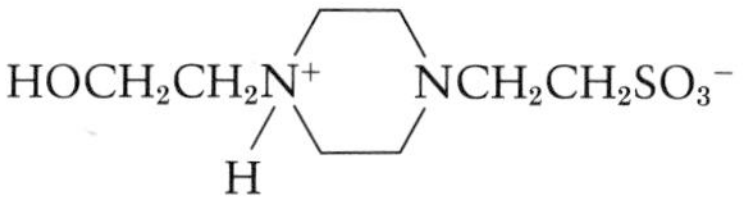

$$HOCH_2CH_2\overset{+}{N} \quad NCH_2CH_2SO_3^- \qquad HOCH_2CH_2N \quad NCH_2CH_2SO_3^-$$
$$\underset{H}{|}$$

N-Tris(hydroxymethyl)methylglycine [TRICINE]

| $T\overline{R}IC\overset{+}{I}NE·H$ (*Zwitterionic form*) $(HOCH_2)_3C\overset{+}{N}H_2CH_2COO^-$ | $\rightleftharpoons$ | TRICINE (*anionic form*) $(HOCH_2)_3CNHCH_2COO^-$ | 8.15 | |

in nature, which means that it can function as an acid or base. This is illustrated in the reactions below for a general system and for one of Good's buffers. *Amino acids*—to be discussed in the next chapter—are also zwitterionic in nature.

$$\overset{+}{HA}H \underset{+H^+}{\overset{+OH^-}{\rightleftharpoons}} H\overset{-}{A} \underset{+H^+}{\overset{+OH^-}{\rightleftharpoons}} A^-$$

Zwitterionic species

$$HOCH_2CH_2\overset{+}{\underset{H}{N}} \quad N CH_2CH_2SO_3^-$$

$+OH^-$ // $+H^+$ N-2-*Hydroxyethylpiperazine*-
                                    N'-2-*ethane*sulfonate
                                    $\overset{+}{H}\cdot H\overset{-}{E}PES$ $+OH^-$ // $+H^+$ $pK_{a_2} = 7.55$

$$HOCH_2CH_2\overset{+}{\underset{H}{N}} \quad N CH_2CH_2SO_3H$$

$$HOCH_2CH_2N \quad N CH_2CH_2SO_3^-$$

HEPES

pK_{a_1} *is very low; sulfonic*
group is strongly acidic

Their study also provides an illustration of why buffer selection can be critical in biological research. One of the *in vitro* biological assays that was used to test the effectiveness of the various buffer systems was the oxidation of succinate to fumarate by mitochondria. The oxidative metabolism of mitochondria will be discussed in greater detail in Chap. 14. At this point a brief summary will suffice. Initially, succinic acid dehydrogenase (an enzyme present in mitochondria) removes the equivalent of two hydrogen atoms (two protons and two electrons) from succinate (a dianion at pH 7). By a complex process involving several different participants that are present in mitochondria, the protons and electrons are finally accepted by oxygen to produce water.

$$^-OOCCH_2CH_2COO^- \xrightarrow{\text{mitochondria}} {}^-OOCCH{=}CHCOO^- + 2H^+ + 2e$$
$$\text{succinate} \qquad\qquad\qquad\qquad \text{fumarate}$$

Then

$$2H^+ + 2e + \tfrac{1}{2}O_2 \xrightarrow{\text{mitochondria}} H_2O$$

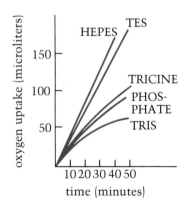

FIGURE 3–2 A study of the mitochondrial oxidation of succinate in the presence of different buffers (pH 7.4, 0.05 *M*, 20 °C). Mitochondrial activity was assayed by measuring oxygen consumption. (Data taken with permission from N. E. Good, G. D. Winget, W. Winter, T. N. Connolly, S. Izawa, and R. M. M. Singh, *Biochemistry*, **5**, 467–477 (1966).)

The basis of the assay was to determine the activity of the mitochondrial particles (isolated from beans) in the different buffer systems by monitoring the rate of oxygen consumption in the presence of succinate. The system was buffered at pH 7.4 at a concentration of 0.05 *M*. The temperature was controlled at 20 °C. Figure 3–2 summarizes the results of the experiment for five different buffers.

Immediately obvious from Fig. 3–2 is the fact that oxygen uptake (that is, mitochondrial activity) was observed to be different with each buffer that was assayed. It is also interesting to note that the traditional biological buffers, phosphate and TRIS, were inferior to three of the newly described buffers. In fact, HEPES and TES were markedly superior, with as yet unexplained beneficial effects. Similar results were obtained with other types of assay systems, such as cell-free protein synthesis. All of this points to the fact that the selection of buffers should not be routine. The presence of low biological activity or the complete absence of activity in an *in vivo* or *in vitro* system should not be hastily assumed, since such results may be attributable to undesirable buffer effects.

Preparation of Buffer Solutions

Buffers can be prepared in either of two ways: (1) both components of the conjugate acid-base pair can be weighed out separately to yield the desired ratio and then dissolved in water; or (2) both components can be obtained from a prescribed amount of only one component, with the second being formed by the addition of a specified amount of strong acid or strong base. The tools needed to prepare both types are a knowledge of the conjugate acid-base system involved, the pK_a for the system, and the Henderson-Hasselbalch equation. An example of each method is given below.

Type 1. Both components weighed out separately.

EXAMPLE. The purpose is to prepare 1 liter of a 0.5 M phosphate buffer at pH 7.5. Assume the availability of H_3PO_4, KH_2PO_4, and K_3PO_4. How would the buffer be prepared?

Solution

STEP 1

The first step is always to determine what the principal components of the buffer system will be. This is no problem with a monoprotic acid. With a diprotic or polyprotic system, however, the components can vary depending on the desired pH. In this instance pH 7.5 is specified. The equilibrium system will thus be determined by selecting the ionization having the pK_a value closest to the desired pH. Reference to a compilation of ionization data for weak acids would show that the desired system in this case is $H_2PO_4^-/HPO_4^{2-}$, with a pK_{a_2} value of 7.21. The $H_3PO_4/H_2PO_4^-$ system has a pK_a of 2.12 and the HPO_4^{2-}/PO_4^{3-} system has a pK_{a_3} of 12.3. These are too low and too high, respectively, and neither would be an effective buffer at pH 7.5. Having established this, we then write the equilibrium equation and identify the conjugate acid-base pair.

$$H_2PO_4^- \quad \rightleftarrows \quad HPO_4^{2-} \quad + H^+ \qquad pK_{a_2} = 7.21$$
$$\text{Bronsted acid} \qquad \text{Bronsted base}$$

STEP 2

Calculate the desired ratio of the acid-base pair from the Henderson-Hasselbalch equation:

$$pH = pK_{a_2} + \log \frac{[HPO_4^{2-}]}{[H_2PO_4^-]}$$

$$\log \frac{[HPO_4^{2-}]}{[H_2PO_4^-]} = 7.5 - 7.21 = 0.29$$

$$\frac{[HPO_4^{2-}]}{[H_2PO_4^-]} = \text{antilog} (0.29) = 1.95$$

Thus, the ratio desired is 1.95 parts of HPO_4^{2-} to 1 part of $H_2PO_4^-$. Since this represents a total of 2.95 parts, we can calculate directly the percentages of each component.

$$\% \ HPO_4^{2-} = \frac{1.95}{2.95} \times 100 = 66.2$$

$$\% \ H_2PO_4^{-} = \frac{1.00}{2.95} \times 100 = 33.8$$

As a check on the solution to this point, determine whether the ratio is consistent with the desired pH. The pH in this case is on the alkaline side of the pK_a value, and thus there should be a larger concentration of the conjugate base than of the conjugate acid. This is verified by the ratio calculated.

STEP 3

Determine the most feasible means of obtaining the desired components. In this instance, the obvious choice would be to weigh out the desired amount of the potassim salts of the acid-base pair—namely, K_2HPO_4 and KH_2PO_4—which upon dissolution will ionize completely to give both components of the conjugate pair.

STEP 4

Calculate the amount of each material required. Since the total phosphate concentration was specified as 0.5 M and since 1 liter is desired,

number of moles of K_2HPO_4 required/liter = $(0.662)(0.5) = 0.331$

number of moles of KH_2PO_4 required/liter = $(0.338)(0.5) = 0.169$

Finally, the grams of each required is

$(0.331 \ mole)(174.2 \ g/mole) = 57.7$ g for K_2HPO_4

and $\qquad$ $(0.169 \ mole)(136.1 \ g/mole) = 23.0$ g for KH_2PO_4

Gram formula weights:
$K_2HPO_4 = 174.2$ g/mole
$KH_2PO_4 = 136.1$ g/mole

STEP 5

Prepare the buffer. Weigh out 23.0 g of KH_2PO_4 and 57.7 g of K_2HPO_4, and dissolve this in about 750 ml of distilled water. Bring the total volume of the solution to 1 liter with distilled water, check the pH with a pH meter, and adjust if necessary.

Type 2. Both components obtained from the same source. This situation generally presents some problems to students. However, the complexity of the problem is artificial and the key to the solution involves basic chemical principles.

EXAMPLE. We need to prepare 1 liter of 0.1 M TRIS buffer of pH 8.3. Assume the availability of crystalline TRIS, 1 M HCl, and 1 M NaOH. How would you proceed? *Note:* In the crystalline state, TRIS exists primarily as the free amine, that is, with $-NH_2$.

Solution

STEP 1
The desired equilibrium is

$$(HOCH_2)_3CNH_3^+ \rightleftarrows (HOCH_2)_3CNH_2 + H^+ \qquad pK_a = 8.3$$
$$\text{Bronsted acid} \qquad \text{Bronsted base}$$
$$\text{protonated} \qquad \text{free amine}$$

STEP 2
Calculate the base-to-acid ratio by use of the Henderson-Hasselbalch equation:

$$pH = pK_a + \log \frac{[\text{free amine}]}{[\text{protonated amine}]}$$

$$8.3 = 8.3 + \log \frac{[-NH_2]}{[-NH_3{}^+]}$$

$$\frac{[-NH_2]}{[-NH_3{}^+]} = \text{antilog } (0.0) = 1.0$$

Thus the solution should contain 50% free amine and 50% protonated species. It should be apparent that this step was not really necessary. All one has to do is recognize that this will always be the situation (50:50) when the pH of the buffer is equal to the pK_a.

STEP 3

Both of the buffer components will be formed from crystalline TRIS. 0.1 mole of crystalline TRIS would be required to yield 1 liter of buffer with a total TRIS concentration of 0.1 M. The next problem is to determine the amount of strong acid that would be required to form the desired composition of the acid-base pair. Since the mixture must contain 50% of the protonated species, 0.05 mole of strong acid will be required. 0.05 mole of the $-NH_3{}^+$ form will be produced and 0.05 mole of the $-NH_2$ form will remain. 0.05 mole of H^+ would be provided by 50 ml of 1 M HCl.

STEP 4

Prepare the buffer. Weigh out 0.1 mole (12.1 g) of crystalline TRIS and dissolve it in approximately 500 ml of water. Add 50 ml of 1 M HCl and mix. Bring the total volume to 1 liter with distilled water, check the pH with a pH meter, and adjust if necessary.

IONIC STRENGTH

A complete description of an ionic solution would consider the types of ions present and the amount of each type. Chemists express this in terms of the *ionic strength* (symbolized as μ), which is defined as follows:

$$\mu = \tfrac{1}{2} \sum_i c_i z_i^2 \tag{3-3}$$

where c_i = molar concentration of the ith ionic species

z_i = electrostatic charge of the ith ionic species

$\sum_i$ = symbol for the summation of all cz^2 terms for each ionic species in solution

Note that the ionic strength is a solution property related to both the concentration of the ions in solution and the ionic nature (i.e., the charge) of the ions. Recall that cellular activity is a function of both.

Generally speaking, the ionic strength of solutions used for cellular studies is approximately 0.15. This value is commonly referred to as the *optimum physiological ionic strength.* It is considered optimal because among other things it maintains a normal water balance in living cells. When whole cells are placed in a solution that does not contain an ionic concentration at least approximately equivalent to the ionic concentration of the intracellular fluid (protoplasm), one of two things (both undesirable) will occur. When the ionic strength of the extracellular fluid is much less than that of the protoplasm, water will enter the cell, causing it to swell and ultimately burst. Alternatively, if the ionic strength of the extracellular fluid is much greater than that of the protoplasm, water will leave the cell,

The intravenous administration of sterile physiological saline (0.9% NaCl with $\mu = 0.154 \, M$) to treat dehydration, to prevent postoperative shock, and to replace fluid lost because of hemorrhage is standard medical practice. This has no effect on the red blood cells since 0.9% NaCl is *isotonic* (having the same salt concentration and hence the same osmotic pressure) with blood. A 5.5% glucose solution can also be used.

causing it to shrink and collapse. In one case the cell is flooded and in the other it is dehydrated.

The utilization of Eq. (3–3) is illustrated below.

EXAMPLE 1. Calculate the ionic strength of physiological saline solution that is 0.154 M NaCl.

Solution:

$$\mu = \tfrac{1}{2} \sum_i c_i z_i^2$$

$$NaCl \rightarrow Na^+ + Cl^-$$
$$0.154\ M \quad 0.154\ M \quad 0.154\ M$$

Thus,

$$\mu = \tfrac{1}{2}[(c_{Na^+}\, z_{Na^+}^2) + (c_{Cl^-}\, z_{Cl^-}^2)]$$

$$= \tfrac{1}{2}[(0.154)(1)^2 + (0.154)(1)^2]$$

$$\mu = 0.154\ M$$

EXAMPLE 2. Calculate the ionic strength of the buffer prepared on p. 77.

Solution: The buffer contains

$$K_2HPO_4 \rightarrow 2K^+ + HPO_4^{2-} \quad \text{and} \quad KH_2PO_4 \rightarrow K^+ + H_2PO_4^-$$
$$0.331\ M \quad 2(0.331\ M) \quad 0.331\ M \qquad 0.169\ M \quad 0.169\ M \quad 0.169\ M$$

Therefore,

$$\mu = \tfrac{1}{2}[c_{K^+} z_{K^+}^2 + c_{H_2PO_4^-} z_{H_2PO_4^-}^2 + c_{HPO_4^{2-}} z_{HPO_4^{2-}}^2]$$

$$\mu = \tfrac{1}{2}[(0.831)(1)^2 + (0.169)(1)^2 + (0.331)(2)^2]$$

Note: Total $c_{K^+} = 2(0.331) + 0.169 = 0.831$

$$\mu = 1.16\ M$$

WATER—THE BIOLOGICAL SOLVENT. HYDROGEN BONDING

Life began in water some 3 billion years ago and continues to be sustained by it. In fact, water is the most abundant material in any living organism, representing approximately two-thirds of the total weight (see p. 9). It is no mere accident that life emerged on our planet. Water was plentiful and it had a low freezing point and a high boiling point. Since the earth was neither constantly extremely cold nor hot, this meant that the beginning life had the opportunity to develop and continue to evolve in a liquid system at a moderate temperature. In the living state the basic role of water is merely to provide a fluid system in which the physicochemical processes of life can occur. Simply put, it is the biological solvent or, if you please, the solvent of the living state. As such, it is truly an essential nutrient for all forms of life. Indeed, water would properly be designated as the most essential nutrient. The following material examines the particular properties of water that are consistent with its solvent function and other biological roles.

Ions will exist only if the solvent in which they are formed prevents their natural recombination. This concept was first stated by Arrhenius in defense of his theory of ionization. He suggested that a proper solvent would very well support the existence of oppositely charged particles by minimizing the force of attraction between them. This capacity of a system to insulate oppositely charged particles from mutual attraction is reflected by

TABLE 3–2 A partial listing of dielectric constants.

Substance	Dielectric constant (D)
Water	80.4
Methanol	33.6
Ethanol	24.3
Ammonia	17.3
Acetic acid	6.15
Chloroform	4.81
Ethyl ether	4.34
Benzene	2.28
Carbon tetrachloride	2.24

the property called the *dielectric constant D*. The relationship of the dielectric constant and the force of attraction (F) between two particles carrying a negative and positive charge $(Q^-$ and Q^+, respectively), and separated by a distance r, is evident from Coulomb's law, which states that

$$F = \frac{Q^+ Q^-}{Dr^2}$$

Note that the force of attraction would be reduced in a medium with a large dielectric constant. The dielectric constants of some liquids are listed in Table 3–2. Observe that water has the highest value of those listed. In fact there are few materials that possess values greater than that for water. In view of the previous comments regarding the ionic character of many of the molecules that occur in living cells, the significance of this should be obvious: *water is very capable of supporting the existence of an ionic environment.*

Water has several other properties that contribute to its biological importance. Due to its *high heat of vaporization* (540 calories/gram), an organism can dissipate a large quantity of heat through the vaporization of small amounts of water. Because of its *high heat capacity* (1 calorie is required to raise the temperature of 1 gram of water 1 degree Celsius), an organism can absorb large amounts of heat without a correspondingly large change in its internal temperature. Both of these properties of water contribute to the maintenance of a relatively constant biotemperature. The *density of liquid water* has a maximum value at 4 °C that *is greater than that for ice.* Consequently, ice floats and preserves an environment capable of supporting the existence of the countless aquatic organisms. Water is a *good conductor of electricity* and thus contributes to the efficient transmission of impulses in nerve tissue, a basically electrical phenomenon.

The degree of all of these properties is unique for water. That is to say, the values are unexpected relative to those of other hydrides of the Group VI elements in the periodic table, namely, H_2S, H_2Se, and H_2Te. This can be explained in terms of the *highly polarized structure* of the water molecule. Because oxygen has a very high electronegativity relative to that of hydrogen, the oxygen atom tends to draw the electrons of each H—O covalent bond to itself. Thus, the electrons are shared unequally. Because of the greater electron density near the oxygen atom, it is designated as having a partial negative charge (δ^-). Conversely, each hydrogen atom having a reduced electron density carries a partial positive charge (δ^+). The net result is that two *permanent dipoles* are established in the molecule.

Electronegativity is a chemical term that refers to the ability of an atom of an element to attract electrons toward its nucleus. Fluorine, the most electronegative element, has an electronegativity value of 4.0. The values of the major elements of which biomolecules are composed are listed below.

Element	Electronegativity
Fluorine (F)	4.0
Oxygen (O)	3.5
Nitrogen (N)	3.0
Carbon (C)	2.5
Sulfur (S)	2.5
Phosphorus (P)	2.1
Hydrogen (H)	2.1

In a covalent bond, the greater the difference in electronegativities of the two atoms, the greater the *polarity of the bond*. As described, the O—H bonds in the water molecule are *polar bonds*. In contrast, bonds involving the same atoms, such as C—C, are without electronegativity difference and thus are without polarity. They are *nonpolar*. Bonds involving two atoms of small electronegativity difference, such as C—H, are so weakly polar that they are also considered as being nonpolar.

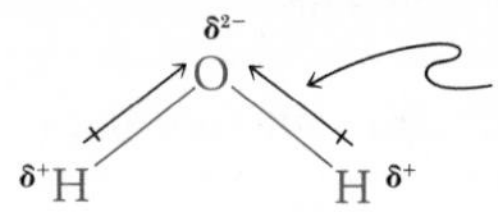

δ^{2-} O

δ^+H   H δ^+

electrons of the covalent O—H *bond are attracted toward the more electronegative oxygen nucleus; this condition is essentially permanent and thus the dipoles* (↔) *are permanent*

This strong dipolar character is directly responsible for the high dielectric constant of water. Positive and negative ions are shielded from mutual interaction by a sheath of water molecules that align themselves around the spherical ions according to electrostatic principles. This engulfment stabilizes the ions in solution and maintains their existence. This phenomenon is specifically referred to as the hydration of solvent water molecules on dissolved solutes. This principle is diagramed below in a two-dimensional representation for a positive and a negative ion. In three dimensions, water molecules would entirely surround the ions.

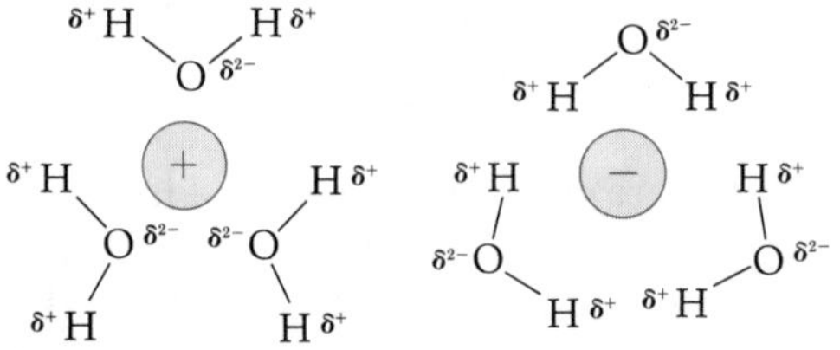

hydration of ⊕ ion
engulfment

hydration of ⊖ ion
by polar solvent molecules

Another extremely important effect of the permanent dipolar character of water is that water molecules can interact with each other by *hydrogen bonding*. A single hydrogen bond is due to the natural electrostatic attraction that one end of a dipole (the δ^- O) has for the end of another dipole (the δ^+ H). This occurs when the two dipoles approach each other very closely. The presence of these permanent dipole-permanent dipole interactions is directly responsible for the high boiling point and the high heat of vaporization of water. The hydrogen bond itself is a weak bond, but the large population of such bonds must be taken into consideration when referring to a given volume of water. For example, 1 milliliter of water contains approximately 3×10^{22} molecules, the majority of which are involved in hydrogen-bond formation in small clusters, with a constant exchange of participating molecules.

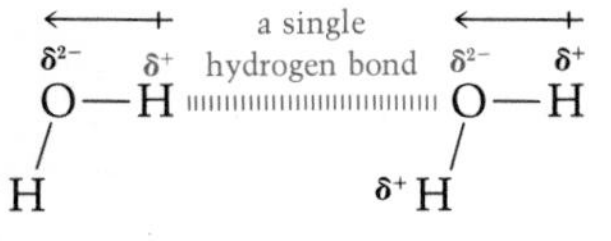

a single
hydrogen bond

δ^{2-} δ^+ δ^{2-} δ^+
O—H ||||||||||||||||||||||||| O—H
δ^+H δ^+H

Hydrogen Bonding in Organic Biomolecules

Hydrogen bonding is not exclusively confined to water. On the contrary, it is common to many systems, with the main requirement being the presence of permanent dipoles. They may occur *intermolecularly* (between different molecules) or *intramolecularly* (within the same molecule). It is preferable that one dipole contain a hydrogen atom with a partial positive charge, and the other dipole contain an oxygen or nitrogen atom with a partial negative charge. For example, the giant protein and nucleic acid molecules are stabilized by hydrogen bonds between dipoles of the following type:

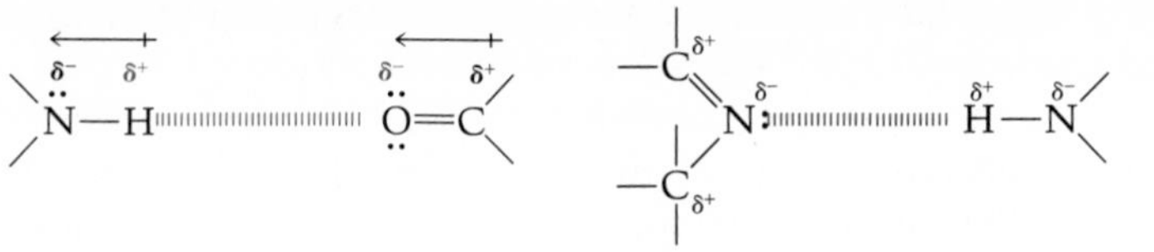

In future chapters, the biological significance of this principle will be illustrated repeatedly. Indeed, we will be left with the inescapable conclusion that hydrogen bonding is a principle that has a tremendous role in the design of nature. In addition to serving as a stabilizing force in the structure of proteins and nucleic acids, hydrogen bonding is also involved in the biochemical expression of genes. That's what you call relevance!

Although the energy of the hydrogen bond is rather small (see Table 3–3), the *presence of several bonds acting cooperatively* does constitute a *considerable stabilizing force*. One further point is that there are good and less good hydrogen bonds, depending on the spatial orientation of the two dipoles. The best situation occurs when both dipoles are coaxial. The bond energy is less if they are coplanar but not coaxial. It is smallest when the dipoles are neither coplanar nor coaxial.

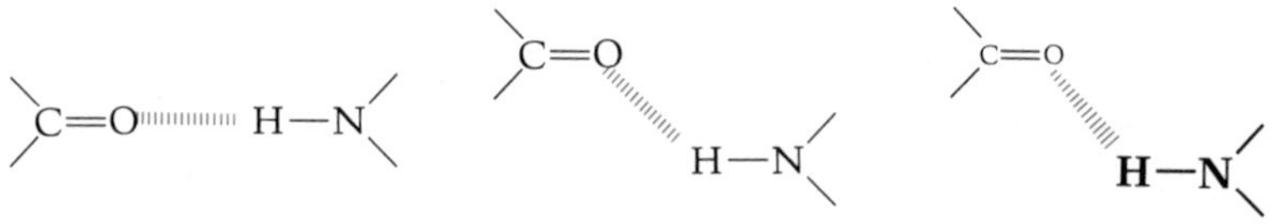

Coaxial
(optimum)

Coplanar;
not coaxial
(intermediate)

neither coplanar
nor coaxial

OTHER NONCOVALENT BONDS OF BIOLOGICAL IMPORTANCE

There are three other types of noncovalent bonding that occur in biomolecules. The easiest to understand is *ionic bonding*, which involves nothing more than a natural electrostatic force of attraction between oppositely charged ionic groupings. An important example is the attraction between the negatively charged carboxylate group ($-COO^-$) and the positively charged protonated amino group ($-NH_3^+$).

$-COO^- \text{ⅲⅲⅲⅲⅲⅲⅲ} H_3\overset{+}{N}-$

Hydrophobic (water hating) *bonding* occurs when nonpolar structures are present in a very polar medium such as water. Let us develop an understanding of this type of bonding by considering the example of oleic acid and water. At room temperature, oleic acid is an oily liquid and very insoluble in water. It is clear from the formula that a large part of oleic acid is

nonpolar (only at C—C and C—H bonds) | slightly polar

in other drawings symbolized as

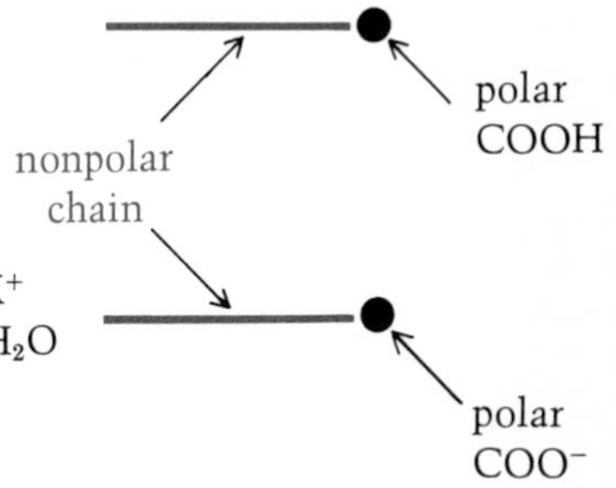

$CH_3CH_2CH_2CH_2CH_2CH_2CH_2CH_2CH=CHCH_2CH_2CH_2CH_2CH_2CH_2CH_2COOH$
oleic acid (protonated form;
water-insoluble oil)

addition of
K^+OH^-

$CH_3CH_2CH_2CH_2CH_2CH_2CH_2CH_2CH=CHCH_2CH_2CH_2CH_2CH_2CH_2CH_2COO^-$ + K^+
+ H_2O

nonpolar | strongly polar

ionic form (water-soluble)

nonpolar
chain

polar
COOH

polar
COO$^-$

extremely nonpolar (hence its water insolubility) but that a small part (the carboxyl group —COOH) has some polar character.

On shaking the two together, a white, cloudy suspension results; in a short time this suspension separates into the original two phases. The suspension consists of spherical particles called *micelles* that are formed under the influence of hydrophobic effects (see diagram below). When oleic acid molecules are dispersed in water, it is argued that a few water molecules orient themselves around each oleic acid molecule. However, the arrangement is incompatible, since water is polar and much of the oleic acid molecule is nonpolar. Rather, the oleic acid molecules will tend to attract each other and expel the water molecules (the hydrophobic effect) from their initial orientation. The mutual attraction of and water expulsion by several oleic acid molecules produces the micelle aggregate, the core of which consists of the mutually attractive nonpolar portion of oleic acid molecules with the polar carboxyl group projecting on the surface. The polar surface of the micelle can interact with water molecules and thus the micelle can exist in water as a hydrated aggregate. In this case it exists for only a few minutes because the interaction of COOH and H_2O groupings is not strong enough for a permanent engulfment. The aggregates deteriorate and eventually two separate layers are reformed.

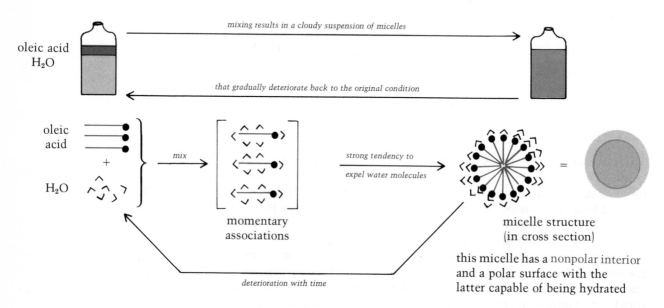

A dramatic event occurs if a small amount of strong base (such as KOH) is added before shaking the oleic acid and water together. Again the cloudy suspension results with only a little shaking; however, once formed it remains. The micelles that form in this case are much more stable. Why? Well, the surface of the aggregate is now ionic and water molecules can form better bonds—bonds that are strong enough to overcome any tendency for the aggregate to deteriorate. The ionic surface is due to the existence of carboxylate ion groups (—COO⁻) that are formed in the presence of the strongly basic potassium hydroxide.

In later chapters you will learn to appreciate that hydrophobic associations are of basic importance to the molecular architecture of proteins, nucleic acids, and biomembranes.

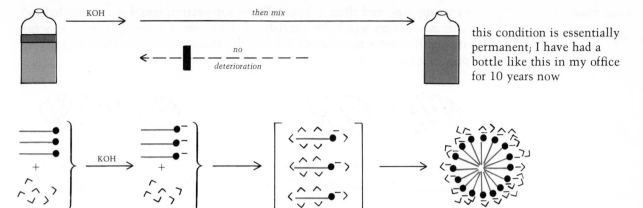

this condition is essentially permanent; I have had a bottle like this in my office for 10 years now

∧ is H₂O

a much more stable micelle with a highly polar surface for a stronger interaction with water

The third type of noncovalent bonding occurring in biomolecules is *van der Waals bonding,* which is similar to hydrogen bonding in that it involves attractive forces among dipoles. However, whereas hydrogen bonds involve permanent dipoles of highly polar covalent bonds such as O—H, van der Waals forces involve induced, nonpermanent dipoles of weakly polar covalent bonds such as C—H. Although weaker than other noncovalent bonds (see Table 3–3), this type of bonding is no less real and in fact is the basis for the mutual association of nonpolar materials for each other, such as in the interior of the micelles we have just described.

A nonpolar bond such as C—H results from the lack of any appreciable difference in the electronegativities of the atoms involved. Thus a naturally permanent dipole does not exist. However, when another atom comes close, a dipole can be induced. Why? Well, when two atoms are in close contact the clouds of negatively charged electrons around each begin to repel each other. Since the electrons are involved in bonding, this repulsion would result in a *distortion of the electron density along the bond axis.* This, in turn, results in a momentary slight electron deficiency on one atom of the bond and a slight excess of electron density on the other atom of the bond. In other words, *the bond has been slightly polarized.* Once polarized, this bond would tend to influence neighboring atoms and induce

TABLE 3–3   **A summary of some bond energies.**

	Type of bond	Approximate energy (kcal/mole)
Strong covalent bonds	H₃C—CH₃	88
	H—H	104
	H₃C—H	104
	H₂C=CH₂	163
	N≡N	226
Weaker noncovalent bonds	Hydrogen bond	5
	Ionic bond	5
	Hydrophobic bond	1–3
	van der Waals bond	1

other dipoles, and then a dipole-dipole interaction could occur (see below). The interaction would be short-lived, however, because the bond is weak and as the atoms move away the van der Waals influence is lost and the dipole is lost.

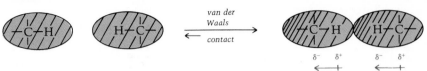

two nonpolar covalent bonds with a uniform electron density around the atoms

at close range a distortion of electron density is induced and it in turn induces a distortion in the other bond; that is, dipoles are induced

As you proceed in your study of biochemistry, I am confident you will appreciate the significance of the subject matter to which this chapter was devoted. Many items that relate to these principles will be encountered; some of these include the isolation and characterization of biomolecules, the ionic character of proteins due to contributions of some of the constituent amino acids, the stabilizing forces of protein and nucleic acid structure, the interaction of protein and lipid molecules in biological membranes, the structural instability of adenosine triphosphate (ATP), the ordered interaction of reactant molecules with the enzyme catalyst, the role of metal-containing proteins, and the transmission of genetic information. To include them in one chapter would likely create more chaos than order in your mind. So now, with your understanding of principles regarding the aqueous ionic environment, let us move on.

LITERATURE

BAILAR, J. C., JR. "Some Coordination Compounds in Biochemistry." *Am. Scientist*, **59,** 586 (1971). A review of the metal ions in life processes.

DOUZOU, P., and P. MAUREL. "Ionic Control of Biochemical Reactions." *Trends Biochem. Sci.*, **2,** 14–17 (1977). A short article on the biological significance of ionic strength.

HUGHES, M. N. *The Inorganic Chemistry of Biological Processes.* New York: John Wiley, 1975. An introduction to the biochemistry of metal ions. Written for chemists.

MONTGOMERY, R., and C. A. SWENSON. *Quantitative Problems in the Biochemical Sciences.* 2nd edition. San Francisco: W. H. Freeman, 1976. Chapters 6, 7, and 8 of this problems book are devoted to ionization and buffers.

PAULING, L. *The Nature of the Chemical Bond.* 3rd edition. Ithaca, New York: Cornell University Press, 1960. A classic description of chemical bonding. Hydrogen bonding is discussed in Chap. 12.

SEGEL, I. H. *Biochemical Calculations.* 2nd edition. New York: John Wiley, 1976. A book on how to solve mathematical problems in general biochemistry that can be used in conjunction with standard textbooks. A large number of problems—solved in detail—are accompanied by descriptive background. Approximately one-third of the book is devoted to ionization, titrations, and buffer systems.

SOBER, H. A., ed. *Handbook of Biochemistry: Selected Data for Molecular Biology.* 3rd edition. Cleveland: The Chemical Rubber Company, 1975. A multivolume, in-depth compilation of evaluated data for those engaged in biochemical research. All of the main types of biologically occurring materials are treated.

WILLIAMS, V. R., W. L. MATTICE, and H. B. WILLIAMS. *Basic Physical Chemistry for the Life Sciences.* 3rd edition. San Francisco-London: W. H. Freeman, 1978. An excellent text of physicochemical principles designed for students in the life sciences. Acid-base equilibria and buffers are treated in Chap. 4.

EXERCISES

3–1. The structure and pertinent physicochemical data for 3-phosphoglyceraldehyde—an intermediate in carbohydrate metabolism—are given below. Identify what ionic species would be present at physiological pH and calculate the approximate percentages of each.

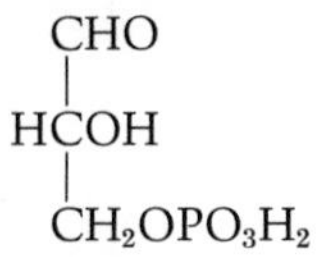

$$
\begin{array}{l}
CHO \\
| \\
HCOH \\
| \\
CH_2OPO_3H_2
\end{array}
\qquad pK_{a_1} = 2.10; \qquad pK_{a_2} = 6.80
$$

3–2. Glycine is a weak diprotic acid. The first ionization of glycine is

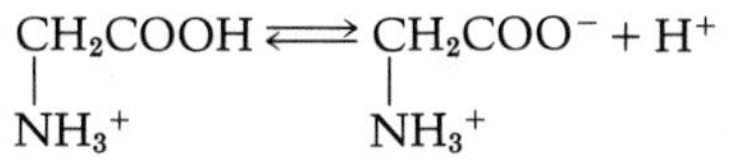

$$
\begin{array}{l}
CH_2COOH \\
| \\
NH_3^+
\end{array}
\rightleftharpoons
\begin{array}{l}
CH_2COO^- + H^+ \\
| \\
NH_3^+
\end{array}
\qquad pK_{a_1} = 2.0
$$

Draw a titration curve that would correspond to this ionization and identify the points on the curve that would correspond to pH values when (a) 90% of the acid species is present, (b) 75% of the acid species is present, and (c) 40% of the base species is present. Do not merely use the scale shown in Fig. 3–1. Calculate each pH value from the Henderson-Hasselbalch equation.

3–3. If a small amount of glycine (see Exercise 3–2) was added to a concentrated buffer solution at pH 2.0, what species of glycine (and the amounts of each) would exist? Explain briefly.

3–4. Compare the structures and pK_a values of glycine (see Exercise 3–2) and acetic acid (see Table 3–1) and describe the effect of the amino group on the acidity of the carboxyl group in glycine relative to the carboxyl-group in acetic acid.

3–5. How would you prepare 1 liter of a 0.01 M citrate buffer at pH 4.55 from monosodium citrate (mol. wt. = 214 g/mole) and disodium citrate (mol. wt. = 236 g/mole)?

3–6. How would you prepare 1 liter of a 0.02 M phosphate buffer at pH 7.6 from crystalline KH_2PO_4 and a solution of 6 M NaOH?

3–7. How would you prepare 1 liter of a 0.05 M HEPES buffer at pH 7.15? At room temperature HEPES is a crystalline solid existing largely in the zwitterionic state (mol. wt = 238). Assume the availability of concentrated hydrochloric acid that is approximately 12 M and solid NaOH.

3–8. Calculate the ionic strength of the buffers prepared in Exercises 3–5 and 3–6.

3–9. From the information given in this chapter, and also given that the molecular weight of hemoglobin is 66,000 g/mole and the volume of blood in normal human adults (2080 ml, male and 1520 ml, female) compute the amount of hemoglobin in grams and pounds in the human adult.

3–10. What is your understanding of the following? (a) hydrogen bonding; (b) hydrophobic bonding.

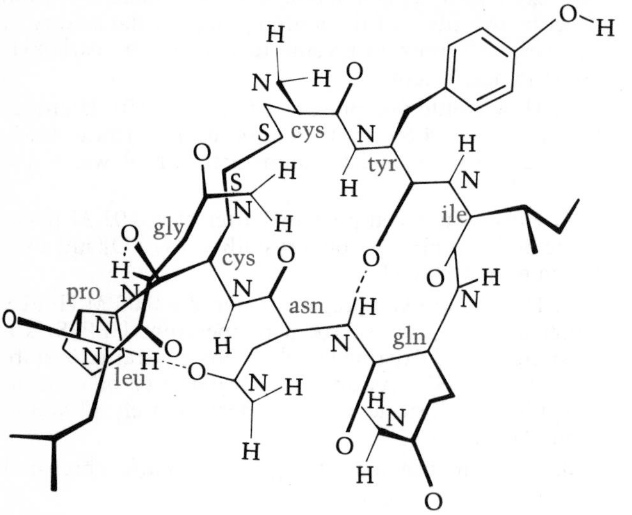

CHAPTER 4

AMINO ACIDS AND PEPTIDES

In the science of biochemistry there is a distinct emphasis on the study of proteins. This is because proteins are involved in a greater number and a greater variety of cellular events than any of the other types of biomolecules. (A listing of the many biological roles of proteins can be found on page 118. In fact, in one way or another you will be learning about proteins throughout most of this course.

Each protein is different from every other protein in terms of its structure and function. We will deal with the reasons for and examples of this individuality beginning with the next chapter. There is also a lot of similarity, the most common aspect of which is that all proteins are polymers, composed of *amino acids*—the monomers that are successively linked to each other by what is called the *peptide bond*. A substance composed of amino acids so linked together is called a *polypeptide* or simply a *peptide*.

Since amino acids are the basic building blocks of all proteins, it is only logical that a knowledge of amino acid biochemistry is needed to understand protein biochemistry. Gradually, however, it will become evident that the biological importance of the amino acids relates to other factors as well, including specific functions involving the metabolism of individual amino acids. For example, methionine functions as a donor of methyl groups; glutamic acid is a key intermediate in the detoxification of ammonia in mammals; glycine is one of the primary biosynthetic precursors of the heme group of hemoglobin; phenylalanine and tyrosine are converted to adrenalin, a substance that you are no doubt familiar with; and the listing could go on and on. These and other considerations will be discussed much later in Chap. 17. The purposes of this chapter are to acquaint you with the identity of the amino acids, to describe some basic principles concerning their chemistry, and to describe some of the important nonprotein peptides that occur in nature.

This chapter will introduce the use of photographs of *space-filling molecular models* to supplement the traditional two-dimensional representations of chemical structure. They are included as visual aids to assist you in developing a mental picture of the three-dimensional world of molecular structure, which will be very helpful in eventually understanding how the molecules of nature function.

OCCURRENCE AND STRUCTURE OF AMINO ACIDS

At the present time there are approximately 300 different amino acids known to occur in nature. Many of these are found only in certain life species and some are found only in one organism. They occur either in the free state as individual molecules dissolved in protoplasm, or covalently linked with other amino acids in peptides and proteins.

Most of the natural amino acids possess the general structure shown below in three equivalent projections.

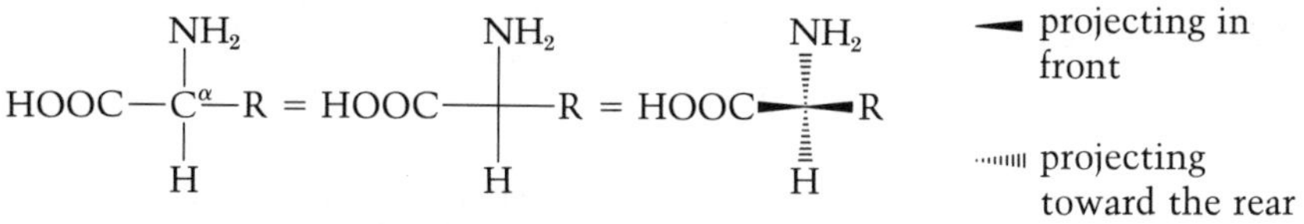

Each represents an *alpha amino acid* wherein the functional amino ($-NH_2$) and carboxyl ($-COOH$) groups are both attached to the alpha (α) carbon.

The structural feature that distinguishes any one amino acid from another is the chemical nature of the R group. It can vary from a single hydrogen atom in glycine to a more complex structure such as the guanidine in arginine.

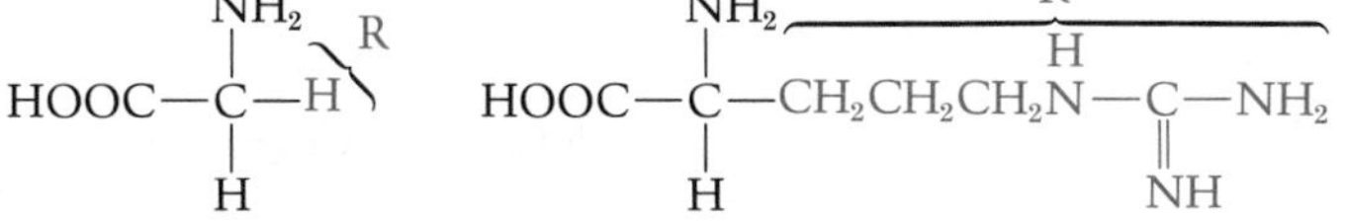

Table 4–1 gives the structure, names, and shorthand abbreviations of some of the important amino acids. The table also includes two common *imino acids* (proline and hydroxyproline), which contain the linkage $\left(\!\!\diagup\!\!\diagdown \text{N}-\text{H}\right)$ instead of the amino group. Some other amino acids not included in Table 4–1 will be encountered later in this and other chapters. The main classification used in the table is based on specific features regarding the *chemical composition* of the R group. Other classifications are also listed. One segre-

TABLE 4–1 Structures, names, abbreviations, and classifications of amino acids.

Classification based on the chemical composition of the R group:

1. Aliphatic
2. Hydroxyl
3. Sulfur
4. Aromatic
5. Acidic (and amides)
6. Basic
7. Imino

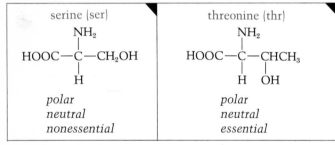

general formula

▼ identifies the 20 amino acids of which most proteins are composed

Aliphatic

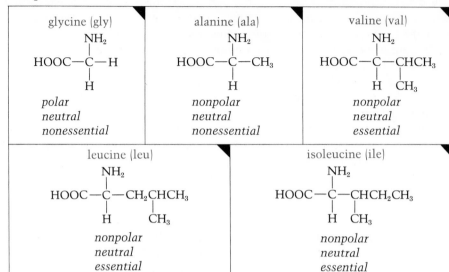

glycine (gly)

HOOC—C—H

polar
neutral
nonessential

alanine (ala)

HOOC—C—CH₃

nonpolar
neutral
nonessential

valine (val)

HOOC—C—CHCH₃

nonpolar
neutral
essential

leucine (leu)

HOOC—C—CH₂CHCH₃

nonpolar
neutral
essential

isoleucine (ile)

HOOC—C—CHCH₂CH₃

nonpolar
neutral
essential

Hydroxyl

serine (ser)

HOOC—C—CH₂OH

polar
neutral
nonessential

threonine (thr)

HOOC—C—CHCH₃

polar
neutral
essential

Sulfur

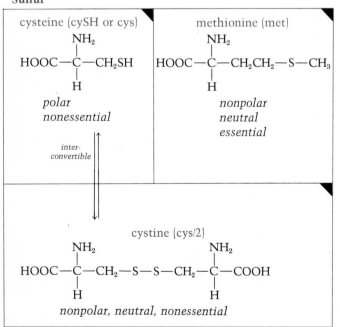

cysteine (cySH or cys)

HOOC—C—CH₂SH

polar
nonessential

methionine (met)

HOOC—C—CH₂CH₂—S—CH₃

nonpolar
neutral
essential

inter-convertible

cystine (cys/2)

HOOC—C—CH₂—S—S—CH₂—C—COOH

nonpolar, neutral, nonessential

Aromatic

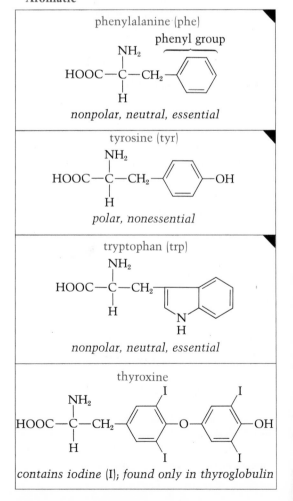

phenylalanine (phe)

phenyl group

HOOC—C—CH₂—

nonpolar, neutral, essential

tyrosine (tyr)

HOOC—C—CH₂—⬡—OH

polar, nonessential

tryptophan (trp)

HOOC—C—CH₂—

nonpolar, neutral, essential

thyroxine

HOOC—C—CH₂—

contains iodine (I); found only in thyroglobulin

TABLE 4–1 (*cont.*)

Imino

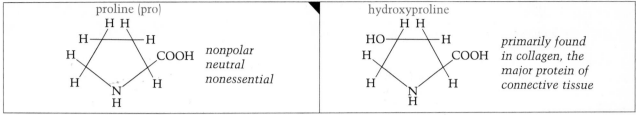

Acidic (and corresponding amides)

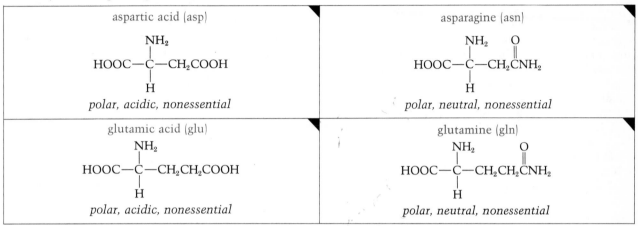

Basic

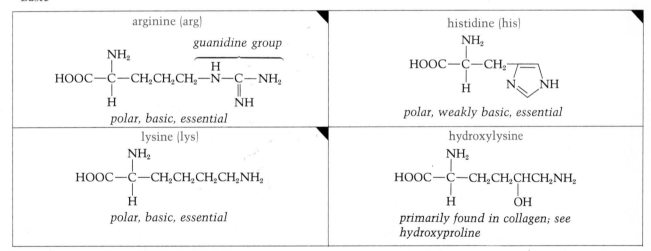

Note: Table 4–1 continues on the next page with three other classifications.

gates the amino acids on the basis of the *polar* or *nonpolar* character of the molecule, a property conferred by the R group; another groups them according to *ionic* character of the R group (see later section); another is a *nutritional* classification based on whether or not they are necessary ingredients in the daily diet of humans. You will find all four classifications useful.

One example of biochemical unity in our biosphere is that, of the large number of amino acids that do exist, only the 24 listed in Table 4–1 are used

TABLE 4–1 *(cont.)*

Other Classifications

Polarity		*Ionic*		*Nutritional (human)*	
(based on R group structure)		*(based on R group structure)*		*(based on dietary requirements)*	
Polar	*Nonpolar*	*Acidic*	*Basic*	*Essential*	*Nonessential*
glycine	alanine	aspartic acid	arginine	(human can't	(humans can
serine	valine	glutamic acid	lysine	produce these)	produce these)
threonine	leucine		*Neutral*	threonine	glycine
cysteine	isoleucine	glycine	serine	methionine	alanine
tyrosine	methionine	alanine	threonine	valine	serine
aspartic acid	phenyl-	valine	asparagine	leucine	cysteine
glutamic acid	alanine	leucine	glutamine	isoleucine	proline
asparagine	tryptophan	isoleucine	proline	phenylalanine	aspartic acid
glutamine	proline	methionine	tryptophan	tryptophan	glutamic acid
arginine		phenylalanine		arginine	asparagine
lysine	many C—C			lysine	glutamine
histidine	and C—H	tyrosine, cysteine, and		histidine	tyrosine
	bonds in	histidine can be considered			
polar C—O,	R group	as weakly basic or acidic			
C—N, and					
O—H bonds					
in R group					

N$^\epsilon$-methyllysine

glycinamide

N-acetylalanine

pyroglutamic acid

by all organisms in the biosynthesis of proteins. Note, however, that thyroxine, hydroxylysine, and hydroxyproline have a limited occurrence, and thus the common pool is reduced to 21. If we consider cystine (containing a disulfide bond) as an oxidized form of cysteine (containing a free sulfhydryl group), the number is further reduced to 20. Various modified forms of these amino acids occur in proteins. About 140 have been identified; some examples (see margin) are N$^\epsilon$-methyllysine, glycinamide, N-acetylalanine, and pyroglutamic acid. The chemical modifications are usually introduced after the parent amino acid has been incorporated into the protein. More will be said about this topic in Chap. 8 (p. 270).

STEREOISOMERISM OF AMINO ACIDS

With the exception of glycine, the alpha carbon atom in amino acids is tetrahedrally attached to four different atoms or groups of atoms. Such a carbon is called a *chiral* center. Because of this structural feature, amino acids can exist in *different stereoisomeric configurations*, distinguished from each other by the spatial orientation of the groups attached to the alpha carbon. For each chiral carbon present there are two different configurations; hence, for a substance with only one such carbon, the number of different stereoisomers is two. (Of the 20 common amino acids, only threonine and isoleucine have more than one chiral carbon.) The two stereoisomers are called L and D configurations in reference to the two nonsuperimposable mirror image structures. The nonsuperimposability of mirror images is a consequence of chirality. The L versus D designation is based on the relationship of the alpha carbon configuration to the known configuration of the two stereoisomers of glyceraldehyde. With the —CHO group oriented

The large grouping at the bottom is intended to represent the R group

$$\begin{array}{c} \text{CHO} \\ \text{HO}\blacktriangleright\text{C}\blacktriangleleft\text{H} \\ \text{CH}_2\text{OH} \end{array}$$

L-glyceraldehyde

$$\begin{array}{c} \text{CHO} \\ \text{H}\blacktriangleright\text{C}\blacktriangleleft\text{OH} \\ \text{CH}_2\text{OH} \end{array}$$

D-glyceraldehyde

$$\begin{array}{c} \text{COOH} \\ \text{H}_2\text{N}\blacktriangleright\text{C}^\alpha\blacktriangleleft\text{H} \\ \text{R} \end{array}$$

L-amino acid
(used for protein biosynthesis)

$$\begin{array}{c} \text{COOH} \\ \text{H}\blacktriangleright\text{C}^\alpha\blacktriangleleft\text{NH}_2 \\ \text{R} \end{array}$$

D-amino acid
(not used for protein biosynthesis)

space-filling model

up and away ↑ and the —CH₂OH group oriented down and away ↓ , the L label of glyceraldehyde designates—by convention—the structure where the —OH group is spatially oriented to the left. The D form of glyceraldehyde corresponds to the structural isomer wherein the —OH group is oriented to the right. The L and D forms of the alpha amino acids are designated similarly in terms of the spatial orientation of the alpha amino group, with the —COOH group oriented up and away and the —R group down and away.

The biological significance of this is that *only L-amino acids are known to occur in proteins*. Although there is no evidence as yet for the occurrence of D-amino acids in proteins, they do occur in many sources—including humans—in both a free state and as a component of other structures. Examples of the latter include the *cell wall material* of bacterial cells (see p. 310 and many *antibiotics* (see p. 111).

$$\begin{array}{c} \text{COOH} \\ \text{H}_2\text{N}-\text{C}^\alpha-\text{H} \\ \text{CH}_3 \end{array}$$

the alpha carbon of alanine is a <u>chiral carbon</u>; the same is true of all other amino acids except glycine →

$$\begin{array}{c} \text{COOH} \\ \text{H}_2\text{N}-\text{C}-\text{H} \\ \text{H} \end{array}$$

which has an achiral carbon

IONIC PROPERTIES OF AMINO ACIDS AND (POLY)PEPTIDES

Depending on pH and in accordance to the principles of acid-base chemistry discussed in Chap. 3, the alpha carboxyl and alpha amino functional groups of an amino acid exist in one of the following combinations, which are interconvertible, as shown.

$$\underbrace{\begin{array}{c}\text{—COOH}\\ \text{—NH}_3{}^+\end{array}}_{\substack{\text{protonated carboxyl}\\ \text{protonated amino}}} \rightleftharpoons \underbrace{\begin{array}{c}\text{—COO}^-\\ \text{—NH}_3{}^+\end{array}}_{\substack{\text{ionized carboxyl}\\ \text{protonated amino}}} \rightleftharpoons \underbrace{\begin{array}{c}\text{—COO}^-\\ \text{—NH}_2\end{array}}_{\substack{\text{ionized carboxyl}\\ \text{free amino}}}$$

First you should recognize that the —COOH/—NH₃⁺ species is fully protonated and hence can be considered as a diprotic Bronsted acid. Thus two ionizations are possible, and they would be described by two separate pK_a values. For most amino acids the pK_{a_1} (for the —COOH group, which is of greater acidity) has a value of about 2 and the pK_{a_2} (for the less acidic —NH₃⁺

93

group) has a value of about 9–10. The stepwise ionizations would be

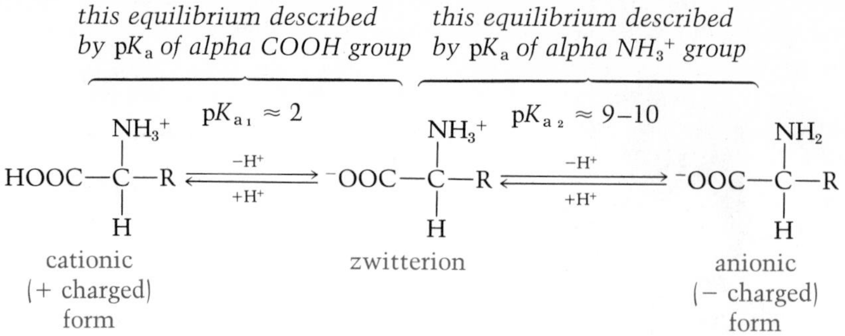

this equilibrium described by pK_a of alpha COOH group   *this equilibrium described by pK_a of alpha NH_3^+ group*

$$HOOC-\underset{\underset{H}{|}}{\overset{\overset{NH_3^+}{|}}{C}}-R \xrightleftharpoons[+H^+]{\overset{pK_{a_1} \approx 2}{-H^+}} {}^-OOC-\underset{\underset{H}{|}}{\overset{\overset{NH_3^+}{|}}{C}}-R \xrightleftharpoons[+H^+]{\overset{pK_{a_2} \approx 9-10}{-H^+}} {}^-OOC-\underset{\underset{H}{|}}{\overset{\overset{NH_2}{|}}{C}}-R$$

cationic
(+ charged)
form

zwitterion

anionic
(− charged)
form

When the pH is very low (in strongly acid solutions), the $-COOH/-NH_3^+$ species would predominate. When the pH is very high (in strongly basic solution), the $-COO^-/-NH_2$ species would predominate. When the pH is near neutral (approximately 7, which is physiological pH), the zwitterionic (see p. 74) $-COO^-/-NH_3^+$ species would predominate.

The preceding is only part of the story. In addition to the contribution of alpha amino and alpha carboxyl groups, the net molecular charge of some amino acids will also reflect the presence of a third ionizable grouping in the R side chain. The amino acids in this category are listed in Table 4–2 along with the name, structure, and ionization reaction (including pK_a value) of the R group.

Due to the presence of the extra carboxyl group, which has a distinctly acidic pK_a value, glutamic acid and aspartic acids are termed *acidic* amino acids. At physiological pH of 7 the side-chain carboxyl will definitely exist in the ionic state, contributing a full negative charge to the net charge of the

TABLE 4–2 Ionization of R groups in amino acids.

Amino acid	Additional R group ionization		Significance
Glutamic acid	$-CH_2CH_2COOH \rightleftharpoons -CH_2CH_2COO^- + H^+$	$pK_a = 3.9$	distinctly acidic R groups
Aspartic acid	$-CH_2COOH \rightleftharpoons -CH_2COO^- + H^+$	$pK_a = 4.3$	
Histidine (imidazole group)	$-CH_2-\text{(imidazole, } HN^+\text{, }NH) \rightleftharpoons -CH_2-\text{(imidazole, } N\text{, }NH) + H^+$	$pK_a = 6.0$	
Cysteine (sulfhydryl group)	$-CH_2SH \rightleftharpoons -CH_2S^- + H^+$	$pK_a = 8.3$	
Tyrosine (aromatic hydroxyl)	$-CH_2-\langle\text{ring}\rangle-OH \rightleftharpoons -CH_2-\langle\text{ring}\rangle-O^- + H^+$	$pK_a = 9.1$	
Lysine	$-(CH_2)_4\overset{+}{N}H_3 \rightleftharpoons -(CH_2)_4NH_2 + H^+$	$pK_a = 10.5$	distinctly basic R groups
Arginine (guanidine group)	$-(CH_2)_3\underset{\overset{\|}{{}^+NH_2}}{\overset{\overset{H}{|}}{N}}CNH_2 \rightleftharpoons -(CH_2)_3\underset{\overset{\|}{NH}}{\overset{\overset{H}{|}}{N}}CNH_2 + H^+$	$pK_a = 12.5$	

for the distinctly acidic and basic acids, the shaded areas identify the species that would exist at pH 7

entire molecule. At the other extreme, with the distinctly basic pK_a values, we have lysine and arginine, which are termed *basic* amino acids. At pH 7 the side chain in each case would definitely contribute a full positive charge to the net charge of the entire molecule. Histidine, cysteine, and tyrosine are not as neatly segregated and could be referred to as weakly basic or weakly acidic substances. All of the other common amino acids listed in Table 4–1 would be classified as *neutral* since their R groups do not ionize and would contribute nothing to the charge. Examples of the ionization steps for a member of each class are shown below.

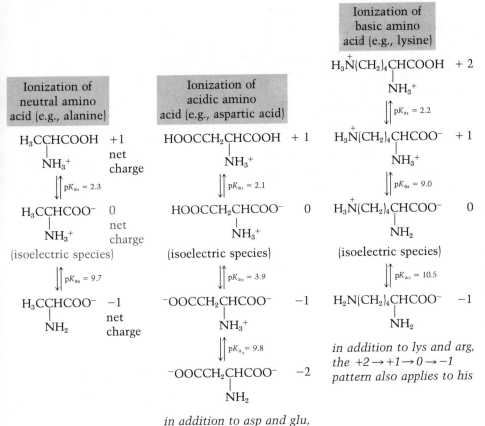

Ionization of neutral amino acid (e.g., alanine)

Ionization of acidic amino acid (e.g., aspartic acid)

Ionization of basic amino acid (e.g., lysine)

in addition to asp and glu, the $+1 \to 0 \to -1 \to -2$ pattern also applies to cySH and tyr

in addition to lys and arg, the $+2 \to +1 \to 0 \to -1$ pattern also applies to his

Acid	α COOH pK_{a_1}	α NH$_3^+$ pK_{a_2}	R pK_{a_3}
	(unless noted otherwise)		
gly	2.34	9.60	
ala	2.34	9.69	
val	2.32	9.62	
leu	2.36	9.60	
ile	2.36	9.68	
ser	2.21	9.15	
thr	2.63	10.43	
cySH	1.71 (1)	10.78 (3)	8.33 (2)
met	2.28	9.21	
phe	1.83	9.13	
tyr	2.20 (1)	10.07 (3)	9.11 (2)
trp	2.38	9.39	
asp	2.09 (1)	9.82 (3)	3.86 (2)
asn	2.02	8.80	
glu	2.19 (1)	9.67 (3)	4.25 (2)
gln	2.17	9.13	
lys	2.18	8.95	10.53
arg	2.17	9.04	12.48
his	1.82 (1)	9.17 (3)	6.0 (2)
pro	1.99	10.6	

neutral

$$pI = \frac{pK_{a_1} + pK_{a_2}}{2}$$

$$pI = \frac{2.34 + 9.69}{2} \text{ (alanine)}$$

$$pI = 6.02$$

acidic

$$pI = \frac{pK_{a_1} + pK_{a_2}}{2}$$

$$pI = \frac{2.09 + 3.86}{2} \text{ (aspartic)}$$

$$pI = 2.98$$

basic

$$pI = \frac{pK_{a_2} + pK_{a_3}}{2}$$

$$pI = \frac{8.95 + 10.53}{2} \text{ (lysine)}$$

$$pI = 9.74$$

In each of these examples, note the designation of the *isoelectric species*, that is, the ionic species *having a net charge of zero*. Also observe that in each instance the isoelectric species is formed between the +1 and −1 species. The pH at which this species would exist nearly exclusively is called the *isoelectric point*. Designated as pI, it is readily calculated as the midpoint between the two pK_a values that apply to the appropriate ionizations on either side of the zero charge species. For the neutral amino acids it is merely the midpoint between pK_{a_1} and pK_{a_2}; for the acidic amino acids, it is the midpoint between pK_{a_1} and pK_{a_2}; for the basic amino acids, it is the midpoint between pK_{a_2} and pK_{a_3}. Sample calculations are given in the margin. The position of the pI value on a titration curve corresponds to the midpoint of the inflection between the two plateau regions for each ionization. The curve in the margin on p. 97 is for a diprotic neutral amino

95

At pH 7 histidine, cysteine, and tyrosine would contribute only partially to the net charge because they would be only partially ionized.

FIGURE 4–1 *Top:* a generalized representation of an abbreviated segment of an amino acid chain. Each rectangle designates an amino acid (AA) containing a neutral (0) or an ionic (+,−) side chain. The horizontal lines signify the peptide bonds involving the α-carboxyl and α-amino groups of adjacent amino acids. The bond itself is not shown in order to emphasize the potential polyionic nature of the chain caused by the presence of free side chains with ionic groups. *Bottom:* a specific segment of a decapeptide chain illustrating the R groups of each amino acid as they would exist at approximately pH 7.

acid. Can you draw similar curves for the acid and basic triprotic amino acids?

The ionic nature of the amino acids, especially that contributed by the side chains, is a physicochemical feature that is extremely relevant to the biochemistry of polypeptides and proteins. As previously stated, the constituent amino acids are successively linked to each other by the covalent peptide bond, which is usually formed through the interaction of the alpha carboxyl and alpha amino groups. Because of this involvement in covalent bonding, these two groups do not then contribute to the net charge. In the case of the R groups we need to consider whether the amino acid is neutral, acidic, or basic. The neutral amino acids would not contribute to the net charge because their R groups lack an ionizable grouping. On the other hand, since their R groups are ionic, the acidic and basic amino acids would contribute to the ionic character of the entire molecule—acidics (glu and asp) contributing a negative charge and basics (lys and arg) contributing a full positive charge. In other words, the ionic character of a polypeptide (i.e., protein) is determined by the ionic side chains that are present. These principles are schematically summarized in Fig. 4–1. This polyionic nature of proteins and biologically active peptides plays a significant role in their overall structure, which in turn controls their biological function. It is also the basis for applying the techniques of ion-exchange chromatography and electrophoresis to the laboratory analysis (next section) of amino acids, peptides, and proteins.

ISOLATION AND SEPARATION OF AMINO ACIDS

One of the most useful techniques for resolving a mixture of amino acids is ion-exchange column chromatography (refer to p. 49 for a review of this

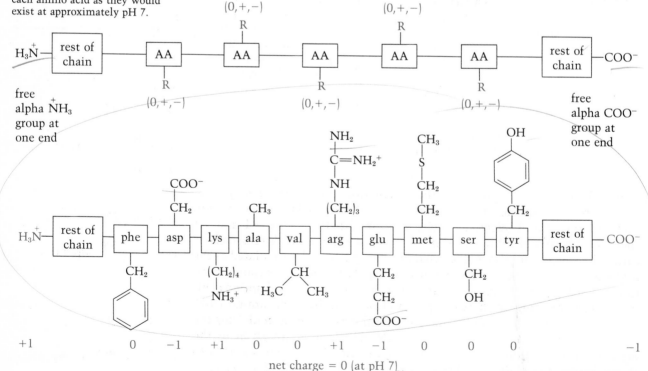

technique). Indeed, with the use of instruments called *amino acid analyzers* (see Fig. 4–2), the composition of very complex mixtures (such as urine and plasma) can be routinely determined and with great sensitivity (nanomole quantities).

After the instrument is prepared for operation, a sample of the mixture is applied to the top of a cation-exchange column. Thereafter, everything is automatic: continual pumping of an elution solvent through the column, continual reaction of the column effluent with ninhydrin (see p. 101), this chapter), and continual monitoring of the absorbance of the ninhydrin-treated effluent with a continual printout of the results on a recorder. Modern instruments are computer programed to handle all operations, including the switching of elution solvents and the quantitative analysis of the recorded data. They also require only a small volume of the sample. A typical pattern was shown earlier in Chap. 2 (see Fig. 2–8) for 50 μl of a deproteinized sample of plasma.

A detailed analysis of why the amino acids elute in a specific order is beyond our scope. The basic idea, however, is that the pH of the eluting buffer will determine the net charge on each amino acid by controlling the ionization of each amino acid. Since the pattern of ionization for each amino acid is uniquely dependent on pH (no two amino acids have exactly the same pK_a values), *each amino acid will have a different net positive charge at the pH of the eluting buffer*. Hence, each amino acid will interact

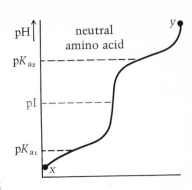

at x: all as $-COOH/-NH_3^+$
at pI: all as $-COO^-/-NH_3^+$
at y: all as $-COO^-/-NH_2$

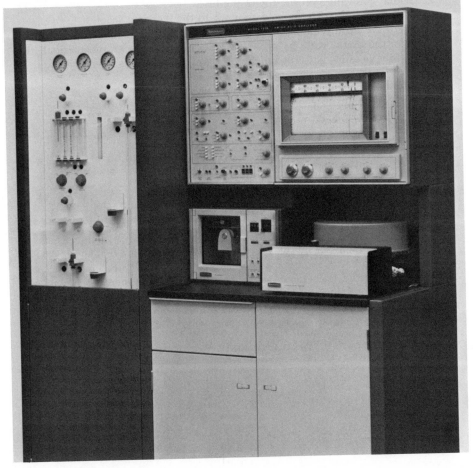

FIGURE 4–2 The amino acid analyzer shown in the photograph can detect 0.1 nanomole of an amino acid in a sample volume as small as 20 microliters. See p. 52 in Chap. 2 for an example of what the instrument can do. (Photograph supplied by Beckman Instruments.)

(exchange) with the cation-exchange resin to different degrees. Differences in the nonpolarity of the R group also result in different degrees of interaction with the solid resin phase and this also contributes to the differential movement of the amino acids.

The paper electrophoresis results (p. 46 in Chap. 2) for a mixture of leucine, lysine, glutamic acid, and histidine are more easily analyzed and worth the space to do so. The analysis involves estimating the net charge of each amino acid at the pH in question—pH 6 in this case. One of various ways this analysis can be done is outlined below. It is based on correlating pH to the ionic composition of an amino acid solution. It considers that each amino acid exists (refer back to p. 95) as one of three groups of ionic species: +1 0 −1; +2 +1 0 −1; and +1 0 −1 −2. The 0 net charge species exists at pH = pI; whenever pH = pK_a (this occurs at the midpoint between +2 and +1, between +1 and 0, between 0 and −1, and between −1 and −2), species of fractional net charge exist corresponding to a 50-50 mixture of each pair, that is, $+1\frac{1}{2}$, $+\frac{1}{2}$, $-\frac{1}{2}$, and $-1\frac{1}{2}$ respectively. Charge estimates can be made as follows.

1. On a straight line draw a lower scale of net charge values from +2 to −2 in equal increments of $\frac{1}{2}$ units with a midpoint value of 0.

$$+2 \quad +1\tfrac{1}{2} \quad +1 \quad +\tfrac{1}{2} \quad 0 \quad -\tfrac{1}{2} \quad -1 \quad -1\tfrac{1}{2} \quad -2 \;\leftarrow\; \text{net charge}$$

2. Display pH values on the top of the line.
   a. Compute the pI value and assign it to the 0 charge species.
   b. For the amino acid in question, select the right assignment of pK_a values for the charge scale and write the values in the appropriate positions; the three possible patterns for the pK_a assignments are shown below.

glu, asp, tyr, cySH → pK_{a_1} pI pK_{a_2} pK_{a_3}
lys, arg, his   → pK_{a_1} pK_{a_2} pI pK_{a_3}
all neutrals   → pK_{a_1} pI pK_{a_2}

$$+2 \quad +1\tfrac{1}{2} \quad +1 \quad +\tfrac{1}{2} \quad 0 \quad -\tfrac{1}{2} \quad -1 \quad -1\tfrac{1}{2} \quad -2$$

$\longleftarrow$ neutral acids $\longrightarrow$

$\longleftarrow$ lys, arg, his $\longrightarrow$

$\longleftarrow$ glu, asp, tyr, cySH $\longrightarrow$

3. Fill in other pH values for full charge species. If the full charge species is flanked on each side by two pK_a values, the pH at which that species exists is the midpoint of the two pK_a values. For the pH of full charge species at the ends of the scale, decrease the first pK_a value and increase the last pK_a value by 1 pH unit.
4. Finally, using the scale of pH values established by writing in the pK_a values, estimate—to the nearest half-unit—the net charge on the amino acid.

For example, the histidine and glutamic evaluations at pH 6 would be

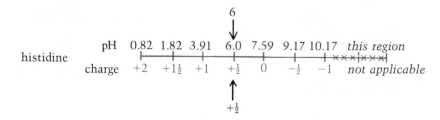

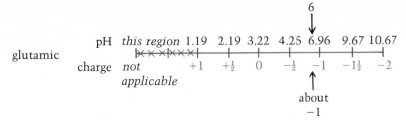

The estimation for leucine would yield 0 charge and for lysine a +1 net charge. Note that the charge analysis is consistent with the results (see p. 46 in Chap. 2) of pH 6 electrophoresis of a mixture of these four amino acids.

The (poly)ionic nature of peptides and proteins is also exploited for their separation, isolation, and characterization by the methods of both ion-exchange chromatography and electrophoresis. A polyacrylamide gel electrophoresis of serum proteins is shown in Chap. 2 (see p. 45). (We will have more to say about the task of protein isolation in the next chapter; see p. 148). Electrophoresis also provides a direct method for the determination of the *isoelectric point of a protein*, the one pH at which the protein carries no net charge, due to the presence of an equal number of individual positive and negative groupings. At the pI the electrophoretic mobility of the protein will be zero. This is determined from a series of mobility measurements at various pHs, as shown in Fig. 4–3.

The value of the isoelectric point is that it can reveal some preliminary information of a general type about the amino acid composition of the protein. For example, a protein with a greater amount of acidic amino acids than basic amino acids will have a pI much less than 7. If the reverse applies, the pI would be much greater than 7. Thus *pepsin*, a digestive protein excreted by cells in the stomach and very rich in glutamic and asparatic acids, has a pI of about 1. On the other hand, proteins associated with nuclear material and very rich in arginine and lysine have a pI of approximately 12. Knowledge of the isoelectric point can also be helpful in designing a scheme for isolating a protein, because the *solubility of a protein is at a minimum at a pH equal to pI*.

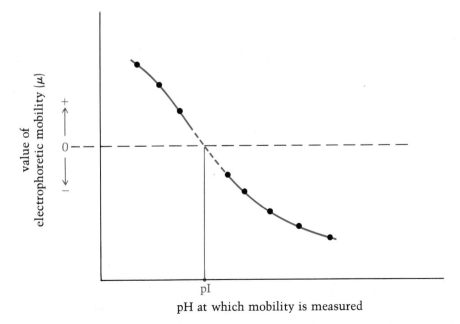

FIGURE 4–3 From a plot of experimentally determined electrophoretic mobilities at different pHs it is possible to extrapolate the data to determine graphically where the curve shows a crossover point at $\mu = 0$. The corresponding pH is the isoelectric point (pI) of the protein.

CHEMICAL PROPERTIES OF AMINO ACIDS

The amino acids are capable of participating in a wide variety of chemical reactions. These include alkylation, arylation, and acylation of the amino group; ester and anhydride formation involving the carboxyl group; and similar reactions involving the reactive groups in the side chains. Over the years a large number of reactions of these specific types, as well as other nonspecific types, have been developed for a variety of purposes such as the identification of amino acids, the chemical synthesis of peptides, the chemical modification of specific amino acids in a polypeptide, and quantitative assays. A few of the important reactions are presented in Table 4–3 along with a brief statement of their applicability. For now the information will not be supplemented further. However, note that most reactions have been cross-referenced to other pertinent sections that illustrate the described application.

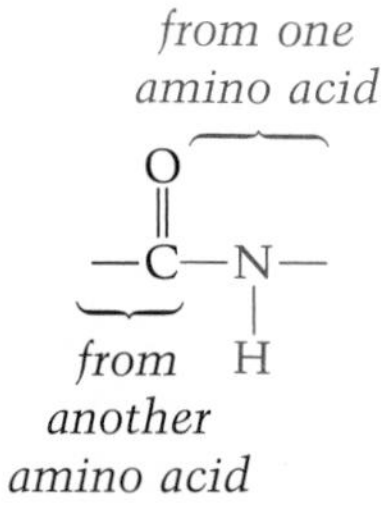

Without question, the most important reaction of biological significance involving the amino acids relates to their ability to react with each other via the alpha carboxyl group of one molecule and the alpha amino group of a second molecule to form a peptide. The amide linkage between two amino acids is called a *peptide bond*. As we will see, this pattern of stepwise extension also applies to both the cellular (biological) and laboratory (chemical) synthesis of polypeptides. However, the way in which both processes occur is much more elaborate than the direct condensation (with loss of H_2O) that is shown here. The primary concern at this point is to indicate the bonding arrangement of the peptide bond, which we will examine in greater detail in Chap. 5 (see p. 131).

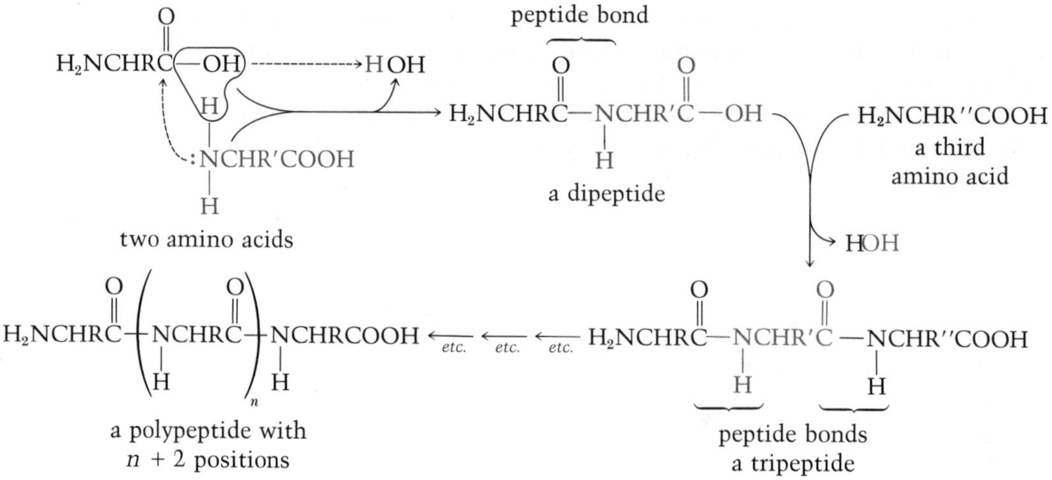

from one amino acid

from another amino acid

peptide bond

PEPTIDES

Terminology, Nomenclature, and Shorthand Notation

In the jargon of biochemistry each position in a peptide is termed an *amino acid residue*, or simply a *residue*. If the peptide contains 2–10 residues, the substance is referred to as an *oligopeptide*. Frequently, the number of residues in oligopeptides is specified by the use of Greek prefixes, as in dipeptide, tripeptide, tetrapeptide, and so on. Ordinarily, a peptide with more

TABLE 4-3   Amino acid reactions (partial listing).

1. Reaction with ninhydrin (used for detection and quantitative measurements of amino acids and peptides)

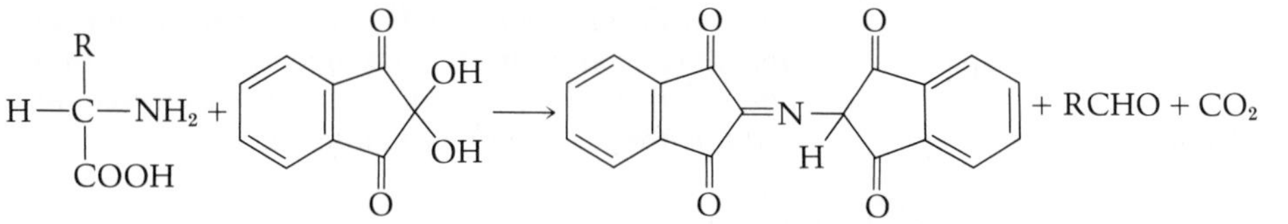

ninhydrin

a substance called Ruheman's Blue having a blue-violet color (λ_{max} = 540 nm); an exception is proline which yields a yellow color (λ_{max} = 440 nm); quantitative measurement based on Beer's law

quantitative determination could also be based on measuring the amount of CO_2 produced

2. Reaction with dinitrofluorobenzene (can be used for detection and quantitative measurements; the primary application is in determining the N-terminal residue of peptide chains; the first reagent to be developed for this purpose; see p. 126).

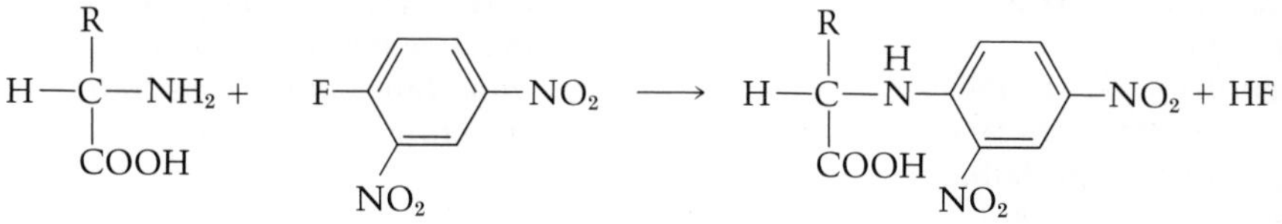

2,4-dinitrofluorobenzene
(Sanger's reagent)

the reagent will also react with certain R groups such as those of lys, arg, cySH, tyr, and his

yellow-colored dinitrophenyl (DNP) derivative; paper or thin-layer chromatography of an unknown DNP-amino acid against standards of known DNP-amino acids provides the basis of identifying the unknown

3. Reaction with phenylisothiocyanate (Edman reaction; useful in identifying amino acids with primary application in determining amino acid sequence of peptide chains from the N-terminus; see p. 126).

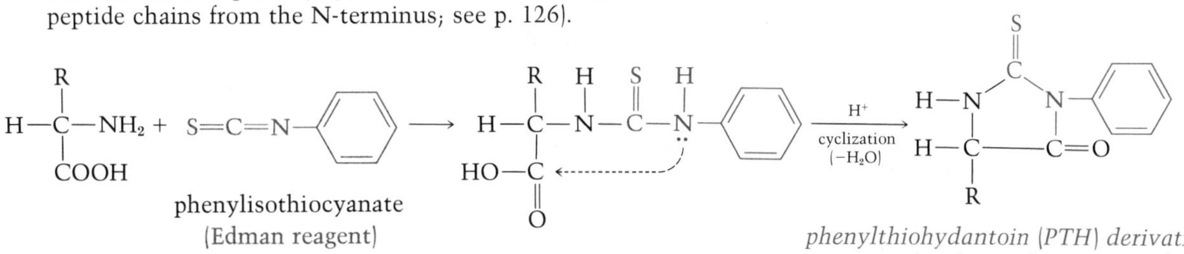

phenylisothiocyanate
(Edman reagent)

phenylthiohydantoin (PTH) derivative; paper or thin-layer chromatography of an unknown PTH-amino acid against standards of known PTH-amino acids provides basis of identifying the unknown

4. Reaction with dansyl chloride (useful in identifying amino acids with the primary application being in determining the N-terminal residue of peptide chains; see p. 126).

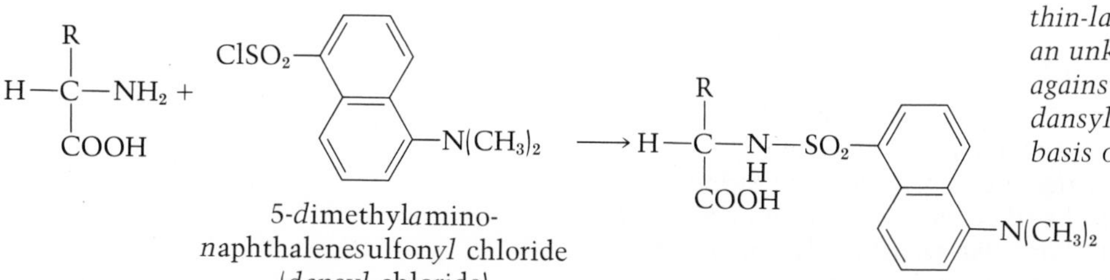

5-dimethylamino-naphthalenesulfonyl chloride
(dansyl chloride)

dansyl derivative; paper or thin-layer chromatography of an unknown dansyl-amino acid against standards of known dansyl-amino acids provides basis of identification

(Table cont. on next page)

TABLE 4-3 (cont.)

5. Reaction with trifluoroacetyl chloride (used for gas-liquid chromatography of amino acids).

$$\underset{\text{COOH}}{\underset{|}{\overset{R}{\overset{|}{H-C-NH_2}}}} + CF_3CCl \longrightarrow \underset{\text{COOH}}{\underset{|}{\overset{R}{\overset{|}{H-C-N-CCF_3}}}} + HCl$$

the trifluoroacetyl derivatives are readily vaporized without decomposition whereas the free amino acids undergo decomposition; consequently, the reaction offers a means of converting amino acids into a form suitable for gas-liquid chromatography

6. Reaction of tyrosine with tetranitromethane (an example of the selective modification of tyrosine residues in a polypeptide).

$$\boxed{P}-CH_2-\langle\rangle-OH + C(NO_2)_4 \longrightarrow \boxed{P}-CH_2-\langle\rangle-OH + {}^{-}C(NO_2)_3 + H^+$$

polypeptide (protein) containing tyrosine

tetranitro-methane (TNM)

nitroformate ion

NO_2

nitrated tyrosine residue of a polypeptide chain; the modified polypeptide may be active or inactive depending on the importance of tyrosine to its structure and function

this is but one example of many different reagents that are used to chemically modify the R groups of amino acid residues in a protein in order to determine their presence and whether or not the amino acid is essential to the structure and function of the protein; under mild conditions the reactivity of TNM is highly selective for tyrosine

7. Reaction of tyrosine with phosphomolybdotungstic acid (basis of quantitative measurement of protein content).

$$\underset{\text{COOH}}{\underset{|}{\overset{NH_2}{\overset{|}{H-C-CH_2-}}}}\langle\rangle-OH \xrightarrow[\text{+Cu}^{2+}\text{/OH}^-]{\text{phosphomolybdotungstic acid}} \text{blue-colored product}$$

this is often referred to as the Lowry method of protein analysis; it is based on detecting tyrosine, with the blue color obeying the Beer's law

8. Reactions to block the amino group (used in chemical synthesis of peptides; see p. 114; two different methods are illustrated).

$$\underset{\text{COOH}}{\underset{|}{\overset{R}{\overset{|}{H-C-NH_2}}}} + N_3-C-O-C(CH_3)_3 \longrightarrow \underset{\text{COOH}}{\underset{|}{\overset{R}{\overset{|}{H-C-N-C-O-C(CH_3)_3}}}}$$

t-butyloxycarbonyl azide (*t*BOC azide)

*t*Boc derivative

$$\underset{\text{COOH}}{\underset{|}{\overset{R}{\overset{|}{H-C-NH_2}}}} + Cl-C-O-CH_2-\langle\rangle \longrightarrow \underset{\text{COOH}}{\underset{|}{\overset{R}{\overset{|}{H-C-N-C-O-CH_2-}}}}\langle\rangle$$

benzylchlorocarbonate

carbobenzoxy(CBZ) derivative

in each instance the alpha amino group of the amino acid is protected and thus not available to participate in peptide bond formation; however, both the tBOC and CBZ groups are sensitive to certain conditions that result in their removal without affecting any peptide bonds

TABLE 4–3 (cont.)

9. Reaction of blocked amino acid with another amino acid (**chemical synthesis** of peptides; see p. 114).

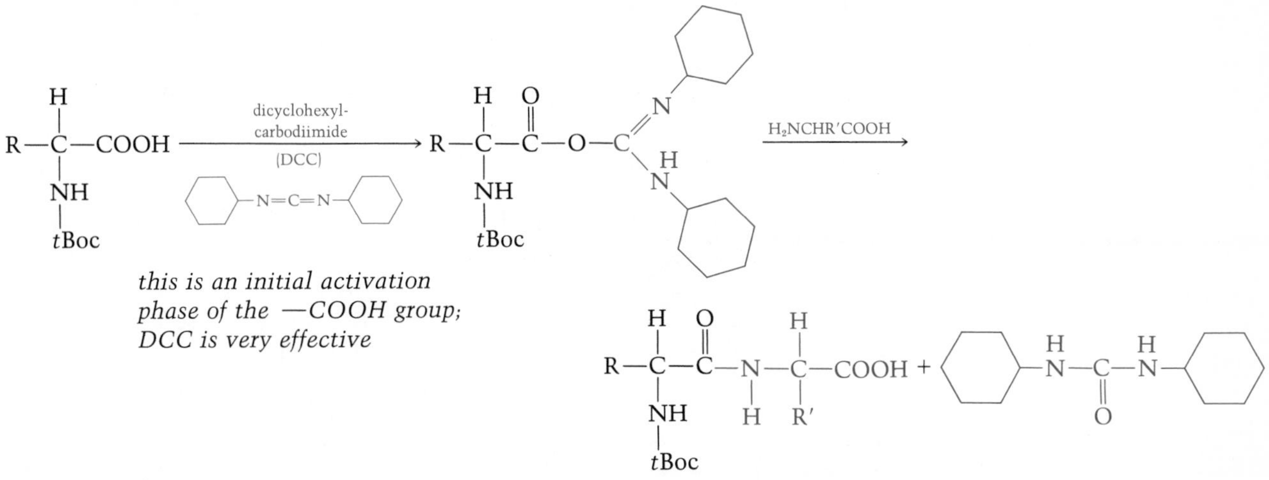

this is an initial activation phase of the —COOH group; DCC is very effective

after isolation this dipeptide in turn could be treated with DCC and a third amino acid to yield a tripeptide which in turn could be used in another cycle, and so on

10. Reaction with adenosine triphosphate — ATP (in all living cells this is the initial enzyme-catalyzed step in the biosynthesis of proteins; see page 262; the process is described as an *activation of the amino acid,* meaning that the amino acyl group of the amino acyl adenylate has a greater reactivity than the free amino acid).

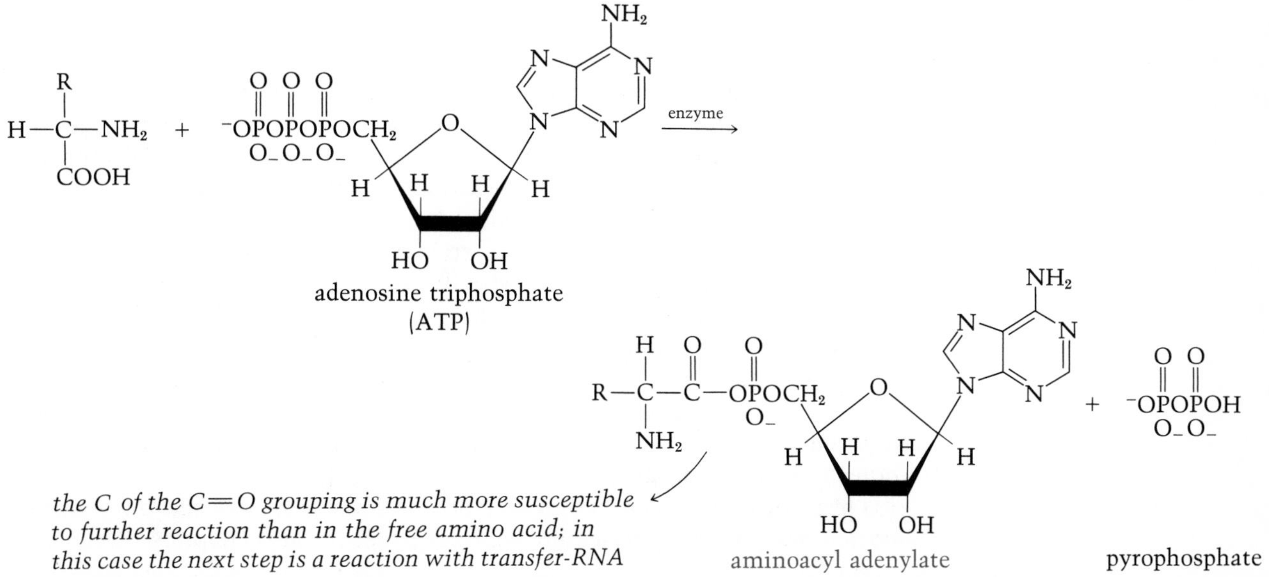

adenosine triphosphate
(ATP)

the C of the C═O grouping is much more susceptible to further reaction than in the free amino acid; in this case the next step is a reaction with transfer-RNA

aminoacyl adenylate

pyrophosphate

than 10 residues is termed a *polypeptide.* Although cyclic and branched peptides do exist, most consist of a *linear, chainlike assembly with two terminal residues.* Since the bond between successive residues involves the α-carboxyl and α-amino groups of adjacent amino acids, it follows that one terminal residue will possess a free amino group and one will have a free carboxyl group. The former is called the *N-terminus residue* and the latter the *C-terminus residue.*

Because the representation of the complete structural formula of a peptide would be cumbersome, shorthand conventions are routinely used. Linear peptides of known sequence (order of residues) are named by beginning at the N-terminus and designating each residue as an acyl substituent of the α-amino group of the succeeding residue. The peptide bond is designated by a dash. If the exact sequence of the peptide or any part of it is unknown, the residues are enclosed by parentheses and the dash is replaced by a comma. An even shorter method of nomenclature is to use the abbreviations (see Table 4–1) to designate each residue. These rules are applied below for the hypothetical octapeptide consisting of two arginine residues, two alanine residues, one glutamic acid residue, one glycine residue, one methionine residue, and one lysine residue. As shown, individual residues are frequently designated numerically, with the N-terminal residue specified as number 1.

Acyl groups are named by dropping the *-ine* or *-ic* ending of the parent name and adding the ending *-yl.*

residue
1   2   3   4   5   6   7   8 ← number
arginyl-alanyl-glycyl-arginyl-glutamyl-alanyl-methionyl-lysine

N-terminus ——————————————————————→ (C-terminus)
(internal residues)

or simply:

arg-ala-gly-arg-glu-ala-met-lys

for when only a partial sequence is known:

1   2,3,4   5   6,7   8
arginyl-(alanyl,arginyl,glycine)-glutamyl-(alanyl,methionyl)-lysine

or simply:

arg-(ala,arg,gly)-glu-(ala,met)-lys

Significance of Amino Acid Sequence

We will begin shortly to describe some naturally occurring oligo- and polypeptides (and then the proteins in the next chapter); however, this is an excellent opportunity to restate a principle of supreme importance regarding their structure and function. As you read this material you will probably ask yourself the question, why are these molecules capable of performing the specific biological functions described for each? The answer to this question is that the functional properties of each peptide are a consequence of *its amino acid sequence.* Certainly, the identity and total number of amino acid residues are important, but it is the *order in which the residues are linked together that is of greatest importance.* Why? Because it is the *sequence of residues that determines the overall three-dimensional shape of the molecule, which in turn is the feature of structure that determines how that molecule will function.* Review the description of *informational molecules* that was given in the Introduction (p. 8). An appreciation of this basic principle is crucial to your study of biochemistry and this justifies my repeating it now and emphasizing it again in later chapters.

Naturally Occurring Nonprotein Peptides

A large number of nonprotein peptides occur naturally and are found in all types of organisms. Structurally, they constitute a very heterogeneous group of materials characterized by size, cyclic chains, branched chains, the presence of D- and L-amino acids, and, in some isolated cases, a unique type of peptide bond. As you might expect in view of the structure-function principle, their physiological functions are also quite varied. This diversity is illustrated in the following specific examples.

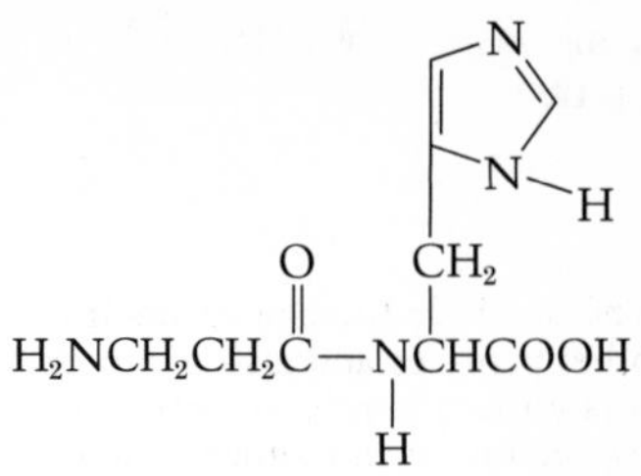

Carnosine and Anserine

Obviously, the smallest possible peptide is a dipeptide. Two examples are *carnosine* and *anserine*, both of which are found in muscle tissue of vertebrates, including human muscle. Both contain *β-alanine*, a structural isomer of α-alanine in which the amino group is on the beta carbon rather than the alpha carbon. (We will see later—p. 357—that β-alanine is also a component of an important vitamin, *pantothenic acid*.) The precise role of these peptides in muscle biochemistry is still unknown—a reminder that even though our knowledge about many things is quite sophisticated, there is a lot we don't know even about simple molecules. It has been suggested, however, that they may function in the buffering of pH in muscle cells.

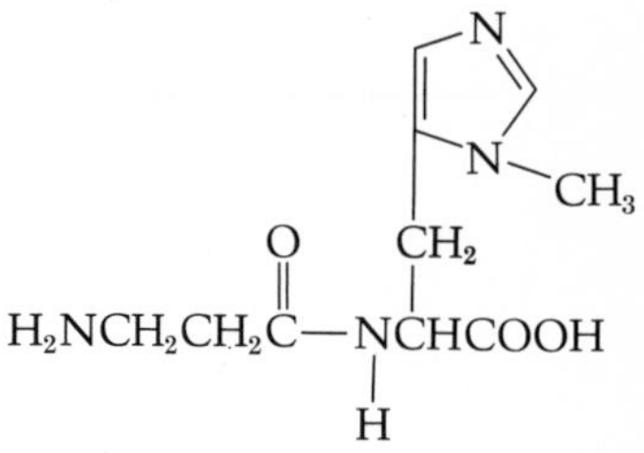

Glutathione

γ-glutamyl-cysteinyl-glycine, a tripeptide commonly called *glutathione*, is universally distributed in animals, plants, and bacteria, and is probably the most abundant simple peptide. The distinguishing structural feature is, of course, the amino acid sequence. A second feature is that the reduced form of the tripeptide contains a functional sulfhydryl (thiol) group (—SH) contributed by cysteine. A third characteristic is the participation of the γ-carboxyl group (rather than the α-carboxyl group) of glutamic acid in the peptide bond. As shown and described below, there are two forms of glutathione—*reduced* and *oxidized*.

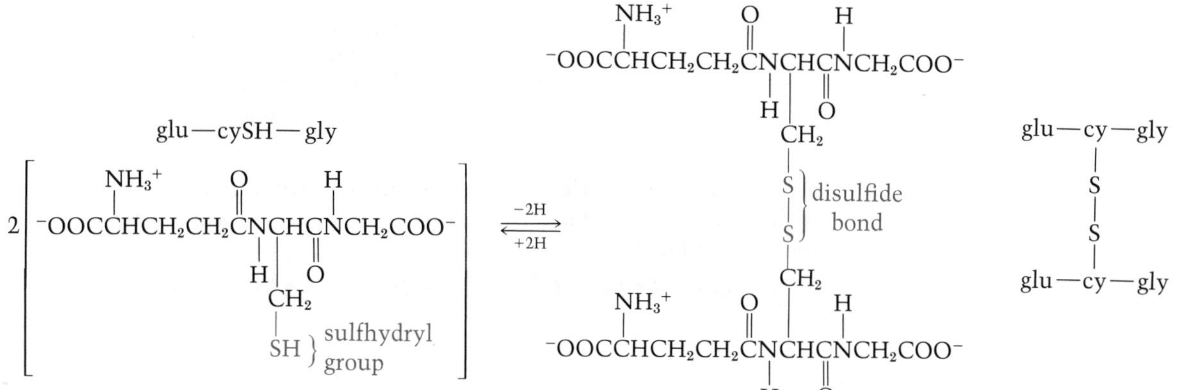

carnosine
(β-alanyl-L-histidine)

anserine
(β-alanyl-N³-methyl-L-histidine)

$$\overset{\beta}{H_2N}CH_2\overset{\alpha}{C}H_2COOH$$

β-alanine

glu—cySH—gly

GSH
(reduced glutathione)

Note: the N-terminal glutamic acid residue is linked to cysteine by its γ-COOH rather than the α-COOH

GSSG
(oxidized glutathione)

glu—cy—gly
|
S
|
S
|
glu—cy—gly

Some of the known physiological functions of glutathione are participation (1) in the transport of amino acids across cell membranes, (2) in the protection of proteins by acting as a scavenger of harmful oxidizing agents that otherwise would oxidize —SH groups in many proteins, and (3) in the maintainance of the iron atom in hemoglobin in its reduced Fe^{2+} state.   Further study may uncover other functions.   For (2) and (3) the reduced thiol form (free —SH) functions as a reducing agent (a source of electrons) that is converted to the oxidized form (a $2 \rightarrow 1$ conversion) consisting of two tripeptide segments connected by an *interchain disulfide bond* (see margin).   The following material describes the protein-protecting role of glutathione.

The disulfide linkage occurs in other peptides and also in proteins. It is either *interchain* (between two chains as in oxidized glutathione) or *intrachain* (if the —S—S— bond involves two cysteine residues within the same chain).

Many proteins are genetically programed to contain free —SH groups contributed by cysteine; however, like any compound containing —SH groups, they are also susceptible to the formation of disulfide linkages.   Although disulfide bonds are found in some proteins, if they are not genetically programed to be present they obviously should not be.   Otherwise the protein loses its native structure, which usually will mean a loss of its biological activity.   It is proposed that reduced glutathione protects the protein —SH groups from oxidation by acting as a scavenger of whatever oxidizing agent is present (O_2 for example).   Thus, the protein is not structurally modified, glutathione is.   The reduced glutathione would then be regenerated by some other reductive process to maintain its optimum level.   Approximately 90% of the nonprotein thiol compounds in mammalian tissues is in the form of reduced glutathione.

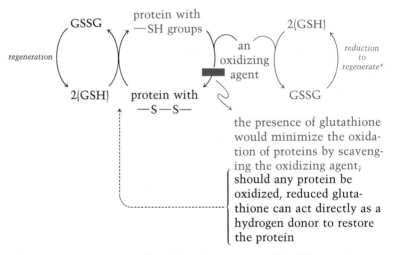

the presence of glutathione would minimize the oxidation of proteins by scavenging the oxidizing agent; should any protein be oxidized, reduced glutathione can act directly as a hydrogen donor to restore the protein

* the regeneration is catalyzed by the enzyme, *glutathione reductase,* with a reduced coenzyme, NADPH(H+) (see p. 359) serving as the source of reducing power:

$$GSSG + NADPH + H^+ \xrightarrow[\text{reductase}]{\text{glutathione}} 2GSH + NADP^+$$

Peptide Hormones

One of the most fascinating aspects of the living state is that organisms are capable of controlling their activities.   This is accomplished by regulating the chemical processes that occur in cells.   There are many facets to this regulation.   In the animal kingdom one aspect involves the production of substances that are discharged by specific producing cells and carried in the circulating fluid to other specific target cells that receive the impact of their effect. These substances are called *hormones* (from the Greek, *hormon*, to rouse or excite). In mammals (including humans) a large number of hormones of many different types with different regulatory properties are produced by various organs. A major group of hormones are peptides (of small to intermediate size) and proteins. Some representatives of this group are described in the following pages.

The endocrine glands are the major hormone-producing organs, hence the formal name for the study of hormones, *endocrinology*. However, hormone production is not restricted to the endocrine system. The scope of modern endocrinology is much broader, including any substance that regulates the activities, growth, and behavior of an organism. The foundation of this speciality (and many others) is biochemistry.

Oxytocin and Vasopressin

The pituitary gland produces several hormones, two of which are cyclic nonapeptides. *Oxytocin* stimulates the contraction of uterine muscle in the pregnant female and the ejection of milk from the mammary glands in lactating females. It is proposed to have no significant influence on the male. *Vasopressin* produces potent antidiuretic effects by stimulating the reabsorption of water by kidney. It also stimulates the contraction of smooth muscle, especially in the blood vessels, thus contributing to the control of blood pressure. *The similar and yet different effects of these two hormones are consistent with their chemical structures* (below), *which are also similar yet different.* By comparing structures, it is obvious that residues 3 and 8 are particularly important, since all other residues are identical. The basic residue (arg) at position 8 is especially important to the vasopressin properties. A similar importance for oxytocin properties results from the nonpolar aliphatic residue (ile) at position 3.

$$
\overset{+}{H_3N}-\overset{1}{cy}-\overset{2}{tyr}-\overset{3}{ile}
$$

N-terminus

all amino acids in the L form

cy—asn—gln $\overset{6\ \ 5\ \ \ 4}{}$

pro—leu—gly(CNH$_2$) $\overset{7\ \ 8\ \ \ 9}{}$

O

human oxytocin

$$
\overset{+}{H_3N}-\overset{1}{cy}-\overset{2}{tyr}-\overset{3}{phe}
$$

cy—asn—gln $\overset{6\ \ 5\ \ \ 4}{}$

pro—arg—gly(CNH$_2$) $\overset{7\ \ 8\ \ \ 9}{}$

O

amide grouping at C-terminus rather than free COOH

human vasopressin

The symbolic representations of peptides as shown above do not indicate anything about the most important aspect of structure—the overall three-dimensional shape. The structure of oxytocin (recently solved) is depicted on p. 108. It is this level of structure, conferred by the ordered interactions of all atoms in the constituent amino acids, that determines the structural and functional individuality of the oxytocin molecule.

Angiotensin

Hypertension (high blood pressure) is one of the major human diseases. Despite years of effort, however, our knowledge of the biochemical events of hypertension is still incomplete. The primary reason for this is that research into the problem has indicated that the biochemistry of the condition is extremely complex.

A major part of the process involves the action of *angiotensin II*, an octapeptide with potent pressor effects (it stimulates constriction of blood

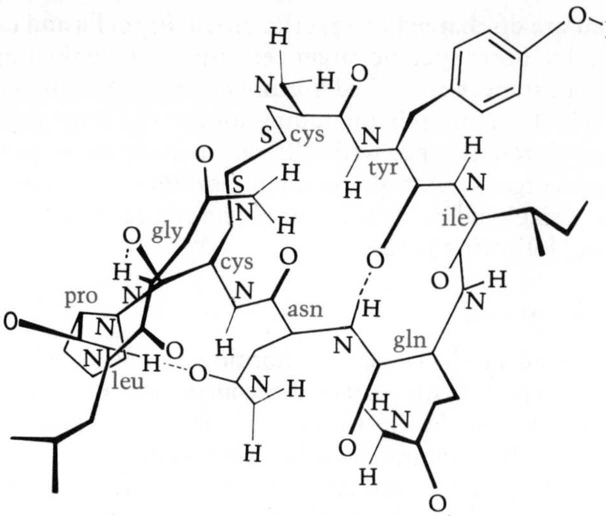

(Structure reproduced from
R. Walter, C. W. Smith, and J. Roy,
Proc. Nat. Acad. Sci. (USA), **73,**
3054 (1976). Drawing generously
supplied by Dr. Roderich Walter.)

skeletal model of oxytocin

vessels, which elevates blood pressure). Angiotensin II is produced by the removal of a dipeptide fragment from a decapeptide precursor called *angiotensin I*, which has much less pressor activity than II. The bulk of this conversion is proposed to occur as blood passes through the lung, which contains high levels of the converting enzyme. The source of angiotensin I is *angiotensinogen*, a plasma protein (polypeptide) that is originally produced in the liver. Angiotensinogen is acted on (in blood) by *renin*, an enzyme that is produced in the kidneys and secreted into blood. The renin cleaves a specific peptide bond in angiotensinogen (the bond between residues 10 and 11), which releases a decapeptide fragment, angiotensin I. In addition to the vasoconstricting effect it has on blood vessels, angiotensin II also acts (see drawing, opposite page) (a) on the central nervous system (brain) to cause thirst and to stimulate the pituitary to increase vasopressin secretion, which in turn causes water retention by the kidneys; and (b) on the adrenal cortex to cause aldosterone (see p. 523) secretion, which acts on the kidneys to cause retention of Na^+. Reasons for the increased production of renin in the kidney and the angiotensinogen in the liver are still unknown.

$$
\overset{1}{\text{asp}}-\text{arg}-\text{val}-\text{tyr}-\overset{5}{\text{ile}}-\text{his}-\text{pro}-\text{phe}-\text{his}-\overset{10}{\text{leu}}
$$

angiotensin I
(weak pressor activity)

converting
enzyme ↓ ↘ his-leu (removed)
(in lung)

$$
\overset{1}{\text{asp}}-\text{arg}-\text{val}-\text{tyr}-\overset{5}{\text{ile}}-\text{his}-\text{pro}-\text{phe}
$$

angiotensin II
(potent pressor activity)

Somatostatin

The hypothalamus produces peptide hormones that in turn act on the pituitary, controlling the production of other hormones by the pituitary (see Table 4–4 on p. 110). One of the hypothalamus hormones inhibits the pituitary's production of human growth hormone. The overall effect is to stay the growth of the body; hence the name of this hypothalamus hormone is

Synthesis of angiotensin II

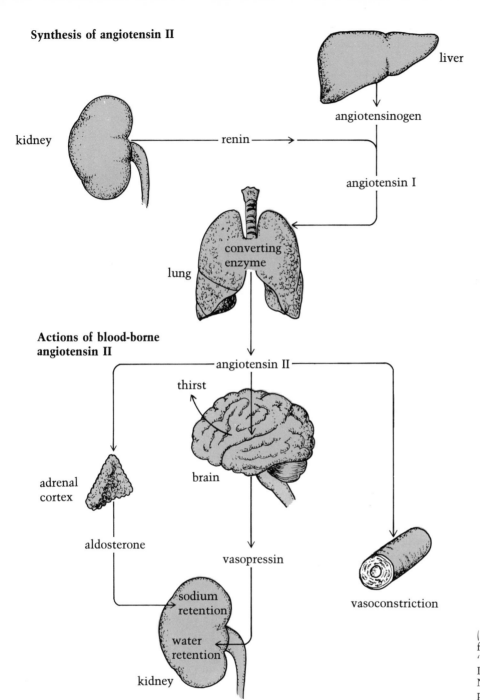

Actions of blood-borne angiotensin II

(Illustration by Nancy Lou Gahan from D. J. Ramsay and W. F. Ganong, "CNS Regulation of Salt and Water Intake," *Hospital Practice*, Vol. 12, No. 3 (March 1977). Reprinted with permission.)

somatostatin. Recent studies have established that somatostatin (14 amino acids) is also produced by the pancreas and has a variety of other regulatory actions in the body. One of these actions is control of the release of insulin and glucagon (see Table 4–4 and also p. 383 in Chap. 12) from the pancreas, which results in the lowering of blood glucose (sugar). This suggests that somatostatin may provide a new approach to the treatment of diabetes. In a remarkable accomplishment of the modern science of genetic

TABLE 4-4 Peptide and protein hormones in mammals. A partial listing.

Substance	No. of amino acid residues (in humans)	Producing site	Affected site	Effects
Gastrin	17	Stomach	Stomach	Stimulates HCl secretion
Secretin	27	Stomach	Pancreas	Stimulates the secretion of water and salts
Glucagon	29	Pancreas	Liver	Stimulates breakdown of glycogen to release glucose (see p. 383)
Calcitonin	32	Thyroid	Bone, kidney	Inhibits release of calcium from bone and stimulates excretion of calcium and phosphorus; see PTH below
Adrenocorticotrophic hormone (ACTH)	39	Pituitary	Adrenal cortex	Stimulates production of adrenal hormones
Insulin	52 (two chains: 21 and 31)	Pancreas	All cells	Controls carbohydrate, fat, and protein metabolism (see pp. 397 and 398)
Parathyroid hormone (PTH)	84	Parathyroid	Bone, kidney	Stimulates release of calcium from bone and inhibits excretion of calcium; see calcitonin
Human growth hormone (HGH)	188	Pituitary	All tissues	Controls many processes
Prolactin	198	Pituitary	Mammary gland	Stimulates milk production
Thyroid stimulating hormone (TSH; thyrotropin)	201 (two chains: 89 and 112)	Pituitary	Thyroid	Stimulates release of thyroxine (see p. 90)
Follicle stimulating hormone (FSH)	202 (two chains: 89 and 113)	Pituitary	Seminiferous tubuler (male); ovary (female)	Stimulates production of sperm and maturation of follicle

The hypothalamus of the brain produces at least three different peptide hormones—*thyrotropin-releasing factor* (TRF), *luteinizing-releasing factor* (LRF), and *growth hormone-inhibiting factor* (GIF; also called *somatostatin*)—each of which acts on the pituitary in a specific way to control hormones produced by the pituitary.

TRF is   pyroglu—his—pro-amide
LRF is   pyroglu—his—trp—ser—tyr—gly—leu—arg—pro—gly-amide
GIF is   shown below

engineering, bacteria have been genetically programed to produce somatostatin in large quantities (see p. 280 in Chap. 8).

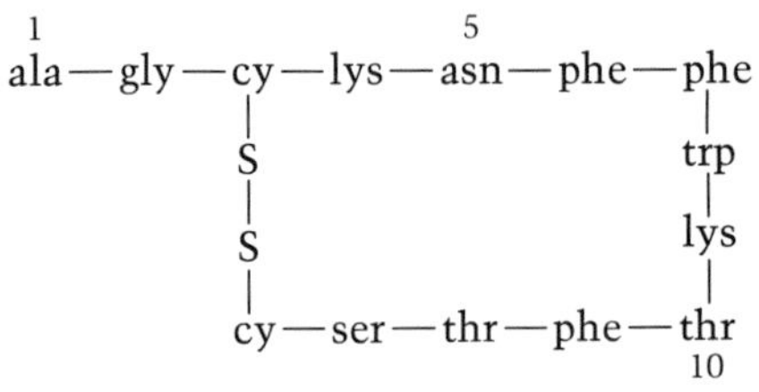

somatostatin

Several other peptide hormones are described in Table 4-4. The list includes some materials that are classified as proteins because of their size. A minimum size of 40–50 amino acid residues for proteins is generally agreed upon.

The regulation of the transmission of impulses in nerve cells involves various types of substances called *neurotransmitters.* In 1976, a new class of compounds (peptide in nature) were isolated; they are proposed to have a neurotransmitter function, specifically in the processing of sensory information dealing with pain and emotional behavior. Because they were originally detected in the brain, the name *enkephalin* (from the Greek, "in the head") has been suggested. Two enkephalins have been isolated, and both are pentapeptides that differ only by one amino acid, that of the C-terminus.

tyr—gly—gly—phe—met
methionine-enkephalin

tyr—gly—gly—phe—leu
leucine-enkephalin

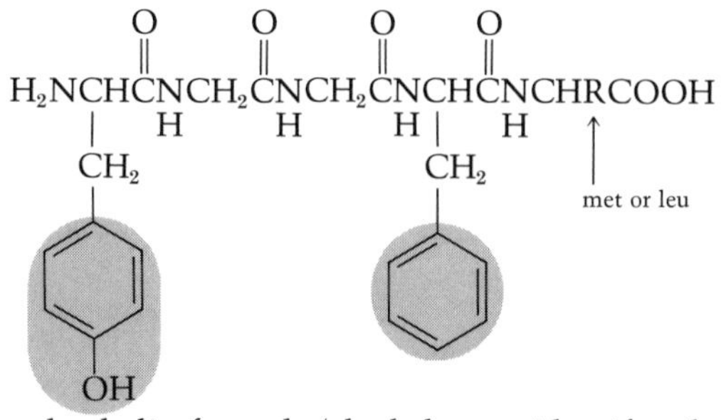

enkephalin formula (shaded areas identify what are proposed to be important structural similarities to opiates, which would explain their similar effects)

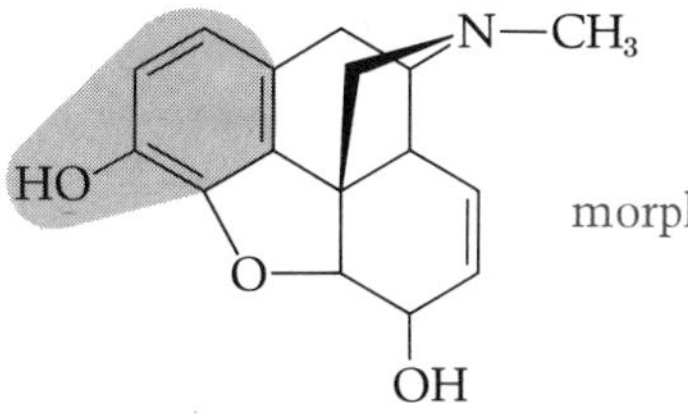

The discovery of the enkephalins has generated considerable excitement because their action mimics that of *morphine* and other opiates, which are used as pain-killers (analgesics) but which unfortunately are addictive. In fact there is evidence that the enkephalins and opiates may act in the same way and possibly at the same membrane receptor site in nerve cells, and that in the presence of opiates the natural production (see margin) and action of the enkephalins is blocked. The cells are then exposed only to opiate action. When the administration of the opiate is stopped, the nerve cells are without both the opiate and the enkephalins—a condition proposed to trigger a sequence of events (yet unknown) that results in withdrawal symptoms, to which the addict responds by taking more of the opiate. In addition to furthering our knowledge of how the brain works, continued research in this area may result in the development of nonaddictive pain killers that are equally as effective as morphine. At present it is thought that the morphinelike effect of enkephalins is due to the two aromatic side-chain groupings of phenylalanine and tyrosine, which may well be the structural characteristics that contribute to the recognition of the specialized receptor sites in the cell membrane. Similar aromatic groupings are present in morphine and phenazocine, which is even more potent than morphine. Note that the greater potency of phenazocine is consistent with a greater similarity to the enkephalin structure.

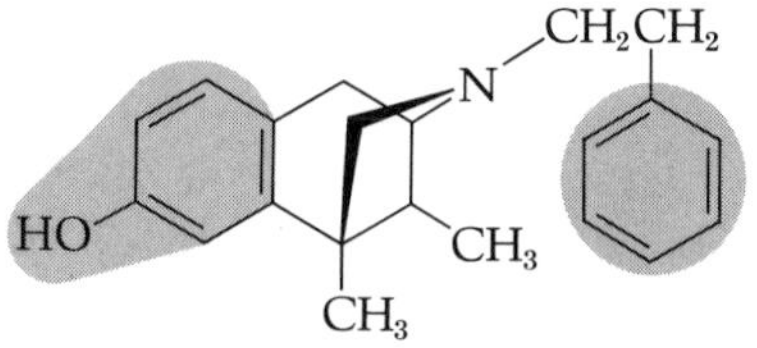

Very little is known at present about the production of the enkephalins. They may be formed as fragments from the cleavage of two larger peptides (produced by the pituitary gland) called *endorphins* (α with 16 residues and β with 31 residues) because they also have analgesic properties mimicking morphine. The endorphins in turn may be produced as fragments from β-*lipotropin,* a protein which is also produced by the pituitary and contains 91 amino acid residues.

Peptide Antibiotics

Many antibiotics are complete peptides or contain a small peptide component as part of the overall structure. This group of materials exhibits a tremendous variety of peptide structure, as illustrated by the structures shown below for *benzyl penicillin, bacitracin, gramicidin, and actinomycin.* The natural biological role of antibiotics is one of self-protection for the microorganisms that produce them. We, of course, use them as chemotherapeutic agents in the fight against disease. In addition, their use in biochemical research has contributed greatly to the understanding of the molecular events of such processes as protein and nucleic acid biosynthesis (see p. 269).

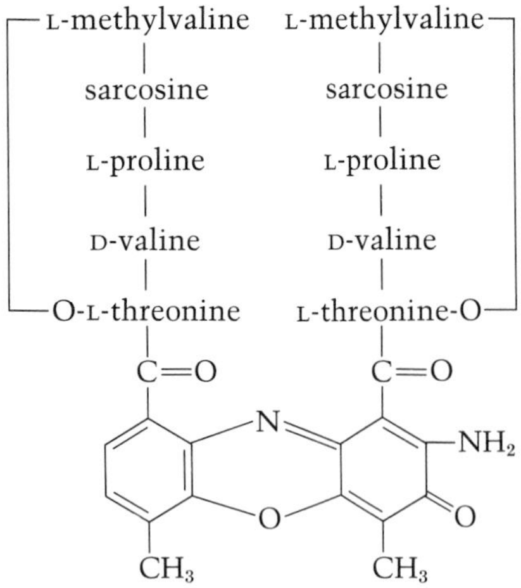

actinomycin D

one of several actinomycins produced by strains of Streptomyces; *see p. 245 for mode of action*

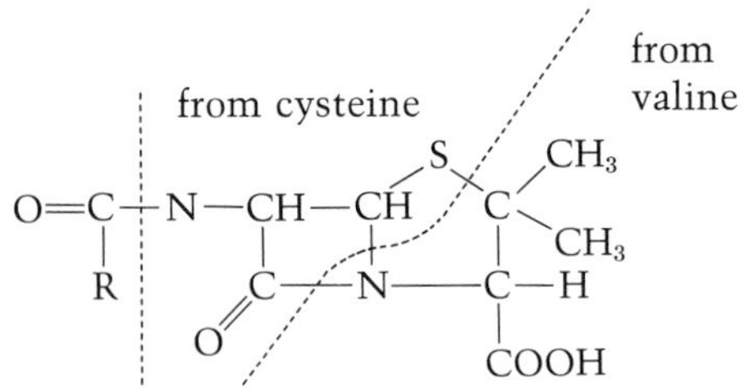

penicillin

produced by Penicillium *molds; most common penicillin has*

$$R = \text{—}CH_2\text{—(benzyl group);}$$

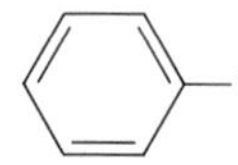

many others are known, both naturally occurring and man made; see p. 311 for mode of action

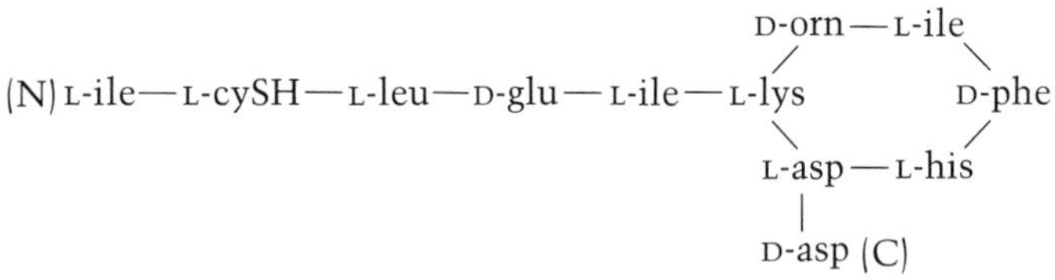

bacitracin A

produced by strains of the bacterium, Bacillus licheniformis; *partially cyclic with* D *and* L *acids including both* D *and* L-*aspartic acid*

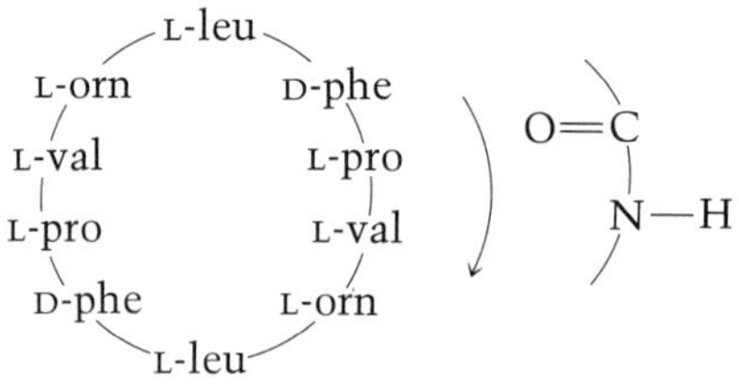

gramicidin S

one of several gramicidins produced by the bacterium, Bacillus brevis; *note the totally cyclic structure and the presence of both* L *and* D *amino acids;* L-*orn is the amino acid, ornithine; see p. 536*

Memory (?) Peptides

The study of how—in molecular terms—knowledge is acquired and stored has been going on for over a half century. Results have been incomplete and sketchy, and claims for this or that process have generated a lot of controversy. A 1970 report by G. Ungar that a specific tetradecapeptide was isolated from the brains of rats that had been trained to have a fear of the dark has fueled the controversy. The peptide was not present in brain extracts obtained from control groups of untrained rats. When the purified peptide was then injected into untrained rats, the result was a significant manifestation of the same trait. The amino acid sequence of this peptide—named *scotophobin,* meaning "dark fear"—is

1   10

ser—asp—asn—asn—glu—gln—gly—lys—ser—gln—gly—gly—gln—tyr

Ungar and others propose that the peptide may be but one of a family of chemical code words of memory that serve for the coding of acquired information in the central nervous system. The proposal implies that the molecular result of the animal's learning to acquire a specific behavior in response to a specific stimulus is the production of a peptide having a specific amino acid sequence. Once the information is acquired—that is, once the molecule has been synthesized—the appropriate message is translated through the central nervous system whenever the same stimulus is received.

This proposal has received some support by the demonstration that untrained rats and goldfish acquire dark-avoidance after injection with a laboratory-prepared, synthetic peptide having the same amino acid sequence as scotophobin. Since the scotophobin work, there have been a few other claims of additional memory peptides being isolated from animals trained for different tasks. For example, the peptide trp-ala-gly-gly-asp-ala-ser-gly-glu is reported to induce sleep. Skepticism still exists, however, and much further research is necessary in this interesting and controversial area.

CHEMICAL SYNTHESIS OF PEPTIDES

The laboratory synthesis of naturally occurring materials of all types has long attracted the attention of organic chemists and biochemists alike. Peptides of known sequence are difficult to synthesize by classical methods. The basic difficulty is that many individual steps are required, and every time a new peptide bond is formed the peptide has to be isolated in pure form for use in the next step. Each step would require blocking the amino group and the R group (needed for the trifunctional amino acids) of the amino acid that is being added in that step. The carboxyl group of this acid would have to be activated in order to have a smooth reaction with the amino group of the peptide that is being lengthened. du Vigneaud's original synthesis of oxytocin required several months of effort.

In 1963, however, Merrifield devised the *solid-phase method* of peptide synthesis, which has had a revolutionary effect. An automated process, it is simpler and much shorter than earlier methods, although the latter are still used. In the Merrifield procedure, the polypeptide chain being formed is supported on small solid beads of a polymeric resin. The beads are only about 50 micrometers in diameter $(50 \times 10^{-6}$ cm), but they are enormous relative to molecular dimension, with about 10^{12} peptide chains being attached to each bead. The original material used by Merrifield (there are now others) was a chloromethylated polystyrene resin.

$$CH_3—O—CH_2Cl + H—\langle\bigcirc\rangle—resin \longrightarrow ClCH_2—\langle\bigcirc\rangle—resin + CH_3OH$$

chloromethyl
methyl ether

polymeric
resin material
with several
phenyl groups;
only one shown
for simplicity

chloromethylated resin;
each resin bead would
contain several sites

(*Note:* a cation-exchange
resin would consist of
—$SO_3^-H^+$ groupings)

The chemically activated resin is then treated in a small reaction vessel with an amino acid containing a blocked amino group. (The t-butyloxycarbonyl (tBOC) group works quite well for amino group protection (see Table 4–3, item 8). If something other than tBOC-chloride were used, any potential reacting group—such OH, NH_2, COOH, or SH—would also be blocked.) The result is the formation of a covalent ester linkage between the amino acid and resin, and thus the first amino acid is referred to as being *anchored to the polymer* (see below). Note that this first amino acid represents the C-terminal residue of the desired peptide chain. The amino blocking group is then removed. (Of course, any reactive functional group on the side chain of the amino acid that was blocked initially would remain blocked to prevent its undesired participation in any further reaction.) The next step would be reaction with the second amino acid (also with a protected amino group and R, if necessary) to yield a dipeptide. Dicyclohexylcarbodiimide (DCC) is employed here to promote condensation (see Table 4–3, item 9). The amino group would be deprotected and the process repeated until the desired number of residues are attached. Then any blocking groups on the side chains are removed, and finally the initial ester bond anchoring the peptide to the resin is cleaved and the released peptide is recovered. Note that a major advantage of the solid-phase method is that the growing polypeptide does not have to be actually isolated after each extension step. Because it is covalently attached to an insoluble polymeric support, the growing polypeptide can be purified by simply extracting out any unused reagent and by-products from the condensation reaction. The polypeptide remains in the reaction vessel.

1. Attachment of C-terminal residue (amino blocked) to resin:

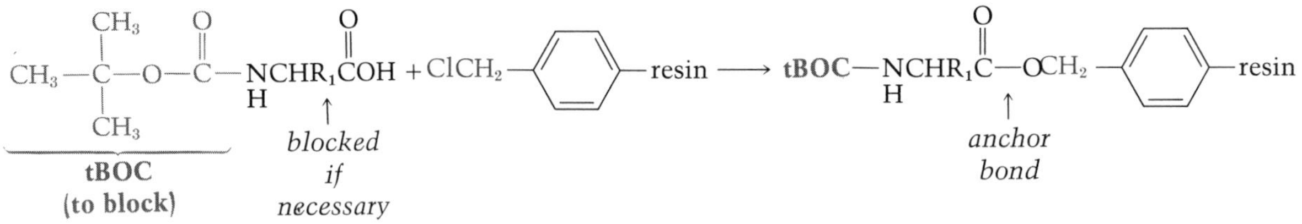

2. Removal of amino blocking group (but not of R) to yield

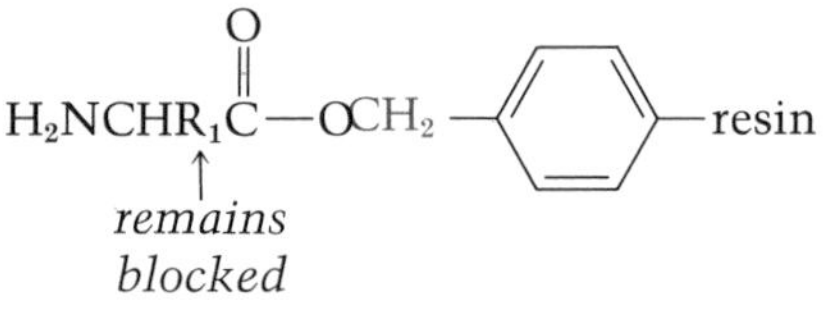

3. Condensation with second amino acid (amino blocked) in presence of dicyclohexylcarbodiimide (DCC):

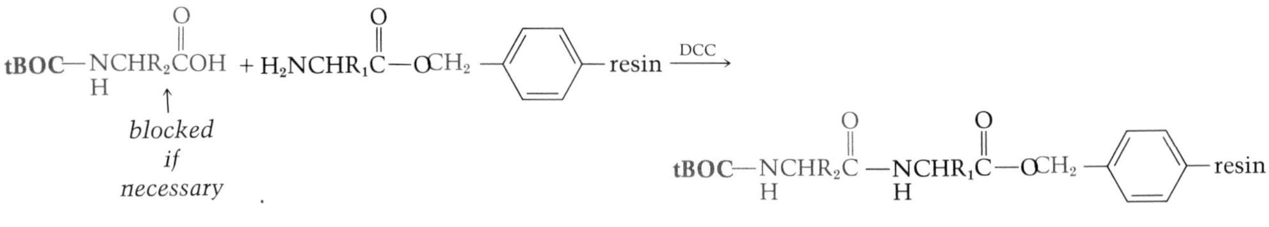

4. Further cycles of steps 2 and 3 until desired peptide is assembled.

5. Removal of any blocking groups in R side chains.

6. Cleavage of anchor bond and recovery of polypeptide product.

Several small peptides have been prepared since this procedure was developed. The first success with large polypeptides occurred in 1969 when Merrifield and coworkers announced the successful synthesis of ribonuclease, an enzyme that contains 124 amino acid residues. The automated process involved 369 different chemical reactions and 11,931 distinct steps, but with continuous operation the synthesis was completed within a few weeks—truly a milestone in synthetic biochemistry. The method has since been used for polypeptides having about 200 residues.

Some Applications

The laboratory synthesis of peptides provides a valuable dimension to biochemical research directed at trying to correlate the biological property of a peptide to any particular residue. For example, the lab synthesis and testing of several variants of oxytocin and vasopressin support the conclusions stated earlier about residues 8 (in oxytocin) and 3 (in vasopressin). Another example is the recent finding that a synthetic met-enkephalin, having D-alanine in position 2 rather than glycine, has much stronger analgesic properties than natural enkephalin. Another example is the synthesis and clinical testing of somatostatin analogs as an insulin substitute.

Another type of use is in the large scale laboratory production of those peptides, such as insulin, that are used as therapeutic agents in treating various diseases but are available (and sometimes then in only very small amounts) only through isolation from natural sources.

Another exciting development regarding the therapeutic use of synthetic peptides has come from the recent report by Votano, Gorecki, and Rich (referenced at end of chapter) that certain synthetic tripeptides, such as phe—phe—arg, may prove to be effective materials in the treatment of individuals with *sickle cell disease* (see p. 130).

LITERATURE

BROWN, M., J. RIVIER, and W. VALE. "Somatostatin: Analogs with Selected Biological Activities." *Science,* **196,** 1467–1469 (1977). A study illustrating the search for a more active peptide hormone by laboratory synthesis of derivatives of the native hormone.

BUMPUS, F. M. "Angiotensin Antagonists in Relation to Hypertension." *Hospital Practice,* **9,** 80–92 (1974). A discussion of the molecular basis for the biological actions of angiotensin II in regard to its amino acid sequence and three-dimensional structure.

CORRIGAN, J. T. "D-Amino Acids in Animals." *Science,* **164,** 142–148 (1969). A review article on the subject.

FRIEDEN, E. and H. LIPNER. *Biochemical Endocrinology of the Vertebrates.* Englewood Cliffs, New Jersey: Prentice-Hall, 1971. A good introduction to this complex topic. Available in paperback.

GOLDSTEIN, A. "Opiate Peptides (Endorphins) in Pituitary and Brain." *Science,* **193,** 1081–1086 (1976). A review article.

MEISTER, A. *Biochemistry of the Amino Acids.* 2nd edition. New York: Academic Press, 1965. An authoritative two-volume work devoted exclusively to the isolation, occurrence, structure, and metabolism of the amino acids. Many references to the original literature.

MEISTER, A., and S. S. TATE. "Glutathione and Related γ-Glutamyl Compounds." In *Annual Review of Bio-chemistry,* **45,** 559–604 (1976). A thorough review article on glutathione biochemistry.

MERRIFIELD, R. B. "The Automatic Synthesis of Proteins." *Scientific American,* **218,** 56–74 (1968). A synopsis of the chemical methodology of solid-phase peptide synthesis, with specific details on the synthesis of insulin.

PAPPENHEIMER, J. R. "The Sleep Factor." *Scientific American,* **235,** 24–29 (1976). Evidence that the brain produces a substance that induces sleep.

SKEGGS, L. T. "Biochemical Relationships of the Renin/Angiotensin System." *Hospital Practice,* **9,** 145–154 (1974). Well-written.

SNYDER, S. H. "Opiate Receptors and Internal Opiates." *Scientific American,* **237,** 44–54 (1977). Enkephalins and endorphins are described.

STEWART, J. M., and J. D. YOUNG. *Solid Phase Peptide Synthesis.* San Francisco: W. H. Freeman, 1969. A detailed description of the technique developed by R. B. Merrifield.

TAGER, H. S., and D. F. STEINER. "Peptide Hormones." *Ann. Rev. Biochem.,* **43,** 509–538 (1974). A review article dealing with structure and biosynthesis.

VOTANO, J. R., M. GORECKI, and A. RICH. "Sickle Hemoglobin Aggregation: A New Class of Inhibitors." *Science,* **196,** 1216–1219 (1977). The study referred to on p. 115 of this chapter.

EXERCISES

4–1. For each of the amino acids given below, write the equilibrium reactions that would apply at a pH corresponding to each of the pK_a values for each amino acid.
(a) alanine (b) glutamic acid (c) arginine
(d) α_1,ϵ-diaminopimelic acid ($pK_{a_1} = 1.8$, $pK_{a_2} = 2.2$, $pK_{a_3} = 8.8$, $pK_{a_4} = 9.9$).

$$HOOCCH(CH_2)_3CHCOOH$$
$$\quad\;\; | \qquad\qquad\;\; |$$
$$\quad NH_2 \qquad\quad\; NH_2$$

4–2. Would the first three amino acids listed in Exercise 4-1 be completely separated from each other by electrophoresis in an electrolytic buffer solution at pH 2, pH 7, or pH 12? Perform each evaluation by estimating the charge of each amino acid as described on pp. 98–99 of this chapter.

4–3. At approximately what pH would each of the amino acids listed in Exercise 4-1 exist almost exclusively in an isoelectric form? Write the structures of the isoelectric species.

4–4. Write the complete structure of the following peptides:
(a) glutamyl-valyl-glycine
(b) leucyl-methionyl-alanine
(c) phenylalanyl-arginyl-tryptophanyl-serine
(d) ile-asn-gly-his-lys-glu-gln-arg

4–5. Classify each of the peptides given in Exercise 4-4 as (a) a basic peptide; (b) an acidic peptide; or (c) a neutral peptide.

4–6. On the basis of your answer to Exercise 4–5 predict which peptide(s) would move most slowly through a column of a strong cation exchanger (see p. 50), using an eluting solvent with a pH of 6. Explain your answer.

4–7. Which of the peptides listed in Exercise 4-5 would move fastest through a gel permeation chromatography column? Assume that (a) the gel material is nonionic, so that only sieving will occur, and (b) the pore size of the swollen gel is slightly larger than molecular dimensions of the largest peptide.

4–8. Draw (see p. 93) all of the stereoisomeric forms of threonine; of isoleucine.

4–9. Classify each of the following amino acids as polar or nonpolar:

(a) $H_2N(CH_2)_3CH(NH_2)COOH$
(b) $CH_3(CH_2)_5CH(NH_2)COOH$

$$\qquad\qquad CHCH_3$$
$$\qquad\qquad ||$$
(c) $HOOCCCH_2CH(NH_2)COOH$

(d) $CH_3CH(NH_2)CH_2CH(NH_2)CH_2COOH$

(e) $CH_3CH=CHCH_2CH_2CH(NH_2)COOH$

(f)

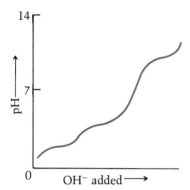

4–10. Would it be possible to separate angiotensin I from angiotensin II by electrophoresis at pH 5? Explain.

4–11. Does the general form of the composite titration curve (see Chap. 2) shown here correspond to the complete titration of glycine, serine, aspartic acid, glutamine, or lysine? Identify the following on the curve: pK_as, buffering regions, pI, and a region where the solution would contain only positively charged molecules of the amino acid.

4–12. The drawings below represent ninhydrin positive zones observed after paper electrophoresis (see p. 46) of a mixture of alanine, aspartic acid, and lysine. Which drawing corresponds to the result observed at pH 1? at pH 12? Identify what each zone corresponds to.

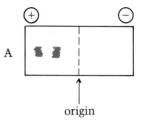

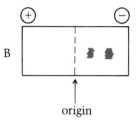

Draw (as shown here) the results expected if the electrophoresis were performed at pH 7.

CHAPTER 5

PROTEINS

Functions of proteins

Catalytic proteins: **enzymes** that accelerate and regulate reactions

Structural proteins: lacking a true dynamic function, these are proteins that confer a molecular foundation to various parts of an organism; examples are collagen, found in the connective tissue of vertebrates (representing about one-third of the total protein in the body) and various proteins in the membranes of all cells.

Contractile proteins: molecules that can reversibly tighten and relax their structural orientation; examples are actin and myosin in muscle tissue and other cells

Natural-defense proteins: examples are the antibodies in the gamma globulin fraction of blood

Digestive proteins: various enzymes present in gastrointestinal secretions; these catalyze the degradation of dietary foodstuffs to smaller substances

In view of the many previous references to the biological importance of proteins, little else need be said at this point. Now let us move on and examine these substances more closely, with a primary focus on their structure. In doing so many facts and principles of tremendous biochemical significance will emerge. Indeed, an understanding of various aspects of protein structure is indispensable to the eventual understanding of much of (if not all of) the biochemical dynamics of a living organism.

We will begin the subject with a consideration of protein classification. In addition to contributing an organization to the subject this also provides an opportunity to (a) provide you with an overview of the extensive involvement of proteins in cellular processes and (b) introduce much of the terminology that is used in reference to the proteins.

CLASSIFICATION

Proteins can be classified in different ways. The list in the margin distinguishes proteins on a *functional basis.* (The listing includes most but not all functions of proteins.) Since all the proteins in an organism are important to sustaining normal life processes, it is not appropriate to rank the various types in any specific hierarchy. However, as mentioned before, those proteins that function as *enzymes* do have a special importance because they are responsible for catalyzing all of the chemical reactions in a cell. This central role of enzymes does not, however, diminish the crucial biological significance of other individual proteins. For example, without insulin the disease of diabetes results; without antibodies our bodies would be defenseless against infection; without hemoglobin the supply of oxygen to cells would be nil; the list could go on and on.

On a *compositional basis* proteins can be divided into two general classes: (1) the *simple proteins*, which yield only α-amino acids on complete hydrolytic degradation; and (2) the *conjugated proteins*, which on degradation yield, in addition to α-amino acids, an organic or inorganic nonpeptide grouping termed a *prosthetic group*. The simple proteins (a blatant misnomer, by the way, since no protein is really a simple molecule in the usual sense of the word) are further subclassified on the basis of solubility and amino acid composition. The conjugated proteins are also subclassified in terms of the chemical nature of the prosthetic group. There is no universal type of bond between the prosthetic group and the polypeptide chain. In some cases the linkage is noncovalent and in others it is covalent. (The table at bottom of the page summarizes this information.)

A third and particularly useful classification is based on *differences in overall three-dimensional structure*. The two divisions in this classification are the *fibrous proteins* and the *globular proteins*. Fibrous protein molecules have a very elongated shape, like that of a thread. They are frequently very large molecules and are often composed of two or more polypeptide chains. They are generally insoluble in water (they are scleroproteins). Collagen is the classic example. Globular proteins, on the other hand, have a much more compact structure due to a highly ordered pattern of folding, bending, and twisting along the polypeptide chain. They are "globlike" with the overall shape ranging from a nearly perfect sphere to an ellipse. The intact protein may consist of a single polypeptide chain, or an aggregate of two or more chains, which may be identical or

Transport proteins: one example is hemoglobin, which transports oxygen in the blood; various others are found in cell membranes and function in the transport of substances across the membrane

Blood proteins: an example is fibrinogen, which participates in the clotting process

Hormonal proteins: proteins such as insulin that regulate cellular activities

Respiratory proteins: an example is the cytochromes, which participate in the transport of electrons to suitable acceptors, such as oxygen in aerobic organisms; they may also be considered as enzymes

Repressor proteins: proteins that regulate the expression of genes in chromosomes

Receptor proteins: various proteins found in cell membranes; these can transmit information to the cell interior after binding with specific substances on the exterior side of the cell

Ribosomal proteins: proteins that in association with RNA yield ribosomes, the multimolecular aggregates that participate in protein biosynthesis

Toxin proteins: proteins in the venom of poisonous reptiles that are responsible for the toxicity to the mammalian nervous system

Vision proteins: an example is the rhodopsins, which participate in the molecular events associated with sight

	Simple proteins			Conjugated proteins	
	Solubility in				
Name	*Water*	*Salt solution*	*Amino acid composition*	*Name*	*Type of prosthetic group*[b]
Albumins	High	High	Nothing distinctive	Lipoproteins	Lipid
Globulins	Low	Variable	Nothing distinctive	Glycoproteins	Carbohydrate
Scleroproteins (such as collagen in the connective tissue of animals)		Insoluble	Many gly, ala, pro	Metalloproteins	Metal ion
				Phosphoproteins	Phosphate
Protamines[a]		Soluble	Many arg	Hemoproteins	Heme group
Histones[a]		Soluble	Many arg and lys	Flavoprotein	Flavin group

[a] These are very basic proteins carrying a large net positive charge at pH 7; the histones are bound to the DNA of chromosomes (see p. 232, Chap. 7).
[b] Two or more prosthetic groups can exist in the same protein.

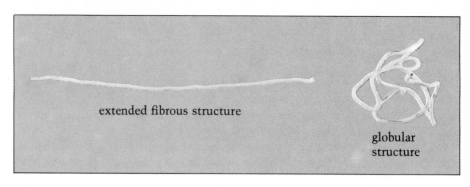

extended fibrous structure

globular
structure

different. They are much more soluble in water than the fibrous proteins. A greater variety and number of proteins are in the globular class; all enzymes are globular proteins.

STRUCTURE—AN INTRODUCTION

It is customary to refer to four levels (aspects) of structure for a protein molecule. They are (1) the **primary level,** which refers to (a) the *identity of the amino acids* that are present, (b) the *relative amount of each amino acid,* and, ultimately, (c) the exact *sequence of the amino acids* in the polypeptide chain(s); if present, the identity and amount of any prosthetic group may also be included at this level; (2) the **secondary level,** which refers to the *geometrical orientation of the polypeptide chain* that serves as the backbone of the polymer; (3) the **tertiary level,** which refers to the complete, *three-dimensional architecture* of the protein, including the orientation of any prosthetic group; and (4) the **quaternary level,** which considers the *noncovalent aggregation of two or more polypeptide chains.* The first three levels apply to all proteins, and the fourth to several.

PRIMARY STRUCTURE

The identity and sequence of monomeric units are the most fundamental structural characteristics of any polymer. The experimental determination of sequence, particularly for a polypeptide, poses a rather formidable problem. For example, the possible number N of different sequences just for a decapeptide containing two residues of each of five different amino acids would be

$$N = \frac{10!}{5(2!)} \approx 360,000$$

The fact that the polypeptide chains of the naturally occurring proteins generally contain varying amounts of the 20 common amino acids and have an average length of 100–150 residues obviously compounds the difficulties of studying this type of polymer. (The number of sequence isomers for a polypeptide of 100 residues that can be generated from the set of 20 amino acids is 20^{100}—a number so large as to be almost beyond comprehension.) Nevertheless, it can be done.

Prior to sequencing, some important preliminary determinations are usually performed.

1. The molecular weight is measured. This can be done by ultracentrifugation, molecular sieve chromatography, or SDS-polyacrylamide gel electrophoresis, as described in Chap. 2 (see pp. 41, 46, and 49). There are also other methods.

2. If it is a conjugated protein, the type and amount of prosthetic group are determined. Before sequencing, the prosthetic group is preferably removed (that is, dissociated) to yield free polypeptide material.

3. The presence of any *interchain* or *intrachain disulfide bonds* (—S—S—) is determined. This type of covalent linkage is not common to all proteins, but is usually critical to those that do possess it. Detection is based on *selective chemical modification techniques*. One approach is to oxidize the disulfide bond; performic acid is a frequently used oxidizing agent. If the product has an increased electrophoretic mobility due to an increased negative charge, one or more disulfide bonds must have been present. Gel-permeation chromatography would also be useful to demonstrate whether the disulfide bond was interchain or intrachain. An alternative tactic is to reduce the disulfide bond. Mercaptoethanol (HSCH₂CH₂OH) or dithiothreitol (*Cleland's reagent*) will do the job. Once again electrophoresis and chromatographic methods would be useful to monitor the results. Without going into the details, suffice it to say that both the oxidative and reductive modifications can be quantified to determine the number of disulfide linkages. If disulfide bonds are present they are usually cleaved and the newly formed peptide (or peptides) is then sequenced.

—S—S— Detection

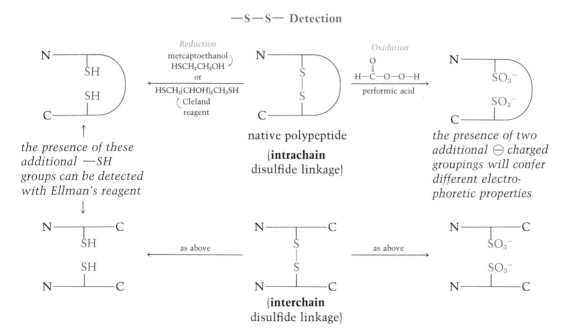

whether by oxidation or reduction two chains will be produced; since the individual chains are smaller than the original protein, they could be detected by molecular sieve chromatography; electrophoresis could also be used

4. Ordinarily the determination of the presence in the native protein of sulf-hydryl (—SH) groups from cysteine is also performed at this time. A very effective, quantitative method is to treat the protein with 5,5'-dithiobis-2-nitrobenzoic acid (*Ellman's reagent*), which selectively reacts with free —SH groups to give a modified protein and the thionitrobenzoate anion. The latter is strongly colored (λ_{max} = 412 nm) and serves as the basis for a quantitative spectrophotometric assay.

—SH Detection

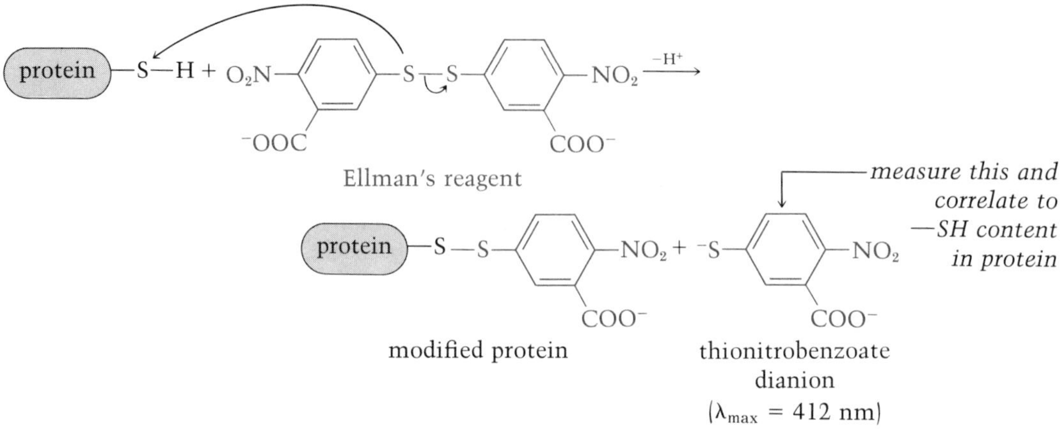

Ellman's reagent

modified protein

thionitrobenzoate
dianion
(λ_{max} = 412 nm)

measure this and
correlate to
—SH content
in protein

5. Proteins having a quaternary level of structure are treated to dissociate the intact aggregate into the individual polypeptide units, which are then isolated and studied separately. Simple but effective ways to dissociate proteins involve changing the pH of the protein solution, treating with a detergent (sodium dodecyl sulfate), mild heating, or repeating cycles of freezing and thawing.

Amino Acid Composition

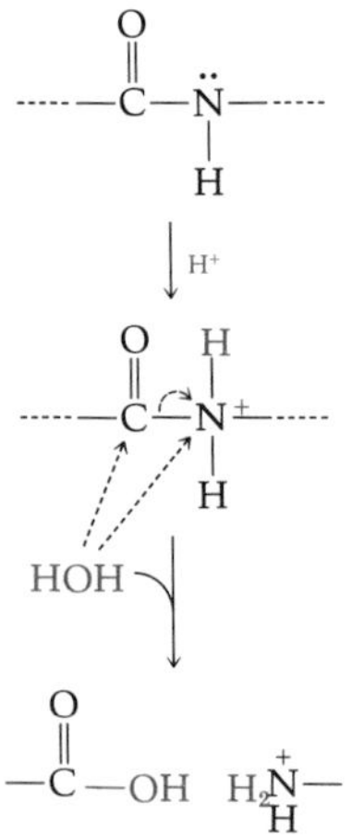

The identity and amount of each amino acid present in a polypeptide is determined by degrading it to a mixture of amino acids and then analyzing the mixture, usually by ion-exchange chromatography. The standard method for degradation is *acid hydrolysis* (usually with 6 N HCl for 12–36 hours at 100–110 °C), which will result in the cleavage of every bond, with no racemization (L → D or D → L) occurring. A serious problem, however, with HCl hydrolysis is that tryptophan, when it is present, is completely destroyed. Recently, the use of methanesulfonic acid in place of HCl has become popular because it does not result in the destruction of tryptophan. *Base hydrolysis* (with 2 N NaOH) can also be used to detect tryptophan but is more destructive of other amino acids. Another disadvantage of base hydrolysis is that extensive racemization occurs. A very nondestructive method involves the *use of enzymes* that occur naturally and catalyze the cleavage of polypeptides. Such enzymes are called *proteolytic* (protein degrading) *enzymes* or simply *proteases* or *peptidases*. *Pronase*, a protease obtained from a microbial source, has been used for this purpose. However, enzyme hydrolysis can result in contamination, since the enzyme, which itself is a polypeptide, can catalyze its own degradation in addition to that of the polypeptide being treated.

Regardless of the method, the protein hydrolyzate (amino acid mixture) is then analyzed qualitatively and quantitatively, generally by use of an amino

mechanism of
acid-catalyzed hydrolysis
of a peptide bond

acid analyzer (see p. 97). The output of the analyzer based on ion-exchange chromatography and ninhydrin detection is the same type of pattern as shown earlier (see Fig. 2–8 on p. 52). The amino acid composition is expressed in the form shown in Table 5–1, p. 124.

Sequence Determination (Terminal Residues)

The actual sequencing generally begins with the determination of the terminal residues.

The C-terminus can be determined by treating the intact polypeptide with *hydrazine*. The hydrazine will react hydrolytically at the carbonyl grouping of each peptide bond, resulting in the formation of acyl hydrazine derivatives of each residue, except the C-terminus. The C-terminus will not be modified, since its α-carboxyl group is not in peptide linkage; thus it can be recovered and identified chromatographically. Another method is treatment of the polypeptide with *carboxypeptidase,* one of a group of diges-

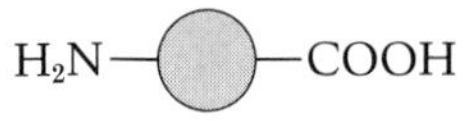

$$H_2N-\bigcirc-COOH$$

(symbolic representation
of amino acid)

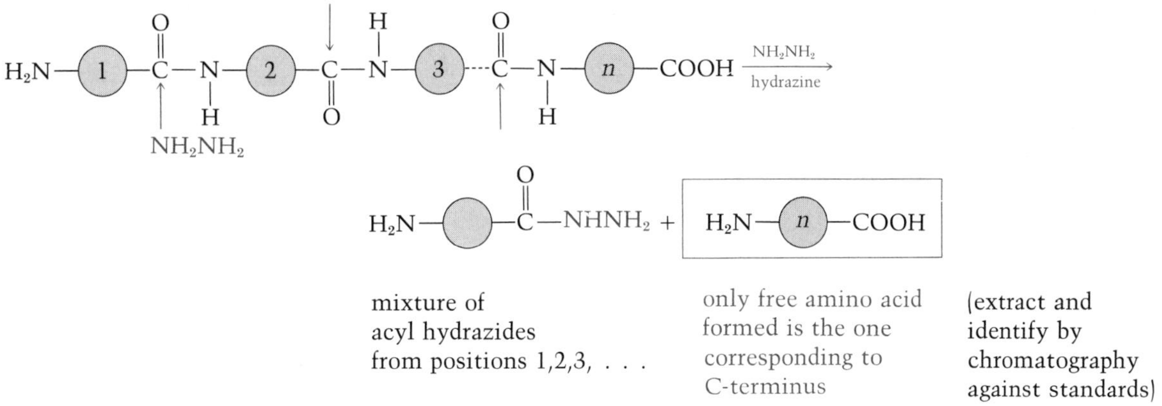

$$H_2N-①-\underset{\underset{NH_2NH_2}{|}}{\overset{\overset{O}{\|}}{C}}-\underset{\underset{H}{|}}{N}-②-\overset{\overset{O}{\|}}{C}-\underset{\underset{O}{\|}}{N}-③\cdots\overset{\overset{O}{\|}}{C}-\underset{\underset{H}{|}}{N}-ⓝ-COOH \xrightarrow[\text{hydrazine}]{NH_2NH_2}$$

$$H_2N-\bigcirc-\overset{\overset{O}{\|}}{C}-\overset{|}{N}HNH_2 + \boxed{H_2N-ⓝ-COOH}$$

mixture of
acyl hydrazides
from positions 1,2,3, . . .

only free amino acid
formed is the one
corresponding to
C-terminus

(extract and
identify by
chromatography
against standards)

tive enzymes obtained from the gastrointestinal secretions of mammals. Carboxypeptidase is relatively specific for removal of C-terminal residues from a polypeptide chain. Difficulties arise, however, because the enzyme does not stop after removing the initial C-terminal residue, but continues to

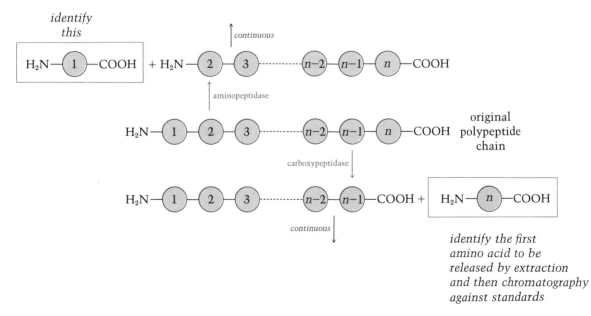

*identify
this*

$$\boxed{H_2N-①-COOH} + H_2N-②-③\cdots\cdots(n\text{-}2)-(n\text{-}1)-ⓝ-COOH$$

↑ *continuous*

↑ aminopeptidase

$$H_2N-①-②-③\cdots\cdots(n\text{-}2)-(n\text{-}1)-ⓝ-COOH$$ original polypeptide chain

carboxypeptidase

$$H_2N-①-②-③\cdots\cdots(n\text{-}2)-(n\text{-}1)-COOH + \boxed{H_2N-ⓝ-COOH}$$

continuous

*identify the first
amino acid to be
released by extraction
and then chromatography
against standards*

TABLE 5-1 Amino acid composition of some proteins. (Values in each column specify the number of residues present in each protein. The relative percentages of each amino acid class are also given.) Note the similarities in the relative occurrence of each class of amino acids in the different proteins. Also note the distribution between nonpolar and polar amino acids. Generally speaking, in most proteins about $\frac{1}{3}$–$\frac{1}{2}$ of the amino acids are nonpolar.

Amino Acid	Class	Alpha chain – human hemoglobin	Beta chain – human hemoglobin	Cytochrome c – human	Glyceraldehyde dehydrogenase	Coat protein – tobacco mosaic virus	Lysozyme – chicken	Human growth hormone	Histone 1 – rabbit and bovine
Glycine	Aliphatic	7	13	13	32	4	12	8	6
Alanine		21	15	6	32	18	12	7	22
Valine		13	18	3	34	10	6	7	2
Leucine		18	18	6	18	11	8	25	5
Isoleucine		0	0	8	21	11	6	8	1
(class %)		42%	44%	35%	41%	35%	34%	29%	48%
Serine	Hydroxyl	11	5	2	19	13	10	18	5
Threonine		9	7	7	22	13	7	10	2
(class %)		14%	8%	9%	12%	17%	13%	15%	9%
Cysteine	Sulfur	1	2	2	4	1	8	4	0
Methionine		7	1	3	9	3	2	3	0
(class %)		6%	2%	5%	4%	3%	8%	4%	0%
Phenylalanine	Aromatic	7	8	3	14	6	3	13	0
Tyrosine		3	3	5	9	7	3	8	1
Tryptophan		1	2	1	3	2	6	1	0
(class %)		8%	9%	9%	8%	10%	9%	12%	1%
Aspartic	Acidic	8	7	3	25	5	8	14	0
Glutamic		4	8	8	13	7	2	20	5
(class %)		9%	10%	11%	12%	8%	8%	18%	7%
Asparagine	Amides	4	6	5	13	12	13	6	2
Glutamine		1	3	2	5	15	3	6	0
(class %)		3%	6%	7%	5%	17%	11%	6%	3%
Lysine	Basic	11	11	18	26	2	6	9	15
Histidine		10	9	3	11	1	1	3	0
Arginine		3	3	2	10	10	11	10	2
(class %)		17%	16%	23%	14%	8%	14%	12%	23%
Proline	Imino	7	7	4	12	8	2	8	7
(class %)		5%	5%	4%	4%	5%	1%	4%	9%
Total Residues		141	146	104	332	156	129	188	75
Nonpolar		74 52%	69 47%	34 33%	143 43%	69 44%	45 35%	72 38%	37 49%
Polar		67 48%	77 53%	70 67%	179 57%	87 56%	84 65%	116 62%	38 51%

124

attack new C-terminal residues every time the chain is shortened. Therefore it is necessary to monitor the rate of release of amino acids. *Aminopeptidase*, another digestive enzyme, but with an opposite site of attack from carboxypeptidase, can be used to determine the N-terminus. It presents the same disadvantage, however, of continuous action.

Because they act only on peptide bonds at the end of a polypeptide, carboxypeptidase and aminopeptidase are called *exopeptidases*.

A much better method of identifying the N-terminus is to treat the polypeptide with *phenyl isothiocyanate (Edman reagent;* see p. 101), which will react with the α amino group of this residue. The modified N-terminal residue is then released by hydrolysis, extracted, and finally treated to yield a rearranged phenylthiohydantoin derivative. The latter can be easily recovered and identified against chromatographic standards. The *dansyl chloride reagent* (see p. 101) can also be used in much the same fashion. Each of these methods is based on the pioneering work of F. Sanger, specifically his use of 2,4-dinitrofluorobenzene in his work on insulin, the first protein to be completely sequenced (in 1955). Each of these methods is diagramed on p. 126. All three are useful, but note that the *Edman procedure is the only one that results in the release of only the modified N-terminal residue without a cleaving of all of the other peptide bonds.*

Sanger received the Nobel Prize for the techniques he developed in his pioneering achievement.

Sequence Determination (Internal Residues)

Determining the sequence of amino acids sandwiched between the terminal residues is, of course, the real problem. Without a lengthy explanation, suffice it to say that the sequence of an average polypeptide chain of a protein cannot feasibly be determined by a systematic removal and identification of one residue at a time from either terminus, although the Edman procedure might appear to offer such a capability from the N-terminus. The limitation is primarily due to the size of the polypeptide. For example, with controlled and mild hydrolytic conditions, the Edman technique can currently be used in this manner on small polypeptides containing 10–20 residues. Beyond that the efficiency of the technique is considerably reduced (recently an instrument was developed that currently permits the automatic identification of amino acid sequence with an acceptable efficiency through 20 Edman cycles). Consequently, the experimental strategy is to *degrade the large polypeptide into several smaller fragments, which are then isolated and sequenced separately with the Edman reagent.* The main requirement, of course, is that the splitting of peptide bonds must occur with some degree of known specificity, so that after the sequence of each small peptide is determined all of the fragments can be positioned in the proper order.

This requirement for specificity of action is best satisfied by the use of biological agents, namely, some of the other digestive proteolytic enzymes isolated from mammals. *Trypsin* and *chymotrypsin* are widely used for this purpose. There are others. Unlike aminopeptidase and carboxypeptidase, which act at the ends of the polypeptide chain, trypsin and chymotrypsin act within the chain and thus are called *endopeptidases* (see p. 186 in Chap. 6 for additional information about all of these enzymes). Neither is characterized by an absolute degree of specificity of peptide bond cleavage, but under carefully controlled conditions they do exhibit some degree of preferential action. Trypsin favors internal peptide bonds that contain a carbonyl group donated by one of the strongly *basic* amino acids (lysine, arginine). Chymotrypsin prefers internal peptide bonds that contain a carbonyl group

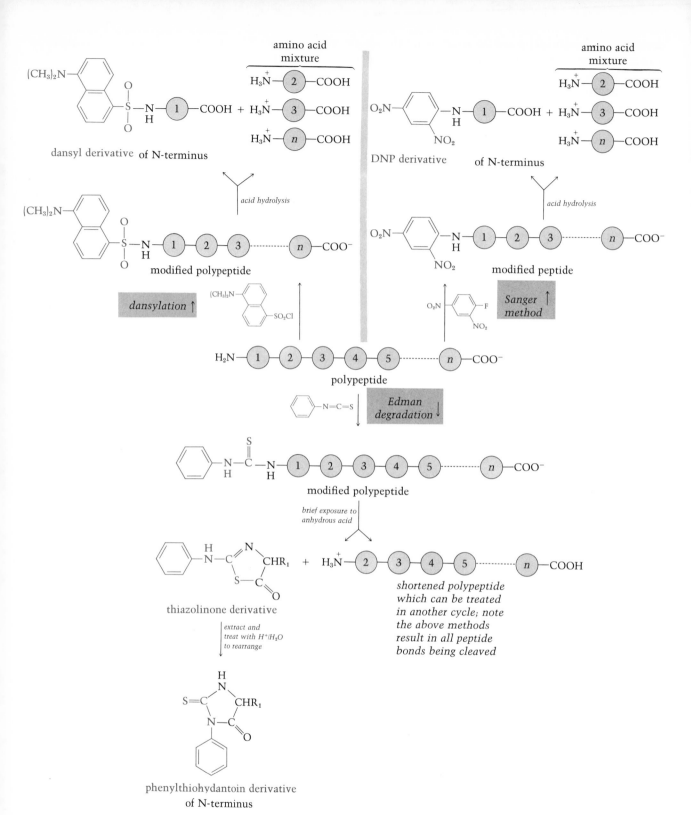

amino acid
mixture

$H_3\overset{+}{N}$—②—COOH

dansyl derivative of N-terminus

$H_3\overset{+}{N}$—③—COOH + $H_3\overset{+}{N}$—③—COOH

$H_3\overset{+}{N}$—ⓝ—COOH

DNP derivative of N-terminus

amino acid
mixture

$H_3\overset{+}{N}$—②—COOH

$H_3\overset{+}{N}$—③—COOH

$H_3\overset{+}{N}$—ⓝ—COOH

acid hydrolysis

modified polypeptide

acid hydrolysis

modified peptide

dansylation ↑

*Sanger
method* ↑

H_2N—①—②—③—④—⑤--------ⓝ—COO^-

polypeptide

*Edman
degradation* ↓

modified polypeptide

*brief exposure to
anhydrous acid*

thiazolinone derivative

$+$ $H_3\overset{+}{N}$—②—③—④—⑤--------ⓝ—COOH

*shortened polypeptide
which can be treated
in another cycle; note
the above methods
result in all peptide
bonds being cleaved*

*extract and
treat with H^+/H_2O
to rearrange*

phenylthiohydantoin derivative
of N-terminus

donated by one of the *aromatic* amino acids (tyrosine, phenylalanine, tryptophan).

$$(N) \cdots CH - \underset{\underset{R}{|}}{\overset{\overset{O}{\|}}{C}} - \underset{H}{\overset{R}{\underset{|}{N}}} - CH \cdots (C) \qquad \cdots CH - \underset{\underset{R}{|}}{\overset{\overset{O}{\|}}{C}} - \underset{H}{\overset{R}{\underset{|}{N}}} - CH \cdots \qquad \cdots CH - \underset{\underset{R}{|}}{\overset{\overset{O}{\|}}{C}} - \underset{H}{\overset{R}{\underset{|}{N}}} - CH \cdots$$

lys
or
arg

    trypsin

phe
or tyr
or trp   chymotrypsin

(met)

   BrCN

(cyanogen bromide)

(in each case the particular amino acids would occupy internal positions rather than being N-terminal)

The use of these biological agents is complemented by a small number of chemical reagents having a known peptide bond specificity. For example, *cyanogen bromide* selectively cleaves a peptide bond only if it contains a carbonyl group donated by methionine. A consequence of the cyanogen bromide treatment is the conversion of the methionine residue to homoserine lactone (see margin).

The isolation and sequencing of small peptides formed only by trypsinolysis, for example, does not solve the problem at all, since the fragments still need to be mapped in the proper order. Thus, the same procedure is repeated with chymotrypsin, with cyanogen bromide, with a combination of both enzymes, or with either enzyme under different conditions, in an attempt to *obtain a large collection of different peptide fragments formed by the cleavage of different peptide bonds within the same polypeptide.* Once all fragments from the different procedures are sequenced, the complete polypeptide chain can be mapped on the basis of positioning the fragments by searching for and aligning overlapping regions. This is more difficult than it appears. Indeed, the task is arduous and generally requires several months depending on the size of the protein and the number of separate chains present in the molecule. The time and effort required have already been reduced by the development of automated peptide sequencers. Even less time will be required when the instrument is perfected to maintain efficiency through more than 20 cycles so that larger fragments can be worked with. There have already been scattered reports that 40–60 cycles have been achieved. The principles of mapping are diagramed on p. 128.

As stated earlier, the first protein to have its amino acid sequence determined was insulin (51 residues). The feat was remarkable in view of the fact that the Edman technique was not available and chromatographic techniques were not as highly refined as they are today. Since then many other proteins and smaller polypeptides from nearly every conceivable source have been successfully sequenced. Improved isolation procedures, the repetitive Edman degradation, improved chromatographic procedures, automatic analyzers, and automatic sequencers have all had a significant impact. The 1972 edition of the *Atlas of Protein Sequence* lists approximately 700 sequences covering 438 different entries. The scope of this type of work is also illustrated by the nearly 2,500 researchers who are referenced in this compilation. The fact that these figures are nearly double those of the 1969 edition reflect the fantastic rate of growth of this area of biochemical

internal methionine

$$\sim N - \underset{\underset{\underset{\underset{\underset{CH_3}{|}}{S}}{\underset{|}{CH_2}}}{\underset{|}{CH_2}}}{\overset{H}{\underset{|}{CH}}} - \underset{O}{\overset{}{\underset{\|}{C}}} - N - CHR - \overset{O}{\overset{\|}{C}} \sim$$

$$\downarrow \text{BrCN(H}_2\text{O)}$$

$$\sim N - \underset{\underset{\underset{CH_2}{|}}{H_2C}}{\overset{}{\underset{|}{HC}}} - \underset{}{\overset{O}{\overset{\|}{C}}} \overset{}{\underset{O}{\diagdown}} + \overset{O}{\underset{H_3\overset{+}{N}CHRC}{\|}} \sim$$

peptide with homoserine lactone as new C-terminus

$+ CH_3SCN$

$+ Br^-$

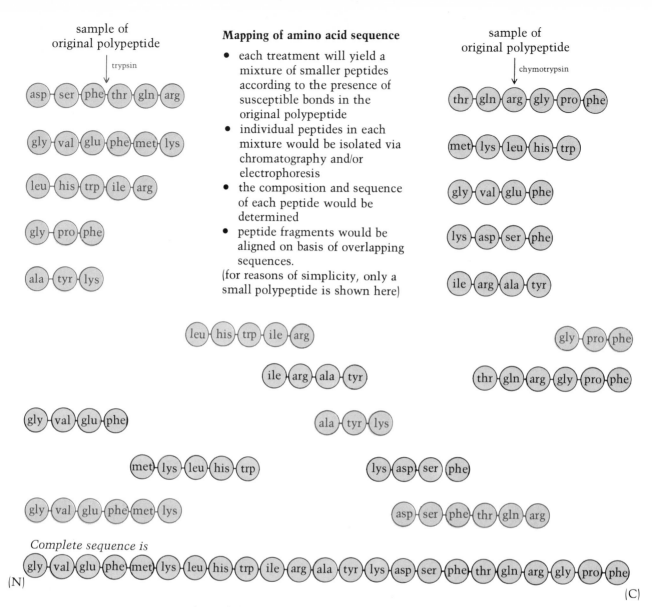

Mapping of amino acid sequence

- each treatment will yield a mixture of smaller peptides according to the presence of susceptible bonds in the original polypeptide
- individual peptides in each mixture would be isolated via chromatography and/or electrophoresis
- the composition and sequence of each peptide would be determined
- peptide fragments would be aligned on basis of overlapping sequences.

(for reasons of simplicity, only a small polypeptide is shown here)

Insulin is a hormone produced by the pancreas (see Table 4–4, p. 110. Cytochrome c is the most studied representative of the cytochrome proteins, which have an important role in the cellular production of energy. The cytochromes are described in Chap. 14.

knowledge. Estimates are that the current number of sequences that are reported in a three month period is greater than the total number of sequences that were known in 1965 when the first edition appeared. The current number of known sequences is probably well over 1,000. The most remarkable of all of these is the recently published sequence of a gamma globulin molecule, with 1320 amino acid residues.

The sequences of insulin and cytochrome c are given in Fig. 5–1. Insulin consists of two chains connected by two interchain disulfide bonds, with one chain also having an intrachain disulfide bond. On the other hand, cytochrome c, a conjugated hemoprotein (see p. 140 for discussion of the heme prosthetic group), consists of a single polypeptide chain with no disulfide bonds.

Some important generalizations regarding amino acid sequence are:

1. There is no single partial sequence or group of partial sequences common to all polypeptides.

					1
man, chimpanzee				gly	—asp—
pigeon				gly	—asp—
fruit fly	gly —val	—pro—ala—	gly	—asp—
yeast	thr —glu—phe—glu—	gln—	gly	—ser —

```
              5                    10                    15
val—glu—lys—gly—lys—lys —ile —phe—ile —met— lys —cySH—ser—gln—cySH—his— thr —
ile —glu—lys—gly—lys—lys —ile —phe—val—gln — lys —cySH—ala—gln—cySH—his— thr —
val—glu—lys—gly—lys—lys —ile —phe—val—gln — arg—cySH—ala—gln—cySH—his— thr —
ala —lys —lys—gly—ala—thr —leu—phe—lys—thr — arg—cySH—glu—leu—cySH—his— thr —

 20              25                    30                    35
val—glu—lys—gly—gly—lys—his—lys—thr—gly— pro—asn—leu—his—gly—leu—phe—
val—glu—lys—gly—gly—lys—his—lys—thr—gly— pro—asn—leu—his—gly—leu—phe—
val—glu—ala—gly—gly—lys—his—lys—val—gly— pro—asn—leu—his—gly—leu—ile —
val—glu—lys—gly—gly—lys—his—lys—val—gly— pro—asn—leu—his—gly—ile —phe—

             40                    45                    50
gly—arg—lys —thr—gly—gln—ala—pro—gly—tyr — ser—tyr—thr—ala —ala—asn—lys—
gly—arg—lys —thr—gly—gln—ala—glu—gly—phe— ser—tyr—thr—asp—ala—asn—lys—
gly—arg—lys —thr—gly—gln—ala—ala —gly—phe— ala—tyr—thr—asn—ala—asn—lys—
gly—arg—his —ser—gly—gln—ala—gln—gly—tyr — ser—tyr—thr—asp—ala—asn—ile —

 55              60                    65                    70
asn—lys—gly—ile —ile —trp—gly —lys —asp—thr— leu —met—glu—tyr—leu—glu—tyr—
asn—lys—gly—ile —thr—trp—gly —lys —asp—thr— leu —met—glu—tyr—leu—glu—tyr—
ala —lys—gly—ile —thr—trp—gln—asp—asp—thr— leu —phe—glu—tyr—leu—glu—tyr—
lys —lys—tyr—val—leu—trp—asp—asp—tyr —tyr— met—ser —glu—tyr—leu—thr—tyr—

             75                    80                    85
pro—lys—lys—tyr—ile—pro—gly —thr—lys—met— ile —phe—val—gly—ile —lys—lys—
pro—lys—lys—tyr—ile—pro—gly —thr—lys—met— ile —phe—ala—gly—ile —lys—lys—
pro—lys—lys—tyr—ile—pro—gly —thr—lys—met— ile —phe—ala—gly—leu—lys—lys—
pro—lys—lys—tyr—ile—pro—gly —thr—lys—met— ala—phe—gly—gly—leu—lys—lys—

             90                    95                    100                   104
lys —glu —glu—arg—ala —asp—leu—ile—ala—tyr—leu—lys—lys—ala—thr  —tyr—glu
lys —ala —glu—arg—ala —asp—leu—ile—ala—tyr—leu—lys—gln—ala—thr  —ala—lys
pro—asn—glu—arg—gly—asp—leu—ile—ala—tyr—leu—lys—ser—ala—thr  —lys
glu —lys —asp—arg—tyr—asp—leu—ile—thr—tyr—leu—lys—ser—ala—cySH—glu
```

FIGURE 5–1 *Top:* comparison of the amino acid sequences of cytochrome *c* from five different sources. Although of slightly different lengths, it is clear that these five molecules have an extensive amount of sequence homology. The 11 residues from 70 to 80 are the same in all cytochrome *c* molecules from 38 different sources studied so far. *Bottom:* human insulin.

2. Every possible combination of two successive amino acids has been detected.

3. Proteins having different functions have widely different sequences. This should be obvious to you.

4. Proteins of similar function have similar sequences but to varying degrees—usually slight, but sometimes extensive. The comparison of sequences for such proteins has contributed to a determination of what part (or parts) of the polypeptide is (are) crucial to its overall structure and function. Comparative information of this sort also helps in tracing the evolution of genes.

129

5. The same proteins performing the same function but occurring in different life species will usually have extensive similarities in sequence. A remarkable example of very extensive similarity is provided by cytochrome *c*, for which the sequence of the single polypeptide chain has been determined from 38 different organisms, ranging from yeast to primates, that cover a 1.2 billion year period of evolution. Among this group the chain size ranges only from 104 to 112. Of greater significance is that 35 of the residues are identical *in all molecules*. Eleven of these 35 represent a continuous segment that is identical *in all molecules*. Another 23 sites involve differences in very similar amino acids (for example, serine vs. threonine, lysine vs. arginine, and aspartic vs. glutamic) and thus they are virtually identical. Radical changes occur in only about six positions. It is evident that the extent of sequence similarity is closely correlated with evolutionary relationships, with less variation common to a protein in organisms of the same phylogenetic class. For example, humans and chimpanzees—both primates—have identical cytochrome *c* molecules.

6. The same protein performing the same function and isolated from members of the same species will almost always have the same exact sequence. When a variation does occur the overall function of the protein may or may not be affected. (See the next section for a description of hemoglobin variants in humans.)

Biological Significance of Amino Acid Sequence

At present there is overwhelming evidence to support the following statement: *The higher orders of protein structure, in particular the overall conformation, and the biological activity of the protein are both intimately dependent upon, and in fact controlled by, the amino acid sequence.* A particularly convincing proof was supplied by the chemical synthesis of the enzyme *ribonuclease* by Merrifield. After the polypeptide was assembled with a sequence identical to the known sequence for native ribonuclease, the synthetic enzyme had the same biological activity as the native enzyme. This strongly suggests that the synthetic product did assume an active three-dimensional conformation, similar to that of the native enzyme, solely on the basis of positioning the 124 amino acid residues in the right order. There are many other similar results with other synthetic polypeptides.

Evidence of a different sort is provided by *sickle cell anemia*, a hereditary disease of the red blood cells characterized by a low oxygen-binding capacity. The malfunction is attributable to the presence of abnormal hemoglobin molecules (HbS) in the sickle cells, so named because of the characteristically abnormal shape they assume. The malfunction of the hemoglobin molecule has been linked to a difference in a single position in the amino acid sequence of the beta (β) polypeptide chain. (An intact hemoglobin molecule consists of two copies of the β chain and two copies of another slightly different polypeptide called the alpha (α) chain; see pp. 124 and 146.) The β chain of normal adult hemoglobin (HbA) contains glutamic acid in position 6, whereas the β chain of HbS contains valine. All other β chain residues are identical and the α chains are completely identical. Despite this small difference (involving a total of two residues out of 574), the HbS molecule doesn't work right.

The ability to bind oxygen is impaired because apparently the HbS molecules tend to stick to each other, forming larger aggregates. This formation of HbS multimolecular aggregates in the red blood cells contributes to their deformation into a sickled shape. In addition to the reduced amount of oxygen-binding ability, the deformed cells can get trapped in narrow blood vessels, which further retards the delivery of any oxygen that is bound. In addition, the sickle cells also rupture easily, hence the anemia. All of this happens because valine (neutral and nonpolar) rather than glutamic acid (negatively charged and polar) is in position 6 of the two β chains. Obviously, the presence of a neutral, nonpolar R group at position 6 in the β chain has disastrous consequences. Putting it another way, the presence of a negatively charged, polar R group in this position is indispensable to the normal structure and function of Hb. A partial listing of other abnormal hemoglobin variants appears in the margin.

It would be incorrect, however, to conclude at this point that every amino acid residue in every protein is indispensable to the normal structure and function of the protein. This simply is not the case. For example, many hemoglobin sequence variants have been documented that are not functionally abnormal. The amino acid residues involved in these would thus be considered dispensable in the sense that substitutions can occur without any effect on the overall molecule. This distinction between *indispensable* (essential) and *dispensable* (nonessential) residues will be amplified further in our discussions of enzymes in the next chapter (see p. 171).

SECONDARY STRUCTURE

The Peptide Bond

A description of the spatial orientation of polypeptide chains properly begins with an examination of the structural characteristics of the linkage that repeats itself along the chain, that is, the peptide bond. Earlier we described it only in the most general way as an amide linkage between the carboxyl group of one amino acid and the amino group of another. However, there is much more to it than that.

Our knowledge of the spatial arrangement of the atoms of the peptide bond and the resultant peptide chain is due primarily to the X-ray diffraction analyses of simple peptides by L. Pauling and R. Corey in the early 1950s. The results of their studies are shown in Fig. 5–2, which illustrates the geometrical features and dimensions of the peptide bond and a segment of a fully extended peptide chain of L-amino acids. The length of 7.23 Å corresponds to the repeat distance between any two alpha carbons in the chain that immediately flank one complete amino acid residue. Based on unstrained bond angles and bond distances, the length of 7.23 Å represents the maximum allowable repeat distance. The significant features of the peptide bond are listed below, and all are represented by the drawings and photographs of Fig. 5–2.

1. *The four atoms of the peptide bond and the two attached alpha carbons are all in the same spatial plane.* The H and R groups on the alpha carbons are projected out of this common plane.

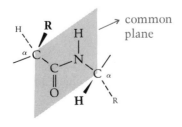

common plane

Some other abnormal hemoglobins

Hb	Alteration (site)
C	glu → lys (β^6)
D	glu → gln (β^{121})
E	glu → lys (β^{26})
G	asn → lys (α^{68})
Q	asp → his (α^{75})
M	his → tyr

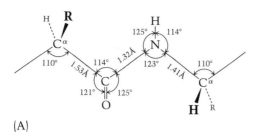

(A)

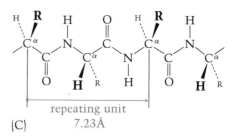

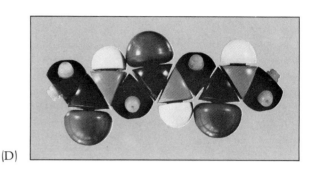

repeating unit
7.23Å

(C)

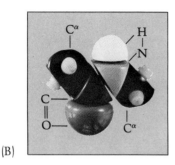

(B)

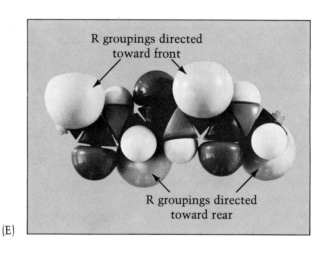

(D)

FIGURE 5-2 *Panel A:* dimensions of the peptide bond. *Panel B:* the symbolic diagram in Panel A shown as a space-filling model. *Panel C:* drawing of a segment of a fully extended polypeptide chain of L-amino acids. *Panel D:* a space-filling model of the drawing in C. The studs correspond to bonds linking H and R groups to alpha carbons. *Panel E:* repeat of panel D showing the H and R groups on the alpha carbons.

R groupings directed
toward front

R groupings directed
toward rear

(E)

2. The geometrical orientation of the *O and H atoms* in the peptide bond is *trans.* In addition, the geometrical orientation of the *two alpha carbon atoms*, relative to the peptide bond to which they are attached, is also *trans.*

3. As a consequence of the spatial orientation in (2), and given the L configuration of each residue, the *R groups* on each of the alpha carbon atoms are arranged in a repeating *trans* fashion. (R ▶ indicates one side of plane and R---indicates the other side.)

4. The C—N bond distance of 1.32 Å for the amide grouping (that is, the peptide bond) is intermediate in length between that of a double covalent bond (1.21 Å) and that of a single covalent bond (1.47 Å). Thus, the intermediate value of its bond length would indicate that it does have *some double-bond (pi-bond) character.* This can be explained in terms of either a resonating or tautomeric structure.

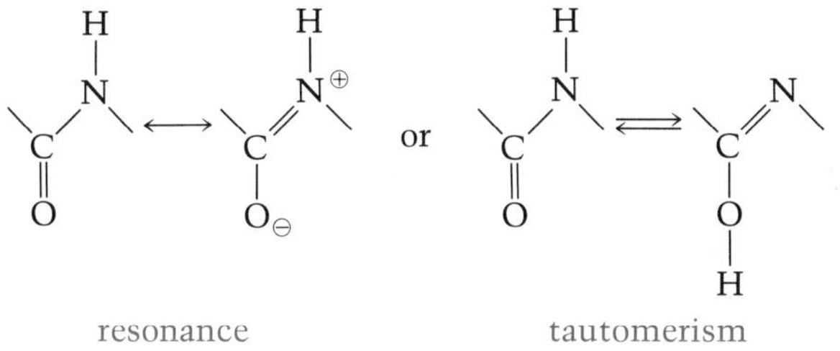

resonance   tautomerism

The primary consequence of this pi-bond character is that there is *restricted rotation around the C—N bond axis*. However, each unit of the planar peptide linkage does contain two bonds involving the alpha carbon atom that are capable of free rotation. Both are sigma bonds without any significant pi character. They are the C^α—N and C^α—C bonds. The notations ϕ (for C^α—N) and ψ (for C^α—C) are used to identify the specific rotational values for each bond. In the fully extended orientation the ϕ and ψ values are both considered to be 0°, and when viewed from the N-terminus, any clockwise rotation through 180° is assigned a + value and any counterclockwise rotation through 180° is assigned a − value.

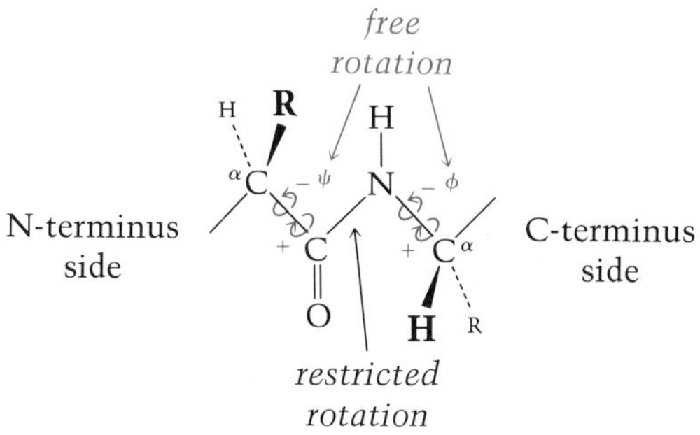

Orientation in Naturally Occurring Proteins

The particular type of orientation assumed by a polypeptide chain is the result of the pattern of free rotation around the bonds of the chain involving the alpha carbon atoms. Three major types of orientation are found in naturally occurring polypeptide chains: (1) *helical*, (2) *sheet*, and (3) *random*. The helical and sheet forms are ordered arrangements, while the random forms are just that, random arrangements. An ordered arrangement means the presence of a distinctly recognizable orientation that is conferred to the chain whenever the ϕ degree of rotation is identical for all the C^α—N bonds and the ψ degree of rotation is identical for all the C^α—C bonds.

In polypeptide chains composed of L-amino acids, the most common helical orientation is called the *right-handed alpha helix*, where each $\phi = -48°$ and each $\psi = -57°$. Why is this specific ϕ,ψ combination so special? Simply because a very stable arrangement results. And why is it stable? First, because there is little or no crowding of atoms, particularly among the R side chain groups of the amino acids. Second, and of greater importance, because the C=O dipole and the N—H dipole of neighboring peptide bonds are optimally oriented (they are nearly coaxial—see p. 83) for maximum dipole-dipole interaction, thereby producing an extensive network of *intrachain cooperative hydrogen bonding* that holds the α helix intact.

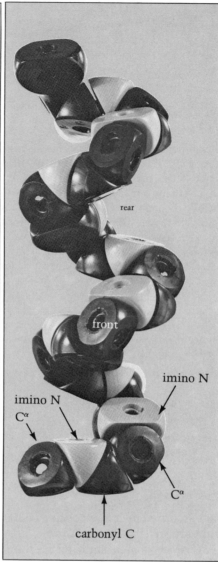

studs for
R groups

5.4 Å
per
turn

rear

front

imino N

imino N

C^α

C^α

carbonyl C

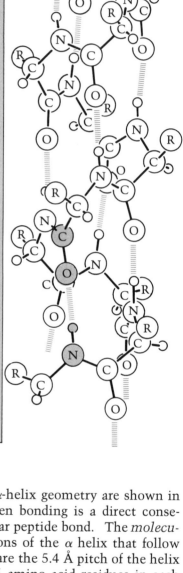

FIGURE 5–3 The α-helix conformation of a polypeptide chain. *Right:* a stick model. *Center:* backbone of α-helix. *Left:* photograph of space-filling model with all C=O, N—H, and C^α—H linkages inserted and corresponding to the segment at the right. The R groups of the alpha carbons are shown as studs so as not to obstruct the view of the hydrogen bonds that exist between successive turns of the helix. (Drawing of the stick model reprinted from Linus Pauling, *The Nature of the Chemical Bond.* Copyright 1939 and 1940 by Cornell University. Third edition © 1960 Cornell University. Used by permission of Cornell University Press.)

A sketch and space-filling models of the α-helix geometry are shown in Fig. 5–3. Note that the intrachain hydrogen bonding is a direct consequence of the *trans* orientation in the coplanar peptide bond. The *molecular fitness* is remarkable. Other specifications of the α helix that follow directly from the specified ϕ,ψ combination are the 5.4 Å pitch of the helix (one complete turn) and the presence of 3.6 amino acid residues in each complete turn. When L-amino acids are present, the right-handed helix is energetically favored over a left-handed helix due to the outward projection of the amino acid R groupings away from the center axis; this minimizes steric hindrance (crowding).

The percentage of α-helix content in globular proteins is quite variable, ranging from 0% (none at all) to values approaching 80–90%. (A polypeptide chain could never contain 100% α helicity and be globular. It would then be a fibrous polypeptide.) Moreover, when the α helix is present, it may be found in just one segment of the polypeptide chains or in two or more separated segments. Two factors that would definitely interrupt a helical orientation are (1) the presence of proline, the cyclic structure of which naturally puts a bend in the polypeptide backbone, and (2) the presence of localized electrostatic forces of repulsion due either to a cluster of positively charged R groups from lysine and arginine or to a cluster of negatively charged R groups from glutamic and aspartic acids. The nonhelical portions of globular polypeptides can be in either a sheet orientation or a random orientation.

The *sheet orientation is also stabilized by cooperative hydrogen bonding* among the same dipoles, but in this case a completely different pattern emerges. Specifically, the bonding occurs on an *interchain basis* among two or more different chains (in fibrous proteins) or on an *intrachain basis* between different segments of the same chain (in globular polypeptide chains). In either case, two types of sheet structure are possible, depending on the alignment of atoms in the different chains or segments. (See Fig. 5–4.) If both chains are aligned in the same direction from one terminus to the other, the arrangement is termed the *parallel sheet.* If the two are aligned in opposite direction, the arrangement is termed the *antiparallel sheet.* Although both types occur, the antiparallel alignment (with each $\phi = -140°$ and each $\psi = +135°$) is more stable because the C=O and N—H dipoles are oriented for maximal interaction, that is, they are coaxial.

In regard to the occurrence of random orientation in polypeptides of globular proteins it should be noted that this does not mean that the chain assumes any orientation. Rather, because of the ordered interactions involving R groups and the possible presence of a disulfide bond(s), a *particular random arrangement is preferred.* That is to say, each of the different ϕ values and each of the different ψ values are preferred. Thus, in the context of the highly ordered native conformation of a globular protein, any nonhelical or nonsheet segment of the polypeptide chain could be referred to as a being in a highly ordered random orientation. If this seems contradictory, think about it a little. The word random means only an absence of identical ϕ values and identical ψ values *in succession.* This point should be clearer after reading the next section.

To summarize:

Type of protein	Orientation of polypeptide chain(s)	
Fibrous		Exclusively helical
	or	exclusively sheet
Globular		Part helix, part sheet, and part random
	or	part helix and part random
	or	part sheet and part random
	or	entirely random

TERTIARY STRUCTURE

It will become very clear that a knowledge of the three-dimensional structure of a biomolecule is extremely useful in understanding its function. This is particularly true for the proteins and nucleic acids. First, let us consider how one can experimentally determine the general shape of a protein.

The α-helix orientation was first reported in 1953 by Pauling and Corey in their study on synthetic polypeptides. They received the Nobel Prize for their achievement.

Orientation	Degrees of rotation	
	C^α—N bonds ϕ	C^α—C bonds ψ
α-helix (right-handed)	−48	−57
sheet (antiparallel)	−140	+135
sheet (parallel)	−119	+113

The segment shown as a helix in Fig. 5–3 is shown here fully extended; how many complete amino acid residues are present?

FIGURE 5–4   The drawing depicts the hydrogen bonding of the pleated sheet configuration between two peptide segments aligned in an antiparallel orientation.   The space-filling model shows a group of interlocking pleated sheets characteristic of many fibrous proteins, including silk.   The section shown consists of 44 residues and 7 polypeptide strands.   (Photograph supplied by The Ealing Corporation, Natick, Mass.)

If the diffusion constant D of the protein has been determined, it is possible to determine the general overall shape of the molecule by calculating the *dissymmetry ratio* f/f_0.   As shown below the calculation also requires an accurate value of the molecular weight M and the partial specific volume $\overline{V}$.

$$\frac{f}{f_0} = \left(\frac{R^3 T^3}{162\ \pi^2 M \overline{V} N^2 D^3 \eta^3}\right)^{1/3} \qquad (N \text{ is Avogadro's number})$$

The interpretation of the value of f/f_0 becomes quite simple if we describe what it is.   It is a ratio of frictional coefficients; specifically a ratio of the frictional coefficient of the protein molecule in its actual shape (f) to the frictional coefficient of the same molecule if it were a perfect sphere (f_0).   Thus, the ratio is an *index of deviation of the molecular shape from a perfect sphere*.   If it is a perfect sphere, $f/f_0 = 1$.   If it is not, $f/f_0 > 1$; and the greater the value then the less spherical and more elliptical or rodlike the molecule.   The values for some proteins are listed in Table 5–2.

Given the following information needed to derive this equation for f/f_0:

1. for any particle

$$f = \frac{RT}{D}$$

2. Stokes's law governing the flow of a spherical particle of radius r in a liquid medium of viscosity η is
$$f_0 = 6\pi\eta r N$$

3. volume of sphere of radius r is

$$v = \frac{4}{3}\pi r^3$$

4. the volume of spherical molecule in terms of partial specific volume $\overline{V}$ and molecular weight M is

$$v = \overline{V}\left(\frac{M}{N}\right)$$

5. can you derive the equation?

TABLE 5–2   **Dissymmetry ratios of proteins.**

Protein	f/f_0	Molecular[a] weight
Enolase	1.00	63,300
Insulin	1.07	12,650
Myoglobin (horse heart)	1.10	17,000
Lactate dehydrogenase (beef heart)	1.13	133,000
Hemoglobin (human)	1.16	66,000
Cytochrome c (beef heart)	1.19	13,400
Albumin (human serum)	1.29	68,500
Gamma globulin (human)	1.51	153,000
Fibrinogen[b] (human)	2.34	340,000
Myosin[b] (cod)	3.63	500,000

[a] Note the lack of any correlation between size and shape.
[b] Both of these proteins have an extensive amount of fibril character.

Sheet conformation

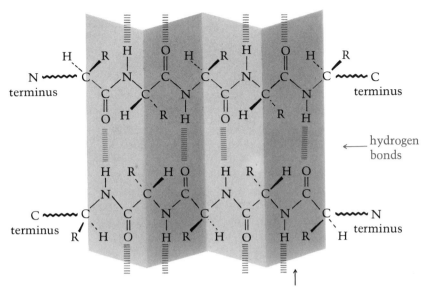

N terminus

C terminus

← hydrogen bonds

all atoms in face of pleat are coplanar

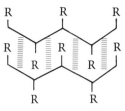

In fibrous proteins this side-by-side alignment involves separate chains (see photo below); in globular proteins it can occur between different regions of the same chain (see Fig. 5–6)

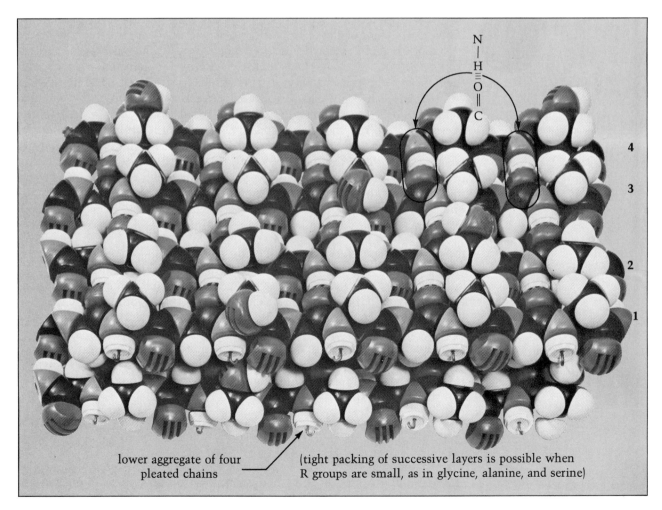

lower aggregate of four pleated chains → (tight packing of successive layers is possible when R groups are small, as in glycine, alanine, and serine)

137

A consideration of more specific features regarding the three-dimensional shape of proteins is best done in the context of differentiating between fibrous and globular proteins.

Fibrous Proteins

Basically the three-dimensional structure of fibrous proteins has been described in the previous section. The *polypeptide chains are entirely in a sheet or helical conformation.* The intact protein, however, is not a single polypeptide chain but rather an aggregate of two or more polypeptide chains. The fibroin proteins in silk, for example, are composed of several sheet chains stacked side by side, as well as above and below each other, in an antiparallel pattern (refer to Fig. 5–4 on p. 137). On the other hand, the α keratin protein fibers of wool, hair, feather, horns, and nails are composed of coiled helical chains, giving a larger, multichain helical structure, which in turn is bundled or coiled together to give a super structure. The various levels of structure are analogous to those found in a rope.

One of the most understood fibrous proteins is *collagen,* which, as the principle constituent of connective tissue, is the major structural protein in all animal life. One estimate of its abundance on earth is 2 trillion pounds. As shown on p. 139, each of the macro collagen fibers found in the connective tissue molecule is about 3000 Å in length, and has an approximate molecular weight of about 300,000. The structure of the collagen molecule is that of a coiled-coil, specifically a *triple helix.* (This is sometimes termed *tropocollagen.*) Each of three identical chains is a left-handed helix, but they are coiled together to give a right-handed triple helix. The helical orientation of each chain is not of the α variety. It is much more extended than the α helix. Of course, the α helix is ruled out anyway, because of the presence of proline and hydroxyproline. Indeed, the amino acid composition and sequence are the likely reasons for the entire triple-helix structure. Nearly one-third of the residues are glycine, another one-quarter are either proline or hydroxyproline, and the rest are made up of small amounts of most of the other amino acids, including some hydroxylysine. The sequence is somewhat repetitive, with glycine found in almost every third position and tripeptide segments of gly—X—pro, gly—X—hypro, and gly—pro—hypro being very common. The $\overset{1\ 2\ 3}{--\text{gly}}\overset{1\ 2\ 3}{--\text{gly}}\overset{1\ 2\ 3}{--\text{gly}}\overset{1\ 2\ 3}{--\text{gly}}$ pattern is proposed to be an absolute requirement for forming the triple helix.

The coiled-coil structure of the triple helix is stabilized by extensive interchain hydrogen bonding and in addition there are some covalent cross-links between chains. Thus, the *structure is rigid and inflexible.* The packing of collagen molecules into the macro collagen fibers, which may also involve covalent cross-links between molecules, confers additional rigidity and inflexibility. The structure of collagen—all the way down to its amino acid composition and sequence—is thus well suited for its biological role as a *tough structural protein.*

Globular Proteins

There is no systematic way to describe the three-dimensional structure of globular polypeptides. Let us then get right into a specific example and use it as a model system representative in many (but not all) respects of other

Levels of collagen structure

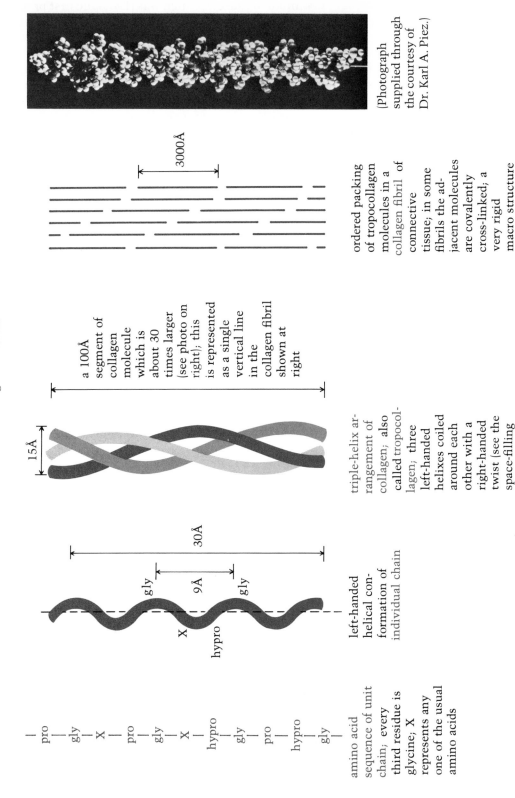

amino acid sequence of unit chain; every third residue is glycine; X represents any one of the usual amino acids

— pro — gly — X — pro — gly — X — hypro — gly — pro — hypro — gly —

left-handed helical conformation of individual chain

30Å
9Å
gly   gly
X   hypro

triple-helix arrangement of collagen; also called tropocollagen; three left-handed helixes coiled around each other with a right-handed twist (see the space-filling model on the right)

15Å

a 100Å segment of collagen molecule which is about 30 times larger (see photo on right); this is represented as a single vertical line in the collagen fibril shown at right

ordered packing of tropocollagen molecules in a collagen fibril of connective tissue; in some fibrils the adjacent molecules are covalently cross-linked; a very rigid macro structure

3000Å

(Photograph supplied through the courtesy of Dr. Karl A. Piez.)

139

For their pioneering developments in applying X-ray diffraction to protein structure determinations, Kendrew and Perutz were awarded the Nobel Prize in 1962.

globular polypeptides. Our example will be *myoglobin*, the oxygen-carrying protein of mammalian muscle, which was (1960) the first protein to have its complete structure unravelled—the brilliant and arduous accomplishment of J. C. Kendrew, M. F. Perutz, and coworkers. There have been several other successes since then and the work continues.

Before considering its three-dimensional structure, let us pause to consider the primary level of myoglobin structure. The protein consists of a single polypeptide chain, containing 153 amino acid residues, to which there is coordinated one prosthetic group, namely, an *iron-containing tetrapyrrole group* called *heme*. Accordingly, myoglobin is termed a hemoprotein. As the name implies, the heme group is also present in hemoglobin. Since heme is present in several other proteins as well, let us now give it a preliminary examination, with further consideration to come in Chap. 14 (see p. 454).

pyrrole

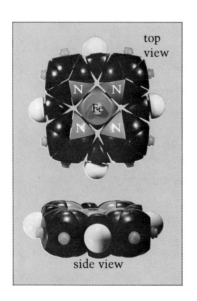

top view

side view

heme

The heme grouping is a bulky planar ring system composed of four *pyrrole* units connected by methenyl ($=CH—$) bridges. The iron atom, with a coordination number of 6, is located in the core of the tetrapyrrole ring and is complexed to each of the four pyrrole nitrogens. In both myoglobin and hemoglobin, the iron is generally in the ferrous state ($+2$). There is strong evidence that the heme grouping is complexed to the polypeptide chain (globin) through two specific histidine residues occupying the fifth and sixth coordination sites of iron. In hemoglobin, four such heme-polypeptide complexes exist (see p. 146). When the proteins are oxygenated, the oxygen, having displaced one of the histidine groups, is transported as a ligand complexed to the sixth coordination site of iron. Various hemes are differentiated by the identity and arrangement of the eight side chains attached to the four pyrrole units (see p. 456). The groups shown above are characteristic of myoglobin and hemoglobin.

The structure of myoglobin was constructed by a sophisticated analysis of X-ray diffraction patterns. One such pattern is shown in Panel A of Fig. 5–5. Several of these patterns were obtained from the same sample studied at different angles. With the indispensable aid of a computer, the intensity of diffraction at each point, as well as the spacings between each point, can be calculated. The final studies of Kendrew and coworkers involved the determination of approximately 25,000 reflections, permitting the measurement of electron density at roughly 250,000 points in the molecule (obviously not a weekend experiment). This provided a resolution of 1.4 Å, which was sufficient to localize nearly every atom in the molecule. These data then served as a basis for the construction of a three-dimensional electron density contour map (Fig. 5–5, Panel B), which was then translated into a molecular model (Fig. 5–5, Panels E and F).

The message conveyed by the display of myoglobin structure regarding the molecular architecture of globular proteins is reinforced by the drawings below of the solved structures of carbonic anhydrase and carboxypeptidase. The message should be obvious, namely, that in contrast to the highly extended structure of a fibrous protein a globular protein does indeed have a more compact structure due to an extensive pattern of folding, bending, and twisting along the polypeptide chain. However you view this—be it with fascination, with excitement, or even with a sensitivity to what could be termed the beauty of biomolecules—realize two things: (1) that the seemingly tortuous three-dimensional arrangement of atoms (the word *conformation* is usually used) in a globular protein does represent a highly ordered, highly specific, and preferred conformation for that protein because of its

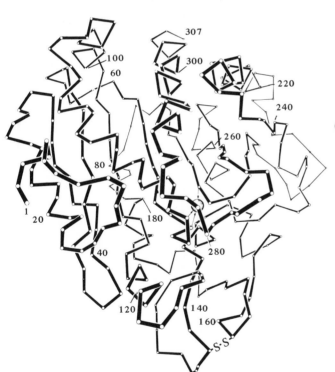

carboxypeptidase

(Sphere in center represents Zn^{2+}; for more details about this protein, see p. 181. Reproduced with permission from W. N. Lipscomb, "Structure and Mechanism in the Enzymatic Activity of Carboxypeptidase A and Relations to Chemical Sequence," *Accts. Chem. Res.*, **3**, 81–89 (1970). Drawing generously supplied by Dr. Lipscomb.)

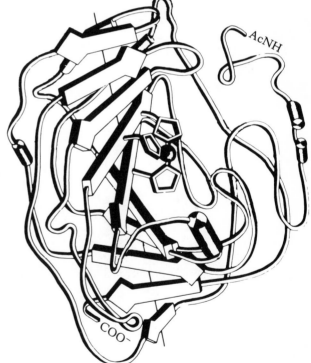

carbonic anhydrase

(Arrows signify antiparallel sheet segments and cylinders signify alpha helix segments; sphere in center represents Zn^{2+} coordinated to histidine R groups. Reproduced with permission from "Crystal Structure of Human Erythrocyte Carbonic Anhydrase C," Cold Spring Harbor Symposium on Quantitative Biology, **36**, 221–231 (1971). Drawing generously supplied by Dr. K. K. Kannan.)

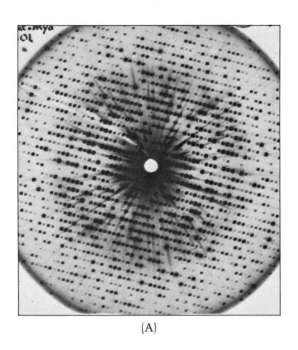

(A)

FIGURE 5–5 Myoglobin structure. *Panel A:* X-ray diffraction photograph of a myoglobin crystal. The calculation of the intensity of each reflection (dots) is used as the basis for constructing a three-dimensional electron density distribution, which is most conveniently represented as a series of contour maps plotted on parallel transparent sheets as shown in *Panel B.* The intense doughnut-shaped area in the center right of Panel B corresponds to an α-helix segment of the polypeptide chain viewed from one end. A similar but much clearer view of the same thing is shown in the upper left of *Panel C,* which is an enlarged photo of a three-dimensional electron density distribution contour map. Panel C also shows a side view of the planar heme grouping. Also visible are the linkages of the heme grouping to the imidazole side chain of a histidine residue (at seven o'clock) and to a water residue (at one o'clock). *Panel D* diagrammatically summarizes the extent of α helical conformation along the polypeptide chain. *Panel E* shows a model of the twists and folds of the polypeptide chain (sausage-shaped) and the orientation of the heme grouping (dark disc at top). *Panel F* is a skeleton model of the myoglobin molecule derived from the contour maps. The white cord follows the course of the polypeptide chain; the heme grouping is located at the upper center with the iron atom (gray sphere) associated with a histidine residue at the left and a water molecule to the right (white sphere). The many projections represent the side

(B)

end view of
helical segment

chains of the amino acid residues. (Panels A, B, C, E, and F taken with permission from J. C. Kendrew, "Myoglobin and the Structure of Proteins," *Science,* **139,** 1259–1266 (1963). Photographs kindly supplied by J. C. Kendrew.)

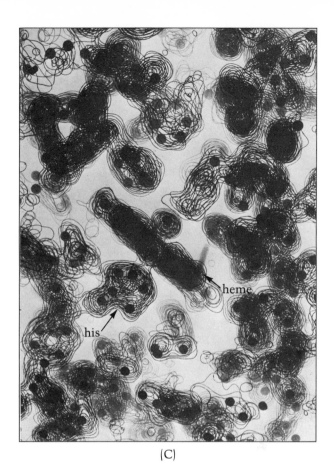

(C)

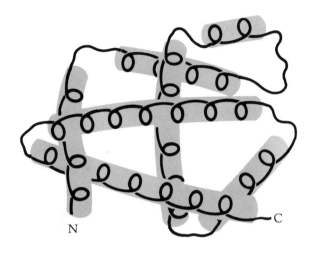

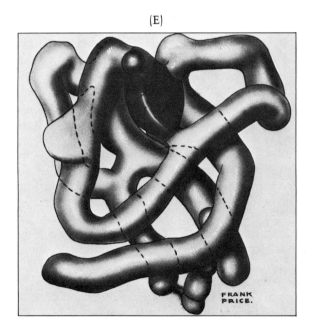

helical segments (70–80% of molecule)

(D)

(E)

(F)

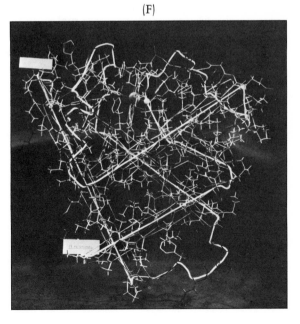

unique amino acid sequence, and (2) that this native conformation—as it is usually termed—represents the ultimate *structural individuality* of the protein, which in turn is primarily responsible for the *functional individuality* of the protein.

The native conformation of a protein is preferred basically because it represents—among countless arrangements—a *conformation of minimum energy*, or in other words, a *conformation of maximum stability* (see the following paragraph). It is incorrect, however, to conclude that every molecule of a protein has exactly the same conformation, into which it is frozen forevermore. It is more likely that a protein has a small number of very closely related and easily interconvertible conformations of minimum energy. Indeed, the principle of interconvertible conformations of oligomeric proteins (see next section) is one of the hallmarks of modern biochemistry. More will be said about this later.

The specification that the tertiary structure of globular proteins is an *ordered conformation* of a severely constrained polypeptide chain obviously dictates a necessity for *stabilizing forces*. In several cases the presence of one or more disulfide bridges in the primary structure fulfills this function. The covalent disulfide linkage is not the only stabilizing force, however, and in fact is not present at all in several proteins. On the whole, stabilization is mediated largely by *noncovalent interactions* (see p. 83) primarily among atoms of the amino acid side chains. The primary types are (a) *electrostatic forces of attraction*, occurring between side chains possessing oppositely charged ionic groups such as lysine, arginine, glutamic acid, and aspartic acid; (b) *nonpeptide bond hydrogen bonding*, such as that occurring between tyrosine and glutamic acid residues; (c) *peptide bond hydrogen bonding*, occurring in helical regions and between different pleated-sheet segments of the chain; and (d) *hydrophobic interactions*, occurring among the nonpolar side chains of leu, val, ile, ala, phe, and the nonpolar acids. Figure 5–6 illustrates each type.

Among the various types, the most important are thought to be the hydrophobic interactions. These are generally localized within the *interior* of the molecule and are responsible for performing the function of stabilizing the internal core of the protein molecule, which is virtually devoid of water. The electrostatic forces of ionic attraction are generally localized at or near the *surface* of the molecule along with other polar R groups. The shielding of most uncharged nonpolar residues in the interior of the molecule, coupled with the surface exposure of most charged and uncharged polar residues, also contributes to the solubility of the protein in water, a polar solvent. Such an arrangement appears to be the *one general feature that applies to all globular polypeptides* (proteins). Recall that this arrangement is similar to that of a micelle (p. 84). Accordingly, you may find it useful to think of a globular protein as a molecular micelle.

Before concluding this section, one further point merits attention, namely, that the native conformation of a globular protein molecule is a highly delicate state and much less rigid than that of fibrous proteins. In other words, the globular conformation is more easily subject to alteration—a feature which has significance for both the handling of globular proteins during their isolation and the way they operate in a cell. The concern about handling is discussed later in this chapter (p. 148). An understanding of the significance of alterations in conformation as part of normal cellular function will be developed gradually.

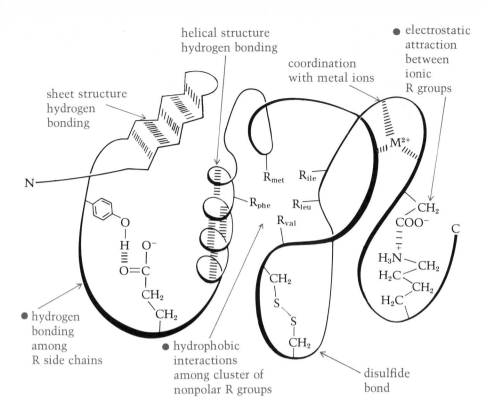

FIGURE 5–6 A diagramatic representation of stabilizing forces in globular proteins. Only one of each type of bond is shown. The frequency of occurrence of each type varies from protein to protein. The sheet and helical structures, disulfide bonds, and metal ion coordination can all be absent, all present, or any combination can be present. The other types of interactions (identified by a •) are found in *all* globular proteins.

QUATERNARY STRUCTURE

This level of structure refers to those proteins existing as *aggregates of two or more polypeptide chains that are linked to each other solely* by *noncovalent forces of attraction.* Covalent interactions such as interchain disulfide bonds are excluded. Proteins of this type are referred to as *oligomers,* with dimers, trimers, and tetramers being the most common, but larger varieties do exist. The nature of the interactions that stabilize the aggregate are certainly electrostatic bonds and hydrogen bonds among side chains localized near the surface of each chain. Hydrophobic forces are occasionally involved but they are less predominant, since the majority of nonpolar residues are generally localized within the inner core of each globular unit, and thus most would not be available for significant surface interactions.

The most studied example of this type of protein is hemoglobin (Fig. 5–7), which is a heterogeneous tetramer consisting of two identical alpha (α) chains and two identical beta (β) chains. Each chain is associated with a heme prosthetic group and is very similar to the myoglobin molecule. Recently, Perutz culminated his years of study on hemoglobin structure with a determination of all the residues that are in contact in both oxygenated and deoxygenated hemoglobin.

The heterogeneous oligomeric proteins have been clearly established as being associated with intracellular regulation. This dimension of this class of proteins is based on the fact that the intact protein is able to assume different conformations wherein the relative orientation of each of the chains is shifted. An important consequence of this is that the activity of the protein is increased or decreased. That is, **different conformations = different levels of activity.** This phenomeon, called *allosterism,* is thoroughly exam-

2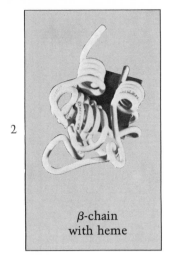

β-chain
with heme

+ 2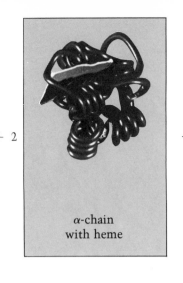

α-chain
with heme

⟶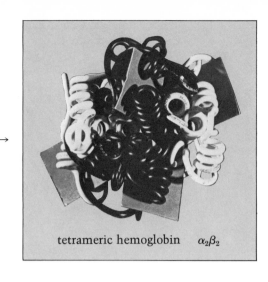

tetrameric hemoglobin $\alpha_2\beta_2$

FIGURE 5–7 A representation of the formation of hemoglobin from its constituent units. The α and β chains were constructed from electrical wire (white and black) and approximate the conformations established by X-ray diffraction. The square piece of cardboard represents a planar heme unit complexed to each chain. The overall structure of each α and β chain, including a considerable relationship in amino acid sequence, is similar to myoglobin, as represented in Panel E of Fig. 5–5. Three dimensionally, the four chains are clustered in a tetrahedral orientation. A comparison of the amino acid sequences of myoglobin and the hemoglobin chains from various sources has established an evolutionary pattern of both proteins; they both originate from a common ancestral polypeptide. This approach of mapping evolution by analyzing amino acid sequences of proteins common to various types of organisms may prove to be the most definitive method of establishing phylogenetic relationships. The premise of this work is that higher lifeforms are the result of a progressive evolution of molecules (see the article by Dickerson referred to at the end of the chapter for more detailed information).

ined in the next chapter (see p. 189). For now, we merely identify the association of regulation with oligomeric proteins.

Before dropping the subject, however, we should note that the myoglobin and hemoglobin molecules do offer the opportunity to illustrate (Fig. 5–8) that an oligomeric structure can affect biological function. Although the function of both proteins involves the binding to O_2, and furthermore, although each has the same heme unit complexed to very similar polypeptide chains, it is clear from the different shapes of the saturation curves in Fig. 5–8 that each protein, prior to saturation, is characterized by a different dependency on the concentration of oxygen. Otherwise, the curves would be identical. Indeed, if either the α or β chain of hemoglobin is examined in the same way, and compared to myoglobin (itself also a single chain), the saturation curves do, in fact, have the same shape. Moreover, the same result is obtained with a completely homogeneous hemoglobin molecule having four alpha chains (α_4) or four beta chains (β_4). Thus, the *heterogeneous oligomeric state* does confer some distinction to both structure and function. The legend to Fig. 5–8 identifies the distinction in function as a *cooperative* effect, a principle about which more will be said later in this chapter.

Hybrid Oligomeric Proteins

A special aspect of certain enzyme proteins of the oligomeric type is their existence in *hybridized forms* called *isoenzymes* or *isozymes*. For example, the enzyme lactate dehydrogenase (LDH) is a tetramer that can contain all possible combinations of two different polypeptide chains H and M, so designated because of their preponderance in either heart (H) or muscle (M) preparations of the enzyme. The five possible tetrameric species are HHHH, HHHM, HHMM, HMMM, and MMMM. All have the same enzymatic function (see p. 390) but to a different degree. Because each hybridized form contains varying levels of two monomeric polypeptide units that differ in amino acid composition, an LDH preparation can be electrophoretically resolved to reveal the presence and amount of each component. This type of analysis of the LDH isoenzymes in blood has proven to be a valuable clinical tool in contributing to the diagnosis of many respiratory and heart

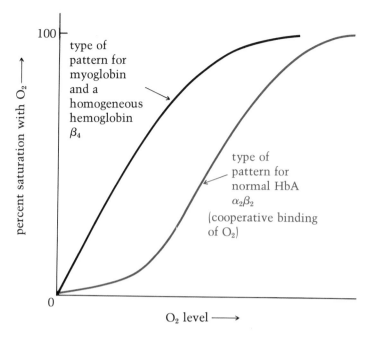

percent saturation with O_2 ⟶

100 —

type of
pattern for
myoglobin
and a
homogeneous
hemoglobin
β_4

type of
pattern for
normal HbA
$\alpha_2\beta_2$
(cooperative binding
of O_2)

0

O_2 level ⟶

FIGURE 5–8 Oxygen saturation curves of myoglobin, normal hemoglobin, and an abnormal hemoglobin. The different curves correspond to the differences in protein structure. The S-shaped (sigmoid) curve for normal HbA typifies what is termed a **cooperative** pattern, meaning that the binding of the first few molecules of O_2 stimulates a more favorable binding of additional O_2 molecules. (Cooperative binding is described in more detail on pp. 152–153 and also in the next chapter.) The curves also state that saturated HbA will release its O_2 in response to a drop in O_2 level more quickly than will myoglobin or the abnormal Hb.

diseases. The blood levels of the isoenzyme forms of creatine phospho-kinase (CPK) provide a particularly good indicator to evaluate heart damage. These clinical topics are discussed more fully in Appendix I.

DENATURATION OF PROTEINS

In addition to being highly ordered and flexible, the native conformation of a protein is a very delicate state, subject to alteration by chemical and/or physical agents without any change in the amino acid sequence. This loss of native conformation is called *denaturation.* Depending on whether or not the protein is oligomeric, whether or not the loss of conformation is partial or complete, and whether or not biological activity is lost, the term has a variety of meanings. These are diagramatically summarized below. Note

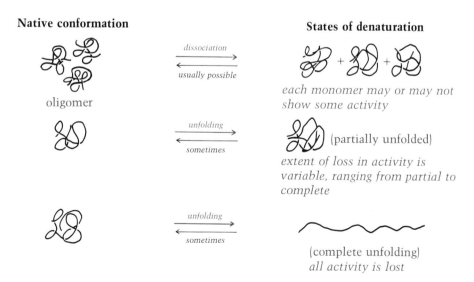

Native conformation

States of denaturation

oligomer

dissociation

usually possible

each monomer may or may not show some activity

unfolding

sometimes

(partially unfolded)

extent of loss in activity is variable, ranging from partial to complete

unfolding

sometimes

(complete unfolding)
all activity is lost

147

that in certain instances the process is reversible; both native conformation and biological activity can be restored. The self-restoring ability of a completely unfolded globular polypeptide chain is remarkable when one considers that it could refold in a countless number of ways. Observations of this sort provide further evidence that the secondary and tertiary levels of protein structure are predetermined by the amino acid sequence.

The biochemist must make every effort to prevent denaturation (particularly of the type that results in the unfolding of a chain) in order to insure successful studies into the activity of a protein. Accordingly, most of the handling is done at reduced temperatures to avoid thermal denaturation, buffers are employed to maintain the natural polyionic character of the protein, and physical trauma such as shaking is avoided or kept to a minimum. Laboratory investigations dealing with the isolation, purification, and analysis of proteins are truly representative of the art of biochemical analysis.

In certain instances the experimental objective may require a deliberate denaturation (partial or complete) of a protein. Some of the common denaturing agents are listed in the margin. Mercaptoethanol will work if disulfide bonds contribute to the native conformation (p. 145). Guanidine hydrochloride interferes with all types of noncovalent interactions. Urea functions by interferring with hydrogen bonds. In any case the *chain unfolds and/or the aggregate dissociates because stabilizing forces are disrupted.*

Protein denaturing agents

---

high temperatures
low pH and high pH
mercaptoethanol[a]
guanidine hydrochloride (6 *M*)
urea (6–8 *M*)
vigorous stirring or shaking
detergents (e.g., sodium dodecyl
    sulfate, SDS)

---

[a] For proteins containing —SH groups this would have a protective effect.

ISOLATION OF PURE PROTEINS

Isolation and purification are necessary prerequisites for the study of a protein, whether it be to characterize its structure or its biological function. Despite the relatively advanced state of biochemical methodology, these tasks are easier said than done. Several different steps are generally required, with each designed to eliminate a greater portion of unwanted material from the fraction of the previous step, until ultimately the desired protein is obtained in pure, uncontaminated form. Since there is no universally applicable procedure for every protein, the choice of each step must be determined solely on the basis of trial and error. A good procedure is one that recovers the maximum amount of desired protein in a pure state. Generally speaking, if the investigator can accomplish this objective with four to six steps, she or he should be pleased. As you might suspect, each step of the procedure requires analyses on each fraction for the total protein concentration and for the total activity of the protein being isolated. Dividing the total activity of a fraction by its total protein content yields the *specific activity* of that fraction (see note in margin).

A typical isolation scheme is illustrated by the data given in Table 5–3 showing the isolation of a protein from bacterial cells. (The protein in question is partially responsible for the movement of simple sugars such as galactose and glucose across the bacterial cell membrane. The proposed function of the protein is to bind with the sugar and then, in conjunction with other membrane components, to transport the sugar to the inside of the cell.) As indicated, the isolation procedure consists of six separate steps. As in any protein isolation procedure, the first step involves rupture of the intact cells to yield a cell-free extract (see p. 33, Chap. 1). The procedure used in this case was *osmotic shock*. The treatment with protamine does

The term *specific activity* is used extensively in biochemistry. It expresses the *amount of activity of a specific substance per unit weight of a mixture containing the substance, or per unit weight of the pure substance.* The increase in specific activity during the isolation of a protein represents a continual enrichment of the total protein in a particular fraction with respect to the protein being isolated. When purity is achieved, further fractionation should yield an unchanged specific activity unless the protein has lost some activity due to denaturation in handling or in storage.

TABLE 5-3 **Isolation of galactose-binding protein from** *Escherichia coli.* **(Data taken with permission; see legend to Fig. 5-9.)**

Steps of isolation	Total protein[a] (mg)	Total binding activity[b] (units)	Specific activity (units/mg)	% yield[c]
1. Cell-free extract (osmotic shock)	2600	940	0.39	100
2. Protamine precipitation	2352	798	0.34	85
3. Ammonium sulfate precipitation	1664	728	0.43	75
4. DEAE-cellulose chromatography	432	488	1.15	52
5. Hydroxyapatite chromatography	46	322	7.0	34
6. DEAE-Sephadex chromatography	18	208	12.0	22

[a] Protein measured by the Lowry method.
[b] Activity of protein was assayed by placing protein fraction in a semipermeable tubing and dialyzing for several hours against a solution containing a known amount of ^{14}C-galactose. The amount of radioactive galactose bound to the protein was calculated by subtracting the radioactivity of the dialyzate from that in the tubing. Radioactivity measurements were made by liquid scintillation. One unit of binding activity represents the binding of 1×10^{-9} mole of galactose.
[c] (Total activity of any step/total activity of initial extract) $\times$ 100.

not seem to have accomplished very much in terms of protein enrichment. What does occur is not represented by the data of Table 5-3. Protamine is a very basic protein (it has a high arginine content) and carries a large net positive charge. Therefore, it is a good precipitating agent of substances such as nucleic acids, which possess a large net negative charge. This is the role of protamine in this procedure—it removes about 95% of the nucleic acid present in the cell-free extract. The third step consists of an *ammonium sulfate fractionation*, a tactic commonly utilized in protein isolation. Addition of ammonium sulfate to protein solutions decreases the solubility of proteins in that solution. The effect, generally called the "salting-out" phenomenon, is dependent on the concentration of ammonium sulfate added and the nature of the protein. The last three steps are column-chromatographic methods (see Chap. 2), based on anion-exchange (DEAE-cellulose), adsorption (hydroxyapatite), and anion-exchange plus molecular-sieve (DEAE-Sephadex).

The course of the galactose-binding protein isolation is readily apparent from Fig. 5-9. Note the gradual elimination of protein until one band is observed in the disc polyacrylamide gel electrophoresis of the last fraction, suggesting that the sample is pure. The presence of only one symmetrical peak in the analytical ultracentrifuge pattern confirms the purity. The molecular weight of the binding protein computed from the latter is 35,000.

Some other tactics that can be used to achieve protein isolation by selective enrichment include *precipitation with organic solvents* such as methyl alcohol or acetone, *mild heating to change solubilities,* and *isoelectric precipitation,* which is based on the fact that a protein has a minimum solubility at a pH corresponding to its isoelectric point.

Affinity Chromatography

All of the various methods we have described so far that are utilized in the purification of biological substances are dependent upon differences in physical and chemical properties among the substances to be separated. For example, with the proteins, differences in size are exploited in gel-permeation chromatography, differences in net charge at a given pH are exploited in electrophoresis and ion-exchange chromatography, differences in solubility are exploited in ammonium sulfate fractionation and isoelectric precipitation, and differences in sedimentation are exploited in high-speed and ultra-

FIGURE 5–9 *Right:* photograph of stained polyacrylamide gels showing the progress of the isolation for galactose-binding protein. Tube 1, electrophoresis of the original cell-free extract; tube 2, electrophoresis of the ammonium sulfate fraction; tube 3, electrophoresis of the DEAE-Sephadex fraction. *Left:* photograph of the optical pattern observed in the analytical ultracentrifuge of boundary movement for the DEAE-Sephadex fraction. (Reproduced with permission from H. Anraku, "Transport of Sugars and Amino Acids in Bacteria," *J. Biol. Chem.,* **243,** 3116–3122; 3123–3127 (1968). Photographs kindly supplied by Y. Anraku.)

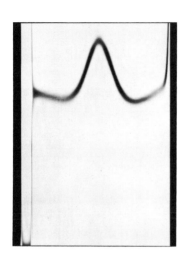

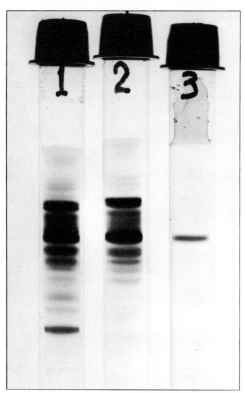

centrifugation. Moreover, as we have just seen, a combination of several of these methods is usually required to achieve total isolation.

In 1968 C. B. Anfinsen, P. Cuatrecasas, and coworkers announced the development of a new and revolutionary chromatographic technique for protein isolation that takes advantage of the *highly specific binding* that occurs between a specific protein P and another substance S, which usually is the substance that the protein interacts with *in vivo.* The technique, called *affinity chromatography,* is a column-chromatographic method with the column filled with a solid polymeric stationary phase (similar to that used in gel-permeation chromatography) to which the substance S is covalently attached. As a mixture of proteins moves through the column, *the only protein that will have its movement retarded is the one P that binds to substance S. All other proteins lacking this specific affinity move right on through the column.* Consider the following example.

One of the proteins found in white blood cells (leukocytes) is a Vitamin B_{12}-binding protein. It is believed that this protein is secreted into the plasma where it functions in the transport of Vitamin B_{12} to body tissues (primarily to the liver). An added advantage of the protein binding of Vitamin B_{12} is that none of the vitamin is lost by excretion in the urine. Until recently the study of this protein was difficult because it could not be obtained in good yields by conventional purification techniques. The results of an affinity chromatography method described in Table 5–4 resolved the difficulty. After devising a way of attaching Vitamin B_{12} to the stationary phase material (trade name Sepharose), the chromatographic analysis of a soluble mixture of leukocyte proteins resulted in nearly a 10,000-fold purification of the B_{12}-binding protein in greater than 90% yield. All in one step and the product was pure!

TABLE 5-4 Purification of Vitamin B_{12}-binding protein from human leukocytes by affinity chromatography.

Step	Volume (ml)	Total activity (ng B_{12} bound)	Total protein (mg)	Specific activity (ng B_{12} bound per mg protein)	Purification	% yield (see Table 5-3)
Soluble fraction obtained from ultrasonic disruption of white blood cells	625	22,100	6,250	3.54 _A_	1 _A/A_	100
Concentrated washings of column after chromatography on Vitamin B_{12}-Sepharose	2.4	20,400	0.585	34,900 _B_	9,820 _B/A_	92.3

SOURCE: Data taken with permission of the copyright owner, The American Society of Biological Chemists, Inc. from Allen, R. H. and P. W. Majerus, *J. Biol. Chem.*, **247**, 7702 (1972).

PROTEIN BINDING

For most proteins (P) the *initial event is a binding* with another substance, which we will generally refer to as a *ligand* (L). With few exceptions the interaction between P and L occurs at a precise location(s) on the surface of the protein molecule called a *binding site(s)*. As implied, sometimes a protein contains more than one site per molecule. For a protein molecule with n sites capable of binding with n ligand molecules, the binding can be represented simply as

$$P + nL \rightleftarrows PL_n \qquad \text{where } n = 1, 2, 3, \ldots$$

for which we can write

$$K = \frac{[PL_n]}{[P][L]^n}$$

where K is the equilibrium *binding constant* or *association constant*. Of interest to the biochemist are (1) the number of binding sites per protein molecule, that is, the value of n, (2) the strength of the binding, that is, the value of K, (3) the specificity of the binding, and (4) when two or more binding sites exist, whether there is any interaction between them.

Various methods can be used for precise quantitative measurements of binding. Although they differ in approach, each method involves measuring the saturation function of the protein (symbolized as ν) at various concentrations of ligand. The symbol ν stands for the ratio of concentration of bound ligand $[L]_b$ to concentration of total protein $[P]_t$. The limits of ν are 0 and n. When $\nu = 0$, no ligand is bound; when $\nu = n$, the maximum amount of ligand is bound and the protein is referred to as 100% saturated or fully saturated; when $\nu = \frac{1}{2}n$ the protein is 50% saturated or half-saturated.

$$\nu = \frac{[L]_b}{[P]_t}$$

It can be shown that the relationship of ν to ligand concentration is expressed by the equation

$$\nu = \frac{nK[L]^c}{1 + K[L]^c}$$

where n and K are as defined above and c, which is related to n, is a measure of the interaction between sites (see below).

151

In the technique of *membrane ultrafiltration* a known amount of free P and free L (free meaning unbound) are mixed together in a compartment constructed of porous membrane material. The pore size is of molecular dimensions but small enough to prevent passage of the high molecular weight protein and of course any ligand that is bound to the protein. Only free ligand will filter through. The diagram below depicts a conical membrane compartment supported in a centrifuge tube. After letting the P and L mixture stand until equilibrium is reached (P + L $\rightleftarrows$ PL), the system can be centrifuged to displace the unbound ligand. The amount of free ligand $[L]_f$ is then measured. Precise and accurate measurements can be made by using radioactively labeled ligand. Since you know the initial concentration of ligand $[L]_i$, the $[L]_b$ is merely $[L]_i - [L]_f$. Several such measurements with various $[L]_i$ values at the same [P] value provide the necessary data to plot ν (that is, $[L]_b/[P]_t$) versus $[L]_i$.

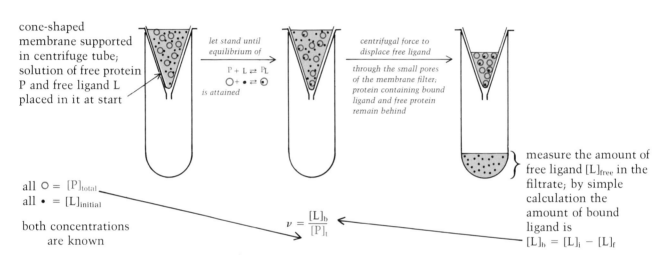

cone-shaped membrane supported in centrifuge tube; solution of free protein P and free ligand L placed in it at start

let stand until equilibrium of

P + L $\rightleftarrows$ PL
○+ ● $\rightleftarrows$ ◉

is attained

centrifugal force to displace free ligand

through the small pores of the membrane filter; protein containing bound ligand and free protein remain behind

} measure the amount of free ligand $[L]_{\text{free}}$ in the filtrate; by simple calculation the amount of bound ligand is
$[L]_b = [L]_i - [L]_f$

all ○ = $[P]_{\text{total}}$
all ● = $[L]_{\text{initial}}$

both concentrations are known

$$\nu = \frac{[L]_b}{[P]_t}$$

Depending on the nature of the binding, a plot of ν versus [L] will generate either a *hyperbolic* or *sigmoid* pattern. A hyperbolic pattern (graph A1 below) means there is only one binding site or there are two or more binding sites that do not interact. A sigmoid pattern (B1 below) means there are two or more binding sites that do interact with each other. For single or multiple noninteracting sites the values of n and K can be determined (A2) by a *Scatchard plot* of the same data when $\nu/[L]$ is plotted against ν. These coordinates yield a straight line where the slope equals $-K$ and extrapolated intercepts correspond to nK and n. In the case of interacting sites, a Scatchard plot does not yield a straight line (B2) but one portion of the curve can still be extrapolated to yield a value for n. The K value and—more importantly—the c value can be determined by a *Hill plot*. The n value (obtained from a Scatchard plot) is used with the same ν and [L] data where the $\log(\nu/n - \nu)$ is plotted against $\log[L]$. This yields a straight line (B3) with slope equal to c and the intercept on the y axis equal to $\log K$.

Interacting Sites—Cooperativity

Having distinguished between noninteracting and interacting multiple sites, let us now examine what the terms mean. The meaning is quite simple. When a protein molecule contains two or more separate binding

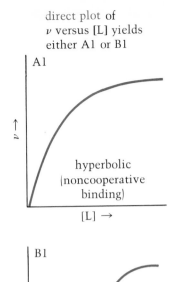

direct plot of
v versus [L] yields
either A1 or B1

A1

$v \uparrow$

[L] →

hyperbolic
(noncooperative
binding)

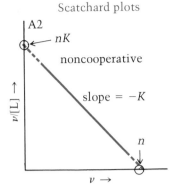

Scatchard plots

A2

nK

noncooperative

slope = $-K$

n

$v/[L] \uparrow$

v →

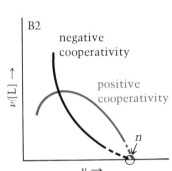

B2

negative
cooperativity

positive
cooperativity

n

$v/[L] \uparrow$

v →

B1

$v \uparrow$

sigmoid
(cooperative
binding)

[L] →

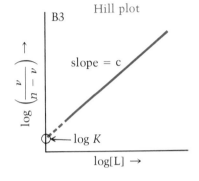

Hill plot

B3

slope = c

log K

$\log\left(\dfrac{v}{n-v}\right) \uparrow$

log[L] →

*Scatchard and Hill
plots are based on
algebraic manipula-
tions of the general
binding equation; the
data are the same as
those used in the
construction of direct
plots.*

For a protein with a single binding site ($n = 1$) A1 and A2 will always be observed.

For a protein with two or more binding sites ($n = 2, 3, 4 \ldots$) that do not interact with each other, A1 and A2 will be observed.

For a protein with two or more binding sites that do interact with each other, B1, B2, and B3 will be observed. Depending on the type of interaction (positive versus negative cooperativity; see text), a Scatchard plot will yield either of two possible patterns.

sites for the same ligand and the event of binding at one site does not influence the same event at any other site, the sites are noninteracting. That is, they function independently of each other. Obviously then, interacting sites behave in just the opposite fashion in that the *binding of ligand to one site influences the same event at a second site,* which in turn influences the same event at a third site, and so on. The influence may be favorable, meaning that the same event occurs more readily, or it may be unfavorable, with the same event occurring less readily. In either case, *what happens at one site influences what happens at another site in the same molecule.* This phenomenon of interacting sites is called *cooperativity.* If the initial binding of ligand promotes the binding of more ligand at other sites in the molecule, the protein is said to exhibit *positive cooperativity.* If further binding is hindered, the protein exhibits *negative cooperativity.*

The determination of cooperativity can be made directly from the previously described plots. If a direct plot yields a sigmoid curve, one can conclude that the protein does exhibit the cooperative effect. A downward curvature of the Scatchard plot (see B2) signifies positive cooperativity and an upward curvature signifies negative cooperativity. The c value (also called the *Hill coefficient*) determined from the Hill plot provides still another way to distinguish between the two types. A c value less than 1 signifies negative cooperativity and a value greater than 1 signifies positive cooperativity. When compared to the value of n, the Hill coefficient—which by the way is never greater than the value of n—can also provide an estimation of how cooperative the sites are.

To summarize:

> Protein with single binding site: $n = 1$, $c = 1$
>
> Protein with multiple binding sites: $n = 2, 3, 4, \ldots$
>
> noncooperativity (noninteracting sites that function independently)
>
> $c = 1$
>
> **cooperativity** (interacting sites, that is, they function interdependently)
>
> $c < 1$ for negative cooperativity; event at one site hinders the same event at another site
>
> $c > 1$ for positive cooperativity; event at one site promotes the same event at another site.

The effect of an event occurring at one place in a protein molecule upon the extent to which the same event occurs at another site in the same molecule is a fascinating topic indeed. Moreover, it is a topic of great biological importance, particularly in regard to those proteins that function as enzymes. In the next chapter we will see how this phenomenon provides a foundation for understanding how *some enzymes can undergo transitions in activity* and hence *regulate the biochemistry of what goes on in a living cell.*

Earlier in this chapter we encountered cooperativity in our discussion of O_2 binding to hemoglobin. At that time we related the cooperative effect to the oligomeric nature of hemoglobin. At this point it should be obvious that *only oligomers would be capable of providing multiple binding sites for the same ligand.* Remember also the distinction between homogeneous and heterogeneous oligomers (see Fig. 5–8).

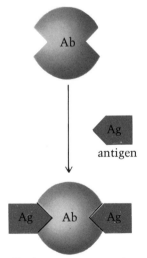

antibody molecule with two binding sites

antigen

antibody–antigen complex

(the binding is very specific and is proposed to involve noncovalent interactions)

ANTIBODIES

Even though it doesn't fit the logical progression of material so far, this seems to be the best spot to include some descriptive material on antibodies.

The gamma globulin fraction of blood serum proteins has been known for many years to contain the glycoproteins, called *antibodies,* that are responsible for combating disease by inactivating alien substances called *antigens.* The phenomenon is referred to as the *immune response.* As shown here the basic explanation is that a free antigen is removed when it is bound by a specific antibody. The distinguishing feature of this process is the remarkable specificity of the antibody molecule in discriminating among antigens that may have very similar structures. Without this natural form of immunization the body would be doomed.

Although this principle of immunity has been known since the early part of this century, little was learned about the nature of antibody molecules until the late 1950s. Just two of the problems were the large size of an antibody molecule and the tremendous heterogenity of the gamma globulin fraction, the latter making the isolation of large amounts of just one antibody very difficult. The discovery in 1959 of ways to cleave the intact antibody molecule into a set of specific smaller fragments, which can then be isolated from each other by chromatography and electrophoresis, resolved the problem of size. The discovery at about the same time that individuals afflicted with the disease called multiple myeloma contained a much more homogeneous gamma globulin serum fraction, due to a marked increase in production of one particular immunoglobulin, resolved the other problem. So it was that the 1960s proved very productive in the study of antibody structure. This culminated in 1969 with a determination of the complete amino acid sequence of a human immunoglobulin by G. Edelman and coworkers.

The immunoglobulin studied by the Edelman group was of the gamma G type (γG), which happens to be the most prevalent type in the antibody fraction of serum. Lesser amounts of four other types (γA, γM, γD, and γE) are also present (see Table 5–5). Regardless of type, the unit of structure is that of two copies of each of two different polypeptide chains, called the *light* (L) and *heavy* (H) chains because of size differences. The molecular weights of the light and heavy chains are on the order of 25,000 and 50,000 respectively. As shown below a total of four interchain disulfide bonds confers a forklike pattern to the overall arrangement. The Edelman study also revealed the presence of 12 additional disulfide bonds of the intrachain variety. Both patterns of disulfide bonding are proposed to be general features of immunoglobulin structure.

The most striking feature of structure is the presence of two different kinds of regions in the amino acid sequence—*variable* (V) regions and *constant* (C) regions. The notion of variable and constant regions had been proposed prior to the completion of the γG sequence on the basis of sequence determinations on polypeptide chains that are secreted in large amounts in the urine of those with multiple myeloma disease. (These polypeptides were subsequently established to represent the light chain of immunoglobulins.) Several of these polypeptides (called Bence-Jones proteins after the English physician who first observed their presence) were obtained from dif-

TABLE 5–5 **A comparison of human immunoglobulins.**

Main type[a]	Molecular weight (g/mole)	Relative abundance in serum (%)	Carbohydrate content (%)
γG[b]	150,000	71	3
γA	180,000–500,000	22	8
γM	950,000	7	12
γD	186,000	very small	?
γE	200,000	very small	11

[a] Subclasses and sub-subclasses of each type are proposed to exist, differing among other things in the size of the heavy chain.
[b] Exists only as H_2L_2 (see diagram); the same may also be true of γD and γE and one type of γA; other γAs appear to be dimers or trimers of the H_2L_2 unit structure; all γMs appear to be pentamers of the H_2L_2 unit structure.

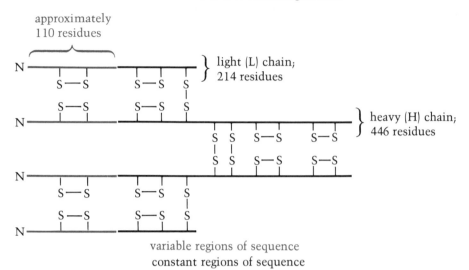

approximately
110 residues

N ——————————————————— } light (L) chain;
S—S   S—S   S   214 residues

N ——————————————————— } heavy (H) chain;
S—S   S—S   S   446 residues

S S S—S   S—S

N ———————————————————
S S S—S   S—S

N ———————————————————
S—S   S—S   S

N ———————————————————
S—S   S—S   S

variable regions of sequence
constant regions of sequence

ferent patients and all were shown to have almost the same amino acid sequence for about half the chain length from the C-terminal end, with only
one residue being different. The remaining half (that is, the variable
region), however, showed about 35 differences in sequence from one
Bence-Jones protein to another. The complete γG sequence, however,
strongly suggests that the variable-constant principle also extends to the
heavy chain as well. (See the article by Edelman cited at the end of this
chapter for how this was established.)

The commonality of antibody function among all immunoglobulins coupled with the specificity of antigen binding of individual immunoglobulins
obviously fits remarkably well with the constant and variable regions of
amino acid sequences. Although the complete three-dimensional structure
of γG is not yet available, it is proposed that the variable regions are clustered together in two separate groupings, which would be consistent with
the binding of two antigen molecules to one antibody molecule. That is,
the individual specificity of an antibody in recognizing a particular antigen
is determined by the variable regions of the γG structure. No doubt, as research continues, the nature of the binding sites will be described with
greater clarity and the exact function of the associated carbohydrate variety,
which is not yet understood too well, will be determined.

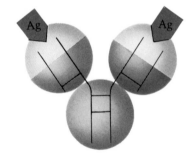

☐ variable regions
■ constant regions

LITERATURE

ANFINSEN, C. B. "Principles That Govern the Folding of
Protein Chains." *Science,* **181,** 223–230 (1973). A review article on the cooperative interactions of amino
acid side chains resulting in the formation of the native
conformation of proteins.

ANFINSEN, C. B., M. L. ANSON, K. BAILEY, J. T. EDSALL,
and F. M. T. RICHARDS, eds. *Advances in Protein Chemistry.* New York: Academic Press. A multivolume reference work (annual volumes) begun in 1944 and devoted to the proteins. Advanced reading.

BORNSTEIN, P. "The Biosynthesis of Collagen." *Ann.
Rev. Biochem.,* **43,** 567–604 (1974). A review article.

CAPRA, J. D., and A. B. EDMUNDSON. "The Antibody
Combining Site." *Scientific American,* **236,** 50–62
(1977).

CERAMI, A., and C. M. PETERSON. "Cyanate and
Sickle-Cell Disease." *Scientific American,* **232,** 45–50
(1975). A description of sickle cell anemia, the abnormal sickle-cell hemoglobin, and a hopeful method of
correcting the low oxygen-binding properties of HbS.

CUATRECASAS, P., M. WILCHEK, and C. B. ANFINSEN. "Selective Enzyme Purification by Affinity Chromatography." *Proc. Nat. Acad. Sci., U.S.*, **61**, 636–643 (1968). Original research paper describing the development of this method.

DAYHOFF, M. O., ed. *Atlas of Protein Sequence and Structure.* Vol. 5. Washington, D.C.: The National Biomedical Research Foundation, 1972. A collection of known amino acid sequences of peptides and proteins with a comparison of sequences of the same molecule isolated from different sources. Some coverage of nucleic acid sequences. Recent supplements have also been prepared.

DICKERSON, R. E., and I. GEIS. *The Structure and Action of Proteins.* New York: Harper & Row, Publishers, 1969. A lucid introduction to the structure and function of proteins. The book contains a wealth of imaginative and highly informative drawings, many of which are available in color slides.

DICKERSON, R. E. "The Structure and History of an Ancient Protein." *Scientific American*, **226**, 58–72 (1972). Discussion of the structure of cytochrome *c* and its evolution as a molecule over 1.2 billion years. The amino acid sequence from 38 different species is examined. Interesting discussion on some specifics regarding the relationship of function to amino acid sequence and three-dimensional structure.

EDELMAN, G. M. "The Structure and Function of Antibodies." *Scientific American*, **223**, 34–42 (1970). A description of how the gamma globulin molecule was sequenced and how antibodies function.

GROSS, J. "Collagen." *Scientific American*, **204**, 120–130 (1961).

HIRS, C. H. W., and S. N. TIMASHEFF, eds. *Enzyme Structure.* Vols. 11, 25, 27, 47, 48, and 49 of *Methods in Enzymology.* New York: Academic Press, 1967, 1972, 1973, 1978, 1978, and 1978. Part of a multivolume work devoted to practical aspects of biochemical studies, particularly those dealing with the isolation and assay of enzymes. Volumes 11 and 47 contain much information on the techniques available for the study of primary level of protein structure, such as determination of amino acid composition; end-group analysis; separation of polypeptide subunits; cleavage of disulfide bonds; separation of peptides; and sequence determination.

KENDREW, J. C. "Myoglobin and the Structure of Proteins." *Science*, **139**, 1259–1266 (1963). A paper adapted from the author's address on accepting the Nobel Prize in chemistry in 1962. Emphasis is given to the use of X-ray crystallography in deciphering protein structures. The treatment is nonmathematical and suitable for beginning students. Several figures from this paper are reproduced in this chapter.

KLOTZ, I. M., N. R. LANGERMAN, and D. W. DARNALL. "Quaternary Structure of Proteins." *Annual Review of Biochemistry*, **39**, 25–62 (1970). A review article summarizing the subunit aspect of protein structure. This article contains the most complete listing (through 1969) of proteins composed of two or more polypeptide chains.

MOSBACH, K. "Enzymes Bound to Artificial Matrixes." *Scientific American*, **224**, 26–33 (1971). Discussion of the binding of proteins to inert polymers and the significance to affinity chromatography and to uses in industry and medicine.

NEURATH, H., and R. L. HILL, eds. *The Proteins.* 3rd edition. New York: Academic Press, 1975. A planned eight-volume reference work dealing with the isolation, composition, structure, and function of proteins.

NOLAN, C., and E. MARGOLIASH. "Comparative Aspects of Primary Structure of Proteins." *Ann. Rev. Biochem.*, **38**, 727–790 (1968). An excellent and thorough review article discussing comparative studies of the amino acid sequences of proteins as they relate to evolutionary, developmental, and genetic mechanisms, as well as to the structure-function principle.

PAULING, L., R. B. COREY, and H. R. BRANSON. "The Structure of Proteins: Two Hydrogen-Bonded Helical Configurations of the Polypeptide Chain." *Proc. Nat. Acad. Sci., U.S.*, **37**, 205–211 (1951). The original paper describing the nature of the alpha helix.

PERUTZ, M. F. "The Hemoglobin Molecule." *Scientific American*, **211**, 64–76 (1964). A description of the three-dimensional structure of the hemoglobin molecule by the primary investigator.

TANFORD, C. "The Hydrophobic Effect and the Organization of Living Matter." *Science*, **200**, 1012–1018 (1978). A review article on the importance of hydrophobic interactions in the assembly and organization of membranes and proteins.

WOLD, F. *Macromolecules—Structure and Function.* Englewood Cliffs, N.J.: Prentice-Hall, Inc., 1971. Chapter 2 deals with protein binding. Good treatment of this and many other topics dealing with proteins in general, enzymes, membranes, nucleic acids, and metabolic control mechanisms. Available in paperback.

ZUCKERHANDL, E. "The Evolution of Hemoglobin." *Scientific American*, **212**, 110–118 (1965). An article comparing the amino acid sequences of the alpha and beta chains of hemoglobin molecules from different species and showing how this provides a basis of establishing evolutionary relationships among organisms on a chemical level, in terms of the evolution of a molecule common to these organisms.

5–1. Verify that there is a total of 67 nonpolar amino acids present in the α chain of human hemoglobin (see Table 5–1).

5–2. For each of the peptides given below, indicate which bonds would be cleaved by the action of (a) trypsin; (b) chymotrypsin; (c) cyanogen bromide; (d) hydrazine; (e) aminopeptidase; and (f) dilute hydrochloric acid and heat. (Assume that the specificity of trypsin and chymotrypsin is limited to text description.)

(a) arg—lys—gly—ala—ser—asp—asp—arg—ala—ser—cySH
(b) glu—tyr—lys—met—lys—phe—gly—val—thr—met—leu—val
(c) phe—trp—lys—tyr—ile—arg—val—ile—val—trp—glu

5–3. How many peptide fragments would be obtained from the treatment of cytochrome *c* from yeast with (a) trypsin, and (b) chymotrypsin, and from the treatment of insulin with (a) and (b)? (Assume that the specificity of each enzyme is ideally limited to text description.)

5–4. A pentapeptide obtained from treatment of a protein with trypsin was shown to contain arginine, aspartic acid, leucine, serine, and tyrosine. To determine the amino acid sequence, the peptide was cycled through the Edman degradation procedure three times. The composition of the peptide remaining after each cycling was as follows:
After Cycle 1: arginine, aspartic acid, leucine, serine
After Cycle 2: arginine, aspartic acid, serine
After Cycle 3: arginine, serine
What is the sequence of the pentapeptide?

5–5. What information, not given in this chapter, would you need to predict whether the total length of a polypeptide chain existing entirely in the α helix conformation should be greater than or less than if it existed in the pleated-sheet conformation? Explain.

5–6. Summarize the nature of each type of bond listed below and its role in protein structure.
(a) peptide bond
(b) disulfide bond
(c) hydrogen bond
(d) electrostatic bond
(e) hydrophobic bond

5–7. Explain why the carbonyl oxygen and the imino hydrogen atoms contributed by the same peptide bond do not enter into hydrogen bond formation with each other.

5–8. How many (a) intrachain hydrogen bonds, and (b) intact amino acid residues, are depicted in the drawing of the α helix segment shown in Fig. 5–3?

5–9. A pure protein is treated with performic acid. You have proved that a reaction has occurred by showing that the untreated protein is eluted faster than the treated protein from the same molecular-sieve column. However, polyacrylamide electrophoresis of the product mixture after performic acid treatment reveals only one very distinct and sharp zone. What can you conclude from this information regarding the structure of the protein?

5–10. Which hemoglobin variant would you predict might be less functional than normal adult hemoglobin HbA? Explain the basis of your answer.

Hemoglobin variant	Description of difference relative to HbA
1	ala rather than val in position 67 of β chain
2	asp rather than his in position 50 of α chain

5–11. Which of the following is most correct in regard to a fibrous polypeptide chain?
(a) all of the φ angles of rotation are identical
(b) all of the ψ angles of rotation are different
(c) all of the φ and ψ angles of rotation are different
(d) all of the φ and ψ angles of rotation are identical and φ = ψ
(e) only some of the φ and ψ angles of rotation are different

5–12. If a mixture containing (1) the coat protein of tobacco mosaic virus, (2) lysozyme, and (3) cytochrome *c* were applied to the top of a polyacrylamide gel and then analyzed by electrophoresis at pH 6, which of the patterns shown here would best represent the appearance of the gel after staining to detect the positions of the proteins? (The data given in Table 5–1 will be helpful in solving this problem. Assume that each acidic and basic residue would contribute a +1 and −1 charge, respectively, to each intact protein.)

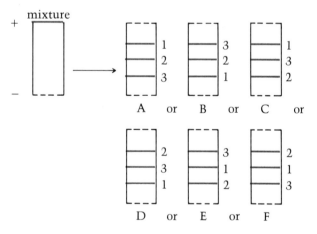

5–13. If a mixture containing proteins A, B, and C were analyzed by gel-permeation column chromatography, which of the following elution profiles would best represent the differential movement of A, B, and C through the column? (Given: Molecular weights of A, B, and C are 150,000, 75,000, and 65,000, respectively. The swollen gel granules had an exclusion limit of approximately 100,000.)

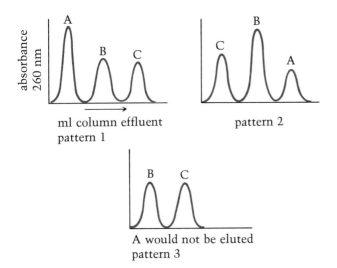

absorbance 260 nm

ml column effluent

pattern 1

pattern 2

A would not be eluted

pattern 3

5–14. Two pure proteins (A and B), each with a molecular weight of 60,000, underwent dissociation in the presence of urea. Sedimentation analysis in the ultracentrifuge showed that the urea-treated A sample gave two separate boundaries, neither of which corresponded to the original untreated A protein. The urea-treated B sample, on the other hand, gave only one boundary, but it likewise did not correspond to the original untreated B sample. What conclusions concerning the structures of proteins A and B can you draw from these statements?

5–15. The complete hydrolysis of an unknown nonapeptide revealed the presence of glutamic acid, two valine, glycine, two lysine, tyrosine, threonine, and phenylalanine residues. The first amino acid to be detected as a phenylthiohydantoin derivative on Edman degradation of the peptide was glutamic acid. The only amino acid detected after treating the peptide with hydrazine was threonine. Treatment of the peptide with trypsin and chymotrypsin gave three fragments in each case: T1, T2, T3 and C1, C2, C3, respectively. None of the trypsin fragments was identical to the chymotrypsin fragments. C2 and T2 proved to be dipeptides; C1 and T1 were tripeptides; and C3 and T3 were tetrapeptides. Hydrolysis of C3 followed by paper chromatography revealed only three ninhydrin-sensitive spots. The N-terminal residue of T3 was shown to be phenylalanine and the C-terminus threonine. The N-terminus of C1 was glycine and the C-terminus was the same as in T3. C2 was shown to contain tyrosine and glutamic acid. T1 was composed of lysine, tyrosine, and glutamic acid. The N-terminus of T2 was valine and the N-terminus of C3 was lysine. At basic pH the C3 fragment migrated with a net charge of +2. Use all of this information to construct as much of the sequence for the original nonapeptide as the data permit. (Assume that the specificity of trypsin and chymotrypsin is limited to text description.)

5–16. You have just isolated a fragment (F) that results from the treatment of a polypeptide chain with cyanogen bromide. Now you proceed to sequence fragment F. Your results are as follows.

(a) Treatment of F with trypsin was negative (i.e., no reaction)

(b) Treatment of F with chymotrypsin gave two fragments (C1 and C2); C2 did not contain any aromatic acids

(c) Paper electrophoresis of F and the two chymotrypsin fragments gave the results shown in the drawings; colored areas represent results after spraying the papers with ninhydrin

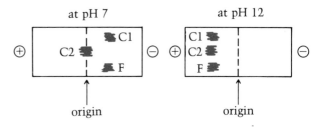

at pH 7   at pH 12

origin   origin

(d) Only three ninhydrin-positive fractions could be collected in cation-exchange chromatography of an acid hydrolyzate of the C1 fragment. One of these fractions was very difficult to elute off the column with the initial eluting buffer of pH 4. This slow moving substance did come off the column only after the eluting buffer was switched to one with a higher pH.

(e) An analysis similar to (d) of the C2 fragment also gave only three ninhydrin-positive fractions, but all were eluted without resorting to a high pH buffer

(f) C2 required three cycles of the Edman reaction before all peptide bonds were broken; the phenylthiohydantoin derivative that was extracted after the third cycle was identified as PTH-valine; the results of the first two cycles were ambiguous

(g) C1 required only two cycles of the Edman degradation; once again the result of the first cycle was ambiguous; the derivative in the second cycle was clearly that of asparagine

(h) Dansylation of C2 yielded dansyl-alanine

What can you conclude about the sequence of the original cyanogen bromide fragment F? (*Note:* The information is insufficient to determine the exact sequence of the entire fragment. Reasonable either/or assignments can be made, however, for uncertainties.)

5–17. A research biochemist wanted to determine the best way to store a protein, having already designed an efficient procedure for its isolation. She took a portion of a solution containing 50 mg of the pure protein obtained from the last step and placed it in a refrigerator for 2 weeks at 4 °C. When analyzed after 2 weeks this protein solution exhibited 360 units of activity. An equal portion of the same protein solution was lyophilized (that is, freeze-dried) and the powdered protein was stored at −20 °C. When analyzed 2 weeks later (after redissolving) a total of 310 units of activity were measured. A third portion of the solution containing 60 mg of the pure protein was placed directly in a freezer operating at −20 °C. When analyzed 2 weeks later (after thawing) a total of 390 units of activity were measured. What storage condition did the biochemist probably decide to use (at least after a 2-week storage period)? Why do you suppose the question is qualified with the notation in parentheses?

159

5–18. An enzyme was purified to constant specific activity; various other tests also supported the isolation of a pure protein. Polyacrylamide gel electrophoresis of a sample from the last step of the isolation procedure gave a stained gel slab with one major zone and two minor zones. Furthermore, the protein associated with each of the three zones catalyzed the same reaction. What can you conclude?

5–19. The results of a binding analysis of protein P for ligand L are summarized below. In each assay the protein concentration was 0.57 mM. Evaluate the values of n and K and whether the protein exhibits cooperativity. If there is cooperativity, verify the type—that is, positive or negative—by using a Hill plot to evaluate the Hill coefficient for a comparison to the value of n. (*Note:* In the plots shown on p. 153 the [L] that is plotted is the initial concentration of ligand.)

5–20. The results of a binding analysis of protein P for ligand L are summarized below. In each assay the protein concentration was 0.40 mM. Evaluate the values of n and K and whether the protein exhibits cooperativity. If there is cooperativity, verify the type—that is, positive or negative—by using a Hill plot to evaluate the Hill coefficient for a comparison to the value of n. (See note in Exercise 5-19.)

Initial concentration of ligand (mM)	Concentration of free ligand after equilibration with protein (mM)
0.570	0.466
0.970	0.800
1.51	1.28
2.27	1.99
3.43	3.09
5.38	4.98
8.90	8.45

Initial concentration of ligand (mM)	Concentration of bound ligand after equilibration with protein (mM)
0.750	0.040
1.00	0.120
1.25	0.240
1.67	0.360
2.31	0.480
3.75	0.600
6.08	0.680
10.00	0.720

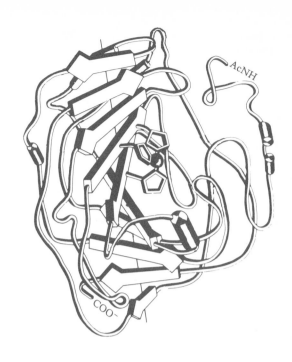

CHAPTER 6

ENZYMES

The most distinguishing feature of reactions that occur in a living cell is the participation of protein catalysts called *enzymes*. As with any catalyst, the basic function of an enzyme is to *increase the rate of a reaction*. Enzymes, however, have three unequaled characteristics. First, they are the most *efficient* catalysts known, with very small (micromolar) quantities of an enzyme able to accelerate a reaction at an extremely fast rate. In fact, most cellular reactions occur about a million times faster than they would in the absence of enzymes, some even faster. Second, the majority of enzymes are distinguished by a *specificity* of action in that virtually every conversion of a reactant (that is, a *substrate*) to a product is catalyzed by a preferred enzyme. In fact, several enzymes exhibit absolute specificity, meaning that they act on only one substrate to yield only one product. The third and perhaps most remarkable characteristic is that the actions of many enzymes are *regulated*, meaning that they are capable of changing from a state of low activity to one of high activity and vice versa. Such changes are controlled by both hormone and nonhormone substances. Collectively, they comprise an elaborate system by which organisms can control all of their activities. Gradually you will appreciate that the individuality of a living cell is due in large part to the unique set of enzymes that it is genetically programed to produce. If even one is missing or defective, the results can be disastrous.

The term "enzyme" was first used by Friedrich Wilhelm Kuhne in 1878. He was attempting to unify the terminology then used to refer to the catalytically active substances (or forces, as some thought) that were proposed to exist in living organisms. Gradually the term was accepted and is now (since about 1940) used almost exclusively in all languages. The word is from the Greek εν ζυμη, meaning "in yeast" but does not, of course, apply only to catalysts from yeast.

161

The first enzyme to be isolated (in 1923) was *urease*. Since then, approximately 2,500 different enzymes have been isolated from all types of organisms. Several of these have been crystallized and about 30 have had their complete three-dimensional structure determined by X-ray diffraction. From this extensive effort we have learned a lot about how enzymes operate but our knowledge is far from being complete. In fact, there is still no precise molecular explanation for the remarkable catalytic efficiency of enzymes.

In this chapter we will examine the basic principles of enzyme action. Particular attention will be given to relating how the catalytic function of an enzyme is related to its chemical structure, and to analyzing how certain enzymes are regulated at the cellular level. Both subjects, however, will be preceded by a study of the development of the classical Michaelis-Menten theory of enzyme kinetics, which has guided the study of enzymes for approximately 50 years.

ENZYME NOMENCLATURE

Some enzymes have seemingly nondescript names such as trypsin, pepsin, renin, and lysozyme. For the most part, however, enzymes are named with an -ASE ending, on the basis of the type of reaction they catalyze and the identity of the substrates involved. For example, the enzyme catalyzing the decarboxylation of histidine would be named *histidine decarboxylase;* another catalyzing the removal of two hydrogen atoms from ethyl alcohol to yield acetaldehyde would be named *alcohol dehydrogenase.* As more en-

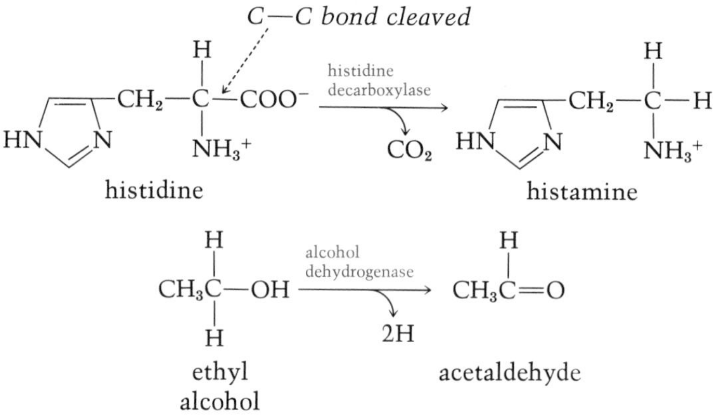

zymes of various functions were discovered, other names were formed: *oxidases, oxygenases, kinases, thiokinases, mutases, transaldolases, transketolases, phosphorylases, phosphatases,* and others. Don't get alarmed. Gradually you will become familiar with most types.

A more systematic approach with new names was first suggested in 1965 and revised in 1972. This system categorizes all enzymes into six main classes on the basis of the general type of reaction they catalyze. The main classes and the type of chemistry they participate in are listed in Table 6–1. The system is designed to zero in on the specific identity of each enzyme by dividing each main class into subclasses and sub-subclasses. By using a numbering system throughout the scheme, each enzyme can be assigned a numerical code, such as 2.1.3.4, where the first number specifies the main

TABLE 6–1   The main enzyme classes according to the International Enzyme Commission. A partial breakdown of class 4 is given in the margin to illustrate the indexing of histidine decarboxylase.

Main class	Type of reaction catalyzed
1. Oxidoreductases	Oxidation-reduction reactions of all types
2. Transferases	Transfer of an intact group of atoms from a donor to an acceptor molecule
3. Hydrolases	The hydrolytic (H_2O participates) cleavage of bonds
4. Lyases	The cleavage of bonds by means other than hydrolysis or oxidation
5. Isomerases	Interconversion of various isomers
6. Ligases	Bond formation due to the condensation of two different substances, with energy provided by ATP

4. Lyases
  4.1 Carbon-Carbon Lyases
    (cleavage of C—C bond)
    4.1.1 Carboxy-lyases
      (cleavage of C—COO$^-$
      bond)
      4.1.1.22 histidine carboxy-
        lyase
        (cleavage of C—CCO$^-$
        bond in histidine)
    4.1.2. aldehyde-lyases
  4.2 Carbon-oxygen lyases
    (cleavage of C—O bond)
  4.3 Carbon-nitrogen lyases
    (cleavage of C—N bonds)
  4.4 Carbon-sulfur lyases
    (cleavage of C—S bonds)
  4.5 Carbon-halogen lyases
  4.6 Phosphorus-oxygen lyases

class, the second and third numbers correspond to specific subclasses and sub-subclasses, and the final number represents the serial listing of the enzyme in its sub-subclass. For example, histidine decarboxylase (traditional name) is identified as histidine carboxy-lyase, 4.1.1.22; alcohol dehydrogenase as alcohol:NAD oxidoreductase, 1.1.1.1.

The use of this modern classification is required in most of the professional chemical and biochemical research journals and scientific abstract services. However, the older system is still widely practiced, mostly in monographs and textbooks. Despite its limitations, the older system has more pedagogical value in some respects, and thus it will be employed in this book.

COFACTOR (COENZYME)-DEPENDENT ENZYMES (VITAMINS)

All enzymes are globular proteins, with each enzyme having a specific function because of its specific globular structure. However, the optimum activity of many enzymes (but not all) depends on the cooperation of non-protein substances called *cofactors*. The molecular partnership of protein-cofactor is termed a *holoenzyme* and exhibits maximal catalytic activity. The protein component stripped of its cofactor is termed an *apoenzyme* and exhibits very low activity, frequently none at all.

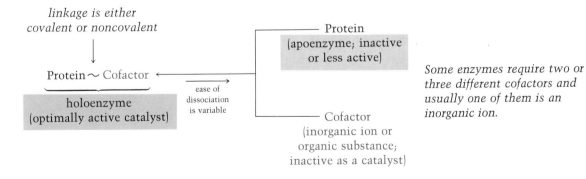

Some enzymes require two or three different cofactors and usually one of them is an inorganic ion.

There are two categories—the *inorganic cofactors*, which include several simple inorganic ions such as Zn^{2+}, Mg^{2+}, Mn^{2+}, Fe^{2+}, Cu^{2+}, K^+, and Na^+, and the *organic cofactors*, which consist of about a dozen substances of diverse structure. The organic cofactors are usually called *coenzymes*. The cofactor participation of inorganic ions represents (in part) the reason why these materials are essential nutrients for every organism (refer back to p. 10. Coenzymes (that is, the organic cofactors) have a special significance to animal (including human) nutrition because they are produced from *vitamins* or are one and the same with a vitamin. For example, the vitamin riboflavin is ingested and converted to either of two cofactors—flavin adenine dinucleotide (FAD) or flavin mononucleotide (FMN). Vitamin K after ingestion is unchanged and used directly as Vitamin K. The reason animals require a daily dietary supply of the vitamins is simple—animals are not able to produce them, whereas plants and bacteria can. Table 6–2 gives the names and vitamin relationship of several major coenzymes and a brief statement of their function. Note that this description is given in terms of the general or specific type of reaction in which the coenzymes participate.

TABLE 6–2 **Coenzymes: name, function, and vitamin relationship. The inclusion of the nucleoside triphosphates as coenzymes is not a common practice.**

Coenzyme	Type of reaction	Group transferred	Vitamin precursor[a]
Nicotinamide adenine dinucleotide (NAD$^+$)	Oxidation-reduction	H (electrons)	Niacin
Nicotinamide adenine dinucleotide phosphate (NADP$^+$)	Oxidation–reduction	H (electrons)	Niacin
Flavin adenine dinucleotide (FAD); flavin mononucleotide (FMN)	Oxidation–reduction	H (electrons)	Riboflavin
Coenzyme Q	Oxidation–reduction	H (electrons)	—
Cytochrome heme groups	Oxidation–reduction	Electrons	—
Coenzyme A	Activation and transfer of acyl groups	$R-\overset{\overset{\text{O}}{\|}}{C}-$	Pantothenic acid
Lipoic acid	Acyl group transfer	$R-\overset{\overset{\text{O}}{\|}}{C}-$	Lipoic acid
Thiamine pyrophosphate	Acyl group transfer	$R-\overset{\overset{\text{O}}{\|}}{C}-$	Thiamine
Biotin	CO_2 fixation	CO_2	Biotin
Pyridoxal phosphate	Transamination of amino acids and other reactions	$-NH_2$	Pyridoxal
Tetrahydrofolic acid	Metabolism of one-carbon fragments	$-CH_3$; $-CH_2-$; or $-CHO$	Folic acid
Cobamide coenzymes	Specialized (see p. 552)		B$_{12}$
Nucleoside triphosphates: adenosine triphosphate (ATP)	Phosphorylation and $-\overset{\overset{\text{O}}{\|}}{C}-$ activation	$-OPO_3$; $-AMP$	—
uridine triphosphate (UTP)	Biosynthesis and interconversion of carbohydrates	glycosyl	—
cytidine triphosphate (CTP)	Phospholipid biosynthesis	Conjugated glyceride	—

[a] All the substances listed constitute what is generally termed the group of B vitamins.

Once again, don't get alarmed. It is not necessary to understand all of these reactions at this time. Throughout succeeding chapters we will describe each coenzyme as the need arises.

The type of association between cofactor and enzyme varies. In some cases they exist separately and become bound to each other only during the course of the reaction. In other cases they are always bound together and sometimes very firmly by covalent bonding.

Generally speaking the role of a cofactor is either (a) to activate the protein by changing its three-dimensional structure, thus providing a shape that maximizes the binding and interaction of the enzyme with its substrate, or (b) to actually participate in the overall reaction as another substrate. The organic coenzymes operate primarily according to (b). The chemistry of this participation is usually described in terms of the coenzyme acting as a donor or acceptor of a particular chemical grouping relative to the other substrate(s). The grouping may be CO_2, a methyl $-CH_3$ group, an amino $-NH_2$ group, or electrons, to name just a few. Accordingly, the coenzymes are sometimes labeled as *group transfer agents*. The diagrams below should help to reinforce the meaning of group transfer. One example

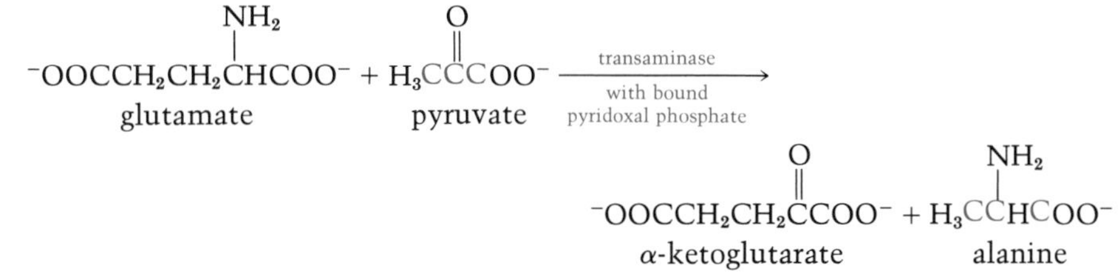

$$\overset{NH_2}{\underset{|}{^-OOCCH_2CH_2CHCOO^-}} + \overset{O}{\underset{||}{H_3CCCOO^-}} \xrightarrow[\substack{\text{with bound} \\ \text{pyridoxal phosphate}}]{\text{transaminase}}$$

$$\overset{O}{\underset{||}{^-OOCCH_2CH_2CCOO^-}} + \overset{NH_2}{\underset{|}{H_3CCHCOO^-}}$$

glutamate pyruvate

α-ketoglutarate alanine

this amino (NH_2) group transfer reaction occurs in two phases:

$$E-PyrP \xrightarrow{\quad glu(NH_2) \quad} E-PyrP-NH_2 \xrightarrow{\quad pyruvate \quad} E-PyrP$$

α-ketoglutarate ala(NH_2)

coenzyme is acceptor *coenzyme is donor*

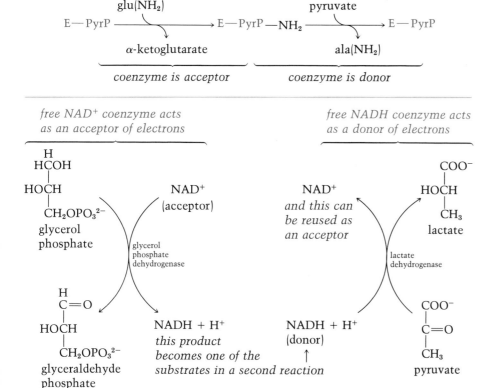

free NAD⁺ coenzyme acts as an acceptor of electrons

free NADH coenzyme acts as a donor of electrons

$$\begin{array}{l} H \\ HCOH \\ | \\ HOCH \\ | \\ CH_2OPO_3{}^{2-} \end{array}$$
glycerol phosphate

NAD⁺ (acceptor)

glycerol phosphate dehydrogenase

$$\begin{array}{l} H \\ C=O \\ | \\ HOCH \\ | \\ CH_2OPO_3{}^{2-} \end{array}$$
glyceraldehyde phosphate

NADH + H⁺
this product becomes one of the substrates in a second reaction

NAD⁺
and this can be reused as an acceptor

$$\begin{array}{l} COO^- \\ | \\ HOCH \\ | \\ CH_3 \end{array}$$
lactate

lactate dehydrogenase

NADH + H⁺ (donor)

$$\begin{array}{l} COO^- \\ | \\ C=O \\ | \\ CH_3 \end{array}$$
pyruvate

illustrates an amino group transfer within the same reaction involving a pyridoxal phosphate (PyrP) dependent enzyme. The other depicts a transfer of electrons (as hydrogen) between two separate reactions, which use different enzymes but have the same coenzyme, namely, nicotinamide adenine dinucleotide (NAD). Structures are not necessary for an appreciation of the principle. The NAD example shows NAD^+ yielding NADH in one reaction and NADH yielding NAD^+ in a second reaction. The two separate reactions are thus linked to each other by a common participant, serving as a product in one and a reactant in the other. Such reactions are referred to as being *coupled reactions*.

BASIC PRINCIPLES OF CHEMICAL KINETICS

The study of *rates of reaction* (how fast they occur) is called *chemical kinetics*. The ultimate objective of a kinetic analysis is to understand how a reaction occurs, that is, to understand the path taken in the conversion of reactants to products. Specifically, how many distinct steps are involved? What is the chemical nature of each step? Which is the slowest occurring step and thus the step that will limit the rate of the overall reaction? A description of a reaction in these terms is a description of the *mechanism* of the reaction. A lot of useful information is provided by knowledge of reaction mechanisms. For one thing, a solved mechanism for one reaction can help in solving the mechanism for other similar reactions. Another benefit is the possibility that the reaction can be controlled, and made to occur faster or slower by manipulating the reaction conditions or by adding another substance. The study and development of various antibiotics is just one example of this.

Although the theory and mathematical language of modern chemical kinetics is sophisticated and complicated, there are some simple basic principles.

REACTION ORDER

At constant temperature, the rate of any reaction is expressed in terms of the existing concentration of the reactants. The degree of dependence of rate on reactant concentration is called the *kinetic order* of the reaction. In this context the overall reaction may be termed *zero order*, *first order*, and so on.

The kinetic order of a reaction is determined experimentally and represents a fitting of the experimental data to the rate equation for the reaction in question. As an example, for the reaction A → B we can write a general mathematical expression defining the rate of reaction (or velocity of reaction), v, in terms of the disappearance of reactant A with respect to time, $-d[A]/dt$, or in terms of the formation of product B with respect to time, $+d[B]/dt$, as follows:

$$v = -\frac{d[A]}{dt} = +\frac{d[B]}{dt} = k_r[A]^n$$

Here k_r corresponds to the rate constant, which has a fixed value for the system under a specified set of conditions, with the only variable being the

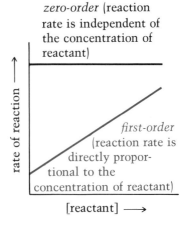

FIGURE 6–1 A graphical representation of zero-order and first-order chemical kinetics.

zero-order (reaction rate is independent of the concentration of reactant)

rate of reaction ⟶

first-order (reaction rate is directly proportional to the concentration of reactant)

[reactant] ⟶

concentration of reactant. The exponent n corresponds to the kinetic (reaction) order. Thus, by determining the reaction rate with different concentrations of A, the value of n can be determined. Once n is known, k_r can be calculated.

By assigning values to n, the meaning of reaction order becomes apparent and the resultant conclusions are obvious.

when $n = 0$ (zero order): $v = k_r[A]^0 = k_r$
when $n = 1$ (first order): $v = k_r[A]^1 = k_r[A]$ or $v \propto [A]$

From these relationships we can see a *zero-order reaction is one for which the reaction rate is independent of reactant concentration*. That is, the reaction rate remains constant regardless of reactant concentration. On the other hand, a *first-order reaction is one for which the reaction rate is directly proportional to the concentration of reactant* raised to the first exponential power. Thus, a twofold increase in reactant concentration will result in a twofold increase in reaction velocity, and so on. These conclusions are depicted graphically in Fig. 6–1. These basic principles of chemical kinetics will be most useful in developing the basic principles of enzyme kinetics.

Transition-state Theory and Catalysis

Any chemical reaction represents a transition from one state (reactants) to another (products). However, the progress of the reaction, energetically speaking, does not proceed directly from reactants to products. On the contrary, modern kinetic theory proposes that the formation of products proceeds through the formation of a *transition state*, corresponding to an *activated (higher-energy) state* of the reactant(s), as shown in Fig. 6–2. It follows then that the velocity of the conversion will be governed by the ease with which this transition state is achieved. The easier it is formed, the greater the velocity.

To achieve the transition state, reactant molecules must (a) acquire sufficient energy to overcome the energy barrier and (b) contact each other in a spatial orientation that will contribute to a productive reaction. The energy difference between the transition state and the initial state of reactants is called the *energy of activation* E_{act}. In the context of transition state theory, a catalyst functions by enhancing the production of the transition state. The simplest explanation for this is that *a catalyst reduces the energy of activation* and does so without affecting the net energetics of the overall reaction (see Fig. 6–3). Countless experimental measurements of a lower E_{act} in the presence of a catalyst support this. Precisely how and why this happens is not yet clear. Generally speaking, however, less energy is required because the *catalyst provides an alternate low-energy path* by which the reaction occurs. This explanation applies to all catalysts including enzymes. However, there is an extra special feature that applies to enzyme catalysis—the remarkable ability of the enzyme to bind and orient the reacting molecules in such a way as to maximize the occurrence of a productive interaction (see (b) above). This occurs at a specific location on the surface of the enzyme, called the *active site*, which we will discuss more thoroughly in later sections.

L. Michaelis and M. L. Menten (1913) pioneered in the kinetic study of enzyme-catalyzed reactions and in the development of explanations as to

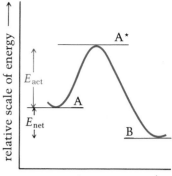

some parameter representative of the progress of reaction

⟶

FIGURE 6–2 An energy profile diagram of a hypothetical reaction, A → B. The finite difference between the energy levels of the ground state of reactants and the excited (transition) state of reactants is the energy of activation, E_{act}, for A. The illustration depicts an energy-yielding reaction, with the net output of energy, E_{net}, corresponding to the difference between the ground states of A and B.

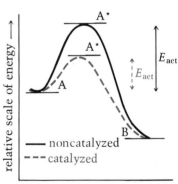

some parameter representative of the progress of reaction

⟶

FIGURE 6–3 Energy profile diagrams of the reaction A → B, illustrating the *smaller energy of activation in the presence of a catalyst*. The presence of the catalyst has *no effect* on the net energy of the reaction.

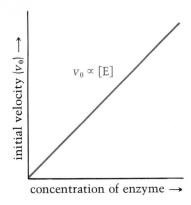

FIGURE 6–4 Kinetics of an enzyme-catalyzed reaction: constant substrate concentration and variable enzyme concentration.

Because of several variables—frequently quite different from one system to another and particularly so with enzyme-catalyzed reactions—that cause alterations in the reaction rate with time, the initial velocity, that is, the rate soon after the reaction has started, is the most accurate measurement of enzyme activity.

how enzymes participate. Their ideas were later confirmed by Briggs and J. B. S. Haldane (1925) who arrived at the same conclusion using a different approach. The general theory proposed has guided enzyme kinetics ever since, although there have been many embellishments of the original.

Initial studies were made with invertase, an enzyme that catalyzes the hydrolysis of sucrose. Data were collected on changes in the *initial velocity* v_0 of the reaction related to separate changes in concentration of enzyme (invertase) and of substrate (sucrose). The results were quite different (see

$$\text{sucrose} \xrightarrow[+\text{H}_2\text{O}]{\text{invertase}} \text{glucose} + \text{fructose}$$

Figs. 6–4 and 6–5). When the substrate concentration was held constant and the amount of enzyme was varied, a linear increase in velocity was observed with increasing concentration of enzyme present; in experiments of the reverse type, when the enzyme concentration was held constant and the amount of substrate was varied, a nonlinear hyperbolic relationship between velocity and substrate concentration was observed.

Both of these relationships have since been found to have a general applicability to enzyme-catalyzed reactions. In terms of classical chemical kinetics, the two distinct slopes of the *hyperbolic plot* (Fig. 6–5) correspond to first-order and zero-order kinetics, respectively. Michaelis and Menten described this as a rate transition from a *substrate-dependent phase* to a *substrate-independent phase*. This fundamental conclusion was a key factor in the formulation of their theory for enzyme action.

In retrospect, the theory of Michaelis and Menten seems remarkably simple. They proposed that the enzyme E reversibly combined with the substrate S to form an *intermediate complex of enzyme and substrate* ES, which then decomposed to yield products P and the free enzyme in its original form. (Since initial velocities were measured, the possibility of ES formation from E and P was neglected because very little P would be available in early stages of the reaction. However, the process can occur.) Kinetically speaking, each reaction is defined by a specific rate constant, designated below as k_1, k_2, and k_3.

$$\text{E} + \text{S} \underset{k_2}{\overset{k_1}{\rightleftharpoons}} \text{ES} \xrightarrow{k_3} \text{E} + \text{P}$$

It was reasoned that, if the proposal were valid, a mathematical equation of state could be derived that would be consistent with the empirical data represented by Figs. 6–4 and 6–5. In developing the rate equation, several other assumptions were made in addition to the postulated formation of an intermediate complex. The most important are the following:

1. A *steady-state equilibrium* is attained very rapidly. Under the steady-state condition, which basically corresponds to a balance of all the reactions in a living organism, the rate of substrate disappearance, $-d[\text{S}]/dt$, is equalled by the rate of product formation, $+d[\text{P}]/dt$. Alternatively, the rate of formation of the intermediate complex, $+d[\text{ES}]/dt$, and the rate of its disappearance, $-d[\text{ES}]/dt$, are balanced. Thus,

$$+\frac{d[\text{ES}]}{dt} = -\frac{d[\text{ES}]}{dt}$$

or
$$k_1[\text{E}][\text{S}] = k_2[\text{ES}] + k_3[\text{ES}]$$

2. The concentration of *total enzyme* $[\text{E}_t]$ is the sum of the enzyme combined with substrate [ES] and the free enzyme $[\text{E}_f]$ not so complexed:

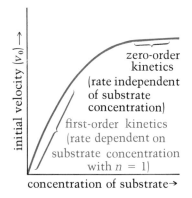

FIGURE 6–5 Kinetics of enzyme-catalyzed reaction: constant enzyme concentration and variable substrate concentration. The hyperbolic pattern typifies classical Michaelis-Menten kinetics.

$$[E_t] = [ES] + [E_f]$$

3. The *rate-limiting step* (that is, the slowest step) is the decomposition of the enzyme-substrate complex, and therefore, the initial velocity v_0 can be expressed in terms of ES concentration:

overall v_0 is the v_0 for $ES \xrightarrow{k_3} E + P$

thus $v_0 = k_3[ES]$

4. A corollary to point 3 is that the *maximum initial velocity V_{max}* will be attained when the concentration of ES reaches a maximum. This will occur when all of the available enzyme is complexed with substrate, that is, when $[E_f] = 0$. This condition is termed *saturation of the enzyme with substrate*. When $[E_f] = 0$, however, $[ES] = [E_t]$, and thus the maximum velocity will be directly proportional to the total enzyme concentration:

maximum $v_0 = V_{max} = k_3[ES]_{max} = k_3[E]_t$

Given these conditions, the development of the rate equation relating v_0 and $[S]$ is straightforward. Beginning with the equation based on the steady-state assumption (see item (1)), but designating $[E]$ more specifically as $[E_f]$ and then solving for $[ES]$, we obtain

$$[ES] = \frac{k_1}{k_2 + k_3}[E_f][S] \tag{6-1}$$

Now, since the proportionality constants of $v_0 \propto [ES]$ and $V_{max} \propto [E_t]$ are the same, it follows that

$$\frac{v_0}{V_{max}} = \frac{[ES]}{[E_t]} \quad \text{and} \quad [ES] = \frac{v_0}{V_{max}}[E_t]$$

Substituting the value of $[ES]$ in this relationship into Eq. (6-1) yields

$$\frac{v_0}{V_{max}} = \frac{k_1}{k_2 + k_3} \frac{[E_f][S]}{[E_t]} \tag{6-2}$$

The multiple rate constant term can be treated as one constant and was originally manipulated in that fashion by Michaelis and Menten as follows:

$$\frac{k_2 + k_3}{k_1} \equiv K_m$$

Termed the *Michaelis-Menten constant*, K_m has a special significance to enzyme kinetics that will be discussed shortly. This substitution into Eq. (6-2), coupled with a substitution for $[E_f]$ and a slight rearrangement, yields

$$v_0 = \frac{V_{max}}{K_m}\left(\frac{[E_t][S]}{[E_t]} - \frac{[ES][S]}{[E_t]}\right) = \frac{V_{max}}{K_m}\left([S] - \frac{[ES][S]}{[E_t]}\right) \tag{6-3}$$

A second substitution in Eq. (6-3) for $[ES]$ according to $[ES] = (v_0/V_{max})[E_t]$ yields

$$v_0 = \frac{V_{max}}{K_m}\left([S] - \frac{v_0[E_t][S]}{V_{max}[E_t]}\right) = \frac{V_{max}}{K_m}\left([S] - \frac{v_0[S]}{V_{max}}\right) \tag{6-4}$$

Which finally, by collecting terms and solving for v_0, gives

$$v_0 = \frac{V_{max}[S]}{K_m + [S]}$$

This last statement is the form of the classical *Michaelis-Menten kinetic equation* corresponding to that of a rectangular hyperbola (Fig. 6-6), with v_0

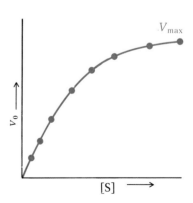

FIGURE 6-6 Graphical representation of Michaelis-Menten equation. See also Fig. 6-7.

169

and [S] as coordinates and with the constant V_{max} as the asymptotic, maximum value of v_0. (Asymptotic means that on the graph the value of V_{max} would be approached but never attained.) The equation is also consistent with the plot of v_0 versus enzyme concentration in Fig. 6–4.

E + S → ES and ES → E + P: Active Site Events
(A Preliminary Consideration)

The enzyme-substrate complex ES *is a real chemical species;* a few have actually been isolated. The formation of the ES species is the result of the *binding of S to E* and as such this is a perfect example of what we referred to earlier (p. 151)—the initial event of any protein associated with a dynamic function is one of binding. *Without binding nothing happens.* The next event of ES → E + P is the *chemistry of breaking and forming bonds to yield products.* It is this chemical conversion that the enzyme catalyzes. However, the catalytic action of the enzyme does not exclude the binding event (see note in diagram). In fact, the binding event is often the sole basis for the specificity of enzyme action.

$$E + S \xrightarrow{\text{binding}} ES \xrightarrow[\substack{\text{product} \\ \text{formation}}]{\text{chemistry of}} E + P$$

although catalysis is primarily dependent on events occurring in this phase, the initial event of binding does contribute to the events of catalysis by specifically positioning S and rendering it more susceptible to the action of the catalytic residues

---

Note the similarity of the Michaelis-Menten equation

$$v_0 = \frac{V_{max}[S]}{K_m + [S]} = \frac{\text{constant}[S]}{\text{constant} + [S]}$$

to the equation defining the relationship of ligand concentration to protein binding discussed earlier in Chap. 5 (p. 151).

$$\nu = \frac{nK[L]^c}{1 + K[L]^c}$$

or

$$\nu = \frac{n[L]^c}{(1/K) + [L]} = \frac{\text{constant}[L]^c}{\text{constant} + [L]^c}$$

Do you think the identical relationship of both v_0 and ν to ligand concentration (substrate is a ligand) is a mere coincidence?

The surface of each enzyme contains at least one specific location, called the *active site,* where the events of binding and chemical conversion occur.

The active site consists of a spatially ordered cluster of a few specific amino acid R groups (and possibly a cofactor), some of which participate in the binding of substrate while others participate in the chemistry of product formation. (Note: some groups may serve a dual function.)

The R groups comprising the active site do not occupy adjacent positions along the polypeptide chain. To the contrary, most of the residues occupy distant positions along the chain and are brought close together by virtue of the many bends, folds, and twists of the polypeptide backbone. These *bind-*

ing residues and *catalytic residues* are obviously *essential to the activity of the enzyme.*

Although not directly involved in active site events, each enzyme contains another group of equally essential amino acid residues. These are the crucial *structural residues,* which through their ordered interactions contribute to the formation and stabilization of the ordered conformation of the entire molecule and thus are responsible for forming the active site. A cofactor may also help in this regard. A severe change in the identity of any one of these essential residues can greatly affect enzyme activity; in fact, a total loss in activity can result. Those residues not involved in a role of structure, binding or catalysis would be *nonessential residues.* The number of essential versus nonessential residues varies from enzyme to enzyme.

Although there has been a continuing advancement in the field of enzyme kinetics, the classical Michaelis-Menten theory serves as the nucleus for contemporary theories and still retains in its own right a general validity and utility in the study of enzymes.

Other Aspects of Michaelis-Menten Theory

Before leaving the subject of the Michaelis-Menten theory of enzyme action, two additional items merit consideration.

The first relates to the oversimplification of the Michaelis-Menten suggestion that only two steps are involved in product formation. A more realistic representation would be as follows, in which it is indicated that several intermediates may actually be involved and that the product is formed in association with the enzyme. Furthermore, for reactions involving two or more substrates the intermediate complex would involve a multiple association.

after the initial binding of
substrate there may be stages of
substrate modification and
product development

$$E + S \rightleftharpoons ES \rightleftharpoons ES' \rightleftharpoons EP' \rightleftharpoons EP \rightleftharpoons E + P$$

ES of Michaelis-Menten theory refers to all of this

for a two-substrate reaction there
are a number of possibilities such as

$$E + S_1 \rightleftharpoons ES_1 \xrightarrow{S_2} ES_1S_2 \rightleftharpoons ES_1'S_2 \rightleftharpoons ES_1'S_2' \rightleftharpoons EP_1'S_2' \rightleftharpoons EP_1P_2 \rightleftharpoons E + P_1 + P_2$$

The second item concerns the application of the Michaelis-Menten kinetic analysis, which is *valid only when the concentration of one substrate is being varied.* Thus, for those reactions that involve the interaction of two or more substrates, all except one must be present in large enough initial concentrations so that their concentration never becomes limiting. Thus, for a two-substrate reaction one can measure the K_m of the enzyme towards either substrate by varying the concentration of one in the presence of an excess concentration of the other. (The same thinking applies to a three-substrate reaction.) In such instances, the K_m values are not necessarily identical. For example, the results shown in Table 6–3 (p. 174) for the Michaelis-Menten analysis of brain hexokinase were obtained under conditions where the concentrations of ATP and Mg^{2+} were in excess.

Measurement of K_m and V_{max}

The values of K_m and V_{max} are determined by measuring the initial velocity of the reaction at different substrate concentrations but using the same amount of enzyme in each measurement. A direct plot (see Fig. 6–7A) of v_0 versus [S] permits an obvious estimation of V_{max} from the plateau region of the hyperbolic curve. Remember, this is only an estimation because the maximum of a rectangular hyperbola is an asymptotic value. K_m is then easily determined, since $K_m = [S]$ when $v_0 = \frac{1}{2}V_{max}$. However, since the V_{max} value is only an estimation, the subsequent determination of the K_m value is also only an estimation.

The limitation of a graphical analysis of a hyperbolic plot can be overcome by plotting the same kinetic data in different ways. The most popular alternative approach is to plot $1/v_0$ versus $1/[S]$, which is called a *Lineweaver-Burk* plot (see Fig. 6–7B). This is based on an algebraic rearrangement of the original Michaelis-Menten hyperbolic rate equation to yield a straight-line rate equation. The rearrangement involves taking the reciprocal of both sides, then multiplying both sides by V_{max}, and finally solving for $1/v_0$ to yield

$$\frac{1}{v_0} = \frac{K_m}{V_{max}}\left(\frac{1}{[S]}\right) + \frac{1}{V_{max}}$$

This equation has the straight-line form of $y = mx + b$ between two variables (y and x) where m is the slope of the line and b is the intercept of the line on the y axis. In the Lineweaver-Burk equation, $y = 1/v_0$, $x = 1/[S]$, $m = K_m/V_{max}$, and $b = 1/V_{max}$. By first determining the value of $1/V_{max}$ (that is, the b value) by extrapolation through the $1/v_0$ axis, K_m can be evaluated from the slope of the line. Alternatively, K_m can be directly determined by further extrapolation through the $1/[S]$ axis where the intercept is equal to $-1/K_m$.

Significance of V_{max}

As the term implies, the maximum velocity is an expression of the efficiency of enzyme operation. To compare the catalytic efficiency of different enzymes, however, it is necessary to first express the V_{max} in terms of the same molar amount of each enzyme. This conversion of V_{max} yields a value that is called the *molecular activity* (or *turnover number*) of the enzyme and represents the moles of substrate reacted per mole of enzyme per unit time (usually 1 minute or second).

Consider the following example of carbonic anhydrase, an important Zn^{2+}-containing enzyme in blood (see p. 72 in Chap. 3) that catalyzes the reaction

$$CO_2 + H_2O \underset{\text{carbonic anhydrase}}{\rightleftharpoons} H_2CO_3$$

A Michaelis-Menten kinetic assay under optimum conditions (pH about 7; T of 25–37 °C) shows that 1 μg of enzyme exhibits a V_{max} of 1.2×10^{-3}

Results[a] of a kinetic assay on the enzyme, acid phosphatase. Reaction rate v_0 is expressed as increase of absorbance A at 405 nm per minute, reflecting the rate of appearance of the product formed in the reaction examined. See Fig. 6–7 for analysis of data.

[S] millimolar	v_0 ΔA_{405}/min
0.50	0.075
0.75	0.090
2.00	0.152
4.00	0.196
6.00	0.210
8.20	0.214
10.0	0.230

[a] Data collected by students in biochemistry lab at John Carroll University.

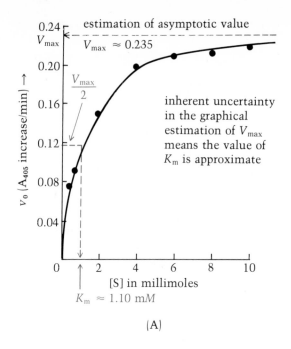

(A)

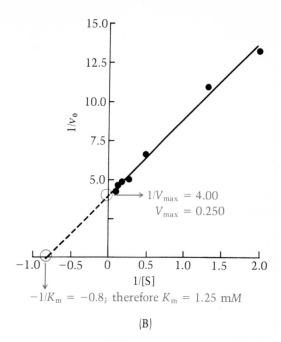

(B)

FIGURE 6–7 Evaluation of the experimental kinetic data listed in the margin. (A) Michaelis-Menten plot; (B) Lineweaver-Burk plot with the straight line extended to intercept both the $1/v_0$ and $1/[S]$ axes.

moles of CO_2 reacted per minute. The molecular weight of carbonic anhydrase is 30,000 g/mole and hence 1 μg of enzyme represents 0.000001/30,000 mole of enzyme, that is, 3.33×10^{-11} mole. Thus

$$\text{molecular activity} = \frac{V_{max}}{\text{moles of E present}} = \frac{1.2 \times 10^{-3} \text{ moles } CO_2 \text{ reacted/min}}{3.33 \times 10^{-11} \text{ mole enzyme}}$$

$$= 36 \times 10^6 \text{ moles } CO_2 \text{ reacted/min/mole of enzyme}$$

In other words, 1 molecule of carbonic anhydrase will catalyze the reaction of 36 million molecules of CO_2 in 1 minute. That's fast! Carbonic anhydrase represents one of the fastest working enzymes. The molecular activity of most enzymes is in the range of 50–1,000,000 with many clustered in the vicinity of 1,000–10,000. Even these values represent fast catalysis when one considers that most organic reactions in the absence of a catalyst require several minutes or several hours even when heated to higher temperatures.

Two other expressions of enzyme activity are the *enzyme unit* and *specific activity*. One enzyme unit corresponds to the number of moles, millimoles, or micromoles of substrate reacted per minute. The specific activity is merely the number of enzyme units per milligram of protein. It has recently been suggested that the traditional usage of the enzyme unit be replaced by the term *katal*, with 1 katal of activity defined as representing the transformation of 1 mole of substrate per second.

Significance of K_m

Assigning K_m the value of $(k_2 + k_3)/k_1$ may not be too informative to you. A more meaningful description is that the K_m represents the *amount of substrate required to bind with one-half of the total amount of enzyme present in solution*. This condition is termed *half (50%) saturation* and would cor-

Question: A solution containing 24 milligrams of enzyme converts 12 millimoles of substrate to product in 2 minutes. (a) How many units of activity are present in solution? (b) How many katals of activity does this represent? (c) What is the specific activity of the solution?

Answers: (a) 6 enzyme units, that is, 6 mmole/min
(b) 0.0001 katal
(c) specific activity = 0.25 units/min

respond to an initial velocity that is one-half the maximum velocity, that is, $v_0 = \frac{1}{2}V_{max}$.

At 50% saturation, where $v_0 = \frac{1}{2}V_{max}$, the Michaelis-Menten equation states that

$$\frac{V_{max}}{2} = \frac{V_{max}\,[S]}{K_m + [S]}$$

which reduces to

$$K_m + [S] = 2[S]$$

or

$$K_m = [S]$$

Thus the K_m value is expressed in units of concentration such as molarity M or millimolarity mM.

This meaning of K_m implies that the measurement of K_m values can be used *to evaluate the specificity of action of a given enzyme toward similar substrates*. To illustrate this, we will use some data (Table 6–3) collected from assays of hexokinase, an enzyme that catalyzes the conversion of simple sugars to phosphoesters (see p. 295). A comparison of K_m values reveals that the amount of glucose required for 50% saturation is 1,000 times less than that required for 50% saturation with allose. The interpretation is that the enzyme has an easier reaction path with glucose than with allose. This may be due to an easier binding of the sugar to yield ES, or an easier conversion of ES to products, or both may be easier. In other words, *the enzyme has a preference* for glucose; in still other words, the *enzyme is more specific in its action* toward glucose. Inspection of the two structures suggests that the spatial orientation of merely one hydroxyl (OH) group on one carbon atom (number 3) is critical. On the other hand, the sugar mannose shows a K_m value very close to the glucose value, suggesting that hexokinase will act on either one with about equal efficiency. One can conclude then that the spatial orientation of the OH group at carbon atom 2 is not as critical to enzyme action.

This same principle can be illustrated another way based on the definition of $K_m = (k_2 + k_3)/k_1$ that was given earlier. If we assume that k_2 is much greater than k_3, the value of K_m would then be approximated by k_2/k_1. This

The evaluation of substrate specificity follows the general rule: the lower the K_m value—the better (more preferred) is the substrate.

TABLE 6–3 K_m values of hexokinase (from brain).

Reaction is   sugar + ATP $\xrightarrow[Mg^{2+}]{hexokinase}$ sugar phosphate + ADP

with the sugar, glucose,	with the sugar, allose,	with the sugar, mannose,
^{1}CHO	^{1}CHO	^{1}CHO
H—^{2}C—OH	H—^{2}C—OH	HO—^{2}C—H
HO—^{3}C—H	H—^{3}C—OH	HO—^{3}C—H
H—^{4}C—OH	H—^{4}C—OH	H—^{4}C—OH
H—^{5}C—OH	H—^{5}C—OH	H—^{5}C—OH
^{6}CH$_2$OH	^{6}CH$_2$OH	^{6}CH$_2$OH
as substrate,	as substrate,	as substrate,
$K_m = 8 \times 10^{-6}\ M$	$K_m = 8 \times 10^{-3}\ M = 8000 \times 10^{-6}\ M$	$K_m = 5 \times 10^{-6}\ M$

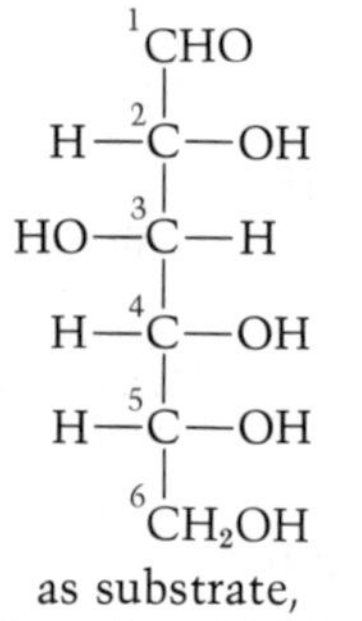

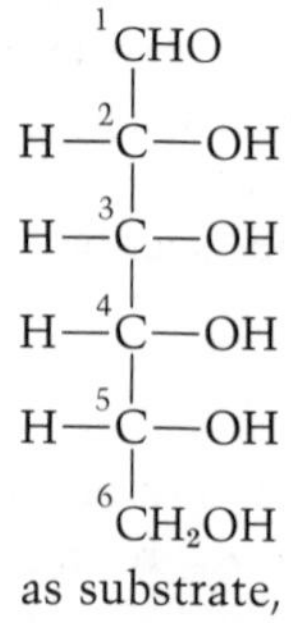

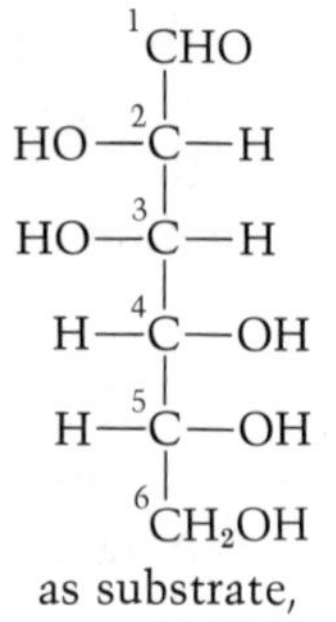

assumption provides a basis to focus only on the events of $E + S \rightleftarrows ES$. As shown below k_2/k_1 would correspond directly to a measure of the *dissociation constant of ES* or inversely to a measure of the *affinity constant of ES*. Thus, a low K_m value corresponds to a low tendency for ES to dissociate into E and S and conversely to a high tendency for E and S to form ES.

Note: The correlation of lower K_m values with a preferred substrate is not as valid when different enzymes are compared, particularly when the type of substrates and reactions involved are not similar.

at equilibrium for

$$E + S \underset{k_2}{\overset{k_1}{\rightleftharpoons}} ES$$

$$\text{rate}_{\text{forward}} = k_1[E][S]$$
$$\text{rate}_{\text{reverse}} = k_2[ES]$$

and hence

$$K_{eq} = K_{affinity} = \frac{[ES]}{[E][S]} = \frac{k_1}{k_2}$$

at equilibrium for

$$ES \underset{k_1}{\overset{k_2}{\rightleftharpoons}} E + S$$

$$\text{rate}_{\text{forward}} = k_2[ES]$$
$$\text{rate}_{\text{reverse}} = k_1[E][S]$$

and hence

$$K_{eq} = K_{dissociation} = \frac{[E][S]}{[ES]} = \frac{k_2}{k_1}$$

$$K_m \approx \frac{k_2}{k_1} = \frac{1}{K_{affinity}} = K_{dissociation}$$

ENZYME INHIBITION

Earlier it was pointed out that the optimum activity of some enzymes requires the participation of a metal ion cofactor. This is one example of enzyme activation, with the metal ion functioning as an *activator*. Now we will examine the action of substances that decrease the activity of enzymes and hence are called enzyme *inhibitors*. The principles of activation and inhibition as they apply to proteins in general and to enzymes in particular will be discussed more thoroughly later in this chapter; many examples will be presented in later chapters.

In addition to being a naturally occurring phenomenon, the inhibition of enzymes is also important in two other respects: (a) it provides information helpful in understanding how an enzyme operates, in identifying amino acid residues essential for catalytic activity, and in further clarifying aspects of any specificity of action; and (b) it aids in the understanding and further development of various antibiotics and other chemotherapeutic drugs and of toxic materials such as insecticides. The following discussion focuses on the two most common types of inhibition—*competitive* inhibition and *noncompetitive* inhibition.

The sources cited at the end of the chapter can be referred to for coverage of other types of inhibitors.

Competitive Inhibition

Competitive inhibitors are substances that have a *structure similar to that of the natural substrate*; when both inhibitor (I) and substrate (S) are together in the presence of an enzyme, *they compete with each other to bind at the active site of the enzyme*. Thus two types of complex can form: EI and ES. It should occur to you that in this instance an EIS ternary complex would not be produced since I and S cannot occupy the same site simultaneously. The formation of an enzyme-inhibitor EI complex reduces the

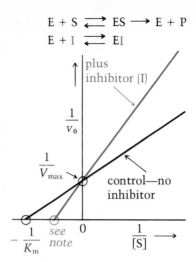

$$E + S \rightleftharpoons ES \longrightarrow E + P$$
$$E + I \rightleftharpoons EI$$

plus inhibitor (I)

$\frac{1}{v_0}$

$\frac{1}{V_{max}}$

control—no inhibitor

$-\frac{1}{K_m}$ *see note*   0 $\frac{1}{[S]}$ →

FIGURE 6–8 Lineweaver-Burk plot of kinetics of competitive inhibition.

In the presence of a competitive inhibitor of concentration [I], it can be shown that the K_m value is increased by a factor equal to

$$1 + \frac{[I]}{K_I}$$

where K_I is the inhibitor dissociation constant of EI and is a measure of the affinity of E for I in the same sense as K_m is a measure of the affinity of E for S (see p. 175). Thus, the intercept on 1/[S] axis in presence of inhibitor equals

$$-\frac{1}{K_m \left(1 + \frac{[I]}{K_I}\right)}$$

The mode of action of *puromycin*, another chemotherapeutic drug that operates as a competitive inhibitor, is described on p. 270.

population of free enzyme molecules that are available for interaction with the natural substrate, which in turn results in a reduction in the amount of substrate converted to product in a given time. A competitive inhibitor normally combines reversibly with the enzyme at its active site and is not converted to any product(s). Hence, competitive inhibition *can be reversed or minimized by merely increasing the concentration of the substrate,* with the greater population of substrate molecules competitively favoring the formation of a larger percentage of the normal ES intermediate complex.

The Michaelis-Menten kinetics that serve as a means of experimentally establishing a substance as a competitive inhibitor are depicted in the Lineweaver-Burk plot of Fig. 6–8. Notice that in the presence of a competitive inhibitor the normal V_{max} is unchanged whereas the K_m is increased. If the molecules do continually compete for the same active site, then a greater amount of substrate should be required for half-saturation (higher K_m). When the system is saturated, however, the maximum velocity should be unaffected, just as though no inhibitor were present.

One of the most classical examples of competitive inhibition is shown by *succinic acid dehydrogenase,* an enzyme of the citric acid cycle (p. 422) that catalyzes the conversion of succinic acid to furmaric acid. The enzyme is dependent on the coenzyme participation of *flavin adenine dinucleotide* (FAD) as a hydrogen acceptor.

$$^{-}OOC-CH_2-CH_2-COO^{-} + FAD \xrightarrow[\text{dehydrogenase}]{\text{succinic acid}} {}^{-}OOC-\overset{H}{\underset{H}{C}}=C-COO^{-} + FADH_2$$

succinate   fumarate

$$^{-}OOC-CH_2-COO^{-}$$ (competitive inhibitor of succinate)
malonate

A variety of competitive inhibitors of this enzyme are known, but the most potent is malonic acid, the next lower methylene homologue of succinic acid. Specifically, if only 2% of the molecules present are malonic acid, there is a 50% inhibition in the rate of production of fumarate.

Many chemotherapeutic drugs function as competitive inhibitors. For example, several of the *sulfa drugs,* used to combat microbial infections in humans, are structurally related to *para-aminobenzoic acid* (PABA), a vital precursor in the microbial biosynthesis of *folic acid,* which in turn is converted to *tetrahydrofolic acid* (see p. 550, Chap. 17), an extremely important coenzyme for several enzymes. A few of these enzymes catalyze crucial steps in the biosynthesis of purine and pyrimidine nucleotides, which in turn are used in the biosynthesis of the nucleic acids RNA and DNA. When the sulfa drug is administered, the immediate effect is the inhibition of the enzyme that catalyzes the PABA-incorporating step in the production of folic acid. This results in a decreased cellular production of tetrahydrofolic acid, which in turn reduces the production of the purine and pyrimidine nucleotides, which in turn will limit the production of nucleic acids. The eventual result of all this is that the organism dies. The selective action of the drugs on the infectious organism is due to the fact that, although humans are critically dependent on folic acid, they do not have the biochemical capacity to synthesize this material from PABA and other precursors. Rather, humans depend on an external dietary source and/or an in-

several enzymatic steps involved with one enzyme catalyzing the incorporation of p-aminobenzoic acid

para-aminobenzoic acid (PABA)

$$H_2N-\langle\bigcirc\rangle-COOH$$

precursors $\rightarrow \rightarrow \rightarrow +\!\!\downarrow\!\!\rightarrow \rightarrow \rightarrow \rightarrow$ folic acid $\rightarrow \rightarrow$ tetrahydrofolic acid

competitive inhibition

$$H_2N-\langle\bigcirc\rangle-SONH_2$$

sulfanilamide
(a sulfa drug)

an essential coenzyme for metabolism of purines and pyrimidines;
(see pages 559 and 563)

ternal supply from the noninfectious intestinal bacteria. One of the dangers of sulfa drug therapy is that an excessive amount can annihilate the intestinal bacteria, resulting in a loss of their many symbiotic life-sustaining functions.

Noncompetitive Inhibition

Inhibitors of this type interact with the enzyme in a variety of ways. The binding may be reversible or irreversible, may occur at the active site or at some other region on the surface of the enzyme protein, and may or may not prevent the binding of substrate. In any case the resultant complex is generally inactive, and the *effect cannot be reversed by merely increasing the ratio of substrate to inhibitor.* This is just the opposite of competitive inhibition. The Michaelis-Menten kinetic characteristics of classical noncompetitive inhibition are shown in Fig. 6–9. Note the reverse effect on the K_m and V_{max} relative to competitive inhibition. In noncompetitive inhibition (classical type) the K_m is unchanged and the V_{max} is decreased. The molecular explanation of the poisonous character of many substances such as cyanide, carbon monoxide, Hg^{2+}, Pb^{2+}, and arsenicals lies in their potent noncompetitive inhibition of certain enzymes. Inhibitor action may result in formation of (a) an inactive EI binary complex with I binding occurring at some specific location, perhaps even at the active site, or (b) an inactive IES ternary complex. In either case dissociation of the complex is not affected by merely adding more substrate.

The use of certain noncompetitive inhibitors can provide valuable information concerning the catalytic nature of enzymes. For example, *p*-chloromercurobenzoate and diisopropylfluorophosphate covalently and irreversibly attach to cysteine sulfhydryl groups and serine hydroxyl groups, respectively. Thus, if inhibition is detected with either material, this is strong experimental evidence for the essential participation of the R groups of cysteine or serine. (See the diagram at the top of the next page.)

Physical Factors

Because enzymes are proteins, their catalytic activity is very sensitive to temperature and pH, with deactivation common in both the high and low extremes of each parameter. The effect may be reversible or irreversible. Of course, denaturation of the protein conformation (see p. 148, Chap. 5) is common to each condition except low temperature. (Indeed, solutions of

$$E + S \rightleftharpoons ES \longrightarrow E + P$$
$$E + I \rightleftharpoons EI$$
$$\text{or } E + S + I \rightleftharpoons SEI$$

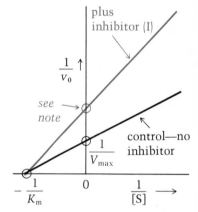

FIGURE 6–9 Lineweaver-Burk plot of kinetics of classical (or pure) noncompetitive inhibition. (Reversible complex formation and only V_{max} affected.)

In the presence of a noncompetitive inhibitor of concentration [I], it can be shown that the $1/V_{max}$ intercept value is increased by a factor of

$$1 + \frac{[I]}{K_I}$$

where K_I is the inhibitor dissociation constant reflecting the affinity of E for I. Thus, the intercept on $1/v_0$ axis in presence of inhibitor equals

$$\left(1 + \frac{[I]}{K_I}\right)\frac{1}{V_{max}}$$

177

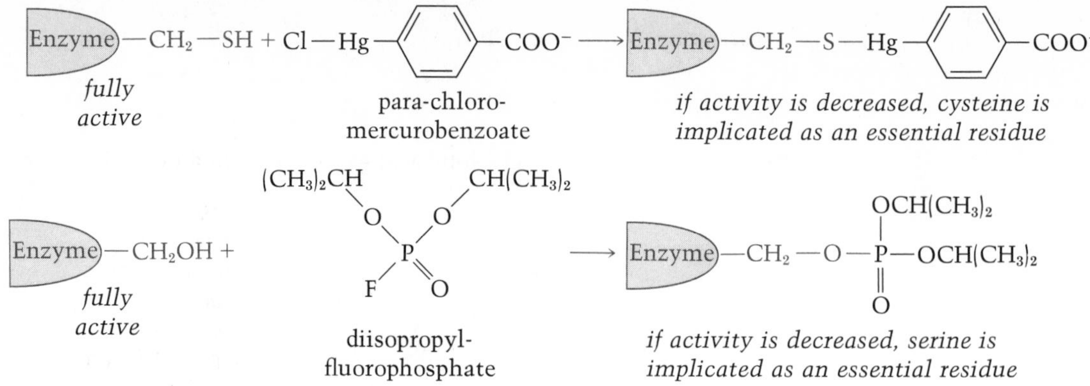

Enzyme—CH$_2$—SH + Cl—Hg—⟨⟩—COO$^-$ → Enzyme—CH$_2$—S—Hg—⟨⟩—COO$^-$

fully active

para-chloro-mercurobenzoate

if activity is decreased, cysteine is implicated as an essential residue

Enzyme—CH$_2$OH +

$(CH_3)_2CH$ $\quad$ $CH(CH_3)_2$

diisopropyl-fluorophosphate

→ Enzyme—CH$_2$—O—P(—OCH(CH$_3$)$_2$)(OCH(CH$_3$)$_2$)=O

if activity is decreased, serine is implicated as an essential residue

most enzymes can be stored at refrigerator temperatures or lower for considerable periods of time without any appreciable loss of activity.) Consistent with the nature of the physiological environment, the optimum temperature, optimum pH, and optimum ionic strength associated with maximum activity generally lie in the ranges of 35–40 °C, pH 6–8, and $\mu = 0.15$ (see p. 79).

ENZYME SPECIFICITY

The preference that enzymes display toward certain substrates is called *enzyme specificity*. In some instances *absolute* specificity is observed, with the enzyme acting on only one substrate. Most enzymes, however, exhibit a *relative* specificity, meaning they have a broader but still limited preference for a small number of chemically related materials. Even in this case, however, the reaction rates and the K_m values will frequently differ due to the preference of an enzyme for a certain member or members of the group. For example, hexokinase will catalyze the ATP-dependent phosphorylation of a large number of hexoses, but it achieves a maximum rate with glucose (see p. 174, this chapter).

The basis of enzyme specificity is the existence of an *active site in the enzyme molecule consisting of a spatially ordered constellation of a small number of amino acid residues*. Remember, the active site location will encompass binding and catalytic residues that are maintained in an optimum spatial orientation because of the interaction of the structural residues that determine and stabilize the foldings of the polypeptide chain.

One of the original theories to account for the ordered formation of the active site-substrate complex was the *lock-and-key* hypothesis of Emil Fischer (1894). The crux of this suggestion is that the recognition of the substrate involves a precise, one-of-a-kind, structural compatibility between the active site of the enzyme and substrate. Obviously this idea is quite applicable to those enzymes that are known to exhibit absolute specificity. However, the major limitation of the lock-key hypothesis is the implication that the entire conformation of the enzyme is rigid. Thus it does not account for the relative specificities of enzymes.

It is now firmly established that most proteins (particularly enzymes) are not rigid structures. Rather, they are capable of undergoing slight and subtle changes in conformation that have dramatic effects on their chemical activity. This concept was first proposed by D. Koshland about 20 years ago

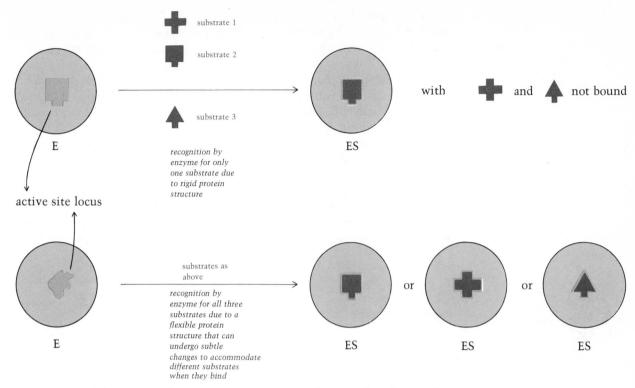

active site locus

recognition by enzyme for only one substrate due to rigid protein structure

recognition by enzyme for all three substrates due to a flexible protein structure that can undergo subtle changes to accommodate different substrates when they bind

in his *induced-fit theory* of enzyme action. According to this theory the initial binding interaction of the substrate initiates subtle alterations in the protein's conformation, producing the optimum orientation of the active site residues for maximum binding with the substrate. With enzymes that are cofactor dependent, this change may be induced by the initial binding of the cofactor, thus rendering the binding of the substrate more likely. The induced-fit proposal more adequately explains the high degree of relative specificity exhibited by several enzymes since similar but different substrates could initiate the development of the active site. The principle applies to other aspects of enzyme action that we will discuss later.

A very special aspect of catalytic specificity exhibited by several (but not all) enzymes is typified by glycerol kinase (see below), an enzyme that cata-

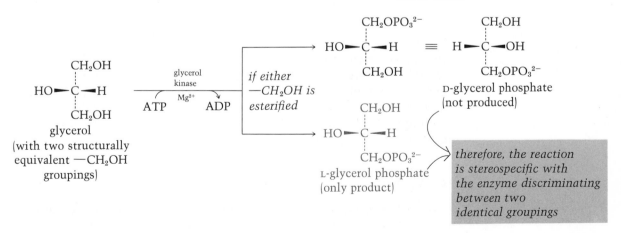

two equivalent projections of the same molecule

if either —CH$_2$OH is esterified

D-glycerol phosphate (not produced)

L-glycerol phosphate (only product)

therefore, the reaction is stereospecific with the enzyme discriminating between two identical groupings

glycerol (with two structurally equivalent —CH$_2$OH groupings)

lyzes the transfer of a phosphate group from ATP to glycerol to yield glycerol phosphate. However, it infallibly produces only one of two possible stereoisomers, namely, the L isomer. This represents an example of *stereospecific catalysis*.

This remarkable property of an enzyme is due to the *asymmetry of the active site locus*. This means that the active site locus is composed of amino acid side chains (R groups) in a three-dimensional arrangement that is *highly ordered but without any pattern of symmetry*. In other words, the active site cannot be split in any plane into two equal halves. Ogston displayed a sharp insight when he proposed this several years ago before much was known about active sites. He specifically proposed that stereospecificity could be explained by a one-of-a-kind three-point attachment of the symmetrical substrate to the asymmetrical active site, followed then by a specific transformation of the bound substrate through the participation of specific catalytic residues (see diagram below). Thus *the asymmetry of the*

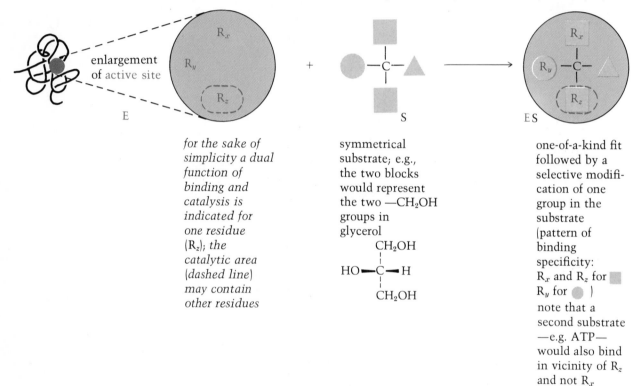

enlargement
of active site

E

S

E S

for the sake of simplicity a dual function of binding and catalysis is indicated for one residue (R$_z$); *the catalytic area* (*dashed line*) *may contain other residues*

symmetrical substrate; e.g., the two blocks would represent the two —CH$_2$OH groups in glycerol

$$\text{CH}_2\text{OH}$$
$$\text{HO} \longleftarrow \text{C} \longrightarrow \text{H}$$
$$\text{CH}_2\text{OH}$$

one-of-a-kind fit followed by a selective modification of one group in the substrate (pattern of binding specificity: R$_x$ and R$_z$ for ▨ R$_y$ for ●) note that a second substrate —e.g. ATP— would also bind in vicinity of R$_z$ and not R$_x$

binding and catalytic residues at the active site determines the stereospecific modification of a symmeterical substrate. In other words, the discrimination between structurally identical groupings in a substrate is due to the microenvironment of the active site, which recognizes the groupings as being different by in effect recognizing only one of them after the substrate is bound. The recognition is based, of course, on the locality of the catalytic residues of the active site.

ACTIVE-SITE EVENTS—A CLOSER LOOK

There is no single, all-inclusive mechanism of action to account for how enzymes operate in the processes of E + S → ES → E + P. Each enzyme is unique in its precise mode of operation, although it is now clear that enzymes having a similar function do exhibit similarities in both molecular structure and chemical action. Despite this individuality, all enzymes operate according to the general principle that there exists a *highly ordered set of chemical interactions between the bound substrate and the R groups of the amino acid residues of the active site.*

The type of chemistry involved at the active site is not mysterious in the sense that a whole new set of chemical principles are needed to explain what happens. On the contrary, classical principles of reaction chemistry are certainly involved. The difficulty in solving the mechanism of action for an enzyme exists primarily because the mechanism is complex in nature, involving several events at the active site that are highly coordinated. Nevertheless, it is possible to generalize about the chemistry of catalysis. For example, the formation of an active enzyme-substrate complex may involve covalent interactions that might contribute to the establishment of bond strains within the substrate (*bond-strain catalysis*). The active-site residues may promote a reaction by acting as donors or acceptors of protons or electrons (*acid-base catalysis*). Still another explanation is that the enzyme's role is merely to provide a centralized and optimally oriented location for the interaction of the substrates, including coenzymes if they are involved (*proximity or orientation catalysis*).

Theoreticians tell us that none of these explanations by themselves can account for the tremendous efficiency of enzymes as catalysts. Thus, the mode of operation of most enzymes probably involves a combination of any two or all three of these factors, plus others that are not yet understood. One thing is fairly certain—*the amino acids that function as catalytic residues are those with reactive side chains;* these are cysteine, serine, threonine, glutamic acid, aspartic acid, lysine, arginine, tyrosine, and histidine. All of these, as well as exposed hydrophobic residues, could also serve as binding residues.

Despite the difficulty of the work, the study of enzyme mechanisms is carried on extensively and several mechanisms have been solved. We will examine two well-understood enzymes to exemplify and reinforce many principles we have covered about enzymes in particular and proteins in general. The enzymes described are *carboxypeptidase* and *chymotrypsin.*

Carboxypeptidase A (CPA; see margin) is a zinc-containing enzyme consisting of 307 amino acid residues in a single polypeptide chain (its molecular weight is approximately 34,000) with one disulfide bond. A drawing of the chain folding is shown in Fig. 6–10. Years of study into its mode of action by many researchers using several methods, including the solution of the three-dimensional structure by X-ray diffraction, has provided the following major facts.

Carboxypeptidase A (CPA) is active toward all polypeptides that do *not* have arginine, lysine, or proline as the C-terminus. Animals also produce carboxypeptidase B (CPB), which acts only if the C-terminus is arginine or lysine. After studying the mechanism of action of CPA, what predictions could you make about the likely mechanism of action of CPB?

1. The zinc ion (Zn^{2+}) is essential for activity.
2. Two separate histidine residues (69 and 196) and a glutamic acid residue (72) are responsible for holding the Zn^{2+} in its position to interact optimally with the substrate.
3. An arginine residue (145) engages in an important binding interaction with the substrate.
4. A tyrosine residue (248), in conjunction with the Zn^{2+}, interacts with the substrate to sensitize the bond to be broken in the substrate.
5. A deep pocket of hydrophobic residues is at the surface to bind the substrate by binding the R group of the C-terminal residue.

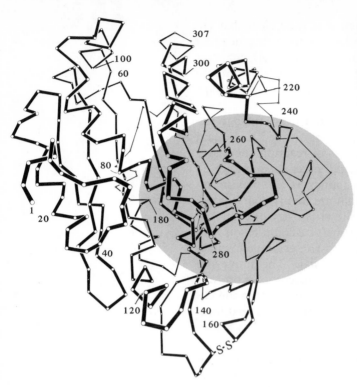

(A) polypeptide chain conformation of carboxypeptidase A; sphere is Zn²⁺; shaded area corresponds to photo at right

(B) space-filling model of the region corresponding to the shaded area in the drawing at the left; this region includes the active site

(D) same region as shown at right, with a model substrate in the binding pocket

(C) close-up view of the active site location consisting of a deep pocket, coordinated Zn²⁺, and other crucial residues

FIGURE 6-10 A study of carboxypeptidase A structure and action. (A)-(D) Polypeptide chain conformation of carboxypeptidase A. (E) The coordinated acid-base catalysis of peptide bond cleavage by water and the binding of the R group deep in the pocket. (Panel A reproduced with permission from W. N. Lipscomb, "Structure and Mechanism in the Enzymatic Activity of Carboxypeptidase A and Relations to Chemical Sequence," *Chem. Res.*, **3**, 81–89 (1970). Drawing generously supplied by Dr. Lipscomb. Photographs for panels B, C, and D supplied through the courtesy of Dr. John Sebastian.

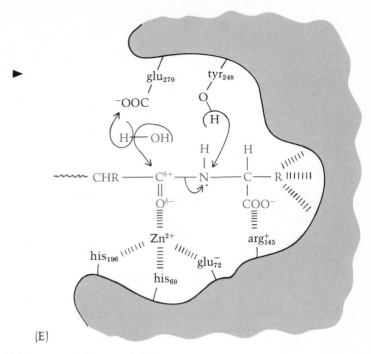

(E)

The display of drawings and models in Figure 6–10 summarize one interpretation of these interactions. The key chemical events in the breaking of the C—N bond are the increased polarity of the O=C bond due to the electron-attracting influence of the Zn^{2+} (it functions as a Lewis acid) and the proton-donating function of tyrosine 248 (it functions as a Bronsted acid). These are accompanied by the attack of water under the influence of glutamic-270 to complete the hydrolysis of the C-terminal peptide bond.

The electrostatic binding of the negative —COO^- group of substrate through the positively charged arginine 145 and the existence of a hydrophobic binding pocket provide a basis for understanding why this enzyme acts on the C-terminus of a polypeptide (rather than the N-terminus) and displays optimum activity when the C-terminal residue has a hydrophobic R group. In examining Fig. 6–10 you want to keep in mind that all of these events occur in a highly ordered and highly efficient, concerted fashion and do so very quickly. You should also be amazed by the intricate workings of enzymes.

183

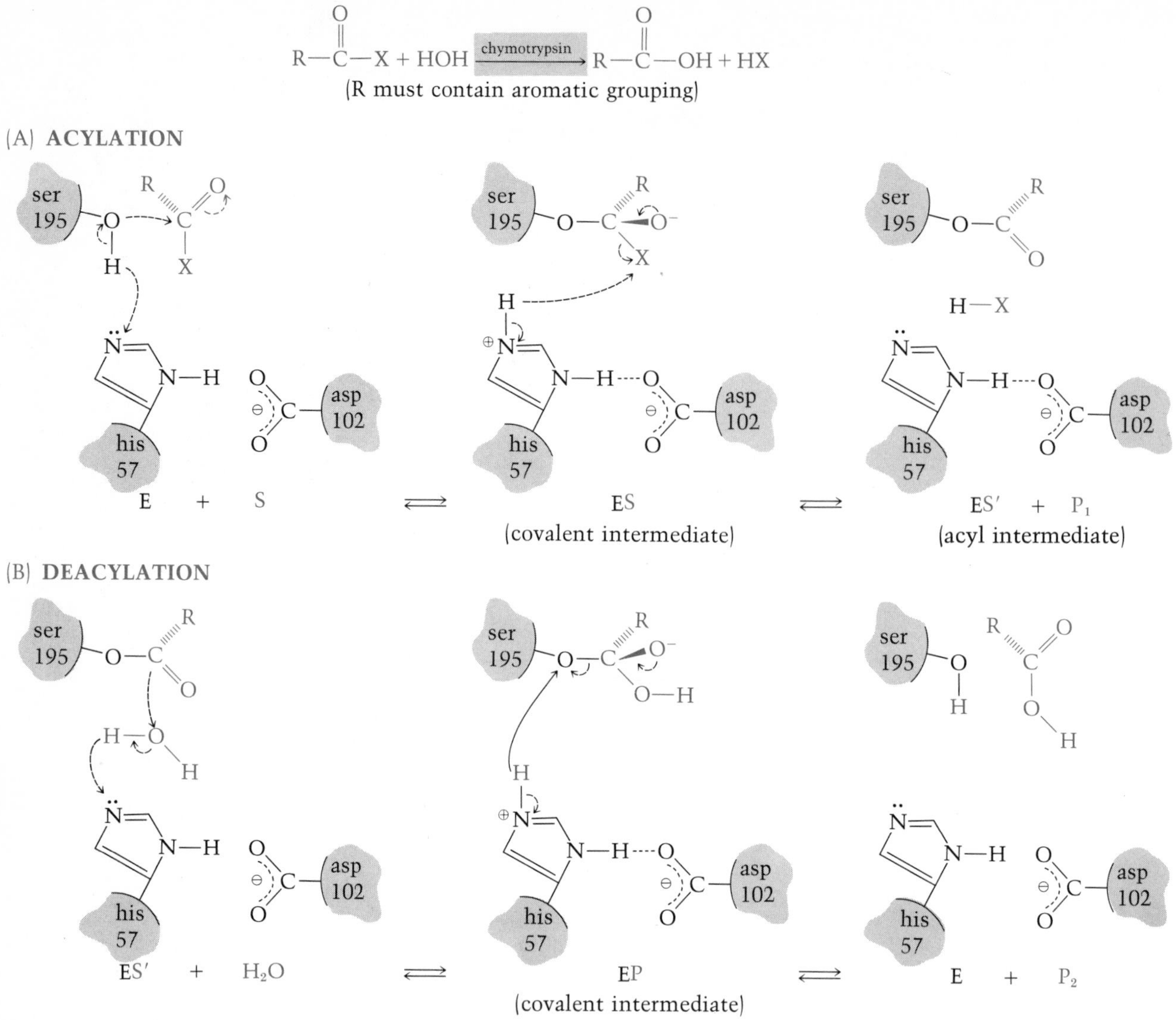

$$R-\overset{\overset{\displaystyle O}{\|}}{C}-X + HOH \xrightarrow{\text{chymotrypsin}} R-\overset{\overset{\displaystyle O}{\|}}{C}-OH + HX$$

(R must contain aromatic grouping)

(A) **ACYLATION**

ser 195 / O—H ⋯ R—C(=O)—X ⟶ ... N: N—H ⋯ O=C—asp 102 / O⁻ ⋯ his 57

E + S ⇌ ES (covalent intermediate) ⇌ ES′ + P₁ (acyl intermediate)

(B) **DEACYLATION**

ser 195 / O—C(=O)—R ... H—O—H ... N: N—H ⋯ O=C—asp 102 / his 57

ES′ + H₂O ⇌ EP (covalent intermediate) ⇌ E + P₂

FIGURE 6–11 A study of chymotrypsin structure and action. *Opposite page:* drawing of the three-dimensional conformation of polypeptide chain. (From B. W. Matthews et al.,"3-D Structure of Tosyl-α-Chymotrypsin," *Nature,* **214,** 652 (1967). Drawing generously supplied by Dr. D. M. Blow.) *Above:* detailed summary of the covalent, acid-base catalysis. The crucial binding interactions for the R group of the substrate are not shown.

Chymotrypsin catalyzes (see Fig. 6–11) the hydrolysis of internal peptide bonds in a polypeptide substrate with a preference toward bonds where aromatic amino acids donate the C=O group. Note that chymotrypsin action involves a completely different type of mechanism than carboxypeptidase: a covalent intermediate is formed involving a reactive —OH group of serine 195, which is made highly reactive under the acid-base influence of histidine 57 and aspartic 102. Once again we have a specific illustration of a localized cluster of catalytic residues that operate in concert to promote a chemical transformation of a substrate.

The proposed mechanism of chymotrypsin catalysis consists of two steps. The first step depicts an acylation of the ser 195 residue, promoted by the abstraction of a H⁺ by a N atom of the imidazole group of his 57. Then under hydrogen bonding influence of asp 102 the H⁺ from his 57 is donated back to the covalent adduct associated with the cleavage of the C—X bond. Thus the substrate has been modified and now it has to be released. This

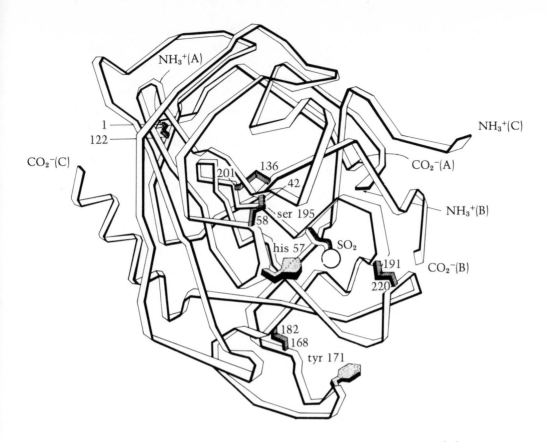

occurs in the second step (a deacylation) and involves participation of the second substrate, H_2O. The his 57 again abstracts a H^+, but this time from water, promoting the formation of a new C—O bond involving the acyl group carbon. Finally, the H^+ from his 57 is donated back to the O of ser 195 with the breaking of the O—C bond and release of the product.

Note that this is a perfect example of the point made earlier (see p. 171) about the oversimplification of the ES notation in Michaelis-Menten theory.

The point of all of this was to provide a basis for an appreciation of what is involved in the representation of $E + S \rightarrow ES \rightarrow E + P$. Clearly there is more than meets the eye. Nevertheless, close examination of these examples should help your understanding. Memorization is not necessary to gain this understanding.

Countless numbers of researchers are engaged in the study of how enzymes operate. In addition to contributing to the basic knowledge of enzyme action in general, these efforts also result in the development of such things as improved antibiotics, herbicides, and insecticides. By knowing precisely how an enzyme operates, it is easier to design agents that would be selective and efficient inhibitors of that enzyme.

ZYMOGENS

Several proteins are originally synthesized in functionally inactive forms, and only after secretion from the cell do they undergo a further modification in structure to yield an active form. The inactive precursor is called a *zymogen* or *proenzyme*. Protein-digesting enzymes, blood-clotting proteins, and some hormone proteins are three important examples of this phenomenon.

zymogen
(inactive protein)

for each zymogen a specific modification in structure is involved

↓

active form

185

The protein-digesting enzymes *pepsin, trypsin, chymotrypsin,* and *carboxypeptidase* (and a few others) catalyze the breakdown of ingested protein to the constituent amino acids, which are then absorbed from the intestine into the bloodstream. The inactive precursors—*pepsinogen, trypsinogen, chymotrypsinogen,* and *procarboxypeptidase*—are originally synthesized in the pancreas. It might occur to you that this inactivity is crucial to the well-being of the cells that synthesize these digestive proteins. If the pancreas cells produced the enzymes in an active form within the cell, this would represent a potential self-destruct situation, since any of the other proteins in the cell would be the immediate targets of their action.

In each case the zymogen conversion involves the modification of its primary level of structure with a resultant alteration in its three-dimensional structure. In the very acidic contents of the stomach, a fragment containing 42 amino acids is removed from the N-terminus of pepsinogen by the action of H^+. The pepsin that is formed can itself then act on pepsinogen to form additional pepsin. In the small intestine, trypsinogen is converted to trypsin by the removal of a hexapeptide fragment from the N-terminus. The process is catalyzed by *enterokinase,* another enzyme secreted in small amounts by the intestinal mucosa, and then also by trypsin itself. The action of enterokinase to produce active trypsin is crucial because trypsin is also responsible for the activation of chymotrypsinogen and procarboxypeptidase.

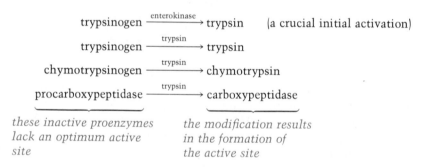

The reason for the original inactivity of the proenzymes is that the optimum orientation of all of the active site residues does not exist in the proenzyme conformation. The binding site residues are not properly aligned and/or the catalytic residues are not properly aligned. When the proenzyme structure is modified by the cleavage of a specific peptide bond or bonds, which may involve the excision of a small polypeptide fragment, a new set of R group interactions occurs throughout the molecule to produce a different conformation. The new conformation contains the proper alignment of all binding and catalytic residues necessary for optimal activity. The precise details of some of these conversions that have been experimentally determined are described in the paper by Stroud cited at the end of this chapter. The phenomenon is a powerful illustration of (a) the requirement of an optimal active site orientation for maximal enzyme action and (b) the dependence of the overall conformation and function of proteins on the amino acid sequence of that protein.

Similar events apply to the clotting of blood. The last step in blood clotting is the conversion of the soluble blood protein *fibrinogen* to protein *fibrin.* The fibrin molecules then undergo end-to-end and side-to-side aggregation yielding insoluble strands of fibrin, which eventually form the clot. The fibrinogen → fibrin conversion is catalyzed by the action of *thrombin*, which cleaves four specific peptide bonds in fibrinogen. Thrombin in turn is produced from the inactive zymogen *prothrombin*. The prothrombin activation requires Ca^{2+} and other blood proteins, called *clotting factors*, which participate in an elaborate cascade of activation steps. One of these proteins, designated *Factor VIII*, is called the *antihemophilic factor* because its absence in the blood is the cause of the common type of hemophilia.

prothrombin $\xrightarrow[\text{several other proteins}]{\text{Ca}^{2+} \text{ and}}$ thrombin   then   fibrinogen $\xrightarrow{\text{thrombin}}$ fibrin
(inactive)   (active   (soluble and   (strong
                                      enzyme)   no tendency   tendency to
                                                              to aggregate)   aggregate)

Another example of this original inactivity in hormone proteins is provided by the conversion of *proinsulin* to *insulin*. (See note in margin for further information on insulin.) The modification involves the action of trypsin, which catalyzes the cleavage of two specific peptide bonds in proinsulin, itself a single polypeptide chain (84 residues) consisting of three intrachain —S—S— bonds. The trypsin action results in the removal of a polypeptide fragment of 33 residues, leaving behind the insulin molecule (two polypeptide chains connected by two interchain —S—S— bonds).

Insulin biosynthesis begins with the assembly of a *preproinsulin* polypeptide containing 107 amino acids. Then at some stage during synthesis, a fragment of 23 amino acids is trimmed from the N terminus and synthesis continues to yield proinsulin

(see p. 129 for sequence)

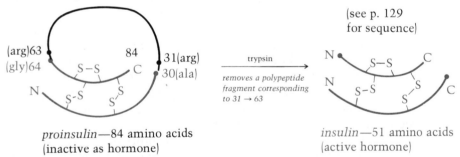

proinsulin—84 amino acids
(inactive as hormone)

insulin—51 amino acids
(active hormone)

Recently, it has been discovered that proinsulin itself is also formed from the trimming of a larger precursor called *preproinsulin*. The preproinsulin molecule has an extension of 23 amino acids on the N-terminus of proinsulin. It is proposed that the presence of this N-terminal fragment during synthesis of the protein represents the way in which proteins to be packaged in Golgi bodies and excreted from the cell are "labeled" as being different from those proteins that are to remain inside the cell.

NATURAL REGULATION OF BIOCATALYSIS

In a previous section we discussed how the sulfa drugs, certain antibiotics, and a host of other chemical agents, all of which are natural or synthetic, can inhibit the progress of enzyme-catalyzed reactions via competitive or

Insulin is a relatively small protein (mol. wt. 6,000) produced by the pancreas and secreted into the circulatory system for transport throughout the body. As a hormone, insulin is known to regulate various processes in various target tissues. The most well-known effect of insulin is its maintenance of normal glucose blood levels by stimulating both the uptake of glucose by cells and its subsequent incorporation into glycogen (see p. 397 in Chap. 12). Elevated levels of blood glucose associated with an impairment of carbohydrate metabolism, plus increased levels of fat and protein metabolism, characterize the disease diabetes mellitus, generally called simply diabetes. The onset of these metabolic abnormalities is due either to the inability of the pancreas to produce insulin or to its production of less than normal amounts. The ultimate effect of the untreated condition is death. However, regular insulin injections can maintain the normal condition. It is estimated that there are about 1.25 million insulin-dependent diabetics in the United States.

187

noncompetitive effects. In effect, their use as chemotherapeutic agents permits the control of enzyme activity or, more generally, the control of metabolism, with the desired objective being the selective destruction of the viability of the infectious organism.

Within the past ten or fifteen years it has become increasingly apparent that metabolic control is also a natural phenomenon common to all types of organisms. The bulk of this regulation is likewise exerted at the level of the enzyme-catalyzed reaction via materials that are normally present and in fact are actually produced within the cell itself. The natural regulatory mechanisms involve effects that not only inhibit, but also enhance enzymatic activity.

General Principles of Active-Inactive Conversions of Proteins *In Vivo*

The preceding discussion of zymogen activation is but one particular example of how a living organism is able to control its processes. We will now discuss in more general terms other tactics that contribute to this capability of biological self-control or self-regulation. *This regulation is achieved mainly by controlling the interconversion of active and inactive forms of proteins*, most notably those proteins that operate as enzymes. You will eventually learn that by the active ⇄ inactive interconversion of a few crucial enzymes in each main area of metabolism a living cell is able to control all of its metabolic processes in a most efficient manner. Remember, the essence of a living cell is in major part a result of all of chemical reactions occurring in the cell, most of which are enzyme-catalyzed. Through regulation of enzyme activity, the reactions themselves are regulated.

Protein activity is regulated by (a) *covalent modification*, (b) *association-dissociation*, (c) *competitive inhibition* (d) *secondary-site effects*, and (e) *gene repression.* (*Note:* The listing is not a ranking of importance.) The first four modes apply to a protein that has already been synthesized in the cell; in these cases the active-inactive transition occurs rapidly. On the other hand, gene repression, which occurs more slowly, regulates the biosynthesis of a protein—an obvious way of regulating a protein-dependent process, since there can be no activity if there is no protein. Gene repression will be examined in Chapter 8.

Covalent modification means that the protein undergoes a change in activity by virtue of the covalent attachment (or detachment) of a chemical grouping from one or more of the amino acid side chains. Frequently, the grouping is phosphate ($-OPO_3^{2-}$) and the side chain involved is that of serine ($-CH_2OH$). Note the reversibility of the process and that enzymes

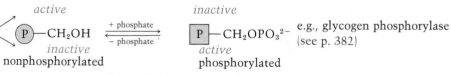

both situations are known but the lower one is more common

| active | | inactive |

P$-CH_2OH$ $\underset{- \text{phosphate}}{\overset{+ \text{phosphate}}{\rightleftharpoons}}$ P$-CH_2OPO_3^{2-}$ e.g., glycogen phosphorylase (see p. 382)

inactive   *active*

nonphosphorylated   phosphorylated

(the ⚪ ⇌ ⬜ symbolism implies that the covalent modification is associated with a change in conformation of the intact protein; occasionally this type of transition is also accompanied by another type of transition, namely, dissociation-association, as discussed below) are involved. Zymogen conversions are also representative of this principle, with the covalent modification involving the enzyme-catalyzed cleavage of a small number of peptide bonds. However, zymogen conversions are not reversible.

An active-inactive conversion of other proteins involves the association-dissociation of the units that comprise an oligomeric state, which is almost always the active form. This type of interconversion is sometimes triggered by a covalent modification (see above) or by the binding of some ligand.

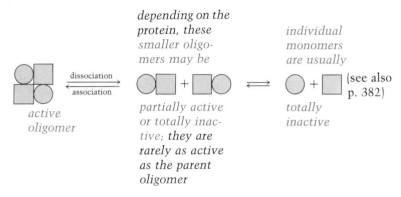

depending on the protein, these smaller oligomers may be

individual monomers are usually

active oligomer

partially active or totally inactive; they are rarely as active as the parent oligomer

totally inactive

(see also p. 382)

There are some proteins that are subject to competitive inhibition by substances that occur naturally. For example, a reaction that requires oxidized NAD^+ might be competitively inhibited by reduced $NADH$; one that requires ATP might be competitively inhibited by ADP or AMP (see p. 210, next chapter); and some enzymes are competitively inhibited by one of the products that are actually produced in the reaction. Examples of all types will be encountered in future chapters.

Regulation via secondary-site effects is examined separately and in greater detail in the next section. It is a most important topic.

Regulation Via Secondary Site Effects—Allosterism

Every scientific discipline characteristically has certain key discoveries that contributed greatly to its advancement. In modern biochemistry a key discovery was that of the interconversion of active and inactive states of a protein by secondary-site effects. This probably represents the major type of biological control mechanism. The phenomenon is described by the *theory of allosterism*, and the resultant effects are termed *allosteric effects*.

Experimental Background. The theory of allosterism (to be defined later) was in part an outgrowth of investigations in the early 1960s designed to determine the mechanism of *feedback inhibition*. The latter term was coined to refer to a specific regulatory phenomenon where an initial enzyme of a multistep biosynthetic reaction sequence is susceptible to a reduction in its catalytic activity by the natural end product of the pathway. The inhibited enzyme decreases the rate of production of an early intermediate (the step is referred to as being *blocked*), which in turn limits the extent to which subsequent steps can occur. The entire sequence is then blocked. An organism benefits immensely from feedback inhibition, because through it the organism can avoid the wasteful production of metabolic end products under physiological conditions in which their biosynthesis is not needed in

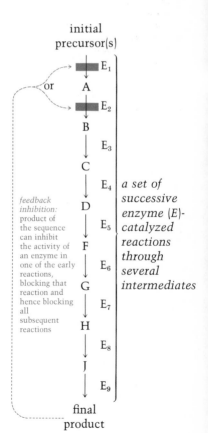

initial precursor(s)

feedback inhibition: product of the sequence can inhibit the activity of an enzyme in one of the early reactions, blocking that reaction and hence blocking all subsequent reactions

a set of successive enzyme (E)-catalyzed reactions through several intermediates

final product

large amounts. Thus, an organism can utilize all of its important basic resources such as carbon, nitrogen, sulfur, phosphorus, and energy in a most efficient manner.

The first systematic study of feedback inhibition was performed by J. C. Gerhart and A. B. Pardee in 1962. The enzyme studied was *aspartate transcarbamylase* (ATCase), which catalyzes the first of a series of reactions in the biosynthesis of the pyrimidine nucleotides, which in turn are used for the biosynthesis of the nucleic acids RNA and DNA. The specific reaction in question is shown in Fig. 6–12. As depicted, ATCase is subject to a strong feedback control by the end product, cytidine triphosphate (CTP). How does this happen?

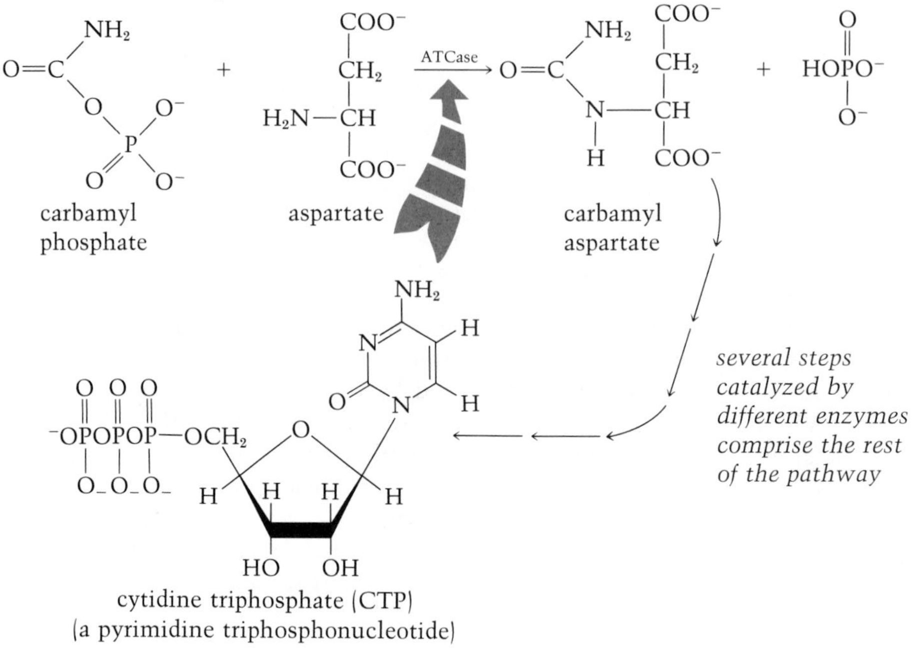

FIGURE 6–12 A schematic diagram of the feedback inhibition of pyrimidine biosynthesis in *Escherichia coli*. The heavy broken arrow indicates feedback inhibition. The succession of smaller arrows represents a subsequent sequence of intermediate enzyme-catalyzed reactions between the initial reaction and the terminal product, CTP.

The first indication that the enzyme operates a little differently than others was provided by a straightforward Michaelis-Menten kinetic study. A plot of v_0 values measured at different concentrations of aspartic acid (a constant level of carbamyl phosphate was maintained) did not yield the classical hyperbola. Rather, the curve (Fig. 6–13) was *sigmoid* in character (S-shaped), showing a very slow increase in v_0 at low concentrations of substrate, followed by a transition to a greater increase in v_0 at higher substrate levels until a maximum velocity was reached. The sigmoid curve of a Michaelis-Menten analysis is now generally regarded as the distinctive kinetic characteristic of an allosteric enzyme. The transition in the two distinct phases of the relationship between v_0 and [S] is referred to as a *cooperative effect* and the phenomenon called *cooperativity*. Briefly, it means that the presence of low levels of substrate exerts an influence that alters the response of the enzyme to higher levels of substrate. In the case we are examining, the influence is a favorable one—that is, the change in v_0 relative to [S] increases—and so it is called *positive cooperativity*. The principle of cooperativity was discussed earlier in Chap. 5 in the description of protein binding. We will say more about this later.

When the same kinetic measurements were made in the presence of

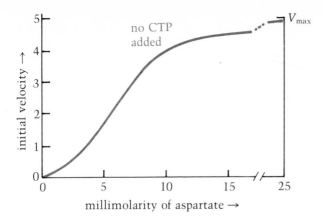

this curve is satisfied by the Hill equation

$$v_0 = \frac{V_{max}[S]^c}{K_m + [S]^c}$$

where c is the Hill coefficient (see p. 154) indicating the type of cooperativity.

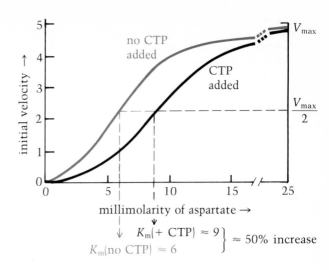

$$\left. \begin{array}{l} K_m(+ \text{ CTP}) \approx 9 \\ K_m(\text{no CTP}) \approx 6 \end{array} \right\} \approx 50\% \text{ increase}$$

FIGURE 6–13 *Left:* Michaelis-Menten kinetics of aspartate trans-carbamylase (ATCase). Aspartate concentration varied as indicated; carbamyl phosphate maintained at constant level; nothing else added. The curve is distinctly sigmoid. *Right:* the color curve is the same as shown before. The black curve represents the kinetics when the enzyme activity is measured in the presence of a small amount of CTP, cytidine triphosphate. (Data taken with permission from J. C. Gerhart and A. B. Pardee, "The Enzymology of Control by Feedback Inhibition," *J. Biol. Chem.*, **237**, 891–96 (1962).)

added CTP, the same sigmoid relationship was observed, but it was more severe (see Fig. 6–13). A graphical evaluation of the Michaelis-Menten plots for the control (no CTP inhibitor added) versus the inhibited systems *suggests competitive inhibition* since the presence of CTP did not alter the V_{max} but did increase the K_m by 50%. *However, it is difficult to explain how CTP could function as a competitive inhibitor of aspartate, since they are not at all similar in their structures* (see Fig. 6–12). The inhibition must be of a noncompetitive type and a special case at that.

To explain the unusual inhibition kinetics Gerhart and Pardee proposed that there are *separate sites* in the enzyme molecule; one selectively binds with the normal substrates, aspartate and carbamyl phosphate (this would be the active site), and the other selectively binds with CTP. Furthermore, it was suggested that the binding of CTP is independent of the binding of aspartate. Finally, it was argued that, if this is so, it might be possible to selectively disrupt one site (that is, selectively denature a part of the protein) without damaging the other. For example, if the CTP binding site were disrupted and the active site were not, then the enzyme would be less sensitive to inhibition by CTP, but would still possess enzymatic activity. If, however, both substrate and inhibitor used the same site for binding, or if the two sites were mutually dependent on each other, the separation of inhibition and normal enzymatic activity would not be possible. One of the denaturing methods used mild heating.

The data of Fig. 6–14 are clear proof that the idea of two sites was correct and that the inhibition site can be (at least in ATCase) selectively destroyed. Note especially in Fig. 6–14 that there was no difference observed in the response of the heat-treated enzyme to the presence or absence of CTP, but the enzyme was still active. Similar results were observed when the enzyme was treated with other denaturants. Thus it can be concluded that the *binding site of the inhibitor was largely separate from the site for the normal substrates.* A closer comparison of the treated and untreated systems indicates, however, that the sites are not entirely independent,

FIGURE 6–14 Evidence for separate but interdependent sites in aspartate transcarbamylase. (Same data source as cited in Fig. 6–13.)

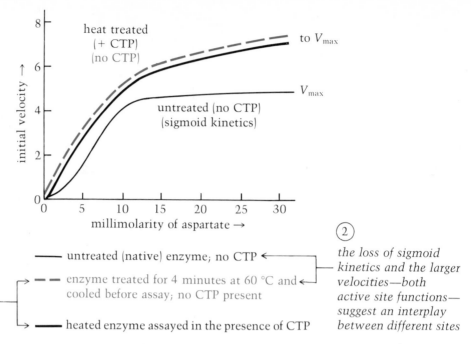

① the lack of any difference strongly suggests that the CTP binding site is absent in the treated enzyme; the active site is still functional, however.

untreated (native) enzyme; no CTP ←

- - - enzyme treated for 4 minutes at 60 °C and ← cooled before assay; no CTP present

heated enzyme assayed in the presence of CTP

② the loss of sigmoid kinetics and the larger velocities—both active site functions— suggest an interplay between different sites

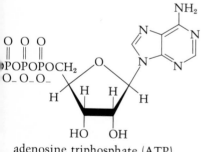

cytidine triphosphate (CTP) a pyrimidine nucleotide; allosteric inhibitor of ATCase

adenosine triphosphate (ATP) a purine nucleotide; allosteric activator of ATCase

since the maximum velocity differed. In fact, the increase in velocity in the treated systems was nearly twofold. Finally, note that the kinetics of the treated systems were no longer characterized by the sigmoid curve common to the native enzyme. Thus, although the data suggest that the sites are independent in the sense that they are defined as separate and distinct regions of the enzyme surface, they also indicate that the *activity of the aspartate (substrate) binding site is profoundly affected by disruption of the inhibitor site.* This intramolecular communication between different locations is mediated by changes in the structure of the protein. (We will discuss this later.)

It so happens that ATCase exhibits still another interesting response in the presence of ATP (adenosine triphosphate). ATP is a purine triphosphonucleotide and although it is distinctly different than CTP, there is a close structural similarity between the two (see margin). The effect of ATP on ATCase is shown in Fig. 6–15. Note the reduced K_m value, representing less substrate required for half-saturation, and the near restoration of the classical hyperbolic kinetics, representing a reduced need for substrate cooperativity. These effects constitute *activation.* In view of the close structural similarity between CTP and ATP it was proposed that the same site could bind either substance. Further studies confirmed this thinking. To restate: ATCase is an allosteric protein that binds either of two similar materials at the same site; however, completely opposite effects result depending on which material is bound, and this in turn triggers completely opposite effects on enzymatic activity. This is not unique to ATCase nor is it applicable to all allosteric proteins.

Before moving on with a further analysis of allosterism, let us summarize the primary conclusions of Gerhart and Pardee.

1. Aspartate transcarbamylase is representative of a group of enzymes that possess more than one binding site. In addition to the primary active site,

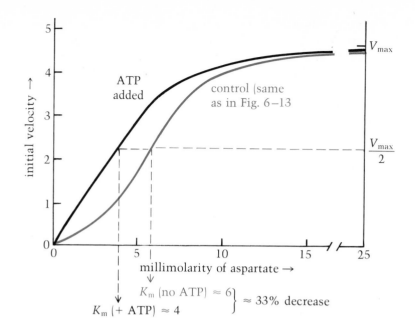

FIGURE 6–15 A comparison of aspartate transcarbamylase kinetics in the presence and absence of adenosine triphosphate (ATP). The greater velocities at low levels of substrate and the indication of no further need of the substrate cooperative effect (loss of sigmoid character) typifies allosteric activation. (Same data source as cited in Fig. 6–13.)

which binds the natural substrates, there also exist, in the same molecule, separate secondary sites, which can bind other substances that function as inhibitors and/or activators.

2. Both the primary and secondary sites can be occupied simultaneously.

3. The secondary sites are not necessarily specific in their action and different materials can bind, resulting in different effects.

4. The binding at the secondary sites may cause a subsequent shift in the structure of the protein molecule, which in turn can affect the catalytic activity of the primary active site.

5. The secondary site effects constitute the basis of feedback control mechanisms that serve as an effective means of regulating metabolism.

Molecular Explanations of Allosteric Effects. Allosteric means *other* (allo-) *spatial conformations or sites* (-steric). Allosterism refers to the *regulation of protein activity by substances* (called allosteric effectors) *that bind to a secondary site and cause the protein molecule to undergo a change in activity.*

The first molecular insight into allosteric effects was made in 1963 by Monod, Changeux, and Jacob, who proposed that proteins subject to allosteric effects have an *oligomeric* structure, composed of identical units called *protomers,* with each protomer containing only one binding site corresponding to each possible effector. For example, a dimeric enzyme (A in margin) that binds only substrate and is not acted on by any inhibitor or activator would contain only two sites—one active site in each protomer. If a dimeric enzyme (B in margin) is subject to an allosteric inhibitor but no allosteric activator, it would contain four sites—one active site and one inhibitor binding site in each protomer. If a trimeric enzyme (C in margin) is subject to both allosteric inhibition and allosteric activation with the inhibitor and activator binding at different locations, the molecule would contain a total of nine sites—one active site, one inhibitor binding site, and one activator binding site in each of the three protomers.

Question: if the allosteric enzyme D in the margin is subject to activation, how many total sites would exist?

protomer units

A

● is active site in each protomer; no inhibitors or activators; total sites = 2

B

● is active site, ▲ is inhibitor site, in each protomer; no activator site; total sites = 4

C

● is active site, ▲ is inhibitor site, ■ is activator site; total sites = 9

D

193

They further proposed that the allosteric effects are achieved according to a *two-state model*. The central idea is that an allosteric protein, prior to any binding with substrate S, activator A, or inhibitor I, *exists in two different conformational states that are in equilibrium with each other*. One is an *active conformation* and the other is an *inactive conformation*. In terms of an enzyme, the active conformation would contain an optimum active site for substrate(s) in each protomer, whereas the inactive conformation would not contain an optimum active site. If the enzyme were subject to activation, the active conformation would also possess optimum binding sites for the activator. If the enzyme were subject to inhibition, the inactive conformation would possess optimum binding sites for the inhibitor.

The model further states that regardless of the number of types of allosteric effectors, the *binding of an effector to its preferred conformation will cause some molecules in the nonpreferred conformation to convert into molecules having the preferred conformation for that effector*. The influence exerted by the effector is said to be *indirect*, meaning that the inhibitor—for example—causes the conversion of an active molecule to an inactive molecule without binding directly to the active molecule. It (the inhibitor) binds to the inactive state. Likewise by binding to the active molecule, an activator causes an inactive molecule to change to an active molecule. The idea is schematically summarized in Fig. 6–16 for a dimeric molecule. To illustrate the principle of cooperativity, the initial equilibrium assumes six inactive molecules and two active molecules. For the sake of simplicity, the effects of an inhibitor and activator are diagramed separately.

After close study of Fig. 6–16 you should understand the basic message. Allosteric enzyme regulation is achieved by the binding of an effector, which causes an interconversion of active and inactive molecules, which in turn determines the total number of optimum active sites that are available. *An increased number of active sites means an increase in enzyme activity; a decreased number of active sites means a decrease in enzyme activity.*

As an alternative explanation of cooperative allosteric effects, Koshland has applied the induced-fit principle (see p. 179) to account for the *direct influence of an effector on a single state*. A schematic summary of this *induced-fit model* is shown below for cooperative binding of substrate to a dimeric enzyme. The key feature is that the first substrate molecule induces

Positive cooperativity of substrate binding. The need for this influence is accentuated by an inhibitor and diminished by an activator

"inactive state"
(less active)

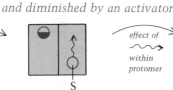

"active state"
(more active)

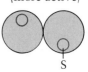

binding of substrate S

effect of ⤳ *within protomer*

effect of ⤳ *on other protomer*

initial state of dimeric enzyme with two active sites that are partially accessible (half open) to substrate

◐ to ○ transition induced by binding of S, which in turn sets off (⤳) other R group interactions throughout the entire protomer

because protomers are associated by noncovalent R group interactions, any conformational change in one will induce a change in the other

now the second site is more accessible to bind additional S

To begin with, assume that a dimeric protein possesses a 6:2 ratio of inactive:active conformers. This situation means that there are initially a total of four active sites available for binding substrate.

When a small level of substrate S is added, it binds to an active conformer. The two-state model proposes that the initial equilibrium of unbound conformers would respond to produce more active conformers. For simplicity, let us assume that one inactive conformer is converted to one active conformer. The final result would be

The noteworthy feature is that there are now five active sites available for further binding of substrate. In other words, the initial binding of a little S causes a transition that increases the population of optimum active sites for additional S. This is the *cooperative effect.*

An allosteric activator A would operate similarly by binding to its preferred site in an active conformer and causing an inactive → active transition. Using the same assumption as above, the initial population of four active sites would be increased to 6. The activator effect would tend to diminish the need for a cooperativity of substrate binding. This effect was depicted in Fig. 6–15 for the ATP influence on aspartate transcarbamylase. The sigmoid nature of the curve was decreased.

An allosteric inhibitor I would operate similarly, but in reverse, by binding to its preferred site in an inactive conformer and causing an active → inactive transition. Again with the same assumption, the initial population of four active sites would be reduced to two. The inhibitor effect would tend to increase the need for a cooperativity of substrate binding. This effect was depicted in Fig. 6–13 for the CTP influence on aspartate transcarbamylase. The sigmoid nature of the curve is increased.

FIGURE 6–16 A simplified, schematic summary of all allosteric effects according to the two-state model. Small circles (● and ○) correspond to substrate sites; small squares (■ and □) correspond to activator sites; small triangles (▲ and △) correspond to inhibitor sites. Open symbols represent optimum ("good") sites and solid symbols represent nonoptimum ("bad") sites. In other words, a set of open and closed symbols (for example, ● and ○ for substrate) identifies the same site, a substrate site, but these sites will have different three-dimensional conformations and hence different activities. The large, shaded rectangles and spheres represent different three-dimensional structures of the same polypeptide chain in the dimeric protein.

a change in one active site to accommodate its binding; this event then triggers other interactions within this protomer that are ultimately transmitted to the other protomer, causing the second active site to assume a more favorable orientation for the binding of the second S molecule. The "message" is in effect transmitted by what can be called a domino-type interaction of R groups. First, the binding of S causes a slight reorientation of a few R groups at the site, which in turn interact with other R groups causing them to assume different orientations; these cause others to shift, and so on, until the entire structure of the protomer exists in a new three-dimensional state, that is, a new conformation. The new shape of the affected protomer can "relay the message" to the partner protomer in the same way—by triggering new R group interactions in the partner, the consequence of which is a better active site in the second protomer of the same molecule to which S bound initially.

The effect of an activator according to the induced-fit model is diagramed below, and below that, the effect of an inhibitor.

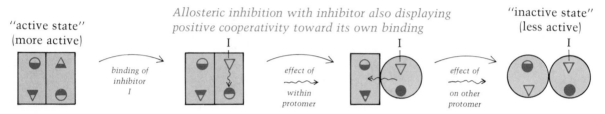

Allosteric inhibition with inhibitor also displaying positive cooperativity toward its own binding

"active state" (more active)

"inactive state" (less active)

two active sites and two inhibitor sites; all shown as partially accessible

I binding induces
(a) a ◖ to ● change at active site, thus inhibiting S binding in the same protomer
(b) and a ▼ to ▽ change of the other inhibitor site, which favors additional I binding; this may also be accompanied by some influence on the second active site; if not, the binding of I to the second ▽ will certainly change the second S site to ●

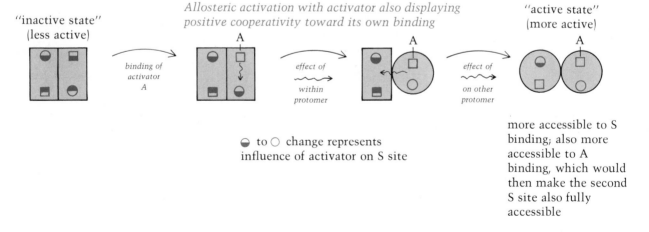

Allosteric activation with activator also displaying positive cooperativity toward its own binding

"inactive state" (less active)

"active state" (more active)

◖ to ○ change represents influence of activator on S site

more accessible to S binding; also more accessible to A binding, which would then make the second S site also fully accessible

Each in its own way, both the indirect, two-state model and the direct, induced-fit model can account for positive cooperativity. However, only the induced-fit model is capable of explaining *negative cooperativity* (see p. 153), where the initial binding of an effector diminishes the further binding

of the same effector. The principle is diagramed below for the negative
cooperativity of an activator.

*Allosteric activation with activator displaying
negative cooperativity toward its own binding*

binding of
activator
A

negative cooperativity activates (or inhibits if
I is ligand) the enzyme partially; on the other
hand, when A or I bind with positive
cooperativity, the corresponding effects can
approach 100%—see previous diagrams

further A binding is not
favored; thus, the
second S site cannot be
further influenced by A

Concluding Comments and a Final Look at Hemoglobin

Although it is possible to establish which of the two models may apply to a
particular allosteric protein, we will not pursue this point, nor will we describe other models (and there are others). The important thing to understand is that each model is explaining the same phenomenon, namely, the
allosteric regulation of *protein* activity. The use of italics is to emphasize
that allosteric regulation is not restricted to oligomeric enzymes, but rather
is applicable to oligomeric proteins in general. For example, the tetrameric
hemoglobin molecule exhibits a strong cooperative effect of oxygen binding,
with the third and fourth oxygen molecule binding much more readily than
the first (see Fig. 5–8).

In the case of hemoglobin, a mechanism of elegant details has been proposed utilizing the principles of the induced-fit model. Key information
was supplied by X-ray diffraction of the entire three-dimensional structures
of both oxygenated hemoglobin (active) and deoxygenated hemoglobin (inactive). See Fig. 6–17. An examination of the many details involving the R
group interactions is beyond our scope. However, the key event (for which
there is extensive evidence) is proposed to be the initial binding of one or
two O_2 molecules to two subunits (say the two α subunits), causing the iron
atom to which they bind to be pulled further into the plane of the heme
grouping. It is this movement of iron that triggers the other R group interactions. These interactions ultimately provide greater accessibility of
the remaining hemes for O_2 binding. (Study Fig. 6–17 carefully.)

In addition to O_2 cooperativity, hemoglobin is also responsive to another
substance in red blood cells, namely, *2,3-diphosphoglycerate* (DPG; see
margin for structure). DPG has the effect of *reducing the affinity of hemoglobin for oxygen*. It has been established that the DPG effect has great
physiological significance since it promotes the release of bound oxygen to
respiring tissues. The single binding site for DPG exists in the central cav-

$$Hb(O_2)_4 + DPG \rightleftharpoons Hb(DPG) + 4O_2$$

→ *at respiring
tissues, DPG
displaces O_2*

← *at lungs, O_2
displaces
DPG*

2,3-diphosphoglycerate
DPG

symbolized as

in Fig. 6-17

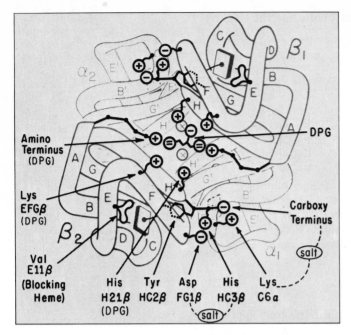
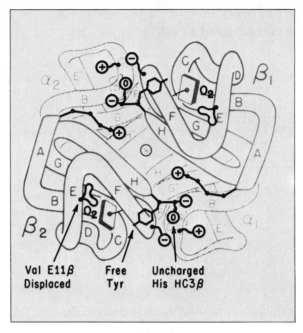

FIGURE 6–17   The three-dimensional structure of hemoglobin (from a horse) viewed with the β subunits closer to the viewer. *Left:* deoxyhemoglobin; *right:* oxyhemoglobin.   Heme groups are shown as small squares.   Boldface arrows point to residues whose R groups undergo subtle but important changes in their position when molecule is oxygenated. A ⊕ indicates a basic side chain; a ⊖ indicates an acidic side chain.   The conformational shift throughout the intact molecule (in this view) is movement of the two β chains toward one another by approximately 7 Å until the two small ⊗s in the center of the diagram (see left) coincide (see right). There are many details to the mechanism.   Note in particular, however, that this shift results in a valine side chain near the heme in each β subunit being reoriented away from the heme groups increasing their accessibility for O_2 binding.   (Reproduced with permission from "X-ray Studies of Protein Mechanisms," *Ann. Rev. Biochem.,* **41,** 815 (1972).   Drawings generously supplied by Richard E. Dickerson.)

ity in the deoxyhemoglobin molecule.   Consistent with the highly negatively charged state of DPG, the binding site is composed of six basic residues (positively charged) contributed by the two β subunits.   Thus, DPG decreases O_2 affinity by cross-linking the β subunits and pushing them away from each other, which in turn contributes to a less accessible orientation of the heme groupings.

LITERATURE (see also Chap. 5 references)

AULT, A.   "An Introduction to Enzyme Kinetics."   *J. Chem. Ed.,* **51,** 381–386 (1974).

BERNHARD, S.   *The Structure and Function of Enzymes.* New York: W. A. Benjamin, Inc., 1968.   An introductory treatment of enzymes with emphasis on kinetics and reaction mechanisms.

BLOW, D. M.   "Structure and Mechanism of Chymotrypsin."   *Accounts Chem. Res.,* **9,** 145–152 (1976).   A review article.

BLOW, D. M., and T. A. STEITZ.   "X-Ray Diffraction Studies of Enzymes."   *Annual Review of Biochemistry,* **39,** 63–100 (1970).   A good review of structural details of seven enzymes in terms of the mechanism and specificity of their catalytic action.

BOYER, P. D., H. LARDY, and K. MYRBACK, eds. *The Enzymes.* 2nd and 3rd editions. New York: Academic Press. A valuable multivolume work covering most aspects of biocatalysis with emphasis on reaction types. Volume 1 (second edition, 1960) is devoted to fundamentals of enzyme catalysis. Volumes 1 and 2 (third edition, 1970) contain the latest information on general principles, including structure, control, kinetics, and mechanism.

CHAN, S. J., and D. F. STEINER. "Preproinsulin, a New Precursor in Insulin Biosynthesis." *Trends Biochem. Sci.,* **2,** 254–252 (1977). A short review article.

DAWES, E. A. *Quantitative Problems in Biochemistry.* Baltimore: Williams and Wilkins, 1967. A problem-oriented textbook covering various physicobiochemical subjects. Chapters 5 and 6 are devoted to reaction kinetics and enzyme kinetics.

DIXON, M., and E. C. WEBB. *Enzymes.* 2nd edition. New York: Academic Press, 1964. A textbook dealing exclusively with enzymes. Although somewhat out of date, it is still an excellent reference for a detailed coverage of general principles.

GUTFREUND, H. *Enzymes: Physical Principles.* New York: John Wiley, 1972. A short book covering many aspects of enzymes from the standpoint of physical chemistry. Available in paperback.

KOSHLAND, D. E. "Correlation of Structure and Function in Enzyme Action." *Science,* **142,** 1533–1541 (1963). A review article defining the nature of the catalytic process from the standpoint of protein structure. Included is a definitive summary of the induced-fit theory, of which the author is the chief proponent.

KOSHLAND, D. E., and K. E. NEET. "The Catalytic and Regulatory Properties of Enzymes." *Ann. Rev. Biochem.,* **38,** 359–410 (1968). An excellent review article.

KOSHLAND, D. E. "Protein Shape and Biological Control." *Scientific American,* **230,** 52–64 (1973). Excellent article discussing allosterism, with an explanation according to the principles of induced-fit theory.

LIPSCOMB, W. N. "Structure and Mechanism in the Enzymatic Activity of Carboxypeptidase A and Relations to Chemical Sequence." *Accounts Chem. Res.,* **3,** 81–89 (1970). A review article.

MONOD, J., J. P. CHANGEUX, and F. JACOB. "Allosteric Proteins and Cellular Control Systems." *J. Mol. Biol.,* **6,** 306–329 (1963). The original article describing the theory of allosterism.

MONOD, J., J. WYMAN, and J. P. CHANGEUX. "On the Nature of Allosteric Interactions: A Plausible Model," *J. Mol. Biol.,* **12,** 88–118 (1965). Original article explaining the formulation of the two-state equilibrium model for enzyme regulation.

NORD, F. F., ed. *Advances in Enzymology.* New York: John Wiley-Interscience. A publication composed of annual volumes (since 1942) devoted to reviewing progress in enzymology. The articles, written by authorities, deal with general and specific subjects. This is an extremely useful reference work for researchers, teachers, students, and writers of biochemistry textbooks. Stadtman's article on regulatory mechanisms, cited below, is a particularly good example.

PISZKIEWICZ, D. *Kinetics of Chemical and Enzyme-catalyzed Reactions.* New York: Oxford University Press, 1977. A short book on reactions kinetics. Available in paperback.

SEGAL, H. L. "Enzymatic Interconversion of Active and Inactive Forms of Enzymes." *Science,* **180,** 25–32 (1973). A review article.

SEGEL, I. H. *Enzyme Kinetics.* New York: John Wiley, 1975. A thorough treatment of many subjects. A condensed version is included in Segel's problems book, which is referenced at the end of Chap. 3.

STADTMAN, E. R., "Allosteric Regulation of Enzyme Activity," *Advances in Enzymology,* **28,** 41–154 (1966). A thorough review of regulatory mechanisms.

STROUD, R. M. "A Family of Protein-Cutting Proteins." *Scientific American,* **231,** 74–88 (1974). An excellent article discussing the structure and function of serine-containing proteolytic enzymes, including chymotrypsin.

References for ATCase

CHANGEUX, J. P., J. C. GERHART, and H. K. SCHACHMAN. "Allosteric Interactions in Aspartate Transcarbamylase: Binding of Specific Ligands to the Native Enzyme and its Isolated Subunits." *Biochemistry,* **7,** 531–538 (1968). A research article describing the binding properties of catalytic and regulatory subunits and the first proposed model of the quaternary structure of the native enzyme.

GERHART, J. C. "A Discussion of the Regulatory Properties of Asparatate Transcarbamylase from *Escherichia coli,*" in *Current Topics in Cellular Regulation,* B. L. Horecker and E. R. Stadtman, eds. **2,** 276–326 (1970). A thorough review article.

GERHART, J. C., and PARDEE, A. B. "The Enzymology of Control by Feedback Inhibition." *J. Biol. Chem.,* **237,** 891–896 (1962). The first study providing evidence for the existence of secondary sites in enzymes and their role in the modulation of catalytic activity.

MARKUS, G., D. K. MCCLINTOCK, and J. B. BUSSEL. "Conformational Changes in Aspartate Transcarbamylase: A Functional Model for Allosteric Behavior." *J. Biol. Chem.,* **246,** 767–771 (1971). A research paper offering the most recent proposal on the structural organization of ATCase and its relationship to the allosteric properties of the enzyme.

ROSENBUSCH, J. P., and K. WEBER. "Subunit Structure of Aspartate Transcarbamylase from *Escherichia coli.*" *J. Biol. Chem.,* **246,** 1644–1657 (1971). Experimental evidence for a hexameric structure of the native protein.

WARREN, S. G., B. F. P. EDWARDS, D. R. EVANS, D. C. WILEY, and W. N. LIPSCOMB. "Aspartate Transcarbamylase from *Escherichia coli:* Electron Density at 5.5Å Resolution." *Proc. Nat. Acad. Sci. USA,* **70,** 1117–1121 (1973). A description of the three-dimensional structure of the hexameric protein.

6–1. From the final form of the Michaelis-Menten equation given on p. 169, prove that K_m is equal to the substrate concentration when the enzyme is 50% saturated.

6–2. The enzyme glutamic acid dehydrogenase catalyzes the reaction:

$$\underset{\text{L-glutamate}}{\overset{\overset{\displaystyle NH_3^+}{\displaystyle |}}{{}^-OOCCH_2CH_2CHCOO^-}} + NAD^+ \longrightarrow$$

$$\underset{\alpha\text{-ketoglutarate}}{\overset{\overset{\displaystyle O}{\displaystyle \|}}{{}^-OOCCH_2CH_2CCOO^-}} + NH_3 + NADH + H^+$$

The dependence of initial velocity on the concentration of L-glutamate is given in the table below. The initial concentration of NAD^+ was held constant in each case. Calculate the K_m and V_{max} of the enzyme by the classical Michaelis-Menten plot and also by the Lineweaver-Burk method, and compare the values. (*Note:* The reaction velocity is expressed in terms of the rate of change in the absorbance at 360 nm. This is a measure of the rate of product formation, specifically NADH formation. See p. 57, Chap. 2, for a review of the absorbance of NAD^+ and NADH.)

L-glutamate concentration (millimolar)	Initial velocity (change in absorbance at 360 nm/min)
1.68	0.172
3.33	0.250
5.00	0.286
6.67	0.303
10.0	0.334
20.0	0.384

6–3. For the system described in Exercise 6-2, what is the value of the dissociation constant for the dehydrogenase and L-glutamate? What is the value for the affinity constant?

6–4. Glycogen is a polymeric carbohydrate composed of several glucose residues. Glycogen synthetase is the enzyme responsible for lengthening the glycogen molecule according to the following reaction. The structures of

$$\text{UDP-glucose} + (glucose)_n \rightarrow (glucose)_{n+1} + \text{UDP}$$
$$\text{acceptor} = \text{chain lengthened}$$

UDP-glucose (the metabolically active substrate form of glucose in this reaction) and glycogen will be considered in Chap. 9. The data below can, however, be analyzed without this knowledge. The results of two separate studies on the catalytic action of glycogen synthetase are given under conditions where the concentration of

UDP-glucose was varied and the concentration of the acceptor polyglucose was held constant. In one case no further additions were made, while in the other, ATP was added (2.5 mM). From a Lineweaver-Burk graphical analysis, determine the apparent type of inhibition by ATP on the activity of the enzyme.

	Initial velocity $\mu moles\ (glucose)_{n+1}\ formed/min$	
UDP-glucose (millimolar)	No additions	ATP added (2.5 mM)
0.8	10.0	2.0
1.4	12.5	3.3
3.3	22.0	6.7
5.0	25.0	10.0

6–5. In the presence of a saturating amount of substrate, 1.4 milligrams of an enzyme results in a velocity of 6.2 millimoles of product formed per minute. If the molecular weight of the enzyme is 52,500 grams/moles, what is the molecular activity of the enzyme?

6–6. The catalytic activity of most enzymes is a sensitive function of pH. A plot (activity versus pH) of the data given below will clearly illustrate the degree of this sensitivity. What is the optimum pH for the enzyme in question?

units of enzyme activity	24.8	33.0	66.7	56.2	41.3	27.5
pH	7.2	7.6	8.0	8.6	8.8	9.0

6–7. After a preincubation with *p*-chloromercurobenzoate, the binding of an enzyme with its substrate was no different from that of the untreated enzyme, but the catalytic activity of the treated enzyme was found to be 40% less. What conclusion can be drawn from this type of observation?

6–8. As exercises in algebra (for a change of pace) perform the following:
(a) convert Eq. (6–4) to the Michaelis-Menten equation
(b) convert the Michaelis-Menten equation to the Lineweaver-Burk equation

6–9. Given the preferred specificity of carboxypeptidase B (see margin on p. 181) and the proposed mechanism of action for carboxypeptidase A (p. 183), propose a likely mechanism of action for carboxypeptidase B.

6–10. The relationship of enzyme activity to the ionic strength of the medium is often similar in form to the relationship of activity to temperature. Given this statement, which of the following curves would best depict the relationship of activity to ionic strength?

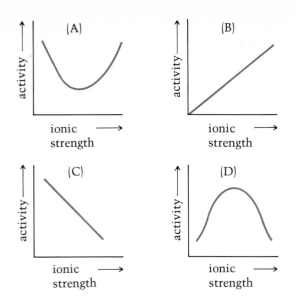

(A) activity — ionic strength →

(B) activity — ionic strength →

(C) activity — ionic strength →

(D) activity — ionic strength →

centration of glucose-6-phosphate was constant and that of $NADP^+$ was varied. (Rose bengal is an organic dye that has been used successfully as a probe into the mechanism of action of certain enzymes. Your analysis of the data will partially indicate why this is so. Note the extremely low concentration of rose bengal used in these studies.) Briefly discuss the data in view of the fact that rose bengal forms strong complexes with imidazole-containing substances.

6–13. For the system described in Exercise 6–12, calculate the K_I for rose bengal.

6–14. Phosphofructokinase is an enzyme, widespread in nature, that catalyzes the reaction given below.

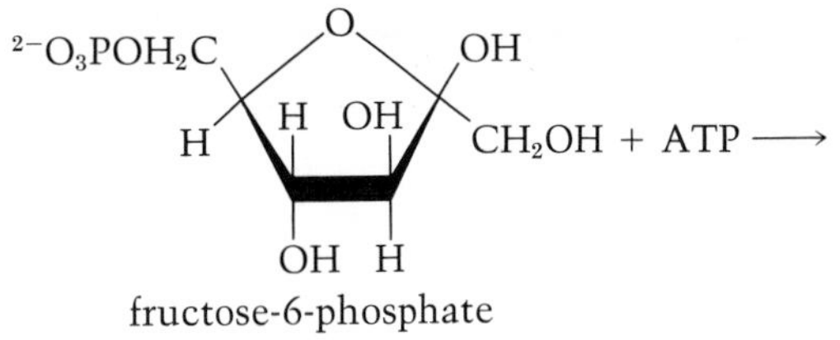

$^{2-}O_3POH_2C$ ... O ... OH

H H OH CH_2OH + ATP ⟶

OH H

fructose-6-phosphate
(F6P)

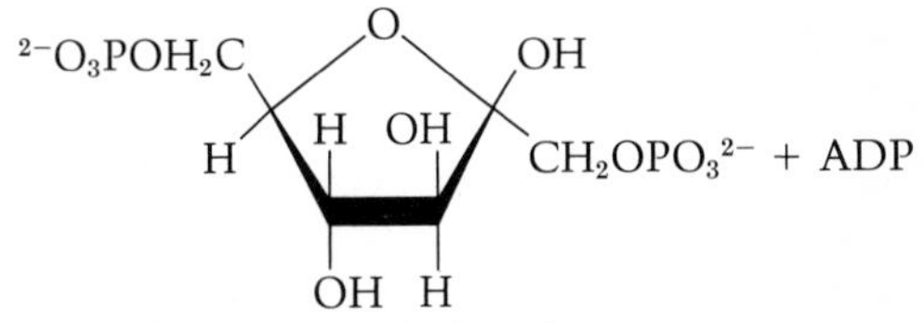

$^{2-}O_3POH_2C$ ... O ... OH

H H OH $CH_2OPO_3^{2-}$ + ADP

OH H

fructose-1,6-diphosphate
(FDP)

6–11. An enzyme is competitively inhibited to different degrees by the same concentration of three separate inhibitors, I_1, I_2, and I_3. The inhibitor dissociation constants are $K_{I_1} = 0.1$ mM, $K_{I_2} = 0.01$ mM, and $K_{I_3} = 1.0$ mM. Which inhibitor would probably produce the greatest inhibition? the least inhibition?

6–12. The enzyme 6-phosphogluconate dehydrogenase catalyzes the oxidative decarboxylation of 6-phosphogluconate according to the following reaction (see Chap. 12, p. 404):

6-phosphogluconate + $NADP^+$ →
ribulose-5-phosphate + CO_2 + NADPH + H^+

From the following kinetic data, determine what effect rose bengal has on the activity of the dehydrogenase. The data were obtained under conditions where the con-

Under conditions where the concentration of ATP was held constant, the dependency of the initial velocity on the concentration of fructose-6-phosphate as measured at two different pHs (6.9 and 8.2) is represented by the data on the next page. Plot the data according to the classical Michaelis-Menten method and interpret what effect the pH has on the activity of the enzyme.

6–15. What type of Michaelis-Menten plot would be obtained if the concentration of CTP were increased above that specified in the legend of Fig. 6-12? Explain the reason for this in terms of the two-state model for allosteric effects.

6–16. According to the principles of the induced-fit model, draw a series of diagrams as shown on p. 196 explaining the following: a dimeric allosteric protein has separate substrate, inhibitor, and activator sites; the inhibitor displays negative cooperativity toward itself; the inhibitor inhibits both substrate and activator binding.

| | Initial velocity (units of enzyme activity/min) | |
Concentration of $NADP^+$ (10^{-5} M)	No additions	+ rose bengal (1.5×10^{-5} M)
2	2.12	1.11
3	2.70	1.47
4	2.94	1.73
6	3.33	2.32
10	3.85	2.78

Fructose-6-phosphate (millimolar)	0.3	0.5	0.8	1.0	1.3	1.5	1.7	2.5	3.5
Initial velocity (micromoles F6P reacted/min) pH 6.9	1.0	2.7	3.0	12	22	46	68	77	80

Fructose-6-phosphate (millimolar)	0.02	0.04	0.05	0.07	0.1	0.15	0.25	0.40
Initial velocity (micromoles F6P reacted/min) pH 8.2	35	47	60	68	80	90	92	95

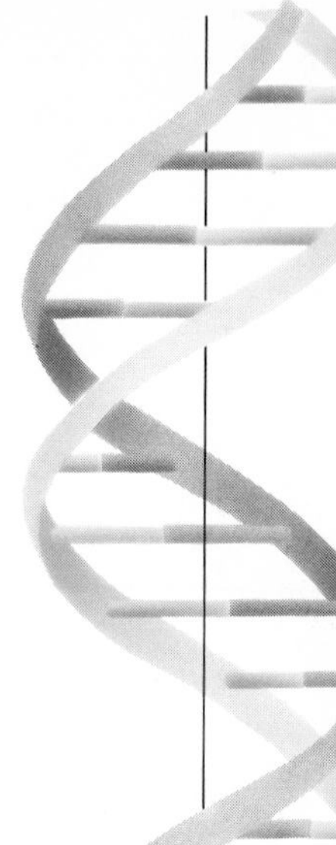

CHAPTER 7

NUCLEOTIDES AND NUCLEIC ACIDS

The two types of nucleic acids found in every living organism are *ribonu-cleic acid (RNA)* and *deoxyribonucleic acid (DNA)*. Viruses, on the other hand, contain only one type, either RNA or DNA. The biological functions of nucleic acids include the *storage, replication, and transmission of genetic information*. In short, they are the molecules that determine what a living cell is and what a living cell does. An important role, to say the least.

All nucleic acids are large polymeric substances, but their sizes cover a wide spectrum. *Transfer-RNA*, a particular type of RNA and the smallest known nucleic acid, has a molecular weight of approximately 25,000, whereas individual molecules of DNA, which constitute some of the largest single molecules yet known, range in molecular weight from 1,000,000 to 1,000,000,000!

Both RNA and DNA are composed of monomeric units called *nucleo-tides*, and hence a nucleic acid is also a *polynucleotide*. A single nucleotide consists of three chemical parts—inorganic phosphate, a simple sugar, and what is called a nitrogen base—covalently attached to each other in the fol-lowing order: phosphate—sugar—nitrogen base. (P, S, and NB in the dia-gram.) In a nucleic acid, successive nucleotides are linked together via

In the 1860s F. Miescher isolated an acidic substance from cell nu-clei that he termed nuclein and later nucleic acid. The first revo-lutionary discovery occurred in the 1940s when Avery, MacLeod, and McCarty established that nucleic acid material—specifically DNA—was responsible for carrying hered-itary information. When the solu-tion for the molecule structure of DNA was reported in 1953 by Watson and Crick, a new era in biology began.

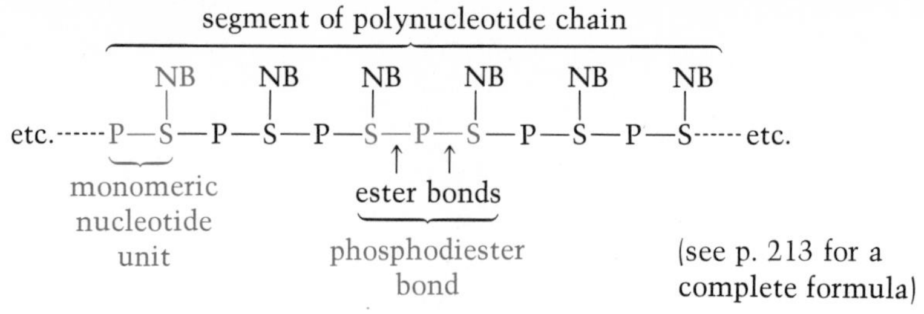

segment of polynucleotide chain

```
            NB     NB     NB     NB     NB     NB
            |      |      |      |      |      |
etc.------P—S—P—S—P—S—P—S—P—S—P—S------etc.
        ‿‿‿                ↑    ↑
     monomeric           ester bonds
     nucleotide      ‿‿‿‿‿‿‿‿‿
        unit          phosphodiester        (see p. 213 for a
                          bond                complete formula)
```

phosphodiester bonds between the sugar and phosphate parts of adjacent nucleotides. The nitrogen bases are not involved in any covalent linkages other than their attachment to the sugar-phosphate backbone.

It will become clear that it is the *sequence of nitrogen bases* along the invariant sugar-phosphate backbone that *constitutes the unique structural and functional individuality of DNA and RNA molecules.* In fact, the entire genetic language of DNA is contained in the sequence of nitrogen bases.

Although there is still much to be learned, progress (particularly in the past 15–20 years) in understanding the biochemistry of DNA and RNA has been nothing short of remarkable. The most fantastic development has been the recent development of the laboratory technology to modify—by design—the composition of genes in DNA. The age of *genetic engineering* has begun.

NUCLEOTIDES

As we did in the discussion of the amino acids and proteins, we must begin any coherent discussion of the nucleic acids with a consideration of the monomeric nucleotides. A further reason for study is that aside from their occurrence in nucleic acids, some nucleotides are of extreme importance in other processes. We will begin by considering each of the distinct parts and how they are linked together.

Ribose and Deoxyribose

The names of RNA and DNA are derived from the type of sugar in the nucleotide monomers. D-*Ribose* is present in ribonucleic acid, whereas 2-D-*deoxyribose* is present in deoxyribonucleic acid. Each sugar contains five carbon atoms (thus they are called pentoses) and functional hydroxyl (—OH) groups. 2-Deoxyribose contains one fewer hydroxyl group, specifically at the carbon-2 position (deoxy—without hydroxyl). The structure of each sugar is shown below in both linear and cyclic formulas, the latter referred to as Haworth representations. The cyclic form more correctly depicts the real structure of the sugar in solution, which is nearly that of a planar, five-membered ring. (See p. 292 in Chap. 9 for additional information regarding the cyclic structures of the simple sugars.) The accompanying systematic names of the sugars contain a -*furanose* suffix because of the structural similarity to *furan*, itself a five-membered, oxygen-containing, cyclic molecule (see margin).

The inclusion of the Greek letter β in the systematic name refers to a subtle but significant structural feature. Specifically, it designates *one of two possible orientations of the hydroxyl group at carbon-1.* As viewed in the

furan

Linear structures	Cyclic structures	Haworth representation

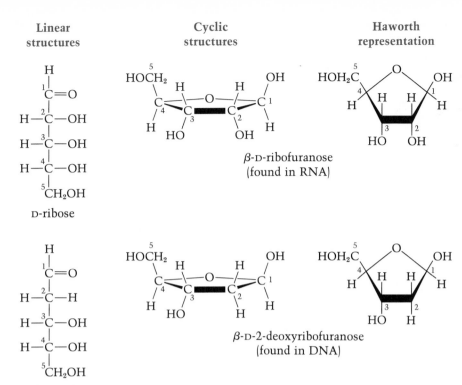

β-D-ribofuranose
(found in RNA)

D-ribose

β-D-2-deoxyribofuranose
(found in DNA)

D-2-deoxyribose

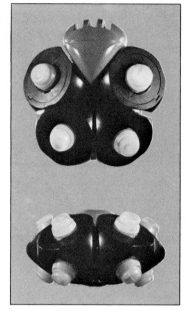

Haworth structure, the β (beta) orientation has the C¹—OH directed upwards from the plane of the ring. The other orientation, corresponding to a downward direction of the C¹—OH, is designated α (alpha) (see margin.) This difference represents two different stereoisomeric forms of ribofuranose and deoxyribofuranose, and all other simple sugars for that matter. The significance of this aspect of structure for the carbohydrates in general will be considered in a later chapter. For now, it is significant for us to note that *only the β forms* of D-ribofuranose and D-deoxyribofuranose are found in RNA and DNA.

α-D-ribofuranose
(not found in RNA; same for deoxyribose and DNA)

Nitrogen Bases (Purines and Pyrimidines)

The nitrogen bases found most frequently in RNA and DNA are *adenine* (A), *guanine* (G), *cytosine* (C), *thymine* (T), and *uracil* (U). A and G are substituted *purines;* C, T, and U are substituted *pyrimidines.* Their structures, systematic names, and occurrence in nucleic acids are given below. The possibility of *tautomeric forms* of each base (with the exception of adenine) is also indicated. In this regard, it is important to note that the *keto isomer predominates* in nature. Another distinguishing structural feature of these molecules evident from the space-filling models is the *coplanarity of all atoms in the ring* and of any atom immediately attached to the ring.

Examples of some nitrogen bases occurring less frequently in nature are shown below. Some, such as *5-methylcytosine,* are minor components of DNA, while others, such as *4-thiouracil* and *dihydrouracil,* are just two of several minor components of RNA, and specifically of transfer-RNA. Others of synthetic rather than natural origin are exemplified by *5-bromouracil* (a potent mutagenic agent) and *6-mercaptopurine* (an anti-

Space-filling models appear on p. 206.

205

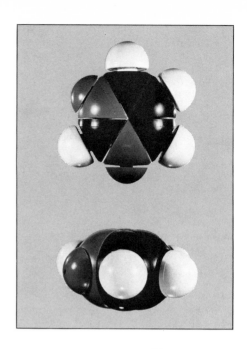

top
views

purines
and
pyrimidines
are
flat
molecules

side
views

purine

pyrimidine

Major purines and pyrimidines of nucleic acids

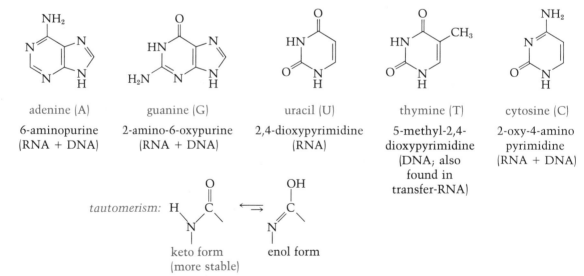

adenine (A)

6-aminopurine
(RNA + DNA)

guanine (G)

2-amino-6-oxypurine
(RNA + DNA)

uracil (U)

2,4-dioxypyrimidine
(RNA)

thymine (T)

5-methyl-2,4-
dioxypyrimidine
(DNA; also
found in
transfer-RNA)

cytosine (C)

2-oxy-4-amino
pyrimidine
(RNA + DNA)

tautomerism:

keto form
(more stable)

enol form

tumor agent).   Unlike the amino acids, the nitrogen bases do not occur in
nature in the free state to any great extent.   Rather, they are found largely in
covalent linkage with ribose or deoxyribose as nucleosides and nucleotides.

Due to their heterocyclic, aromatic nature, the purines and pyrimidines
characteristically *absorb energy in the ultraviolet (UV) region* of the electromagnetic spectrum.   Although each substance has a unique absorption
spectrum, each shows maximum absorption at or close to 260 nm (see Fig.

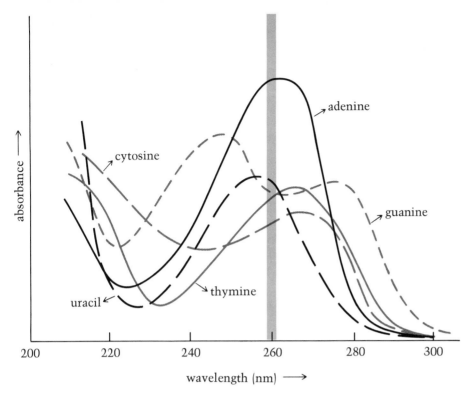

FIGURE 7–1 Ultraviolet absorption spectra of the major purines and pyrimidines. Absorption measurements at 260 nm are used for detection and quantitative determination of nucleosides, nucleotides, and nucleic acids. This characteristic absorption is also the basis for viewing living tissue in ultraviolet microscopy, which permits the observation of the cellular location of RNA and DNA, and for the mutagenic effect of UV radiation (see p. 261). The spectra in the figure were obtained with solutions having the same concentration and a pH of 7. As you can see, adenine is the strongest absorber and cytosine the weakest.

7–1). Accordingly, nucleic acids also absorb UV light—a property that is commonly used for their detection and quantitative measurement. It is also the basis for the mutagenic effect of UV radiation (see p. 261).

Other nitrogen bases

5-methyl-cytosine (DNA)   4-thiouracil (in some transfer-RNAs)   dihydrouracil (in some transfer-RNAs)   6-mercaptopurine   5-bromouracil

Phosphate

In the vast majority of cases when phosphorus occurs in an organic biomolecule it is present as part of (a) a mono- or diester linkage, (b) a mono- or dianhydride linkage, or (c) both ester and anhydride linkages. We will encounter each type in the next few pages. (See structures, p. 208, top.)

Nucleoside Structure

The sugar-nitrogen base portion of a nucleotide is called a *nucleoside*. The covalent linkage usually (see below) involves the C^1 atom of the sugar and the N^1 atom of a pyrimidine or the N^9 atom of a purine. According to carbo-

monoester linkage 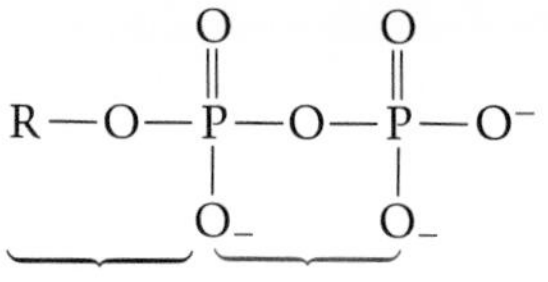   ester   anhydride 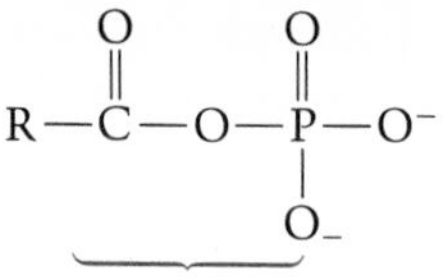   mixed anhydride

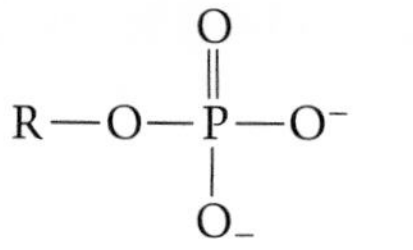

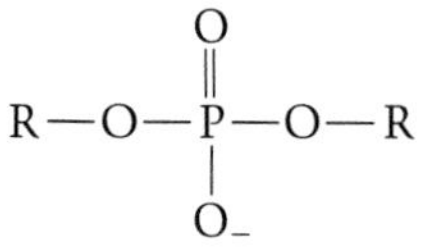

diester linkage 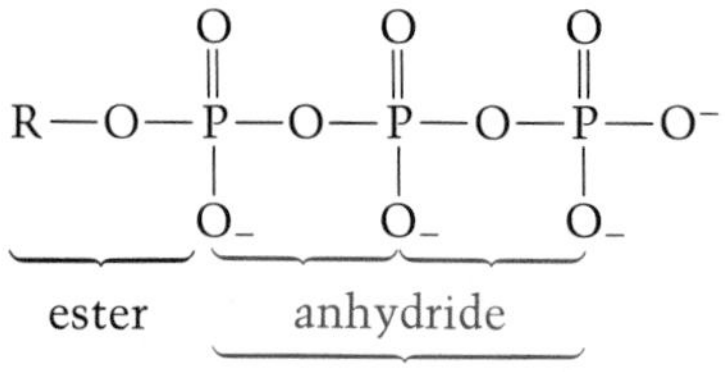   ester   anhydride

triphosphoanhydride

hydrate nomenclature, the linkage is called a *glycoside bond.* To avoid confusion in numbering, the atoms of the carbohydrate unit are differentiated by a prime superscript. The accepted trivial names of the *common ri-*

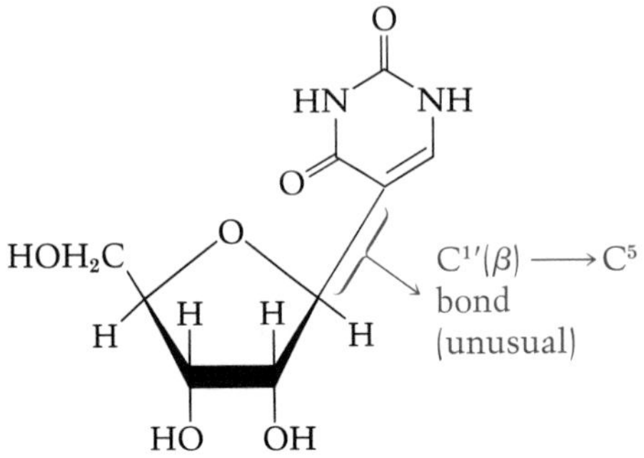

$C^{1'}(\beta) \longrightarrow C^5$ bond (unusual)

This is an important exception to the usual $C^{1'}$—N bond in nucleosides; pseudouridine is found in transfer-RNA molecules

5-β-D-ribofuranosyluracil
(also called *pseudouridine*)

bonucleosides are *adenosine*, *guanosine*, *uridine*, and *cytidine*. The *common deoxyribonucleosides* are *deoxyadenosine*, *deoxyguanosine*, *deoxycytidine*, and *thymidine*. (Thymidine is not prefixed because of its infrequent occurrence as an N-riboside; it would be named ribosylthymine.) On a more systematic basis, the nucleosides are named as β-D-ribosyl or β-D-2'-deoxyribosyl derivatives of the purine or pyrimidine, as indicated below.

A pyrimidine ribonucleoside

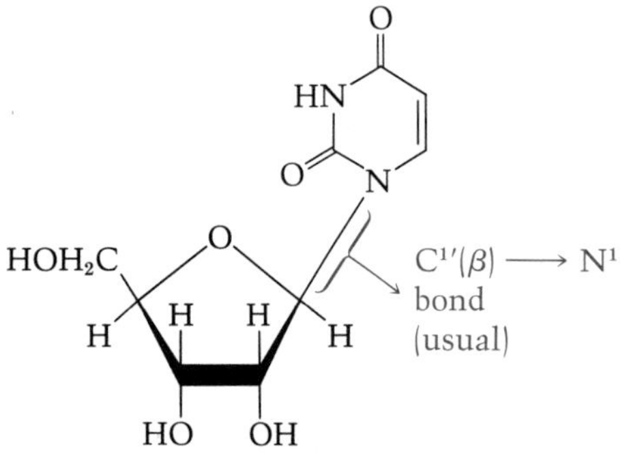

$C^{1'}(\beta) \longrightarrow N^1$ bond (usual)

uridine
1-β-D-ribofuranosyluracil

A purine deoxyribonucleoside

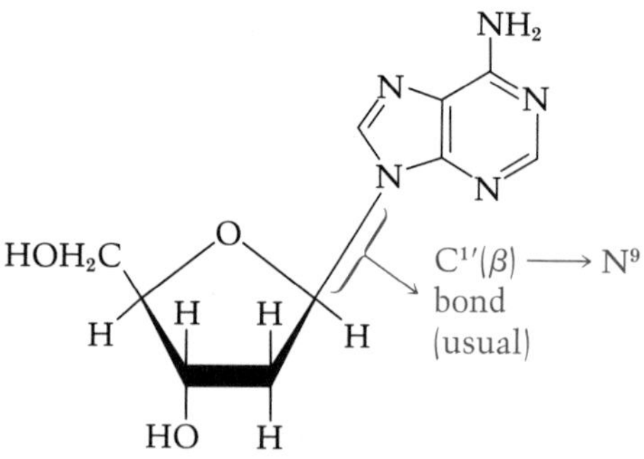

$C^{1'}(\beta) \longrightarrow N^9$ bond (usual)

deoxyadenosine
9-β-D-2'-deoxyribofuranosyladenine

Although we have described nucleosides as part of the nucleotide structure, some free nucleosides exist. Several have valuable clinical use. For example, a *Streptomyces* organism excretes *puromycin*, a potent antibiotic that operates by functioning as an inhibitor of protein biosynthesis (see p. 270). Various microorganisms produce *arabinosyl cytosine (ara-C)* and *arabinosyl adenine (ara-A)*, which contain the pentose β-D-arabinose rather than ribose. Both substances have attracted a lot of attention because they have proven to be successful as potent antiviral and antifungal agents without having any adverse effect on the host cell. The use of ara-A, ara-C, and related materials as effective treatments in the cure of various crippling and fatal diseases of viral origin, including some cancers, is an exciting prospect. The mode of action of ara-A and ara-C is based on their inhibition of DNA biosynthesis (see p. 255 in Chap. 8).

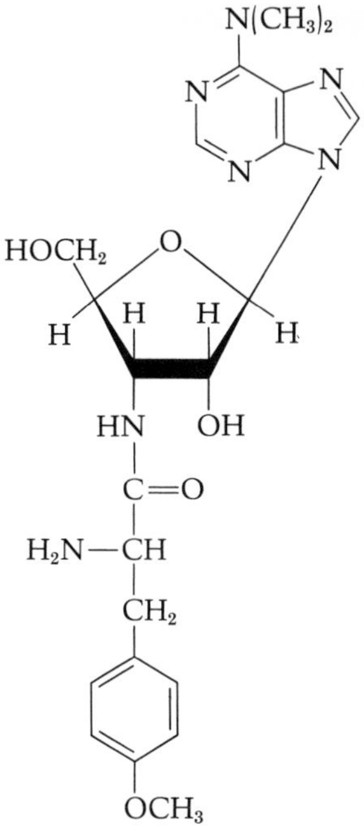

puromycin

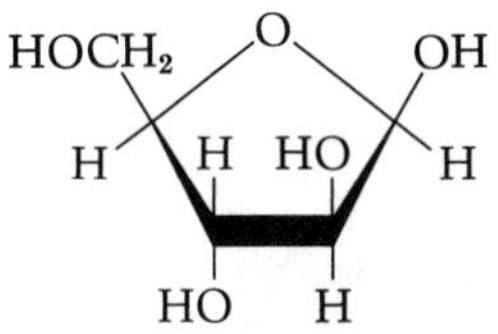

β-D-arabinose
(furanose form)

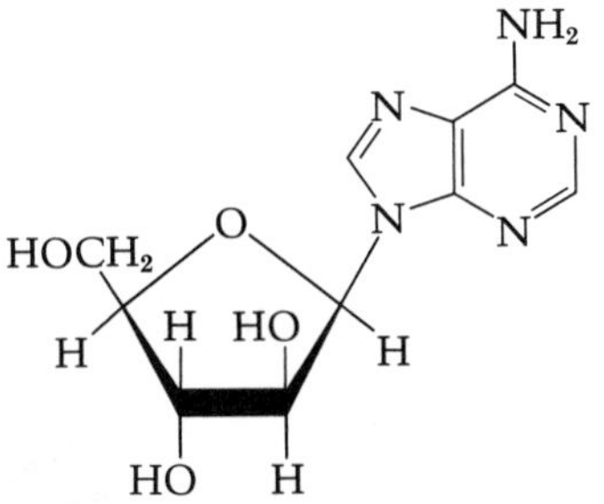

9-β-D-arabinofuranosyladenine
ara-A

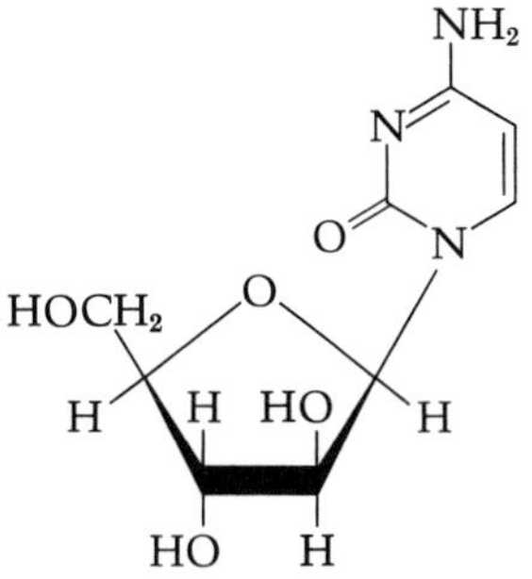

1-β-D-arabinofuranosylcytidine
ara-C

Nucleotide Structure

A nucleotide consists of a phosphate group in ester linkage to the sugar unit of a nucleoside. Usually the phosphate bond involves the 5′ position on the pentose. Depending on the identity of the pentose, all nucleotides can be categorized as either ribonucleotides or deoxyribonucleotides. Then, according to the number of phosphate residues present, nucleosides can be further subclassified as *monophospho*nucleotides, *diphospho*nucleotides, or *triphospho*nucleotides. In diphospho- and triphosphonucleotides the phosphate groups are covalently attached in linear sequence by phosphoanhydride bonds. All types have a universal occurrence. (See p. 210, top.)

Specific nucleotides are named in a variety of ways. One method considers them as phosphate esters. Thus, the adenosine family would consist of adenosine-5′-monophosphate (5′-AMP), adenosine-5′-diphosphate (5′-ADP), and adenosine-5′-triphosphate (5′-ATP), with the expression in parentheses being an obvious shorthand notation. The deoxy counterparts would be named deoxyadenosine-5′-monophosphate (5′-dAMP), and so on. Alternatively, the monophosphonucleotides are sometimes named as acyl acids of the parent nucleoside due to the presence of the acidic phosphate group. Some examples are *adenylic acid, deoxyadenylic acid, uridylic acid, thymidylic acid,* and so on.

Cyclic Nucleotides

Adenosine-3′,5′-cyclic monophosphate (cyclic-AMP or just cAMP) is a universally occurring nucleotide of immense biological importance. Some have called it the "wonder-molecule." It was first isolated in 1959 by E.

**mono-, di-, triphosphonucleotides (5′)
of adenosine**

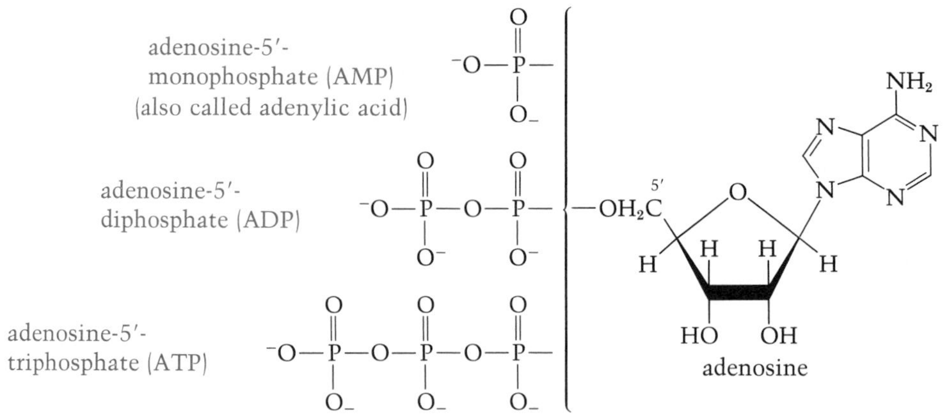

adenosine-5′-
monophosphate (AMP)
(also called adenylic acid)

adenosine-5′-
diphosphate (ADP)

adenosine-5′-
triphosphate (ATP)

adenosine

Sutherland (Nobel Prize, 1971) and coworkers as part of investigations into the mechanism of action of certain hormones, such as adrenalin, in regulating carbohydrate metabolism. On the basis of their study they proposed that the immediate action of adrenalin and many other hormones is to activate the enzyme that is responsible for the production of cAMP. In turn the

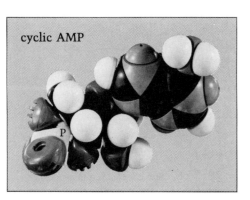

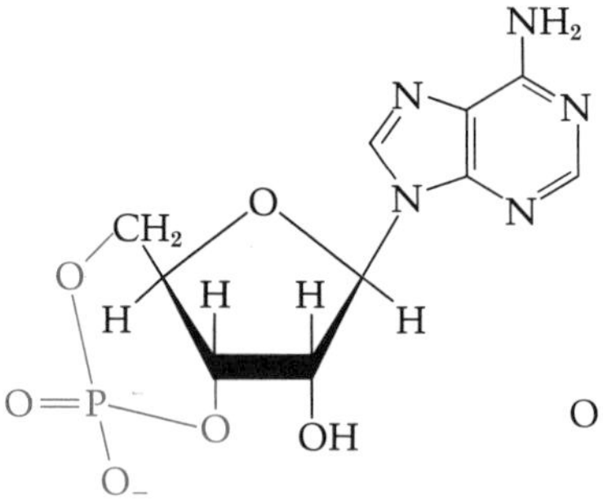

cyclic AMP

adenosine-3′,5′-cyclic
monophosphate
3′,5′-cyclic AMP

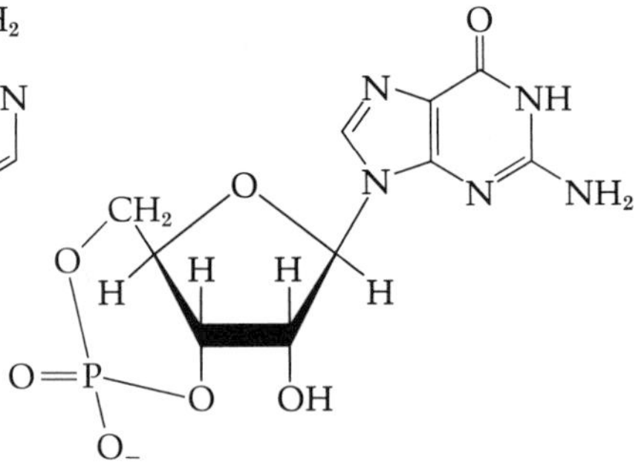

guanosine-3′,5′-cyclic
monophosphate
3′,5′-cyclic GMP

cAMP acts to control the activity of other enzymes, frequently by an allosteric activation. In other words, the cAMP was proposed to act as an *intermediate regulator* of the primary hormone regulator. Sutherland labeled this role of cAMP as that of a *secondary messenger*. Since then cyclic AMP has been implicated in many diverse processes in all types of organisms and cells.

Approximately 10 years later, N. Goldberg and J. Larner discovered the natural occurrence of another cyclic nucleotide, *guanosine-3',5'-cyclic monophosphate* (cyclic GMP), in several types of mammalian tissues. Its presence had gone undetected because of the attention to cAMP and also because its level in cells is only about one one-hundredth that of cAMP. Since then cGMP has been detected in many cells of animal, plant, and microbial origin. Although the biological role of cGMP has not been categorically established, evidence is mounting that it functions in the same fashion as cAMP, but *in reverse*. That is to say, cAMP functions primarily as an activator of enzymes, cGMP as a deactivator of the enzymes. In fact, in some cases the two materials exert their opposite effects on the same enzyme. It has been further suggested that the intracellular balance of cAMP and cGMP levels may have some clinical application in human medicine if certain disease states are related to the presence of improper balances and if it is possible to restore these balances to normal. In Chap. 12 we will have more to say about the cyclic nucleotides, and their specific involvement in carbohydrate metabolism.

The biochemistry of cyclic nucleotides must now include a third substance as a result of the recent isolation of cytidine-3',5'-cyclic monophosphate (cyclic CMP).

The cyclic nucleotides are formed from the corresponding triphosphonucleotides by the enzymes *adenylate cyclase* and *guanylate cyclase*. Both enzymes are *localized in the cell membrane*. *Cyclic phosphodiesterase*, a cytoplasmic enzyme, acts to catalyze the degradation of the cyclic diester on the 3' side, yielding the corresponding 5'-monophosphonucleotides.

$$5'\text{-ATP} \xrightarrow[\text{cyclase} \searrow]{\text{adenylate}} 3',5'\text{-cAMP} \xrightarrow[\text{phosphodiesterase}]{\overset{\text{H}_2\text{O}}{\underset{\text{cyclic}}{\frown}}} 5'\text{-AMP}$$
$$\text{PP}_i$$

$$5'\text{-GTP} \xrightarrow[\text{cyclase} \searrow]{\text{guanylate}} 3',5'\text{-cGMP} \xrightarrow[\text{phosphodiesterase}]{\overset{\text{H}_2\text{O}}{\underset{\text{cyclic}}{\frown}}} 5'\text{-GMP}$$
$$\text{PP}_i$$

ATP AND BIOENERGETICS—AN INTRODUCTION

In addition to their being utilized in nucleic acid biosynthesis (see the next chapter), the triphosphonucleotides participate in many enzyme-catalyzed reactions involved in the metabolism of all types of compounds. In some cases their involvement is rather specific. For example, CTP participates in phospholipid biosynthesis (see p. 518), and UTP functions in the biosynthesis and interconversions of various carbohydrates (see p. 297).

While all of these functions are important to normal cellular processes, there is one *central role of ATP that is of extreme importance: in all living cells ATP functions as "a receiving and shipping department of chemical energy"* (at the molecular level, of course). A preliminary and compact explanation of this function is summarized in the diagram below, which shows the interconversion of ATP and ADP. Chemically speaking, the nature of these conversions is quite simple: a phosphoanhydride bond is

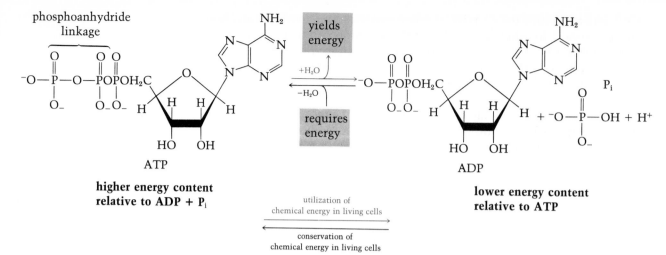

phosphoanhydride linkage

ATP

higher energy content relative to ADP + P$_i$

yields energy

requires energy

utilization of chemical energy in living cells

conservation of chemical energy in living cells

ADP

lower energy content relative to ATP

Note: ATP is also used as an energy source in reactions wherein AMP and PP$_i$ are produced rather than ADP and P$_i$. The same principles apply except that *in vivo* ATP is not directly reformed from AMP. The return route is

AMP $\longrightarrow$ ADP $\longrightarrow$ ATP. The fate of pyrophosphate (PP$_i$) is: PP$_i$ $\xrightarrow{\text{H}_2\text{O}}$ 2P$_i$. This yields additional energy (see p. 258). To summarize then,

$$H_2O + ATP \rightleftharpoons ADP + P_i$$
and
$$\uparrow +ATP \quad \uparrow +H_2O$$
$$H_2O + ATP \longrightarrow AMP + PP_i$$

broken during hydrolysis (ATP degradation) and formed during phosphorylation (ATP synthesis). This, however, is only half of the description. The other half involves the energetics of the two processes. In this regard, the simple fact is that ATP is a much less stable molecule relative to ADP and P$_i$. In other words, *ATP represents a higher energy state than ADP and P$_i$*. Thus, the ATP $\rightarrow$ ADP + P$_i$ conversion *is accompanied by a release of energy* and the ADP + P$_i$ $\rightarrow$ ATP conversion *requires a utilization of energy*. A similar description also applies to the hydrolysis of ATP to AMP and PP$_i$ (pyrophosphate).

An explanation of why ATP represents a higher energy state relative to ADP and P$_i$ is given on p. 354.

Simply put, the ATP $\rightleftharpoons$ ADP + P$_i$ (or AMP + PP$_i$) interconversions comprise the molecular basis for the flow of chemical energy within all living cells. When a cell degrades high energy foodstuffs such as carbohydrate, the energy released is used to form ATP. When a cell uses energy (for a process such as biosynthesis, muscle contraction, or transport of substances across membranes), ATP is degraded to provide the energy. The ATP-ADP-AMP system and other aspects of bioenergetics are examined in more detail in Chap. 14.

NUCLEIC ACIDS

Covalent Backbone of Polynucleotides

Polypeptides are composed of amino acids linked together by peptide bonds. Polynucleotides are composed of *5'-monophosphonucleotides linked together by phosphodiester bonds*. Since the diester bond involves the 3' and 5' positions of the pentose unit in adjacent nucleotides, it is designated as a 3' $\rightarrow$ 5' phosphodiester bond.

A complete formula representation of a pentanucleotide chain is shown below. From left to right the structure progresses from the *5′ terminus* (with a phosphate group) to the *3′ terminus* (with a free 3′—OH group). The large polyribonucleotide chains (RNA) and polydeoxyribonucleotide chains (DNA) are termed nucleic acids because of the presence of several mildly acidic phosphate groups along the chain. At pH 7 the phosphate group would be nearly completely ionized ($pK_a \approx 6$), and so the nucleic acids are *polyanionic* (having many negative changes) in nature.

A pentanucleotide structure illustrating 3′ → 5′ phosphodiester bonds between adjacent nucleotides. As indicated, each phosphate group would be ionized at about pH 7

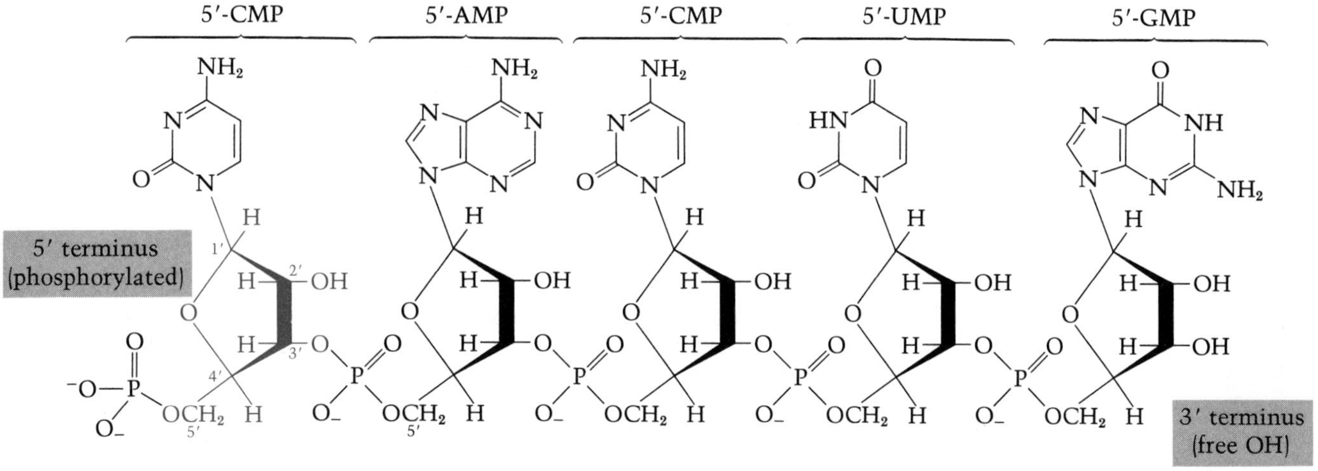

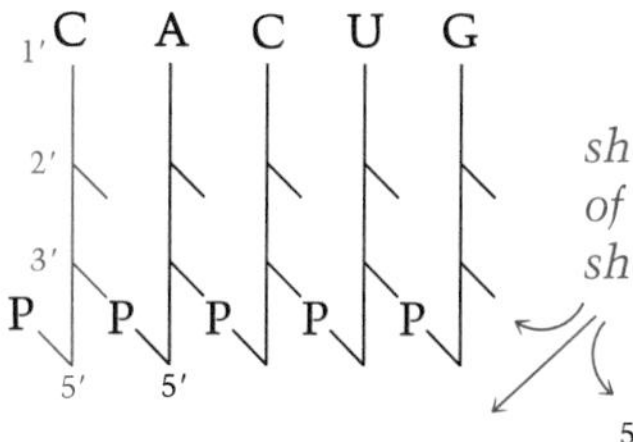

shorthand representations of the pentanucleotide shown above

p C p A p C p U p G   or simply $\overset{5'}{\text{pC}}$—A—C—U—$\overset{3'}{\text{G}}$ or   pCACUG

For obvious reasons shorthand designations of polynucleotide chains are in common use, with each base being identified by a single letter code (N in general, or specifically A, G, C, T, U). One convention utilizes a vertical line and diagonal slashes to represent the sugar unit, with phosphodiester bonds being represented by the letter P between the 3′ and 5′ slashes. An even simpler representation uses pN to symbolize a 5′-monophosphonucleotide, with Np designating 3′-nucleotide. A polynucleotide is then shown as a succession of pN notations reading left-to-right from the 5′ terminus to the 3′ terminus. The simplest representation uses just the single letter code with a p to identify only the phosphorylated 5′ terminus.

Isolation of Nucleic Acids

The isolation of nucleic acids begins with an extraction from its cellular or viral source. If the source is cellular, the extraction is done on a particular subcellular fraction of disrupted cells—the nuclei, mitochondria, chloroplasts, ribosomes, or soluble fraction (see Chap. 1). Some of the properties

DNA properties

- insoluble in dilute solutions of NaCl
- soluble in concentrated solutions of NaCl
- insoluble in alcohol
- can be dissociated from protein by treatment with a detergent or phenol

(table cont. on p. 214)

- soluble in dilute solutions of NaCl
- insoluble in alcohol
- can be dissociated from protein by treatment with a detergent or phenol

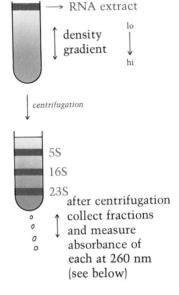

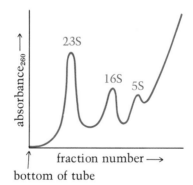

FIGURE 7–2 A typical density gradient ultracentrifugation (rate zonal) pattern of RNA extracted from intact ribosomes of *Escherichia coli*. The absorbance pattern of the tube contents after centrifugation illustrates the presence of three different RNAs, distinguished by different sizes: 23S (heavy), 16S (intermediate), and 5S (light).

of RNA and DNA that are exploited in these procedures are listed in the margin.

If the extract contains different species of nucleic acid, the second problem is one of resolving the individual components. The most frequently used technique for this is density gradient ultracentrifugation (see p. 42, Chap. 2). A separation of three major RNA components obtained from bacterial ribosomes is diagramed in Fig. 7–2 and illustrates the application. Once a particular nucleic acid component is isolated, the same method may be used to confirm purity and for additional characterization studies such as the determination of molecular weight. However, the techniques of ion-exchange, adsorption, and molecular-sieve chromatography and electrophoresis can also be used. Isopycnic density-gradient centrifugation is a particularly powerful tool for studies on DNA preparations (see p. 226, this chapter).

A very effective method to achieve the isolation of a specific polynucleotide chain is the technique of *hybridization*. We will defer discussion of this method until later (see p. 225) since it is based on a principle of structure that we have not yet examined.

Primary Structure—Nucleotide Composition

The nucleotide composition (also referred to as base composition and expressed as %A, %G, %C, and %T or %U) of a nucleic acid can be determined by routine procedures involving (a) complete degradation of the nucleic acid into a mixture of its constituent nucleotides and (b) an analysis of the mixture by chromatography—usually an ion-exchange column method. RNA can be degraded by treatment with sodium hydroxide (base hydrolysis) or by using enzymes that could catalyze the hydrolysis of all phosphodiester bonds. RNA is susceptible to both treatments but DNA is resistant to base hydrolysis. An examination of the mechanism of OH^- action shown below reveals the basis for the DNA resistance, namely, the cyclic intermediate cannot be formed because of the absence of a hydroxyl group on the 2′ position of deoxyribose.

The base composition of DNA can also be determined by two other reliable and relatively simple methods. One method involves the heating of DNA and the other involves ultracentrifugal sedimentation. Both are discussed later in this chapter (see p. 226) after DNA structure has been described.

The identity and relative amounts of purine and pyrimidine bases are fundamental properties of nucleic acid structure. In fact, in the case of DNA some important generalizations regarding base composition data contributed greatly to the discovery of the molecular structure of DNA. These generalizations (*Chargaff's rules*) are itemized in the next section (see p. 221). However, base composition does not define the ultimate individuality of nucleic acids. For example, DNA preparations from very diverse sources may have a very similar base composition (see Table 7–2, page 220). The *uniqueness of a nucleic acid* is contained in the *sequence of bases* along the polynucleotide backbone.

Before progressing into the subject of base sequence, let us first consider the enzymatic degradation of nucleic acids. Enzymes that catalyze the hydrolysis of nucleic acids are called *nucleases* (they are sometimes also called *phosphodiesterases*), with *exonucleases* catalyzing the sequential hydrol-

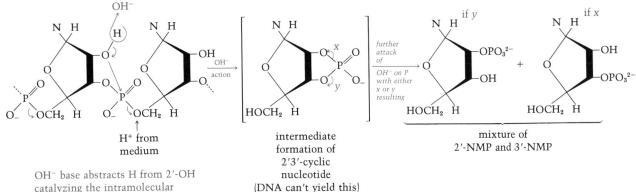

OH⁻ base abstracts H from 2'-OH catalyzing the intramolecular attack of P by 2' oxygen and promoting cleavage of P—O bond

intermediate formation of 2'3'-cyclic nucleotide (DNA can't yield this)

further attack of OH⁻ on P with either x or y resulting

mixture of 2'-NMP and 3'-NMP

(as shown, the action of OH⁻ results in cleavage of the P—O bond involving the 5' oxygen and is called *b*-type cleavage; cleavage of the O—P bond involving the 3' oxygen is called *a*-type cleavage; the cleavage pattern occurring with enzymes is characteristic of the enzyme, with some yielding *a*-cleavage and others *b*-cleavage)

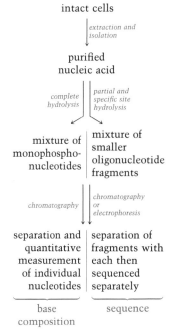

ysis of terminal diester bonds and *endonucleases* acting on internal diester bonds. The total hydrolysis of a nucleic acid can be achieved by using a mixture of an endonuclease and an exonuclease.

Elements of specificity for individual nucleases, varying from one to another (see Table 7–1), include the ability (a) to act on either RNA, DNA, or both; (b) to act on single-stranded and/or double-stranded nucleic acids; (c) to produce *a*-type or *b*-type cleavage; (d) to recognize either purine sites or pyrimidine sites—this is called *base specificity*; and (e) in the case of special endodeoxyribonucleases, to recognize certain sequences—this is called *sequence specificity*.

The enzymes in (e) that act only at sites in DNA where there is a specific base sequence and then only at specific bonds in or near that sequence are called *restriction enzymes* (see margin). The elegant specificity of action of the restriction enzymes has been used to develop ingenious laboratory procedures for sequencing DNA and *to excise intact genes from one DNA chromosome and then splice them into another DNA chromosome* (see p. 280 in the next chapter.)

The degradation of DNA and RNA by nuclease enzymes is of course a natural event. Processes involving nucleases include such things as digestion of dietary nucleic acids, the normal cycle of cellular turnover of nucleic acids, genetic recombination, and the defense against infection from foreign agents such as viruses. On the basis of successful clinical trials, the Soviet Union recently approved the use of pancreatic ribonuclease as a new antiviral drug for the treatment of tick-borne encephalitis. This application renews the hope for more widespread testing and development of enzymes as curative therapeutic agents for various diseases.

Primary Structure—Base Sequence

The first achievement in sequencing a complete nucleic acid molecule—a small transfer-RNA—was reported by Holley and coworkers in 1965. The

Restriction enzymes were first discovered (in 1970) in the bacterium *E. coli*, and since then have been isolated from various other prokaryotes. They have been termed restriction enzymes because of their proposed function in protecting the parent cell from infection by a bacterial virus. They do this by preferentially degrading the viral DNA chromosome; hence the expression of the viral chromosome and resultant viral multiplication are restricted. Although searched for, no restriction enzymes have yet been detected in eukaryotic cells.

TABLE 7–1 Functional characterization of some nucleases.

Enzyme	Substrate	Mode of attack	Specificity	Cleavage sites
1. Pancreatic ribonuclease	RNA	Endo (*b* type)	Linkage between pyrimidine (Py) and a nonspecific base (X) on the 3′ side; phosphodiester bond cleaved at 5′ position	. . Py p X p . .
2. Takadiastase T_1 (isolated from fungi)	RNA	Endo (*b* type)	Linkage between guanine (G) and nonspecific base (X) on 3′ side; phosphodiester bond cleaved at 5′ position	. . G p X p . .
3. Micrococcal nuclease	RNA, DNA	Endo (*b* type)	Linkage between adenine (A) and nonspecific base (X) on 5′ side; phosphodiester bond cleaved at 5′ position	. . X p A p . .
4. Snake venom nuclease	RNA, DNA	Exo (*a* type)	Attack starts from 3′ hydroxyl end with phosphodiester bond cleaved at 3′ position; no base specificity	. . X p Y p Z (3′-OH)
5. Spleen nuclease	RNA, DNA	Exo (*b* type)	Attack starts from and requires a free 5′ hydroxyl terminus; phosphodiester bond cleaved at 5′ position; no base specificity; *5′-OH generated by initial treatment with a 5′-monoesterase,* which cleaves a terminal phosphate bond	[5′-OH]X p Y p Z . .

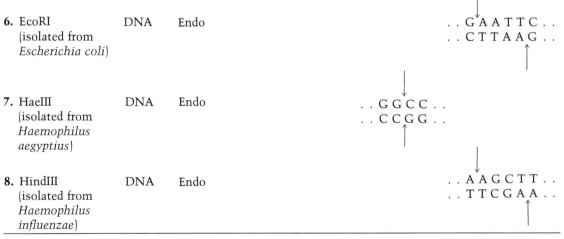

arrows designate sites of action by enzymes listed (numbers) in table

The following are restriction enzymes (arrows identify the two cleavage sites in each sequence). The special feature of restriction sites is described on p. 241; can you deduce what it is from these examples?

6. EcoRI (isolated from *Escherichia coli*)	DNA	Endo		. . G A A T T C . . . . C T T A A G . .
7. HaeIII (isolated from *Haemophilus aegyptius*)	DNA	Endo		. . G G C C . . . . C C G G . .
8. HindIII (isolated from *Haemophilus influenzae*)	DNA	Endo		. . A A G C T T . . . . T T C G A A . .

procedure resembled the strategy of determining the amino acid sequence of a polypeptide discussed in Chap. 5 (see p. 125). The nucleic acid is partially degraded into smaller oligonucleotide fragments by treatment with an endonuclease—preferably one that exhibits some base specificity. The fragments are then isolated (by ion-exchange and/or molecular-sieve chromatography and/or electrophoresis), analyzed for base composition, and sequenced separately. The sequencing is done with exonucleases that catalyze the one-by-one removal of nucleotides from the end of the fragment. The first transfer-RNA molecule sequenced by Holley had 77 nucleotides (see p. 228 in this chapter). Since then, about 80 other transfer-RNA molecules have been sequenced. There have been numerous other successes—the most remarkable of which is the report (1976) of the complete sequencing of the entire RNA chromosome of the MS2 bacterial virus consisting of 3,569 nucleotides! The sequencing of such a massive chain is made possible by using highly specific endonucleases to produce a limited number of manageable fragments and by using a few other tricks we will not go into. The point is that what many thought would be a near impossible task—the sequencing of large RNA and DNA molecules—can now be done.

Until recently the sequencing of DNA was a formidable task. In fact, the approach was to use enzymes to prepare an RNA copy of the DNA and then to sequence the RNA copy. Now three very efficient techniques are available to sequence DNA directly. One developed in 1975 by Sanger (the same Sanger who 25 years earlier pioneered the methodology for determining amino acid sequences) has been used to determine the sequence of the single-stranded DNA chromosome of the ϕX174 bacterial virus—one of the smallest DNA molecules known—which consists of 5,375 nucleotides! Spectacular results, indeed. In 1977 two more techniques were described, one developed by W. Gilbert and A. Maxam, and the other by the Sanger group. Since limitations of space prevent a description of each procedure, we will examine only the Gilbert-Maxam method, even though it is best-suited for DNA fragments containing 100–200 nucleotides. It truly exemplifies a pinnacle of scientific ingenuity.

The Gilbert-Maxam procedure begins by labeling the 5' end of a single strand of DNA with radioactive phosphate (^{32}P). If the original DNA is double-stranded, both 5' ends are labeled and the strands are separated and processed separately. The ^{32}P labeling is accomplished enzymatically by (a)

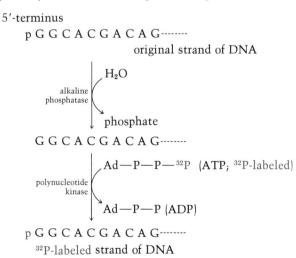

first using *alkaline phosphatase* to remove the existing phosphate from the 5′ end and (b) then treating the dephosphorylated DNA with *polynucleotide kinase* and ^{32}P-labeled ATP to replace the phosphate (now a ^{32}P) on the 5′ end.

The labeled DNA is then divided into four batches with each batch receiving a different chemical treatment. Each treatment is designed to alter chemically the purine and pyrimidine bases in a specific way—and a different way each time—so as to sensitize the DNA to further cleavage at specific points. Each treatment is described below and schematically summarized in Fig. 7–3.

For the oligonucleotide

pCGCAC

methylation and cleavage (signified by arrows) would occur at either the G or the A site

$_p$C G C A C $_p$C G C A C

to yield the labeled fragments

pC   pCGC

But, because the specific chemistry occurs better (heavy arrow) at a G location than at an A location, more pentanucleotide molecules will yield this pattern and hence there will be more pC fragments than pCGC fragments present in the mixture.

1. *Purine cleavage: guanine enhanced—adenine suppressed* $(G > A)$. Treating the ^{32}P-labeled DNA with dimethyl sulfate methylates N atoms in both A and G bases, but is more complete with G than with A. This sensitizes the glycoside bonds between the methylated bases and the sugar to easy cleavage by mild heating at neutral pH (7). After the methylated bases are removed, the phosphoester bonds involving the free sugar are preferentially cleaved by heating in the presence of NaOH. Since the original methylation occurs more readily at G than at A, more G than A sites are cleaved. The net result of all this is *a mixture of oligonucleotide fragments all of different lengths, some of which are labeled with* ^{32}P. By taking advantage of the differences in size, the mixture can be analyzed by polyacrylamide gel electrophoresis to identify the pieces containing the ^{32}P label. The pattern of resolution is recorded by exposing the gel to a piece of X-ray film. Bands of the gel containing ^{32}P will appear as darkened zones, intense for large amounts of ^{32}P and lighter for small amounts of ^{32}P. The larger amount of ^{32}P will be present in the intense zones containing fragments that result from G-site cleavage since more G-sensitive sites result in the methylation. Fainter zones contain less ^{32}P because the fragments arise from A-site cleavage, which is suppressed. See the accompanying diagram. The developed film is called an autoradiograph of the gel.

2. *Purine cleavage: adenine enhanced—guanine suppressed* $(A > G)$. The methylation is the same as above. However, by then gently treating the methylated DNA with dilute HCl and at 0 °C rather than by heating, methylated adenine bases are removed much more completely than methylated guanines. The cleavage of bonds involving the free sugar is the same as above. The mixture of fragments is then also analyzed by polyacrylamide gel electrophoresis. The major fragments (intense zones) in the mixture would be produced from cleavage at A sites and minor fragments (lighter zones) from G sites. This is just the reverse of the result in (1).

3. *Pyrimidine cleavage: both cytosine and thymine* $(C + T)$. The ^{32}P-labeled DNA is treated (with hydrazine and then piperidine) to eliminate selectively C and T bases from the sugars to which they are attached and also to cleave the phosphoester bonds involving the free sugars that are generated. The final product mixture contains fragments of different lengths corresponding to cleavage at both cytosine and thymine sites. The mixture is analyzed by gel electrophoresis and all zones display equal intensity.

4. *Pyrimidine cleavage:—only cytosine* (C). The same treatment as in (3) is used except NaCl is present, which prevents hydrazine reaction with thymine bases. Only C sites are modified, removed, and cleaved. Hence, the mixture contains fragments that arise only from C site cleavage.

The complete sequence of the original DNA is then "read" directly from the patterns of the electrophoretic resolution of the labeled fragments on each gel. As shown in Fig. 7–3 the patterns of each gel are aligned side by side and then read as follows. A faint band on A > G pattern coupled with an

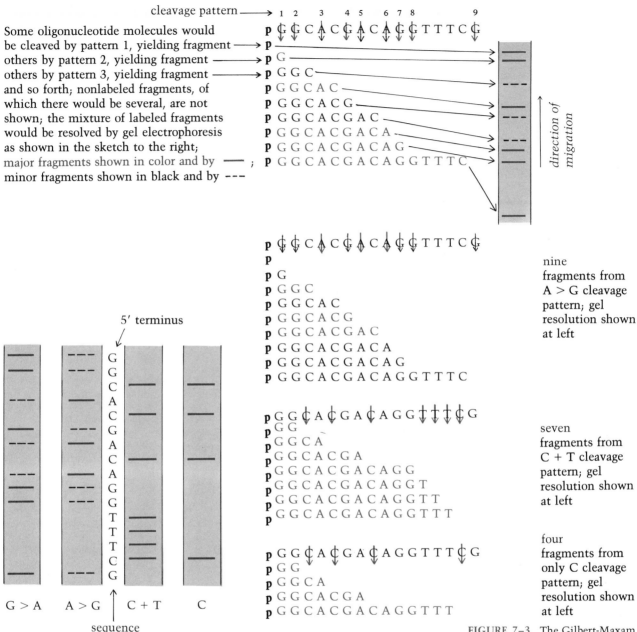

Some oligonucleotide molecules would be cleaved by pattern 1, yielding fragment → others by pattern 2, yielding fragment → others by pattern 3, yielding fragment → and so forth; nonlabeled fragments, of which there would be several, are not shown; the mixture of labeled fragments would be resolved by gel electrophoresis as shown in the sketch to the right; major fragments shown in color and by ——; minor fragments shown in black and by ---

cleavage pattern

p G G C A C G A C A G G T T T C G

p
p G
p G G C
p G G C A C
p G G C A C G
p G G C A C G A C
p G G C A C G A C A
p G G C A C G A C A G
p G G C A C G A C A G G T T T C

direction of migration

5' terminus

p G G C A C G A C A G G T T T C G
p
p G
p G G C
p G G C A C
p G G C A C G
p G G C A C G A C
p G G C A C G A C A
p G G C A C G A C A G
p G G C A C G A C A G G T T T C

nine fragments from A > G cleavage pattern; gel resolution shown at left

p G G C A C G A C A G G T T T C G
p G G
p G G C A
p G G C A C G A
p G G C A C G A C A G G
p G G C A C G A C A G G T
p G G C A C G A C A G G T T
p G G C A C G A C A G G T T T

seven fragments from C + T cleavage pattern; gel resolution shown at left

p G G C A C G A C A G G T T T C G
p G G
p G G C A
p G G C A C G A
p G G C A C G A C A G G T T T

four fragments from only C cleavage pattern; gel resolution shown at left

G > A   A > G   C + T   C

G
G
C
A
C
G
A
C
A
G
G
T
T
T
C
G

sequence corresponding to gel patterns

FIGURE 7–3 The Gilbert-Maxam technique of sequencing DNA requires four separate steps. Each step involves chemical modification of a purine or pyrimidine base; removal of the modified base and cleavage at the vacated site (see text). Heavy arrows identify major sites in the two steps that involve purine bases. Color identifies the major fragments in each mixture that would yield intense bands. As shown below, after the gel patterns are grouped together the DNA sequence can be "read" directly. (The letter **p** represents [32]P.)

intense band in the G > A pattern is read as G. Just the reverse is read as A. An intense band on the C + T pattern, is read as C or T, but if the same band is missing on the C pattern, the base is T. Ingenious, to say the least!

Secondary and Tertiary Structure—DNA

In the April, 1953 issue of *Nature*, J. D. Watson and F. H. C. Crick used only about 1,000 words to describe the three-dimensional structure of DNA. They proposed (see Fig. 7–4) that *DNA consists of two right-handed, helical*

polynucleotide chains, each coiled around a central axis but running in opposite directions. This double helix is stabilized by interchain hydrogen bonding between specific pairs of nitrogen bases, which are projected from each chain towards the core of the helix and perpendicular to the axis. Their brilliant proposal has been proven to be correct and is generally recognized as the single achievement that has revolutionized twentieth-century biological science. It was revolutionary because for the first time it was possible to express biological phenomena such as the storage, replication, and expression of genetic information in the precise terms of molecular structure. A new term—*molecular biology*—was also popularized as a label for this type of research. However, molecular biology is not a distinct discipline from biochemistry, but rather a specific area of biochemistry focusing on the relationship of molecular structure to biological phenomena with a particular focus on nucleic acids, proteins, and genetic phenomena.

The genius of Watson and Crick was in their interpretation of data available to them, and the construction of a model consistent with the data and their interpretation. In addition, they happened to be at the right place at the right time. The critical empirical data were provided by E. Chargaff's earlier studies on the base composition of many DNA samples and by the brilliant X-ray diffraction studies on crystalline DNA being performed at that time by M. Wilkins and R. Franklin at nearby King's College of London. The X-ray analysis revealed a distinct symmetry of structure, which was interpreted as helical.

The studies of Chargaff revealed some remarkable generalizations about the base composition (Table 7–2) of DNA. These data provided the impor-

FIGURE 7–4 A diagramatic representation of DNA according to Watson and Crick. The two ribbons symbolize the two phosphate-sugar chains, and the horizontal rods the pairs of hydrogen-bonded nitrogen bases holding the chain together. The vertical line marks the central axis. (Taken with permission from J. D. Watson and F. H. C. Crick, "Genetical Implications of the Structure of Deoxyribonucleic Acid," *Nature*, **171**, No. 4361, 964–967 (1953).)

TABLE 7–2 **Distribution of purines and pyrimidines in deoxyribonucleic acids (a partial listing).**

Biological source	%				Ratios		
	A	G	C[a]	T	A/T	G/C[a]	purine / pyrimidine
Viruses							
Lambda (*E. coli*)	26.0	23.8	24.3	25.8	1.01	0.98	0.99
Adenovirus (human)	24.5	25.2	24.8	25.5	0.96	1.02	0.99
Fowlpox	32.3	18.0	17.2	32.6	0.99	1.05	1.01
Bacteria							
Escherichia coli	23.8	26.0	26.4	23.8	1.00	0.98	0.99
Bacillus subtilis	28.9	21.0	21.4	28.7	1.01	0.98	1.00
Pseudomonas fluorescens	18.2	33.0	30.0	18.8	0.97	1.10	1.05
Plants							
Carrot	26.7	23.1	23.2	26.9	0.99	1.00	0.99
Tobacco leaf	29.7	19.8	20.0	30.4	0.98	0.99	0.98
Peanut	32.1	17.6	18.0	32.2	1.00	0.98	0.99
Animals							
Frog	26.3	23.5	23.8	26.4	1.00	0.99	0.99
Chick embryo	28.9	23.7	21.2	26.2	1.10	1.12	1.11
Chicken (liver)	30.3	22.0	19.7	28.0	1.08	1.12	1.10
Rat (liver)	28.6	21.4	21.5	28.4	1.01	1.00	1.00
Human (liver)	30.3	19.5	19.9	30.3	1.00	0.98	0.99
Human (spleen)	28.1	24.7	21.1	26.1	1.08	1.11	1.12
Human (thymus)	29.8	20.2	18.2	31.8	0.94	1.11	1.02

[a] Includes any contribution from 5-methylcytosine (see p. 205)

tant clues regarding the number of chains in the helix and their arrangement.

1. The number of purine bases (A + G) was balanced by the number of pyrimidine bases (T + C); that is, the ratio of purines to pyrimidines was approximately one (Pu/Py = 1.0).
2. The number of adenine residues was balanced by the number of thymine residues; that is, the ratio of adenine to thymine was approximately one (A/T = 1.0).
3. The number of guanine residues was balanced by the number of cytosine residues; that is, the ratio of guanine to cytosine was approximately one (G/C = 1.0).
4. The sum of adenine and cytosine residues was balanced by the sum of guanine and thymine residues; that is, (A + C) = (G + T).
5. Each of these relationships was found to be true of all DNA samples examined, regardless of biological source.

The *novel feature* of the double-helix conformation *is the interchain hydrogen bonding between nitrogen bases on different chains directly opposite each other, with a purine always bonded to a pyrimidine.* More specifically, *adenine is always bonded to thymine, A═══T (hence the A/T ratio of 1.0) and guanine is always bonded to cytosine, G═══C (hence the G/C ratio of 1.0).* This bonding arrangement is in agreement with the base composition data and is also consistent with the structures of the nitrogen bases. In fact, it represents an optimum spatial fit. Accordingly, the A═══T and G═══C pairings are termed <u>*complementary base pairs.*</u> Line drawings and space-filling models of each pair are given in Fig. 7–5. The latter are especially informative in that they clearly illustrate the structural com-

For their contribution, Watson, Crick, and Wilkins were awarded the Nobel Prize in 1962. An extremely interesting and controversial narrative account of this period in scientific history and of the personalities involved is given by Watson himself in his book, *The Double Helix.* (A subtitle to Watson's account could very well be "The Saga of Rosalind Franklin." Read the book and discover what I mean.)

FIGURE 7–5 Purine-pyrimidine base pairs in deoxyribonucleic acids.

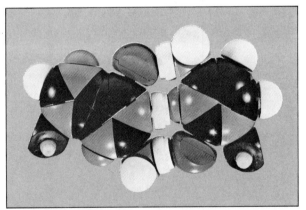

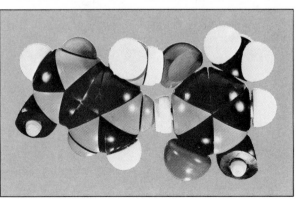

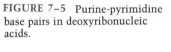

Guanine ┄┄┄┄┄ Cytosine
(three hydrogen bonds)

≈11Å

Adenine ┄┄┄┄┄ Thymine
(two hydrogen bonds)

≈11Å

patibility of the planar bases in forming linear hydrogen bonds. Without question, this is one of the most striking examples of the molecular logic that exists in nature. If you conclude that the bond strength of a G≡≡≡C pair with three hydrogen bonds should be greater than that of the A===T pair with two hydrogen bonds, you are correct.

This pattern of base-pairing exists along the entire length of the duplex. Hence the entire base sequence of one strand is complementary to the other strand and there is an extensive network of cooperative hydrogen bonding. A two-dimensional representation is shown at the top of p. 223.

The specification that the two helical chains run in opposite directions is most important, because this provides the proper alignment of bases for maximizing hydrogen bond formation. In structural jargon, this feature is

FIGURE 7–6 A schematic representation of DNA double helix, including significant molecular dimensions. These dimensions of DNA structure correspond to *crystalline* DNA prepared for X-ray analysis. Specifically, they represent the so-called Watson-Crick B type of DNA crystals prepared under particular conditions. Humidity is one of the critical conditions affecting the extent of crystal hydration. The B type crystal is obtained under high humidity. At reduced humidity the A conformation of DNA is obtained. It has 11 base pairs per turn with each pair spaced about 2.6 Å apart and oriented at about 20° to the sugar-phosphate backbone. There is now some evidence (see the Griffith reference at the end of chapter) that in solution the base pairs of DNA are spaced 2.9 Å apart with 10.5 base pairs per turn. More study is needed, however, to determine the DNA structure in solution.

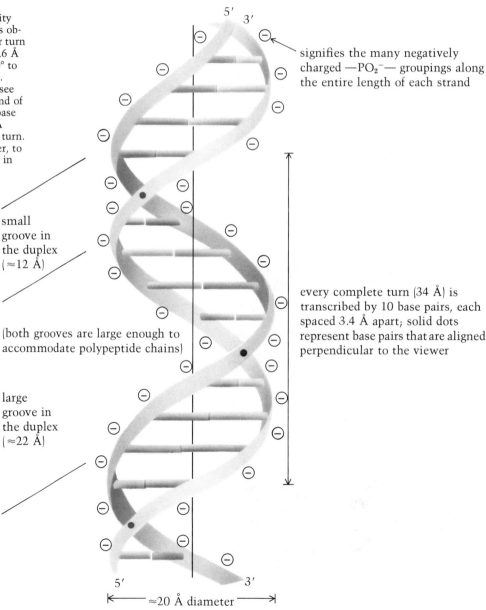

signifies the many negatively charged —PO₂⁻— groupings along the entire length of each strand

small groove in the duplex (≈12 Å)

(both grooves are large enough to accommodate polypeptide chains)

large groove in the duplex (≈22 Å)

every complete turn (34 Å) is transcribed by 10 base pairs, each spaced 3.4 Å apart; solid dots represent base pairs that are aligned perpendicular to the viewer

≈20 Å diameter

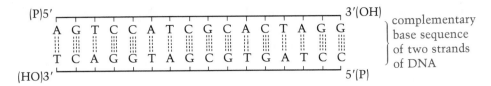

(P)5′ ——————————————————————————— 3′(OH)

A G T C C A T C G C A C T A G G

T C A G G T A G C G T G A T C C

(HO)3′ ——————————————————————————— 5′(P)

} complementary base sequence of two strands of DNA

referred to as *opposite polarity.* Since the propagation of one chain is 3′ → 5′ and the other is 5′ → 3′, each terminus will consist of a 3′-OH and 5′-phosphate.

All aspects of DNA structure are summarized in Figs. 7–6 and 7–7; Fig. 7–8 reveals that DNA molecules are large enough (20 Å diameter) to be observed by electron microscopy. The micrograph shows the *circular* double helix that is isolated from bacteria and viruses. It has been argued that all

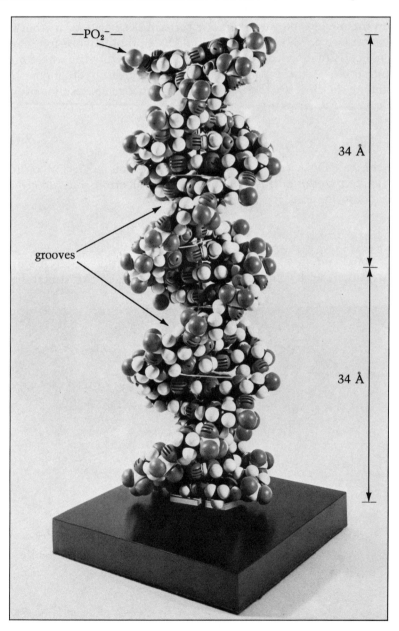

—PO₂⁻—

grooves

34 Å

34 Å

FIGURE 7–7 A space-filling model of a segment (two complete turns) of a DNA duplex molecule. (Photograph supplied through the courtesy of The Ealing Corporation, Natick, Mass.)

223

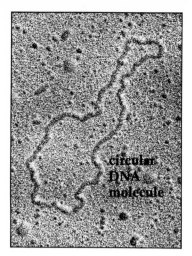

FIGURE 7–8 An electron micrograph of a DNA molecule obtained from *Escherichia coli*. This particular specimen is *plasmid DNA*, a small circular duplex molecule found in some bacteria in addition to the major chromosome. Plasmids carry supplementary genes, most significant of which are genes that confer antibiotic resistance. Plasmids are also used as carrier molecules in genetic engineering experiments (see p. 280 in next chapter). (Micrograph supplied by Stanley N. Cohen, Department of Medicine, Stanford University.)

DNA molecules are circular and that the linear duplex is the result of shearing during isolation procedures. An exception to the Watson-Crick structure is represented by some viral DNAs, such as that found in the φX174 bacterial virus, which are circular, *single-stranded* molecules (see p. 256, Chap. 8).

To summarize: DNA is a double-helix fibril consisting of two complementary, right-handed, helical, polynucleotide chains of opposite polarity intertwined around the same axis and held together by internal purine-pyrimidine hydrogen-bonded base pairs projected within the central core from a sugar-phosphate backbone. The polyanionic character of the molecule permits an additional mode of stabilization via electrostatic interaction with inorganic counter-cations such as Mg^{2+}, or with basic proteins such as the *histones* (see p. 119), which contain large numbers of positively charged amino acid side chains. This association with basic proteins applies to the DNA in the nucleus of the eukaryotic cells and is discussed further later in this chapter (see p. 232). Mitochondrial DNA and the DNA in prokaryotic microbial cells that lack a nucleus are not associated with protein but with inorganic ions. A third stabilizing force arises from hydrophobic interactions among the closely packed (see Fig. 7–7) nitrogen bases that are stacked in the core of the structure.

In Vivo Structure of DNA

Inside the cell DNA exists not as an extended molecule but rather in a *super-twisted conformation*. Evidence from electron microscopy of this "native conformation" is shown in Fig. 7–9.

Denaturation of DNA

At this point you should realize that the denaturation of DNA would involve treatment with some chemical or physical agent that destabilizes the

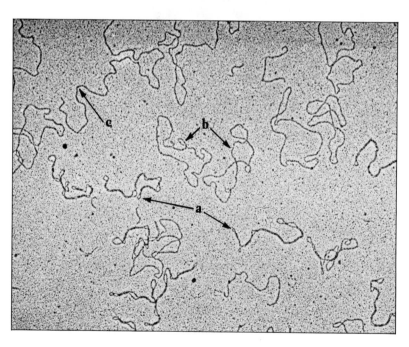

FIGURE 7–9 The circular φX174 DNA molecule in (a) super-twisted conformations. (b) Various relaxed conformations, and (c) linear fragments produced on shearing. (Micrograph prepared and supplied by Dr. Jack Griffith, Stanford University.)

double-stranded, helical conformation and produces a partially or completely unfolded state.

A most effective method of disrupting DNA is to heat it. The input of thermal energy exceeds the stabilizing forces of attraction due to hydrogen bonds and hydrophobic bonds causing to duplex to unwind—partially, and then completely. This process has been termed *melting*. The course of

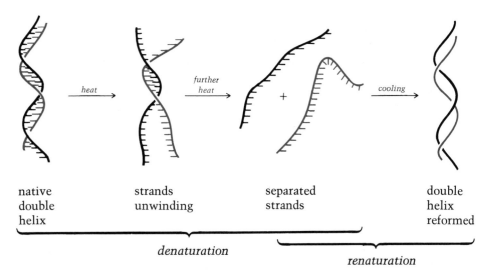

native double helix

strands unwinding

separated strands

double helix reformed

denaturation

renaturation

thermal denaturation of DNA is depicted in Fig. 7–10. Called a *melting curve*, the graph depicts a plot of absorbance at 260 nm versus temperature. If you understand DNA structure, with a little thought you should be able to justify the use of these coordinates. If you argue that denatured DNA should exhibit a greater absorbance than native DNA because the purine and pyrimidine bases would be more exposed for interaction with the incident radiation, you are correct. This increase in absorption is called *hyperchromicity*. Accordingly, the important region of the curve is the steep sigmoid area that corresponds to the temperature range in which the complete transition in conformation occurs. The midpoint of this region is termed the *transition* or *melting temperature* (T_m) and is a distinctly reproducible physical property of a DNA molecule under a specific set of conditions. It is significant to note that T_m values are usually quite high, ranging from 85–100 °C. By comparison, proteins are disrupted at much lower temperatures. This is not too surprising, since the actual number of stabilizing forces per typical protein molecule is much less than that for a DNA molecule. Thus, we can conclude that the presence of thousands of hydrogen bonds and thousands of hydrophobic (apolar) interactions does, in fact, contribute considerable stability to the compact double helix conformation of DNA.

A final note: Under very controlled conditions the process of melting is reversible when the temperature is lowered. Certainly this *self-association* (renaturation) is convincing evidence for the significance of complementary base-pairing in the double helix. The inherent tendency for self-association between complementary polynucleotide strands is the basis of an elegant technique called *hybridization*, which is illustrated in the next chapter (see p. 238).

Melting temperatures of DNA have proven extremely useful and reliable

The entire curve would be shifted to the right for a DNA with higher GC content; to the left for a DNA with a lower GC content

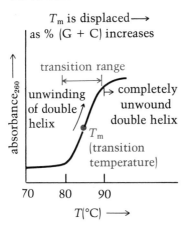

FIGURE 7–10 A typical melting curve profile of DNA, depicting the hyperchromic effect observed on heating.

for determining base composition. On the basis of T_m measurements on many DNA molecules whose base composition had been determined by another method, empirical relationships, such as the following, have been derived:

$$T_m = 69.3 + 0.41(\%G + \%C)$$

when conducted in a solution containing 0.2 M Na$^+$

This linear relationship is depicted in Fig. 7–11.

Another reliable method for determining base composition is to measure the *buoyant density* ρ_B of DNA via isopycnic density gradient centrifugation. The buoyant density refers to the suspension of DNA at a point in a cesium chloride density gradient where the densities of DNA and CsCl are equal. As the following empirical equation illustrates, there is also a linear correlation between DNA density and G + C content:

Both techniques are good examples of exploiting a molecular property to good advantage in order to learn something about the intact molecule.

$$\rho_B = 1.660 + \frac{0.098(\%G + \%C)}{100}$$

(see also Fig. 7–11)

RNA Structure

In *E. coli* the relative amounts of each type of RNA are about 2% mRNA, 16% tRNA, and 82% rRNA. Similar values—as rough approximations—would apply to all cells in general.

Living cells contain (see margin) three types of RNA: *messenger-RNA* (mRNA), *transfer-RNA* (tRNA), and *ribosomal-RNA* (rRNA). All three are single-stranded molecules. Each participates in the elaborate process of protein biosynthesis where the sequence of amino acids is determined by the nucleotide sequence in DNA. Before examining the structure of these different RNAs, let us first outline their participation.

In a process called *transcription*, the nucleotide sequence of a gene locus in one strand of DNA is copied into an RNA molecule. This is called messenger-RNA because, obviously, it now carries the genetic instructions originally present in DNA. In the process called *translation*, the mRNA then binds to ribosomes, which are particles composed of protein and RNA—hence the name ribosomal-RNA. It is here that the mRNA sequence of nucleotides governs the sequence of positioning amino acids during polypeptide biosynthesis. Individual amino acids are "carried" to the ribosome site by transfer-RNA molecules, which then also "read" the message in mRNA. The message is contained in successive groups of three nucleotides in mRNA called *triplet codons*. Each tRNA contains a unique trinucleotide sequence called an *anticodon*, complementary in sequence to a codon. Hence, the message is read by a *codon-anticodon recognition* between mRNA and tRNA, with the ribosomes providing the location where the two come together. Details of transcription, translation, and the genetic code will be examined in the next chapter.

$$\text{DNA} \longrightarrow \text{messenger RNA} \xrightarrow[\text{amino acids}]{\overset{\textit{participation of}}{\textit{ribosomes and transfer RNA}}} \text{polypeptide}$$

transcription   translation

Messenger-RNA. A living cell is capable of producing hundreds to thousands of different mRNA molecules. Each is a single-stranded molecule of variable length existing basically as an open chain without any precise pattern of intrachain folding. In other words, mRNA lacks an ordered three-dimensional structure. The most distinct structural feature of each mRNA molecule is its unique sequence of nucleotides. Each successive set of three nucleotides (a codon) provides the information for alignment of amino acids in polypeptide biosynthesis; for example, the codon UUU specifies phenyl-alanine, AUA isoleucine, GAU aspartic acid, and so forth. Hence the sequence <u>UUU</u><u>AUA</u><u>GAU</u> (read in the order underlined) would specify the tri-peptide segment phe-ile-asp. The entire genetic code is presented on p. 273.

Although prokaryotes and eukaryotes utilize the same genetic code, it has been recently established that eukaryotic mRNA is different in at least three other respects. Specifically, in the mRNA in eukaryotic cells (a) a pol-yadenylate (poly(A)) fragment of about 150–200 nucleotides long is at-tached to the 3'-end of mRNA, (b) Gppp is present at the 5'-end as a "cap," and (c) two or three nucleotides at the 5'-end are methylated. Items (b) and (c) are illustrated below. Why? Well, we don't yet know for sure. One

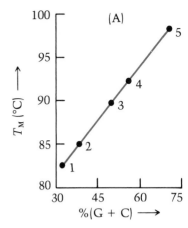

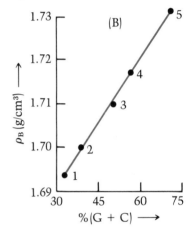

FIGURE 7–11 *Panel A:* relationship of melting temperature to GC content of DNA. *Panel B:* relationship of buoyant density to GC content of DNA. DNAs: *Staphylococcus aureus* (1), calf thymus (2), *Escherichia coli* (3), *Brucella abortus* (4), and *Streptomyces griseus* (5). (Data taken from the *Handbook of Biochemistry*, Cleveland, Ohio: CRC Press, 1970.)

suggestion regarding the poly(A) fragment is that it may assist in the transport of the mRNA to the cytoplasm from the nucleus where it is assembled. Another suggestion is that the poly(A) fragment may help protect mRNA from a rapid destruction by intracellular proteases. The Gppp cap and methylated bases at the 5'-end are proposed to be essential for the ordered positioning of mRNA with ribosomes in the initial step of protein bio-synthesis. None of these features is present in the mRNA of prokaryotic cells (see diagram, p. 228, top). The significance of "leader sequences" be-fore the message and "trailer sequences" after the message are discussed in the next chapter (see p. 244).

Transfer-RNA. As previously described, transfer-RNA molecules are so-named because they carry the amino acids—as amino acyl-tRNA adducts (see p. 103 in Chap. 4 and p. 262 in next chapter)—to the ribosome site for assembly into polypeptides. Each cell contains several different tRNA mol-ecules, perhaps, as many as 60. They are the smallest nucleic acid (mol. wt. about 25,000), consisting of approximately 73–93 nucleotides. Because of the small size and ease of isolation, a lot has been learned about tRNA struc-ture. In fact, the complete three-dimensional structure has recently been solved.

Solutions of purified tRNA display a significant hyperchromic effect

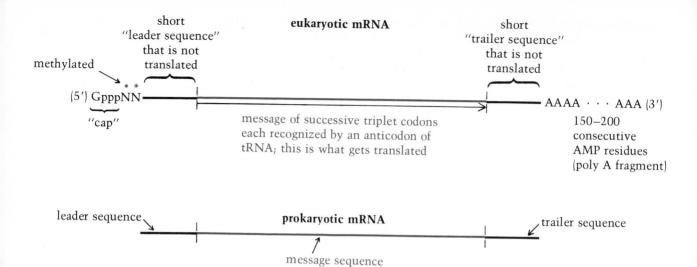

when heated, a strong indication that the native molecular structure is highly ordered. The first structural insight was provided in 1965 with the solution of the complete nucleotide sequence for a tRNA molecule. The sequence revealed that the polynucleotide chain could assume a variety of ordered arrangements by *folding back on itself* on the basis of *intrachain base-pairing* involving GC and AU recognitions. Intrachain folding would yield some double-stranded character and perhaps double-stranded regions would assume a double-helix orientation. This would explain the hyperchromic effect of heating tRNA, since bases in these regions would be buried in the native conformation and would become more exposed as the strand unravels.

FIGURE 7–12 Two-dimensional representation of the cloverleaf model of transfer-RNA structure. This arrangement has been proved to be consistent with all known sequences of tRNA molecules, but is not an accurate representation of the native structure (see Fig. 7–13).

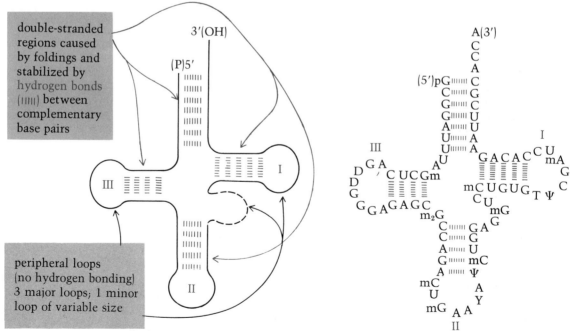

phenylalanine tRNA (from yeast)
76 nucleotides

Of the various folding patterns suggested, the one that has proven to be consistent with all tRNA sequences is the so-called *cloverleaf structure* (see Fig. 7–12), which consists of four distinct double-stranded segments forming three major loops, one minor loop of variable size, and an open arm that includes the two ends of the polynucleotide chain. This two-dimensional cloverleaf representation has guided tRNA research, and in doing so has been very useful in attempting to correlate the functions of tRNA to its structure. However, the question has always been, what is the native, three-dimensional structure of tRNA?

A. Rich and S. Kim in 1974 answered the question by using X-ray diffraction to solve the three-dimensional structure (see Fig. 7–13). The molecule is L-shaped, consisting of two distinct stems (each having partial double-helical character due to complementary base-pairing) that are aligned almost perpendicular to each other. What do you find in the folded conformation? Three major loops and an open stem! Thus, the cloverleaf pattern is a valid two-dimensional representation.

Another distinctive feature of tRNA structure is the presence of a variety of atypical (or modified) nucleosides in addition to the usual ones of A, G, C,

FIGURE 7–13 Photograph of a molecular wire model of the phenylalanine transfer-RNA of yeast. The view is approximately perpendicular to the molecule. Note the partial double-helix character of the two arms. An upside-down view from the rear reveals the L shape. (Reproduced with permission from S. H. Kim, et al., "Three-Dimensional Structure for Yeast Phenylalanine Transfer RNA," *Science*, **185,** 435–439 (1974). Copyright 1974 by the American Association for the Advancement of Science. See also F. L. Suddath, et al., *Nature*, **248,** 20–24 (1974). Photograph supplied by S. H. Kim.) A photograph of a space-filling model of tRNA appears on the title page at the front of the book.

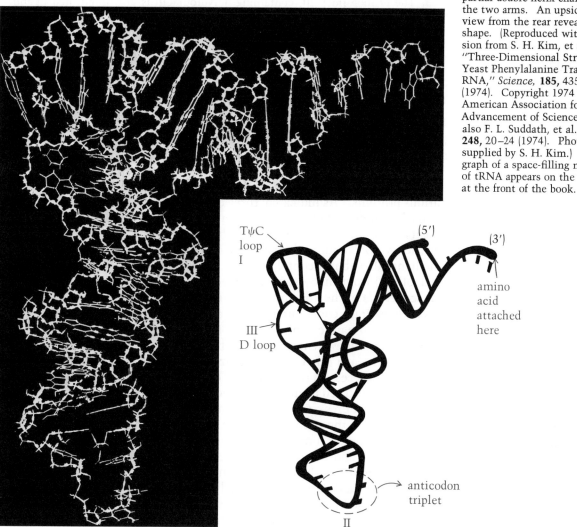

TψC loop I

III D loop

(5')

(3')

amino acid attached here

anticodon triplet

II

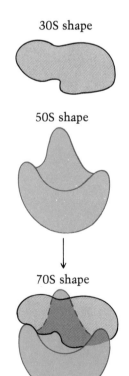

N,N-dimethylguanosine (m₂G)

1-methylguanosine (mG)

N⁶-isopentyladenosine (iA)

30S shape

50S shape

70S shape

and U. To date, more than 50 modified nucleosides have been detected. Some were identified earlier (p. 207), and a few more are shown in the margin. Many of these result from methylation in the base position, but several others involve more complex modification. Some of these modified nucleosides—such as *ribosylthymidine* (T), *pseudouridine* (ψ), and *dihydrouridine* (D)—are found in tRNAs of most organisms. Others are found only in prokaryotes, others only in eukaryotes, and others only in specific organisms. In any event, it is clear that—at least in some cases—the presence of modified nucleosides in particular regions (and in some instances in a particular position) of the tRNA molecule is responsible for the structure of tRNA and also for many of the specific functions of the tRNA molecule.

Some of the established assignments of specific functions to specific regions of tRNA structure are listed below and in Fig. 7–13. They are discussed in more detail in the next chapter.

1. The 3′ end of the open stem (with a —CCA sequence common to all tRNAs) is the site for the covalent attachment of the amino acid to tRNA.
2. The TψC loop (loop I) is associated with the binding of the aminoacyl-tRNA species to ribosomes.
3. The D loop (loop III, containing dihydrouracil) is also implicated in the binding of aminoacyl-tRNA to ribosomes.
4. The anticodon is always contained in loop II. The presence of a modified nucleoside adjacent to the anticodon is also universal. The anticodon loop is also associated with ribosome binding.

Ribosomal-RNA

Ribosomes are multimolecular aggregates of protein (about 60%) and RNA (about 40%); obviously the RNA is called *ribosomal-RNA*. An intact ribosome (70S in bacteria) is a complex of *two subunits*, one heavy (50S) and one light (30S). As shown in Fig. 7–14, the intact complex can be dissociated to yield the subunits, which in turn can be dissociated to yield rRNA and various proteins. As indicated, three different sizes of rRNA can be isolated from ribosomes. They are easily separable from each other, for example by density gradient ultracentrifugation (refer back to p. 214 in this chapter). The heavy 50S subunit contains two RNAs—a 23S rRNA and a 5S rRNA. The light 30S subunit contains only a 16S rRNA. 5S rRNA molecules (120 nucleotides) from many sources have been completely sequenced and partial sequences of the 16S and 23S have been reported. There are many unsolved questions concerning rRNA, including whether or not there is a highly ordered native conformation of each type.

There is much evidence that in each subunit RNA and protein molecules interact with each other in an organized pattern. One indication of this is provided by successful *reconstitution* experiments wherein active ribosome subunits are restored from their component parts previously isolated by dissociation procedures. Of course, the *self-assembly of ribosomes* is a natural event as well as the terminal stage in the formation in the cell. Investigations are currently underway to determine the pattern of interactions between RNA and protein and between protein and protein. Despite the inherent difficulty of the task, some information is already available. For example, the overall three-dimensional shape of a ribosome is proposed (at present, anyway) to look something like the drawing in the margin. The point to note is that the boxes used in Fig. 7–14 are highly idealized representations of the real structure.

230

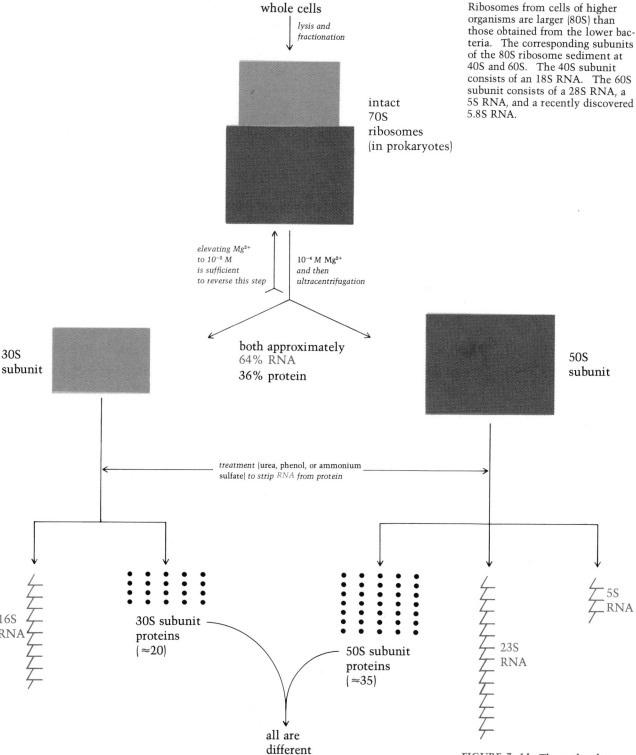

whole cells

*lysis and
fractionation*

intact
70S
ribosomes
(in prokaryotes)

Ribosomes from cells of higher
organisms are larger (80S) than
those obtained from the lower bacteria. The corresponding subunits
of the 80S ribosome sediment at
40S and 60S. The 40S subunit
consists of an 18S RNA. The 60S
subunit consists of a 28S RNA, a
5S RNA, and a recently discovered
5.8S RNA.

*elevating Mg²⁺
to 10⁻² M
is sufficient
to reverse this step*

*10⁻⁴ M Mg²⁺
and then
ultracentrifugation*

30S
subunit

both approximately
64% RNA
36% protein

50S
subunit

*treatment (urea, phenol, or ammonium
sulfate) to strip RNA from protein*

16S
RNA

30S subunit
proteins
(≈20)

50S subunit
proteins
(≈35)

23S
RNA

5S
RNA

all are
different

FIGURE 7–14 The molecular
composition of ribosomes. Once
completely dissociated, the component parts can be mixed together
to restore the intact subunits,
which in turn reassociate to yield
the intact ribosome.

Double-Stranded RNA. As with DNA, there are some RNA molecules that
are composed of two separate strands in a double helix arrangement.
Double-stranded RNA is found primarily in some viruses, where it serves as
the viral chromosome carrying all the genetic instructions for the replica-

tion of the virus. However, it is also found in very small amounts in normal eukaryotic cells (including mammalian cells). Although the function(s) of double-stranded RNA in eukaryotic cells is still obscure, there is evidence that one role is its stimulation of a special group of proteins collectively referred to as *interferons*. The name, interferon, is related to the proposed role of these substances, which is to combat viral infection by interfering with the processes of viral multiplication. This role of interferons has been debated for over a decade. It is an exciting area of current research, and is potentially of enormous benefit in the treatment of certain viral diseases, including some associated with cancer. One possibility is to use synthetic double-stranded RNAs as antiviral drugs.

CHROMATIN AND VIRUSES—OTHER NUCLEOPROTEIN AGGREGATES

The association of protein molecules with nucleic acid molecules involving ordered interactions between nucleic acid and protein is called a *nucleoprotein aggregate*. Ribosomes are just one example. Two others are chromatin and viruses.

Chromatin refers to the nuclear DNA that exists not as free DNA but as a complex in association primarily with proteins. (In electron micrographs (see p. 23) the chromatin is seen as electron dense regions irregularly distributed throughout the nucleus.) The proteins of chromatin are of two general classes. *Histone proteins*—of which there are five principal types: H1, H2A, H2B, H3, and H4—are characterized by a high content of the basic amino acids lysine and arginine (see margin). All other chromatin proteins—of which there are many types—are classified as *nonhistone proteins* because they lack the high lysine and arginine content and are more "averagelike" in their amino acid composition.

Because of their highly polycationic nature (they have many ⊕ charges), histones are "tailor-made" to interact with the highly polyanionic nature (many ⊖ charges), of the sugar-phosphate backbone of the DNA double helix. In addition to contributing to the stabilization of DNA structure, there is strong evidence that histones also function in regulating the expression of genes in DNA. How this occurs is not yet resolved and much further study is needed. However, recent studies have established that the histone-DNA interactions follow a regular pattern in which successive clusters of histones are bound to sections of DNA separated by short spacer

Lysine and arginine content of histone proteins

Histone[a]	% lys	% arg
H1	24.8	2.6
H2A	10.9	9.3
H2B	16.0	6.4
H3	9.6	13.3
H4	10.8	13.7

[a] Data are for histones isolated from calf thymus; most eukaryotic cells contain the same types.

Note that $\frac{1}{4}$–$\frac{1}{3}$ of amino acids are highly basic.

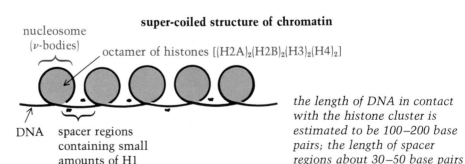

super-coiled structure of chromatin

nucleosome (ν-bodies)

octamer of histones [(H2A)₂(H2B)₂(H3)₂(H4)₂]

DNA spacer regions containing small amounts of H1 and nonhistone proteins

the length of DNA in contact with the histone cluster is estimated to be 100–200 base pairs; the length of spacer regions about 30–50 base pairs

regions of DNA. Each histone cluster is proposed to be an octamer consisting of two molecules each of H2A, H2B, H3, and H4 and it binds with stretches of DNA containing about 100–200 base pairs. These aggregates have been termed *nucleosomes* or *ν*-bodies (nu-bodies). The H1 protein is proposed to be bound to DNA in the spacer regions, which are about 30–50 base pairs long. When viewed under an electron microscope, this pattern of chromatin structure looks like beads on a string. As shown on p. 232 (bottom), if the DNA is wrapped around the histone cluster, a *super-coiled structure* of chromatin results. A super-coiled structure of DNA would help explain, among other things, how all of the DNA in the nucleus would fit in the nucleus.

Viruses

It is probably common knowledge to you that a virus is not considered as a true life form but rather as a parasite that depends on its infection of a host organism for a replicative existence. What you may not realize is that most viruses are nucleoprotein particles consisting only of several proteins and a single nucleic acid molecule, with the nucleic acid molecule, which serves as the viral chromosome, encapsulated by a protein sheath. Chemically speaking, there are two classes of viruses: those that contain an RNA chromosome and those that contain a DNA chromosome. Depending on the virus, the nucleic acid may be single-stranded or double-stranded and be either linear or circular. Physically speaking, the viruses comprise a heterogeneous group having various sizes (they are all small) and shapes. The unique shape of a virus results from the geometrical pattern of the nucleic acid and protein association. Some are rods (see margin), some are spheres, some are hexagonal (see margin), some are octagonal, and so on. Biologically speaking, we have animal, plant, and bacterial viruses, and several types of each of these.

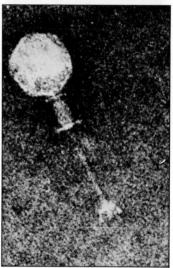

On a practical level the causal relationship of viruses to many diseases has of course guided research in virology for quite some time. In addition, however, from the standpoint of basic research, viruses have proven to be immensely valuable as simple *model systems that can be used to study the complex biochemistry of gene duplication and expression.*

LITERATURE

ADAMS, R. L. P., R. H. BURDON, A. M. CAMPBELL, and R. M. S. SMELLIE. *The Biochemistry of Nucleic Acids.* 8th edition. New York: Academic Press, 1976. An excellent introductory source.

CRICK, F. H. C. "The Structure of the Hereditary Material," *Scientific American,* **191,** 54–61 (1954). The first teaching article to be published on the structure of DNA.

FIDDES, J. C. "The Nucleotide Sequence of a Viral DNA." *Scientific American,* **237,** 54–67 (1977). A description of Sanger's "plus-and-minus" method for sequencing DNA.

FRAENKEL-CONRAT, H. *The Chemistry and Biology of Viruses.* New York: Academic Press, 1969. An excellent introduction to the principal facts known about the structure and function of bacterial, plant, and animal viruses. Several hundred references are given to the original literature.

FRAENKEL-CONRAT, H. *Design and Function at the Threshold of Life: The Viruses.* New York: Academic Press, 1962. A somewhat dated but still very informative introduction at the basic level. Available in paperback.

GOLDBERG, M. A. "Cyclic Nucleotides and Cell Function," *Hospital Practice,* **9,** 127–142 (1974). An excellent, introductory-level review article summarizing the regulatory functions of cyclic AMP and cyclic GMP. Good diagramatic illustrations of principles. Primary focus on mammalian cells.

GOLDBERG, N. D., and M. K. HADDOX. "Cyclic GMP Metabolism and Involvement in Biological Regulation." In *Ann. Rev. Biochem.,* **46,** 823–896 (1977). A thorough review article.

GREENGARD, P. and G. A. ROBINSON, eds. *Advances in Cyclic Nucleotide Research.* New York: Raven Press. Yearly publication reviewing various aspects of cyclic nucleotides.

GRIFFITH, J. D. "DNA Structure: Evidence from Electron Microscopy." *Science,* **201,** 525–527 (1978). The structure of DNA in solution is suggested to be significantly different than the Watson-Crick B conformation.

GROSSMAN, L., and K. MOLDAVE, eds. "Nucleic Acids." In *Methods in Enzymology,* Vol. 12, S. P. Colowick and N. O. Kaplan, eds. New York: Academic Press, 1968. A collection of papers from many contributors, dealing with practical aspects in the study of nucleosides, nucleotides, and nucleic acids.

HOLLEY, R. W. "The Nucleotide Sequence of a Nucleic Acid." *Scientific American,* **214,** 30–39 (1966). A description of how the first nucleic acid molecule, a species of transfer-RNA, had its primary structure determined.

HORNE, R. W. "The Structure of Viruses." *Scientific American,* **208,** 48–56 (1963). A description of the structure of various types of viral particles as revealed by high-resolution electron microscopy.

KORNBERG, R. P. "The Structure of Chromatin." In *Ann. Rev. Biochem.,* **46,** 931–954 (1977). A review article.

KURLAND, C. G. "The Structure and Function of the Bacterial Ribosome." In *Ann. Rev. Biochem.,* **46,** 173–200 (1977). A review article.

MAXAM, A. M., and W. GILBERT. "A New Method for Sequencing DNA." *Proc. Nat. Acad. Sci. USA,* **74,** 560–564 (1977). The original research paper.

NOMURA, M. "Ribosomes." *Scientific American,* **221,** 28–35 (1969). An account of the anatomy of a ribosome with particular attention given to the ribosomal proteins.

NOMURA, M. "The Assembly of Bacterial Ribosomes." *Science,* **179,** 864–873 (1973). A review article describing the molecular anatomy of ribosomes, the pattern in which they may be assembled, and their cellular function.

RICH, A. "The Three-dimensional Structure of Transfer RNA." *Scientific American,* **238,** 52–73 (1978). An excellent article.

RICH, A., and U. L. RAJBHANDRY. "Transfer RNA: Molecular Structure, Sequence, and Properties." In *Ann. Rev. Biochem.,* **45,** 805–860 (1976). A thorough review article.

ROBISON, G. A., R. W. BUTCHER, and E. W. SUTHERLAND. "Cyclic AMP." *Annual Review of Biochemistry,* **38,** 149–174 (1968). A review article emphasizing the biochemistry of cyclic AMP in mammalian organisms.

SANGER, F., et al. "Nucleotide Sequence of Bacteriophage ϕX174 DNA." *Nature,* **265,** 687–695 (1977). The original research paper.

STEIN, J. S. and L. J. KLEINSMITH. "Chromosomal Proteins and Gene Regulation." *Scientific American,* **232,** 46–57 (1975). Discussion of the suggested role of histones and nonhistone chromosomal proteins as regulatory elements for the expression of DNA genetic information in nuclei of higher organisms.

WATSON, J. D. *The Double Helix.* New York: Atheneum, 1968. An interesting and revealing narrative account of the events and persons associated with the discovery of DNA structure. Became a best seller.

EXERCISES

7–1. Draw the structural formula for each of the following substances.
(a) guanosine-3'-monophosphate
(b) deoxyadenosine-5'-diphosphate
(c) 5'-dADP
(d) 5'-dTMP
(e) thymidine
(f) cytosine-5'-triphosphate
(g) . . . pUpGp . . .
(h) guanylic acid

7–2. The values of the molar extinction coefficients (that is, the value of ϵ for a 1 M solution at the wavelength of maximum absorption in the UV region) for the major purines and pyrimidines are 6,100, 7,900, 8,200, and 13,400 for those with a single absorption maximum, and 8,200 and 10,700 for the base with two absorption maxima. Using other information given in this chapter, assign the ϵ values to the appropriate nitrogen bases.

7–3. Shown below is an oligonucleotide segment of a ribonucleic acid molecule. What type of cleavage pattern (if any) would result in this segment by treating the RNA with
(a) sodium hydroxide
(b) pancreatic ribonuclease
(c) micrococcal nuclease
(d) spleen nuclease after 5'-monoesterase
(e) takadiastase

pGpGpCpUpApCpGpUpApGpApUpCpAp

7–4. Sketch the four patterns of radioactive zones expected if the entire p C A C . . . chain in the duplex shown below were subjected to the Gilbert-Maxam sequencing technique. What would the patterns look like if the other chain were sequenced?

p C A C T T A C T T G T T A A C C G C C
G T G A A T G A A C A A T T G G C G G p

7–5. A purified DNA preparation was found to contain 30.4% adenine and 19.6% cytosine. The adenine-thymine ratio was 0.98 and the guanine-cytosine ratio was 0.97. Calculate the amount of guanine and thymine in this DNA and also the ratio of purine bases to pyrimidine bases.

7–6. Draw a representation of the complementary base pair found in RNA that would correspond to the adenine-thymine pair in DNA.

7–7. The length of the chromosome of *Escherichia coli* cells has been determined by electron microscopy to be approximately 1.2 millimeters. If it is assumed that the chromosome consists of a single DNA molecule, how many complementary base pairs are present in the chromosome? (1 angstrom = 1×10^{-8} centimeters.)

7–8. You build a space-filling model (6 feet long) of a DNA segment and place it on its side. Then a small flashlight is positioned at one end so that the light shines into the core of the model. How much light might you see from the other end of the model? Explain. (Look closely at Fig. 7–6.) Assuming the model represents the structure of DNA in aqueous salt solution, how much water is probably present in the core of the DNA molecule?

7–9. A pure DNA preparation was shown to have a T_m of 85 °C. What is the percent of A + T in this DNA? (Assume the measurement was made on a solution of the DNA containing 0.2 M Na$^+$.)

7–10. Estimate the buoyant density of the DNA referred to in Exercise 7–9.

7–11. How would you account for the fact that the T_m of DNA preparations increases with increasing amounts of guanine and cytosine?

7–12. According to the L-shaped structure of transfer-RNA, what maintains the relative spatial orientations of loops I and III?

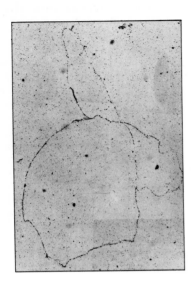

CHAPTER 8

BIOSYNTHESIS OF NUCLEIC ACIDS AND PROTEINS

Living cells must do several things in order to carry out normal life processes and remain alive. Of central importance are the events of nucleic acid and protein biosynthesis. Collectively they are responsible for the duplication and expression of genetic information and the study of these activities comprises the area of *biochemical genetics*. The duplication of chromosomes is necessary for the passage of genetic information during cell division to yield identical daughter cells. The expression of chromosomes is necessary to produce all of the proteins that determine the biochemical individuality of the living cell. A popular way of summarizing the flow of information is the so-called *central dogma of molecular biology*, which can be diagramatically represented as shown at the top of p. 237.

The *replication* of DNA is the biosynthesis of an exact copy of a DNA molecule from the monomeric deoxyribonucleotides. Genetic information is preserved by duplicating the nucleotide sequence of the parent DNA. The molecular expression of the genes in DNA involves two distinct phases. First, in the process called *transcription*, the genetic information contained in the nitrogen base sequence of DNA is rewritten as a complementary base sequence of RNA. That is to say, the genetic specificity originally in a

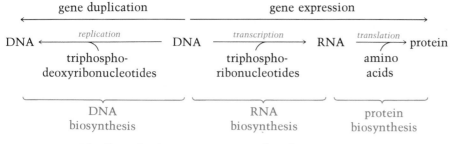

(the flow of information associated with some viruses
involves RNA→RNA and RNA→DNA)

DNA molecule is copied in the form of an RNA molecule—be it transfer-RNA, ribosomal-RNA, or messenger-RNA. Then, in the process called *translation*, the genetic information now contained in messenger-RNA is utilized to direct the assembly of proteins. In other words, the original language of genetic specificity and function, that is, the sequence of nucleotides in a nucleic acid, is converted into the language of protein specificity and function, that is, the sequence of amino acids in a polypeptide chain.

In addition to describing the biochemistry of these processes, in this chapter we will also examine the nature of the genetic code, cancer-related viruses, and the new technology of genetic engineering. The overall design of this book prohibits a thorough coverage of these complex events. However, most of the important highlights are included.

RNA BIOSYNTHESIS—GENE TRANSCRIPTION

Three types of genes in DNA are transcribed into complementary RNA molecules—transfer-RNA genes, ribosomal-RNA genes, and messenger-RNA genes—with messenger-RNA carrying the code for the assembly of a polypeptide with a specific amino acid sequence. Although each RNA is different, they are all produced in the same way.

The gene → mRNA → polypeptide hypothesis was first proposed in 1961 by F. Jacob and J. Monod. The existence of mRNA was demonstrated experimentally shortly thereafter in the same year.

Enzymology of Transcription

The enzyme responsible for transcription (RNA biosynthesis) is called *DNA-dependent RNA polymerase* (sometimes referred to as *transcriptase*). It was originally discovered in rat liver by S. Weiss and coworkers in 1959, and since then has been detected in virtually all types of living cells. The enzyme catalyzes the sequential polymerization of triphosphonucleotides with the instructions for nucleotide insertion provided by DNA, which functions as a *template*. Important characteristics of the enzyme's mode of action are described below.

$$\left. \begin{array}{c} n\text{UTP} + n\text{CTP} \\ + \ n\text{ATP} + n\text{GTP} \end{array} \right\} \xrightarrow[\text{(Mg}^{2+}\text{ and DNA required)}]{\substack{\text{DNA-dependent–} \\ \text{RNA polymerase}}} \cdots (pU)_n(pC)_n(pA)_n(pG)_n \cdots + 4n\text{PP}_i$$

mixture of
5'-triphosphoribonucleotides

polyribonucleotide
(RNA copy of DNA)

Chain growth is in the 5' → 3' direction. Polymerization involves the condensation of triphosphonucleotides via phosphodiester bonds. This could occur in either of two ways: (a) by 5′ → 3′ synthesis, if the chain is extended from the 5′ end, or (b) by 3′ → 5′ synthesis, if the chain is extended from the 3′ end. The chemistry catalyzed by DNA-dependent–RNA polymerase (and all other related enzymes) is *exclusively 5′ → 3′ synthesis*.

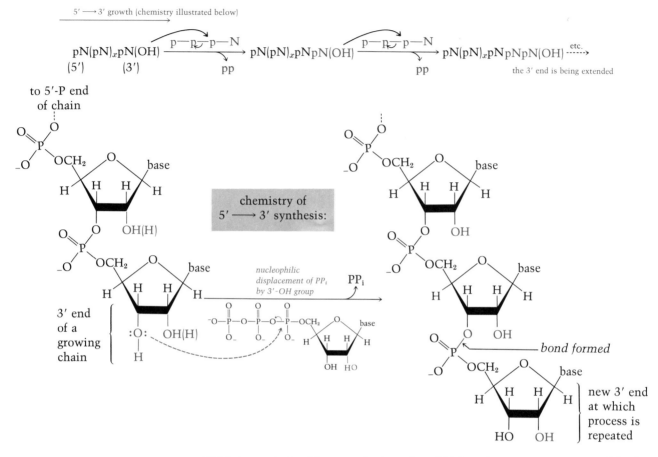

DNA is copied. The conclusion that RNA polymerase is responsible for the *in vivo* transcription of DNA is based upon much evidence obtained from *in vitro* studies demonstrating that the RNA product is complementary in base sequence to the DNA used. That is to say, the DNA actually *functions as a template, which is copied during synthesis*.

The most convincing evidence of DNA-copying by RNA polymerase has been provided by *hybridization* experiments (see p. 225, Chap. 7 and below). In the hybridization technique, a mixture containing both RNA and DNA is heated quickly and then cooled slowly. On heating, the DNA duplex unwinds to yield disordered polynucleotide strands. RNA molecules will likewise lose whatever conformational structure they possess. On slow cooling, the disordered chains can align with each other via complementary hydrogen bonding of base pairs to form stable duplexes. Thus, if this were done with a pure DNA solution, the DNA duplex would be destroyed and then reformed. If, on the other hand, we were to start with a mixture of DNA and RNA to yield a mixture of denatured DNA and RNA strands, and if the RNA had a base sequence complementary to that of one of the DNA strands, then slow cooling should also result in the formation of a few *DNA-RNA hybrid duplexes*. Ultracentrifugation, ultrafiltration, or gel

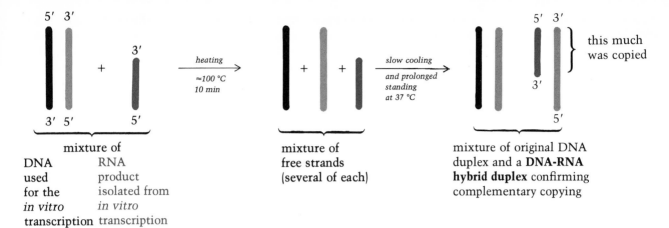

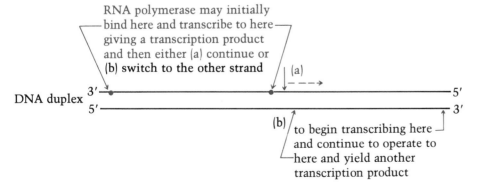

DNA **RNA**	
used product	
for the isolated from	
in vitro *in vitro*	
transcription transcription	

chromatography could then be used to isolate the DNA-RNA hybrids. This is precisely what was observed in studies of the hybridization between the product RNA and primer DNA of the RNA polymerase reaction, confirming that the RNA was a complementary copy of the DNA.

Strands can be started. The operation of RNA polymerase begins with the binding of the first NTP (usually ATP); then the second NTP binds, and then the active site residues catalyze the formation of the first phosphodiester bond, with a PP_i unit from the second NTP being released. The dinucleotide and the DNA template remain attached to the enzyme but there is displacement of each to allow for the third NTP unit to be positioned. The second phosphodiester bond is then formed and the process continues. Because RNA polymerase can start the synthesis by positioning the first two NTP units it is said to be *self-priming*. This is an important aspect of RNA polymerase action not shared by DNA polymerase enzymes, which operate differently. (We will learn more about this later (see p. 252).)

One strand copying—but either strand. When a single RNA polymerase molecule binds to DNA, it uses only one strand of the duplex as a template until transcription of that gene segment is complete. As shown below, however, the other strand can be transcribed (a) by the same polymerase molecule after it finishes the segment it started, or (b) by another polymerase molecule.

Structure of RNA polymerase. The following description applies to the most thoroughly studied transcription enzyme, that of *E. coli.* RNA polymerase is a large oligomeric enzyme (mol. wt. ≈ 500,000) composed of four different types of subunits labeled as α, β, β', and σ (sigma). The subunits are present in relative proportions of 2:1:1:1. Thus the intact enzyme is a pentamer: $\alpha_2\beta\beta'\sigma$. Tightly bound Zn^{2+} is also present. Enzymes from other sources are similar in both structure and mode of operation, but there are also significant differences (see p. 245).

Stages of Transcription

The DNA → RNA process involves a sequence of successive events, two of which involve *recognition signals*. Each step is described below and the entire sequence is diagramatically summarized in Fig. 8–1.

1. *Binding* of the polymerase to the DNA template is of course the first event. Obviously, this binding must occur at the start of the gene segment. This recognition is the function of the *sigma* (σ) *subunit*. The start of a gene (or group of genes) corresponds to a nontranscribed segment of nucleotides that is called a *promoter locus*. The promoter locus is itself a gene in the sense

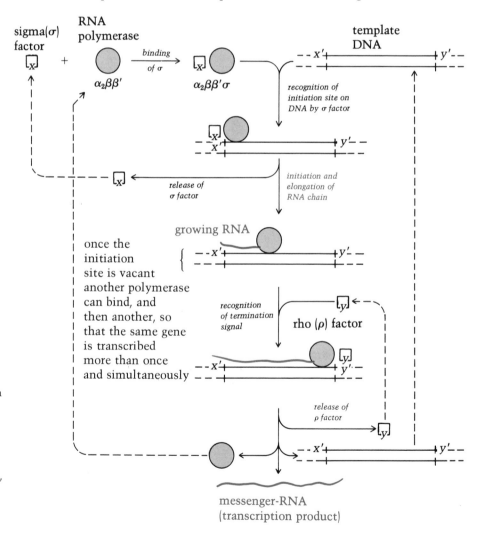

FIGURE 8–1 A diagramatic summary of DNA transcription by RNA polymerase. There is much evidence that the transcription sequence can begin with another RNA polymerase before the first is completed. In fact, it appears that several RNA products are assembled simultaneously, with each in a different stage of development on separate RNA polymerase molecules. There is nothing particularly mysterious about this since the initiation binding site is vacated as a result of the relative movement of the polymerase and the DNA template. Once vacated, another RNA polymerase would bind and proceed with transcribing, and then another would, and then another. See Fig. 8–10 for electron microscopic evidence of multiple transcriptions. The diagram here shows only one transcription sequence, for the sake of simplicity.

x = binding site in σ factor recognizing initiation site (x') in DNA
y = binding site in ρ factor recognizing termination site (y') in DNA

(- - - → indicates recycling of participants)

that it carries information. The information it carries is a *nucleotide sequence that is recognized as a binding site by the σ factor* (see next section).

2. *Initiation* of RNA assembly begins with the binding of ATP (or GTP), followed then by the positioning of a second nucleotide (NTP) according to template instructions from the DNA that is being copied. Phosphodiester bond formation yields pppApN(3'-OH).

3. *Elongation* of the RNA chain proceeds according to 5' → 3' chemistry, with DNA providing instructions for the sequential insertion of each nucleotide. During elongation the double helix conformation of the template DNA must unfold to expose the bases on the strand being copied. The unfolding involves the participation of other proteins, which we will not discuss further.

4. *Termination* occurs at the end of the transcription unit when another non-transcribed segment—the *terminator* locus—is reached. The recognition is similar to that of the σ factor and protomer. The protein responsible for recognizing and binding to the terminator locus has been isolated. It is called the *rho (ρ) factor* (see next section).

5. *Release* of the polymerase and the assembled RNA product completes one cycle.

Protein-specific Recognition of DNA Sites

The protein-specific recognition of "start" and "stop" sites (promoter and terminator sequences) guarantees the transcription of intact segments of genetic information. Obviously without such signals and their faithful recognition the readout of genetic information would be completely random and disordered. The RNA polymerase would bind at any site on DNA and begin transcribing to any other site. Thus, transcription could start within one gene and stop within any other gene, resulting in a "garbled" product.

How this recognition occurs is not yet known with certainty. One suggestion for promoter recognition is the presence of an AT-rich region of base pairs flanked on each side by GC-rich regions. Another suggestion (which is proposed to have a general application to all types of recognition sites in DNA) is that the individuality of the DNA sites is contained in *sequences of inverted symmetry* in the two strands of DNA. Such regions as referred to as *palindromes*. A palindrome is a word (or sentence) that spells (or reads) the same way in both directions. For example:

$$
\begin{array}{lll}
\text{word palindromes:} & \overrightarrow{\underleftarrow{\text{madam}}} \quad \overrightarrow{\underleftarrow{\text{tut}}} \quad \overrightarrow{\underleftarrow{\text{toot}}}
\end{array}
$$

a sentence palindrome:

$$\overset{1\quad 2\quad 3\quad 4\quad 5\quad 6\qquad\quad 7}{\text{a man, a plan, a canal, Panama}}\underset{7\qquad\quad 6\quad\; 5\quad 4\quad 3\; 2\; 1}{}$$

In duplex DNA a palindrome nucleotide sequence would be represented by

a contiguous palindrome:

3'------C A G C G G A A T T C A T G C C------5'
5'------G T C G C C T T A A G T A C G G------3'

a spaced palindrome:

There is no doubt that palindrome sequences exist and no doubt that they serve as specific sites that are recognized by specific proteins. For example, the continuous sequence shown above represents the recognition site of the EcoRI *restriction enzyme*. Indeed, the specificity of all restriction enzymes is due to their recognition of palindrome sites. Examine the other examples given in Table 7–1 on p. 216. The example of the spaced sequence shown above corresponds to a 26 base-pair segment of a promoter gene isolated and sequenced from the *E. coli* chromosome.

The recognition of a palindrome sequence by a specific protein may occur either (a) because the palindrome sequence is longitudinally symmetrical, or (b) because the palindrome region may snap back on itself to form a cruciform (a hairpinlike distortion) of the DNA duplex. Either structure could be recognized by a dimeric protein having twofold symmetry.

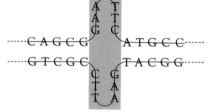

——C A G C G G A A T T C A T G C C——   *"snap-back"*   ——C A G C G⟍ ⟋A T G C C——

——G T C G C C T T A A G T A C G G——   ⇌   ——G T C G C⟋ ⟍T A C G G——

fully extended duplex segment with the
EcoRI palindrome site

cruciform structure
(hairpin loop distortion)

*because each strand of the
palindrome sequence consists of
two complementary halves*

Modification of RNA—Post-Transcription Events

In many instances the initial transcription products are not identical to the functionally competent, mature forms of RNA, but rather are precursors that must be converted to the mature forms through various molecular alterations known as *processing*. The modes of processing include: (a) *nucleolytic events* in which specific phosphodiester bonds are cleaved, resulting in the *trimming* of larger precursors; (b) *terminal additions of nucleotides*; and (c) *specific modifications of nucleotides*. All types of cellular RNAs are processed but each in a characteristic manner, with both similarities and differences between eukaryotic and prokaryotic sources.

Transfer-RNA and Ribosomal-RNA. The initial transcription products of transfer-RNA genes and ribosomal-RNA genes are not the final tRNA and rRNA molecules but are larger precursors that are first timmed to size. The trimming is done by exo- and endonuclease enzymes operating at specific cleavage sites. The complex series of steps that intervene between precursor to product is illustrated on p. 243. This diagram depicts a dimeric tRNA-precursor where two tRNAs are contained in the same precursor. The two-from-one pattern does not apply to all tRNAs. Depending on the specific tRNA, sometimes the precursor is monomeric (one-from-one) and sometimes it is multimeric (three-or-more-from-one). However, the removal and replacement of small oligonucleotide fragments at the 3′ terminus to yield the -CCA sequence is universal to the biosynthesis of all tRNAs.

Transfer-RNA molecules also undergo further processing (done mostly at

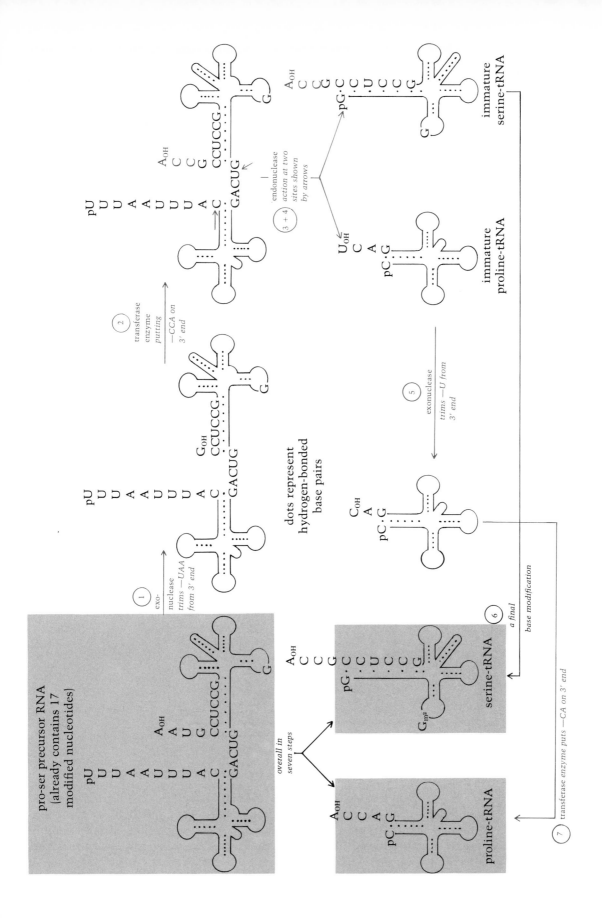

the level of the larger precursor), with various enzymes catalyzing the modification of a few crucial nitrogen bases. These modifications (see p. 230 in
Chap. 7) contribute to the structural and functional individuality of tRNA
molecules. After trimming, ribosomal-RNA molecules do not undergo any
further processing except for their ordered association with ribosomal proteins to yield intact ribosomes.

Messenger-RNA. In prokaryotic cells the mRNA transcript product is used
without any further processing. In eukaryotic cells, however, most mRNA
molecules undergo further processing, namely, (a) the "capping" of the 5'
terminus with Gppp, (b) the methylation of a few residues at the 5' end, and
(c) the attachment of a poly(A) segment at the 3' terminus. At present the
order in which these alterations occur is still uncertain. The significance of
these modifications was mentioned in the previous chapter (see p. 227).

The newest discovery regarding mRNA biosynthesis in eukaryotic cells is
quite remarkable. Recently it was discovered that a number of genes in
DNA contain nucleotide sequences (containing as many as 500 base pairs)
that are not found in the messenger RNAs corresponding to the genes. As
yet it is not known why these sequences (they have been called *intervening*
or *spacer* sequences) are present. The interesting thing is that they are
transcribed in the initial formation of mRNA but are then selectively ex-
cised to yield the final mRNA. Since the intervening sequences are located
within the initial transcript, it is obvious that the excision and the rejoining
of the two RNA pieces must be performed with great precision.

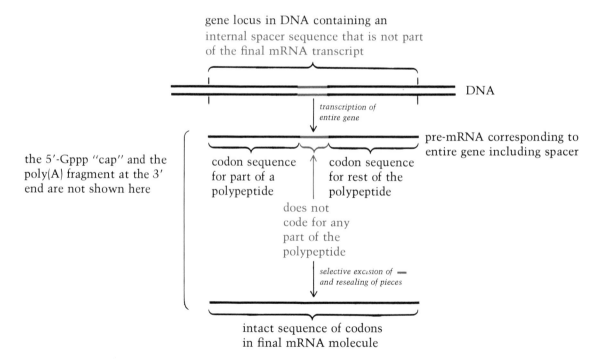

gene locus in DNA containing an
internal spacer sequence that is not part
of the final mRNA transcript

DNA

*transcription of
entire gene*

pre-mRNA corresponding to
entire gene including spacer

the 5'-Gppp "cap" and the
poly(A) fragment at the 3'
end are not shown here

codon sequence
for part of a
polypeptide

codon sequence
for rest of the
polypeptide

does not
code for any
part of the
polypeptide

selective excision of —
and resealing of pieces

intact sequence of codons
in final mRNA molecule

As far as is known, prokaryotic mRNAs have no caps, have no poly(A)
fragments, and they are total copies of their genes with no missing segments. Thus it appears that eukaryotic gene expression is more complicated than originally predicted on the basis of knowing how it occurs in bacteria.

About 30 different antibiotics have had their mode of action linked to an inhibition of the transcription process. This inhibition can occur one of two ways. *Rifampicin* is typical of a *direct* inhibitory action of the RNA polymerase. Specifically, rifampicin binds to the β subunit of RNA polymerase. The inhibition is severe, with nanomolar quantities of the antibiotic resulting in a 50% loss in polymerase activity (*E. coli*). However, eukaryotic polymerases are relatively resistant to rifampicin, indicating that significant differences between the transcription enzyme in eukaryotic and prokaryotic cells do exist. From a chemotherapeutic standpoint in the treatment of disease (in humans, for instance) this distinction is rather important, since it is desirable to utilize drugs that will preferentially attack the infectious organism and not harm the cells of the host organism.

rifampicin
(binds with RNA polymerase
preventing the binding of NTPs
to the polymerase)

Actinomycin D inhibits transcription by a different and less discriminating mechanism than rifampicin. Actinomycin functions by binding firmly with the DNA template. This means that continued action of RNA polymerase is prevented since the enzyme's movement on the template is impeded. In other words, it has an *indirect* inhibitory action on enzyme activity. The binding is strong, involving (1) the wedging of the phenoxazone portion of the actinomycin molecule in the core of the duplex between adjacent nitrogen base residues and (2) the association of the two cyclic peptide units with the strands of DNA forming one of the grooves.

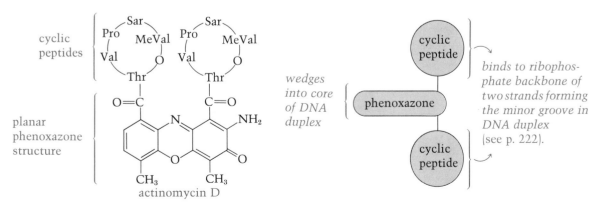

binds to ribophosphate backbone of two strands forming the minor groove in DNA duplex (see p. 222).

245

Natural Control of Transcription

Another important aspect of the transcription process is that it is subject to control by the cell itself through a phenomenon called *induction-repression*. This mode of control accounts in part for how a living organism can regulate itself at the level of gene expression by the regulation of protein biosynthesis. We will examine this subject later in the chapter (see p. 277).

Reverse Transcription (RNA is template; DNA is product)

One of the most important findings of recent years was the discovery (in 1970 by H. Termin, S. Mizutani, and D. Baltimore) of an *RNA-dependent DNA polymerase*. In effect, the enzyme is a *reverse transcriptase* in that it is capable of *forming DNA from an RNA template*. In other words, the central dogma of the flow of information as originally enunciated by Crick in 1958 has proved to be an oversimplification. The process is reversible, at least between DNA and RNA; that is, DNA $\rightleftarrows$ RNA.

On an applied level, a more significant aspect of this enzyme is that it was originally detected in an RNA tumor virus of animal origin, and since then has been detected in several other animal RNA viruses. This has prompted the suggestion that the ability of these viruses to transform normal cells into cancer cells may result from the transcription of viral-RNA: it would first be transcribed into a DNA molecule by its own reverse transcriptase, which becomes active on infection of a host cell; this viral-DNA *might* then be incorporated into the chromosome of the normal host cell, which *might* transform it into a tumor cell. The uncertainty implied by the italicized words is significant at least to human cells. Progress is being made in linking viruses to human cancer, but as of yet there is no decisive evidence that a virus is the direct cause of transforming normal human cells into cancer cells.

Research in this area with human cancers has been very controversial. Various reports announcing the isolation of cancer-related viruses of human origin have appeared in recent years but they proved to be false and have been retracted. In 1974, however, C. McGrath and M. Rich announced the isolation of a virus from cancerous human breast tissue. The virus could not be traced to having originated from contamination during laboratory culture of the breast tissue over a long term. It was a human virus of human origin. Then, more recently, in 1975, R. E. Gallagher and R. C. Gallo also announced a similar success with the isolation of a virus from leukocytes cultured from a patient with acute myelogenous leukemia. This virus also was characterized as being of human origin. Both were RNA viruses and both contained reverse transcriptase activity. These findings offer the hope that certain types of cancer—in this case, those induced by RNA viruses—will be understood and possibly prove to be susceptible to chemotherapeutic treatment. If a causative viral function proves to be valid, then it should be feasible to find substances that would inhibit the activity of the viral reverse transcriptase and thus prevent the multiplication of the virus and additional transformation of normal cells.

DNA BIOSYNTHESIS—GENE REPLICATION

General Principle: Semiconservative Replication

In the late 1950s M. Meselson and F. W. Stahl in a brilliantly conceived experiment demonstrated that in dividing *E. coli* cells the bacterial chromo-

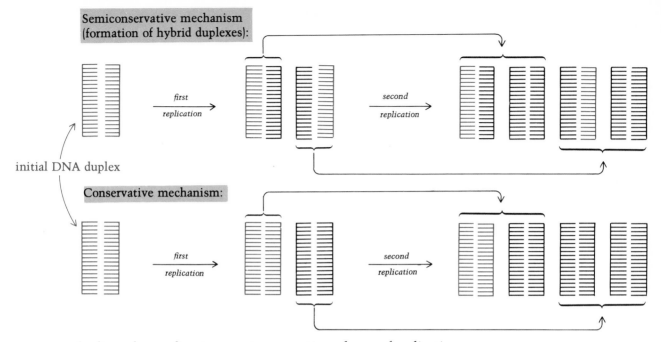

Semiconservative mechanism
(formation of hybrid duplexes):

initial DNA duplex

first replication

second replication

Conservative mechanism:

first replication

second replication

FIGURE 8–2 A comparison of two possible mechanisms for DNA replication.

some was duplicated according to a *semiconservative* scheme of replication. Subsequent studies verified the original finding with *E. coli* and established that the same scheme applies to all types of organisms.

According to the semiconservative mechanism, each strand of the original DNA molecule acts separately as a template to direct the formation of its complement and then combines with it (see Fig. 8–2). Each of the two new double-helix molecules would thus consist of a newly synthesized polynucleotide chain combined with its complementary chain from the original molecule. In other words, each newly produced DNA would be a *hybrid* of old and new. This is distinct from a *conservative* scheme of replication wherein each chain of the original DNA molecule would direct the formation of a new complementary chain, but then the two new chains would combine with each other. No hybrids of old and new chains would be produced.

The experiment was based on attempting to prove or disprove the formation of hybrid duplexes and thus to prove one of the two possible schemes. To do this (see Fig. 8–3), *E. coli* cells were first grown in a medium containing $^{15}NH_4Cl$ as the sole nitrogen source. ^{15}N is a *stable, heavy isotope* of the most abundant natural form of nitrogen, ^{14}N. Consequently, after growth in such a medium, all nitrogen-containing compounds will eventually be "labeled" with ^{15}N. The DNA of the cells will be extensively labeled, because each molecule will contain several thousand ^{15}N atoms in the many purine and pyrimidine nitrogen bases. Because of this, ^{15}N-DNA will have a greater density (it will be heavier) than ^{14}N-DNA or any hybrid of DNA containing a portion of ^{14}N and a portion of ^{15}N, and should be separable by density gradient ultracentrifugation.

After the cells were labeled in ^{15}N medium, they were washed and transferred to a similar medium containing only $^{14}NH_4Cl$, and growth was continued for several generations. Every time the population doubled, that is, with the production of every new generation of cells, a sample was removed and the *DNA assayed for its content of ^{14}N and ^{15}N*. Ultracentrifugal analysis of the DNA obtained after one generation revealed only one zone in the

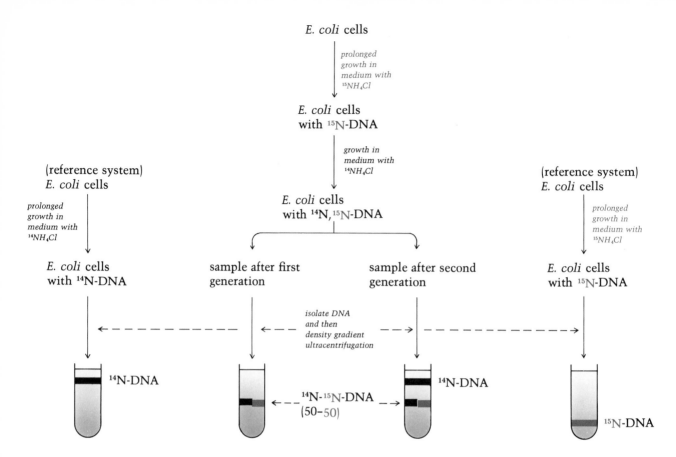

FIGURE 8-3 Diagramatic summary of Meselson-Stahl experiment proving the semiconservative mechanism of DNA replication.

tube, which was 50% lighter than the completely ^{15}N-DNA and 50% heavier than the completely ^{14}N-DNA. No other bands were present in significant amounts. This clearly indicated that after one doubling all of the DNA molecules *were 50–50 hybrids* of ^{14}N and ^{15}N. Furthermore, the DNA obtained after the second generation displayed two bands. One was equivalent to the band obtained after the first generation and the second was equivalent to ^{14}N-DNA. Moreover, both were present in equal amounts. None of the samples had a band equivalent to all ^{15}N-DNA. All of these data proved that the formation of hybrid DNAs could have arisen only by a semiconservative scheme of replication.

General Principle: Simultaneous Synthesis of Both Strands from a Replication Origin

The next significant insight into the molecular nature of chromosome replication was provided by J. Cairns and coworkers in 1963, who succeeded in *observing* the chromosome of *E. coli* during stages of its replication. This feat was accomplished by growing *E. coli* cells in a medium containing tritium (^{3}H)-labeled thymidine deoxyribonucleotide. The intent was that the nucleotide would enter the cell and be utilized in the production of DNA as it was replicated. Obviously the DNA would then be radioactively labeled due to the thymine positions being ^{3}H-labeled. After the addition, cells were removed at various times and the DNA was extracted. The DNA extract was then spread on a piece of X-ray film to detect the weak β radiation emitted by the ³H atoms of the thymine bases. After prolonged periods of exposure, the developed film showed a track of silver grains corresponding to the labeled DNA. An idealized drawing of such an image is shown below (left side). This image corresponds to an intact duplex (double helix) obtained after one generation of ³H-thymine-labeled bases. Since one of the strands of the duplex would be uniformly labeled, the track of silver grains would be of uniform intensity. According to the semiconservative pattern of replication, two types of images would be expected after a second round of cell division and DNA replication. Each image would be of uniform intensity but one image would be stronger because both strands would be ³H-labeled.

The type of silver grain patterns observed *during* the second generation period (right side) was truly remarkable. It showed two circular regions, with part of one having about twice the silver grain intensity of the other. The more intense part of the image was interpreted as representing the copying of the labeled strand of the first generation duplex, which would contain twice as much ³H-thymine, since both strands now contain the ³H-thymine label. The less intense regions represent (a) the second duplex being synthesized as the unlabeled strand of the original duplex is replicated, and (b) the part of the original first generation duplex that had not yet been replicated.

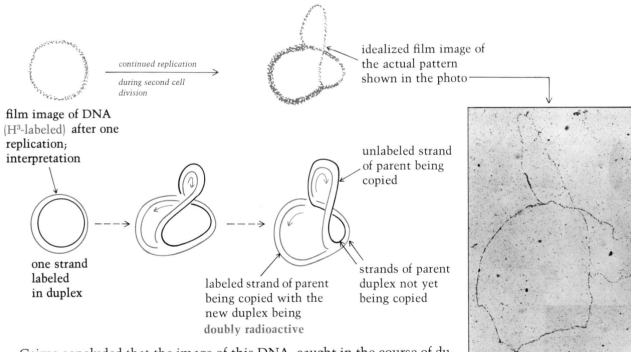

continued replication

during second cell division

idealized film image of the actual pattern shown in the photo

film image of DNA (H³-labeled) after one replication; interpretation

unlabeled strand of parent being copied

one strand labeled in duplex

labeled strand of parent being copied with the new duplex being **doubly radioactive**

strands of parent duplex not yet being copied

Cairns concluded that the image of this DNA, caught in the course of duplication, must mean not only that replication was *semiconservative* but also that the synthesis of both strands was *simultaneous*. Cairns went on to propose a model of replication with the key feature being that synthesis proceeds from a certain region on the duplex, which was called the origin. At the origin it was proposed that some sort of swivel mechanism would operate to unwind a short segment of the duplex, exposing both strands for simultaneous replication. From this point each strand could then be synthesized in the same direction (*unidirectional synthesis*) or in opposite

(Reproduced with permission from *Cold Spring Harbor Symposium in Quantitative Biology,* **28:** 43 (1963); photograph supplied through the courtesy of John Cairns.)

"fork"

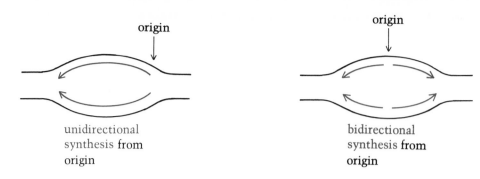

origin

unidirectional synthesis from origin

origin

bidirectional synthesis from origin

directions (*bidirectional synthesis*) progressing away from the point of origin. With a continuing action of the swivel mechanism, the replication fork would continue to be displaced along the original duplex until both strands (and thus two new duplexes) were synthesized.

The simultaneous synthesis of both strands in a duplex of opposite polarity (see margin) would seem to demand that the enzyme (or enzymes) that participates in the synthesis of DNA be capable of forming phosphodiester bonds in both the $5' \rightarrow 3'$ direction and the $3' \rightarrow 5$ direction. (*Note:* To yield a duplex of opposite polarity, the strand being assembled from the $3' \rightarrow 5'$ strand in the parent duplex must be assembled in the $5' \rightarrow 3'$ direction. On the other hand, the strand being assembled from the $5' \rightarrow 3'$ strand in the parent duplex must be assembled in the $3' \rightarrow 5'$ direction.) As we proceed now to consider the enzymology of DNA biosynthesis you will want to keep this item in mind.

Recall: the two strands in DNA are aligned in opposite directions.

5' ——————— 3'
3' ——————— 5'

Enzymology of DNA Replication

The replication of DNA is accomplished by *DNA-dependent DNA polymerase.* As shown below, the action of this enzyme is similar in a general way to the DNA-dependent RNA polymerase of transcription (see p. 237). Each enzyme (a) utilizes triphosphonucleotides as substrates, (b) requires Mg^{2+} for optimal activity, (c) requires DNA, which is copied, and (d) catalyzes the formation of a new chain *only in the $5' \rightarrow 3'$ direction.* All DNA polymerases isolated to date from all types of sources have the same characteristics.

$$\left. \begin{array}{l} n\text{dTTP} + n\text{dCTP} \\ + n\text{dATP} + n\text{dGTP} \end{array} \right\} \xrightarrow[\text{(Mg}^{2+}\text{ and DNA required)}]{\substack{\text{DNA-dependent} \\ \text{DNA polymerase}}} \cdots (\text{pdT})_n(\text{pdC})_n(\text{pdA})_n(\text{pdG})_n \cdots + 4n\text{PP}_i$$

mixture of
5'-triphosphodeoxyribonucleotides

polydeoxyribonucleotide
(DNA copy of DNA)

The initial isolation and characterization of DNA polymerase from *E. coli* was accomplished in 1958 by Arthur Kornberg (Nobel Prize, 1959) and coworkers. Since then, Kornberg and other investigators have unraveled many details about this extremely complex multienzyme process as it occurs *in vivo.* The summary of DNA polymerase action shown here is simplified.

DNA is copied—chemical proof. After its isolation (see margin) it was necessary to prove that DNA polymerase produced a copy of DNA. Base composition studies showing the A, T, G, and C content of product DNA to be identical with the primer DNA were supportive but not conclusive evidence. A most convincing proof was obtained by an ingenious experimental strategy, developed by the Kornberg group, called *nearest-neighbor analysis.* As the language implies, this technique permits the measurement of the relative frequency with which each nitrogen base appears adjacent to itself and to each of the other three bases in a polynucleotide chain.

Since there are four nitrogen bases, this means that there are 16 (4^2) possible dinucleotide sequences in question.

The technique of nearest-neighbor analysis is performed in four separate phases, with each phase consisting of the following steps: (a) DNA polymerase forms DNA with one of the triphosphonucleotides being radioactively labeled with ^{32}P; (b) the labeled DNA product is recovered and hydrolyzed completely by the action of a 5'-nuclease; and (c) the hydrolyzate mixture of nucleotides is then analyzed to determine the relative amounts of each labeled nucleotide. It may seem foolish to form DNA in one step and degrade it in another, but the pattern of formation of the phosphodiester bonds is different than the cleavage of the bonds. To be specific, DNA polymerase forms a bond between phosphate and the 3'-OH of the sugar (*a*-type formation), whereas the 5'-nuclease cleaves the bond between phosphate and the 5'-OH of the sugar (*b*-type cleavage). The result (see Fig. 8–4) is that the phosphorus atom originally contributed by a substrate

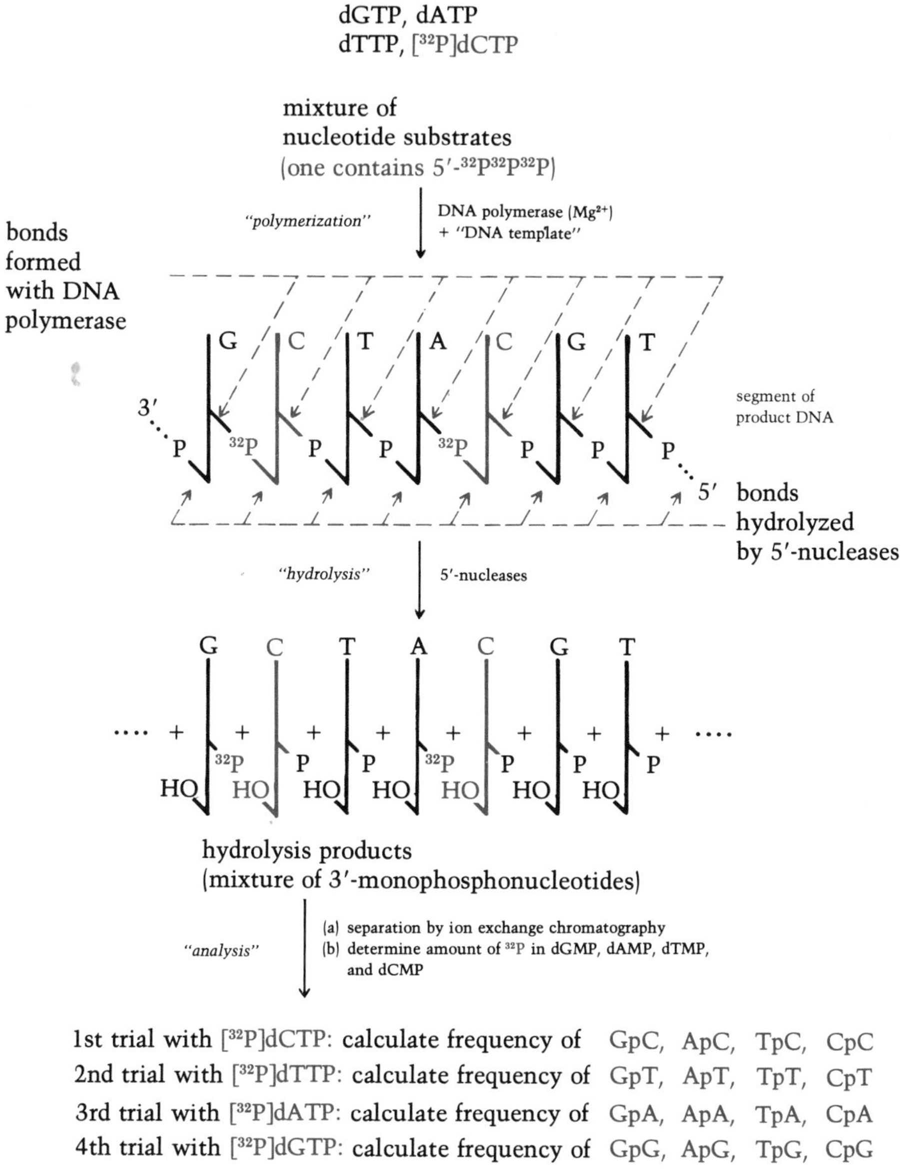

1st trial with [^{32}P]dCTP: calculate frequency of GpC, ApC, TpC, CpC
2nd trial with [^{32}P]dTTP: calculate frequency of GpT, ApT, TpT, CpT
3rd trial with [^{32}P]dATP: calculate frequency of GpA, ApA, TpA, CpA
4th trial with [^{32}P]dGTP: calculate frequency of GpG, ApG, TpG, CpG

FIGURE 8–4 A summary of the experimental approach of nearest-neighbor analysis.

nucleotide in the formation of phosphodiester bonds by DNA polymerase is transferred to its nearest neighbor on the 5' side by the action of the 5'-nucleases. The use of radioactive phosphorus serves as the marker to detect and measure the frequency of such transfers when *only one* of the four triphosphonucleotide substrates is labeled. For example, if dTTP, dATP, dGTP, and (^{32}P)dCTP were used as substrates, hydrolysis of the product DNA would yield specific amounts of 3'-(^{32}P)dTMP, 3'-(^{32}P)dAMP, 3'-(^{32}P)dGMP, and 3'-(^{32}P)dCMP. The amount of radioactivity associated with each of the 3'-monophosphonucleotide fractions would be governed by the frequency of occurrence along the chain of the following dinucleotide sequences: TpC, ApC, GpC, and CpC. The repetition of this experiment under the same conditions with the same template DNA, but with a different labeled triphosphonucleotide each time, would permit the frequency of all 16 sequences to be measured. The calculations are not shown.

Determining the 16 dinucleotide frequencies of the product DNA proves nothing about the fidelity of the enzyme in copying the original DNA. But what was done then was to use the first product DNA as the template DNA and redetermine the dinucleotide frequencies again in the second-generation product DNA. These turned out to be nearly identical and thus it could be concluded that DNA polymerase can indeed use DNA as a template and do so very faithfully.

In addition, the nearest-neighbor analysis also provided the first direct experimental proof of the opposite polarity (that is, antiparallel orientation of strands) of the DNA duplex. This conclusion was based on the fact that the frequency of occurrence of each of the 16 dinucleotide sequences in the $5' \rightarrow 3'$ direction was identical to the frequency of occurrence of its complementary dinucleotide sequence in the opposite $3' \rightarrow 5'$ direction, rather than the same $5' \rightarrow 3'$ direction. For example (see margin), the frequency of occurrence of TpG was equal to that of CpA and not ApC; the occurrence of GpA was equal to that of TpC and not CpT; and so on.

$(5')$TpG$(3')$ →

$(5')$ApC$(3')$ →

ApC would be the dinucleotide complement of TpG if the two strands were oriented in the same direction

$(5')$TpG$(3')$ →

$(3')$ApC$(5')$ ←

CpA would be the dinucleotide complement of TpG if the two strands were oriented in opposite directions

DNA is copied—biological proof. In 1967 M. Goulian, Kornberg, and R. L. Sinsheimer achieved the ultimate proof for the fidelity of the copying ability of DNA polymerase. They used the enzyme *in vitro* to produce a copy of an intact viral chromosome and then demonstrated that the synthetic product was biologically active. The chromosome chosen was the single-stranded circular DNA of ϕX174 (a tiny bacterial virus), which consists of nearly 5,400 nucleotides forming nine different genes. The synthetic chromosome infected *E. coli* cells, underwent replication in the cells, and intact ϕX174 viruses were produced. The ingenious strategy of the experiment is described later in this chapter (pp. 255–257).

A 3'-OH primer terminus is required. DNA polymerase is incapable of copying a DNA template when the template exists as an intact duplex (linear or circular). The reason is that, unlike the RNA polymerase enzyme of transcription, *DNA polymerase cannot start the assembly of a new chain.* Rather, it requires a 3'-OH primer terminus in one strand and then extends the length of this primer strand according to the template instructions of the other strand. The drawings below illustrate this requirement.

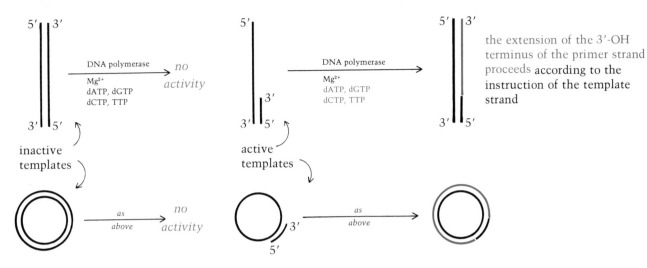

the extension of the 3'-OH terminus of the primer strand proceeds according to the instruction of the template strand

The *In Vivo* Process—Other Enzymes Participate

The discovery and characterization of DNA polymerase did not provide an immediate understanding of how DNA replication occurs inside a cell. In fact, the *in vivo* participation of DNA polymerase was seriously challenged. One of the major arguments against the enzyme was that it is capable of building a chain only in the $5' \rightarrow 3'$ direction, yet the Cairns study indicated that DNA synthesis proceeds in both the $5' \rightarrow 3'$ and $3' \rightarrow 5'$ directions.

The resolution of this directionality problem was aided by the work of R. Okazaki, who used certain growth conditions that slow down the replication of *E. coli*. Okazaki found that much of the newly formed DNA in these cells could be isolated as short pieces of DNA sedimenting at 10S–12S, corresponding to polynucleotide chain lengths in the range of 1,000–2,000 residues. This was interpreted to mean that DNA was not assembled continuously along the entire length of each strand, but rather *in a discontinuous piece-by-piece manner with all segments eventually being joined together.* Immediately it was proposed that the sealing of segments to each other was accomplished by another enzyme called *DNA ligase.* DNA ligase was known to exist, but until this time its biological role had not been determined. Further studies, showing that Okazaki fragments (as they are called) accumulate in mutant *E. coli* cells deficient in DNA ligase activity, provided conclusive proof of this aspect of DNA replication. How this fits in with the exclusive $5' \rightarrow 3'$ directionality of DNA polymerase will be evident shortly.

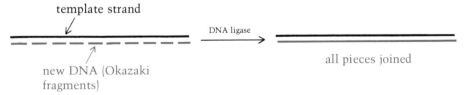

template strand

new DNA (Okazaki fragments)

all pieces joined

Another apparent problem with DNA polymerase action is in its inability to use an intact DNA duplex as template. How then can it start the replication process *in vivo*? Well, it is now certain that DNA polymerase does not start the process. Rather, other enzymes are first involved to generate a 3'-OH primer location and only then does the action shift to DNA polymerase.

253

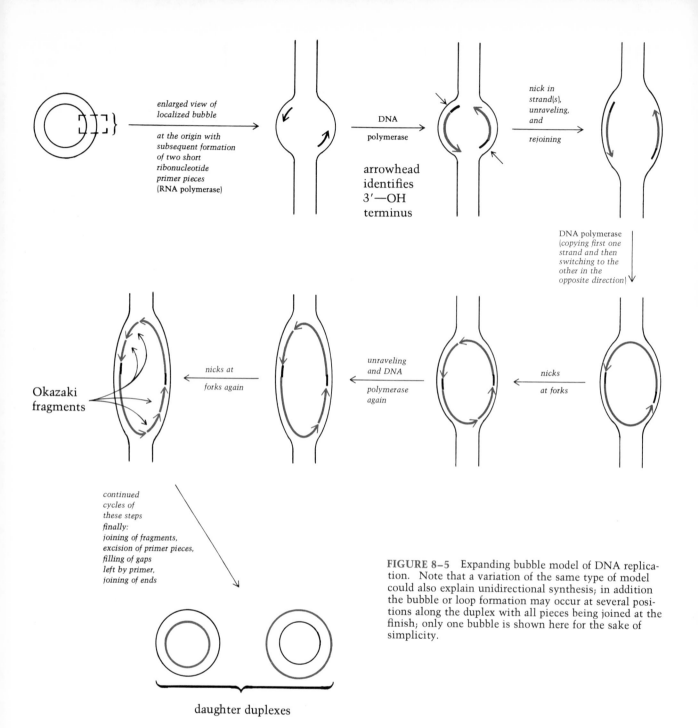

enlarged view of
localized bubble

at the origin with
subsequent formation
of two short
ribonucleotide
primer pieces
(RNA polymerase)

DNA
polymerase

arrowhead
identifies
3'—OH
terminus

nick in
strand(s),
unraveling,
and
rejoining

DNA polymerase
(copying first one
strand and then
switching to the
other in the
opposite direction)

nicks
at forks

unraveling
and DNA
polymerase
again

nicks at
forks again

Okazaki
fragments

continued
cycles of
these steps
finally:
joining of fragments,
excision of primer pieces,
filling of gaps
left by primer,
joining of ends

FIGURE 8–5 Expanding bubble model of DNA replication. Note that a variation of the same type of model
could also explain unidirectional synthesis; in addition
the bubble or loop formation may occur at several positions along the duplex with all pieces being joined at the
finish; only one bubble is shown here for the sake of
simplicity.

daughter duplexes

The general highlights of one model for DNA replication in *E. coli* are
summarized in Fig. 8–5. It is not the final word on the *E. coli* process nor is
it totally applicable to all types of cells. The scheme is bidirectional (see
margin) in that new synthesis proceeds in opposite directions from a replication origin. For the sake of simplicity only a single origin site is indicated,
although several may exist. Initially the duplex is deformed to expose each
strand. This deformation (Cairns called it a bubble) would involve the participation of an enzyme that "nicks" one of the strands (a single phospho-

diester bond is cleaved), causing the duplex to undergo a localized unwinding, and then the same or another enzyme would seal the nick. The next step requires the action of *RNA polymerase to assemble two small primer segments*. DNA polymerase would then extend the primer segments according to template instructions of each strand until the deformed region was filled. Another sequence of nicking, unraveling, and rejoining of strands would enlarge the location in each direction and DNA polymerase action would resume—advancing toward each fork. At the fork, DNA polymerase would then *switch strands* and begin using a new template, stopping when it runs out of template. The new strands would then be nicked at precisely the point when the switch occurred and yet another stage of deforming the parent duplex would begin. Continued cycles would eventually replicate the entire parent duplex. The replication process would be completed by (a) the action of *DNA ligase* to join all the new DNA pieces (these are the Okazaki fragments), (b) the excision of the RNA primer segments, and (c) the filling of the gaps left after the primer segments are removed. *Note: The exclusive 5' → 3' capability of DNA polymerase is sufficient to copy both strands because of its remarkable ability to copy a segment of one strand and then switch to the other strand for further template instructions.*

In *E. coli* there are three different DNA polymerases, referred to as pol I, pol II, and pol III (the numbers reflect the order of discovery). Although present in the least amount (about 10 molecules per cell), pol III is believed to be the major *replicating enzyme*. The primary roles of pol I are proposed to be in (a) filling the gaps that result from excision of the RNA primer segments and (b) acting as a *repair enzyme* for DNA when it undergoes damage due to ultraviolet radiation (see below) or the internal action of various endo- and exonucleases in the cell.

The antiviral drugs, ara-A and ara-C (see p. 209), operate by interfering with DNA replication. The precise mode of action is not yet established but it is known that the ara-A and ara-C nucleosides are first converted in the cell to the triphosphonucleotides ara-ATP and ara-CTP. The inhibition of the DNA replication process may involve competition with the normal positioning of dATP and dCTP. Insertion of ara-AMP and ara-CMP may also prevent continued extension of new DNA.

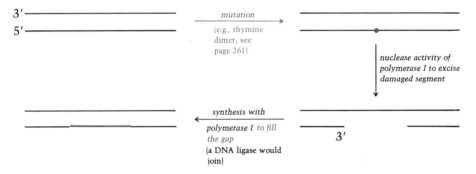

Multiple polymerases also occur in eukaryotic cells. To date three distinct cellular enzymes, called DNA polymerase α, β, and γ, have been discovered. A fourth enzyme has also been detected in mitochondria. The *in vivo* roles of α, β, and γ are not yet known, and this includes the question of which is the primary replicating enzyme. In eukaryotic cells, the replication of DNA may also involve participation of the histone and nonhistone proteins (see p. 232) that are present in the chromatin complex.

In Vitro **Synthesis of a Biologically Active DNA**

Further proof that this DNA polymerase is responsible for DNA replication was obtained in a remarkable experiment performed by M. Goulian, Kornberg, and R. L. Sinsheimer in 1967. The objective of the experiment was to

With approximately 5,500 nucleotides arranged in specific sequence, the φX174 viral DNA would appear to be a massive template. In comparison to other chromosomal DNAs, however, the size of φX174 DNA is quite small. Most other viral chromosomes are larger. On the cellular level, the fact that the single chromosome of an *E. coli* cell contains approximately 10,000,000 nucleotides further indicates the smallness of the φX174 DNA. The DNA from higher organisms would, of course, be even larger.

nonreplicative form of φX174 DNA

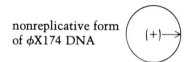

replicative form of φX174 DNA

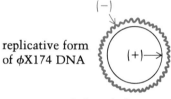

(−) circle **is** complementary to (+) circle

deoxybromouridine triphosphate (dBUTP)

O

HN—Br

O—N

deoxyribose-P-P-P

deoxythymidine triphosphate (dTTP)

O

HN—CH₃

O—N

deoxyribose-P-P-P

duplicate *in vitro* an intact DNA molecule obtained from a known biological source, and then to test whether the synthetic copy was biologically active. The DNA chosen as a template was the single chromosome of a tiny bacterial virus, designated φX174.

Nature aided considerably in the design of the experiment, because it so happens that the φX174 chromosome exists in both a *nonreplicative* state and a *replicative* state. The former is a circular molecule consisting of only one polynucleotide strand (+). That is, it is single stranded. The replicative state, however, consists of two intertwined complementary strands (+ and −). That is, it is double stranded. Thus, the objective was first to use the (+) strand as a template to make a synthetic (−) strand, and then to use the (−) strand to make a synthetic (+) strand. If the DNA polymerase were truly competent, then the synthetic (+) strand should be identical to the original template (+) strand. In other words, the synthetic (+) strand should carry all of the genetic information of φX174, and hence be biologically active φX174 viral DNA.

To test this, a brilliant technique was devised, which in part required the closure of the newly synthesized DNA molecule to give an intact circular (−) strand and then an intact circular (+) strand. This was done enzymatically by use of a *joining enzyme,* called *DNA ligase,* which was also isolated from *E. coli.* The closure to yield an intact circular duplex was then followed by controlled treatment of the duplex product with DNAase (an endonuclease). This was done to produce a nick in the strand that was being copied in each phase. Since the action of DNAase is not selective for either of the strands, and since some of the duplex molecules would escape the DNAase action, a mixture of products was obtained. Mild exposure to heat was then used to unravel the nicked strands from the duplexes. Under the conditions used, the intact duplex remained unchanged. The intact circular synthetic strand was then isolated by density gradient ultracentrifugation. The separation of strands was made possible by using *deoxybromouridine triphosphate* (dBUTP) in place of deoxythymidine triphosphate in the first polymerization step. Bromouracil is a structural analog of thymine that will enter into complementary base pairing with an adenine residue. Thus, during assembly, wherever a thymine would have been inserted, bromouracil was inserted instead. The product DNA is then labeled, because bromouracil is much *heavier* than thymine (the Br atom is heavier than the—CH₃ grouping). Consequently, the presence of several bromouracil residues would produce a DNA strand of greater density that could be separated from those strands not containing bromouracil. The pattern of the experiment is diagrammed in Fig. 8–6. The final result of the experiment was that the synthetic (+) strand infected *E. coli* cells, and resulted in the production of additional (+) strands that were identical to the native (+) strand used as a template in the first place. In other words, the product was indeed biologically active.

Chemical Synthesis of DNA

It is now possible to synthesize a DNA duplex of known base-pair sequence in the laboratory. The techniques used are primarily those of organic synthesis but enzymes are also used to seal certain bonds. The basic strategy (see p. 258, top) involves the stepwise assembly of small oligonucleotide seg-

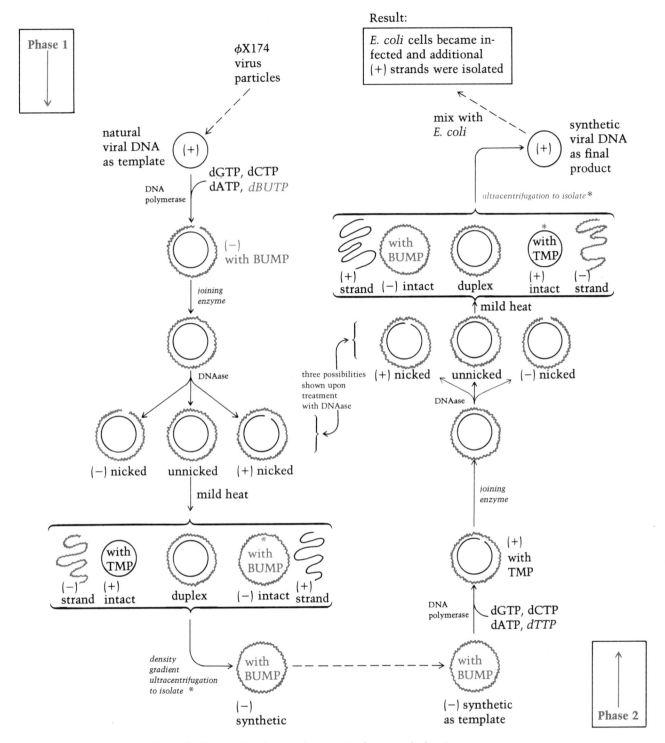

FIGURE 8–6 A diagramatic summary of the *in vitro* synthesis of a biologically active DNA.

ments (e.g., 12–20 nucleotides), part duplex and part single-stranded. Segments are spliced together by the technique of hybridization and then joined together by using DNA ligase. The same strategy would be used to extend the duplex even further, until finally the terminal segments are inserted.

After nearly 10 years of effort G. Khorana and coworkers have recently completed the synthesis of an intact 207 base-pair duplex corresponding to

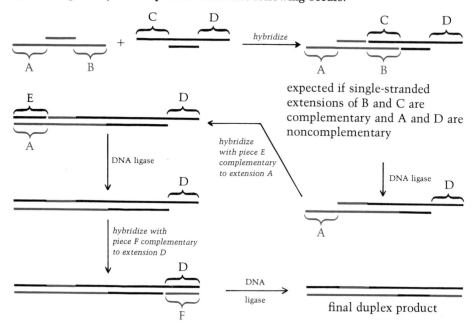

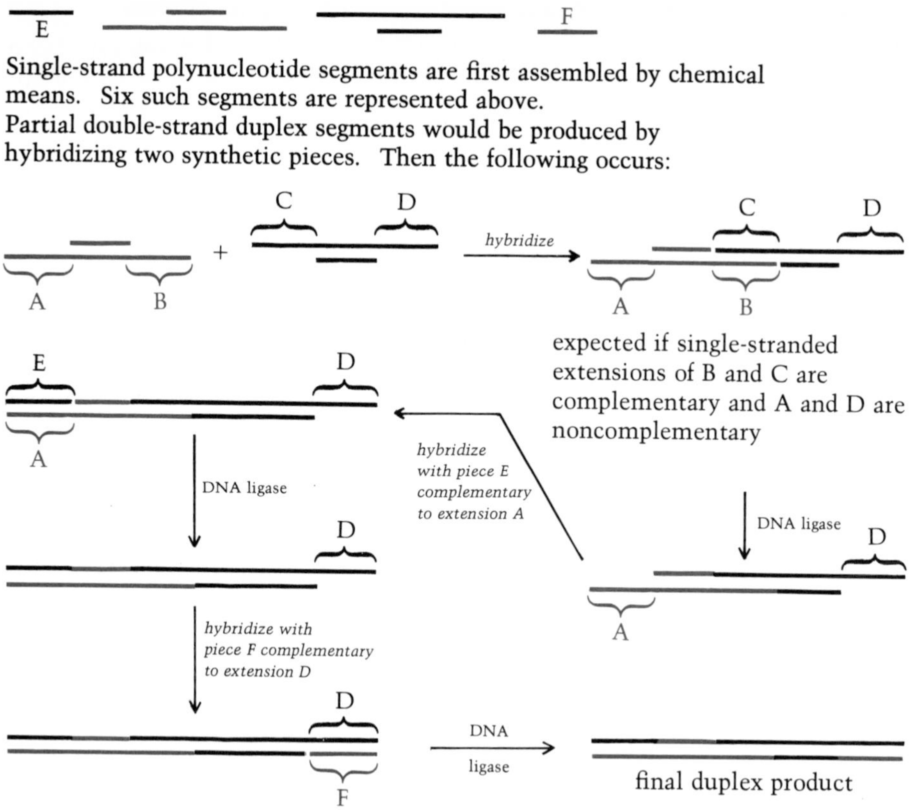

E
F
Single-strand polynucleotide segments are first assembled by chemical means. Six such segments are represented above.
Partial double-strand duplex segments would be produced by hybridizing two synthetic pieces. Then the following occurs:
C
D
A
B
+
hybridize
C
D
A
B
expected if single-stranded extensions of B and C are complementary and A and D are noncomplementary
E
D
A
hybridize with piece E complementary to extension A
DNA ligase
D
DNA ligase
D
A
hybridize with piece F complementary to extension D
D
F
DNA ligase
final duplex product

the gene for a transfer-RNA precursor molecule. The overall strategy required the synthesis of 39 separate oligonucleotide fragments. Further experiments established that the synthetic gene was correctly transcribed *in vivo* by RNA polymerase to yield transfer-RNA. *The synthetic gene was biologically active.* This paves the way for future research, which might involve (a) the fine-level study of how specific proteins can recognize and bind to specific base-pair sequences and control the readout of genetic information, or (b) a consideration of the possibility of transplanting synthetic genes into chromosomes that lack the gene. An example of the latter is discussed later in the chapter (see p. 283).

METABOLIC FATE OF PYROPHOSPHATE

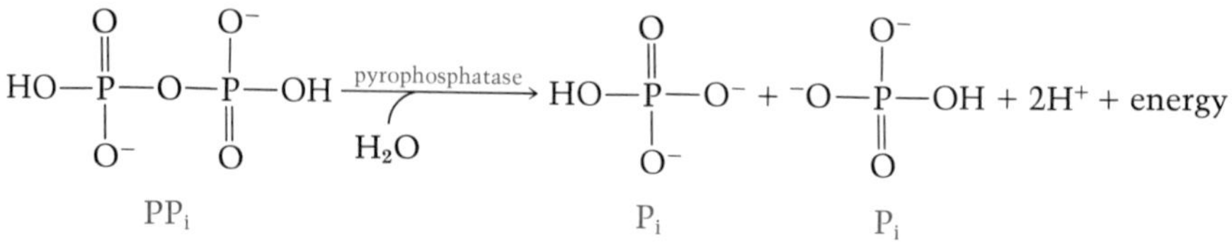

In the preceding sections, we have seen that the enzymes that function in replication and transcription require triphosphonucleotides as substrates. This results in the cleavage of the triphosphoanhydride linkage, and for each added nucleotide, one unit of inorganic pyrophosphate (PP_i) is produced. What is the metabolic fate of this pyrophosphate, and is there any significance associated with it? The presence of *pyrophosphatase* in virtually all

pyrophosphatase
H₂O
+ 2H⁺ + energy
PPᵢ
Pᵢ
Pᵢ

types of cells answers the first part of the question. This enzyme catalyzes the hydrolysis of inorganic pyrophosphate to two units of inorganic ortho-phosphate (P_i). The latter, of course, can then be utilized in various phosphorylation reactions.

It is also significant to note that the hydrolysis of pyrophosphate is energy-yielding. It can be argued then that this final energy-yielding hydrolysis would also help in displacing the overall process of nucleic acid biosynthesis, which is energy-requiring, to completion. This argument explains then (at least in part) why NTPs rather than NDPs are the preferred substrates of RNA and DNA polymerases. The integration between energy-yielding and energy-requiring reactions is discussed more fully in Chap. 11.

MUTATIONS

The common understanding of the word *mutation* is that it is *a change in one or more hereditary characteristics of an organism due to an alteration in one or more genes of a chromosome.* A fact not so commonly understood is that the reason for the alteration in molecular terms is a change of the nitrogen base sequence. In fact *all that is required is a change of a single base-pair.* Depending on the way in which the change occurs, the final consequence may or may not be seriously damaging to the organism. The failure to produce a particular protein (or transfer-RNA or ribosomal-RNA) molecule or the production of a functionally defective molecule will usually have serious consequence and such a mutation is called *lethal.* Otherwise, the mutation is said to be of a *silent* type.

A base-pair change can occur in any one of three ways—by *substitution, addition,* or *deletion.* Of the three, substitution mutations are the least likely to produce a lethal effect (see note in margin), since only one triplet of the coding sequence is changed (see diagram, p. 260) and thus only a single change in amino acid sequence would result. Unless the single triplet change results in a substitution that would affect a crucial amino acid residue involved in structure and/or function, the protein will still be normal. This type of alteration is also called a *transitional* mutation, meaning that a GC pair is changed to an AT pair or vice versa.

A base-pair addition or deletion modifies all coding triplets beyond the site of change and is called a *frame-shift* mutation. You should readily appreciate that either event would result (see diagram, p. 260) in a drastic alteration of amino acid sequence, and hence the protein will most likely be seriously defective, totally inactive, or even be (though it is only a slim chance) a new protein with a new function.

A mutation can occur as a natural spontaneous process at a very low frequency or it can be induced by various mutagenic agents and have a high frequency of occurrence. One of the most powerful mutagens is *nitrous acid* (HNO_2). It works (see Fig. 8–7) by causing a chemical modification in DNA of either cytosine, adenine, or guanine, which in turn changes the base-pairing recognition by the modified base. The result is a base-pair substitution. In each case, the modification is an oxidative deamination with a

$\diagdown$C—NH$_2$ grouping being converted to a $\diagdown$C=O group. As an example, suppose that cytosine is converted to uracil. After two cycles of DNA bio-

At the present time, about 275 different hemoglobin variants have been discovered and the great majority of them involve single amino acid substitutions in either the α or β chains. Only four of these variants are established as seriously defective in that they are associated with pathological manifestations; one of these is HbS, the sickle cell hemoglobin.

259

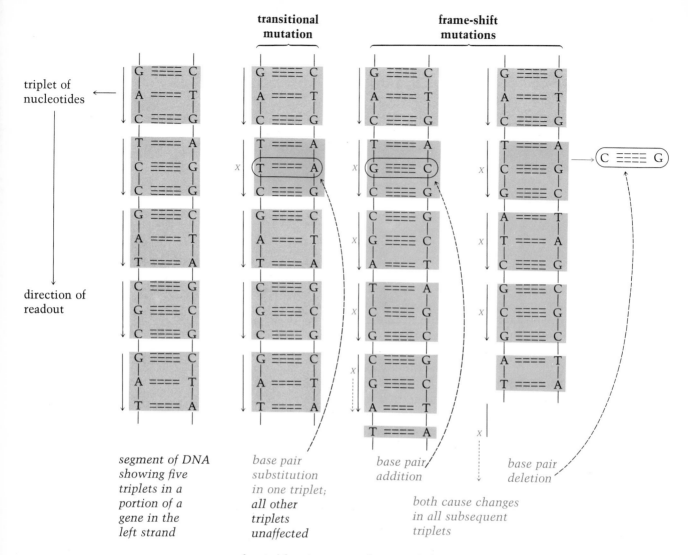

triplet of nucleotides →

direction of readout

segment of DNA showing five triplets in a portion of a gene in the left strand

base pair substitution in one triplet; all other triplets unaffected

base pair addition

base pair deletion

both cause changes in all subsequent triplets

synthesis (that is, two replications) the final effect would be that the original CG pair would become a TA pair—a CG → TA transition. The reason for this is that after C becomes U it prefers to pair with A (the complement of U), which in turn then pairs with T. Similar events apply to AT → GC and GC → TA transitions (see Fig. 8–7).

The biochemistry of nitrous acid is relevant to the concern over nitrite and nitrate levels in certain meats processed for human consumption. Nitrate (NO_3^-) and nitrite (NO_2^-) can be converted to nitrous acid in the stomach, where the nitrous acid can in turn react with various secondary amines to produce *nitrosamines*—potent mutagens that may then play an important role in inducing cancer in the body.

Mutations can also be caused by certain physical agents, with the most common being *ultraviolet (UV) radiation*. The current view is that the mutagenicity of UV light is primarily due to the formation of *thymine dimers* in DNA. As shown below, the dimer is formed between two adjacent thymine bases. Although cells have the ability to repair this type of chromosome damage (see p. 255 in this chapter), the frequency of mistakes (and hence, mutations) that occur when several lesions have to be replaced is high.

Earth organisms are protected from the harmful solar ultraviolet emission by the ozone layer in the outer atmosphere.

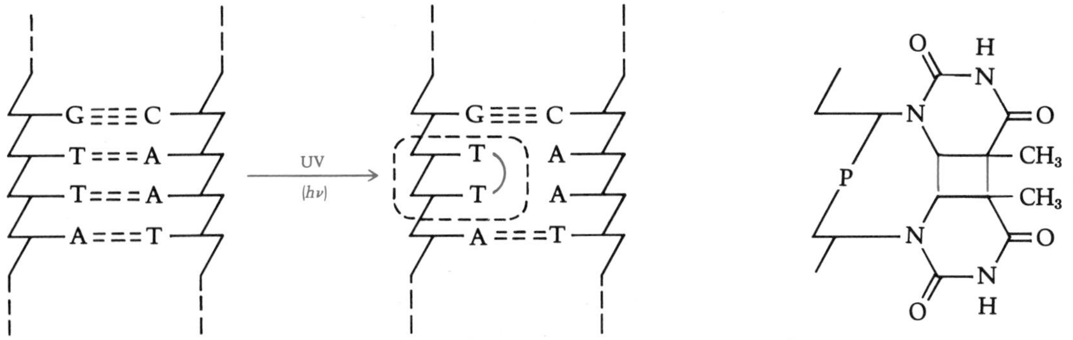
G
C
T
A
T
A
A
T
UV
(hν)
G
C
T
A
T
A
A
T
O
H
N
N
O
CH3
P
CH3
N
O
N
O
H

formation of thymine dimer in one strand
(structure at right)

POLYPEPTIDE (PROTEIN) BIOSYNTHESIS—GENE TRANSLATION

The assembly of a polypeptide chain from its constituent amino acids is one of the most fascinating life processes. It is also a very complex process. Merely listing the substances that participate is sufficient indication of the complexity. The list includes amino acids, ATP, GTP, transfer-RNAs, aminoacyl-tRNA synthetases, ribosomes, messenger-RNA, K^+, peptide synthetase, and various noncatalytic proteins. Although our current understanding of the process is extensive, various aspects are still under active in-

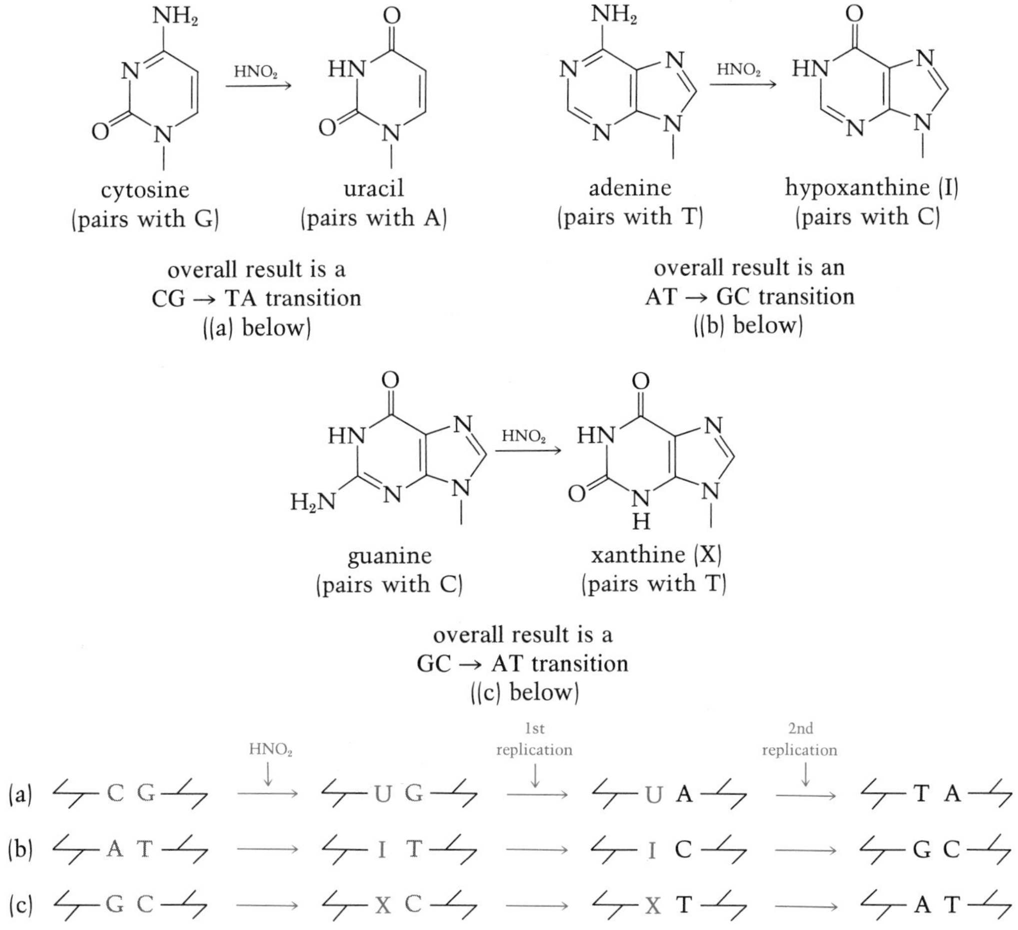
cytosine
(pairs with G)
uracil
(pairs with A)
adenine
(pairs with T)
hypoxanthine (I)
(pairs with C)
overall result is a
CG → TA transition
((a) below)
overall result is an
AT → GC transition
((b) below)
guanine
(pairs with C)
xanthine (X)
(pairs with T)
overall result is a
GC → AT transition
((c) below)
HNO2
1st replication
2nd replication
(a) C G → U G → U A → T A
(b) A T → I T → I C → G C
(c) G C → X C → X T → A T

FIGURE 8–7 A summary of the mutagenic effect of HNO_2.

vestigation. The following description of the important highlights is necessarily brief.

It is useful to consider the process as occurring in four stages, namely:

1. *Activation* and *selection* of amino acids (ATP-dependent).
2. *Initiation* of polypeptide chain formation (GTP-dependent).
3. *Elongation* of the polypeptide chain (GTP-dependent).
4. *Termination* of polypeptide chain formation.

Activation and selection.   In all types of cells the first event of translation is the enzyme-catalyzed conversion of each amino acid to an *aminoacyl-tRNA* species. This accomplishes two things: (1) the reactivity of the amino acid for peptide bond formation is enhanced (activation), and (2) the amino acid is matched with a specific transfer-RNA (selection). The enzyme is termed an *aminoacyl-tRNA synthetase* and operates in two steps (see below). Mg^{2+} is required for optimal activity. The involvement of pyrophosphatase is also shown because it can assist the overall process.

Step 1: $\qquad$ amino acid + ATP $\xrightarrow[\text{synthetase (Mg}^{2+})]{\text{aminoacyl-tRNA}}$ aminoacyl-AMP + PP_i

Step 2: $\qquad$ aminoacyl-AMP + tRNA $\xrightarrow[\text{synthetase (Mg}^{2+})]{\text{aminoacyl-tRNA}}$ aminoacyl-tRNA + AMP

1 + 2: $\quad$ amino acid + ATP + tRNA $\xrightarrow{\text{synthetase}}$ aminoacyl-tRNA + AMP + PP_i

Then: $\qquad\qquad\qquad\qquad\qquad\qquad$ $PP_i \xrightarrow{\text{pyrophosphatase}} 2P_i$

Overall: $\quad$ amino acid + ATP + tRNA $\xrightarrow[\text{pyrophosphatase}]{\text{synthetase}}$ aminoacyl-tRNA + AMP + $2P_i$

The formation of an aminocyl-AMP species in step 1 represents the initial activation of the amino acid. Since ATP is consumed, the reaction would be labeled as *energy-requiring*. The carbonyl carbon of the carboxyl group in the free amino acid becomes (see below) a carbonyl carbon of a mixed anhydride in aminoacyl-AMP, which increases its reactivity. In step 2 a terminal OH of transfer-RNA attacks the carbonyl carbon (displacing AMP) to yield the aminoacyl-tRNA product. Since this is an ester, the increased reactivity of the aminoacyl carbonyl group is conserved.

In addition to the chemistry it catalyzes, the synthetase enzyme is also

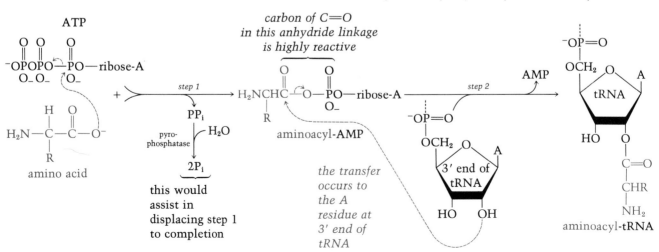

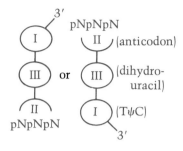

in this section on translation the transfer-RNA structure will be symbolized in a shorthand representation as follows:

a review of
transfer-RNA structure

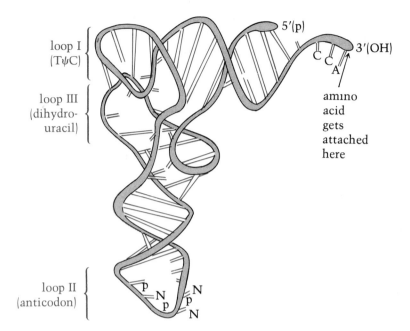

loop I
(TΨC)

loop III
(dihydro-
uracil)

loop II
(anticodon)

5'(p)

3'(OH)

amino
acid
gets
attached
here

and aminoacyl-tRNA as

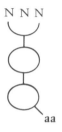

distinct in one other important way. It is *highly specific in its recognition of both amino acid and transfer-RNA.* One might expect then that a living cell would consist of several synthetase enzmes, each having different recognition properties. In fact 20 different synthetases are believed to exist—one for each of the 20 amino acids. The number of different tRNA molecules per cell is even greater—there are perhaps as many as 60—since most of the amino acids can be matched with at least two different tRNA molecules and some with three.

The basis of this enzyme specificity is, of course, the highly ordered recognition of the active site for a particular combination of amino acid and transfer-RNA. A logical question in this regard is, what is the feature of tRNA structure that is recognized for binding by the synthetase? Each amino acid, of course, has its own unique structure, conferred by the side chain. With transfer-RNA, the selection appears to be based on the ability of the synthetase to recognize (and hence prefer for binding) unique features of the open stem region.

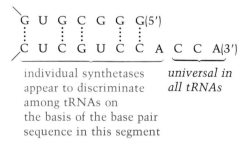

individual synthetases
appear to discriminate
among tRNAs on
the basis of the base pair
sequence in this segment

*universal in
all tRNAs*

Initiation. Polypeptide chains are assembled one amino acid at a time from the N terminus to the C terminus. Obviously then, this stepwise process has a beginning, a middle, and a end. It is the beginning that concerns us here.

It should seem logical to you that the assembly of a polypeptide chain with a specific sequence of amino acid residues would involve specific "start" and "end" signals. In prokaryotic cells (e.g., *E. coli*) it is known that polypeptide assembly begins with the positioning of *N-formylmethionine-tRNA* (fmet-tRNA). Thus, N-formylmethionine becomes the first residue (N terminus) of the chain. A short time after assembly the formylmethionine residue is then removed.

Conclusive proof for this molecular signal was provided by discoveries that *E. coli* contains (a) two transfer-RNAs for methionine, although the genetic code predicts that only one is necessary (see p. 273), and (b) a *formyltransferase* enzyme, which catalyzes the formylation (see margin) of only one of the corresponding met-tRNA complexes. Symbolically, the two tRNAs are designated as tRNA^{fmet} and tRNA^{met} and the corresponding aminoacyl-tRNA complexes as fmet-tRNA^{fmet} and met-tRNA^{met}. Although the nucleotide sequences of tRNA^{fmet} and tRNA^{met} are different, it is significant that they both contain the same anticodon, UAC. Hence, they both should have the ability to recognize the same mRNA codon, AUG, and so they do. While it seems that this would tend to create some confusion in the translation process, there is clear evidence that such is not the case at all. Rather, the UAC anticodon of fmet-tRNA^{fmet} recognizes the AUG codon only when the latter is at the beginning of the genetic message in mRNA. When AUG is localized within the message, it is recognized by nonformylated met-tRNA^{met}. Accordingly, the mRNA codon, AUG, is termed the genetic *initiation signal*, and the fmet-tRNA^{fmet} complex is termed the *initiator*. Amazing!

Formylation is the attachment of a formyl —$\underset{H}{C}$=O grouping. The transfer is made (in this case to the amino group of the amino acid) from a formylated form of the *coenzyme tetrahydrofolic acid* (FH$_4$). The origin of the formyl-FH$_4$ species is discussed on p. 551 in Chap. 17.

The initial positioning of the fmet-tRNA^{fmet} in relation to the ribosome is very involved. First, under the influence of a noncatalytic protein called *initiation factor III* or simply IF$_3$, an intact 70S ribosome dissociates to yield the 30S and 50S subunits; then mRNA binds to the 30S subunit; then the fmet-tRNA^{fmet} binds; and finally the 50S reassociates. GTP and at least two other noncatalytic proteins (IF$_1$ and IF$_2$) participate in various stages as initiation factors. The net result is the formation of a 70S-mRNA-fmet-tRNA complex.

Note that the fmet-tRNA^{fmet}, via its TψC loop, is initially bound to a specific location (the P site) on the surface of the 50S subunit. A similar neighbor site (the A site) is proposed to accommodate an incoming aminoacyl-tRNA (see the discussion of elongation). A third 50S site is involved in catalyzing peptide bond formation.

In the mitochondrion of eukaryotic cells the same N-formylmethionine

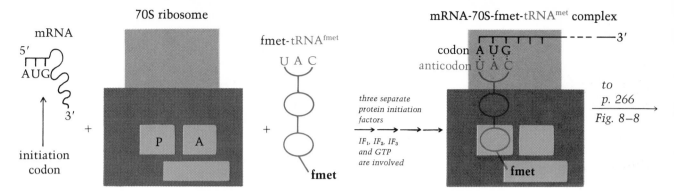

Note 1: The AUG initiating codon is shown at the 5' end of mRNA and thus the others follow in sequence toward the 3' end. In other words, the genetic code of mRNA is read in the 5' → 3' direction and not the other way.

Note 2: In eukaryotic cells the formation of the initiation complex requires the Gppp "cap" on the 5' end of mRNA (see p. 227, Chap. 7).

initiation mechanism operates and this provides further evidence to support the proposal that the mitochondrion evolved from prokaryotic bacteria (see p. 26). However, in the cytoplasm of eukaryotes a different mechanism is known to operate. Methionine is still the first residue positioned, but formylation is not necessary. Eukaryotic cells contain two tRNA^{met} molecules, with one to position internal methionine and the other (the initiator) to position methionine initially at the N-terminus. The primary structural distinction between the two is that the initiator tRNA^{met}, unlike all other tRNAs, *does not have a TψC sequence in loop I.* This is only one difference in the specific details of translation as it occurs in prokaryotes and eukaryotes. However, the overall strategy (as described here for prokaryotes) is the same in both types of cells.

Elongation. After formation of the initiation complex, each subsequent amino acid is positioned in a repetitive cycle of three steps:

1. The binding of aminoacyl-tRNA, directed by a specific codon-anticodon recognition.
2. Peptide bond formation.
3. Translocation.

Each step of the cycle is described below and diagramatically summarized in Fig. 8–8 for the initial formation of a dipeptide after one cycle and the ultimate formation after *n* cycles of a complete polypeptide with *n* residues.

1. Beginning with the 70S-mRNA-fmet-tRNA complex, a second aminoacyl-tRNA (aa$_2$-tRNA^{aa2}) is positioned according to recognition of codon-anticodon base-pairing. As shown, the incoming aminoacyl-tRNA becomes anchored at a second 50S binding site (A) very close to the P site already occupied by fmet-tRNA.
2. A third site on the 50S subunit contains *peptide synthetase,* which catalyzes (K$^+$ is required for optimal activity) the formation of a peptide bond between the two aminoacyl groups. The chemistry is straightforward (see p. 267), involving the nucleophilic attack of the N atom in the free amino group of the newly positioned aminoacyl-tRNA at the carbonyl carbon of the N-formylmethionine grouping. Note that the newly formed peptide is now attached to the transfer-RNA that was positioned in the first phase.
3. The final step involves the displacement of free tRNA^{fmet} from the ribosome

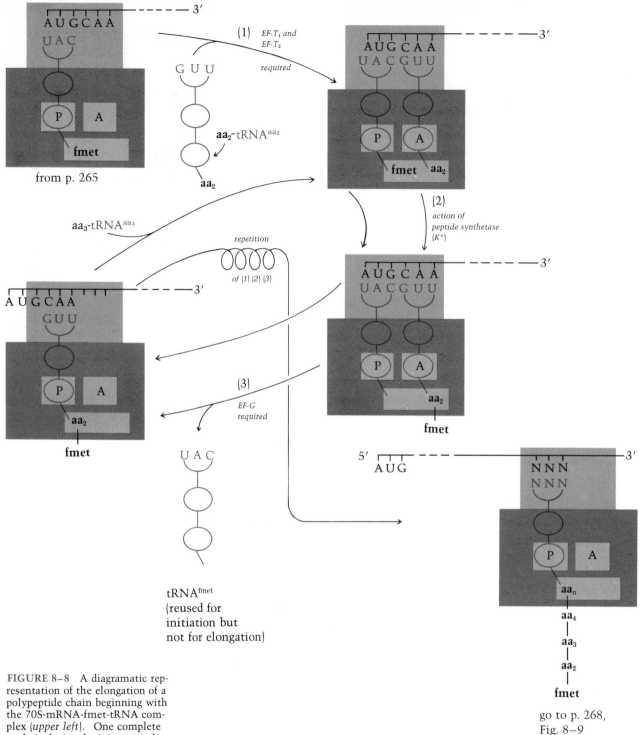

FIGURE 8–8 A diagramatic representation of the elongation of a polypeptide chain beginning with the 70S-mRNA-fmet-tRNA complex (*upper left*). One complete cycle is depicted, giving an indication of the successive cycles that result in the eventual assembly of the entire polypeptide, which is still attached to the ribosome via a tRNA (*lower right*). Steps 1, 2, and 3 are coded to the text description.

go to p. 268, Fig. 8–9

surface and a displacement (i.e., a translocation) of fmet-aa$_2$-tRNA^{aa_2} to the vacated P site. Since this dipeptidyl-tRNA is still engaged in base-pairing with messenger-RNA, the net effect also involves a movement of the messenger-RNA relative to the ribosome surface. At this point step 1 would be repeated with the binding at the vacated A site by the next

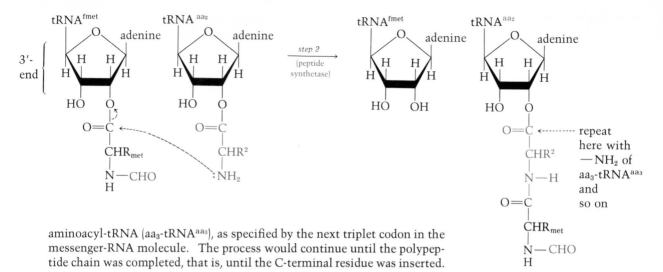

aminoacyl-tRNA (aa₃-tRNA^aa₃), as specified by the next triplet codon in the messenger-RNA molecule. The process would continue until the polypeptide chain was completed, that is, until the C-terminal residue was inserted.

As indicated in Fig. 8–8 steps 1 and 3 of each cycle also involve the participation of other proteins and GTP. Referred to as *elongation factors*, the proteins have essential roles in promoting the binding of the incoming aminoacyl-tRNA and the displacement of the newly extended peptidyl-tRNA.

Termination. Translation begins with a "start signal" of an AUG codon recognized by UAC anticodon in fmet-tRNA^fmet. Obviously, the process should consist of a "stop signal" as well and so it does. In fact, there are (at least in prokaryotes) three separate mRNA codons representing *termination signals:* UAA, UAG, and UGA. However, the termination codons are not recognized by a specific aminoacyl-tRNA but by soluble proteins called *release factors.* In bacteria and yeasts there are at least two such release factors, designated R_1 and R_2 (R for release). The R_1 and R_2 proteins are capable of selectively binding with the termination codons at the surface of the 30S subunit, thus preventing the addition of more aminoacyl-tRNAs to the 50S subunit. They are called release factors because, besides performing the function just described, they also promote the release of the completed polypeptide chain from the tRNA to which the polypeptide is attached (see Fig. 8–9). This is also accompanied by the displacement of free tRNA and mRNA from the ribosome. Thus the mRNA and 70S ribosome are available for use over again. The last step of polypeptide biosynthesis is the enzymatic removal of the formylmethionine residue (and possibly one or two more residues) from the N-terminus.

Simultaneous Synthesis of Polypeptides on Polysomes

In order to focus attention of the different steps of translation, our preceding discussions and the accompanying diagrams were limited to the growth of a single polypeptide on a single ribosome attached to a single mRNA. The *in vivo* process is much more efficient, however, with a single messenger-RNA generally being translated simultaneously by more than one ribosome. In other words, a single mRNA is attached to a cluster of ribosomes, termed a *polysomal complex,* with each individual ribosome engaged in the assembly of a polypeptide chain. Of course, each of the ribosomal-associated stages of translation (initiation, elongation, and termination) would apply to each

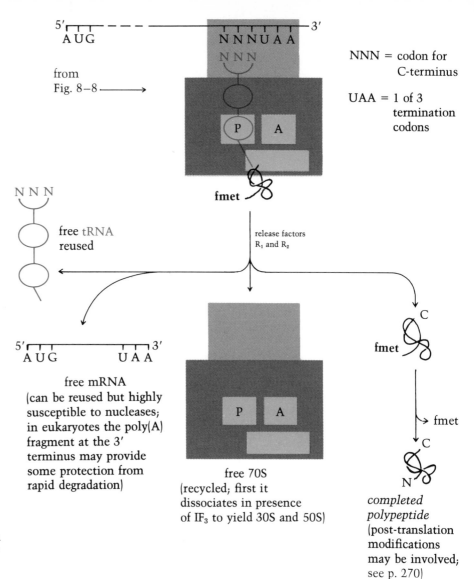

FIGURE 8–9 A diagramatic representation summarizing the termination and release of a fully assembled polypeptide chain from the mRNA-70S-polypeptidyl-tRNA complex. The precise mode of action of the release factors is not yet known.

ribosome and to its growing polypeptide. The advantage conferred on the cell by this arrangement is that several copies of the polypeptide can be made before the mRNA undergoes degradation.

The size of polysomal complexes varies widely, but is generally a function of the length of the mRNA molecule. Extremely large mRNAs, consisting of a few thousand nucleotides, may be complexed to as many as 50–100 ribosomes. Generally, however, polysomal clusters are comprised of 3–20 ribosomes.

It is said that seeing is believing. Accordingly, the most convincing evidence of the mRNA-polysomal complex has been supplied by electron microscopy. The electron micrograph in Fig. 8–10 is but one example of the remarkable success in this area. A polysomal cluster of about 12 ribosomes (heavy black bodies) in association with messenger-RNA (thin thread that appears to connect the ribosomes) is clearly visible at the extreme right. Progressively smaller polysome clusters, also in association with mRNA,

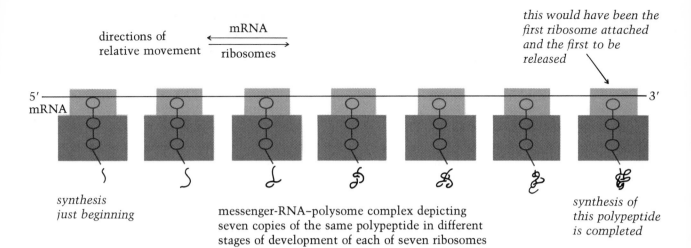

directions of relative movement

mRNA ←

ribosomes →

this would have been the first ribosome attached and the first to be released

5′ ——————————————————————————————————————— 3′

mRNA

synthesis just beginning

messenger-RNA–polysome complex depicting seven copies of the same polypeptide in different stages of development of each of seven ribosomes

synthesis of this polypeptide is completed

are evident towards the left. The long thread running across the field *is DNA.* In addition, where a ribosome is in close proximity to the DNA, note the presence of a smaller granule that is less electron dense than the ribosome. Presumably, this represents a molecule of RNA polymerase attached to the DNA template. Thus, the entire image is a view of *both transcription and translation* and obviously indicates that the *two processes are tightly coupled in both space and time.* In other words, soon after the transcription of messenger-RNA is begun and before the transcription is completed, its translation begins. Then as the size of the messenger-RNA increases with continued transcription, more ribosomes become associated with it. Note how the polysome cluster gets larger as transcription proceeds along the chromosome from left to right in the field.

Inhibition of Protein Biosynthesis by Antibiotics

Several antibiotics exert their effect by inhibiting protein biosynthesis. Depending on the antibiotic, however, the mode of action is quite variable, with different antibiotics inhibiting different steps in the overall translation

FIGURE 8–10 Electron microscopic evidence of genes in action. This micrograph is interpreted as depicting DNA, RNA polymerase, messenger-RNA, and ribosomes caught in the acts of transcription and translation. The specimen was obtained from a cell-free extract (produced by osmotic shock) of rapidly dividing *E. coli.* See text for further description. (Reproduced, with permission, from "Electron Microscopy of Genetic Activity" by B. A. Hamkalo and O. L. Miller, *Annual Review of Biochemistry,* Volume 42. Copyright © 1973 by Annual Reviews, Inc. All rights reserved. Photograph generously supplied by Dr. O. L. Miller.)

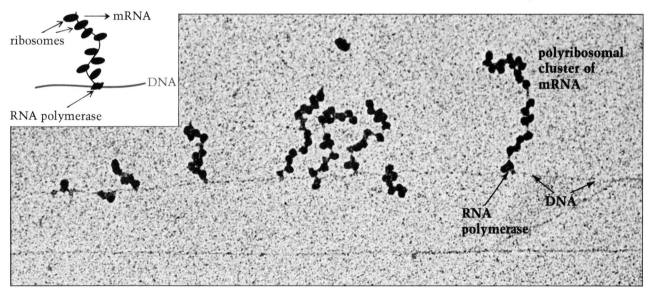

ribosomes → mRNA

DNA

RNA polymerase

polyribosomal cluster of mRNA

RNA polymerase

DNA

$$O_2N-\underset{\underset{OH}{|}}{\overset{\overset{CH_2OH}{|}}{\underset{}{C}}}HCHN\underset{H}{\overset{}{-}}C\overset{O}{\underset{CHCl_2}{}}$$

chloramphenicol
(trade name: Chloromycetin)

process in different ways. One of the most potent is *chloramphenicol*, which acts by binding to the aminoacyl-tRNA binding site (the A site) of the 50S subunit, preventing further binding of aminoacyl-tRNAs. *Streptomycin* binds to a specific ribosomal protein of the 30S subunit. Presumably this changes the conformation of the ribosome, causing a less favorable recognition between the codons of mRNA and the anticodons of aminoacyl-tRNAs. In addition to the fact that the translation process becomes sluggish, this also results in many errors in the positioning of aminoacyl-tRNAs, yielding defective polypeptides. That is to say, the genetic instructions in mRNA are not read correctly. *Puromycin* inhibits translation by interrupting the elongation stage, yielding incomplete polypeptides. You may recall (p. 209, Chap. 7) that puromycin is classified as a nucleoside. The special feature of the nucleoside structure of puromycin is that it is capable of being recognized by the aminoacyl-tRNA binding site of the 50S subunit as resembling the 3'-terminus of an aminoacyl-tRNA. In other words, puromycin functions as a competitive inhibitor. When it becomes attached to the 50S subunit (an event that can occur at any time), the elongation of the growing polypeptide proceeds no further. The partially completed polypeptide chain then links with puromycin to form a puromycin-peptide species, which is then displaced from the ribosome. It is without function unless normal assembly has proceeded far enough to still yield an active polypeptide chain.

A detailed analysis of the mode of action of these and other antibiotics that inhibit protein biosynthesis is beyond the scope of this book. However, it should be noted that research in this area is quite extensive and a considerable amount of information is available. At the clinical level this research has definite implications, since the knowledge of the precise mechanism of action of antibiotics offers a basis for the development of modified ones, with even better clinical properties, such as quicker action and fewer undesirable side effects. The ideal situation is to inhibit preferentially the process in the infectious cells, with little or no effect in the host cells. Remember, although the general mechanism of the translation process that we have just described appears to operate universally in prokaryotic and eukaryotic cells alike, there are differences, and some, such as those found in the initiation stage, are very distinct. In terms of our current discussion these differences are evidenced by the fact that certain of the antibiotics differ in their ability to inhibit protein synthesis in either prokaryotic or eukaryotic cells. Chloramphenicol, for example, inhibits the assembly by 70S ribosomes in prokaryotic cells, but is without appreciable effect on the process mediated by 80S ribosomes in eukaryotic cells.

Post-translational Events

The biosynthesis of many proteins involves more than the condensation of amino acids into a polypeptide chain of specific sequence, size, and shape. Depending on the protein, one or more of the following modifications to the chain may also occur: (1) disulfide bond formation, (2) attachment of cofactors and coenzymes, (3) attachment of prosthetic groups, particularly the addition of carbohydrate groupings, which yields glycoproteins (see p. 310), (4) cleavage of one or more specific peptide bonds to convert a precursor polypeptide to the final product (remember, for example, the preproinsulin → proinsulin → insulin conversions), (5) chemical modification—such as formylation, acylation, amidation, methylation, phosphorylation, hydroxylation, iodination, and carboxylation—of specific amino acid residues, and (6) association of monomers to yield oligomers.

The general rationale for most of these processes is fairly obvious in that (a) they contribute directly to the structure and hence to the activity of the protein, or (b) they are directly involved in the activity of the protein and less important as structural factors. The least understood alterations are those referred to in (5). So far, about 150 amino acid modifications have been identified and the reason for many remain as yet undiscovered. The topic was recently reviewed by Uy and Wold in an article cited at the end of the chapter.

THE GENETIC CODE

Coding Assignments

That genetic information flows from DNA through a messenger-RNA molecule was first proposed in 1961 by two brilliant French biochemists, F. Jacob and J. Monod. Four years later the "language" of the genetic code was solved. For their separate contributions, M. Nirenberg, H. G. Khorana, and R. Holley were joint recipients of the Nobel Prize in 1968. (Jacob and Monod received the Prize earlier for their messenger hypothesis.) In their solution they (a) proved the messenger-RNA hypothesis; (b) proved that the code was *triplet* in nature, meaning that each amino acid is programed in mRNA by a set of three successive bases, called a codon; (c) established that the mRNA code is "read" by complementary recognition of anticodon triplets in tRNA; and (d) made major contributions in the identification of the amino acid-coding assignments for all 64 codons (see margin).

Given only four different bases in DNA, which originally contains the code, a triplet coding ratio would permit 64 different trinucleotide sequences ($4 \times 4 \times 4$). If the code was doublet, as it once may have been before the emergence of bacteria, 16 different codons (4×4) would be possible.

A crucial development in probing the nature of the genetic code was the demonstration by Nirenberg that polypeptide synthesis could occur *in vitro* using soluble cell-free systems to which synthetic messengers (made via laboratory synthesis) were added. For example, when *homo*polyribonucleo-

$$
\begin{array}{l}
\text{synthetic} \\
\text{messenger-RNA}
\end{array} +
\begin{array}{l}
\text{cell-free} \\
\text{system}
\end{array} + \text{ATP} + \text{GTP} +
\begin{array}{l}
\text{mixture of 20} \\
\text{radioactive} \\
\text{amino acids}
\end{array} \longrightarrow
\begin{array}{l}
\text{polypeptide containing} \\
\text{the radioactive amino} \\
\text{acids corresponding} \\
\text{to the coding} \\
\text{instructions of the} \\
\text{synthetic messenger}
\end{array}
$$

tides (all one nucleotide) were used as the messenger, homopolypeptides (all one amino acid) were produced: a poly(U) messenger gave polyphenylalanine; poly(A) gave polylysine; poly(C) gave polyproline; poly(G) gave polyglycine. Although these observations conclusively proved the messenger RNA hypothesis, they did not completely resolve the problem of whether the nature of the genetic code was doublet, triplet, or an even higher coding ratio.

Conclusive proof of the triplet code was eventually achieved by Khorana and coworkers in 1964. These workers also utilized the approach of cell-free synthesis of polypeptides but in the presence of synthetic *hetero*polyribonucleotide messengers of *known sequence*. The unique contribution of the Khorana group was the development of the chemical procedures to prepare such RNAs in the laboratory. As an example of their findings, consider

271

the following. One of the RNAs they prepared was a poly(UG) with 50% U and 50% G and with a sequence of UGUGUGUGUGUGUGUGUG. . . . Now, if the genetic code were read in groups of three successive bases, we would predict that when this type of synthetic RNA was used as messenger, it should be read as UGU-GUG-UGU-GUG-UGU-GUG. . . , and the resultant polypeptide should consist of only *two amino acids in alternating sequence.* This is exactly what was found. Specifically, the polypeptide consisted of cysteine and valine (cys-val-cys-val-cys-val . . .). Therefore, codons had to be translated as triplets and in this case specifically UGU = cys and GUG = val. If they were translated as doublets, the polypeptide would have been homogeneous, containing only one amino acid. (Why?) Thus, the results with homogeneous messengers meant that UUU = phe, AAA = lys, CCC = pro, and GGG = gly.

Several other coding assignments were similarly established. However, the yeoman work of elucidating codon sequence was done in the most direct way possible by Nirenberg, who chemically synthesized all 64 possible codons and then assayed for the ability of each one to direct the *binding* of specific aminoacyl-tRNAs to ribosomes, again in a cell-free system (see Fig. 8–11) supplemented with a mixture of radioactive amino acids (all 20). The basis of this assay was that each synthetic codon would be recognized by its complementary anticodon in a specific tRNA, to which was attached a specific amino acid. Because no intact messenger-RNA is present, there would be no polypeptide synthesis. What does occur is that the base-paired codon-aminoacyl-tRNA complex binds to the ribosomes to yield a ternary complex of ribosome-codon-aminoacyl-tRNA. Ultrafiltration was then used to isolate the complex. The identity of the radioactive amino acid was established by dissociating the complex into its component parts and chromatographing the extract against amino acid standards. Thus, a coding as-

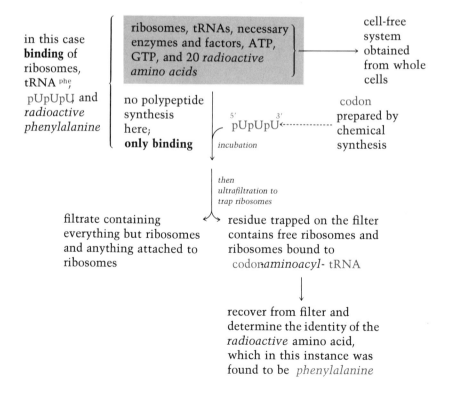

FIGURE 8–11 A summary of Nirenberg's binding assay of synthetic codons.

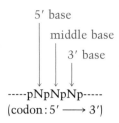

```
        5' base
         |
         | middle base
         |    |
         |    | 3' base
         |    |  |
         ↓    ↓  ↓
  -----pNpNpNp-----
   (codon: 5' ——→ 3')
```

base at 5' end of codon ↓	middle base of codon →				base at 3' end of codon ↓
	U	C	A	G	
U	phe-UUU	ser-UCU	tyr-UAU	cys-UGU	U
	phe-UUC	ser-UCC	tyr-UAC	cys-UGC	C
	leu-UUA	ser-UCA	UAA	UGA	A
	leu-UUG	ser-UCG	UAG	trp-UGG	G
C	leu-CUU	pro-CCU	his-CAU	arg-CGU	U
	leu-CUC	pro-CCC	his-CAC	arg-CGC	C
	leu-CUA	pro-CCA	gln-CAA	arg-CGA	A
	leu-CUG	pro-CCG	gln-CAG	arg-CGG	G
A	ile-AUU	thr-ACU	asn-AAU	ser-AGU	U
	ile-AUC	thr-ACC	asn-AAC	ser-AGC	C
	ile-AUA	thr-ACA	lys-AAA	arg-AGA	A
	met-AUG (and initiation)	thr-ACG	lys-AAG	arg-AGG	G
G	val-GUU	ala-GCU	asp-GAU	gly-GGU	U
	val-GUC	ala-GCC	asp-GAC	gly-GGC	C
	val-GUA	ala-GCA	glu-GAA	gly-GGA	A
	val-GUG	ala-GCG	glu-GAG	gly-GGG	G

UAA, UGA
UAG } termination

FIGURE 8–12 A summary of the 64 triplet codons of the genetic code and their known coding assignments.

signment for each synthetic triplet could be made directly. With this technique, nearly 50 codon sequences were unequivocally assigned to code for certain amino acids. Ambiguities in the assays of the others prevented complete success, although probable assignments were made in several cases. Eventually, with improvements in the binding assay, and with the work of other investigators using different approaches, including *in vivo* studies with mutant organisms, the entire genetic code has been defined. Figure 8–12 summarizes this knowledge in tabular form.

Degeneracy of the Genetic Code

Two distinctive features of the genetic code are clearly evident. First, *all triplets have a known function,* with 61 coding for amino acids and the other three coding for termination of polypeptide chain formation. Second, the code is grossly *degenerate,* meaning that several amino acids are coded for by more than one codon. Indeed, with the notable exceptions of methionine and tryptophan, which have only one codon, the phenomenon of degeneracy applies to all of the amino acids. In fact, three amino acids (arg, ser, and leu) show sixfold degeneracy (six codons). It is customarily argued, and rightly so, that the degeneracy of the code confers definite selective advantages on living organisms. The advantages are that certain errors in DNA replication, DNA transcription, and RNA translation can occur without a corresponding change in the genetic information or in its expression. Consider the codon GCU, for example; it arises from the transcription of CGA in DNA and codes for alanine. If this were the only code for alanine, then any alteration during replication or transcription of the CGA sequence would change the information. However, because of the fourfold degeneracy that exists in the third position, only an error involving the first two bases would change the information.

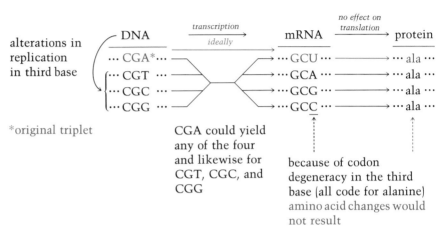

alterations in
replication
in third base

*original triplet

CGA could yield
any of the four
and likewise for
CGT, CGC, and
CGG

because of codon
degeneracy in the third
base (all code for alanine)
amino acid changes would
not result

Close inspection of Fig. 8–12 reveals that in all cases of twofold, threefold, and fourfold generacy the variation involves the *third* base of the triplets. The same relationship applies to the three cases of sixfold degeneracy if you consider the six codons as one group of four and a second group of two. On still closer inspection, note that where only twofold degeneracy exists, the pattern of the third base change always involves a purine for a purine or a pyrimidine for a pyrimidine: $NNPu_a$ and $NNPu_b$, or $NNPy_a$ and $NNPy_b$. Both characteristics indicate then that the genetic code may originally have been a doublet one that eventually evolved into a triplet.

Further inspection of the genetic code reveals that it is characterized by another type of degeneracy. To illustrate this, the sets of degenerate codons for alanine, valine, leucine, and isoleucine are listed below.

alanine	valine	leucine	isoleucine
GCU	GUU	CUU	AUU
GCA	GUA	CUA	AUA
GCC	GUC	CUC	AUC
GCG	GUG	CUG	
		UUA	
		UUG	

We have already noted that a code for alanine will survive errors in DNA replication and DNA transcription involving the third base position because of the four-codon degeneracy for alanine. If errors involve the first and second base, the code for alanine is lost. *However*, certain changes in the first or second base could occur *without losing the code for the type of amino acid* that alanine represents. Look at the columns of codons. GCU and GUU (differing in the second base position) code for alanine and valine, respectively, but each is an aliphatic, nonpolar amino acid. Likewise, CUU and AUU (differing in the first base position) code for leucine and isoleucine, respectively, but again both are aliphatic nonpolar amino acids. Indeed, any one of the 17 codons will code for an aliphatic, nonpolar amino acid. This sort of pattern also applies to other groups of amino acids although not to the same extent. But don't let this mislead you; most changes in the first and second base position, or both, would result in significant alterations in the code (see Exercises 8–7 and 8–9).

Universality of the Code

Even though there are some specific differences, all living cells and viruses use the same basic strategy for the translation process. In fact, the same codons with basically the same coding assignments are used and hence the genetic code is said to be *universal*. This strongly suggests then that the general operation of the molecular apparatus responsible for gene expression (protein biosynthesis) was probably already perfected with the evolution of photosynthetic bacteria some 3,000,000,000 years ago, and that it has remained essentially unchanged since then. I submit this to you as the single most illustrative example of biochemical unity in our biosphere.

New Discoveries—Overlapping Codons and the Reading of Codons in Different Frames

The chromosome of the φX174 bacterial virus is a single-stranded molecule, consisting of 5,375 nucleotides, which codes for the production for nine different proteins: A, B, C, D, E, F, G, H, J. However, various genetic and physical studies established that 5,375 nucleotides do not contain enough information to code for all these proteins. The successful sequencing in 1977 of the entire chromosome explained the discrepancy. It was found that at four different places the *same stretch of DNA can code for different functions*, meaning that transcription and translation can occur in different reading frames. As shown below, genes A and B are both coded in the same region and likewise genes D and E. Although not indicated in the drawing, it has also been established that some of the adjacent genes use initiation codons that overlap with termination codons of the proceding gene. Overlaps of this type occur at the A/C junction and the D/J junction.

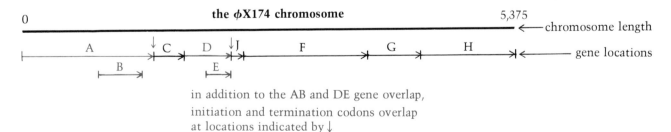

in addition to the AB and DE gene overlap,
initiation and termination codons overlap
at locations indicated by ↓

The multifunctional character of the φX174 chromosome may be a unique property displayed by only a few very small DNA viruses. However, overlapping initiation and termination codons have also been detected in *E. coli*, and thus the possibility exists that some of these tactics of compressing information into DNA may be of more general significance.

CODON-ANTICODON RECOGNITION—DEGENERACY AND "WOBBLE"

The key to the fidelity of the stepwise assembly of a polypeptide chain is the recognition of a $5' \to 3'$ codon sequence in messenger-RNA by a $3' \to 5'$ anticodon sequence in transfer-RNA. The basis of the recognition is, of course, complementary base-pairing, which is optimized when oligonucleotide sequences are aligned with opposite strand polarity.

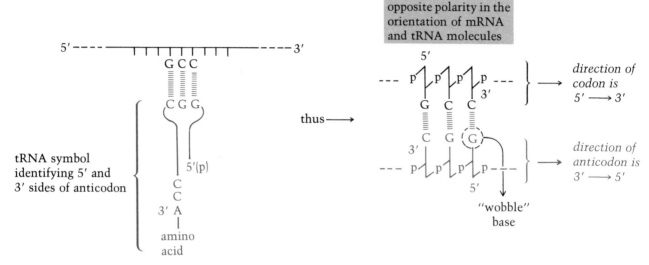

tRNA symbol identifying 5' and 3' sides of anticodon

opposite polarity in the orientation of mRNA and tRNA molecules

thus ⟶

direction of codon is $5' \longrightarrow 3'$

direction of anticodon is $3' \longrightarrow 5'$

"wobble" base

Although some tRNAs do recognize and bind with only one specific codon, others have the ability to do so with two or three different codons. The biological significance of this is related to the degeneracy of the genetic code and was first expressed by Francis Crick in what he termed the *wobble hypothesis* of codon-anticodon recognition.

"Wobble" refers to the ability of a base in the anticodon to assume slightly different spatial positions; the other two bases exist in a more fixed orientation. For reasons we will not pursue, Crick argued that the "wobble position" is located at the 5' end of the anticodon. It was further argued that *wobble would permit the same anticodon to recognize more than one codon because novel base-pairs could be produced* between the 5' end of the anticodon and the 3' end of the codon. The base-pairs are novel in the sense that they do not follow the classical pairings of A-T, G-C, and A-U known to exist in DNA and RNA.

On the basis of theory and model-building it was further proposed that wobble would only be possible with certain bases, and furthermore, that novel base-pairings would be restricted to certain combinations. The wobble bases are U, G, and hypoxanthine (designated as I for inosine, the ribonucleoside of hypoxanthine). Wobble is not permitted with A and C. Table 8–1 summarizes the base-pairing possibilities predicted by Crick—three if the base at the wobble position is I and two if the base of the wobble position is G or U.

Having described its basic meaning, let us now briefly examine how the wobble concept is related to the degeneracy of the genetic code. As a specific example, consider the sequence of yeast alanine-tRNA, which has the anticodon sequence (3')CGI(5'). Since it is called alanine-tRNA, it obviously recognizes an alanine codon. It is known (Fig. 8–12), however, that

TABLE 8–1 Base-pair combinations predicted by the wobble hypothesis (I is hypoxanthine).

Base at the 5' end of anticodon	Base at the 3' end of codon
I	A, C, or U
G	C or U
U	A or G
A[a]	U
C[a]	G

[a] Wobble does not permit any novel combinations with these bases when they are in the anticodon.

Hypoxanthine was included in the original analysis by Crick because it had been detected in the anticodon of some tRNAs that had been sequenced at the time.

the genetic code exhibits a fourfold degeneracy for alanine, with the four codons being GCU, GCC, GCA, and GCG (all $5' \rightarrow 3'$). This degeneracy could be accounted for if four different alanine-tRNAs existed in the cell; but, if the wobble concept applies, then it is possible that all four codons could be recognized by a minimum of two different tRNAs. Furthermore the tRNA described above could recognize three of the four degenerate codons. Why? Well, the 5' base of the anticodon is I, which means that the CGI anticodon could recognize GCU, GCC, and GCA, but not GCG. It so happens that this is precisely the pattern observed in binding studies made with purified alanine-tRNA and synthetic preparations of these four codons. Although a second yeast alanine-tRNA containing an anticodon sequence complementary to the GCG codon has not yet been found, this does not negate the agreement of the properties of the known alanine-tRNA with the wobble concept. There are many other examples that substantiate this principle.

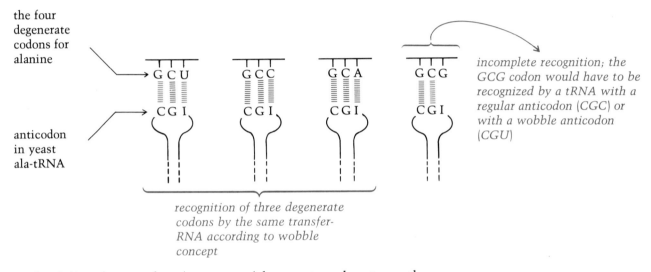

the four degenerate codons for alanine

anticodon in yeast ala-tRNA

recognition of three degenerate codons by the same transfer-RNA according to wobble concept

incomplete recognition; the GCG codon would have to be recognized by a tRNA with a regular anticodon (CGC) or with a wobble anticodon (CGU)

The ability of two or three (never more) degenerate codons to use the same tRNA means a cell does not have to produce 64 unique tRNA molecules and is also another safety valve that can help minimize errors in the readout of genetic information.

REGULATION OF PROTEIN BIOSYNTHESIS

The size of the DNA chromosome in *E. coli* is sufficient to code for approximately 3,000 proteins in addition to ribosomal-RNA and transfer-RNA products. However, only about 1,000 of these proteins are produced at any one time on a regular basis. In the 46 human chromosomes the potential for protein production is roughly estimated to be 10–100 times greater, but again not all of the human proteins are regularly produced. Clearly this means that living cells have the ability to control protein biosynthesis, with several proteins being produced only in response to specific conditions. To explain how the control mechanism operates, F. Jacob and J. Monod proposed (in 1961, with later refinements) the *induction-repression theory* of gene regulation. Their ideas have been proven correct by various genetic and biochemical studies.

277

The basic idea for the induction of the gene expression was originally formulated to explain events typified by the ability of *E. coli* to degrade the sugar lactose, which, for reasons explained in the next chapter, is termed a β-galactoside. *E. coli* is able to utilize this sugar as a carbon and energy source by first degrading it into its constituent parts—galactose and glucose. The hydrolytic cleavage is catalyzed by the enzyme, β-galactosidase. However, the enzyme is produced only when lactose (or a similar β-galactoside) is present in the growth medium, and when the supply of lactose is exhausted, the enzyme is no longer produced. Accordingly, lactose is referred to as an *inducer* and β-galactosidase as an *inducible enzyme*.

Jacob, Monod, and other workers established that the expression of a *structural gene* (that is, a gene coding for the assembly of a protein involved in cell structure or cell metabolism) was controlled by the expression of a *regulatory gene*. The basic strategy proposed was that the regulatory gene directed the formation of a *repressor* substance, the function of which was to block the expression of a structural gene. In the presence of the inducer, the repressor was rendered inactive as a result of being bound to the inducer, and hence the influence of regulatory gene on structural gene was short-circuited, permitting the structural gene to be expressed.

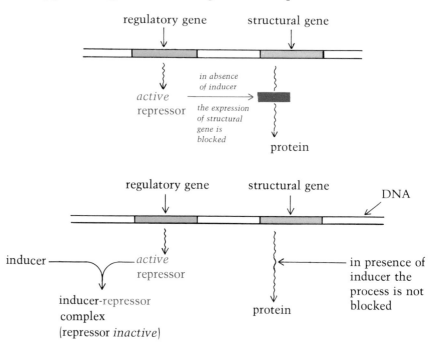

Further research established that (a) the repressor is a protein molecule, (b) the repressor protein—in its active state—functions by binding to a specific DNA segment called the *operator gene*, and (c) the binding of repressor to the operator locus prevents the binding of RNA polymerase to the *promoter gene*. As diagrammed in Fig. 8–13, the operator O and promoter P genes are adjacent DNA segments at the beginning of the structural gene sequence. The entire sequence of promoter, operator, and structural gene(s) is called an *operon*. Note that the regulatory R gene need not be adjacent to the operon it controls, but can be anywhere on the chromosome or even in a different chromosome. Figure 8–13 presents a general summary of these events. Note that the operon under control may consist of more than one structural gene (SG).

In the presence of inducer the promoter locus is accessible for the binding of RNA polymerase. In the case of a multifunctional operon, one intact mRNA is transcribed and it in turn is translated in segments to yield three different polypeptides.

The segment of *E. coli* DNA that comprises the entire
promoter-operator region of the lactose operon has
been isolated and its sequence determined (see Dickson,
et al., cited at the end of this chapter). The total length
is 122 base pairs, with the operator locus comprising
about 40($\frac{1}{3}$) and the promoter comprising about 80($\frac{2}{3}$)

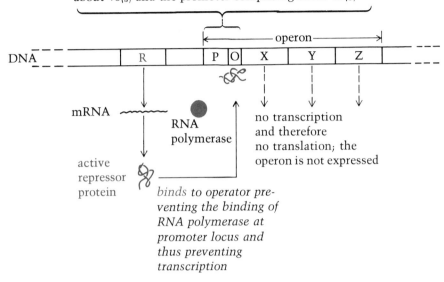

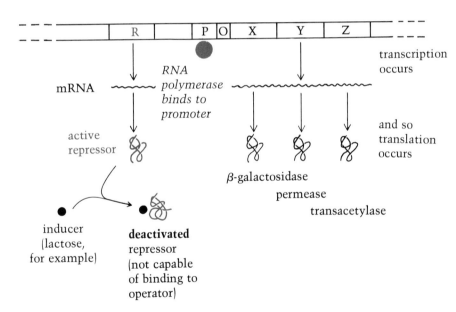

FIGURE 8–13 A diagramatic
summary of gene control by induction. *Upper half:* in the absence
of inducer, protein biosynthesis is
inhibited by the action of the repressor. *Lower half:* in the presence of inducer, the repressor is
deactivated and the inhibition of
protein biosynthesis is removed.
As indicated, it is proposed that
the operon may and usually does
consist of more than one structural gene.

There are two important features of this control process: (1) The biosynthesis of protein is achieved by regulation of messenger-RNA biosynthesis. In other words, it is the *process of transcription that is under
direct control*—the control of the translation process is consequential. (2)
The controlling event is the *interconversion of active and inactive states of
the repressor protein* under the influence of the inducer. How this occurs is
not at all mysterious; it represents an *example of allosterism.*

Other operons are known to be under control by still other processes with
most being variations of the induction mode. In gene *repression,* for ex-

ample, the repressor protein is originally synthesized in an inactive state and thus in the absence of inducer the operon is expressed. The binding of inducer yields an active repressor complex and the expression of the operon is inhibited. Because of its coordinate action in activating the repressor, the inducer in this case has been termed a *corepressor*. A most remarkable example of this action is the repression of the histidine operon, which codes for the production of 10 different enzymes, all responsible for the biosynthesis of histidine, with histidine itself functioning as the inducer (corepressor).

Biological Significance

It is not difficult to appreciate that the significance of induction and repression as control systems is that they allow an organism *to adapt to a changing environment.* Such a capability is definitely advantageous in the context of metabolic economy. There is no reason why a cell should produce the enzymes necessary for the entry and hydrolysis of a β-galactoside if the latter is not present in the growth medium, or produce the 10 enzymes for the biosynthesis of histidine if the intracellular concentration of histidine is sufficient to sustain normal metabolism.

It should be noted that this control mechanism is not confined to bacteria, but is common to higher organisms, including humans. However, the extent to which it may operate and its consequences are largely unknown. To what extent is it related to natural defense mechanisms? To what extent is it related to cellular differentiation? Is it related to infectious diseases? Is it related to cancer? Is it related to the aging process? Is it related to genetic diseases? In future years it is likely that we will have answers to such questions and to many others. We have now completed an analysis of two of the major molecular mechanisms with which all living cells are endowed. One is the allosteric regulation of protein biosynthesis, and the other is the allosteric regulation of activity (Chap. 6) after a protein is assembled.

$$\text{DNA (genes)} \xrightarrow{\text{\textit{transcription}}} \text{mRNA} \xrightarrow{\text{\textit{translation}}} \text{protein}$$

*regulation of
protein synthesis
(and hence activity)
at this level by
repressor-inducer
and repressor-
corepressor
allosteric effects*

*after synthesis,
protein activity can
be regulated
directly by
allosteric effects*

RECOMBINANT DNA—GENETIC ENGINEERING

The occurrence—as a natural event—of the transfer of genetic information from one DNA chromosome to another has attracted the interest of scientist and non-scientist alike ever since the work of Darwin and Mendel. The molecular events associated with genetic recombination are not yet completely understood, although it is certain that the process requires the operation of specific enzymes—endonucleases, polymerases, and ligases.

In recent years, biochemists have learned how to conduct genetic recombinations in the laboratory. This newest development in genetic engineering, called *recombinant DNA technology*, has stirred considerable controversy over the past four years because it represents a double-edged sword. On the one hand, it promises enormous practical benefit to humanity in such areas as medicine and agriculture, and also provides new approaches for probing the biochemistry of gene regulation. On the other hand, however, the opponents of this technology argue that it could by accident or by reckless or devious application result in considerable harm to all animal and plant life.

Recognizing the ominous nature of this technology, the scientists who pioneered in its development took the initiative and announced their concern to the rest of the scientific community and the general public. This resulted in a brief self-imposed moratorium on this type of research while a conference was convened to discuss the merits of the research and what—if any—guidelines might be proposed to conduct further research. After extensive debate (at times heated), guidelines were established and adopted by the National Institutes of Health (NIH), the federal agency that would oversee their application. Research has now resumed under these guidelines and already fantastic results have been reported (see below).

The issue of DNA research is still not settled and at the time of this writing, the United States Congress was still considering whether federal legislation is necessary to construct a framework for this research that will both foster its further development and attend to the important social and ethical implications of this revolutionary new tool.

What is all the fuss about? Well, in its basic sense recombinant DNA technology allows the implantation of an intact piece of genetic information, isolated from an organism or even chemically synthesized in the laboratory, into another host organism where the implanted DNA can be both replicated and expressed. The ability to carry out such transfer is made possible by exploiting (a) the remarkable specificity of action of *restriction enzymes* (see page 216) and (b) the existence in many bacterial cells of a second DNA chromosome molecule, called a *plasmid* factor, which can be employed as a vector to carry the new DNA segment and as a replicating vehicle to clone the implanted DNA segment.

A plasmid is an intact circular DNA duplex found in many bacteria in addition to the main chromosome. In other words, a plasmid is a physically distinct extrachromosomal element. Other physical features are that they are usually much smaller in size than the main chromosome and that multiple copies of a plasmid (as many as 20) are frequently found in the same cell. Despite their physical separation from the main chromosome, plasmids are capable of self-replication and do carry genetic information. In many instances they carry the function of antibiotic resistance; that is, they contain a gene that codes for a protein that can degrade an antibiotic, rendering it harmless to the parent cell. Because they exist separately and are small in size, plasmids can be easily *isolated in intact form*. After isolation, a restriction enzyme is used to "cut" the plasmid in order to insert a "foreign" piece of DNA obtained from another source by the use of the same restriction enzyme. After the ends of this *hybrid plasmid* (a recombinant DNA) are sealed, it can be inserted intact into another cell.

A summary of the recombinant plasmid technology is given in Fig. 8–14. *Note that the key to the splicing of "foreign" DNA into plasmid DNA is the existence in each of "sticky ends" (single-stranded flaps that have complementary sequences) generated by the action of the restriction enzyme.*

Specific successes of recombinant plasmid technology include (1) the transfer of a variety of antibiotic resistance from one bacteria to another, (2) the transfer of a ribosomal-RNA gene from frog cells to bacterial cells, (3)

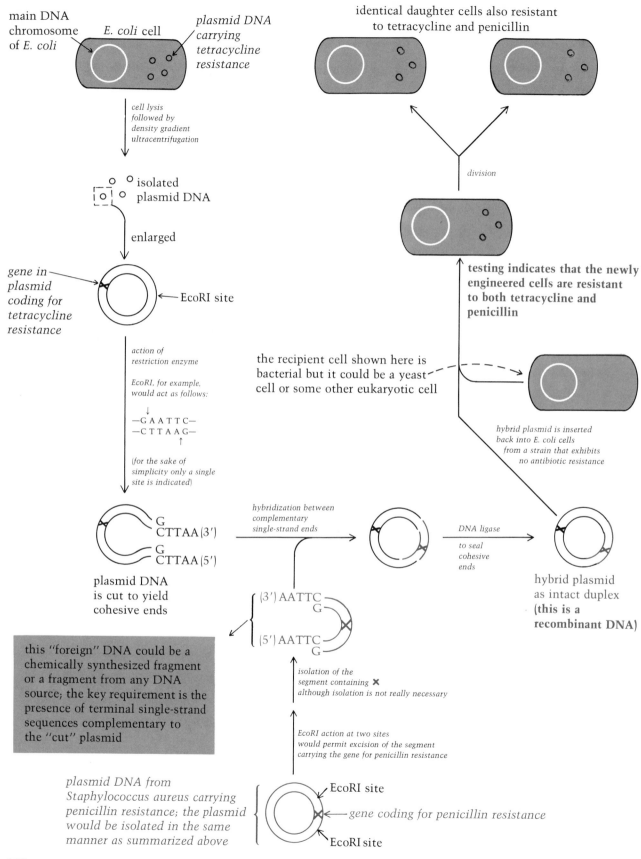

main DNA chromosome of *E. coli*

E. coli cell

plasmid DNA carrying tetracycline resistance

cell lysis followed by density gradient ultracentrifugation

isolated plasmid DNA

enlarged

gene in plasmid coding for tetracycline resistance

EcoRI site

action of restriction enzyme

EcoRI, for example, would act as follows:

↓
—G A A T T C—
—C T T A A G—
↑

(for the sake of simplicity only a single site is indicated)

G
CTTAA (3')
G
CTTAA (5')

plasmid DNA is cut to yield cohesive ends

hybridization between complementary single-strand ends

DNA ligase

to seal cohesive ends

hybrid plasmid as intact duplex **(this is a recombinant DNA)**

this "foreign" DNA could be a chemically synthesized fragment or a fragment from any DNA source; the key requirement is the presence of terminal single-strand sequences complementary to the "cut" plasmid

(3') AATTC—
G
(5') AATTC—
G

isolation of the segment containing ✗ although isolation is not really necessary

EcoRI action at two sites would permit excision of the segment carrying the gene for penicillin resistance

plasmid DNA from Staphylococcus aureus carrying penicillin resistance; the plasmid would be isolated in the same manner as summarized above

EcoRI site

gene coding for penicillin resistance

EcoRI site

identical daughter cells also resistant to tetracycline and penicillin

division

testing indicates that the newly engineered cells are resistant to both tetracycline and penicillin

the recipient cell shown here is bacterial but it could be a yeast cell or some other eukaryotic cell

hybrid plasmid is inserted back into E. coli cells from a strain that exhibits no antibiotic resistance

the insertion of a synthetic gene for insulin into bacterial cells, and (4) the insertion of a synthetic gene coding for the mammalian peptide hormone, somatostatin, into bacterial cells. The somatostatin experiment had the ultimate success—*the bacteria produced somatostatin*! This was the first time that the insertion of a synthetic gene (naturally of eukaryotic origin) resulted in the occurrence of all three phases of the molecular flow of genetic information in prokaryotic cells: the gene was replicated, transcribed, and translated. So now we have a strain of prokaryotic bacteria that is genetically programed to perform a eukaryotic function—it will produce somatostatin. This may prove to be of considerable benefit if it is determined that somatostatin can serve as an effective substitute for insulin in the treatment of diabetes.

W. Gilbert and coworkers have recently (1978) succeeded in engineering the bacterium *E. coli* to manufacture and secrete rat proinsulin. Thus, in a few years the insulin supply for diabetics may be produced by bacteria—a prospect once thought to be merely fantasy.

With future refinements, we can expect to hear of other similar successes (eukaryotic-to-prokaryotic transfers) where bacteria are "taught" to produce insulin and other peptide hormones. This technology will increase the supply of such hormones and at a considerable savings in production cost. In the area of agriculture, it would be a tremendous accomplishment if the gene for nitrogen fixation could be transferred from N₂-fixing bacteria to plants (a prokaryotic-to-eukaryotic transfer). This would mean there would be less need for synthetic fertilizers in farming. In some cases, it might be that the gene would be the desired product. Remember, once a gene is inserted into a plasmid and the recombinant plasmid inserted into a host cell, the gene is duplicated every time the cell divides—and so the original gene inserted is *cloned*. After growing the bacteria, a batch of the gene (plasmid) could be isolated and possibly be transferred to other cells that are defective or missing that gene. The future is exciting!

In another remarkable development Paul Berg and coworkers have succeeded (1978) in transferring the rabbit hemoglobin gene to monkey cells; the monkey cells then produced rabbit hemoglobin. This transfer and its results are immensely significant and have profound implications.

LITERATURE

BALTIMORE, D. "RNA-dependent–DNA Polymerase in Virions of RNA Tumor Viruses"; H. TEMIN and S. MIZUTANI, "RNA Dependent–DNA Polymerase in Virions of Rous Sarcoma Virus." *Nature*, **226**, 1209–1211; 1211–1213 (1970). Two research papers announcing the independent isolation of a reverse transcription enzyme from animal tumor viruses.

BENZER, S. "Fine Structure of a Gene." *Scientific American*, **206**, 70–84 (1962). A synopsis of classical experiments with bacterial viruses that revolutionized our understanding of genes.

CAIRNS, J. "The Bacterial Chromosome." *Scientific American*, **214**, 36–44 (1966). A description of the *in vivo* experiments showing that the two polynucleotide strands of the *E. coli* DNA chromosome are replicated simultaneously, and a proposed mechanism of how this occurs.

CHAMBERLIN, M. J. "The Selectivity of Transcription." *Ann. Rev. Biochem.*, **43**, 1974. A thorough review article of the structure and operation of RNA polymerase with emphasis on the initiation and termination phases.

CLARK, R. F. C., and K. A. MARCKER. "How Proteins Start." *Scientific American*, **218**, 36–42 (1968). Discussion of the discovery of formyl-methionine.

COHEN, S. N. "The Manipulation of Genes." *Scientific American*, **233**, 24–40 (1975). Excellent article describing more fully the gene splicing experiment described in this chapter.

CRICK, F. H. C. "The Genetic Code: II." *Scientific American*, **215**, 55–62 (1966). An updated description of the nature of the genetic code and of how it works.

CRICK, F. "Central Dogma of Molecular Biology." *Nature*, **227**, 561–563 (1970). A restatement and defense of Crick's original theory accounting for the flow of genetic information among DNA, RNA, and protein in view of new discoveries since 1958.

DE CROMBRUGGHE, B., B. CHEN, M. GOTTESMAN, J. PASTAN, H. E. VARMUS, M. EMMER, and R. L. PERLMAN. "Regulation of mRNA Synthesis in a Soluble Cell-free System." *Nat. New Biol.*, **230**, 37–40 (1971). Experimental evidence for a requirement of cyclic-AMP in the transcription of the β-galactoside operon.

DICKSON, R. C., J. ABELSON, W. M. BARNES, and W. S. REZNIKOFF. "Genetic Regulation: The Lac Control

Region." *Science,* **187,** 27–35 (1975). A description of the nucleotide sequence of the promoter-operator region of *E. coli* DNA and a discussion of how the lac operon is regulated.

FIDDES, J. C. "The Nucleotide Sequence of a Viral DNA." *Scientific American,* **237,** 54–67 (1977). A description of the pattern of overlapping genes in the ΦX174 chromosome.

GALLAGHER, R. E., and R. C. GALLO. "Type C RNA Tumor Virus Isolated from Cultured Human Acute Myelogenous Leukemia Cells." *Science,* **187,** 350–353 (1975). See p. 246 of this chapter.

GROBSTEIN, C. "The Recombinant-DNA Debate." *Scientific American,* **237,** 22–33 (1977). The capabilities of scientific technology, the regulations governing research, and the significance to society of genetic engineering are discussed.

HAMKALO, B., and O. L. MILLER. "Electronmicroscopy of Genetic Activity," *Ann. Rev. Biochem.,* **42,** 379–396 (1973).

HANAWALT, P. C., and R. H. HAYNES. "The Repair of DNA." *Scientific American,* **216,** 36–43 (1967). A description of experiments proving that living cells have the ability to repair damaged DNA.

HASELKORN, R., and L. B. ROTHMAN-DENES. "Protein Synthesis." *Ann. Rev. Biochem.,* **42,** 397–438 (1973). Review article of the events of the translation process.

HELINSKI, D. R. "Plasmids as Vehicles for Gene Cloning: Impact on Basic and Applied Research." *Trends Biochem. Sci.,* **3,** 10–15 (1978). A short review article.

HOUSEMAN, D., M. JACOBS-LORENA, U. L. RAJBHANDRY, and H. F. LODISH. "Initiation of Hemoglobin Synthesis by Methionyl-tRNA." *Nature,* **227,** 913–918 (1970).

ITAKURA, K., et al. "Expression in *Escherichia coli* of a Chemically Synthesized Gene for the Hormone Somatostatin," *Science,* **198,** 1056–1063 (1977). A remarkable success in genetic engineering.

JACOB, F., and J. MONOD. "Genetic Regulatory Mechanisms in the synthesis of Proteins." *J. Mol. Biol.,* **3,** 318–356 (1961). The original paper proposing the messenger-RNA hypothesis and the nature of gene control by induction and repression of operons by regulatory genes.

KOLATA, G. B. "Bacterial Genetics: Action at a Distance on DNA." *Research News* article in *Science,* **198,** 41–42 (1977). Brief description of the new discovery that a regulatory protein can bind to one site on DNA and effect the binding of a different regulatory protein at a second DNA site far removed from the first. "Overlapping Genes: More Than Anomalies?" *Research News* article in *Science,* **196,** 1187–1188 (1977). A brief description of the discovery of overlapping genes.

KORNBERG, A. "The Synthesis of DNA." *Scientific American,* **219,** 64–78 (1968). A nontechnical account of this epoch achievement. The article also contains a brief historical survey of the events that led to the discovery of DNA polymerase.

KORNBERG, A. "Active Center of DNA Polymerase." *Science,* **163,** 1410–1418 (1969). A review of the struc-

tural and catalytic properties of DNA polymerase by its discoverer and primary investigator. Emphasis given to polymerase I from *E. coli.*

KORNBERG, A. *DNA Synthesis,* San Francisco: W. H. Freeman, 1974. An excellent presentation of facts and ideas about biochemical aspects of DNA biosynthesis. Polymerases, repair, recombination, restriction, and transcription are covered. Students with a background in general biochemistry can handle the reading. (See also the Scheckman source cited below.)

LEHMAN, I. R. "DNA Ligase: Structure, Mechanism, and Function." *Science,* **186,** 790–797 (1974).

MARX, J. L. "Gene Structure: More Surprising Developments." *Research News* article in *Science,* **199,** 717–518 (1978). A brief review of the discovery of spacer sequences in eukaryotic genes.

MARSHALL, R. E., C. T. CASKEY, and M. NIRENBERG. "Fine Structure of RNA Codewords Recognized by Bacterial, Amphibian, and Mammalian Transfer-RNA." *Science,* **155,** 820–825 (1967). Conclusive evidence for the universality of the basic language of the genetic code on the basis of *in vitro* binding studies with 50 synthetic codons, with a discussion of specific variations in terms of their phylogenetic and evolutionary significance.

MATHEWS, M. B. "Mammalian Messenger RNA." In *Essays in Biochemistry,* P. N. Campbell and F. Dickens, eds., Vol. 9, pp. 59–101. New York: Academic Press, 1973. A good review article; the poly(A) fragment is discussed.

MESELSON, M., and F. W. STAHL. "The Replication of DNA in *Escherichia coli*." *Proc. Nat. Acad. Sci., U.S.,* **44,** 671–682 (1958). The original paper describing the experimental approach used to establish the semiconservative scheme of DNA replication.

NIRENBERG, M. W. "The Genetic Code: I." *Scientific American,* **208,** 80–94 (1963). A description of (a) classical experiments with synthetic oligo- and polyribonucleotides proving that the sequence of bases in an RNA molecule specifies the order of insertion of amino acids during the assembly of a polypeptide chain; (b) the nature of the genetic code; and (c) how many of the triplet codes were determined by laboratory studies.

PERRY, R. P. "Processing of RNA." In *Ann. Rev. Biochem.,* **45,** 605–630 (1976). A thorough review article describing the posttranscription processing of tRNA, rRNA, and mRNA in both prokaryotes and eukaryotes.

PIPER, P. W., and B. F. C. CLARK. "Primary Structure of a Mouse Myeloma Cell Initiator Transfer RNA." *Nature,* **247,** 516–518 (1974) and SIMSEK, M., U. L. RAJBHANDRY, M. BOISNARD, and G. PETRISSANT. "Nucleotide Sequence of Rabbit Liver and Sheep Mammary Gland Cytoplasmic Initiator Transfer RNAs." *Nature,* **147,** 518–520 (1974). The uniqueness of the eukaryotic initiator tRNA is described.

PTHASHNE, M., and W. GILBERT. "Genetic Repressors." *Scientific American,* **222,** 36–44 (1970). Description of the experiments proving that the repressor is a protein, and discussion of repressor action.

RICH, A. "Polyribosomes." *Scientific American,* **209,** 44–53 (1963). A description of experiments and electron microscopic investigations proving that clusters of ribosomes serve as the active sites of protein biosynthesis.

RICHTER, D., V. A. ERDMANN, and M. SPRINZL. "Specific Recognition of GTΨC Loop of tRNA by 50S Ribosomal Subunits from *E. coli*." *Nat. New Biol.*, **246**, 132–135 (1973).

SCHECKMAN, R., A. WEINER, and A. KORNBERG. "Multienzyme Systems of DNA Replication." *Science*, **186**, 987–993 (1974). A review of current ideas regarding the *in vivo* enzymology of DNA biosynthesis. Data are presented supporting the participation of several proteins, emphasizing polymerase III.

SEBASTIAN, J. "Structure and Function of the Yeast RNA Polymerases." *Trends Biochem. Sci.*, **2**, 102–104 (1977). A short review article.

SINSHEIMER, R. L. "Recombinant DNA." In *Ann. Rev. Biochem.*, **46**, 415–438 (1977). A review article of the state of the art through late 1976.

SOBELL, H. M. "How Antimycin Binds to DNA." *Scientific American*, **231**, 82 (August, 1974).

SPRINZL, M., and F. CRAMER. "Accepting Site for Aminoacylation of tRNA^phe from Yeast." *Nat. New Biol.*, **245**, 3–5 (1973). Evidence for the 2′ position as the initial site of aminoacylation at the acceptor stem.

STENT, G. S. *Molecular Genetics—An Introductory Narrative.* San Francisco: W. H. Freeman, 1971. A new textbook designed for undergraduate courses in molecular biology and genetics. The author, an authority, uses an historical approach, tracing the growth of knowledge from important pioneering studies to the present day.

TEMIN, H. M. "The DNA Provirus Hypothesis." *Science*, **192**, 1075–1080 (1976). Nobel Prize lecture describing current thoughts on the significance of reverse transcription and the infectivity of RNA viruses.

UY, R., and F. WOLD. "Posttranslational Covalent Modification of Proteins." *Science*, **198**, 890–896 (1977). A good review article.

WATSON, J. D. "Involvement of RNA in the Synthesis of Proteins." *Science*, **140**, 17–26 (1963). The author's Nobel address.

WATSON, J. D. *Molecular Biology of the Gene.* 3rd edition. New York: W. A. Benjamin, 1976. The best available introductory treatment of the molecular aspects of gene function; it should be in the library of every biological scientist.

WATSON, J. D., and F. H. C. CRICK. "Genetical Implications of the Structure of Deoxyribonucleic Acid." *Nature*, **171**, 964–967 (1953). The original paper by the discovers of DNA structure proposing how a DNA duplex molecule could be replicated.

WEISSBACH, H., and S. OCHOA. "Soluble Factors Required for Eukaryotic Protein Synthesis." In *Annual Review of Biochemistry*, **45**, 191–216 (1976). A thorough review article.

WINTERSBERGER, E. "DNA-dependent DNA Polymerases from Eukaryotes." *Trends Biochem. Sci.*, **2**, 58–60 (1977). A short review article.

YANOFSKY, C. "Gene Structure and Protein Structure." *Scientific American*, **216**, 80–94 (1967). Discussion of the proof for the colinearity of base sequence in DNA and of amino acid sequence in proteins.

EXERCISES

8–1. Assuming that the chromosome of *E. coli* replicates according to the *conservative* scheme, which of the drawings below, numbered 1 through 6, will represent the density gradient ultracentrifugation pattern of the cellular DNA obtained in the Meselson-Stahl experiment from (a) the first generation of cells, that is, after one replication, and (b) the second generation of cells, that is, after two replications? Explain.

8–2. In the Meselson-Stahl experiment proving the *semi-conservative* scheme for the replication of DNA, which of the patterns in Exercise 8–1 would correspond

to the density gradient pattern obtained from the third generation of cells, that is, after three replications? Describe the relative proportions of each type of DNA that would be present in the extract.

8–3. Under certain conditions, *E. coli* cells can multiply quite rapidly. For example, in a nutritionally luxuriant broth the generation time at 37 °C is approximately 30 minutes. Assuming that the same time is required for replication of DNA, calculate and then reflect upon the number of nucleotides that are added in phosphodiester

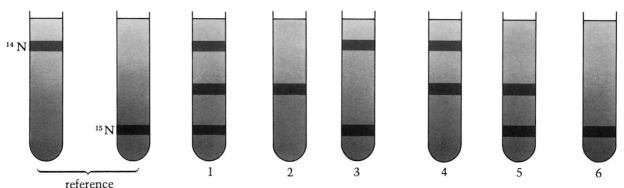

reference   1   2   3   4   5   6

linkage per minute. There are approximately 4.4 million nitrogen base pairs in the single *E. coli* chromosome.

8–4. In the DNA duplex segment below identify a successive sequence of nucleotides that could serve as the site of action of a restriction enzyme.

pCAGTTACTTGTTAACCGCC
GTCAATGAACAATTGGCGGp

8–5. If DNA polymerase were incubated with a mixture of dGTP, dCTP, dTTP, and [³²P]dATP in the presence of a single-stranded primer DNA that possessed in part the sequence of bases given below, how many units of 3'-[³²P]dGMP, 3'-[³²P]dCMP, 3'-[³²P]dTMP, and 3'-[³²P]dAMP would be contributed by the complementary segment in the product DNA after complete hydrolysis by 5'-nucleases?

...pApTpCpTpTpCpGpCpApTpGpCpApTpGpTpCpT...

8–6. Shown below is a hypothetical segment of a DNA molecule. What would be the maximum number of bases potentially susceptible to modification by treatment with nitrous acid? What would be the maximum number of nitrogen base pairs potentially susceptible to a transitional mutation upon treatment with nitrous acid?

8–7. Using the genetic code and the description of the difference in the amino acid sequence in the β chain of sickle cell hemoglobin (see p. 130), prove that a change in a single base pair in DNA can result in a defective protein.

8–8. The polypeptide chain of myoglobin contains 153 amino acid residues. Theoretically then, how many nucleotides would be present in the gene that specifies the configuration of this polypeptide?

8–9. The amino acid sequence of the polypeptide chain of cytochrome *c* is given on page 129. In symbolic fashion, write a nucleotide sequence for the gene of pigeon DNA that would correspond to residues 99 through 104 of the cytochrome *c* chain. Then determine the number of base substitution mutations that could occur in this gene segment without a resultant change in the amino acid sequence.

8–10. On page 276 it was stated that a leucine-tRNA molecule (yeast) having the anticodon (3')GAA is capable of binding with only one of the six degenerate codons assigned to leucine (see Figure 8–12). Given the fact that a second leucine-tRNA (yeast) having the anticodon (3')AAC has been isolated, predict (a) what codon or codons it would bind with, and (b) the *minimum* and the *maximum* number of additional leucine-tRNAs that would be required to recognize the codons that still remain unaccounted for.

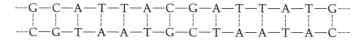

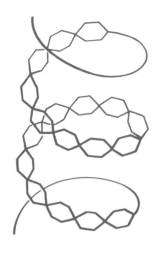

CHAPTER 9

CARBOHYDRATES

Carbohydrate is not a very exact term since it applies to a very large number of materials and includes a wide spectrum of chemical structure and biological function. Without any knowledge of structure it was suggested as a name almost 100 years ago to refer to those naturally occurring substances having a composition according to the formula $-(C \cdot H_2O)_n$, that is, carbon·hydrate. The modern meaning is much more general and includes any substance that satisfies this criterion and many derived substances.

In most organisms carbohydrate material—largely in the form of the simple sugar, *glucose*—is the primary foodstuff, which upon degradation supplies the bulk of the *energy and carbon* required in the biosynthesis of proteins, nucleic acids, lipids, and other carbohydrates. Many of the polymeric carbohydrates, however, have more specific functions, falling mainly into one of two categories—those that have a *structural role* and those that have a *carbon-storage role*. The *celluloses* in plants best exemplify the structural carbohydrates and the *starches* (in plants) and *glycogens* (in animals and bacteria) the carbon-storage carbohydrates. In addition to these pure carbohydrates, there are many substances, such as various *glycoproteins* and *glycolipids*, that are *partially composed* of carbohydrate material and have unique functions of their own. (The prefix *glyco-* is commonly used to designate the presence of carbohydrate.) Other materials serve as cementing substances, as the amorphous, gellike substances that lubricate bone joints, as the protective coatings of bacteria, as the components of bacterial cell walls, as the components of some antibiotics, as the gum secretions that assist in the healing of plant wounds, and as the determinants of the different blood groupings.

Obviously, there is no way that the structure and function of carbohydrates can be thoroughly covered in a single chapter. What we will try to do is introduce important principles that apply to carbohydrates in general and highlight only a few of the specifics mentioned above. The Pigman-Horton source (see end of chapter) is an excellent reference for more information. In basic outline this chapter covers (in sequence) the following classes of carbohydrates: (1) the *monosaccharides* (monomeric units), sometimes referred to as the *simple sugars*, (2) the *oligosaccharides* (two to eight monomers in glycosidic linkage), and (3) the *polysaccharides*, consisting of more than eight monomers, but usually referring to very high molecular weight substances with from 100 to a few thousand monomeric units. (The term *saccharide* (derived from Greek *sakchar*, meaning sugar or sweetness) is related to the characteristic taste of many of the simple carbohydrates.) The chapter concludes with a discussion of (4) the *glycoproteins*.

Our understanding of carbohydrate structure has its origins in the latter part of the nineteenth century with the pioneering studies of Emil Fischer, who was the first to establish the structure of several of the monosaccharides. Fischer's work in carbohydrate chemistry is referred to by many as the *birth of modern organic chemistry*.

Methods of carbohydrate separation and isolation will not be discussed. Suffice it to say that mono- and oligosaccharides can be analyzed by any of the chromatographic methods considered in Chap. 2. Gel-permeation chromatography and ultracentrifugation are quite useful in the analysis of polysaccharides.

MONOSACCHARIDES

Basic Structure and Stereoisomerism

Chemically speaking, the monosaccharides can be described as polyhydroxy aldehydes, polyhydroxy ketones, and derivatives thereof. The derivatives will be discussed later. All simple monosaccharides have the general empirical formula $(CH_2O)_n$, where n is a whole number ranging from 3 to 9. Regardless of carbon number, all monosaccharides can be grouped into one of two general classes: *aldoses* or *ketoses*. The -ose ending is characteristic in carbohydrate nomenclature. The ending *-ulose* is irregularly used to designate a simple ketose. Aldoses contain a functional aldehyde grouping

$$-\overset{\overset{\textstyle O}{\|}}{C}-H,$$ whereas ketoses contain a functional ketone grouping $\diagdown C=O$

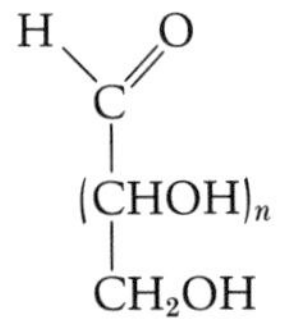

Subclasses are then distinguished on the basis of carbon content according to the following terms: aldotriose, ketotriose, aldotetrose, ketotetrose, and so on. *Glyceraldehyde* is the simplest aldose and *dihydroxyacetone* is the simplest ketose. Each is considered as the parent compound of higher $\diagdown CHOH$ homologues in each class.

The structures of glyceraldehyde and dihydroxyacetone are also distinct in that glyceraldehyde contains an asymmetric (chiral) carbon atom whereas dihydroxyacetone does not. In fact, as revealed by close inspection of the general structures shown above, the absence of an asymmetric center is unique to dihydroxyacetone, with all other classes of simple carbohydrates

H—C=O
|
(CHOH)$_n$
|
CH$_2$OH

polyhydroxy
aldehyde

CH$_2$OH
|
C=O
|
(CHOH)$_n$
|
CH$_2$OH

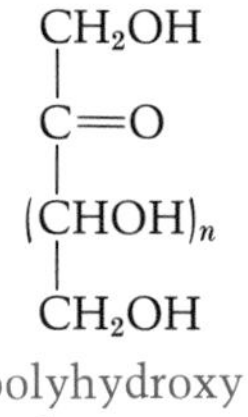

polyhydroxy
ketone
(2-keto)

Aldose family

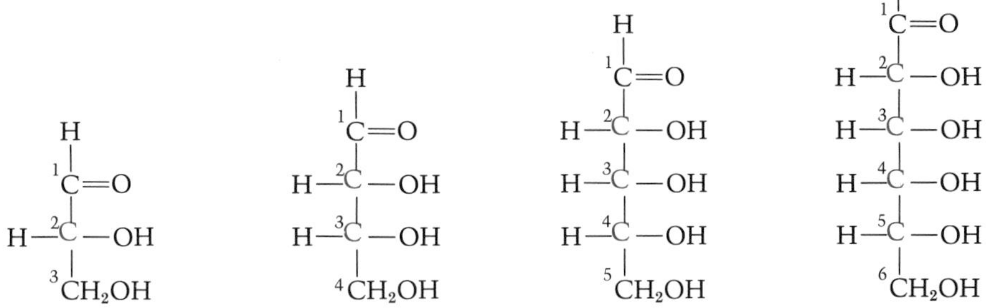

H
$|$
$^1C=O$
$|$
$H-^2C-OH$
$|$
$^3CH_2OH$

$C_3H_6O_3$
aldotriose
(glyceraldehyde)

H
$|$
$^1C=O$
$|$
$H-^2C-OH$
$|$
$H-^3C-OH$
$|$
$^4CH_2OH$

$C_4H_8O_4$
aldotetrose

H
$|$
$^1C=O$
$|$
$H-^2C-OH$
$|$
$H-^3C-OH$
$|$
$H-^4C-OH$
$|$
$^5CH_2OH$

$C_5H_{10}O_5$
aldopentose

H
$|$
$^1C=O$
$|$
$H-^2C-OH$
$|$
$H-^3C-OH$
$|$
$H-^4C-OH$
$|$
$H-^5C-OH$
$|$
$^6CH_2OH$

$C_6H_{12}O_6$
aldohexose

These linear structural formulas are referred to as Fischer projection formulas; the accepted convention of numbering carbons is as shown here; asymmetric carbons are in color.

2-Ketose family

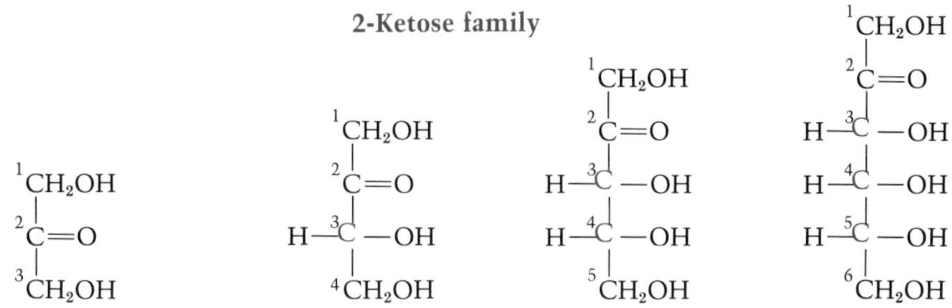

$^1CH_2OH$
$|$
$^2C=O$
$|$
$^3CH_2OH$

ketotriose
(dihydroxyacetone)

$^1CH_2OH$
$|$
$^2C=O$
$|$
$H-^3C-OH$
$|$
$^4CH_2OH$

ketotetrose

$^1CH_2OH$
$|$
$^2C=O$
$|$
$H-^3C-OH$
$|$
$H-^4C-OH$
$|$
$^5CH_2OH$

ketopentose

$^1CH_2OH$
$|$
$^2C=O$
$|$
$H-^3C-OH$
$|$
$H-^4C-OH$
$|$
$H-^5C-OH$
$|$
$^6CH_2OH$

ketohexose

having at least one asymmetric center, the total number being equal to the number of internal $\diagdown$CHOH$\diagup$ groups. Obviously, a progression through the homologous families of aldoses and ketoses will yield greater numbers of isomeric forms with each higher homologue. The number of stereoisomers corresponds to 2^n where n equals the number of asymmetric carbons. For example, an aldohexose with a general formula of $C_6H_{12}O_6$ and four asymmetric carbons, that is, four $\diagdown$CHOH$\diagup$ groups, could exist in any one of 16 possible isomeric forms, with eight D forms and eight L forms.

With D-glyceraldehyde as the parent compound, it is possible to construct a chart (Fig. 9–1) illustrating the structures of all D-aldoses through the aldohexose group. The diagram illustrates the sugars of one homologous family as originating from the next lower homologous group, with the chain extended by the generation of a new $\diagdown$CHOH$\diagup$ at position 2. Note that every time this occurs, the hydroxyl group at C^2 can assume two possible orientations, while all other $\diagdown$CHOH$\diagup$ groupings remain unchanged. The overall configuration of each sugar is fixed by the orientation of the $\diagdown$CHOH$\diagup$ group most distant from the functional aldehyde group. This would be $—C^5$ in hexoses; $—C^4$ in pentoses; and $—C^3$ in tetroses. A similar diagram can be constructed for the L series, beginning with L-glyceraldehyde. Diagrams for the ketoses are also possible (Fig. 9–2, p. 291).

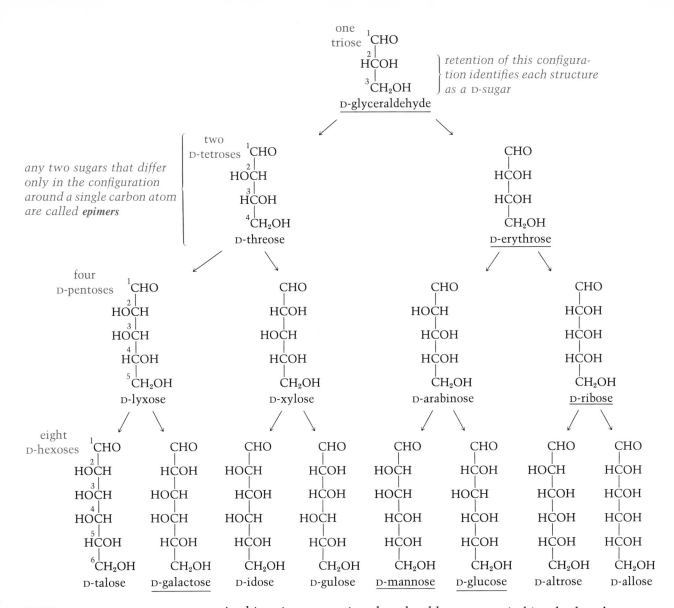

FIGURE 9–1 Structural relationships among D-aldoses. Those that are underlined occur more commonly.

Table 9–1 appears on p. 292.

At this point, a question that should come to mind is whether the D or L form predominates in nature. The question is valid, and yes, one form does predominate—the D configuration. Thus, through the first four homologous classes of aldoses and ketoses, the number of biologically important simple sugars has been reduced from 46 to 24. In addition, only about half of these 24 have what might be called a general importance in that they participate most frequently in the living processes. These are listed in Table 9–1. As either an unmodified sugar or as a derivative, each occurs in the free monomeric state or in oligomeric and/or polymeric structures. The structures of two important sugars not previously given in Figs. 9–1 or 9–2 are also shown: *sedoheptulose*, a seven carbon ketose, and *2-deoxyribose*, an oxidized derivative of ribose found in DNA.

Anomeric Isomers

Consider the following facts. Two separate aqueous solutions containing equal amounts of pure D-glucose may exhibit widely different optical prop-

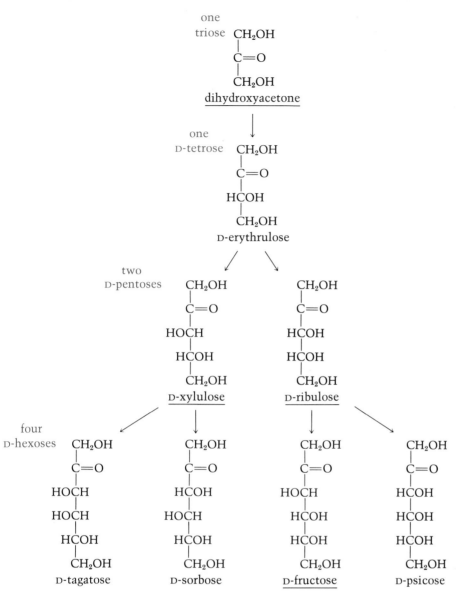

FIGURE 9–2 Structural relationships among D-ketoses. Those that are underlined occur more commonly.

erties when analyzed in a polarimeter. On prolonged standing (quicker if a small amount of acid is present) the specific rotation of each solution would gradually change and ultimately attain the same equilibrium value. The phenomenon, called *mutarotation,* is common to all of the simple monosaccharides.

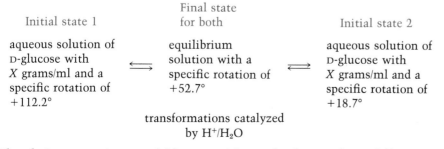

Initial state 1	Final state for both	Initial state 2
aqueous solution of D-glucose with X grams/ml and a specific rotation of $+112.2°$	equilibrium solution with a specific rotation of $+52.7°$	aqueous solution of D-glucose with X grams/ml and a specific rotation of $+18.7°$

transformations catalyzed by H^+/H_2O

The obvious questions are (a) how can identical solutions have different optical properties and (b) how can both solutions attain the same specific rota-

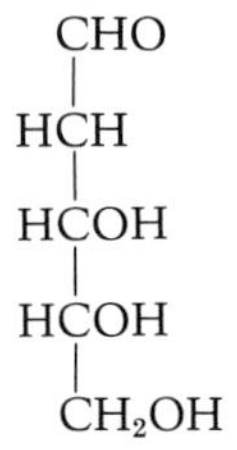

CHO
|
HCH
|
HCOH
|
HCOH
|
CH₂OH

The Fischer structure in the margin:

$$
\begin{array}{c}
CHO \\
|\\
HCH \\
|\\
HCOH \\
|\\
HCOH \\
|\\
CH_2OH
\end{array}
$$

TABLE 9–1 Common monosaccharides.

dihydroxyacetone
D-glyceraldehyde
D-erythrose
D-ribose
D-2-deoxyribose
 (see above)
D-ribulose
D-xylulose
D-glucose (most abundant)
D-galactose
D-mannose
D-fructose
D-sedoheptulose
 (a 2-ketoheptose; see below)

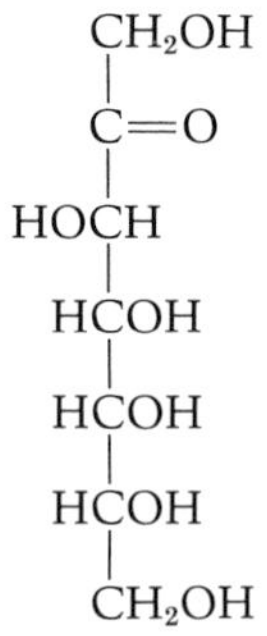

$$
\begin{array}{c}
CH_2OH \\
|\\
C=O \\
|\\
HOCH \\
|\\
HCOH \\
|\\
HCOH \\
|\\
HCOH \\
|\\
CH_2OH
\end{array}
$$

tion? Well, it should occur to you that, in fact, the samples must *not be identical*, and that the observed differences would be most logically explained by proposing that each solution probably contains a *different form* of D-glucose. Furthermore, since the two solutions differ only in their ability to rotate a ray of polarized light, in all probability they contain two *different stereoisomeric forms* of D-glucose. With regard to the second question, we can interpret the data as representative of the interconversion of these two forms to yield an equilibrium mixture containing each isomer in quantities characteristic of D-glucose.

The above description and accompanying interpretation of data are representative of the first encounter with this apparent anomaly about 80 years ago. To account for the existence of two stereoisomeric forms of D-glucose, it was proposed that *in solution very few sugar molecules exist with free aldehyde or ketone functional groups, but exist rather as cyclic hemiacetals or hemiketals, respectively.* As shown (margin, p. 293), the hemiacetal linkage is formed from the condensation of an aldehyde grouping and a hydroxyl grouping. If the interaction is *intra*molecular, as it is in the monosaccharides, the resultant hemiacetal is cyclic. For reasons of chemical stability, five- and six-membered rings are most common. (Generally, aldohexoses form six-membered rings via a C^1-C^5 interaction; ketohexoses form five-membered rings via a C^2-C^5 interaction to yield cyclic hemiketals; aldopentoses form five-membered rings via a C^1-C^4 interaction.

The rationale for this suggestion becomes evident by inspecting the structures shown below, which reveal that the formation of a cyclic hemiacetal generates an additional asymmetric center at the original carbonyl atom. The new asymmetric center is termed the *anomeric carbon.* Two stereoisomers exist because the anomeric hydroxyl group can assume either one of two possible spatial orientations. In the linear Fischer representations the structure with the anomeric hydroxyl group directed to the right is termed the α form, and that with the opposite orientation (—OH to the left) is termed the β form. Anomeric isomers exist for all simple sugars; their existence is an extremely significant structural feature of the carbohydrates. We have already encounted one example in Chap. 7. Can you identify it without checking?

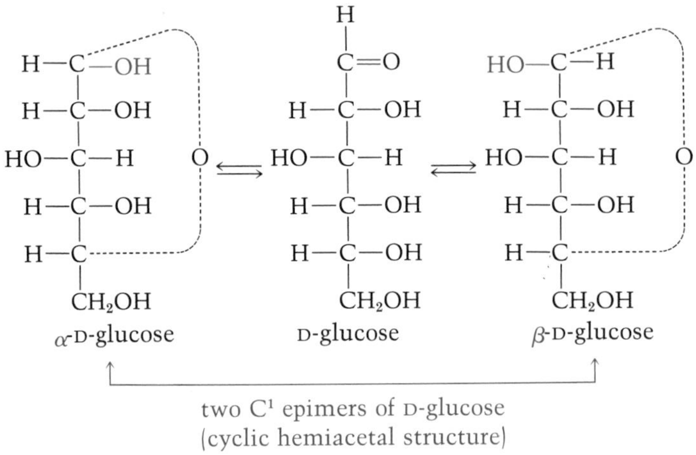

the dashed line represents a distorted bond projecting toward the rear

α-D-glucose   D-glucose   β-D-glucose

two C^1 epimers of D-glucose
(cyclic hemiacetal structure)

Mutarotation is more completely described, then, as the interconversion of the α and β anomers. The conversion is not direct, but is bridged by the noncyclic form containing a free carbonyl group. An equilibrium mixture of unchanging optical properties does not imply 50-50 amounts of each anomer. One form is usually in excess, with the relative amounts being determined by differences in chemical stability.

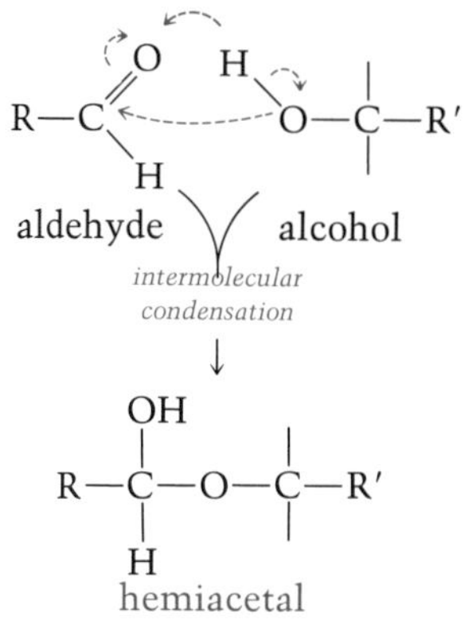

Haworth Projection Formulas

It is quite obvious that a Fischer projection formula of the hemiacetal structure is inaccurate in what it represents. Chemical bonds do not have right-angle bends. In 1929 Haworth suggested a more realistic representation in which both five-membered and six-membered cyclic structures are depicted as planar ring systems with the hydroxyl groups oriented either above or below the plane of the ring. Although it, too, does not really represent the actual three-dimensional structure of a sugar, the Haworth representation has been used for nearly 50 years as an easy-to-draw formula that also permits a quick evaluation of the relative orientations of the —OH groups in the structure. Because of the structural similarity to the organic compounds called *furan* and *pyran*, a five-membered cyclic hemiacetal is labeled a *furanose* and a six-membered hemiacetal ring is called a *pyranose*. As shown below the front edges are usually heavily shaded to confer some three-dimensional character. The simplified Haworth form with carbons omitted from the ring is routinely used.

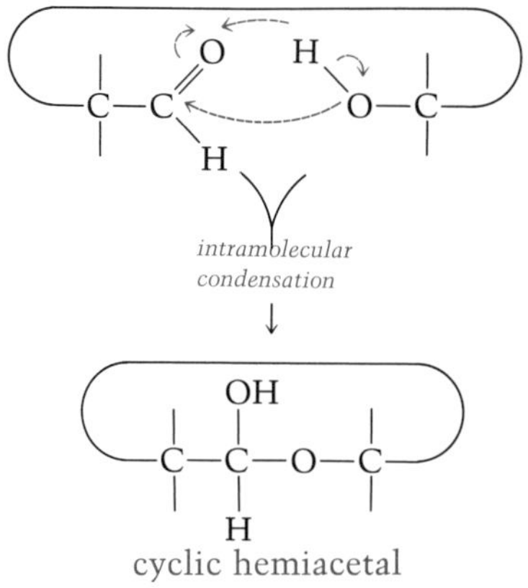

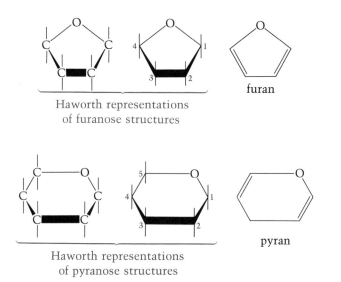

Haworth representations
of furanose structures

furan

Haworth representations
of pyranose structures

pyran

For any D sugar, the conversion of a Fischer formula into a Haworth formula proceeds as follows: (a) anything that is directed to the right in the Fischer structure is given a lower (downward) orientation in the Haworth structure; (b) anything that is directed to the left in the Fischer structure is given an upper (upward) orientation in the Haworth structure; and (c) the terminal —CH₂OH grouping is given an upward orientation in the Haworth structure. (For an L sugar, (a) and (b) are the same, but the terminal —CH₂OH grouping is projected downward.) The structures of α-D-glucopyranose and β-D-fructofuranose (see next page) illustrate the conversion. Note the shorthand form, in which only dashes are used to represent the position of the —OH group and all Hs are omitted.

293

Fischer   complete
Haworth   abbreviated
Haworth

α-D-glucopyranose

β-D-fructofuranose

Conformation of Sugars in Solution

Albeit very useful, the completely planar Haworth structures are still distorted representations of the actual molecule. The furanose ring, although nearly flat, has a slight pucker when seen from the side, as in the ball and stick model below. The preferred conformation of most pyranose rings is nowhere near being flat, but is the *chair* structure that is common to such things as cyclohexane. In the chair conformation the OH groups would exist in either *axial* (vertical) or *equatorial* (nonvertical) positions. To convert from Haworth to the chair conformation is simple; whatever has a downward or upward orientation in the Haworth representation has a similar orientation in the chair conformation.

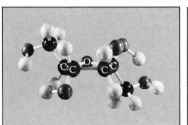

Haworth   chair conformation

There is much more to this subject of sugar conformations that we will not consider. Throughout this and later chapters we will use whatever representation best illustrates the point of our discussion. Usually the Haworth forms will be presented.

Due to the presence of the functional OH groups and the potential for the existence of a free —$\overset{H}{C}$=O or $\overset{\diagdown}{\underset{\diagup}{C}}$=O grouping, sugars can undergo the spectrum of reactions that are common to alcohols, aldehydes, and ketones. Then there are some class reactions that are a consequence of the cyclic polyhydroxy ring structure. All of this amounts to a large number of reactions. However, the scope of our consideration will be limited to only a few of these, with a particular focus on reactions of biological significance.

Mutarotation (Reaction with Acid)

The interconversion of the α and β anomeric forms, which was mentioned earlier, occurs as the hemiacetal ring opens and recloses, yielding the opposite configuration around the anomeric carbon. The catalytic effect of a dilute acid solution is attributed to the protonation of the oxygen atom in the ring. This facilitates an electron rearrangement, which results in the cleavage of the O—C^α bond and formation of a carbonyl (C=O) grouping, after loss of a proton. The reverse of these events, which would begin with the protonation of the oxygen in C=O, would result in ring formation.

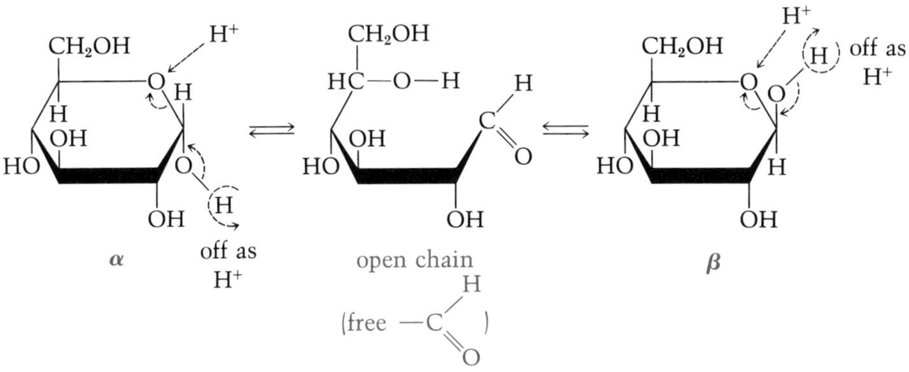

An important aspect of this transformation is that sugars, even in the absence of catalytic amounts of H^+, can exist in the open-chain structure with a free carbonyl grouping. Consequently, a sugar can participate in a reaction—including reactions that occur under physiological conditions—as either the open-chain form or the cyclic form. Thus, depending on the reaction, we will depict the sugar in either structure.

Esterification (Phosphate Esters and Nucleoside Diphospho Sugars)

Alcohols readily form esters when reacted with acids, anhydrides, or acyl halides. The most important type of sugar esters that occur in living cells are (a) the *phosphate esters* (phosphoesters) and (b) the *nucleoside diphosphate esters*.

The biosynthesis of phosphate esters usually occurs in an *ATP-dependent reaction,* catalyzed by a *kinase* enzyme. In this type of reaction ATP, with its very reactive phosphoanhydride linkage, acts as a donor of phosphate.

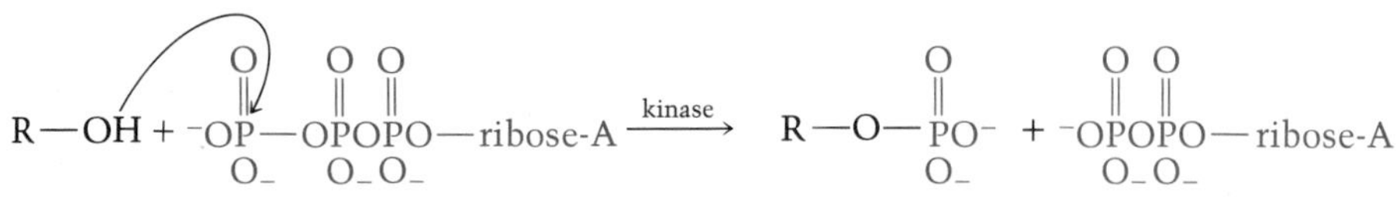

$$R\text{—OH} + {}^-OP\text{—OPOPO—ribose-A} \xrightarrow{\text{kinase}} R\text{—O—PO}^- + {}^-OPOPO\text{—ribose-A}$$

free
sugar   ATP   sugar phosphate   ADP

The most widespread enzyme is *hexokinase,* which has a broad specificity toward the hexose substrate but always yields a sugar-6-phosphate. An important exception to this general rule is represented by galactokinase, which prefers galactose and yields galactose-1-phosphate. The formation of glucose-1-phosphate is presented elsewhere (see p. 379).

Three important diphosphesters are fructose-1,6-diphosphate, ribulose-1,5-diphosphate, and sedoheptulose-1,7-diphosphate. The formation (see below) of fructose-1,6-diphosphate involves hexokinase and a second highly specific and immensely important (see p. 385) enzyme, *phosphofructokinase.* The other two diphosphoesters are produced as part of the CO_2-fixation pathway in photosynthesis (see p. 492 in Chap. 15).

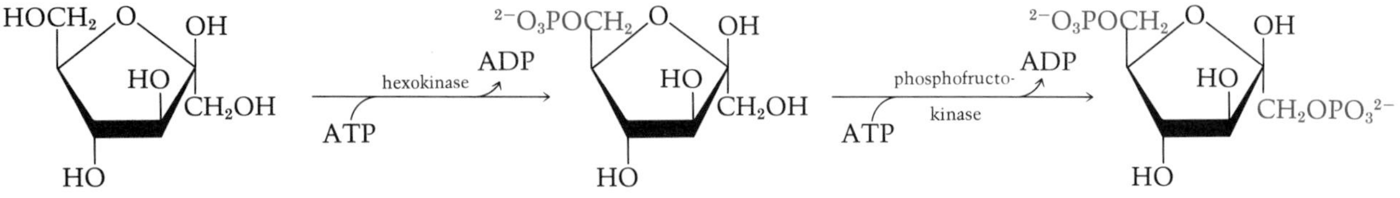

> The name kinase is always synonymous with an ATP-dependent reaction in which a phosphate grouping from ATP is transferred to a substrate.

β-D-fructofuranose
or
β-D-fructose

$\xrightarrow[\text{ATP}]{\text{hexokinase} \quad \text{ADP}}$

β-D-fructofuranose-6-phosphate
or
fructose-6-phosphate (F6P)

$\xrightarrow[\text{ATP}]{\text{phosphofructo-} \atop \text{kinase} \quad \text{ADP}}$

β-D-fructofuranose-1,6-diphosphate
or
fructose-1,6-diphosphate (FDP)

The biological significance of the sugar phosphates is that they represent the *metabolically active form* of sugars. In other words, in most instances when a sugar participates as a substrate in an enzyme-catalyzed reaction, it does so as a phosphate ester. One group of obvious exceptions to this generalization is of course, the kinase-catalyzed reactions, wherein the phosphorylated species are themselves formed from the neutral sugars. There are a few others. In addition to enhancing the reactivity of sugars, phosphorylation confers another benefit to a cell because the cell membrane is not very permeable to the passage of sugar phosphates. In effect, this means that once they are formed within the cell, the sugar phosphates are more or less trapped inside.

One of the most important reactions that the sugar-1-phosphates undergo is with another unit of a nucleoside triphosphate to yield *nucleoside diphospho derivatives.* The general reaction, catalyzed by an enzyme referred to as either a *NuDP-sugar pyrophosphorylase* or a *nucleoidityl transferase* is shown below. Although all types of nucleoside diphospho sugars are found in nature, the *UDP-sugars* appear to be most abundant.

The NuDP-sugars participate in many reactions. Basically, however, all of these reactions can be grouped under two types, involving (a) the biosynthesis of oligomeric and polymeric carbohydrates, and (b) various chemical transformations of simple sugars such as the isomeric interconversion of galactose and glucose. Both types are discussed in Chap. 12 (see p. 392). A few examples of type (b) are given in the next few pages.

> For his many contributions in furthering the understanding of carbohydrate metabolism, and especially for the discovery and elucidation of the role of NuDP-sugars, L. F. Leloir was awarded the Nobel Prize in 1970.

(a) $(\text{glucose})_n + \text{UDP-glucose} \rightarrow (\text{glucose})_{n+1} + \text{UDP}$
(b) $\text{UDP-glucose} \rightleftarrows \text{UDP-galactose}$

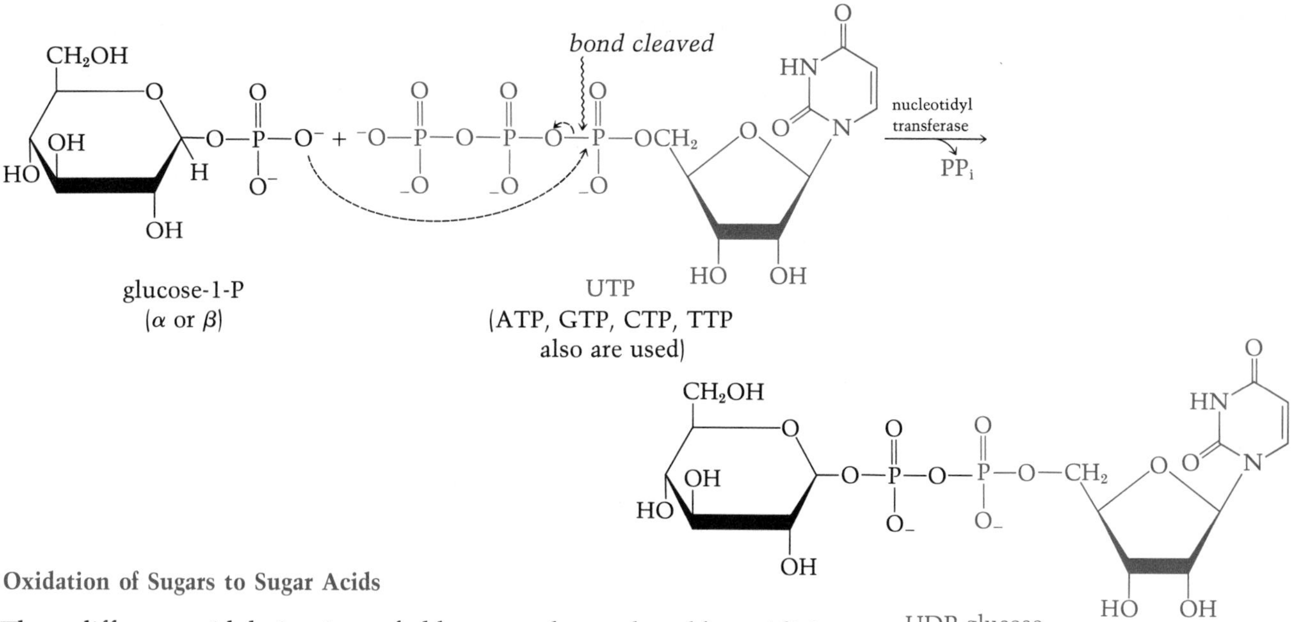

glucose-1-P
(α or β)

UTP
(ATP, GTP, CTP, TTP
also are used)

nucleotidyl
transferase
PP_i

UDP-glucose

Oxidation of Sugars to Sugar Acids

Three different acid derivatives of aldoses can be produced by oxidizing ter-minal groupings to COOH groups. Acids arising from the oxidation of the terminal —CHO group are called *glyconic acids*. If the terminal —CH_2OH group is oxidized, a *glycuronic acid* is produced, and if both terminal groups are oxidized, a *glycaric acid* is produced. The three acids of D-glucose are shown below; also indicated is their tendency to undergo an intramolecular elimination of H_2O to yield six-membered *cyclic lactones*.

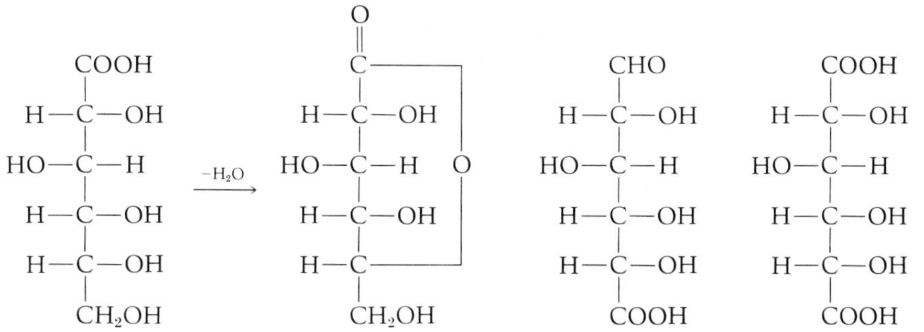

D-gluconic acid   D-gluconolactone   D-glucuronic acid   D-glucaric acid

In plants and most vertebrates—*except* humans, primates, and guinea pigs—an important reaction sequence leads to the biosynthesis of *ascorbic acid.*

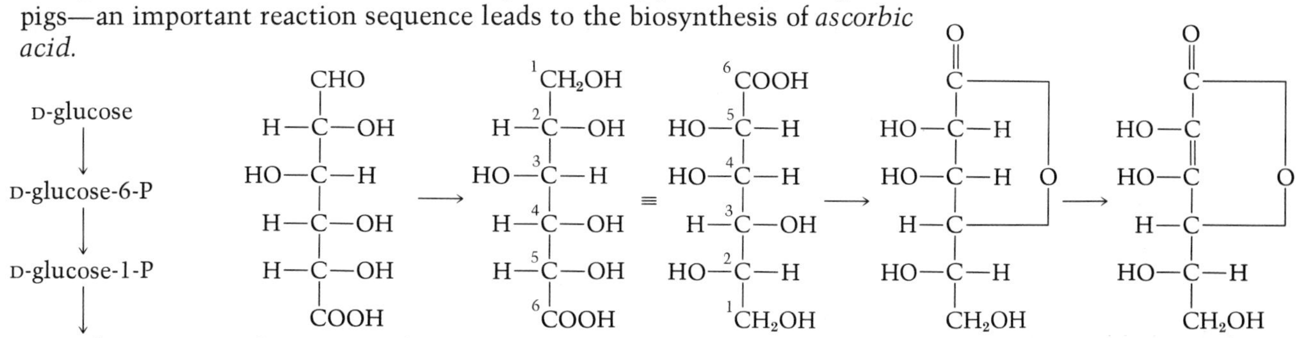

D-glucose → D-glucose-6-P → D-glucose-1-P → UDP-D-glucose ⟶ D-glucuronic acid

L-gulonic acid   L-gulonolactone   L-ascorbic acid (lactone form)

297

Without a dietary supply of ascorbic acid (*Vitamin C*), humans are prone to the disease of *scurvy*. Humans are unable to synthesize ascorbic acid because human cells do not have the enzyme (or a defective form of the enzyme is present) that catalyzes the last step in this process. The precise biochemical function of ascorbic acid is not yet known, but there is evidence that it serves as an *antioxidant*, protecting cells, particularly from the highly toxic action of powerful oxidizing agents that are proposed to be formed from O_2 (see p. 472 in Chap. 14).

Any carbohydrate that can yield a free carbonyl linkage susceptible to further oxidation is termed a *reducing sugar*. Obviously then, a *nonreducing sugar* is one that cannot yield a free carbonyl linkage. This terminology, reducing vs. nonreducing sugar, will be encountered later in the chapter.

The complete oxidation of a sugar results in its total degradation to CO_2 and is accompanied by a release of energy. This, of course, represents the net chemistry of respiration, with carbohydrate as the foodstuff. The oxidative and energy-yielding process of *respiration* is complemented by the reductive, energy-requiring process of *photosynthesis*, which results in the fixation of atmospheric CO_2 and its conversion to carbohydrate. The net chemistry and net energetics of this relationship are shown in the following diagram.

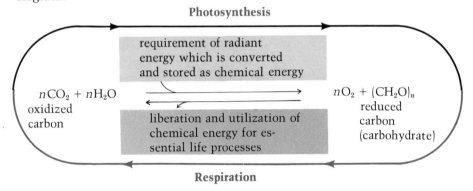

Photosynthesis

requirement of radiant energy which is converted and stored as chemical energy

$nCO_2 + nH_2O$
oxidized carbon

liberation and utilization of chemical energy for essential life processes

$nO_2 + (CH_2O)_n$
reduced carbon (carbohydrate)

Respiration

Reduction of Sugars to Sugar Alcohols and Deoxy Sugars

Two important reduced forms of the sugars are the *polyhydroxyalcohols* and the *deoxy* structures. In cells, the sugar alcohols are usually formed in reactions catalyzed by specific dehydrogenases, using the coenzyme NADH or NADPH as a hydrogen (electron) donor. Some specific examples are shown on p. 299 (top), and a brief statement of biological role is also given.

The formation of deoxy sugars involves the substitution of an OH group by an H atom. The most important biological example is D-*2-deoxyribose*, the "sugar of DNA." The complex details of the D-ribose → D-2-deoxyribose conversion are discussed in Chap. 17 (p. 563).

In all types of organisms, glucose-6-phosphate is—among other things—a precursor in the biosynthesis of *inositol*, a six-membered *cyclic* polyhydroxy alcohol. The most common isomer (there are nine different steroisomers) is *myo-inositol*. In plants a large portion of inositol occurs as a fully phosphorylated form (i.e., a hexophosphoester), which is believed to be a storage form of phosphate in plants. A similar phosphate-storage role is not found in animals. In animals most of the inositol is found in cell membranes as a component of the lipid *phosphatidyl inositol* (see p. 326 in next chapter).

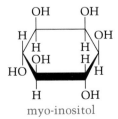
myo-inositol

CH₂OH structures:

$$
\begin{array}{ccccc}
& \text{CH}_2\text{OH} & \text{H}\quad\ \ \text{C}{=}\text{O} & \text{CH}_2\text{OH} & \text{CH}_2\text{OH} \\
& \text{H}-\text{C}-\text{OH} & \text{H}-\text{C}-\text{H} & \text{H}-\text{C}-\text{OH} & \text{H}-\text{C}-\text{OH} \\
\text{CH}_2\text{OH} & \text{H}-\text{C}-\text{OH} & \text{H}-\text{C}-\text{OH} & \text{HO}-\text{C}-\text{H} & \text{HO}-\text{C}-\text{H} \\
\text{H}-\text{C}-\text{OH} & \text{H}-\text{C}-\text{OH} & \text{H}-\text{C}-\text{OH} & \text{H}-\text{C}-\text{OH} & \text{HO}-\text{C}-\text{H} \\
\text{CH}_2\text{OH} & \text{CH}_2\text{OH} & \text{CH}_2\text{OH} & \text{H}-\text{C}-\text{OH} & \text{H}-\text{C}-\text{OH} \\
& & & \text{CH}_2\text{OH} & \text{CH}_2\text{OH}
\end{array}
$$

glycerol	D-ribitol	D-2-deoxyribose	D-glucitol	D-galactitol
(a component of lipids)	(a component of riboflavin and the coenzyme, FAD)	(found in nucleotides of DNA)	also called sorbitol	also called dulcitol
			(various commercial uses, such as sweetening agent; high levels are found in semen)	(cataract formation in lens of eye may be related to an accumulation of this sugar alcohol)

Glycosides

When an alcohol (ROH) reacts with another alcohol (R′OH) the product is an ether (R—O—R′). The sugars are no different in this basic sense, although there are a few distinguishing features. In the first place, a sugar can provide several OH groupings. A very special aspect involves the participation of the anomeric OH in the hemiacetal structure, which yields a "full" cyclic acetal. Since the anomeric position is more susceptible to reaction, it is possible with mild conditions to achieve nearly complete selectivity at this site. In the terminology of carbohydrate structure the full acetal structure is called a *glycoside*, with those derived from pyranoses called *pyranosides*, and those from furanoses called *furanosides*. In either case the newly formed linkage of C^α—OR is called a *glycosidic bond*. Under stronger conditions the other C—OH positions will react to give a completely substituted sugar.

α-D-glucopyranose (hemiacetal) → methyl-α-D-glucopyranoside a glycoside (full acetal) → methyl-2,3,4,6-tetra-O-methyl-α-D-glucopyranoside

The glycosidic linkage is of extreme biological significance, since it represents (neglecting some obscure exceptions) *the covalent bond of all monosaccharide-monosaccharide interactions.* Consequently, the glycoside bond is to the oligo- and polysaccharide carbohydrates what the peptide bond is to oligo- and polypeptides, and what the phosphodiester bond is to oligo- and polynucleotides. The glycoside linkage involves the anomeric hydroxyl group (α or β) of one monosaccharide and any available hydroxyl group in the second monosaccharide. The formation of α(1 → 4) and α(1 → 6) glycosidic linkages between two glucose molecules is diagramed below. Note that the symbolism merely identifies the OH sites that are involved in the bond. Although all combinations (α(1 → 3), α(1 → 2),

299

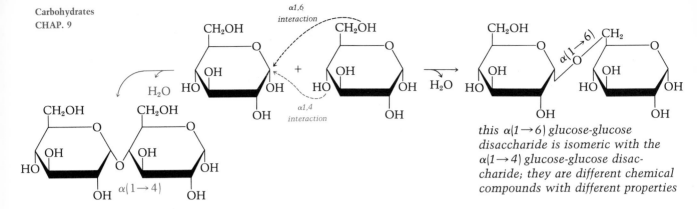

this α(1→6) glucose-glucose disaccharide is isomeric with the α(1→4) glucose-glucose disaccharide; they are different chemical compounds with different properties

β(1 → 4), β(1 → 6), and so on) are found in the naturally occurring oligo- and polysaccharides, *each particular oligosaccharide and polysaccharide contains a specific glycosidic linkage.* Indeed, in some instances this is the primary structural difference between otherwise identical oligomers or polymers (for example, compare the cellulose and amylose structures shown in the section on polysaccharides). With the exception of disaccharides, all linear oligomers or polymers will contain monomeric residues involved in two glycosidic linkages, except for the two residues at each end of the chain, which are involved in only one. Some residues may be involved in three glycosidic bonds, a situation common to *branched* polymers such as glycogen.

Linear polyglucose chain

all glycoside bonds are α(1→4)

nonreducing
end
(no free
anomeric OH)

reducing
end
(free
anomeric OH)

A variety of glycoside materials are found in nature, particularly in the plant kingdom. Several of these have therapeutic uses. The potent heart stimulant *digitoxin* is but one of several examples. Several antibiotics are glycosides; some examples are *erythromycin, streptomycin,* and *puromycin* (see p. 209). You should also recall that a nucleoside contains a C—N glycosidic bond.

AMINO SUGARS, N-ACETYL AMINO SUGARS, AND SIALIC ACID

Many naturally occurring oligo- and polysaccharides contain a monosaccharide unit that consists of a —NH₂ or —NHCCH₃ grouping in place of an —OH group. Such sugars are called *amino sugars* and *N-acetyl amino sugars*, respectively. Hexose derivatives are most common, with the replacement position usually being C^2. *2-amino-*D*-glucose* (D-glucosamine), *2-amino-*D*-galactose* (D-galactosamine), and the corresponding acetylated forms, *N-acetyl-*D*-glucosamine* and *N-acetyl-*D*-galactosamine*, are the most abundant.

The biosynthesis of these derivatives (as the UDP form) originates (see below) with fructose-6-phosphate and the amino acid glutamine, which functions as the amino group donor. Either of the UDP derivatives would then be utilized for the formation of various oligo- and polysaccharides, a few examples of which are cited later in the chapter.

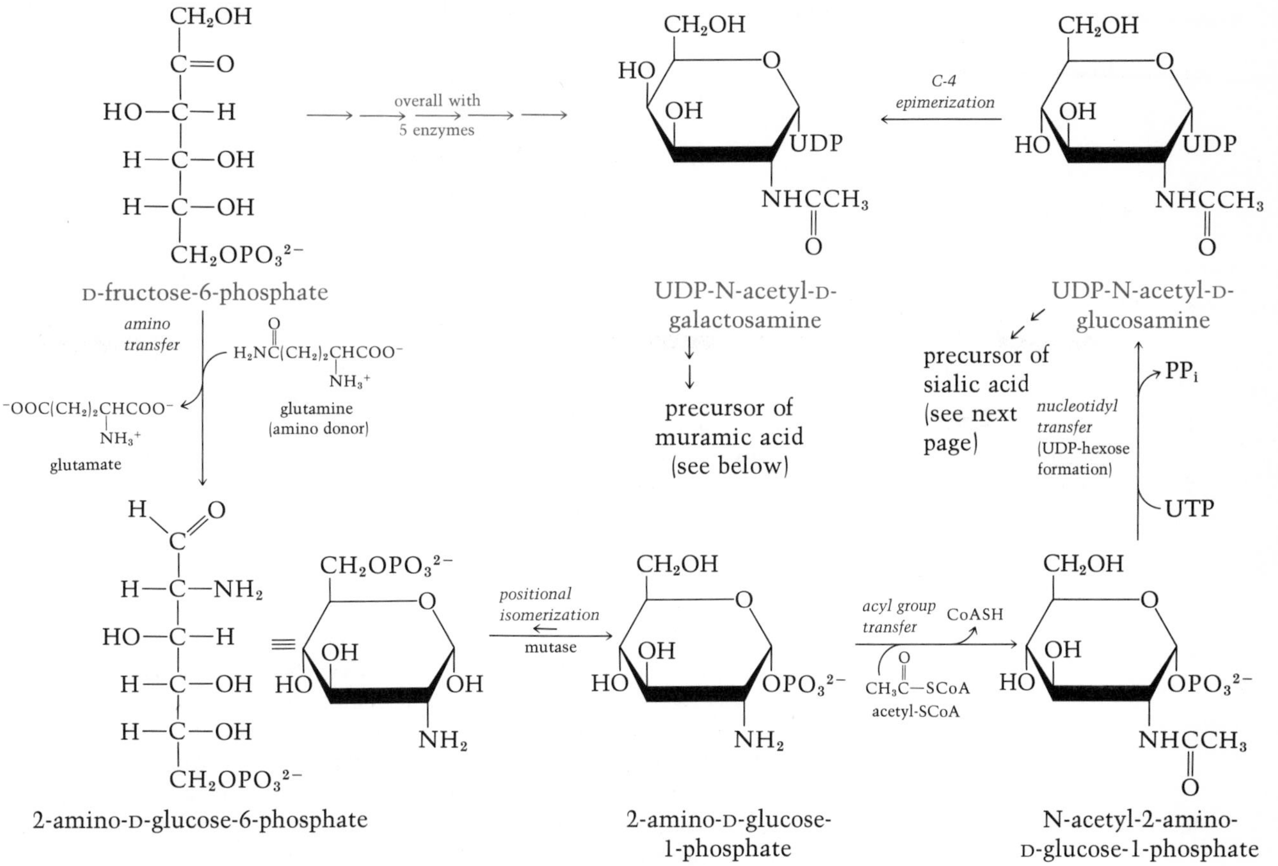

D-fructose-6-phosphate

UDP-N-acetyl-D-galactosamine

precursor of muramic acid (see below)

UDP-N-acetyl-D-glucosamine

C-4 epimerization

precursor of sialic acid (see next page)

nucleotidyl transfer (UDP-hexose formation)

amino transfer

glutamine (amino donor)

glutamate

2-amino-D-glucose-6-phosphate

2-amino-D-glucose-1-phosphate

N-acetyl-2-amino-D-glucose-1-phosphate

positional isomerization mutase

acyl group transfer CoASH

acetyl-SCoA

In bacteria UDP-N-acetyl-D-glucosamine is the precursor of *muramic acid*, a component of bacterial cell walls (see p. 310 in this chapter). Muramic acid is a derivative of N-acetyl-D-glucosamine that contains a *lactic acid* substituent at position C^3. As shown below the lactic acid

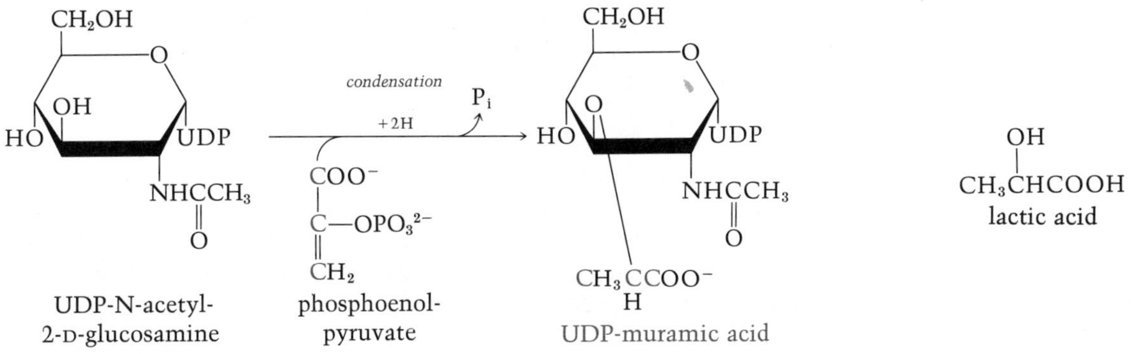

UDP-N-acetyl-2-D-glucosamine

condensation +2H P_i

phosphoenolpyruvate

UDP-muramic acid

lactic acid

$CH_3CHCOOH$

301

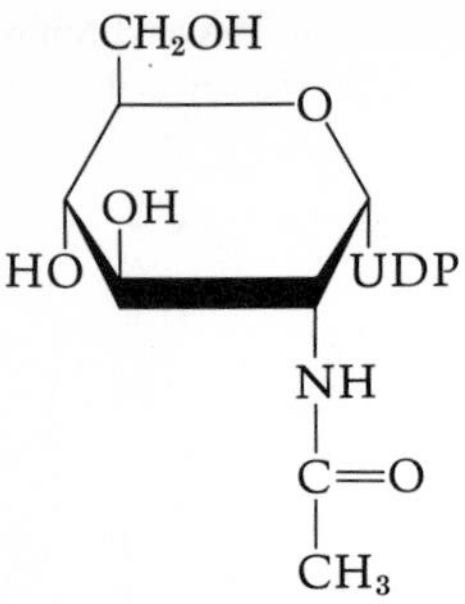

UDP-N-acetyl-2-D-glucosamine

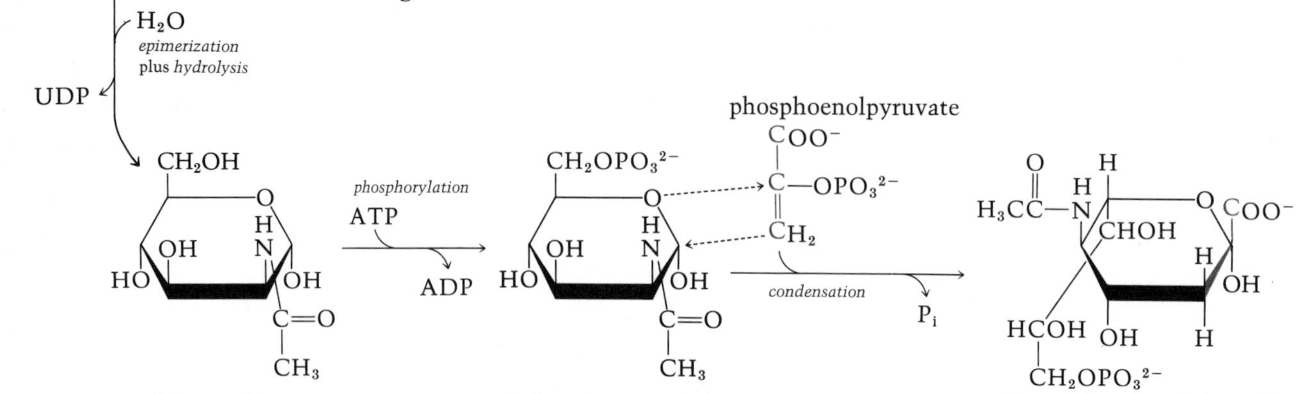

H_2O

epimerization
plus *hydrolysis*

UDP

N-acetyl-2-D-mannosamine

phosphorylation
ATP

ADP

6-phospho-N-acetyl-2-D-mannosamine

phosphoenolpyruvate

condensation

P_i

N-acetyl-neuraminic acid
(sialic acid)

moiety originates from *phosphoenolpyruvate* (see p. 373), which condenses with the C^3—OH position of the carbohydrate substrate.

UDP-N-acetyl-D-glucosamine is also the precursor of *sialic acid* (also called *N-acetyl-neuraminic acid*), which is an important component of (a) various glycoproteins (see p. 312, this chapter) and (b) the gangliosides, which are carbohydrate-containing lipids (see p. 329 next chapter). The overall conversion (see below) begins with an epimerization at the C^2 position of UDP-N-acetyl-D-glucosamine to yield UDP and N-acetyl-D-mannosamine. Then, after a kinase-catalyzed, ATP-dependent phosphorylation, a condensation with phosphoenolpyruvate occurs to give the phosphoester of sialic acid. Note that sialic acid is a nonose, i.e., a nine-carbon sugar.

The discussion of these sugar derivatives provided an opportunity to introduce the fascinating topic of *metabolism*, that is, *cellular reactions*. The various reactions that were presented can provide you with an opportunity to develop a method of study before we formally get into this area. One recommended way is to study a metabolic sequence by identifying the pattern of chemical transformations that occur between the beginning and end of the sequence. In other words, try to understand the chemistry rather than just memorize structures and names. The notation accompanying each conversion was included to focus your attention on this approach. To assist your study, this notation device will also be used in later chapters.

OLIGOSACCHARIDES

Depending on the number of monomeric residues linked to each other via glycosidic bonds, an oligosaccharide is termed a *di*saccharide, a *tri*saccharide, and so on, with an upper limit of eight residues generally accepted as the distinction from polysaccharides. Most oligosaccharides are comprised of hexose sugars. If all residues are identical, the substance is termed a *homogeneous oligomer*. Obviously then, the presence of two or more different types of monomers characterizes a *heterogeneous oligomer*. Both types occur in nature and examples of each are described in this section. Coverage is confined to disaccharides since larger oligomers do not have a widespread occurrence.

Among the many disaccharides of natural origin that occur in the free state, *sucrose* and *lactose* are the most abundant and most important. Both are heterogeneous disaccharides.

Sucrose is composed of α-D-glucose and β-D-fructose linked via the $\alpha1$ of glucose and the $\beta2$ of fructose, that is, via an $\alpha,\beta(1 \rightarrow 2)$ glycosidic linkage. Sucrose is found throughout the plant kingdom, but is most abundant in sugar cane, sugar beets, and maple syrup. It is the primary granulated product obtained from the processing of these materials and is commonly known as table sugar. After being synthesized in the green leaves it is then transported to various other parts of the plant, primarily for storage. When a carbon and energy source are needed, the sucrose is then hydrolyzed to glucose and fructose, which enter the mainstream of metabolism. The same hydrolytic degradation occurs during digestion in the animals that consume plants. This provides one of the major dietary supplies of hexoses for the animal kingdom. Everyone is familiar with its sweetening and flavor-enhancing properties, as well as the fact that an excess intake can be harmful, particularly to maintenance of good teeth. The Haworth formula of sucrose is

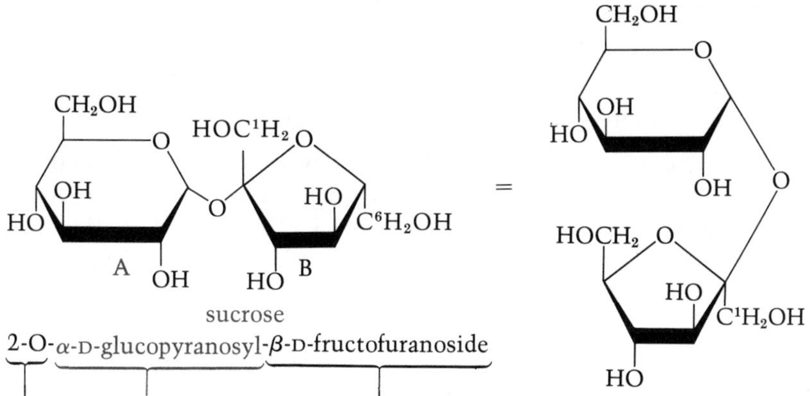

sucrose

2-O-α-D-glucopyranosyl-β-D-fructofuranoside

Note that the $\alpha,\beta(1 \rightarrow 2)$ glycosidic bond in sucrose involves the anomeric hydroxyl of both monomers, which eliminates the possibility of a free aldehyde or ketone grouping. Since a free carbonyl group cannot be formed without breaking the glycosidic bond, sucrose is a *nonreducing* sugar. The biosynthesis of sucrose in plants proceeds by either of two methods, both of which involve the participation of UDP-glucose:

UDP-D-glucose + D-fructose $\rightarrow$ UDP + sucrose   (major route)

UDP-D-glucose + D-fructose-6-phosphate $\rightarrow$ UDP + sucrose phosphate $\rightarrow$ sucrose + P_i

Lactose is composed of β-D-galactose and D-glucose, linked via the $\beta1$ of galactose and position 4 of glucose, that is, via a $\beta(1 \rightarrow 4)$ glycosidic linkage. It is the most abundant carbohydrate material in the milk of mammals (milk is about 5% lactose) and represents the major carbon and energy source for a breast-fed infant. Since the anomeric carbon of the glucose residue is not involved in the glycosidic linkage, the potential for a free aldehyde group does exist, and thus lactose is classified as a *reducing* sugar.

303

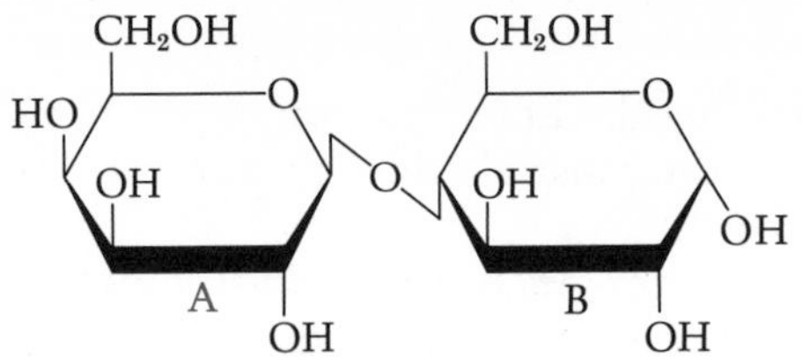

lactose (α form)

4-O-β-galactopyranosyl-α-D-glucopyranose

After its formation from lactose during digestion, the metabolic utilization of D-galactose is preceded by conversion to D-glucose in a multistep, UTP-dependent process. Normal operation of this conversion is critical, since high levels of blood galactose contribute to severe physiological disturbances that can be fatal, especially to infants. The abnormality is a genetic, metabolic disorder involving a malfunctional enzyme in one of the steps in the D-galactose → D-glucose conversion. The condition is termed *galactosemia* (see p. 393 in Chap. 12). The biosynthesis of lactose in the mammary gland proceeds as follows:

$$\text{UDP-D-galactose} + \text{glucose} \rightarrow \text{UDP} + \text{lactose}$$

Cellobiose, maltose, and *isomaltose* are examples of homogeneous disaccharides. Specifically, all three are diglucose molecules. However, they differ in the nature of the glycosidic linkage. Cellobiose $\beta(1 \rightarrow 4)$ is the sole repeating unit in cellulose; maltose $\alpha(1 \rightarrow 4)$ is the sole repeating unit of the amylose fraction of starch; isomaltose $\alpha(1 \rightarrow 6)$, though not a repeating unit, is found in the amylopectin fraction of starch and glycogen. The notion of a repeating unit is developed further in the section on polysaccharides. As the structures below indicate, all three would be *reducing* sugars.

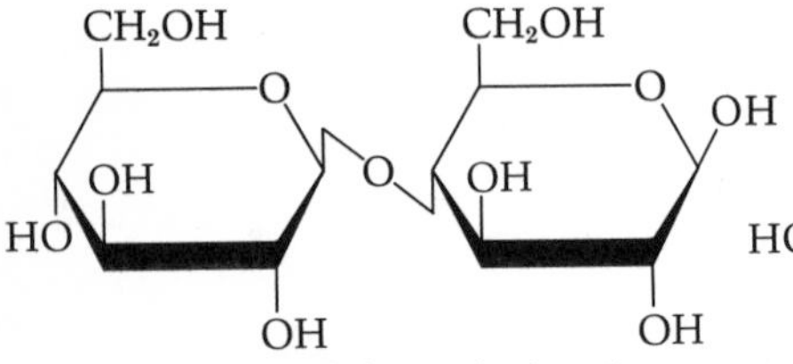

cellobiose (β form)

4-O-β-D-glucopyranosyl-β-D-glucopyranose

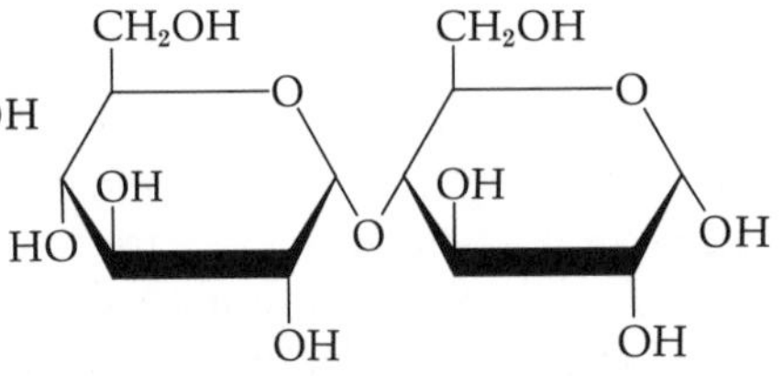

maltose (α form)

4-O-α-D-glucopyranosyl-α-D-glucopyranose

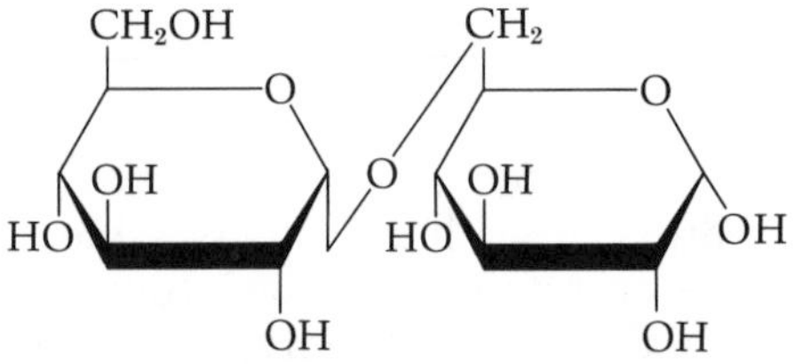

isomaltose (α form)

6-O-α-D-glucopyranosyl-α-D-glucopyranose

POLYSACCHARIDES

The three structural features that essentially determine the functional properties of any given polysaccharide are (a) the identity of the constituent monomers, (b) the nature of the glycosidic linkages between them, and (c), when appropriate, the sequence of residues. Relative to the first point, polysaccharides can be classified as *homopolysaccharides* (all residues identical) or *heteropolysaccharides* (different residues). Unlike polypeptides and polynucleotides, which are composed of a variety of monomers, the heteropolysaccharides are usually composed of only two different monomers. Furthermore, the residue sequence is not random but usually *repetitive*. Polysaccharides exist both in the free state and in conjugation with other materials such as lipids, peptides, and proteins. Generally the conjugation is covalent.

Polysaccharides of natural origin vary greatly in size, ranging up to several thousand monomeric residues. In addition the residue count frequently varies among samples of the same material. Thus, the molecular weight of polysaccharides has limited physicochemical value.

Homopolysaccharides: Celluloses

Cellulose is the most abundant organic compound of natural origin on the face of the earth. It occurs throughout the plant kingdom as a component of the cell wall, and frequently it is the major component. As the primary *structural component* in woody plants, it may be considered the backbone of the vegetable world. In parts of some plants, it is present in almost pure form. For example, the seed hairs of the cotton plant are particularly rich in cellulose, containing 98–99%. Although plants are the major source, cellulose is also produced (in small quantity) by some bacteria and animal organisms.

Microscopic cellulose fibers are aggregates of a variable number of unbranched *polyglucose* chains in parallel alignment with each other. If your eyes had the resolving power of a microscope, the pages these words are printed on would serve as a good visual aid for studying such fibrils. Since the number of glucose residues in an individual chain is also variable, it is unrealistic to refer to molecular weight. All glucose residues are in the β-D configuration and all are linked via $\beta(1 \rightarrow 4)$ glycosidic bonds, as illustrated below. The notion of a *repeating unit* in a polysaccharide chain should be obvious at this point, with β-cellobiose repeating in cellulose.

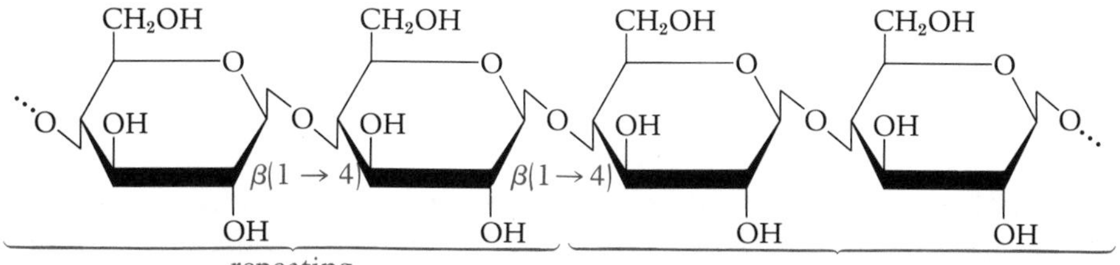

repeating
disaccharide in
cellulose
(β-cellobiose)

the linkage within the repeating unit is $\beta(1 \rightarrow 4)$ and also $\beta(1 \rightarrow 4)$ between successive repeating units

The chemical structure of cellulose is well suited for its biological role as a structural material. The hydrogen-bonding capacity among individual chains is quite high, with each residue contributing three OH groups that could participate. This confers a great degree of strength to the intact fiber and is also the basis of its water insolubility. In the cell walls of plants, these cellulose fibers are densely packed in layers that are further strengthened by the presence of other polysaccharide substances such as *hemicellulose*, *pectin*, and *lignin*, which function as cementing materials. The toughness of this cell wall aggregate should be obvious when you gaze at a 100-foot oak tree standing erect even in a heavy wind. A hemicellulose is a polymer composed primarily of D-xylose; pectin is composed primarily of D-galacturonate; the composition of lignin is still unresolved.

The direct nutritional value of cellulose is virtually nil for the higher animals, with the exception of ruminants. The basic reason for this is that the combined digestive secretions of the mouth, stomach, and intestine do not contain a *cellulase* that would cleave the glycosidic bonds and yield free glucose units. However, cellulases are found in nature, most commonly in ruminants, various insects, snails, fungi, algae, and bacteria. For obvious reasons, such organisms are referred to as being *cellulolytic*. Think about this the next time you see a grazing cow, a moth-eaten piece of clothing, a rotted piece of wood, or a damp, odor-ridden basement.

Chitin

A subtle variation on the structure of cellulose is found in chitin, which is a linear homopolysaccharide consisting of *N-acetyl*-D-*glucosamine* residues linked by $\beta(1 \rightarrow 4)$ bonds. Because it is a close structural relative of cellulose, we might argue that it should have a biological function similar to that of cellulose. Indeed, such is the case, with chitin being the major organic structural component of the exoshells of the invertebrates. It is also found in most fungi, many algae, and some yeasts as a cell wall component. As in cellulose, individual chains are bundled together via hydrogen bonding.

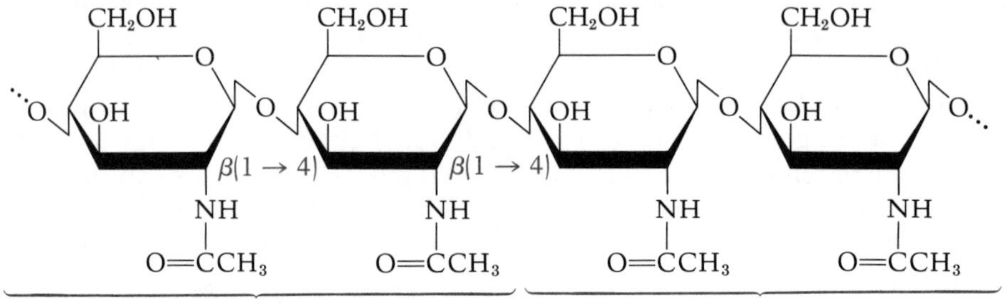

repeating
disaccharide in
chitin

Starches

The starches comprise a group of polyglucose materials of varying size and shape. Occurring exclusively in plants, they exist inside the plant cell as granules dispersed in the cytoplasm. Like the celluloses, all starches are homogeneous, containing only D-glucose residues. Unlike the celluloses, the starches are not single molecules, but are normally mixtures of two structurally distinct polysaccharides. One component is termed *amylose*, the other *amylopectin*. Both are poly-α-D-glucose molecules. Amylose is a *linear* molecule with all residues linked via $\alpha(1 \rightarrow 4)$ bonds. Amylopectin, on the other hand, is a *branched* molecule, a result of the presence of a small number of $\alpha(1 \rightarrow 6)$ linkages at various points along a chain consisting of $\alpha(1 \rightarrow 4)$ linkages. Amylose appears to prefer a helical-coiled conformation. No preferred conformation of amylopectin has been suggested. A diagramatic representation of each material is shown in Fig. 9–3.

The biological role of the starches is that of *food storage* in plants. That is to say, glucose is held in reserve in polymeric form. When a source of both carbon and energy is needed, the starches are released from granules and then degraded by enzymes. Most plants contain two distinct hydrolyzing enzymes, traditionally named α-*amylase* and β-*amylase*. Both attack the amylose and amylopectin fraction at $\alpha(1 \rightarrow 4)$ sites, but in a different pattern. Cleavage with α-amylase is random, occurring at different

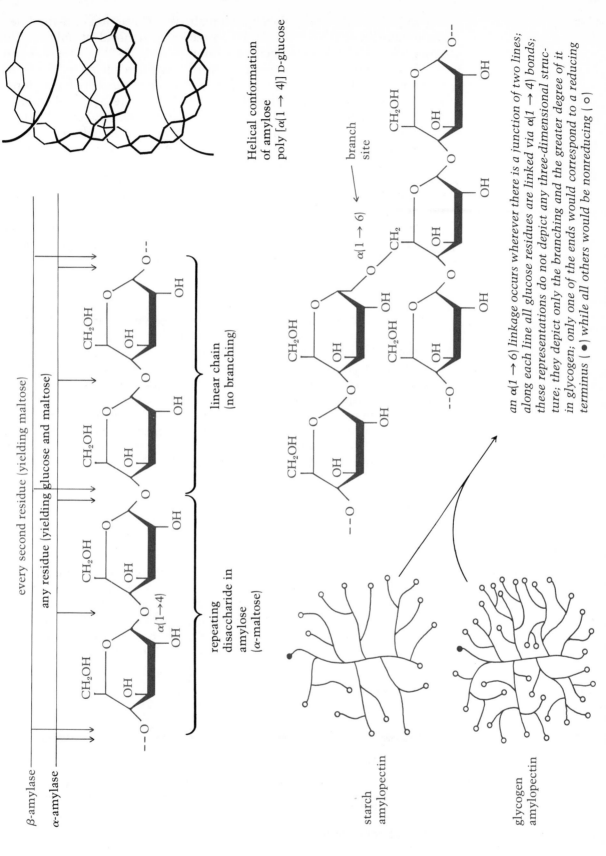

FIGURE 9–3 Diagramatic representations of starch amylose, starch amylopectin, and glycogen amylopectin. The action of α- and β-amylases is also indicated.

β-amylase

every second residue (yielding maltose)

α-amylase

any residue (yielding glucose and maltose)

CH₂OH

OH

OH

α(1→4)

CH₂OH

OH

OH

CH₂OH

OH

OH

CH₂OH

OH

OH

repeating disaccharide in amylose (α-maltose)

linear chain (no branching)

Helical conformation of amylose poly [α(1 → 4)] D-glucose

CH₂OH

OH

α(1 → 6)

CH₂

branch site

CH₂OH

OH

OH

CH₂OH

OH

OH

CH₂OH

OH

OH

an α(1 → 6) linkage occurs wherever there is a junction of two lines; along each line all glucose residues are linked via α(1 → 4) bonds; these representations do not depict any three-dimensional structure; they depict only the branching and the greater degree of it in glycogen; only one of the ends would correspond to a reducing terminus (•) while all others would be nonreducing (○)

starch amylopectin

glycogen amylopectin

loci to yield a mixture of glucose and maltose. The action of β-amylase is more ordered; it is characterized by successive removal of only maltose units, beginning at a nonreducing terminus (see Fig. 9–3). (Note that amylose has a single nonreducing terminus and a single reducing terminus, but that amylopectin has several nonreducing ends and a single reducing terminus.) Neither enzyme is capable of hydrolyzing the $\alpha(1 \rightarrow 6)$ linkages. Thus, whereas the combined action of the two enzymes will completely degrade amylose to glucose and maltose, amylopectin is only partially degraded. However, other catalysts, called *debranching enzymes*, specific for hydrolyzing the $\alpha(1 \rightarrow 6)$ linkage, do exist in nature.

Unlike cellulose, starch is digestible by humans (and most other organisms) due to the presence of *salivary amylase* and *pancreatic amylase* in the digestive secretions. Both enzymes are similar in action to the α-amylase of plants. In combined action with other digestive enzymes (notably maltase and debranching enzymes), starch is completely degraded to α-D-glucose, which is absorbed and metabolized further.

Glycogen

The storage of carbon and energy in a poly-α-D-glucose state is not unique to plants. In animal and bacterial cells the same thing occurs, with the storage function fulfilled by *glycogen*. In higher animals, glycogen granules are most abundant in cells of liver and muscle tissue. Structurally, glycogen is a branched polyglucose molecule identical to the amylopectin fraction of starch in all respects except that glycogen is more highly branched (see Fig. 9–3). Branch points on glycogen amylopectin occur about every 8–10 residues along the $\alpha(1 \rightarrow 4)$ chain. In starch amylopectin, branch points occur about every 25–30 residues. Within the cell, the glycogen molecule is degraded by *glycogen phosphorylase*, an enzyme that sequentially removes one glucose residue at a time, yielding glucose-1-phosphate (see p. 379 in Chap. 12). A debranching enzyme would again operate to cleave $\alpha(1 \rightarrow 6)$ branch points.

Heteropolysaccharides: Hyaluronic Acid

Hyaluronic acid is a heteropolysaccharide of major importance in higher animals. In connective tissue, it is the primary component of *ground-substance*, a gelatinous material filling the extracellular spaces of tissue. Another abundant source is synovial fluid, a viscous packing around bone joints serving as a lubricant and shock absorber. The vitreous humor and umbilical cord are also rich in hyaluronic acid. Since aqueous solutions of

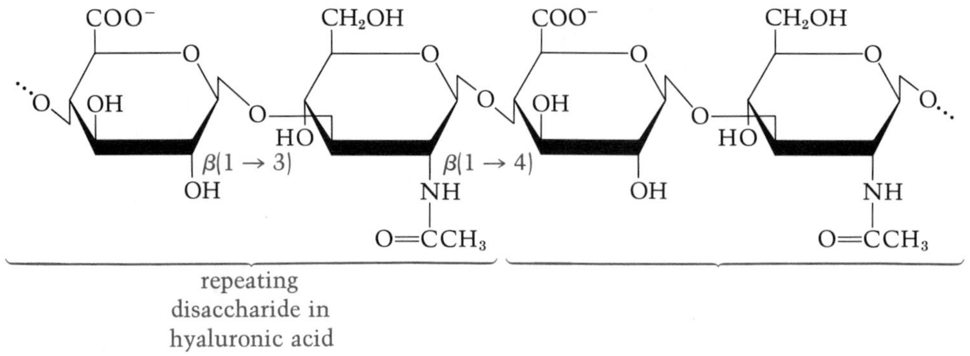

repeating
disaccharide in
hyaluronic acid

this material are gelatinous, it is synonomously termed a *mucopolysaccharide*. Structurally, the molecule is largely a linear polymer of a disaccharide repeating unit, composed of D-*glucuronic acid* and *N-acetyl*-D-*glucosamine* linked together by a $\beta(1 \rightarrow 3)$ bond. Adjacent repeating units are joined by $\beta(1 \rightarrow 4)$ linkages. Thus, $\beta(1 \rightarrow 3)$ and $\beta(1 \rightarrow 4)$ bonds alternate along the chain. The high viscosity of hyaluronic acid is probably in part related to its polyanionic character at physiological pH, which would favor excessive hydration and interchain hydrogen bonding.

Chondroitin Sulfates

Hyaluronic is but one of the mucopolysaccharides that are associated with connective tissue of animals. Another major type is represented by the group of materials known as *chondroitin sulfates*. These are related to hyaluronic acid in that they are also composed of a similar repeating unit consisting of D-*glucuronic acid* and *N-acetyl*-D-*galactosamine* (glucosamine in hyaluronic acid) linked together by $\beta(1 \rightarrow 3)$ bonds. The major distinction is that either the C^4 or C^6 OH group in the galactosamine residue is *sulfated*. The extent and pattern of sulfation in successive disaccharide units distinguishes one type of chrondroitin from another. In any event, the presence of the additional SO_3^- groupings confers an even greater polyanionic character to the chondroitins.

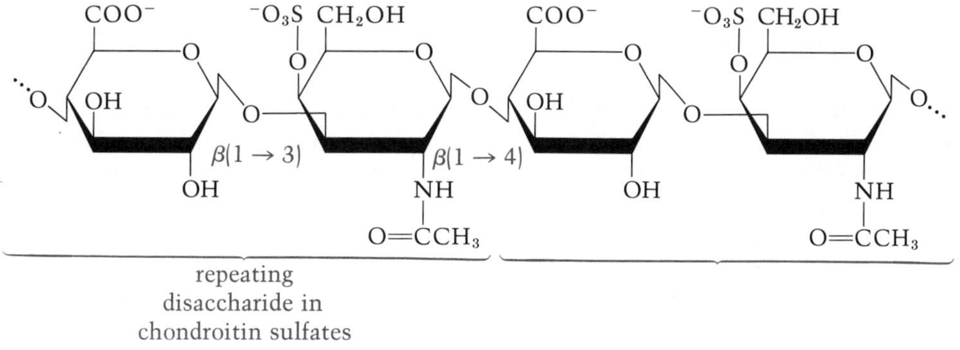

repeating
disaccharide in
chondroitin sulfates

Heparin

Found in many mammalian tissues, *heparin* is another interesting mucopolysaccharide. Most evidence suggests that it is a linear or nearly linear molecule composed of a disaccharide repeating unit of *sulfated glucuronic acid* and *disulfated glucosamine* residues linked via an $\alpha(1 \rightarrow 4)$ bond. Adjacent repeating units are similarly linked. Heparin is a good anticoagulant of blood and is widely used as such in medical practice.

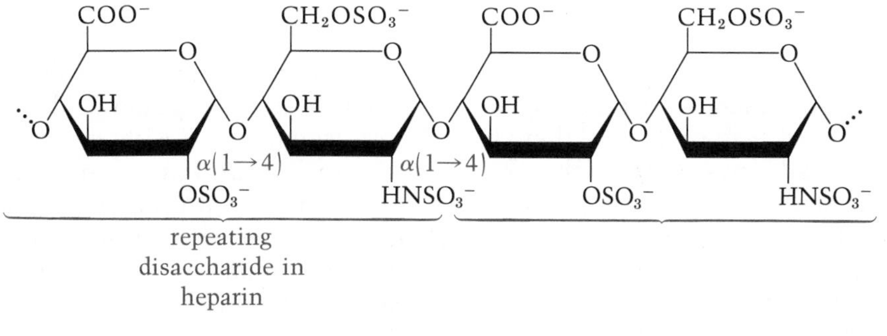

repeating
disaccharide in
heparin

CONJUGATED POLYSACCHARIDES

Bacterial Cell Walls

All unicellular bacteria contain a rigid cell wall. For years it was thought that its only notable function was a protective one, but recently this view has changed considerably and many diverse functions are now associated with the cell wall. In accord with the structure-function theme of modern biochemical research, it was only logical that studies should be initiated to ascertain the chemical structure of this cytological unit. Success was first reported in 1965 by Strominger for the cell wall material of *Staphylococcus aureus*. Since then, cell wall preparations of other organisms have been similarly examined. On a comparative basis, the picture that has emerged is that bacterial cell walls are both heterogeneous and homogeneous in composition. They are heterogeneous in the sense that the total composition of wall preparations differs from one genus to another, with varying levels of peptides, proteins, lipids, and carbohydrates being the major components. They are homogeneous in the sense that all sources possess a similar (but not identical) polymeric unit that acts as the basic structural framework of the wall.

The enzyme *lysozyme* acts on the polysaccharide moiety of the cell wall structure, cleaving specific glycoside linkages and thus fragmenting the intact gridlike structure. The fragments dissociate from the cell surface, leaving the cell (now called a *protoplast*) in a deprotected state that is highly susceptible to rupture. Lysozyme is present in eye tears and the whites of eggs, serving as a defense against bacterial infection. Lysozyme preparations are also used in the laboratory for lysing bacterial cells to obtain cell-free systems for further fractionation (see p. 32, Chap. 1).

The common unit is a network of *polysaccharide chains covalently cross-linked to each other via small polypeptide bridges.* Due to this conjugation of peptide and carbohydrate, the material is conventionally termed a *peptidoglycan*. (Figure 9–4 should be referred to as you read the following descriptive material.) The polysaccharide moiety is heterogeneous, composed of a repeating disaccharide unit of *N-acetyl-D-glucosamine* and *N-acetyl-muramic* acid, with a $\beta(1 \rightarrow 4)$ linkage. Successive units are also attached via $\beta(1 \rightarrow 4)$ linkages. The peptide portion can be considered as consisting of two parts. First, we can identify a *tetrapeptide unit* composed of both D- and L-amino acids and covalently attached to the C^3 lactic acid side chain of the muramic acid residue. In fact, every muramic acid residue in every chain is so characterized. Second, we can distinguish a *pentaglycine unit* that acts as a *bridge* between the terminal alanine residue of one peptide branch on one chain and the L-lysine residue of a second branch attached to a neighboring parallel polysaccharide chain.

As shown in Fig. 9–4, the repetition of such linkages among several chains confers a *grid* pattern on the complete peptidoglycan structure. Considering the extensive, crisscross presence of covalent bonds, the basic *rigidity* of the cell wall should be evident. Thus, we encounter another example of the structural fitness of a biological material relative to its natural function. (Note the reference to the *mode of action of penicillin* in the legend to Fig. 9–4.)

Glycoproteins

Structurally many proteins consist of carbohydrate prosthetic groups (as linear or branched *oligosaccharides*) attached to specific amino acid side chains (see below); hence, they are called *glycoproteins*. Although they occur in all types of organisms, glycoproteins are especially prevalent in the fluids and cells of animals, where they are associated with numerous and various functions (see margin, p. 312). As this information in the margin indicates, glycoproteins occur extensively in the membranes of cells as integral components of the membrane or in association with the membrane as a component of the surface coat.

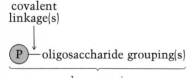

covalent
linkage(s)

P—oligosaccharide grouping(s)

glycoprotein

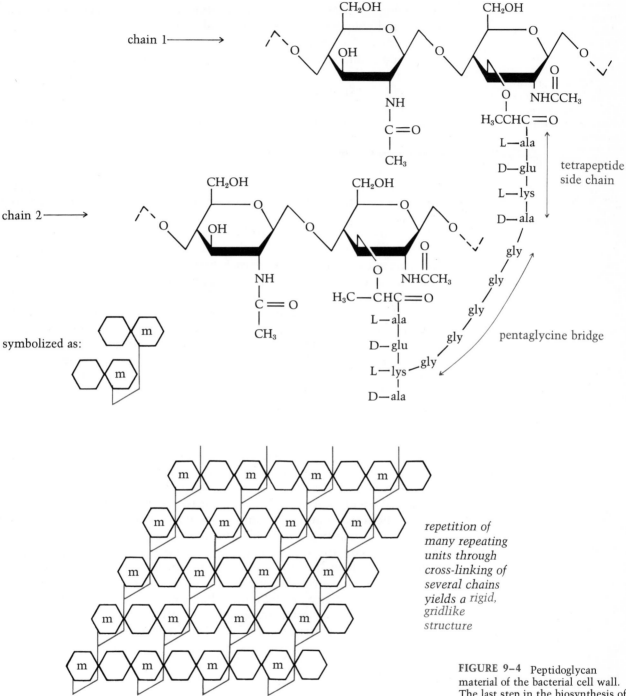

FIGURE 9–4 Peptidoglycan material of the bacterial cell wall. The last step in the biosynthesis of the cell wall is the formation of the pentaglycine bridges; this confers the gridlike pattern. **Penicillin** inhibits this reaction(s) and thus prevents completion of synthesis. Lacking the protection of a normal rigid cell wall, the cell fails to survive.

Despite its universal occurrence, the carbohydrate portion of the glyco-proteins has been the subject of extensive study for only about a decade. From these studies it is now quite clear that the glycoproteins represent one of the most—if not the most—complex group of biomolecules. The diversity of function is one thing, but this is exceeded manyfold by the variability

A listing of glycoproteins according to the type of function associated with a particular glycoprotein

---

Hormones
Antibodies
* Enzymes
* Receptor proteins
* Transport proteins
* Cell adhesion proteins
* Growth control proteins
* Cell recognition proteins
* Proteins conferring blood group
  characteristics
  Proteins conferring structural
  stability to multimolecular
  aggregates

---

* Glycoproteins of these types are usually present in the surface coat and the plasma membrane of cells.

of the carbohydrate component. Indeed, in this respect it appears that each glycoprotein may be unique, meaning that the information content of glycoprotein structure is twofold. The puzzling aspect of this is, why should an identifying carbohydrate moiety be attached to a polypeptide molecule already individualized by its amino acid sequence?

One proposal is that the attachment of sugars to a protein is the identifying chemical label used to tag proteins that are destined to be utilized outside of the cell or in the membranous network of the cell. The reciprocal aspect of this hypothesis is that those proteins to be retained and used in the cytoplasm of the cell are nonglycosylated. Although this hypothesis is still being debated, there is clear evidence that its basic premise is valid, namely, that the *carbohydrate moiety does confer an additional recognition factor to the protein.* In the simplest sense this means that the specific binding of a glycoprotein to another molecule or molecular aggregate may be based on the specific binding of the carbohydrate moiety. Whether the carbohydrate moiety is then further involved in the protein function is still an unresolved question.

Glycoproteins as a group exhibit great differences in their carbohydrate content, which ranges from less than 1% to as high as 80% of total weight. Glycoproteins with 4% or more carbohydrate are sometimes termed *mucoproteins* because they exhibit a very high viscosity. The covalent linkage to the polypeptide is made via a glycosidic bond to the side chain of either *serine, threonine, asparagine,* or *hydroxylysine* (when it is present, as in collagen). The number of oligosaccharide units per protein molecule is variable, but all units in the molecule are usually identical.

The sugars that commonly occur in the oligosaccharide grouping include D-galactose, D-glucose, D-mannose, L-fucose, N-acetyl-D-glucosamine, N-acetyl-D-galactosamine, and N-acetyl-neuraminic acid (sialic acid). In many cases the presence of sialic acid is especially important. The total individuality of the oligosaccharide grouping depends on (a) the composition of sugar residues, including the identification of anomeric configuration of each residue, (b) the sequence of the residues, (c) the pattern of glycosidic linkages within the sequence, and (d) the nature of the linkage to the protein. A few examples of oligosaccharidic groupings are shown in Table 9–2.

The list of glycoprotein functions in the margin reveals that many of these functions relate to "communication"—cell-cell recognition, cell-molecule recognition, organelle-molecule recognition, and molecule-molecule recognition. Clearly, all of these recognition phenomena, especially at the cell-cell level, are important normal processes that occur in all living organisms. Abnormalities of these processes are most probably associated with many diseases, including cancer, and with other processes such as rejection of tissue transplants. Continued research in the area of glycoprotein biochemistry should result in exciting discoveries.

Lectins

Throughout the plant kingdom there exist proteins, called *lectins*, that bind very strongly to glycoproteins. The binding between lectin and glycoprotein involves specific binding sites on the lectin molecule that recognize specific carbohydrate groupings in the glycoprotein. In plants, the lectins may serve as a defense mechanism against invasion of plant cells by foreign

TABLE 9-2 Oligosaccharide groupings in some glycoproteins.[a]

1. A glycoprotein isolated from the membrane of the glomerulus in kidney.

$$\text{GLU} \xrightarrow{\alpha(1 \longrightarrow 2)} \text{GAL} \xrightarrow{\beta} \text{OH of HYLYS side chains}$$

2. One of the blood group-determining (A positive) proteins in pig.

$$\text{GALNAc} \xrightarrow{\alpha(1 \longrightarrow 3)} \text{GAL} \xrightarrow{\beta(1 \longrightarrow 3)} \text{GALNAc} \xrightarrow{\alpha} \text{OH of SER and THR side chains}$$

with $\alpha(1 \longrightarrow 2)$ FUC branch on GAL and $\alpha(2 \longrightarrow 6)$ NAN branch on GALNAc

approximately 500 units/molecule

3. The major glycoprotein present in the membrane of human red blood cells.

$$\text{NAN} \xrightarrow{\alpha(2 \longrightarrow 3)} \text{GAL} \xrightarrow{\beta(1 \longrightarrow 3)} \text{GALNAc} \xrightarrow{\alpha} \text{OH of SER and THR side chains}$$

with $\alpha(2 \longrightarrow 6)$ NAN branch on GALNAc

exactly 16 units/molecule; this protein also contains a single
copy of another oligosaccharide grouping attached to a specific
asparagine residue

4. In one of the human antibodies, namely, immunoglobulin G.

$$\text{NAN} \xrightarrow{\alpha(2 \to 6)} \text{GAL} \xrightarrow{\beta(1 \to 4)} \text{GLNAc} \xrightarrow{\beta(1 \to 2)} \text{MAN} \xrightarrow{\alpha(1 \to 6)}$$

$$\text{NAN} \xrightarrow{\alpha(2 \to 6)} \text{GAL} \xrightarrow{\beta(1 \to 4)} \text{GLNAc} \xrightarrow{\beta(1 \to 2)} \text{MAN} \xrightarrow{\alpha(1 \to 3)}$$

$$\text{MAN} \xrightarrow{\beta(1 \to 4)} \text{GLNAc} \xrightarrow{\beta(1 \to 4)} \text{GLNAc} \longrightarrow \text{NH}_2 \text{ of ASN side chains}$$

with $\beta(1 \to 4)$ GLNAc branch and $\alpha(1 \to 6)$ FUC branch

(Some glycoproteins contain oligosaccharide groups
even more complex than this)

5. A protein (antifreeze protein) isolated from the serum of Antarctic fishes; the intact protein has the
amazing property of lowering the freezing point of water to the same extent as *an equal weight* of NaCl.

$$\text{GAL} \xrightarrow{\beta(1 \to 3)} \text{GALNAc} \xrightarrow{\alpha} \text{OH of THR side chains} \qquad 31 \text{ units/molecule}$$

[a] GAL = D-galactose; GLU = D-glucose; MAN = D-mannose; FUC = L-fucose; GLNAc = N-acetyl-D-2-glucosamine; GALNAc =
N-acetyl-D-2-galactosamine; NAN = acetyl-neuraminic acid (sialic acid); HYLYS = hydroxylysine; ASN = asparagine; SER = serine;
THR = threonine.

cells and/or viruses. Other evidence indicates that lectins may also explain
how plants recognize certain "good" bacteria with which legume plants
must associate for their very survival (the good bacteria being the
nitrogen-fixing bacteria in the soil that grow symbiotically with the plant).
Recent studies have uncovered the occurrence of lectins in several nonplant
organisms, including mammals; here, too, they appear to function in recognition processes.

The most studied lectin is *concanavalin A* (conA) from the jack bean.
The complete three-dimensional structure has been determined and the
binding site events have been identified. For several years conA has proven

to be a useful tool in probing the biochemistry of cell surfaces and in the study of cell agglutination (clumping)—an event caused by the binding of conA. Modified forms of conA have been shown to be successful in restoring normal growth patterns to previously transformed cells; obviously, this has generated a lot of interest in conA specifically and lectins in general. Can they be used in a therapeutic way to curb the growth of tumors? What can we learn about the relationship of cell surface biochemistry to the normal process of growth control? These and many other questions are being probed in current research.

LITERATURE

FLORKIN, M., and E. H. STOTZ, eds. *Comprehensive Biochemistry*. Amsterdam-New York: Elsevier Publishing Company, 1963. An advanced multivolume treatise, not yet complete, planned to cover all aspects of biochemistry. Volume 5 contains several articles on the chemistry of the monosaccharides and their derivatives, oligosaccharides, and polysaccharides. The coverage of polysaccharides is excellent.

GHUYSEN, J. M., J. L. STROMINGER, and D. J. TIPPEN. "Bacterial Cell Walls." In *Comprehensive Biochemistry*, Vol. 26, Part A, M. Florkin and E. H. Stotz, eds., pp. 53–104. Amsterdam-New York: Elsevier, 1968. A review article on the chemical structure of cell walls in bacteria.

KENT, P. W. "Structure and Function of Glycoproteins." In *Essays in Biochemistry*, Vol. 3, P. N. Campbell and G. D. Greville, eds., pp. 105–152. New York: Academic Press, 1967. A review article confined to major glycoproteins of mammalian origin found in serum, blood, connective tissue, and bone.

KORNFELD, R., and S. KORNFELD. "Comparative Aspects of Glycoprotein Structure." In *Annual Review of Biochemistry*, **45**, 217–238 (1976). A review article focusing on the structure of the carbohydrate grouping of glycoproteins from varied sources.

MORRISON, R. T., and R. N. BOYD. *Organic Chemistry*, 3rd edition. Boston: Allyn and Bacon, 1973. Acclaimed as the best organic chemistry text available. Chapters 34 and 35 contain a superb introductory treatment of carbohydrate chemistry.

PIGMAN, W., and D. HORTON, eds. *The Carbohydrates: Chemistry and Biochemistry*. 2nd edition. 3 vols. New York: Academic Press, 1972. An authoritative and complete reference source.

SHARON, N. "The Bacterial Cell Wall." *Scientific American*, **220**, 87–98 (1969). Description of the chemical structure of the cell wall in bacteria and the action of lysozyme and penicillin.

SHARON, N. "Glycoproteins." *Scientific American*, **230**, 78–86 (1974).

WINTERBLUM, P. J., and C. F. PHELPS. "The Significance of Glycosylated Proteins." *Nature*, **236**, 147–151 (1972). A proposal for the revision of the hypothesis mentioned briefly in this chapter on p. 312 and originally formulated by E. H. Hylar (see *J. Theor. Biol.*, **10**, 89 (1966)).

EXERCISES

9–1. Transfer the Fischer projection formula for each of the following monosaccharides into its Haworth representation and then (when appropriate) into the chair conformation.
(a) α-D-gulopyranose
(b) β-D-allopyranose
(c) α-D-galactopyranose-1-phosphate
(d) β-D-sorbofuranose-1,6-diphosphate
(e) α-L-lyxofuranose
(f) the C^4 epimer of (b)

9–2. In sugar X the orientations of C^1, C^2, C^3, and C^4 hydroxyls are axial, equatorial, equatorial, and equatorial, respectively, and the C^{6}H$_2$OH orientation on C^5 is axial. Draw the chair conformation (in the form shown on p. 294). Then convert this to a Haworth representation, a Fischer projection, and a free carbonyl form. Now identify the name of sugar X.

9–3. Draw the Haworth formula for each of the following carbohydrates.
(a) O-α-D-glucopyranosyl-(1 → 1)-α-D-glucopyranoside
(b) O-α-D-galactopyranosyl-(1 → 6)-β-D-glucopyranose
(c) O-α-D-glucopyranosyl-(1 → 1)-α-D-glucosaminopyranoside
(d) O-α-D-galactopyranosyl-(1 → 6)-O-α-D-glucopyranosyl-(1 → 2)-β-D-fructofuranoside

9–4. β-D-Mannose has a specific rotation of −16.3°, and an equilibrium mixture of the α and β isomers of D-mannose has a specific rotation of +14.5°. What conclusion can be drawn with respect to the specific rotation of α-D-mannose relative to the value given for β-D-mannose?

9–5. Give a brief description of the catalytic role of the following enzymes.
(a) hexokinase

(b) α-amylase
(c) β-amylase
(d) UDP-sugar pyrophosphorylase

9–6. Determine the maximum number of different disaccharides that could exist if they were composed only of D-glucose. Classify each as a reducing or nonreducing sugar.

9–7. How many different types of glucosyl residues are present in glycogen? Recall that it is the involvement in bonding that distinguishes each residue.

9–8. Methylene blue is an organic dye capable of existing in an oxidized (blue) and a reduced (colorless) state. Consequently, if the oxidized form of methylene blue were added to a solution containing a reducing sugar such as lactose, a blue to colorless transition would be observed. With this information, predict what would occur if the solution contained any of the disaccharides shown below.

(a)

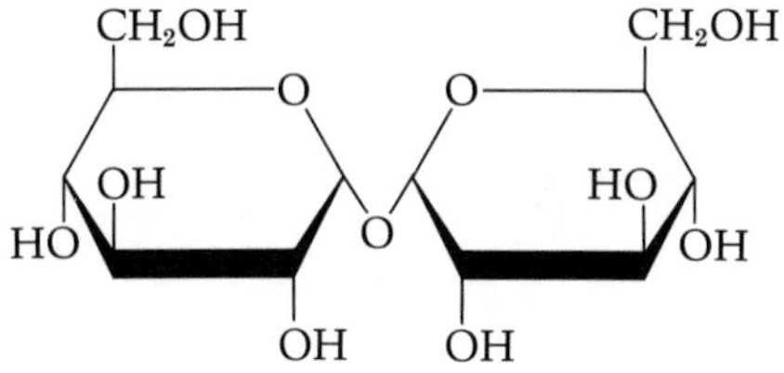

(b)

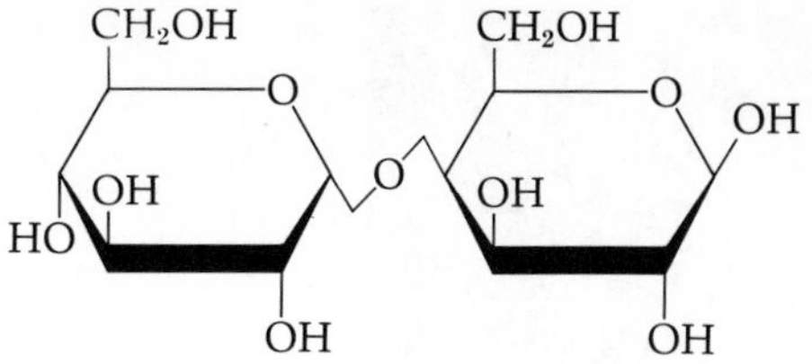

(c)

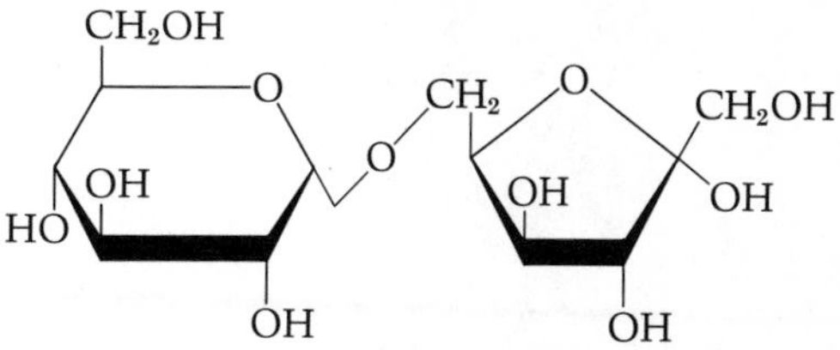

9–9. How would you describe the similarities and differences between the amylopectin fractions of potato starch and liver glycogen?

9–10. Draw a complete structure corresponding to the linkage between the C-terminal residue of the pentaglycine bridge and the L-lysine residue of the tetrapeptide side chain in the peptidoglycan material of the bacterial cell wall.

9–11. Draw the complete structure of the carbohydrate moiety of examples 1, 2, and 4 in Table 9–2.

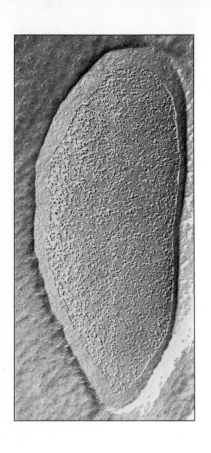

CHAPTER 10

LIPIDS AND BIOMEMBRANES

The term lipid (from the Greek *lipos*, meaning *fat*) refers to any naturally occurring, *nonpolar* substance that is nearly or totally insoluble in water, but is soluble in other nonpolar solvents such as chloroform, carbon disulfide, ether, and hot ethanol. Because of the structural and biofunctional diversity of the lipids, the statement is necessarily rather vague and very generalized.

Stated in general terms, the major biological roles of lipids include serving (a) as *components of membranes*, (b) as a major *storage form of carbon and energy*, (c) as *precursors* of other important substances, (d) as *insulation barriers* to avoid thermal and physical shock, and (e) as *protective coatings* to prevent excessive loss or gain of water and infection. The membrane and storage roles are particularly important. Other more specific roles are associated with specific lipids that function as *vitamins* or *hormones*.

In the definition above, the emphasis on the word nonpolar signifies a common denominator of this heterogeneous group of materials. In several instances, this characteristic of nonpolarity is conferred on the lipid material by the presence of one or more *fatty acids*, which contain long aliphatic hydrocarbon chains. Accordingly, the chapter begins with consideration of the fatty acids.

Thereafter, the organization of the chapter follows a common classifica-

TABLE 10–1 A comparison of simple and compound lipids in terms of their chemical composition; summarized here by the identification of the constituent parts that are liberated on complete hydrolysis.

	Lipid	Constituent parts
Simple lipids	1. Acylglycerols	$\xrightarrow{\text{hydrolysis}}$ glycerol + fatty acid(s)
	2. Waxes	$\longrightarrow$ alcohol + fatty acid (both long chain)
	3. Phosphoacylglycerols	$\longrightarrow$ glycerol + fatty acid(s) + HPO_4^{2-} + (choline or ethanolamine primarily)
Compound lipids	4. Sphingomyelins	$\longrightarrow$ sphingosine + fatty acid + HPO_4^{2-} + choline
	5. Cerebrosides	$\longrightarrow$ sphingosine + fatty acid + simple sugar
	6. Gangliosides	$\longrightarrow$ sphingosine + fatty acid + 2–6 simple sugars (including sialic acid)

Note: 3 and 4 are also referred to as *phospholipids* because of the presence of phosphate
4, 5, and 6 are also referred to as *sphingolipids* because of the presence of sphingosine
5 and 6 are also referred to as *glycolipids* because of the presence of carbohydrate

tion based on structure, which segregates the lipids into three broad groups: (a) *simple* lipids; (b) *compound* lipids; and (c) *derived* lipids. The simple lipids include only those materials that are esters of fatty acids and the trihydroxyalcohol, glycerol, or other long-chain monohydroxyalcohols. The major types are the *acylglycerols* and the *waxes*. The compound lipids include a host of materials that contain other substances in addition to an alcohol and fatty acids. They are so named only because their structures consist of more parts than a simple lipid does. The four major types are *phosphoacylglycerols*, *sphingomyelins*, *cerebrosides*, and *gangliosides*. The derived lipids represent an outright hodgepodge that includes any material that meets the general definition of a lipid, but cannot be neatly classified as a simple or compound lipid. *Steroids*, *carotenoids*, and the *lipid vitamins* are just a few examples. The basis of this classification scheme may seem just as nebulous as the definition of a lipid, but it provides a useful approach for a brief discussion.

The information in Table 10–1 summarizes the composition of the individual simple and compound lipids described in this chapter. Again, note the presence of fatty acids in each type.

FATTY ACIDS

Other than being a traditional reference to its isolation from fats (see p. 324 for definition) the term "fatty acid" has no precise meaning. Nevertheless, it is in routine use. Chemically speaking, a fatty acid is a *long-chain aliphatic carboxylic acid*. Although there are a considerable number and variety of naturally occurring fatty acids, we can simplify matters by providing generalizations concerning those acids that occur most frequently in nature:

(a) Most are *mono*carboxylic acids containing *linear* hydrocarbon chains with an *even* number of carbon atoms, generally in the range of C_{12}–C_{20}; shorter and longer chain acids, branched and cyclic chain acids, and acids of odd-number carbon content do occur but at a much lower frequency.

(b) *Unsaturation is common* but largely confined to the C_{18} and C_{20} acids; when two or more double bonds exist they are almost always separated by a single methylene group, that is,

TABLE 10-2 Naturally occurring fatty acids; a partial listing.[a]

Fatty acid (carbon content)	No. of double bonds	Formula
I. Even-numbered—straight-chain—fully saturated		
Lauric acid (C_{12})	0	$CH_3(CH_2)_{10}COOH$
Myristic acid (C_{14})	0	$CH_3(CH_2)_{12}COOH$
Palmitic acid (C_{16})	0	$CH_3(CH_2)_{14}COOH$
Stearic acid (C_{18})	0	$CH_3(CH_2)_{16}COOH$
Arachidic acid (C_{20})	0	$CH_3(CH_2)_{18}COOH$
II. Even-numbered—straight-chain—unsaturated (all *cis*)		
Palmitoleic acid (C_{16}) Δ^9	1	$CH_3(CH_2)_5CH=CH(CH_2)_7COOH$
Oleic acid (C_{18}) Δ^9	1	$CH_3(CH_2)_7CH=CH(CH_2)_7COOH$
Linoleic acid (C_{18}) $\Delta^{9,12}$	2	$CH_3(CH_2)_4CH=CHCH_2CH=CH(CH_2)_7COOH$
Linolenic acid (C_{18}) $\Delta^{9,12,15}$	3	$CH_3CH_2CH=CHCH_2CH=CHCH_2CH=CH(CH_2)_7COOH$
Arachidonic acid (C_{20}) $\Delta^{5,8,11,14}$	4	$CH_3(CH_2)_4CH=CHCH_2CH=CHCH_2CH=CHCH_2CH=CH(CH_2)_3COOH$
III. Miscellaneous acids (very limited occurrence)		
Ricinoleic acid (C_{18}) Δ^9 (hydroxy-containing)	1	$CH_3(CH_2)_5CHCH_2CH=CH(CH_2)_7COOH$ OH
Tuberculostearic acid (C_{19}) (branched, odd)	0	$CH_3(CH_2)_7CH(CH_2)_8COOH$ CH_3
Lactobacillic acid (C_{19}) (cyclic branch, odd)	0	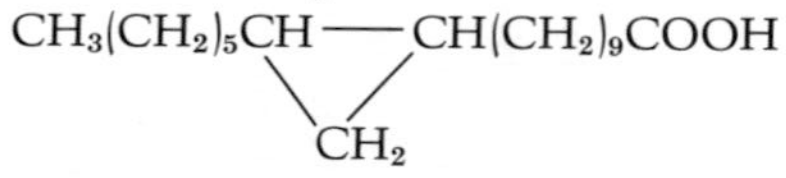 $CH_3(CH_2)_5CH\!-\!\!-\!CH(CH_2)_9COOH$ CH_2

[a] The Δ symbol is used to designate the position of double bonds. The carbon of the COOH grouping is number 1.

—CH=CH—CH$_2$—CH=CH—   and

—CH=CH—CH$_2$—CH=CH—CH$_2$—CH=CH—

(c) In the unsaturated acids, the double bonds are nearly always in the *cis* configuration.

The names and structural formulas of those acids consistent with these generalizations are given in Table 10–2 (I and II). Some exceptions (III) are also listed as a useful frame of reference.

Volumes have been written on the chemistry and biochemistry of fatty acids. We will consider only three general aspects: structural, analytical, and metabolic. If we take stearic acid as an example, the structural principle becomes evident by recalling some basic descriptive chemistry. Suppose you are given the structural formula of stearic acid and asked to predict its water solubility. How would you proceed? Well, you should remember that the dissolution of any solute in any solvent is based on the formation of mutual attractive forces between the two substances, which in turn is controlled by the structural similarity of the solute and solvent. To dust off some classical generalizations: like dissolves like; polar dissolves polar; nonpolar dissolves nonpolar. The exercise is now straightforward. All that is required is an evaluation of the polarity or nonpolarity of stearic acid. This presents no problem. Study the structural formula of stearic acid for a moment.

Most conspicuous are the many C—C bonds (17) and C—H bonds (35). Because of the minimal difference in electronegativity between C and H (refer back to p. 81 in Chap. 3) and because there is obviously none at all between C and C, we deduce that there is a total of 52 extremely nonpolar linkages that completely overshadow the one polar COOH group. We conclude then that stearic acid is a very nonpolar substance and thus very insoluble in water. It should be obvious that this nonpolarity applies to

stearic acid

H—C—C—C—C—C—C—C—C—C—C—C—C—C—C—C—C—C—C—O—H

←———————————————————————————————————→   ←————→
strongly nonpolar hydrocarbon chain of   polar terminal
17 C—C nonpolar bonds + 35 C—H nonpolar bonds   group

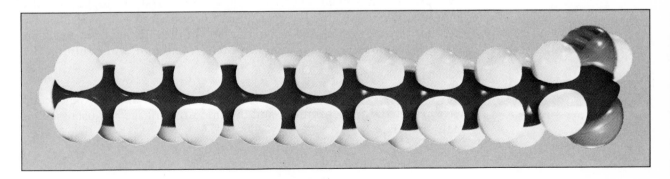

hydrocarbon chains are also symbolized
as shown at the right

/\/\/\/\/\COOH

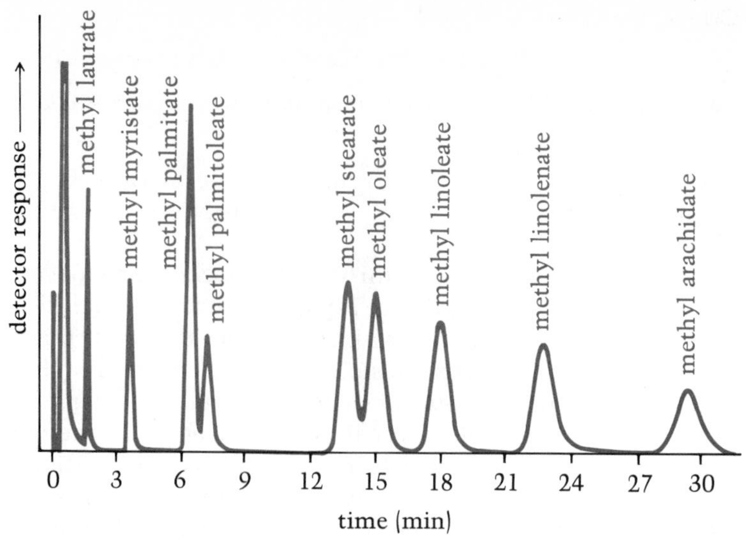

FIGURE 10–1 A representation of a typical separation of a mixture of the volatile methyl esters of fatty acids by gas-liquid chromatography. Note the very distinct separation of the sharp zones corresponding to homologous components. (Elution pattern reproduced with permission of Applied Science Laboratories.)

mixture of fatty acids

RCOOH

R'COOH

R"COOH

$\downarrow$ CH₃OH

RCOOCH₃

R'COOCH₃

R"COOCH₃

mixture of methyl esters

$\downarrow$

analysis via
gas-liquid chromatography

other fatty acids and to any material containing one or more fatty acids. In addition to explaining water insolubility, this is a factor that is most basic to a discussion of lipid function, particularly the ordered complexing of protein and lipids in cellular membranes (see p. 330).

The analytical aspect of the study of fatty acids is noteworthy because it exemplifies the fantastic resolving power and sensitivity of *gas-liquid chromatography* (GLC), the principles of which were previously discussed in Chap. 2 (p. 51). With GLC, fatty acid composition studies of simple or compound lipids are now rapid and routine. After hydrolysis, the reaction mixture is chemically treated to effect the conversion of the nonvolatile, free fatty acids to *volatile ester derivatives*, generally *methyl esters*. The ester mixture is recovered and a small sample is injected into the gas chromatograph. A typical GLC pattern is shown in Fig. 10–1.

Although the details of lipid metabolism will be treated in Chap. 16, one principle, pertinent to the metabolism of fatty acids, is worthy of our brief attention at this point. As with the simple sugars, the fatty acids are converted to a so-called *metabolically active form*. Whereas the simple sugars are processed as sugar phosphates or NuDP-sugars, the fatty acids are metabolized as *acyl esters of coenzyme A*. As shown below, formation of the active species occurs in an ATP-dependent reaction catalyzed by a *thiokinase*. Note that the active group of coenzyme A (symbolized as CoASH) is a free *sulfhydryl group* (—SH), and hence the ester of the fatty acid and coenzyme A is called a *thioester* (*thio-* from the Greek, *theion*, meaning sulfur). The

acyl thioester linkage $\left(\begin{matrix} \overset{\displaystyle O}{\overset{\|}{-C}}-S- \end{matrix}\right)$ is most important and will be encountered repeatedly. Coenzyme A and the thioester linkage are discussed formally in Chap. 11.

$$\underset{\text{fatty acid}}{\overset{O}{\overset{\|}{R C O H}}} + \underset{\text{coenzyme A}}{HSCoA} + ATP \xrightarrow{\text{thiokinase}} \underset{\substack{\text{acyl-SCoA} \\ \text{(a thioester)}}}{\overset{O}{\overset{\|}{R-C}}-SCoA} + AMP + PP_i$$

So-named because they were originally detected in seminal plasma (with the source being the prostate gland), *prostaglandins* are now known to be present in most mammalian tissues. In most cells these are present in extremely small amounts—nanogram quantities (10^{-9} g) or less. The general biological role of prostaglandins is *hormonal*, with specific prostaglandins exerting a regulatory effect on specific processes in specific cells. The many and diverse processes known to be under prostaglandin control are remarkable and the list—already vast and somewhat confusing—is still growing (see margin). Although in most cases precise explanations of the prostaglandin effect are not yet available, two general features have been established: (1) many target cells of prostaglandin control do contain specific *membrane receptors* that bind with individual prostaglandins, and (2) in many cases the ultimate effect is *mediated through cyclic nucleotides* via the secondary messenger mechanism (p. 211), with prostaglandins increasing or decreasing (depending on the tissue) the cellular levels of cyclic-AMP and/or cyclic-GMP.

Physiological processes associated with prostaglandin control

contraction of smooth muscle with a particularly potent effect on uterine muscle (see text)
blood supply (blood pressure)
nerve transmission
development of the inflammatory response
water retention
electrolyte balances
blood clotting (see text)

The basic structure of a prostaglandin is that of a C_{20} monocarboxylic acid containing an internal cyclopentane ring. Individual prostaglandins have one or more double bonds in specific positions and are also oxygenated at specific positions. These features are illustrated below for PGE_2, one of the major prostaglandins (there are five others); note that the biosynthesis is shown as originating from the corresponding C_{20} linear-chain unsaturated fatty acid—hence the reason for discussing prostoglandins in this section. (Prostaglandins of different structures are identified by different notations; for example, PGE_2 is differentiated from PGE_1 or $PGF_{1\alpha}$. The structural variations among different prostaglandins are very minor.) The events of PGE_2 biosynthesis from arachidonic acid are depicted in Fig. 10–2. Note that three enzymes are involved and that one is believed to require glutathione (see p. 105). The significance of other features of Fig. 10–2 are discussed below.

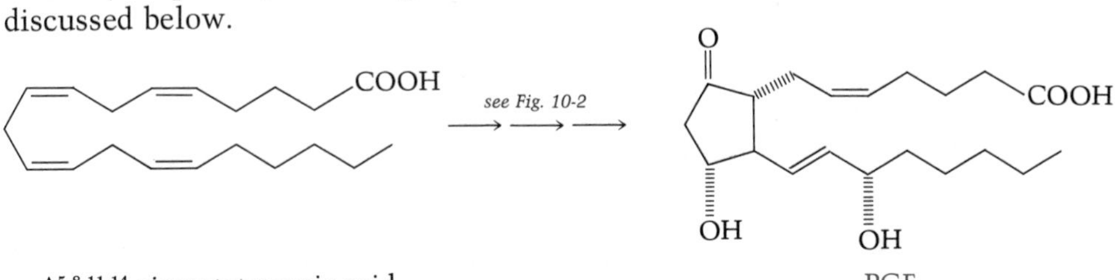

see Fig. 10-2

$\Delta^{5,8,11,14}$-eicosatetraenoic acid
(arachidonic acid)

PGE_2

One reason for the tremendous interest in the prostaglandins is that they have both contraceptive and abortive properties. Although clinical trials are not yet totally complete, reports so far indicate that both effects are achieved efficiently and in some situations without undesirable side effects for the woman. It has been suggested that these applications may represent the greatest breakthrough so far for achieving voluntary regulation of human reproduction.

Prostaglandins are also associated with the therapeutic properties of *aspirin*, which has long been used as an effective analgesic for the relief of minor pain as well as for the treatment of inflammation and fever. Ironically, however, the biochemical bases of these effects were largely unknown until recent studies provided evidence that the effect of aspirin may be

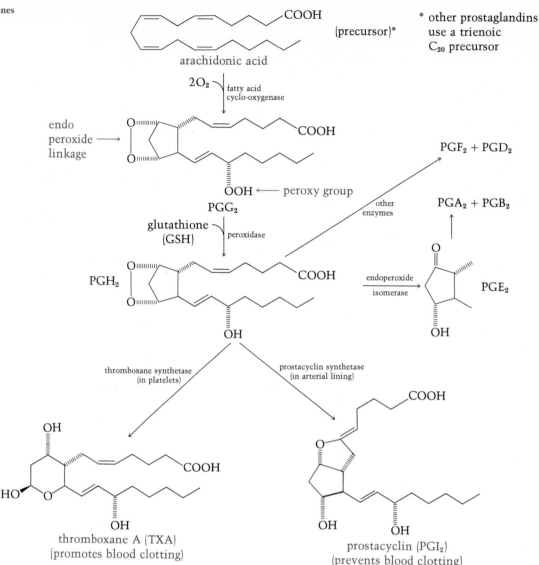

arachidonic acid

(precursor)*

* other prostaglandins
use a trienoic
C_{20} precursor

$2O_2$ — fatty acid cyclo-oxygenase

endo peroxide ⟶ linkage

COOH

OOH ⟵ peroxy group

PGG_2

$PGF_2 + PGD_2$

glutathione (GSH) — peroxidase

other enzymes

$PGA_2 + PGB_2$

PGH_2

COOH

OH

endoperoxide isomerase

PGE_2

OH

thromboxane synthetase (in platelets)

prostacyclin synthetase (in arterial lining)

OH

HO   O

COOH

OH

thromboxane A (TXA)
(promotes blood clotting)

COOH

O

OH   OH

prostacyclin (PGI_2)
(prevents blood clotting)

FIGURE 10–2 A summary of some important aspects of prostaglandin biosynthesis.

linked to its ability to cause the inhibition of the biosynthesis (see Fig. 10–2) of the prostaglandins in various organs. Although the precise role played by the prostaglandins in the normal regulation of body temperature are not yet understood, this breakthrough has opened the door to understanding, so that future research into the mode of action of aspirin and related drugs can now proceed with some direction.

It has recently been established that the biosynthesis of several prostaglandins (refer again to Fig. 10–2) proceeds through *endoperoxide intermediates*, designated as PGG_2 and PGH_2, with specific enzymes in specific tissues then catalyzing the conversion of these intermediates to different classes of prostaglandins. An exciting and important discovery is that in some cells the PGG_2 and PGH_2 endoperoxide intermediates are converted to two other products—*thromboxane* (TXA) and *prostacyclin* (PGI_2). *Thromboxane* formation occurs in blood platelets and *promotes blood clotting* by causing platelets to clump and arteries to constrict. On the other hand, *prostacyclin* formation occurs in cells lining the arteries and veins and *inhibits blood clotting* by inhibiting both the clumping of platelets and the constriction of arteries. This strongly suggests that the relative activities of these two substances may determine whether or not a clot will form. In addition to their role in the normal regulation of blood clotting, an imbalance in their action favoring platelet aggregation may also be one of the initial steps in the formation of atherosclerotic plaques. Thus the findings are of potentially great clinical significance because heart attacks and strokes are often caused by abnormal clot formation. Are certain individuals prone to heart attacks and strokes because of a deficiency (inherited or developed) in the production of prostacyclin? If so, can a synthetic analog of prostacyclin be prepared that could be administered to such individuals to prevent the attack? These questions indicate the importance of research in this area.

Peroxides of Fatty Acids

The enzyme-catalyzed oxidation of arachidonic acid to a hydroperoxide precursor of prostaglandins is an essential step of normal metabolism in mammals. However, in general the *oxidation of unsaturated fatty acids to hydroperoxide products is an undesirable chemical modification* because lipid peroxides can cause damage to other biomolecules, particularly the proteins and nucleic acids. Various studies have linked the formation of lipid peroxides to (a) the process of aging, (b) the toxicity of oxygen (see p. 471 in Chap. 14), (c) the damaging effect on cells by certain types of air pollution, (d) the spoiling of food, and (e) a host of other degenerative biochemical transformations, including some that are drug-induced.

Peroxide formation of a lipid occurs nonenzymatically from the reaction of a strong oxidizing agent—such as hydrogen peroxide, singlet oxygen, or the hydroxy radical (all toxic forms of oxygen that are discussed further in Chap. 14)—with unsaturated fatty acids having two or more double bonds. The reaction is shown below in general and then specifically with linoleic acid.

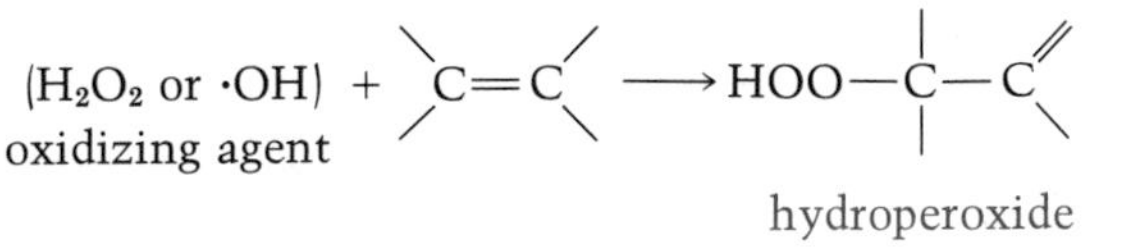

$$(H_2O_2 \text{ or } \cdot OH) + \quad \underset{\text{oxidizing agent}}{}\underset{}{\overset{}{C}}=C \longrightarrow HOO-\overset{|}{C}-C$$

hydroperoxide

For example, with linoleic acid,

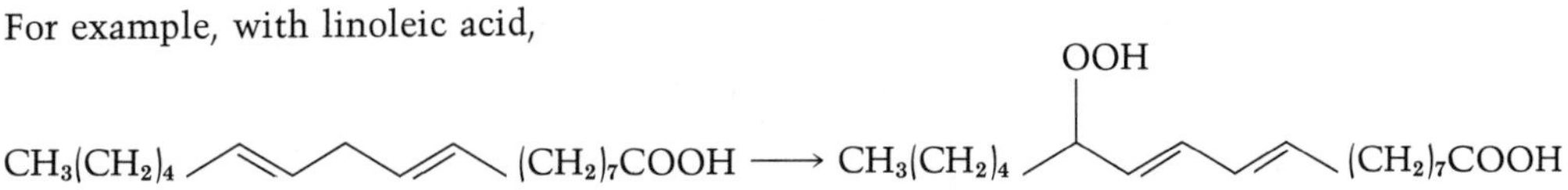

$$CH_3(CH_2)_4 \overset{}{\diagup\!\!\diagdown}(CH_2)_7COOH \longrightarrow CH_3(CH_2)_4 \overset{OOH}{\overset{|}{\diagup\!\!\diagdown}}(CH_2)_7COOH$$

The lipids most susceptible to peroxide formation are the compound lipids (next section) present in membranes. These membrane lipids have an abundance of unsaturated fatty acids. The peroxidation of membrane lipids can in turn cause oxidation of membrane proteins—an alteration that can have damaging effects on the structure and function of the membrane. This damage is attributed primarily to the conversion of the lipid peroxides to peroxide radicals, which are highly reactive oxidizing agents—a conversion that may be mediated by the presence of metal ions such as iron.

$$ROOH \xrightarrow[\text{Fe}^{2+}]{\text{Fe}^{3+}} RO\cdot \quad + OH^-$$

highly reactive
peroxy radical

Detoxification of lipid peroxides is attributed to glutathione, vitamin E (see p. 339 in this chapter), and ascorbic acid. Each of these substances can function as an antioxidant, preventing the undesired action on proteins and nucleic acids. Abnormalities caused by lipid peroxides (or the normal process of aging) could be caused by a defect in the detoxification mechanism or by the production of lipid peroxides in excess of the level that can be efficiently detoxified.

SIMPLE LIPIDS

The simple lipids can be subdivided into two groups: (1) the *neutral acylglycerols* and (2) the *waxes*.

Neutral Acylglycerols

An acylglycerol (also called a glyceride) is an *ester* of the trihydroxyalcohol, *glycerol*, and *fatty acid*(s). Depending on the number of esterified hydroxyl groups, there are *mono*acylglycerols, *di*acylglycerols, and *tri*acylglycerols. The triacyl species is the most abundant in nature. In any case, a simple acylglycerol does not contain any ionic functional groups and hence is said to be a *neutral lipid*. The structures of a mono-, a di-, and a triacylglycerol are shown below. The R groups, which are usually different, designate the acyl side chains of the aliphatic fatty acids. The reason for the predominance of L-acylglycerols is the stereospecificity of glycerol kinase, which converts glycerol only to L-glycerol-1-phosphate, which in turn is the immediate *in vivo* precursor of the acylglycerols (see p. 179, Chap. 6 and p. 518, Chap. 16). Depending on their physical state at room temperature, triacylglycerols are termed neutral *fats* (solids) or neutral *oils* (liquids). Their water insolubility is common knowledge.

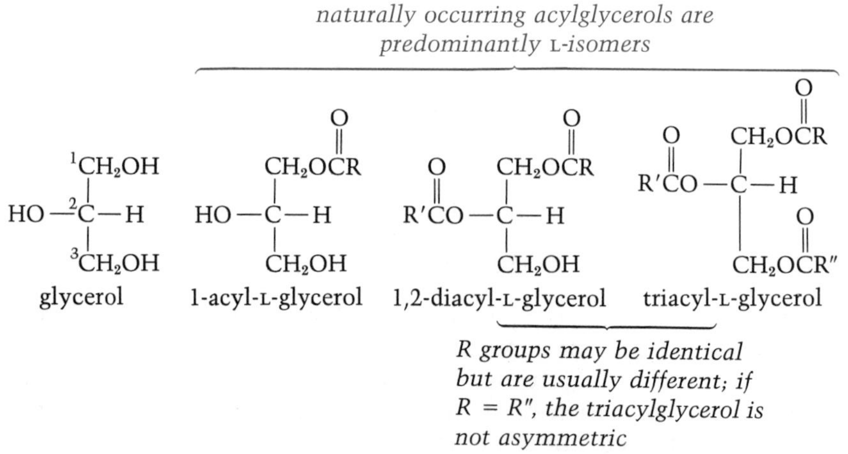

naturally occurring acylglycerols are predominantly L-isomers

$$^1CH_2OH \qquad CH_2O\overset{O}{\overset{\|}{C}}R \qquad \overset{O}{\overset{\|}{R'C}}O \quad CH_2O\overset{O}{\overset{\|}{C}}R \qquad \overset{O}{\overset{\|}{R'C}}O \quad \overset{O}{\overset{\|}{CH_2O\overset{O}{\overset{\|}{C}}R}}$$

HO—^{2}C—H   HO—C—H   R'CO—C—H   R'CO—C—H

$^3CH_2OH$ CH_2OH CH_2OH CH_2OCR''

glycerol   1-acyl-L-glycerol   1,2-diacyl-L-glycerol   triacyl-L-glycerol

R groups may be identical but are usually different; if R = R", the triacylglycerol is not asymmetric

In animals the triacylglycerols fulfill three basic functions: (1) They constitute the so-called fat depots, which are the primary *storage reservoirs of carbon*. In this context, the term "lipid" is synonymous with "fatty acids," and hence, to be more specific, the triacylglycerols are storage forms of the fatty acids. Since the complete oxidative degradation of the fatty acids to CO_2 produces a sizeable amount of useful energy, the fat depots are also termed *energy reserves*. In animal cells they are the major energy reserves.

Although the fat depots are in dynamic equilibrium with all metabolic processes, their degradation is favored under normal conditions when the level of dietary carbohydrate is low. On the other hand, a high level of dietary carbohydrate coupled with a low rate of respiration favors their formation. (2) In the form of lipoprotein particles called *chylomicrons*, the triacylglycerols serve as the means by which ingested fatty acids are *transported* via the lymphatic system and blood for distribution within the animal body. (3) They provide *physical protection* and *thermal insulation* for the various body organs.

There is only one relevant reaction of the simple acylglycerols, namely, their hydrolysis to yield free glycerol and the fatty acids. Chemically, this is readily accomplished by treatment with dilute acid or dilute alkali. When alkali is used, the process is termed *saponification*. In living cells, hydrolysis is achieved via enzymes termed *lipases* (see p. 500, Chap. 16). As you might anticipate in view of our previous discussion on enzymes, the enzymatic hydrolysis is characterized by some degree of specificity, with different lipases acting preferentially on different linkages.

$$
\begin{array}{c}
\overset{O}{\underset{\|}{H_2COCR_1}} \\
R_2COC-H \\
\underset{\|}{O} \\
H_2COCR_3 \\
\underset{\|}{O}
\end{array}
$$

$$K^+OH^-/H_2O \qquad H^+/H_2O$$

glycerol ← ↘ glycerol

$R_1COO^-K^+$	R_1COOH
+	+
$R_2COO^-K^+$	R_2COOH
+	+
$R_3COO^-K^+$	R_3COOH

the term saponification refers to the reaction of a glyceryl ester with potassium (or sodium) hydroxide to produce the potassium salts of the long chain carboxylic acid, which is a soap

Waxes

A wax is also an ester, but is different in that the constituent alcohol and acid *both contain long hydrocarbon chains*. Generally speaking, all waxes are totally insoluble in water. Commercial applications of synthetic and naturally occuring waxes are widespread. In nature the waxes are generally metabolic end products, with the most important biological role being to serve as protective chemical coatings. The surface feathers and skin of animals are coated with a waxy covering that acts as a waterproofing agent. The waxy coating on the leaves and fruits of plants prevents loss of moisture and also minimizes the chance of infection.

general formula of a wax: $R-O-\overset{O}{\underset{\|}{C}}-R'$ (R and R' are long hydrocarbon chains)

$$CH_2-O-\overset{O}{\underset{\|}{C}}$$

EXAMPLE: $CH_3(CH_2)_{28}CH_2-O-\overset{O}{\underset{\|}{C}}(CH_2)_{14}CH_3$

myricyl palmitate
(primary component of beeswax)

COMPOUND LIPIDS

There are three major classes of compound lipids: (1) the *phosphoacylglycerols*, (2) the *sphingomyelins*, and (3) the *glycolipids*. The first two classes are frequently called *phosphatides* or *phospholipids* because of the presence of phosphorus. All types occur exclusively in membranes.

Phosphoacylglycerols (Phosphoglycerides)

Phosphoacylglycerols are the most abundant of the compound lipids. They occur in all types of membranes in all types of cells. Their presence is easily demonstrated by treatment of cells with a chloroform-methanol solvent. Because of their appreciable nonpolar character, the phosphoacylglycerols are readily extracted from the cells with this nonpolar solvent, with the extract then easily analyzed by thin-layer chromatography. The extract is nearly always a mixture of different phosphoacylglycerols. Depending on the type of cell and its physiological state, the number of components and their relative concentrations is quite variable. Despite this diversity, there are what can be called major and minor components. The major ones include *phosphatidyl choline* (also called lecithin), *phosphatidyl ethanolamine* (also called cephalin), and *phosphatidyl glycerol*. The minor ones are *phosphatidyl serine, diphosphatidyl glycerol* (also called cardiolipin), *phosphatidyl inositol,* and the parent compound of all phosphoacylglycerols, *phosphatidic acid.* In view of its precursor role (see p. 518, Chap. 16), our analysis of structure most logically begins with phosphatidic acid.

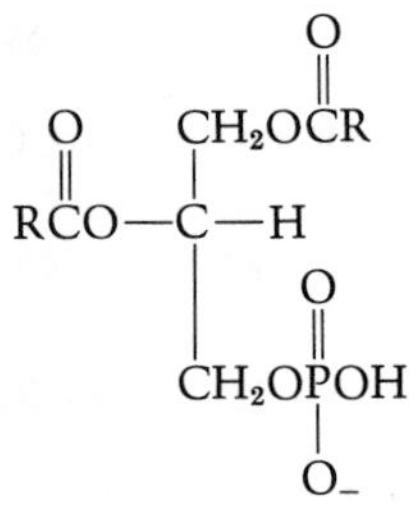

Phosphatidic acid (see margin) consists of a 1,2-diacylglycerol moiety esterified at position 3 to phosphoric acid. Structurally speaking, it is the simplest phosphoacylglycerol. The free acid occurs in nature only in small quantities, with most found in ester linkage as the phosphatidyl component of the other phosphoacylglycerols. As indicated by the names above and illustrated below, the different phosphoacylglycerols are distinguished from one another by the nature of the grouping esterified to phosphatidic acid. The fatty acid composition of the various phosphoglycerides follows no clear pattern. In fact, variations in fatty acid composition occur for each phosphoacylglycerol species. Thus, 10 molecules of phosphatidyl choline obtained from 10 different sources most generally will have 10 different sets of fatty acids esterified with the glycerol moiety. The same even applies to

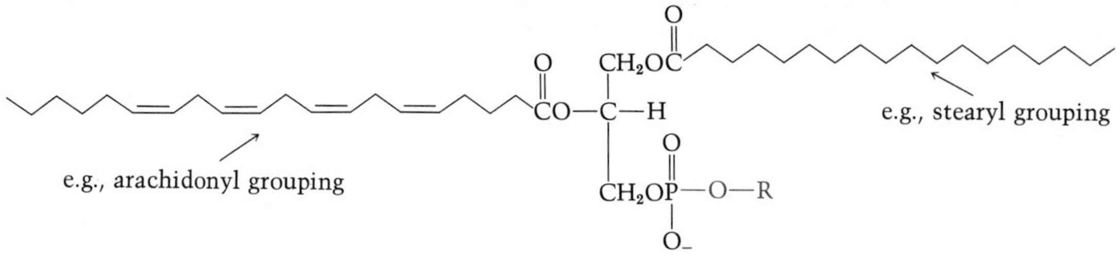

L-phosphatidic acid

e.g., arachidonyl grouping

e.g., stearyl grouping

phosphatidyl esters

identity of (H)O—R

in phosphatidyl ethanolamine

$HO-CH_2CH_2\overset{+}{N}H_3$

in phosphatidyl choline

$HO-CH_2CH_2\overset{+}{N}(CH_3)_3$

in phosphatidyl serine

$HO-CH_2CHCOO^-$
$\qquad\qquad |$
$\qquad\qquad \underset{+}{NH_3}$

in phosphatidyl glycerol

$HO-CH_2CHCH_2OH$
$\qquad\qquad |$
$\qquad\qquad OH$

in diphosphatidyl glycerol

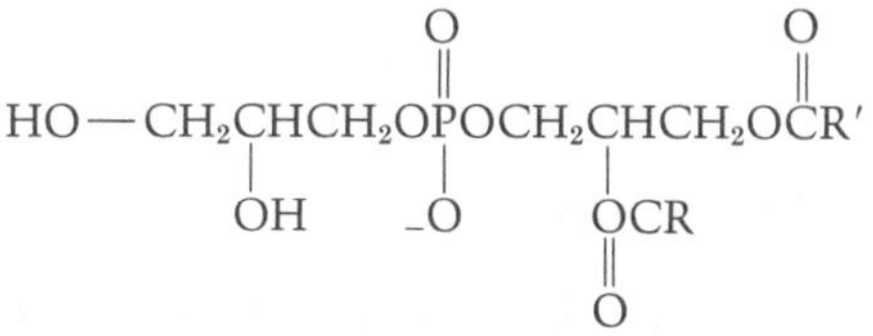

in phosphatidyl inositol

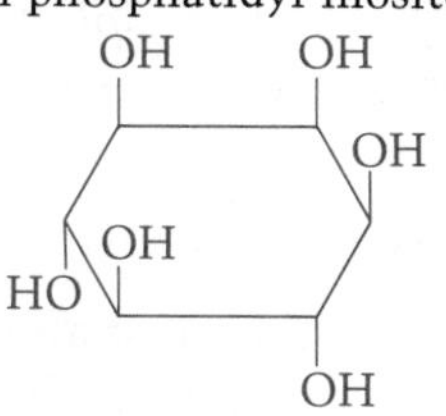

phosphatidyl choline molecules obtained from the same source. In other words, phosphatidyl choline is phosphatidyl choline not because of its fatty acid composition, but because of the presence of choline.

An understanding of the role of the phosphoacylglycerols (as well as the other compound lipids) in biomembranes is founded on an understanding of their molecular structure. In this context, we put aside the distinguishing structural features specified above and instead focus attention on gross structural similarities. In so doing, we note that the phosphoacylglycerols are by their nature *amphipathic* (from the Greek *amphi*, of both sides; *pathos*, feeling) molecules containing distinct regions that have been termed the *hydrophobic, nonpolar "tail"* and the *hydrophilic, polar "head."* The tail region includes the fatty-acid hydrocarbon chain, whereas the head region includes the negatively charged (at physiological pH) phosphate group with its ester component (see structure and photograph). At a minimum, the polarity of the head region is limited to the anionic phosphate group. However, if the alcohol moiety is also charged at physiological pH, as is the case with the quaternary amino grouping of choline (see above), the polarity of the head portion is correspondingly greater. The same would also apply to phosphatidyl ethanolamine and phosphatidyl serine. The discussion in a later section of the molecular ultrastructure of cell membranes will illustrate the biological significance of this physicochemical characteristic.

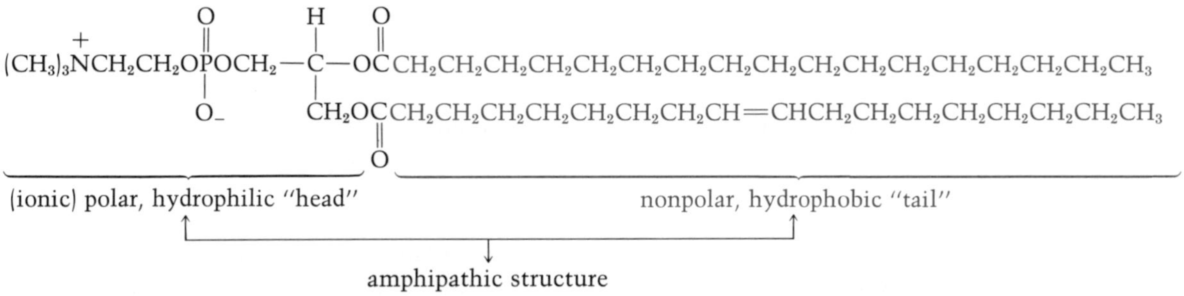

$$(CH_3)_3\overset{+}{N}CH_2CH_2O\overset{\overset{\displaystyle O}{\|}}{\underset{\underset{\displaystyle O_-}{|}}{P}}OCH_2 - \overset{\overset{\displaystyle H}{|}}{C} - O\overset{\overset{\displaystyle O}{\|}}{C}CH_2CH_2CH_2CH_2CH_2CH_2CH_2CH_2CH_2CH_2CH_2CH_2CH_2CH_2CH_2CH_3$$

$$CH_2O\overset{\overset{\displaystyle }{}}{C}CH_2CH_2CH_2CH_2CH_2CH_2CH_2CH=CHCH_2CH_2CH_2CH_2CH_2CH_2CH_2CH_3$$

(ionic) polar, hydrophilic "head"   nonpolar, hydrophobic "tail"

amphipathic structure

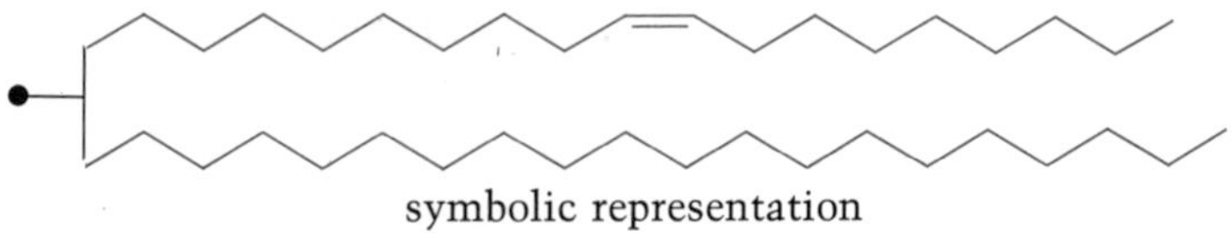

symbolic representation

Phosphonolipids. Recently, a new class of lipid, called *phosphonolipid*, was detected in bovine liver. The term phosphonolipid designates a phospholipid consisting of *phosphonic acid* linked via a C—P bond rather than phosphoric acid linked via an ester C—O—P bond. The liver phosphonolipid discovered was a diacylglyceryl-(2-aminoethyl)phosphonate, which is a phosphono analogue of phosphatidyl ethanolamine. The previous detection of 2-aminoethyl phosphonate (also called ciliatine) in various sources such as sea anemone, goat liver, beef brain, and several human tissues suggests that phosphonolipids may have a widespread occurrence. Their formation and function(s) in cells are at present unknown.

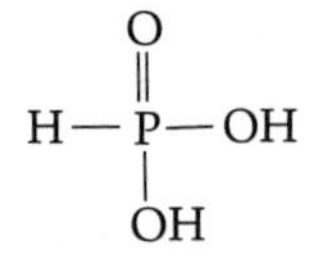

$$H—\overset{\overset{O}{\|}}{\underset{\underset{OH}{|}}{P}}—OH$$

phosphonic acid

$$RC\overset{\overset{O}{\|}}{—}O—\overset{\overset{\displaystyle CH_2O\overset{\overset{O}{\|}}{C}R}{|}}{\underset{\underset{H_2C—\overset{\overset{O^-}{|}}{\underset{\underset{O}{\|}}{P}}—OCH_2CH_2\overset{+}{N}H_3}{|}}{C}}—H$$

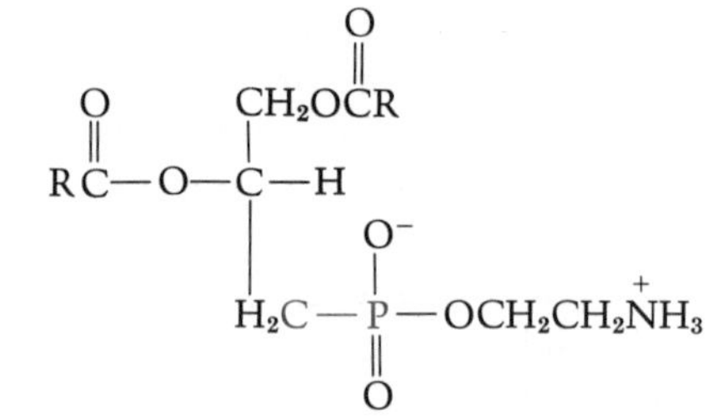

1,2-diacylglyceryl-(2-aminoethyl)phosphonate

$$HO—\overset{\overset{O}{\|}}{\underset{\underset{OH}{|}}{P}}—OH$$

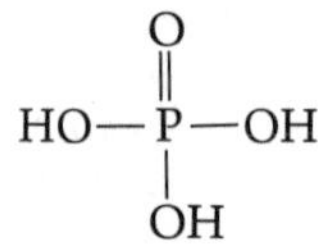

phosphoric acid

Sphingomyelins

The complete hydrolysis of a sphingomyelin yields one fatty acid, choline, phosphoric acid, and *sphingosine*. No glycerol is present. The sphingosines are a family of *long-chain, unsaturated amino alcohols* varying in terms of carbon length (see structure below). Although two OH groups are present, the fatty acid moiety is linked via an amide bond to the lone NH_2 group. This species is often called a *ceramide*. The linkage of a phosphorylcholine moiety to the ceramide yields the complete sphingomyelin. Note that the sphingomyelins would also have a distinct amphipathic character as described above for the phosphoacylglycerols. Sphingomyelins are found in considerable quantities in the cellular membranes of nerve and brain tissue.

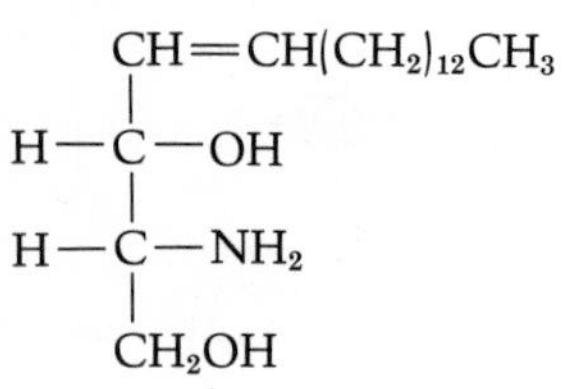

sphingosine

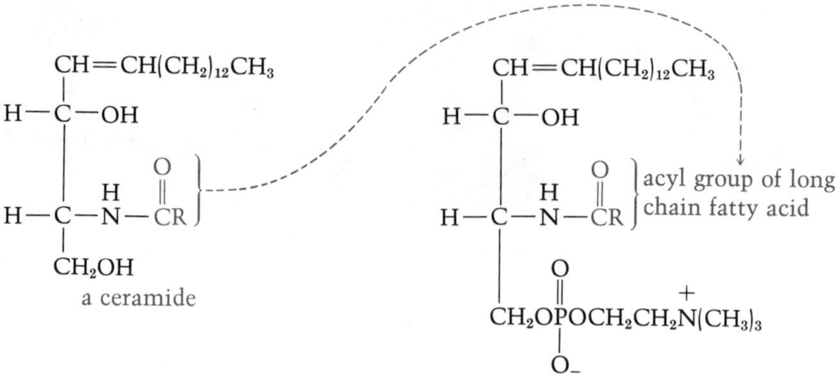

a ceramide

acyl group of long chain fatty acid

a sphingomyelin

Glycolipids

Glycolipids consist of carbohydrate and lipid moieties in covalent linkage. There are two important classes, namely, the *cerebrosides* and the *gangliosides*. Both lack phosphorus.

The complete hydrolysis of a cerebroside yields one or two fatty acids, a sphingosine, and a simple sugar, generally glucose or galactose. Taking into account the presence of the sphingosine moiety, a more precise general term would be *glycosphingolipid*. Like sphingomyelins, the cerebrosides are most abundant in brain and nervous tissue. Cerebrosides are also amphipathic in nature, but to a lesser degree than the phosphoacylglycerols and sphingomyelins due to the presence of the uncharged but polar carbohydrate moiety.

The gangliosides, which are also glycosphingolipids, are derivatives of cerebrosides containing a more complex carbohydrate moiety that includes acetylated derivatives of amino hexoses and *N-acetylneuraminic acid*, also called *sialic acid*; see p. 302, Chap. 9). The exact nature of the oligosaccharide component differentiates one ganglioside from another. The structure below is that of G_{M_1}, one of the more common gangliosides. Others have a smaller oligosaccharide unit or have one or two additional sialyl residues. Gangliosides, like the cerebrosides and sphingomyelins, are found primarily (though not exclusively) in cell membranes of nerve and brain tissue. The role of gangliosides may be more than that of being a structural component of membranes. In other words, they may be directly associated with some dynamic function characteristic of membranes. For example, recent studies suggest that gangliosides may form part of the receptor sites in membranes. *Membrane receptor sites* are individually distinct locations on the surface of the membrane to which specific molecules become attached, triggering a response within the cell. We will have much more to say about these sites on various occasions. Among other roles, there is recent evidence linking membrane gangliosides to the binding of the tetanus toxin and the cholera toxin.

Although sphingomyelins and glycolipids occur ubiquitously, their exact biological role(s) is as yet unknown. It is known, however, that their synthesis and breakdown in higher animals are required for normal life processes, and that defects in their metabolism can result in serious physiological diseases. Four such well-documented conditions in humans are *Gaucher's disease, Fabry's disease, Tay-Sachs disease*, and *Niemann-Pick disease*. In each instance, the problem results from abnormal degradation of glycolipids because defective genes fail to direct the synthesis of specific, required enzymes. Since they are genetic (and thus transferable) diseases manifested by abnormal metabolism, conditions of this sort are frequently termed *inborn errors of metabolism*.

Unfortunately, these four conditions are not the only examples of inborn errors of metabolism. Galactosemia, a disease associated with abnormal galactose metabolism, is described in the next chapter. A few diseases related to abnormalities in the metabolism of amino acids are discussed in Chap. 17. Several others are listed in Appendix II. In each case, research continues in an attempt to establish the precise biochemical basis for the disease, that is, the cause-effect relationship, and possible treatment. There is growing evidence that some symptoms of the lipid diseases are related to alterations of the cell surface membranes.

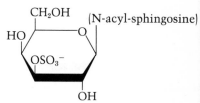

a glucocerebroside

galactocerebrosides *are often sulfated at C-3; intact lipid is called* a **sulfatide** *and represents about 25% of brain cerebrosides*

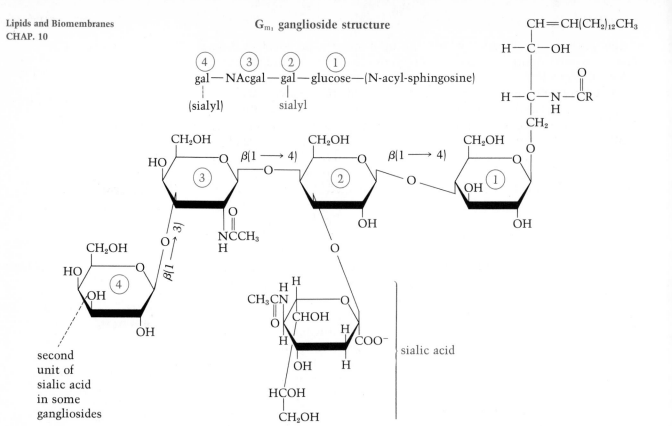

G_{m_1} ganglioside structure

BIOLOGICAL (CELLULAR) MEMBRANES

Any living cell is enclosed by a boundary that confines the interior of the cell and separates it from the external environment. This boundary is called a *membrane*. In addition, in the higher cell forms there is also an elaborate compartmentalization within the cell interior itself; this is represented by the presence of membrane-bound subcellular organelles such as the nucleus and mitochondria, as well as the presence of other subcellular systems, such as the Golgi apparatus and the endoplasmic reticulum, that are entirely membranous in nature.

However, membranes are more than just static boundaries. In fact, they are very active bio(physico)chemical systems responsible for such phenomena as the selective transport of substances into and out of the cell and subcellular compartments, the binding of hormones and other regulatory agents, enzyme-catalyzed reactions, the transmission of electrical impulses, and even the production of ATP. Not every membrane is identical, however, and specific examples of these functions are found in different types of membranes. For example, only the inner mitochondrial membrane is responsible for ATP production and certain hormones bind only at certain membranes. In subsequent chapters various examples of membrane-associated events will be described. The focus in this section will be on the structure of membranes.

Basically membranes are *composed of protein and lipid*, with the relative amount of each varying considerably among different membranes (see Table 10–3). The smaller carbohydrate content of several membranes is contrib-

TABLE 10–3 Chemical composition of some selected cell membranes.[a]

Membrane source	% Protein	% Lipid[b]	% Carbohydrate
Myelin membrane of nerve cells	18	79	3
Liver cells (mouse)	46	54	3
Liver cells (rat)	53	42	5–10
Red blood cells (human)	49	43	8
Nuclear membrane (rat liver cells)	59	35	2–4
Retinal rods (bovine)	51	49	4
Outer mitochondrial membrane	52	48	2–4
Inner mitochondrial membrane	76	24	1–2
Chloroplast lamellae (spinach)	70	30	6
Gram-positive bacteria	75	25	10
Mycoplasma bacteria	58	37	1.5

[a] Data taken from G. Guidotti, "Membrane Proteins," *Annual Review of Biochemistry,* **41,** 731 (1972).
[b] The types of lipids that are present in any one membrane include the phosphoacylglycerols, sphingomyelins, cerebrosides, gangliosides, and cholesterol, in relative proportions particular to the membrane. Most of the lipids are responsible for the permeability properties of the membrane. Others are probably associated with specific membrane functions.

uted in the form of glycoproteins and glycolipids. The molecular ultra-structure resulting from the arrangement of the protein and lipid materials has been the subject of debate and investigation for nearly 40 years, and has been most intensive in recent years. It does appear, however, that there is no one single design of ultrastructure that applies in complete detail to all membranes. The two most popular proposals are (1) the *unit-bilayer* model and (2) the *penetrating-protein* (or *fluid-mosaic*) model. As indicated by the descriptions below, both models are similar in terms of the participation of noncovalent interactions and the orientation of lipid, but they differ in terms of the orientation of protein material

The unit-bilayer model (often referred to as the "classical" model) was first proposed by Danielli and Davson in 1935 and further developed by J. D. Robertson in the 1950s. The distinctive feature is that the membrane is considered to be composed of a *lipid core sandwiched between two protein surfaces.* The lipid core consists of two layers of lipid molecules associated with each other by virtue of hydrophobic associations of their nonpolar "tails." The polar "head" region on each side of the lipid bilayer is then accessible for participation in polar interactions with charged protein molecules. (Refer to the diagram, p. 332.) Since the protein-lipid and lipid-lipid interactions are of the noncovalent variety, the overall structure would not expected to be as rigid and tough as that of a bacterial cell wall and, indeed, it is neither. Rather, it is more flexible and more fragile. The symmetry of the unit-bilayer structure conveyed by the drawing is only apparent, and there is much evidence that *membranes are asymmetric,* with one side of the membrane being different than the other side in terms both of what lipids are present in each region of the bilayer and what proteins are present on the exterior and interior surfaces.

Electron microscopy (see Fig. 10–3), X-ray diffraction, and other experimental studies have guided the development of the unit-bilayer model and also contributed to its validity. Indeed, there is considerable evidence that the myelin membrane of nerve cells, the membrane of retinal rod cells, and the membrane of *Mycoplasma* cells (one of the smallest bacterial organisms known that can cause pulmonary disease in humans) have this arrange-

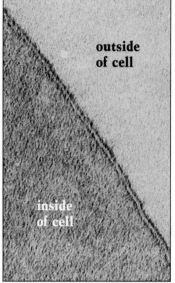

outside of cell

inside of cell

FIGURE 10–3 An electron micrograph of a portion of the cytoplasmic membrane of a red blood cell. The dense granular area at the left represents the inside of the cell, and the lighter gray area at the right represents the outside of the cell. The two electron-dense lines represent the inner and outer protein layers of the membrane, which sandwich the less electron-dense lipid region according to the unit-bilayer model. An extensive area of the erythrocyte membrane may have this structure as well as some segments (not detectable in this type of microscopy) that contain proteins embedded in lipid bilayer. Certain membranes, such as those of nerve cells, appear to consist exclusively of unit-bilayer structure. Magnification: 240,000×. (Photograph generously supplied by J. David Robertson, Department of Anatomy, Duke University.)

Unit-bilayer model of membrane structure

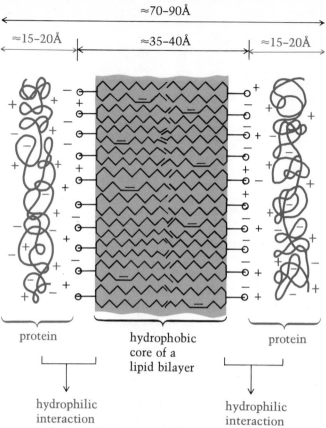

≈70–90Å

≈15–20Å   ≈35–40Å   ≈15–20Å

protein

hydrophobic
core of a
lipid bilayer

protein

hydrophilic
interaction

hydrophilic
interaction

ment. However, the unit-bilayer model is not a universal ultrastructure that applies to all cell membranes. Without itemizing various experimental data with which the unit-bilayer model is not wholly consistent, suffice it to say that these inconsistencies have prompted the proposal of alternative models. One of these models is the *penetrating-protein* (or *fluid mosaic*) model suggested by S. J. Singer and D. Wallach and their associates in 1966.

The fluid-mosaic model retains the proposals of the unit-bilayer model regarding the localization and orientation of lipid in the membrane. The difference is the proposal that *protein is not confined to the two hydrophilic surfaces of the lipid bilayer.* Rather, it is suggested that *proteins are also integrated into the lipid layer,* with some embedding only partially from either side, and others passing through and traversing the entire bilayer. Among other attributes, such a model is consistent with the differentiation that characterizes the ease with which membrane proteins can be extracted. Some are solubilized very easily with mild extraction procedures, while others are difficult to remove even with harsh treatments. Singer classifies the two groups as *peripheral* (surface-oriented) proteins and *integral* (embedded) proteins. The drawings on p. 334 and in Fig. 10–4 depict the features of this model and the photo in Fig. 10–4 offers support from electron microscopy. The photo is the result of the *freeze-etch technique* for examining membrane preparations by electron microscopy; it permits the viewing of a surface that is revealed by a longitudinal cleavage through the lipid bilayer of the membrane.

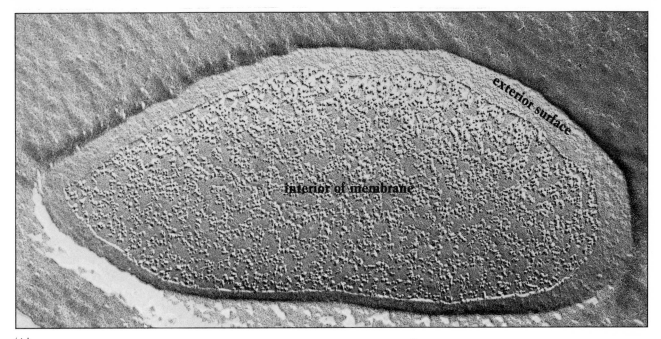

(A)

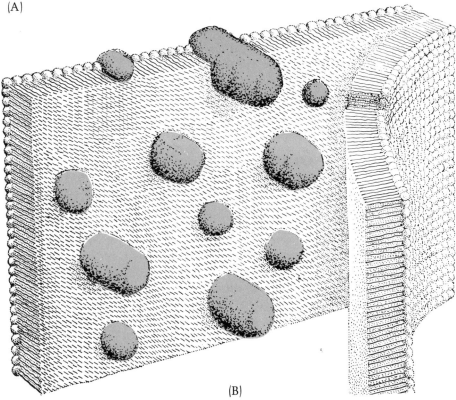

(B)

FIGURE 10–4 (A) Electron microscopic image of a fracture plane through the erythrocyte (red blood cell; human) membrane; obtained by the technique of freeze etching. The oblong body in the center of the field is a red blood cell with about 90% of the surface layer of its membrane removed. The surface layer itself, with a smooth appearance, is visible on the entire perimeter. The inner granular area is interpreted as representing the lipoid interior of the membrane, with many intramembrane protein-containing particles. Magnification: 88,000×. (B) The drawing is intended to depict how these intramembrane proteins would be exposed when the membrane is cleaved so as to peel away the outside layer. (Electron micrograph reproduced with permission from Daniel Branton, Department of Biology, Harvard University. Drawing reproduced with permission from S. J. Singer, "Architecture and Topography of Biologic Membranes," *Hospital Practice*, **8**, 81–90 (May, 1973).)

Recent studies of various sorts, including the isolation of a few membrane proteins that have a structural characteristic consistent with their integration in the lipid bilayer, have supported this model. If a protein is embedded in a highly hydrophobic lipoid region, we would predict that the embedded portion of the protein would have to have a distinctly hydrophobic

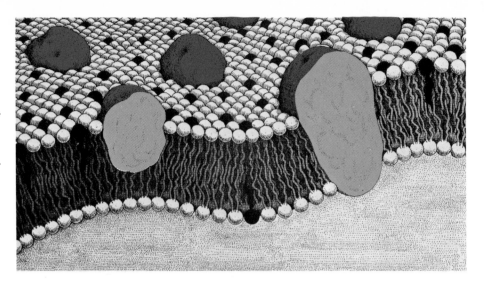

In the penetrating-protein membrane model, proteins are viewed as being predominantly globular and amphipathic, with their hydrophilic ends protruding from the membrane and their hydrophobic ends embedded in the bilayer of lipids (gray) and cholesterol (black). The proteins make up the membrane's "active sites"; some are simply embedded on one or the other side, others pass entirely through the bilayer. (Reproduced with permission from S. J. Singer, "Architecture and Topography of Biologic Membranes," *Hospital Practice*, **8**, 81–90 (May, 1973).)

section in order to coexist with the lipid. Otherwise, it would be expelled. One such amphipathic protein has been isolated from the membrane of red blood cells; it contains a consecutive stretch of about 25 nonpolar amino acid residues.

It is significant to note that the unit-bilayer and penetrating models are not mutually exclusive, and thus both ultrastructures could very well apply to the same membrane. That is to say, in a given membrane, the amount of integrated proteins could vary from 0% to 2%, 10%, 20%, or higher, and when it does occur it could be present in a random fashion or in localized pockets of the membrane. In other words, the lipid-bilayer aspect of ultrastructure may be universal, but the percent of integrated protein will vary from one membrane to another.

It is important to realize, whatever the case, that a *membrane is not a very rigid system* and that some membrane proteins and lipids are mobile molecules. Indeed, some proteins may be capable of transposing their position from one side of the membrane to another and then back again. This displacement may apply to an individual protein molecule or to a multimolecular complex of lipid and protein. In either case this principle of protein mobility within the membrane does have significance for the transport of materials across membranes, a subject to which we will return shortly (see p. 340).

The preceding description could easily have been extended to include a further elaboration on both models and an examination of other models. However, the somewhat simplified discussion has emphasized the most important features of cellular membranes: (a) they are composed of a noncovalent, lipoprotein matrix; (b) they are nonrigid; and (c) some proteins (lipoprotein complexes) participate actively as components of binding sites, or as transport agents, and even as enzymes. (The need for further study was also stressed.) We will have several occasions throughout subsequent chapters to refer to these features, particularly (c).

A striking example of a highly organized and biochemically active membrane system is provided by the mitochondrial membrane (Chap. 14).

DERIVED LIPIDS

As stated previously, the derived lipids, in terms of both structure and function, represent a truly heterogeneous group similar only in their water insolubility. In the following pages, we will briefly consider a few important representatives: the *steroids*, the *carotenoids*, and related *lipid vitamins*.

Steroids

Steroids are found in all organisms, where they are associated with various functions. In humans, for example, they function as sex hormones, as emulsifying agents in lipid digestion, and in the transport of lipids across membranes and through plasma fluids.

All steroids have a similar basic structure consisting of a *fused hydrocarbon ring system*. Due to its structural relationship to the aromatic hydrocarbon *phenanthrene*, the basic structure is called *perhydrocyclopentanophenanthrene* (*perhydro*, meaning completely hydrogenated). The diversity of steroid structure is due to numerous structural variations of the perhydrocyclopentanophenanthrene nucleus, which include varying levels of unsaturation, the presence of ring substituents, and various chemical modifications of the cyclopentano moiety in particular.

The presence of a C_8—C_{10} hydrocarbon side chain at position 17, and of a hydroxyl group at position 3, characterizes a large number of steroids called the *sterols*. The most important member of this family, indeed, the most abundant sterol in the animal kingdom, is *cholesterol*. Cholesterol is a structural component of cell membranes, but it varies in concentration, representing anywhere from 0 to 40% of the total membrane lipid. Because of its fused ring structure, which is not as flexible as an extended hydrocarbon chain, it is proposed that the presence of cholesterol contributes more rigidity (stiffness) to a membrane. Cholesterol is also the primary metabolic precursor of other important steroids, including the bile acids and the sex hormones (p. 522 in Chap. 16); in certain tissues it is also a precursor of Vitamin D (p. 338 in this chapter).

phenanthrene   cyclopentane

perhydrocyclopentano-
phenanthrene

The complete biosynthetic pathway of cholesterol has been determined, as have many of the steps involved with its conversion to other steroids (see Chap. 16).

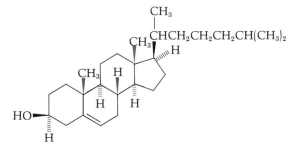

cholesterol

the indicated orientations of the two methyls and three hydrogens result from a specific pattern of ring fusion referred to as trans *fusion; another consequence is that the overall structure of cholesterol is rather* **flat** (see photo)

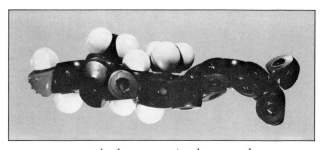

most hydrogens omitted to reveal the coplanarity of fused rings

Three steroids of particular interest are the male sex hormone, *testosterone*, and the female sex hormones, *estradiol* and *progesterone*. In the male testosterone regulates the development of nearly all sex character-

testosterone

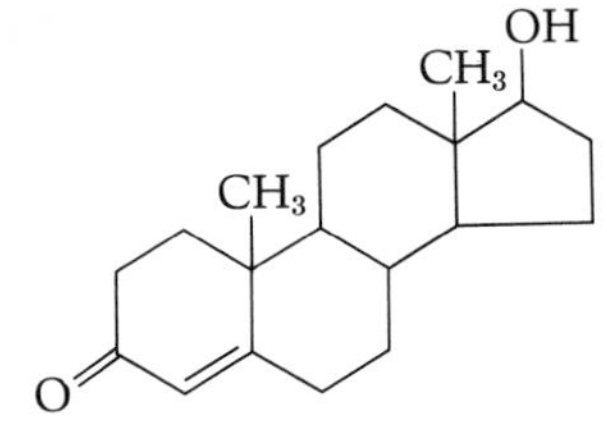

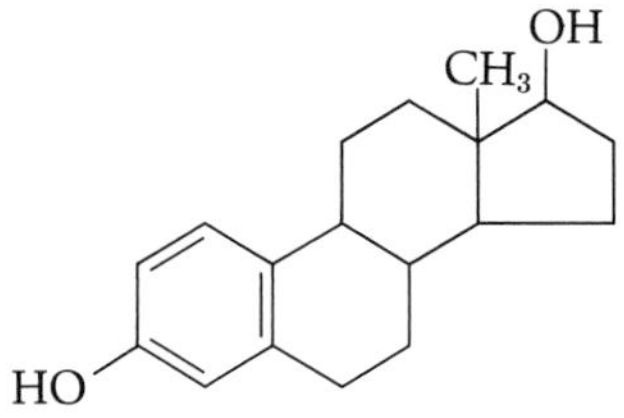

estradiol

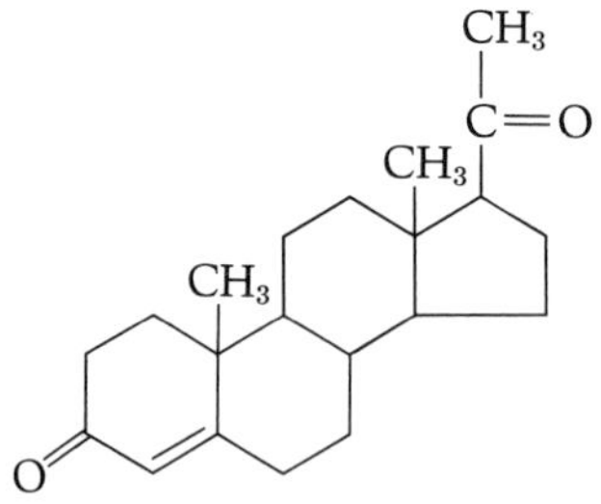

progesterone

istics, the maturation of the sperm, and the activity of the genital organs. In the female estradiol and progesterone, both products of the ovary glands, are largely responsible for regulation of the menstrual cycle. Progesterone is produced only during a certain period of the cycle, most notably after the release of the ovum from the ruptured follicles, at which time it begins to regulate the preparation of the uterine mucosa for the deposition of the fertilized ovum. If fertilization occurs, the production of progesterone continues through pregnancy. If the egg is not fertilized, the level of progesterone drops and production does not resume until the next cycle.

The biochemistry of the hormonal regulation of the menstrual cycle has been extensively studied, with a primary objective being the development of a safe and effective method of fertility control. The pioneering studies were performed in the late 1930s, and the first significant discovery was that daily injections of the natural progesterone inhibited ovulation. In the past 30–35 years, hundreds of steroid preparations, mostly synthetic in nature, have been tested for inhibitory and/or regulatory properties affecting both ovulation and menstruation. Several of these materials are now available in oral "pill" form. Although these steroids have been declared safe by manufacturers and the Food and Drug Administration, our knowledge of possible undesirable effects on both a short- and a long-term basis is not precise or complete. Accordingly, even when conditions warrant their use, they should be employed only under advisement and with caution.

Carotenoids (β-Carotene and Vitamin A)

The carotenoids consist of two main groups, the *carotenes* and the *xanthophylls*. Both types are water-insoluble pigments widely distributed in nature, but they are most abundant in plants and algae. The carotenes are pure hydrocarbons, whereas the xanthophylls are oxygen-containing derivatives. The former are more abundant and only these are considered here.

The most common carotenoid is the carotene β-carotene. As shown below, β-carotene is a C_{40} hydrocarbon consisting of a highly branched, unsaturated chain containing identical substituted ring structures at each end. Virtually all other carotenoids can be considered as variants of this structure. Although the carotenoids have been linked as participants in the

β-carotene $(C_{40}H_{56})$

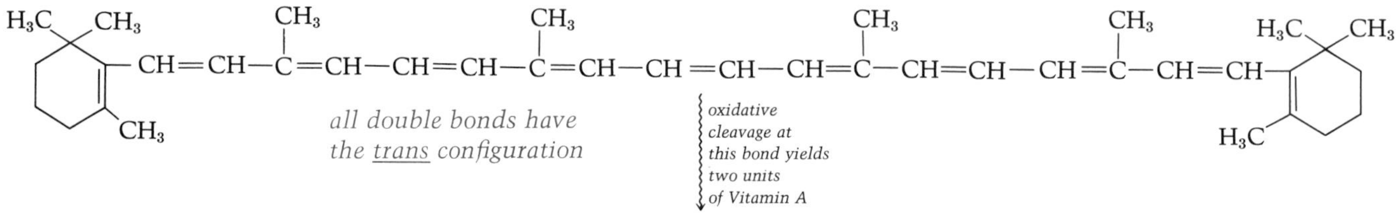

all double bonds have the trans configuration

oxidative cleavage at this bond yields two units of Vitamin A

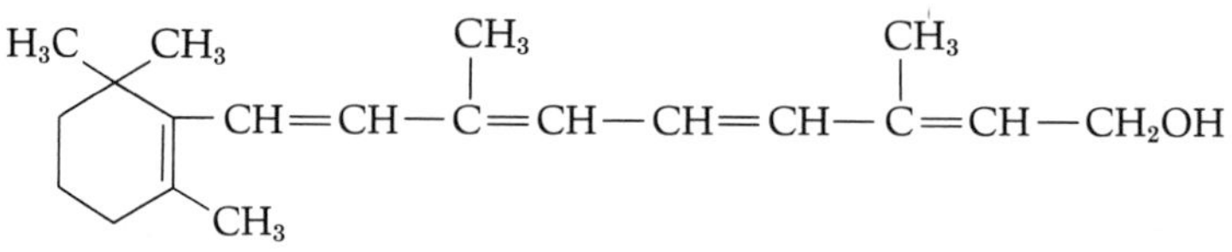

Vitamin A (alcohol form)
"retinol"

336

harnessing of solar radiation in the process of photosynthesis, the exact mechanism of their participation is yet to be resolved. The same is true of other plant and microbial processes known to involve carotenoids. Of great importance is the enzyme-catalyzed oxidative cleavage of β-carotene to *Vitamin A*, which occurs with a stoichiometry of $1 \rightarrow 2$. In animals this conversion represents a chief natural source of Vitamin A.

One instance where the physiological function of Vitamin A is understood on a molecular level is in the retina of the eye, where the reduced, alcohol form of Vitamin A (*retinol*) is enzymatically converted to the oxidized, aldehyde form (*retinal*), which then becomes complexed with different retinal proteins, called *opsins*, and forms the active proteins that function in vision. The complexes of retinal and an opsin protein are the primary *photoreceptors* of incident light in the visual cells, and transmit information to the nervous system by a process not yet clearly understood. Most vertebrates contain two types of visual cells in the retina: (1) *rod cells*, which are dim-light receptors and do not perceive color, and (2) *cone cells*, which are bright-light receptors and also responsible for color vision. In the rod cells, there appears to be only one opsin, and the active lipid-protein receptor complex is called *rhodopsin*. In cone cells, at least three different opsins are known to occur; these are complexed to retinal to constitute a blue-sensitive receptor, a red-sensitive receptor, and a green-sensitive receptor. In the case of rhodopsin, it has been established that the retinal is covalently attached to the protein via the side-chain amino group of a lysine residue.

retinol (all *trans*)

retinal (all *trans*)
(becomes complexed to
opsin proteins)

At one time it was thought that only the all-*trans* form of retinal was present in visual cells. However in subsequent studies a second isomeric form was discovered, namely, 11-*cis*-retinal, in which one of the double bonds has a *cis* orientation. In fact, it is the 11-*cis* species that is known to be complexed to the opsin protein. Another aspect of vision chemistry was uncovered with the discovery that 11-*cis*-retinal, when exposed to light, was converted to the all-*trans* isomer.

In accordance with these findings, the molecular events of vision are proposed to consist of a *cis-trans* isomerization cycle, as shown on p. 338 for the rod pigment. The distinguishing photochemical act is the cleavage of the lipoprotein rhodopsin complex—a complicated process accompanied by the isomerization of retinal. The conversion is not direct, but involves many intermediates. There is reasonable evidence suggesting that one or more of these intermediate steps may be subsequently involved in *generating extremely small electrical potentials that activate the nervous system.* The vision cycle, in terms of the fate of the visual pigment, is completed by the regeneration of 11-*cis*-retinal, which is required for the refor-

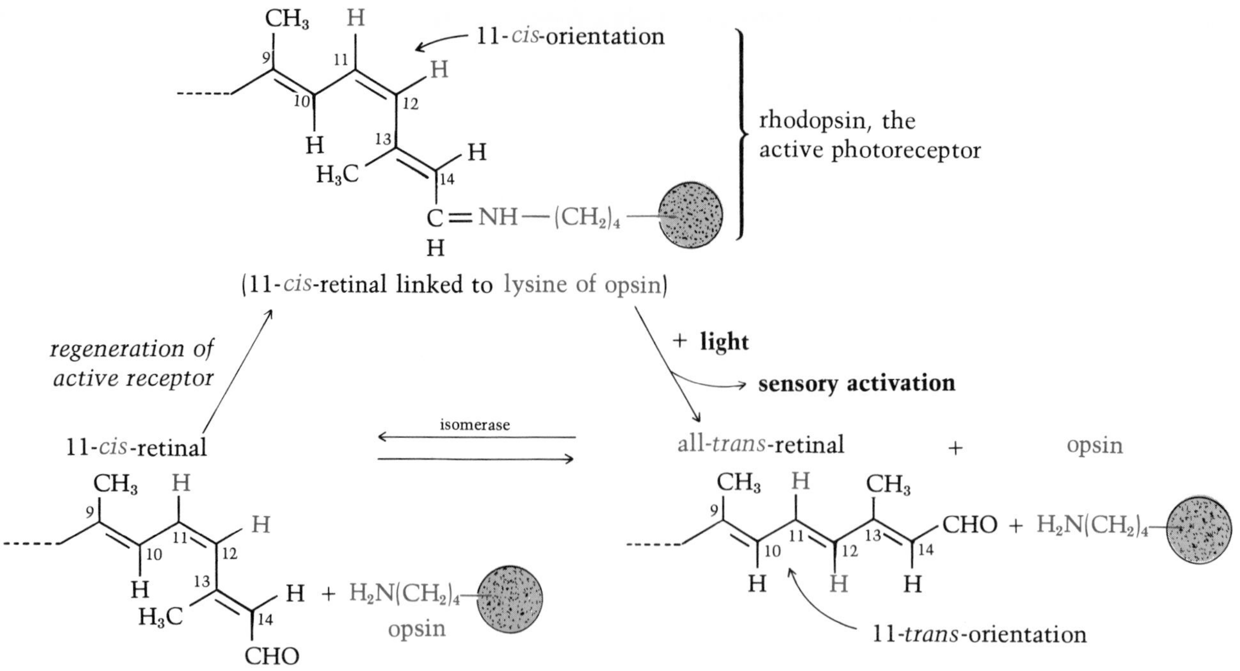

(11-cis-retinal linked to lysine of opsin)

mation of the active rhodopsin pigment. One possible route is a direct enzymatic conversion catalyzed by an isomerase.

Vitamin D

Animals require *Vitamin D* in the normal calcium and phosphorous metabolism necessary for healthy bone and tooth development. A deficiency leads to *rickets*, a disease in which the bones become soft and pliable, producing various deformities.

One of the most interesting aspects of Vitamin D is that it is formed from a sterol precursor by exposure to ultraviolet radiation. One important sterol precursor is 7-dehydrocholesterol (itself produced enzymatically from cholesterol), which yields Vitamin D_3.

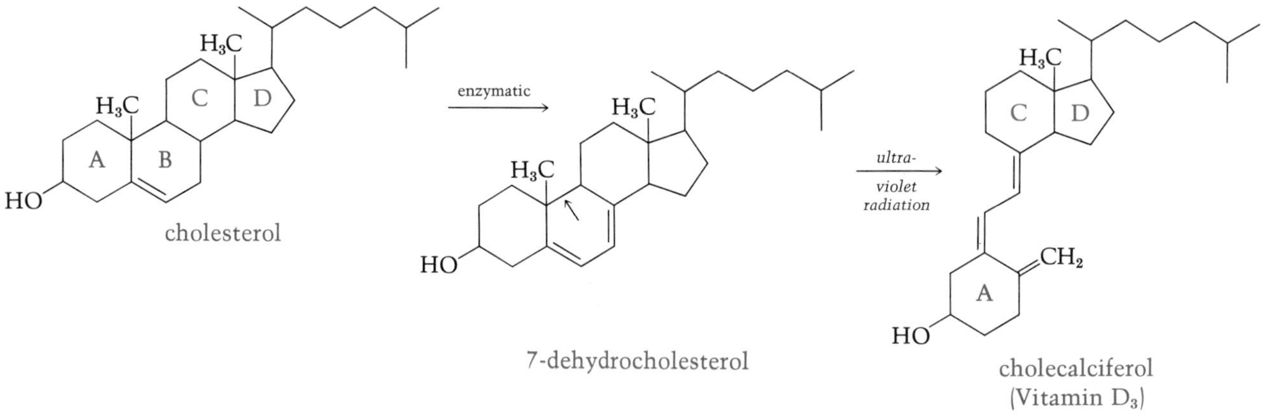

the photochemical cleavage occurs at the bond shown by the arrow; electron rearrangements after the cleavage yield the final product

As implied by the D_3 designation, there are various forms of Vitamin D. The D_3 species (cholecalciferol) is the form present in milk and fish-liver oils, both of which are major dietary sources of the vitamin. The D_3 species can also be synthesized in the skin of animals. Because of this synthesis in skin tissue, adults receiving normal exposure to sunlight require less Vitamin D in the diet than infants. It is now established that after D_3 intake or formation in the skin, the D_3 species (carried in plasma) is converted in the liver and kidney to hydroxylated derivatives, which are even more active. Indeed, there is strong evidence that the major active form in the body is the 1,25-dihydroxy derivative.

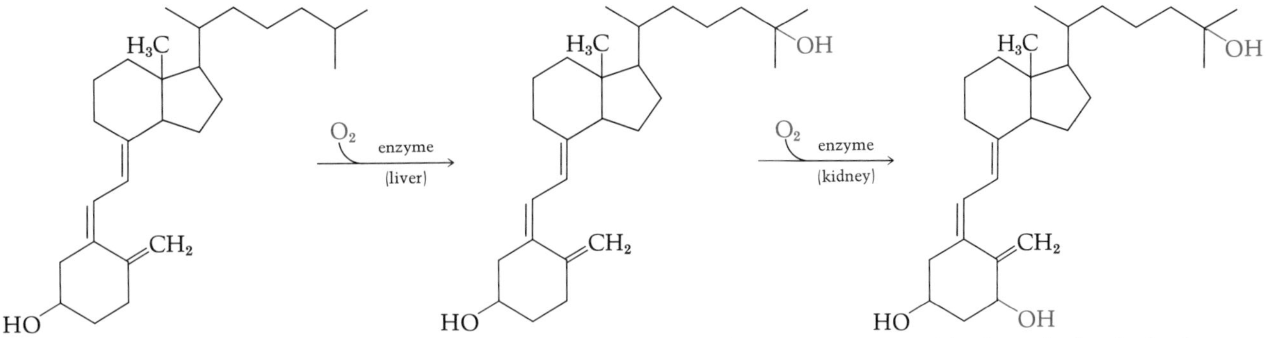

cholecalciferol
(Vitamin D₃)

25-hydroxycholecalciferol

1,25-dihydroxycholecalciferol
(most active at stimulating the intestinal absorption of Ca^{2+}—and phosphate—and the mobilization of Ca^{2+} for bone development)

Vitamin E

A substance required for reproduction in animals is *Vitamin E*, the so-called antisterility or fertility vitamin. The basic structure of the E vitamins (there are different forms) is called *tocopherol* (from the Greek *tokos*, meaning childbirth, and *pherein*, meaning to carry). The most potent form of Vitamin E is α-tocopherol.

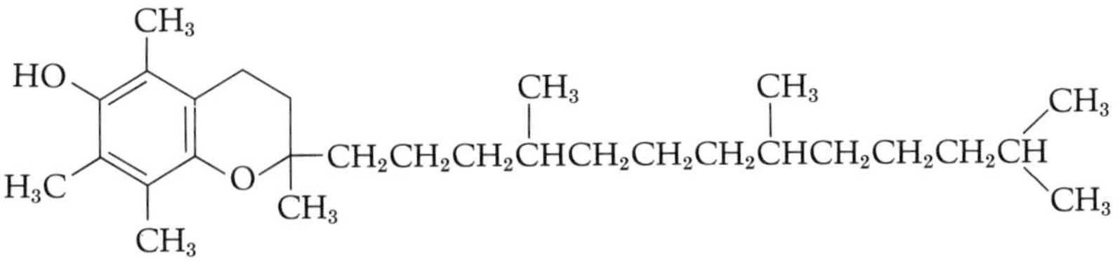

Vitamin E (alpha tocopherol)

Although Vitamin E has proven essential for reproduction in the laboratory rat, there is no conclusive evidence for the same relationship in humans. Of greater significance (in humans and various other animals) is the proposed effect Vitamin E and close derivatives have in maintaining the normal chemical composition and function of muscle tissue. For example, in certain animals a Vitamin E deficient diet results in muscular dystrophy.

In the processing of food, Vitamin E is commonly added because it acts as an antioxidant, preventing the spoilage of foods through oxidation. In living cells, Vitamin E may also function as an antioxidant, along with ascorbic acid (p. 297) and glutathione (p. 105).

Vitamin K

A deficiency of *Vitamin K* slows the rate of blood clotting. When first detected as being necessary for clotting, it was termed in German *Koaglutations-vitamin*; hence the name Vitamin K.

The basic structure of Vitamin K is a *napthoquinone* bicyclic ring system with a long hydrocarbon chain of variable length attached to the quinone ring. The hydrocarbon chain in Vitamin K_2 consists of repeating *isoprene* units. Animals depend on two sources of Vitamin K: (1) diet, particularly one that includes green vegetables, tomatoes, and cheese; and (2) synthesis by bacteria in the intestinal tract.

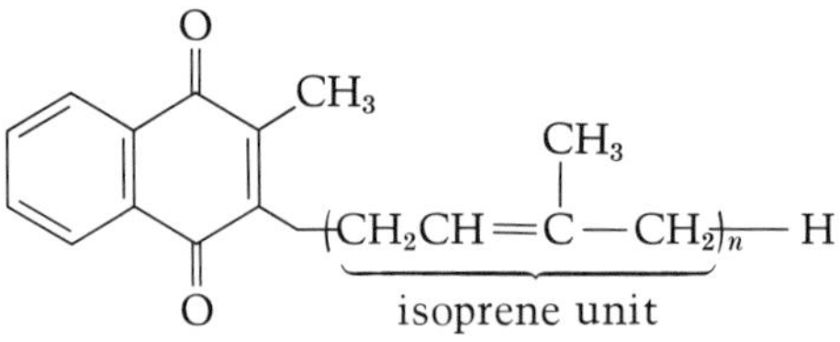

Vitamin K (general structure)
(required for blood clotting)
n is variable but usually < 10

The precise function Vitamin K serves in blood clotting has recently been established. It promotes clotting by participating in the synthesis of a *"complete" prothrombin* molecule, the precursor of thrombin (see p. 187). After the prothrombin molecule is assembled, Vitamin K participates in a post-translational modification. The specific modification is the *carboxylation of several glutamic acid* side chains. The insertion of the additional —COO⁻ sites is necessary for the optimum binding of Ca^{2+}, which activates the conversion of prothrombin to thrombin.

Post-translation modification

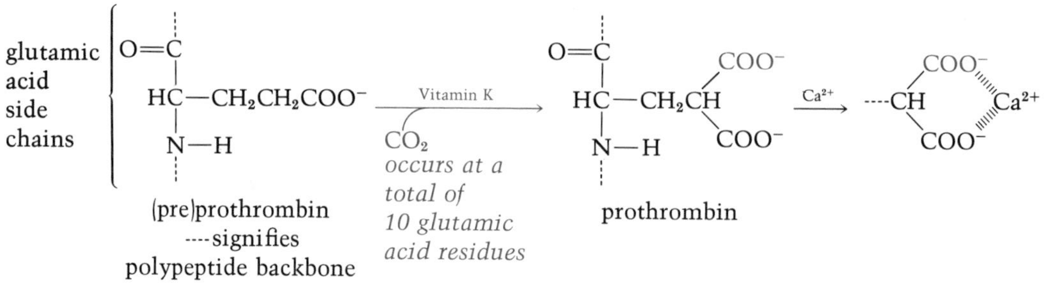

MEMBRANE TRANSPORT

The passage (transport) of substances into and out of cells and also their transport between the cytoplasm and the various subcellular organelles (mitochondria, nucleic, etc.) are determined by membranes. This is a particularly important role of membranes. Obviously, if a biomembrane were a completely impermeable barrier, regions would be totally isolated and nutrients could not enter and products could not leave. On the other hand, if it were a completely permeable partition, any and all substances would be free to move from one region to another. Neither of these extremes applies. Rather, the transport properties of membranes are in between; they are *semipermeable* partitions. Some materials can move across whereas others cannot. Furthermore, different membranes often have different permeability characteristics.

Depending on the substance and the membrane involved, direct (see margin) transport may occur via one of two general processes. One does not require an expenditure of energy by the cell and is referred to as *passive transport*; the second does require energy and is referred to as *active transport*. Further distinctions involve whether transport requires the participation of membrane proteins and whether or not the substance is chemically modified during transport.

Direct transport means that a substance (X) actually moves across a membrane. There are a few examples of *indirect* transport wherein, despite the impermeability of a membrane to X, it can cross the membrane in a different form. Two examples of indirect transport are described elsewhere (see p. 394 and p. 471).

Passive Transport

Some substances pass through a membrane by *simple diffusion*. No energy is necessary and no specific interactions with membrane proteins are involved. The movement may occur right through the lipoprotein matrix or it may occur through pores or channels that exist in the lipoprotein matrix. Water and several lipid-type substances are examples of such substances.

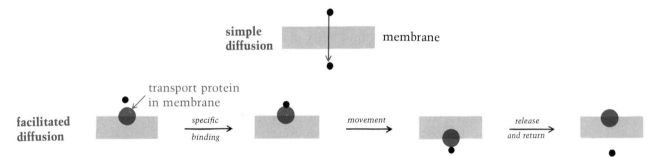

In *facilitated diffusion* the movement requires the participation of a carrier molecule, usually a protein located in the membrane. The *transport protein* binds with the substance on one side of the membrane, moves to the other side of the membrane, releases the substance, and returns to its original position. Although it is simple enough to describe, very little is known about how the movement of protein occurs. Does it prefer to exist on one side of the membrane (and if so, why?) or is it constantly floating between the two membrane surfaces? Is there a conformational change on binding that causes the protein to reorient itself? As yet, these questions remain unanswered.

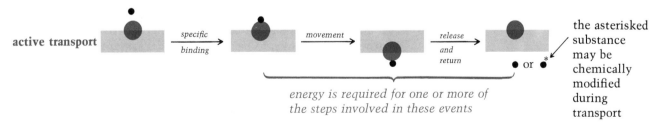

Active Transport

When the passage of a substance requires an expenditure of metabolic energy by the cell, the process is called *active transport*. The energy requirement is generally satisfied by ATP, but other energy-yielding processes are known. In addition to the energy requirement, an active transport mechanism also requires the participation of membrane components (i.e., transport proteins) and in this respect is similar to facilitated transport. As indicated in the diagram below, the energy-requiring step(s) would occur subsequently to the binding action of the carrier molecule.

The preceding diagrams are effective in summarizing general features but are greatly oversimplified, particularly in terms of the involvement of transport proteins. For one thing, more than one membrane component may be involved, that is, the transport system may be multimolecular. In addition, a transport protein may span the entire width of the membrane. In other instances, a hormone may participate, usually by stimulating one of the steps.

One of the most important active transport systems in animal cells is responsible for the movement of both Na^+ and K^+ across cell membranes. Referred to as the *K^+-Na^+ pump*, the system is responsible for maintaining a cell interior that is high in K^+ and low in Na^+ by moving K^+ into the cell and Na^+ out of the cell. Both exchanges with the plasma are against concentration gradients (and are thus energy-requiring) since the normal plasma level of K^+ is lower and that of Na^+ is higher than that in the cytoplasm. This ionic imbalance is a characteristic of normal cells that is important to the regulation of water content in the cell, to protein biosynthesis in the cell, and to the excitability of nerve and muscle cells. The fact that nearly 25% of the total energy expended in the entire animal body is estimated to be used for this purpose is an indication of how important these ionic gradients are to a normal condition.

The operation of this pump involves only a single protein called *Na^+-K^+ ATPase*. There is evidence that this protein spans the entire width of the membrane, providing a binding site for Na^+ on one side and a site for K^+ on the other side. This double binding is followed by (a) a reorientation of the protein carrier and the bound ions, (b) release of the bound ions, and (c) a second reorientation to restore the initial state. It is not known yet whether the hydrolysis of ATP (which is catalyzed by the same protein, hence the name ATPase) is necessary for both (a) and (b), or just for (a) or (b). There is evidence, however, to indicate that the "energy is delivered" via the phosphorylation of the protein, a modification which likely alters its conformation so that (a) and/or (b) occur.

An example of a multimolecular transport process is the *PEP-phosphotransferase system* in *E. coli* and other bacteria. (PEP stands for phosphoenolpyruvate, which, like ATP, is a high-energy phosphorylating molecule. See p. 356 in Chap. 11.) This bacterial phosphotransferase system accounts for the transport of hexoses into the cell as hexose phosphates. Although not present in all bacteria and apparently not present in higher cells either, it has proven to be a useful model system that may have features similar to other transport systems.

At least three different proteins (and possibly four or five) are involved, participating in a two-step reaction sequence (see p. 343). The first step involves the catalyzed (enzyme I) phosphorylation of a heat-stable, low-molecular-weight protein (HPr), with phosphoenolpyruvate acting as a high-energy phosphate donor. In the second step, catalyzed by enzyme II, the sugar is phosphorylated with the phospho-HPr acting as a phospho donor.

Although an idealized oversimplification, the drawing on p. 343 suggests how the transport could be effected. (The drawing is consistent with the finding that HPr and enzyme I are primarily soluble proteins that become at-

$$\text{(1)} \quad {}^-\text{OOC}-\underset{\underset{\text{OPO}_3{}^{2-}}{|}}{\text{C}}=\text{CH}_2 \;+\; \text{HPr} \xrightarrow{\text{enzyme I}} {}^-\text{OOC}-\underset{\underset{\text{O}}{\|}}{\text{C}}-\text{CH}_3 \;+\; {}^{2-}\text{O}_3\text{P}-\text{HPr}$$

phosphoenolpyruvate   pyruvate

$$\text{(2)} \quad {}^{2-}\text{O}_3\text{P}-\text{HPr} + \text{Sugar} \xrightarrow[\substack{\text{(maybe other}\\\text{proteins)}}]{\text{enzyme II}} \text{HPr} + \text{sugar}-\text{phosphate}$$

$$\textit{Net:} \quad {}^-\text{OOC}-\underset{\underset{\text{OPO}_3{}^{2-}}{|}}{\text{C}}=\text{CH}_2 \;+\; \underset{\text{(outside)}}{\text{sugar}} \xrightarrow[\text{HPr}]{\;\text{EI}\;\xrightarrow{}\;\text{EII}\;} {}^-\text{OOC}-\underset{\underset{\text{O}}{\|}}{\text{C}}-\text{CH}_3 \;+\; \underset{\text{(inside)}}{\text{sugar}-\text{phosphate}}$$

PEP   PYR

tached to the membrane only during phosphotransferase activity, whereas the protein component that contains enzyme II is definitely membrane-localized.) The distinctive feature is that the protein II component can assume different positions in the membrane. It can expose its binding site to glucose on the outside face of membrane and then reorient itself so that the bound glucose is directed towards the inside face. The HPr and enzyme I participants would then attach at specific sites of the protein II component. The HPr gets phosphorylated and the HPr-P_i in turn acts as the phosphate donor for glucose, with glucose-6-phosphate released on the inside.

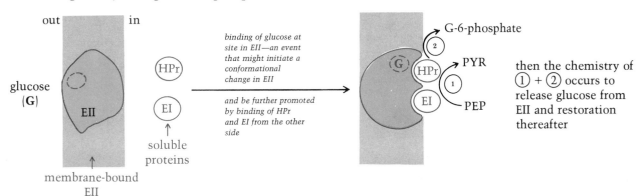

LITERATURE

ANSELL, G. B., J. N. HAWTHORNE, and R. M. C. DAWSON, eds. *Form and Function of Phospholipids.* 2nd edition. New York: American Elsevier, 1973. An authoritative treatise; an extensive compilation of phospholipid composition of membranes is included.

BERGSTROM, S., and B. SAMUELSSON. "Prostaglandins." *Ann. Rev. Biochem.,* **34,** 101–109 (1965). A brief review article.

BRETSCHER, M. S. "Membrane Structure: Some General Principles." *Science,* **181,** 622–629 (1973). A review article.

BLAND, J. "Biochemical Consequences of Lipid Peroxidation." *J. Chem. Ed.,* **55,** 151–155 (1978). A short review article on the formation of lipid peroxides and a survey of their harmful biological effects.

CAPALDI, R. A. "A Dynamic Model of Cell Membranes." *Scientific American,* **230,** 26–34 (1974).

CUATRECASAS, P. "Hormone-Receptor Interactions and the Plasma Membrane." *Hospital Practice,* **9,** 73–80 (July, 1974). Excellent article (at the introductory level) on the regulation of cellular activity initiated by the binding of hormones to membrane receptor sites. Emphasis is given to the effects of insulin.

CUATRECASAS, P. "Membrane Receptors." *Ann. Biochemistry,* **43,** 169–214 (1974). A review of recent progress in the identification, isolation, and purification of various membrane-localized receptors. A focus on insulin receptors.

DEUEL, H. J. *The Lipids.* New York: John Wiley-Interscience, 1951, 1955, 1957. A three-volume treatise on the chemistry and biochemistry of lipids. A useful collection of chemical and physiological information, although the biochemical material is considerably dated.

FISHMAN, P. H., and R. O. BRADY, "Biosynthesis and Function of Gangliosides," *Science,* **194,** 906–915 (1976). A good review article covering (a) the formation of gangliosides, (b) the consequences of impaired formation, and (c) the physiological roles of gangliosides, such as that of mediating the action of cholera toxin.

FLORKIN, M., and E. H. STOLTZ, eds. *Comprehensive Biochemistry*, Vol. 6. Amsterdam-New York: Elsevier Publishing Company, 1965. Review articles on the chemistry of fatty acids, waxes, neutral fats and oils, phospholipids, glycolipids, and sphingolipids.

FOX, C. F. "The Structure of Cell Membranes." *Scientific American*, **226**, 30–38 (1972). An informative article on the roles of proteins in membrane structure and membrane function; includes a discussion of the process of active transport of materials across membranes.

HEFTMANN, E. *Steroid Biochemistry*. New York: Academic Press, 1970. An introductory survey of the steroids and related compounds. Coverage includes structure, metabolism, and physiological properties.

HUBBARD, R., and A. KROPF. "Molecular Isomers in Vision." *Scientific American*, **216**, 64–76 (1967). In addition to a discussion of the basic chemistry of the visual process, this article also contains an excellent introductory treatment of the subject of geometrical (*cis-trans*) isomerism in organic compounds.

JACOBS, S., and P. CUATRECASAS. "The Mobile Receptor Hypothesis for Cell Membrane Receptor Action." *Trends Biochem. Sci.*, **2**, 280–282 (1977). A short review article.

KABACK, H. R. "Transport Studies in Bacterial Membrane Vesicles." *Science*, **186**, 882–892 (1974). A review article of membrane transport focusing on a few bacterial systems that have been characterized.

KEENAN, T. W., and D. J. MOORE. "Mammary Carcinoma: Enzymatic Block in Disialoganglioside Biosynthesis." *Science*, **182**, 935–936 (1973). Referred to on p. 329 in this chapter.

MARCHESI, V. T. "The Structure and Function of a Membrane Protein." *Hospital Practice*, **8**, 76–84 (June, 1973). Introductory level discussion of a major glycoprotein isolated from the membrane of red blood cells. (Original research described in *Proc. Nat. Acad. Sci., USA*, **69**, 1445 (1972).)

MARCHESI, V. T., H. F. FURTHMAYR, and M. TOMITO. "The Red Cell Membrane." In *Ann. Rev. Biochem.*, **45**, 667–698 (1976). A thorough review article.

PETROW, V. "Steroidal Oral Contraceptive Agents." In Vol. 2 of *Essays in Biochemistry*, P. N. Campbell and G. D. Greville, eds., pp. 117–146. New York: Academic Press, 1966. A review article surveying the structures and mechanism of action of natural and synthetic contraceptive steroids.

RAMWELL, D. W., and J. E. SHAW. "Biological Significance of the Prostaglandins." In *Recent Progress in Hormone Research*, Vol. 6, E. B. Astwood, ed. New York: Academic Press, 1970. A recent and informative review article, containing discussions of experimental data.

RAZIN, S., and S. ROTTEM. "Cholesterol in Membranes." *Trends Biochem. Sci.*, **3**, 51–55 (1978). A short review article on the role of cholesterol as a regulator of membrane fluidity.

SINGER, S. J. "The Molecular Organization of Membranes." *Annual Review of Biochemistry*, **43**, 805–833 (1974). A detailed reviw article focusing on the arrangement of proteins in membranes; some coverage of the movement (fluidity) of lipids and proteins in the membrane.

STENFLO, J., and J. W. SUTTIE. "Vitamin K-dependent Formation of γ-Carboxyglutamic Acid." In *Ann. Rev. Biochem.*, **46**, 157–172 (1977). A review article on the activation of prothrombin by Vitamin K.

VANE, J. R. "Inhibition of Prostaglandin Synthesis as a Mechanism of Action for Aspirin-like Drugs." *Nat. New Biol.*, **231**, 232–235 (1971). Experimental evidence relating the therapeutic effects of aspirin and related drugs to the production of prostaglandins in the lung. Two suceeding articles report on similar effects with spleen cells and with human blood platelets.

WALD, G. "Molecular Basis of Visual Excitation." *Science*, **162**, 230–239 (1968). The address delivered by the author on receiving the Nobel Prize in 1967 for his work in the field.

YAMAKAWA, T., and Y. NAGAI. "Glycolipids at the Cell Surface and Their Biological Functions." *Trends Biochem. Sci.*, **3**, 128–131 (1978). A short review article.

EXERCISES

10–1. The melting points of a few fatty acids are listed below. On the basis of these data, explain the characteristic that distinguishes neutral fats and oils.

Acid	Melting point (°C)
linoleic acid	−5.0
linolenic acid	−10.0
myristic acid	53.9
oleic acid	13.4
palmitic acid	63.1
stearic acid	69.6

10–2. What prediction can be made regarding the value of the melting point temperature for (a) palmitoleic acid, (b) lauric acid, (c) arachidic acid, and (d) arachidonic acid? (Refer to the data in the preceding problem.)

10–3. If a mixture consisting of a triglyceride and phosphatidyl choline were analyzed by thin-layer chromatography on silica gel in a chloroform-methanol-water developing solvent, you would observe complete separation, with the R_f of the triglyceride being approximately 1, and that of phosphatidyl choline approximately 0.4. Explain why the R_f values of these two lipids differ so widely.

10–4. Draw the structure of the intact lipid that would correspond to each of the following mixtures of products obtained from the complete hydrolysis of the lipid.
(a) glycerol, palmitic acid, stearic acid, inorganic phosphate
(b) glycerol, palmitoleic acid, oleic acid, ethanolamine, inorganic phosphate
(c) sphingosine, palmitic acid, inorganic phosphate
(d) sphingosine, glucose, oleic acid

10–5. Which would have a greater amphipathic character, a glucocerebroside or a sphingomyelin? Explain.

CHAPTER 11

PRINCIPLES OF BIOENERGETICS

To sustain their existence, living cells depend on an intake of "food" to serve not only as sources of carbon, nitrogen, sulfur, phosphorus, and the other bioelements but also as sources from which useful energy is extracted. Without energy a cell is a nonfunctional machine—incapable of synthesizing the numerous materials necessary for its viability; incapable of active membrane transport; and incapable of various specialized physiological functions such as muscle contraction, nerve transmission, and motility.

In Chap. 7 we had our first encounter with biochemical energetics in describing the general metabolic role of *adenosine triphosphate ATP* (see p. 211), the pre-eminent material in the flow of energy in a cell. To restate that role briefly: ATP formation represents the conservation by a cell of chemical energy released in energy-yielding degradative reactions, and ATP breakdown represents the utilization of energy by a cell in the energy-requiring events of synthesis and other processes. In this chapter we will expand on this theme, focus on ATP in more detail, and also examine other materials that function in the transfer of energy. As you will discover, this topic provides both a necessary background for and a logical progression into the subject of metabolism.

Describing the energetics of any system—be it living or nonliving; organic or inorganic; chemical, physical, or biological—is the domain of a specialized field called *thermodynamics*, where emphasis is placed on *en-*

ergy changes as the system undergoes a transformation from one state to another. In the language of thermodynamics, energy changes can be described in a variety of ways, but the most useful description is given in terms of the *change in free energy.* For chemical systems, the free-energy change is extremely useful, because under the commonly encountered conditions of constant temperature and constant pressure, it provides a valid method of predicting the feasibility of a reaction, as well as being representative of the maximum amount of chemical energy that is potentially available for doing useful work.

PRINCIPLES OF THERMODYNAMICS

Energetics of State Transitions

Any physical or chemical transformation is most directly and most simply described by contrasting the physical and/or chemical properties of the initial and final states of the system. Included among these properties are pressure, temperature, volume, the physical states of the materials, the concentration of each material, and the chemical composition of each material. Two simple examples are given below, one a physical transformation and the other a chemical transformation. (For both we will impose conditions of constant temperature and pressure and, for reasons of simplicity, neglect any changes in volume.) In the former, the nature of the initial and final states is obvious; water is converted from a liquid to a vapor state. The chemical process is also simply described; gaseous propane (an organic alkane) and oxygen react to yield carbon dioxide and water in respective reacting proportions of $1 + 5 \rightarrow 3 + 4$. This of course, represents a complete oxidation (combustion) of propane corresponding to the use of natural gas as a fuel.

$$H_2O(liq) \xrightarrow{\text{vaporization}} H_2O(vap) \qquad (P \text{ and } T \text{ are constant})$$
$$\text{initial state} \qquad\qquad\qquad \text{final state}$$

$$CH_3CH_2CH_3(vap) + 5O_2(vap) \xrightarrow[\substack{\text{(complete}\\ \text{oxidation)}}]{\text{combustion}} 3CO_2(vap) + 4H_2O(vap) \qquad (P \text{ and } T \text{ are constant})$$
$$\text{initial state} \qquad\qquad\qquad\qquad \text{final state}$$

Although both descriptions are valid, common experience tells us they are incomplete. The major shortcoming is that we have neglected to indicate that one transformation requires an input of energy from the surroundings and the other results in an output of energy. (That is to say, we have not included a comparison of the energy levels of the initial and final states.) In the energy-requiring vaporization of water, the initial liquid state is at a lower energy level than the final vapor state, with the difference in energy levels primarily due to a greater amount of molecular motion in water molecules in the vapor state than in the liquid state. Thus, heat energy is converted to kinetic energy of molecular motion. In the combustion of propane the difference in energy between the two states is that there is a greater chemical bonding energy in the reactants than in the products. Thus, in the course of the chemical change, the conversion of the higher energy reactants is accompanied by a release of heat energy. (*Of course, the energetics of the reverse of each of these processes would be just the opposite of what has been described.*)

$$H_2O(liq) + \textit{energy} \longrightarrow H_2O(vap)$$

energy-requiring process with
540 calories required per
gram of water

$$CH_3CH_2CH_3 + 5O_2 \longrightarrow 3CO_2 + 4H_2O + \textit{energy}$$

energy-yielding process with
531 calories produced per
mole of C_3H_8

The intent of these introductory remarks is to emphasize that *a complete physicochemical description of any change in state necessarily includes an analysis of the energetics of the process, which provides a deeper insight into the nature of the transformation and the participating substances.* While this principle applies to all systems, it is particularly appropriate to the study of biochemical transformations in the living state. Indeed, as stated previously, without such an approach our understanding of cellular metabolism would be obscured.

Note: In the SI system the unit of energy is the *joule* J. The interconversion of thermochemical calories and joules is made according to these relationships: 1 calorie = 4.184 joules and 1 kilocalorie = 4.184 kilojoules.

Thermodynamic State Functions

A description in thermodynamic terms of the transition from one state to another can be given in terms of four *state functions,* called *internal energy E,* *enthalpy H,* *entropy S,* and *free energy G.* The internal energy expresses the total energy of a system; the enthalpy expresses the heat content of a system; the entropy expresses the degree of disorderliness of a system; and the free energy expresses the useful energy of a system, that is, energy that is available for conversion to useful work.

Each of these thermodynamic properties is referred to as a *state function* because their values are determined only by the specific condition of the system, that is, the state of the system. In most instances, the actual value of E, H, S, or G is difficult if not impossible to measure, and hence thermodynamics deals primarily with changes (Δ for finite change) in state functions: ΔE, ΔH, ΔS, and ΔG. The universal expression of the difference for the general case of initial state $\rightarrow$ final state is always *final state value minus initial state value* $(E_f - E_i)$. For a chemical transformation of reactants $\rightarrow$ products, the difference would be stated as *product state value minus reactant state value* $(H_p - H_r)$.

Each of the expressions (ΔE, ΔH, ΔS, and ΔG) has a different meaning and application, and hence each has its own utility. For biochemical applications, the free energy change ΔG has greatest utility because it applies under conditions of constant pressure and constant temperature, the conditions under which cellular reactions occur. In addition, the value of the free energy change for a chemical reaction is easily measured in various ways, and it has very useful interpretations. Although we could develop the concept of ΔG in terms of ΔE, ΔH, and ΔS (they are all interrelated; for example, it can be shown that $\Delta G = \Delta H - T\Delta S$), we will bypass this approach and proceed directly to an analysis of what ΔG means.

Free Energy

The thermodynamic function G was appropriately termed free energy because a finite change in its value from one state to another $(G_f - G_i)$ is a measure of the *maximum amount of energy in the system that is potentially available for useful work* when the change occurs under the conditions of constant T and P. In the technical language of thermodynamics, there are several other interpretations of the free energy function. An interpretation of particular usefulness to chemical systems is that the free energy is a thermodynamic property directly *related to the total chemical energy* of the system and hence *to the chemical stability* of the system.

In this context, a high free energy value is representative of a potentially unstable system that under the proper conditions would spontaneously go to a lower level of free energy. In other words, a *negative value* for the change in free energy $(-\Delta G)$ corresponds to an *energy-yielding* reaction that results in a change from an unstable state of high chemical energy content to a more stable state of lower chemical energy content. Such a reaction is termed an *exergonic reaction* and is said to be *thermodynamically favorable*. In contrast, a *positive value* $(+\Delta G)$ corresponds to an *energy-requiring* reaction that results in a change from a stable state of low chemical energy to a more unstable (less stable) state of higher chemical energy. Such a reaction is termed an *endergonic* reaction and is said to be *thermodynamically unfavorable*, that is, it does not have the capacity to occur spontaneously unless energy is supplied.

The condition of $\Delta G = 0$ has a special meaning signifying that the forward and reverse processes of a reversible reaction system (one exergonic and the other endergonic) are occurring at the same rate and the system is at *equilibrium*, at which point there is no tendency to undergo any further net change.

If an exergonic reaction were to occur by itself, the output of chemical energy would be lost, primarily as heat energy. However, if it were to occur in the presence of an endergonic reaction, the output of chemical energy from the exergonic reaction could serve as an input of chemical energy to drive the endergonic process. This type of behavior is termed *coupling*, and represents the basic design of energy flow in living organisms. To understand this point further, it is necessary that we first examine some of the biomolecules that participate in cellular energetics. We will begin such an analysis shortly.

> If the chemical energy of reactants (G_r) is less than that of products (G_p), the reaction would be endergonic, that is, energy-requiring, and the value of ΔG would be positive.
>
>
>
> $$\text{reactants} \longrightarrow \text{products}$$
> $$\text{(initial state)} \qquad \text{(final state)}$$
> $$G_r \qquad\qquad G_p$$
> $$\Delta G = G_p - G_r$$
>
> If the chemical energy of reactants (G_r) is greater than that of products (G_p), the reaction would be exergonic, that is, energy-yielding, and the value of ΔG would be negative.

Meaning and Measurement of the Standard Free Energy Change $(\Delta G°)$

Under conditions of constant temperature and pressure, for a general chemical transformation involving reactants A and B and products C and D,

$$a\text{A} + b\text{B} \rightleftarrows c\text{C} + d\text{D} \qquad \text{(constant } T, P\text{)}$$

it can be shown that the free energy change is given by the classical relationship

$$\Delta G = \Delta G° + RT \ln\left(\frac{[\text{C}]^c[\text{D}]^d}{[\text{A}]^a[\text{B}]^b}\right)$$

where 　[ ] = molar concentration
　　　　R = 1.99 calories/degree/mole (or approximately 2.0)
　　　　T = absolute temperature (degrees Kelvin)

Here ΔG represents the free energy change at any point in the course of the transformation, and $\Delta G°$ corresponds to the *standard free energy change*, which applies only to a particular set of conditions, namely, when all participants are in their *standard states*. (Although the conditions of the standard state are purely arbitrary, there is a universal agreement; for solutes dissolved in solution, the accepted criterion is unit activity $(a = 1.0)$, or approximately 1 M for most materials.) Thus, at a specific temperature and pressure, the value of ΔG will vary with changes in the existing concentrations of all participants, whereas $\Delta G°$ is constant and will change only if the temperature and/or pressure are altered. For these reasons, and because the use of the standard-state condition creates a common denominator for all systems, the standard free-energy change is a much more desirable measurement, permitting a *comparison of the chemical energetics of different systems* at the same temperature and pressure.

A direct method of determining $\Delta G°$ is possible if we recognize that $\Delta G = 0$ when the system is at equilibrium, and hence

$$\Delta G° = -RT \ln \left(\frac{[C]^c [D]^d}{[A]^a [B]^b} \right)$$

But now since the system is at equilibrium, the concentration ratio is actually the equilibrium constant (K_{eq}) of the system, and we can write

$$\Delta G° = -RT \ln (K_{eq}) = -2.3 RT \log (K_{eq})$$

This last equation is one of the most useful relationships of thermodynamics. Its utility is obvious: at any temperature, the $\Delta G°$ can be calculated directly from the equilibrium constant, which in most cases can be determined in the laboratory. Some problems based on the use of this equation are provided at the end of the chapter.

Many reactions, especially those that occur in nature, involve H^+ as a product or reactant, and hence a quotation of their $\Delta G°$ would necessarily mean that the concentration of H^+ would be approximately 1 M (standard condition of unit activity). Since this means that the pH would equal zero, it would certainly be unrealistic to quote the $\Delta G°$ of H^+-dependent reactions in a living cell where the pH is approximately 7. Accordingly, when the equilibrium constants of pH-dependent reactions are measured, the system is studied at a pH of 7 and the resultant standard free energy changes are specified as being so calculated. To avoid confusion with a true $\Delta G°$, the standard free energy changes of H^+-dependent reactions calculated under a non-standard-state condition of $[H^+]$ are symbolized differently, with a $\Delta G°'$ notation being most common.

The meaning of $\Delta G°(\Delta G°')$ values is slightly different from that given previously for ΔG. Most different is the interpretation given to a zero value. When $\Delta G°(\Delta G°') = 0$, this means that there is no appreciable difference between the chemical stability of initial and final states, and under standard conditions the reaction is neither endergonic nor exergonic, but the system is not necessarily at equilibrium. When $\Delta G(\Delta G') = 0$, the system is at equi-

endergonic *(requires energy for completion)*	A + B ⟶ C + D equilibrium favors reactants $K'_{eq} \lll 1$ $\Delta G°' = +$ value stability$_{reactants}$ > stability$_{products}$
exergonic *(yields energy)*	A + B ⟶ C + D equilibrium favors products $K'_{eq} \ggg 1$ $\Delta G°' = -$ value stability$_{reactants}$ < stability$_{products}$

librium. The important point, however, is that the interpretation given to negative and positive values of $\Delta G°(\Delta G°')$ is basically the same as that for $\Delta G(\Delta G')$, with minus (negative) values corresponding to a thermodynamically favorable (exergonic) process and plus (positive) values corresponding to a thermodynamically unfavorable (endergonic) process.

Although the equation $\Delta G° = -RT \ln (K_{eq})$ has a general utility for all chemical reactions, it is possible to prove that for *oxidation-reduction reactions the standard free energy change is given by*

$$\Delta G° = -n\mathscr{F}\mathscr{E}° \quad \text{or} \quad \Delta G°' = -n\mathscr{F}\mathscr{E}°'$$

where n = number of moles of electrons transferred
       $\mathscr{F}$ = Faraday's constant ($\approx$ 23,000 calories/volt)
       $\mathscr{E}°$ = net standard *oxidation-reduction potential* (volt)
       $\mathscr{E}°'$ = $\mathscr{E}°$ at pH 7

The basis of this relationship is that an oxidation-reduction system is capable of doing useful work due to the transfer of electrons from that part of the system undergoing oxidation (loss of electrons) to that undergoing reduction (gain of electrons). Such a system is called an *electrochemical cell,* and the net potential ($\mathscr{E}°$) is merely a measure of the difference between each part of the system that undergoes oxidation and reduction. Although we could continue here with an analysis of the basic principles and thermodynamics of electrochemical cells, the subject is more logically discussed in conjunction with the phenomena of respiratory electron transfer and the cellular production of useful metabolic energy in the form of ATP. Accordingly, we will defer further discussion until we consider these subjects in Chap. 14.

HIGH-ENERGY BIOMOLECULES

The preceding discussion of the general principles of thermodynamics and the free energy function in particular constitutes the first of two phases in the development of the groundwork necessary for a meaningful study of cellular energetics. The rest of the chapter will be devoted to the second phase, wherein we will consider the thermodynamic characteristics of a special group of molecules that participate in the flow of chemical energy within a cell.

After completing the next several chapters, you will discover that without question the single most important substance of cellular energetics is *adenosine triphosphate* (ATP). At this point, however, you need only identify the general features of this concept, as summarized in the following diagram. Two important points are illustrated. On an overall basis, it is stated that

1. chemical energy is conserved via the formation of ATP in association with the energy-yielding degradative reactions of catabolism, and
2. chemical energy is then utilized via the cleavage of ATP for the energy-requiring synthetic reactions of anabolism and other energy-requiring processes such as active transport and muscle contraction.

In this section our objective will be to establish why ATP is so well suited for this central role. In addition to ATP, we will also consider other compounds that fulfill supplementary but equally important roles in cellular energetics. All have one thing in common—a large amount of energy is re-

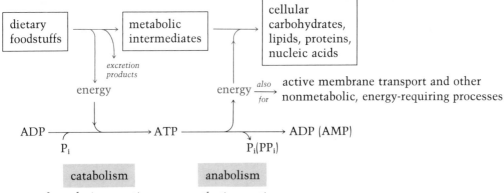

catabolism

degradative reactions

anabolism

synthetic reactions

leased when they undergo a chemical transformation. Accordingly, they are called *high-energy compounds*. It is generally accepted that the term "high-energy" implies that the amount of energy involved is at least 7,000 calories/mole. Being an arbitrary value, there is, of course, nothing sacred about it. It is merely a standard that sets ATP and a few other compounds apart from all other naturally occurring substances. The high-energy label is just a popular way of referring to these compounds as a group and implying their particular importance to the flow of energy in a living cell.

Adenosine Triphosphate

Rather than proceed with a direct description of ATP energetics, let us develop the subject by analyzing a single reaction system that is representative of ATP participation in metabolism, namely, the enzymatic conversion of glutamate and NH_3 to glutamine. This reaction is particularly important in the area of nitrogen metabolism (p. 530), but for the moment its metabolic function is not of interest. At this time we are primarily interested in analyzing the net reaction as a *model* system for learning something about ATP.

$$\underset{\text{glutamate}}{\overset{\displaystyle O \qquad\qquad O}{\text{HOCCH}_2\text{CH}_2\underset{\underset{NH_3^+}{|}}{\text{CHCO}^-}}} + NH_3 + ATP \underset{(Mg^{2+})}{\overset{\text{glutamine}}{\underset{\text{synthetase}}{\rightleftharpoons}}} \underset{\text{glutamine}}{\overset{\displaystyle O \qquad\qquad O}{\text{H}_2\text{NCCH}_2\text{CH}_2\underset{\underset{NH_3^+}{|}}{\text{CHCO}^-}}} + ADP + P_i$$

(Mg²⁺ required as a cofactor for optimal activity)

at pH 7 and 37 °C (310 °K):

$$K'_{eq} = 1.2 \times 10^3$$

therefore $\quad \Delta G^{\circ\prime} = -2.3 \, RT \log K'_{eq}$
$= -2.3 \, (1.99)(310) \log (1.2 \times 10^3)$
$= -4,370$ calories

From the value of $\Delta G^{\circ\prime}$ we conclude that the net process is exergonic, and hence the formation of glutamine from glutamate in the presence of ATP degradation is thermodynamically favorable. However, this conclusion does not in any way suggest what purpose is served by the conversion of ATP to ADP and P_i.

To resolve this question, we can look upon the net process as being composed of two separate reactions:

(1) glutamate + NH$_3$ $\longrightarrow$ glutamine + H$_2$O **(11–1)**

(2) ATP + H$_2$O $\xrightarrow{\text{Mg}^{2+}}$ ADP + P$_i$ **(11–2)**

Net: glutamate + NH$_3$ + ATP $\longrightarrow$ glutamine + ADP + P$_i$ **(11–3)**

This maneuver is conceptually permissible since a *net process can be considered as the sum of individual steps.* Moreover, in this case both of the individual steps can indeed occur separately and in the absence of any enzyme. (As a catalyst, the enzyme only accelerates the reaction. Remember, although it reduces the energy of activation, the enzyme does not alter the overall thermodynamics of the reaction.)

Equilibrium measurements on the glutamate-glutamine interconversion in the absence of ATP, ADP, and P$_i$ yielded the following result: at pH 7 and 37 °C, $K'_{eq} = 3.13 \times 10^{-3}$. Thus, for

$$\text{glutamate} + \text{NH}_3 \to \text{glutamine} + \text{H}_2\text{O}$$

$$\Delta G^{\circ\prime} = -2.3\, RT \log (3.13 \times 10^{-3})$$
$$= +3{,}550 \text{ calories}$$

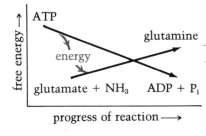

free energy →

ATP

energy

glutamine

glutamate + NH$_3$ ADP + P$_i$

progress of reaction →

Yes, the sign of the free energy change is correct, and the formation of the amide from the free acid is actually an endergonic reaction. In the presence of ATP, however, we have already established that the same reaction is exergonic. With a little thought, a conclusion as to the participation of ATP is inescapable: *the hydrolysis of ATP to ADP and P$_i$ must be sufficiently exergonic to provide the chemical energy needed to mediate the endergonic formation of glutamine.* In other words, the complete reaction involving glutamate, NH$_3$, and ATP can be considered as a coupling of a thermodynamically unfavorable reaction to a thermodynamically favorable reaction, with the latter acting as an energetic driving force of the former. In this particular case, it has been experimentally established that the *transfer of energy* is mediated via the intermediate formation of *phosphoglutamate*, as shown below. In other words, the energy released by the cleavage of a terminal phosphoryl grouping from ATP is conserved in the transfer of the phosphate group. This transfer produces a highly reactive acyl phosphate linkage (see p. 355) in phosphoglutamate in which the C of the acyl group is more susceptible to attack by the :NH$_3$ molecule than the C of the free carboxyl group. The enzyme specificity is responsible for this overall sequence, including the selection of the side-chain carboxyl group over the alpha carboxyl group as the site for phospho transfer.

ATP

$$\text{HOCCH}_2\text{CH}_2\text{CHCO}^- + {}^-\text{OP}\,\text{OPOPO-rib-A} \xrightarrow[\text{E}]{\text{Mg}^{2+}} {}^-\text{OP}-\text{O}-\text{CCH}_2\text{CH}_2\text{CHCO}^- \xrightarrow[\text{E}]{\text{Mg}^{2+}} \text{H}_2\text{N}-\text{CCH}_2\text{CH}_2\text{CHCO}^-$$

:NH$_3$

phosphotransfer

ADP

phosphoglutamate
(a highly reactive anhydride
produced as a reaction
intermediate)

P$_i$

To summarize: In our model system we have established that the $ATP + H_2O \rightarrow ADP + P_i$ conversion is a thermodynamic driving force for an otherwise unfavorable process. Although we have considered a specific chemical transformation of the living state, distinguished by specific thermodynamic values and a specific mechanism, the *theme of ATP participation applies to all ATP-dependent processes.* Many different examples will be encountered in subsequent chapters.

A more precise analysis of ATP energetics necessitates that we establish the value of the free energy change for ATP hydrolysis under physiological conditions. To make this calculation, we need look no further than the glutamate-glutamine reaction system for which the $\Delta G^{\circ\prime}$ values for glutamine formation were given in the presence and absence of ATP. Since changes in thermodynamic functions are dependent solely on the initial and final states, we can express the net standard free energy change as equal to the sum of the free energy changes corresponding to any component parts—an application of Hess's law of summation. Hence,

$$\Delta G^{\circ\prime}_{net} = \Delta G^{\circ\prime}_1 + \Delta G^{\circ\prime}_2$$

where
$$\begin{aligned}
\Delta G^{\circ\prime}_{net} &= -4{,}370 \text{ calories} &&\text{(reaction 11-3)} \\
\Delta G^{\circ\prime}_1 &= +3{,}550 \text{ calories} &&\text{(reaction 11-1)} \\
\Delta G^{\circ\prime}_2 &= ? &&\text{(reaction 11-2)}
\end{aligned}$$

We then solve for $\Delta G^{\circ\prime}_2$:

$$\begin{aligned}
\Delta G^{\circ\prime}_2 &= \Delta G^{\circ\prime}_{net} - \Delta G^{\circ\prime}_1 \\
&= -4{,}370 - (+3{,}550) \\
\Delta G^{\circ\prime}_2 &= -7{,}920 \text{ calories}
\end{aligned}$$

a value at pH 7 and 37 °C that applies to

$$-7{,}920 \text{ for } \longrightarrow$$
$$ATP + H_2O \xrightleftharpoons{Mg^{2+}} ADP + P_i$$
$$+7{,}920 \text{ for } \longleftarrow$$

Over the years there has been considerable debate as to the most accurate value of $\Delta G^{\circ\prime}$ for ATP hydrolysis, with reported values ranging from $-7{,}000$ to $-9{,}000$ calories. The chief difficulty is that the hydrolysis of ATP is quite sensitive to changes in temperature, pH, and Mg^{2+} concentration. (Details can be found in the research paper by Alberty cited at the chapter end. It is advanced reading, however.) The issue is even fuzzier when one considers what the value is in a living cell, where the environment does change, and the concentrations are anything but those that apply to standard conditions. It is estimated that *in vivo* a value of $-11{,}000$ or more may be more realistic. Whatever, we will not bicker over this point but rather make routine use of a $-8{,}000$ figure in all future reference to ATP. (We will also use the same value for the hydrolysis of ATP to AMP and PP_i and also for the hydrolysis of PP_i to $2P_i$.)

To conclude our analysis of ATP, we can ask one final question: *why* is ATP such a thermodynamically unstable compound? Or, to ask the same question differently: why is a large amount of energy required for the formation of ATP from ADP and P_i? In general terms, the answer to each is that the ATP molecule must contain sufficient chemical energy to alleviate structural features that tend to act as destabilizing forces. In particular, two such forces can be identified in the triphosphoanhydride grouping: (1) elec-

this is R below

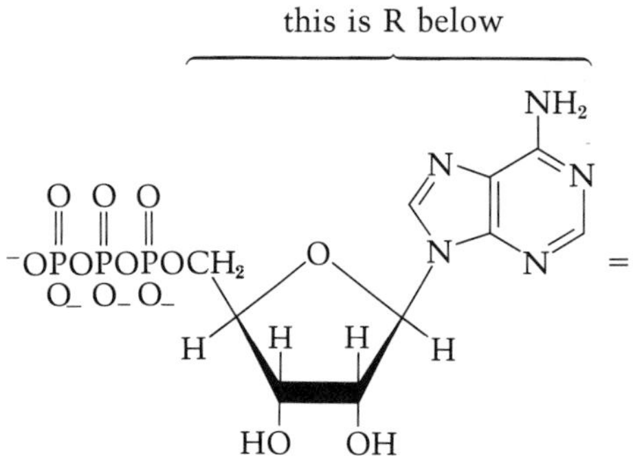

NH_2

$^-OPOPOPOCH_2$

ATP^{-4} (tetraanion at pH 7)

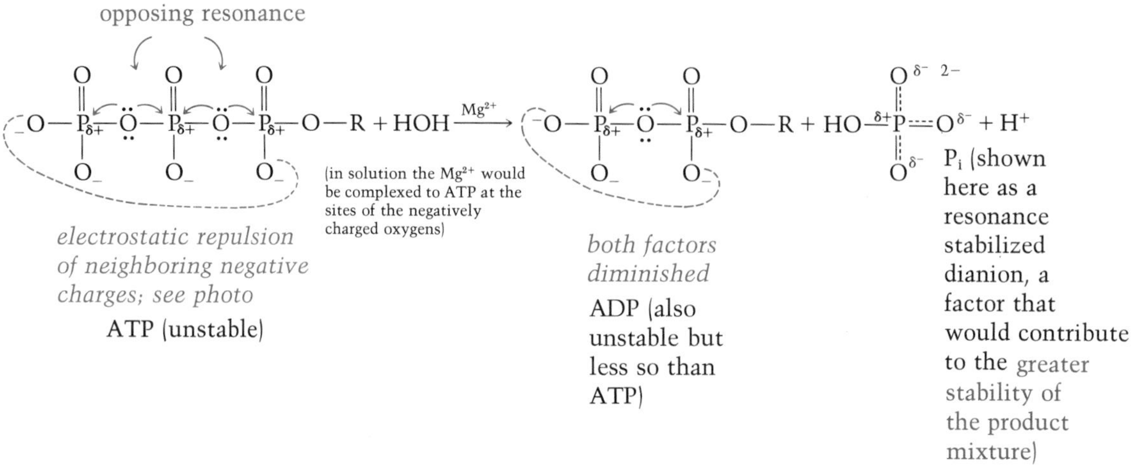

opposing resonance

electrostatic repulsion
of neighboring negative
charges; see photo

ATP (unstable)

(in solution the Mg²⁺ would
be complexed to ATP at the
sites of the negatively
charged oxygens)

both factors
diminished

ADP (also
unstable but
less so than
ATP)

P$_i$ (shown
here as a
resonance
stabilized
dianion, a
factor that
would contribute
to the greater
stability of
the product
mixture)

 trostatic repulsion and (2) opposing resonance. The potential for electrostatic repulsion is readily apparent if we recognize that, at physiological pH, ATP probably exists as a tetraanion due to the ionization of the four —P—O—H bonds. The result is a spatial localization of like negative charges that have a natural tendency to repel each other. This repulsion constitutes a structural strain on the whole molecule, but the —P—O—P— bonds are most affected. The principle of opposing resonance argues that a potential competition will exist between successive phosphorus atoms for the available unshared electrons of the sandwiched oxygen atoms. The competition is established by the presence of a partial positive charge on each P atom resulting from the polarization of each P=O bond. Although the electrostatic stress is probably more significant (particularly under physiological conditions), the important point to realize is that the very existence of ATP is dependent on the presence of sufficient chemical energy within the molecule to overcome these physiocochemical stresses. The energy is not confined to any one particular bond in ATP, but is distributed among several. When ATP is converted to ADP and P$_i$, the stresses are minimized (that is to say, the products are more stable) and energy is released. The hydrolysis is favored not only by this inherent instability of ATP, but also by the stability conferred on the product state by the resonance of the dianion species of inorganic phosphate.

Finally, it should be noted that the energetics of ATP hydrolysis are not significantly dependent on the presence of an adenine grouping. That is to say, all of the purine and pyrimidine triphosphonucleotides would have a very similar $\Delta G^{\circ\prime}$ value for their hydrolysis, so that, thermodynamically speaking, ATP, GTP, UTP, and all of the others are considered to be identical. (Similar arguments would apply to the hydrolysis of ATP → AMP + PP$_i$ and PP$_i$ → 2P$_i$ both of which exhibit energy changes comparable to ATP → ADP + P$_i$.

The list of high-energy compounds is not restricted to nucleoside triphosphates. Other naturally occurring substances that are given the same designation are *acyl phosphates, enoyl phosphates, thioesters, phosphoguanidines,* and the *nicotinamide adenine dinucleotides.* An example of each type, along with a reference to illustrate its participation in metabolism, is given in Table 11–1. Of the substances listed, those with a more general importance are the acylphosphates, the thioesters, and the nicotinamide adenine dinucleotides. Each is described further in the following sections.

Table 11–1 appears on p. 356.

Acyl Phosphates

An acyl phosphate (see margin) is a highly reactive anhydride linkage wherein the carbonyl carbon of the acyl group is particularly susceptible to reaction with an attacking nucleophile. The $\Delta G°'$ value of about −12,000 calories, obtained when the anhydride is hydrolyzed (water is the nucleophile), is clear evidence of this instability. One specific example of biological importance is *1,3-diphosphoglycerate* (see p. 387, next chapter), which serves as a precursor of ATP formation. Another important example of a whole family of acyl phosphates is the *amino acyl-AMP* derivatives of amino acids, which we previously encountered in Chap. 8 (see p. 262; also see p. 103 in Chap. 4) where the reacting nucleophile was a terminal OH group of transfer-RNA. *Note:* The formation of amino acyl-AMP derivatives also illustrates the use of energy in ATP.

acyl phosphate

increased reactivity

A final look at the flow of energy in terms of ATP participation is given below. Gradually you will learn to appreciate that this diagram summarizes in large part the logic of metabolism. It will also become evident to you that the overall metabolism of a cell is regulated in large part by the intracellular concentrations of ATP, ADP, and AMP. D. Atkinson has termed this regulation *energy-charge control.* Energy charge is an expression of the relative concentrations of the three adenylate-containing (AMP-containing) species. As Atkinson has defined it,

adenylate kinase is an important enzyme responsible for recycling AMP; the conversion to ADP uses ATP as a donor of phosphate

AMP + ATP ⟶ 2ADP

$$\text{energy charge} = \frac{[ATP] + \frac{1}{2}[ADP]}{[ATP] + [ADP] + [AMP]}$$

	System	$\Delta G^{\circ\prime}(pH\ 7)$ calories/mole

Hydrolysis of triphosphonucleotides and pyrophosphate:

$$\text{ATP} \xrightarrow[\text{H}_2\text{O}]{} \text{ADP} + \text{P}_i \qquad -8{,}000$$

By comparison, the hydrolysis of phosphoesters is typified by a $\Delta G^{\circ\prime}$ value of about $-3{,}000$ to $-4{,}000$ calories. They are still reactive derivatives of high energy but less so than the materials listed here.

$$\text{ATP} \xrightarrow[\text{H}_2\text{O}]{} \text{AMP} + \text{PP}_i \qquad \text{ca. } -8{,}000$$

$$\underset{\text{PP}_i}{{}^{-}\text{O}-\overset{\overset{\text{O}}{\|}}{\underset{\underset{\text{O}_{-}}{|}}{\text{P}}}-\text{O}-\overset{\overset{\text{O}}{\|}}{\underset{\underset{\text{O}_{-}}{|}}{\text{P}}}-\text{O}^{-}} \xrightarrow[\text{H}_2\text{O}]{} \underset{2\text{P}_i}{2\ \text{HO}-\overset{\overset{\text{O}}{\|}}{\underset{\underset{\text{O}_{-}}{|}}{\text{P}}}-\text{O}_{-}} \qquad \text{ca. } -8{,}000$$

Hydrolysis of acyl phosphates:

$$\text{R}-\overset{\overset{\text{O}}{\|}}{\text{C}}-\text{OPO}_3{}^{2-} \xrightarrow[\text{H}_2\text{O}]{} \text{R}-\text{COO}^- + \text{P}_i \qquad -11{,}800$$

$$\text{1,3-diphosphoglycerate} \xrightarrow[\text{H}_2\text{O}]{} \text{3-phosphoglycerate} + \text{P}_i \quad \text{(see p. 373)}$$

Hydrolysis of enoyl phosphates:

$$\text{RCH}{=}\underset{\underset{\text{OPO}_3{}^{2-}}{|}}{\text{C}}-\text{COO}^- \xrightarrow[\text{H}_2\text{O}]{} \text{RCH}_2-\overset{\overset{}{\underset{\underset{\text{O}}{\|}}{\text{C}}}}{}-\text{COO}^- + \text{P}_i \qquad -14{,}800$$

$$\text{phosphoenolpyruvate} \xrightarrow[\text{H}_2\text{O}]{} \text{pyruvate} + \text{P}_i \quad \text{(see p. 373)}$$

Hydrolysis of acyl thioesters of coenzyme A:

$$\text{R}-\overset{\overset{\text{O}}{\|}}{\text{C}}-\text{SCoA} \xrightarrow[\text{H}_2\text{O}]{} \text{R}-\overset{\overset{\text{O}}{\|}}{\text{C}}-\text{O}^- + \text{HSCoA} \qquad -7{,}370$$

$$\text{acetyl-SCoA} \xrightarrow[\text{H}_2\text{O}]{} \text{acetate} + \text{HSCoA} \quad \text{(see text, this chapter)}$$

Hydrolysis of guanidinium phosphates:

$$\text{R}-\underset{\underset{{}^{+}\text{NH}_2}{\|}}{\text{CH}_2\text{NHCNHPO}_3{}^{2-}} \xrightarrow[\text{H}_2\text{O}]{} \text{RCH}_2\underset{\underset{{}^{+}\text{NH}_2}{\|}}{\text{NHCNH}_2} + \text{P}_i \qquad -10{,}300$$

$$\text{creatine phosphate} \xrightarrow[\text{H}_2\text{O}]{} \text{creatine} + \text{P}_i \quad \text{(see p. 402)}$$

Oxidation of nicotinamide adenine nucleotides:

$$\underset{\text{reduced forms}}{\text{NADH(or NADPH)} + \text{H}^+} \xrightarrow[\frac{1}{2}\text{O}_2]{} \underset{\text{oxidized forms}}{\text{NAD}^+(\text{or NADP}^+) + \text{H}_2\text{O}}$$

$-52{,}600$ (with O_2 as oxidizing agent)

His argument is that living cells are capable of maintaining an energy charge value within the narrow range of about 0.8–0.9 (see margin). When the energy charge falls below 0.8, signaling a drop in the cellular ATP level due to an increased rate of energy consumption, the cell responds by increasing the rate of ATP-yielding metabolic reactions. On the other hand, when the energy charge exceeds 0.9, signaling a rise in the cellular ATP level due to a diminished rate of energy consumption, the cell responds by decreasing the rate ATP-yielding reactions. We will study the regulation of metabolism in these and other terms throughout future chapters.

The maximum value of energy charge is 1.0, a state where [ADP] and [AMP] both equal zero and only ATP exists. A cell in such a state would be saturated with energy. The minimum value of energy charge is 0, a state where [ADP] and [ATP] both equal zero and only AMP exists. A cell in such a state would be starved for energy. In living cells these extremes are avoided by the regulation of metabolism.

Thioesters

Thioesters (see margin), particularly those involving *coenzyme A* (symbolized CoASH or HSCoA), play a very important role in metablism; namely,

they serve as the *metabolically active form of acyl* $\left(R-\overset{\displaystyle O}{\underset{\displaystyle \|}{C}}- \right)$ *groups*. Indeed, the bulk of enzyme reactions involving either the transfer of acyl groups, or the chemical modification of acyl groups, require the acyl group to be in the form of an acyl-SCoA ester. The simplest and most common species is *acetyl-SCoA*.

Although the structure of HSCoA is complex, note that the reactive functional group is the free *sulfhydryl group*. (In those organisms capable of synthesizing coenzyme A, the SH group originates from cysteine—see p. 547, Chap. 17.) Also noteworthy is that part of the HSCoA molecule is composed of *β-alanine* and *pantothenic acid*, the latter being an essential *vitamin* for mammals. As indicated, living cells must expend ATP energy

thioester

$$R-\overset{\displaystyle O}{\underset{\displaystyle \|}{C}}-S-R$$

activated
acyl
group

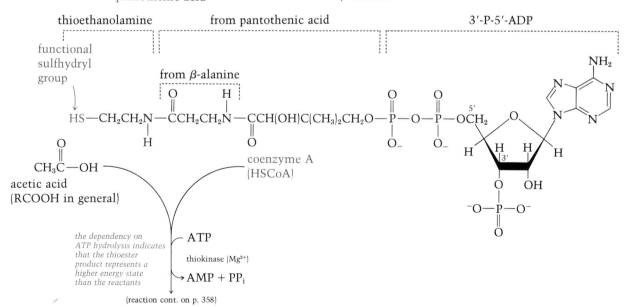

pantothenic acid

β-alanine

coenzyme A
(HSCoA)

acetic acid
(RCOOH in general)

the dependency on ATP hydrolysis indicates that the thioester product represents a higher energy state than the reactants

ATP

thiokinase (Mg²⁺)

AMP + PPᵢ

(reaction cont. on p. 358)

357

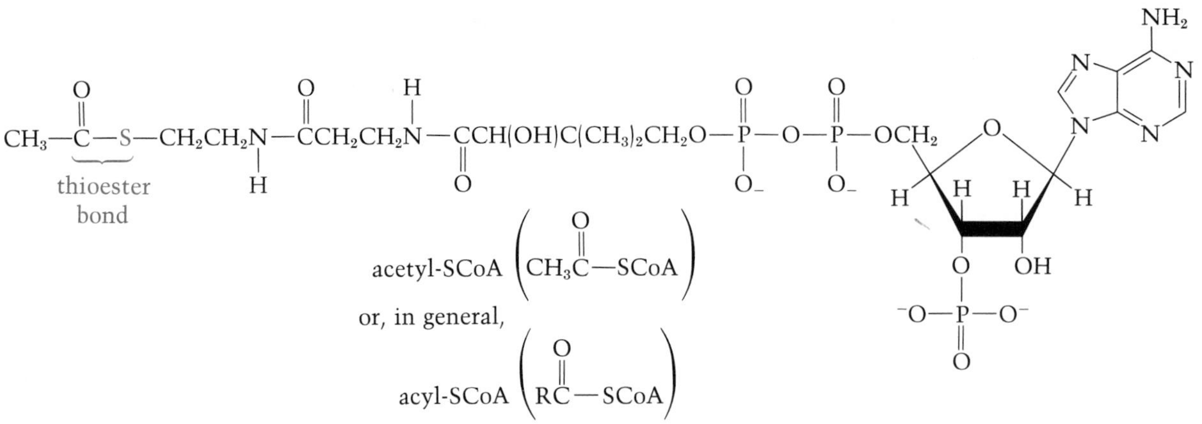

thioester bond

acetyl-SCoA $\left(CH_3\overset{O}{\underset{}{C}}-SCoA\right)$

or, in general,

acyl-SCoA $\left(R\overset{O}{\underset{}{C}}-SCoA\right)$

to produce the thioester CoA-derivative of a free acid—a role of ATP that should make perfect sense to you. The reaction is catalyzed by a *thiokinase* enzyme.

It is logical that the metabolically active form of acyl groups be the thioesters simply because thioesters are more chemically reactive than oxyesters. The sites in the acyl grouping most susceptible to reaction are the *carbonyl carbon* and its *alpha carbon* atom. The carbonyl carbon is more susceptible to a substitution reaction with a nucleophilic (electron-donating) grouping and the alpha carbon is more susceptible to a condensation reaction with an electrophilic (electron-seeking) grouping. (Still other types of reactions will be encountered when we discuss the metabolism of fatty acids.) Two of the important cellular reactions of acyl-SCoA compounds in general, and acetyl-SCoA in particular, are shown below to illus-

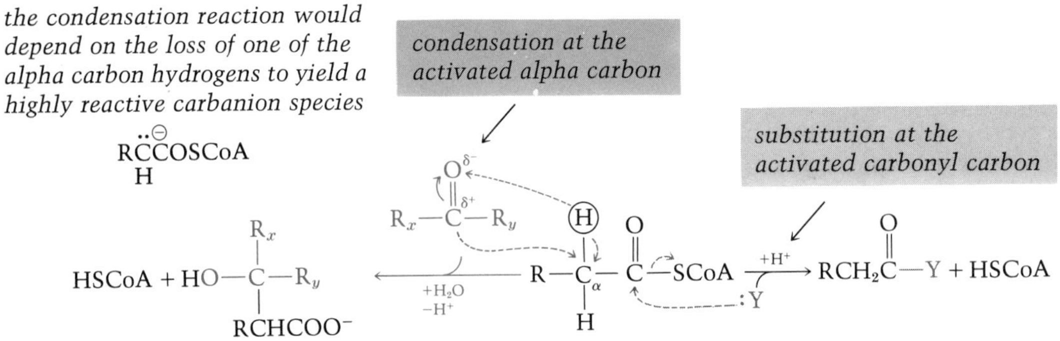

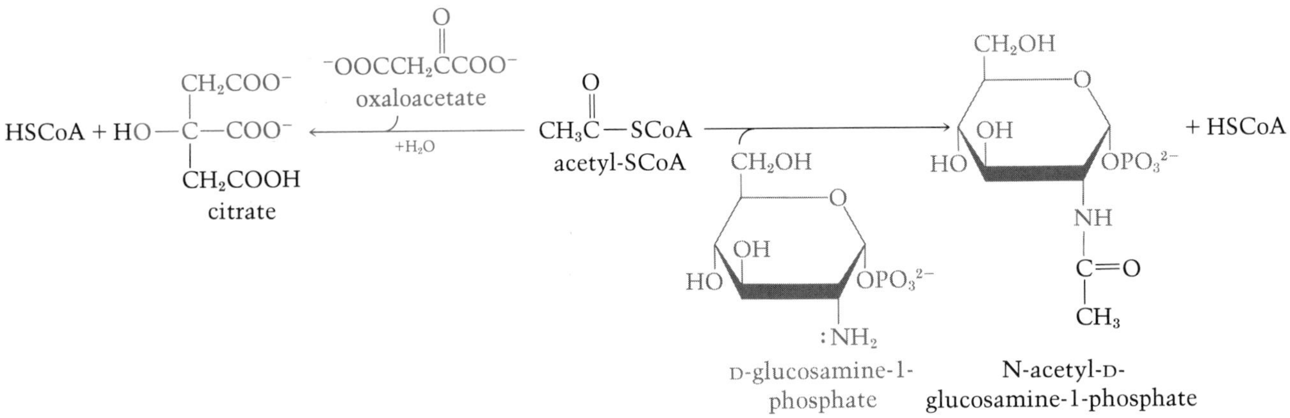

the condensation reaction would depend on the loss of one of the alpha carbon hydrogens to yield a highly reactive carbanion species

trate each type of reaction. For simplicity the enzymes that catalyze each reaction are not shown. The formation of citrate from acetyl-SCoA and oxaloacetate is the first reaction of an extremely vital metabolic pathway, the citric acid cycle (see Chap. 13).

Nicotinamide Dinucleotides

The nicotinamide dinucleotides represent another class of high-energy compounds of special biochemical importance. To understand why they are so classified, let us first examine their structure and metabolic function. Two distinct types exist in nature—*nicotinamide adenine dinucleotide* (symbolized *NAD*) and *nicotinamide adenine dinucleotide phosphate* (symbolized *NADP*). (According to an older and now less used convention, they were termed pyridine nucleotides and called diphosphopyridine nucleotide (DPN) and triphosphopyridine nucleotide (TPN), respectively.) As shown below, the structural difference between NAD and NADP is rather subtle, to say the least; NADP contains an extra phosphate group at $C^{2'}$ of the ribose unit attached to adenine.

Although the line drawings of these materials tend to convey a complex structure, they are not complex at all. Actually, there is only one part of the structure that we haven't yet encountered, namely, the *nicotinamide* moiety, which is a substituted pyridine. The remainder is composed of adenine, β-D-ribose, and phosphate. Moreover, the linkages are familiar. In fact, as the name implies, the molecules can be considered as being composed of two nucleotide units linked together by a phosphodiester bridge (hence, the name dinucleotide). The nicotinamide component is especially important for two reasons. First, nicotinamide is a derivative of the *vitamin nicotinic acid*, sometimes called *niacin*. Since nicotinamide does

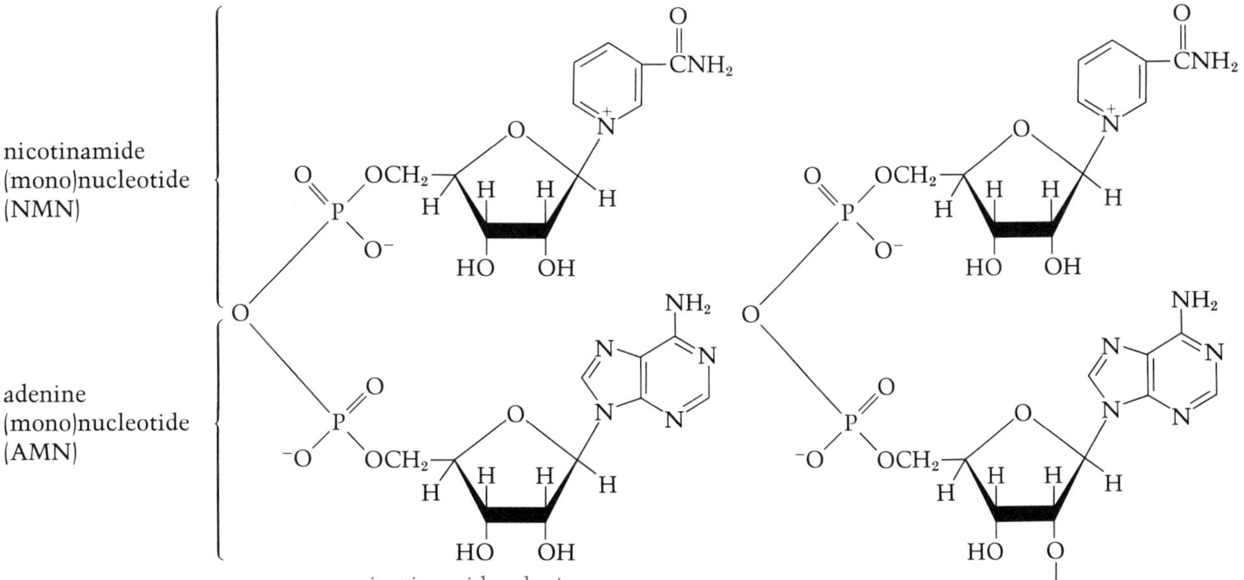

nicotinamide (mono)nucleotide (NMN)

adenine (mono)nucleotide (AMN)

nicotinamide adenine dinucleotide (NAD$^+$)

(see photo)

nicotinamide adenine dinucleotide phosphate (NADP$^+$)

359

not generally occur in the free state but largely in glycosidic linkage with ribose in NAD and NADP, the latter are assumed to be the *metabolically active forms of the vitamin.* Second, and more important, is the fact that the pyridine ring of nicotinamide comprises the active site of NAD and NADP (see below). That is to say, it is the pyridine ring that actually functions in the reversible transfer of electrons and a proton.

On a functional basis, both NAD and NADP are classified as coenzymes for a large number of *dehydrogenase* enzymes; they act as agents in the transfer of hydrogen atoms in oxidation-reduction reactions. Accordingly, two forms of each coenzyme exist, an *oxidized form* and a *reduced form*. The structures given above are of the oxidized species, and are normally symbolized as NAD^+ and $NADP^+$ due to the positive charge on the pyridine N atom; the reduced forms are symbolized as NADH and NADPH. In the presence of a reduced substrate (hydrogen donor) and an appropriate dehydrogenase, the *pyridine ring is reduced* by accepting (always at position 4) the equivalent of one proton (H^+) and two electrons, with the second proton released to the medium. Accordingly, NAD^+ ($NADP^+$) is generally termed an *electron acceptor* (or *hydrogen acceptor*), although the actual group transferred is a hydride ion $H:^-$. In the reverse type of process, NADH (NADPH) would act as an *electron donor* (or *hydrogen donor*).

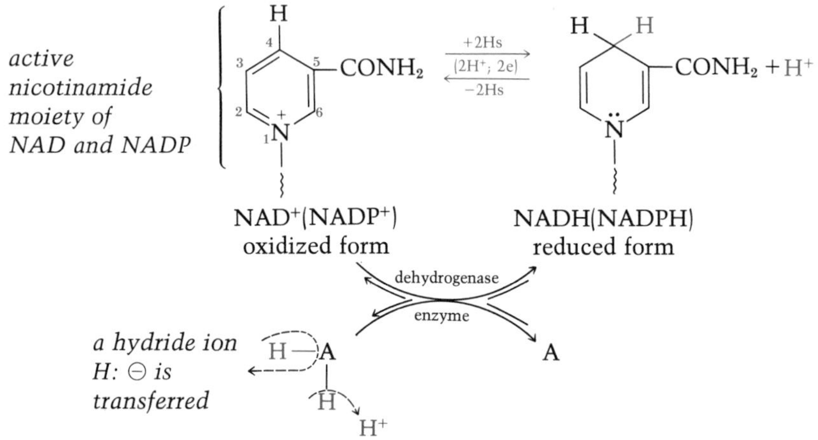

No doubt you are wondering what all this has to do with bioenergetics and high-energy compounds. Well, the explanation is that, once they are produced in the cell, the primary metabolic fate of the reduced forms, especially NADH, is to be reoxidized as the first step in a series of consecutive oxidation-reduction reactions that terminate with the reduction of molecular oxygen. Although this also applies to NADPH, its primary metabolic fate is to serve as a hydrogen donor in many biosynthetic reactions (see p. 367, Chap. 12). The complete process, termed *electron transport*, is most important because the great majority of organisms in the biosphere utilize this operation or a variation of it (using a terminal electron acceptor other than molecular oxygen) as the main source of energy needed for the intracellular formation of ATP. In other words, it constitutes the major means by which a cell converts the chemical energy of exogenously supplied foodstuffs to useful metabolic energy.

Still, why are NADH and NADPH to be considered as high-energy compounds? Because we are deferring discussion of oxidation-reduction potentials until Chap. 14, our approach to this question will necessarily be limited at this point. For now, suffice it to say that whenever an oxidation-reduction reaction is composed of a reducing system and an oxidizing system that are separated by a positive potential difference ($\mathscr{E}^\circ_\text{net}$ is +), the reaction will be exergonic, with the amount of released energy dependent on the magnitude of the potential difference. As a statement of fact (to be discussed later in Chap. 14), under physiological conditions (pH 7; 37 °C), the overall potential difference between the initial reducing half-reaction (NADH $\rightarrow$ NAD$^+$ + 2H$^+$ + 2e) or (NADPH $\rightarrow$ NADP$^+$ + 2H$^+$ + 2e), and the terminal oxidizing half-reaction ($\frac{1}{2}$O$_2$ + 2H$^+$ + 2e $\rightarrow$ H$_2$O), is +1.14 volts. We can appreciate the significance of this number by using the equation on p. 350 of this chapter to determine that $\Delta G^{\circ\prime} \approx -53,000$ calories ($n = 2$ for the transfer of two electrons). Hence, the process is extremely exergonic and we conclude that the *reduced* nicotinamide dinucleotides represent high-energy compounds. The component steps of electron transfer, and the means by which the released energy is conserved by a cell through the coupled formation of ATP, comprise one of the most fascinating biochemical processes in nature. The complete coupled process is termed *oxidative phosphorylation*. Many of the details have been established and will be discussed in Chap. 14.

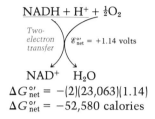

$$\text{NADH} + \text{H}^+ + \tfrac{1}{2}\text{O}_2$$

Two-electron transfer $\quad \mathscr{E}^{\circ\prime}_\text{net} = +1.14 \text{ volts}$

$$\text{NAD}^+ \qquad \text{H}_2\text{O}$$

$$\Delta G^{\circ\prime}_\text{net} = -(2)(23,063)(1.14)$$
$$\Delta G^{\circ\prime}_\text{net} = -52,580 \text{ calories}$$

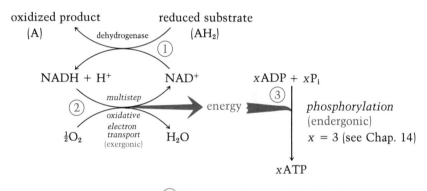

① formation of NADH by dehydrogenation of reduced substrate

② + ③ oxidative phosphorylation

② oxidation of NADH by electron transport, with oxygen as the terminal electron acceptor

③ phosphorylation of ADP to yield ATP

COUPLING PHENOMENON

As previously stated, the *coupling principle states that an energetically unfavorable reaction is driven to completion by an energetically favorable reaction.* This principle is a theme through all of metabolism, explaining (a) the efficient conservation and utilization of metabolic chemical energy, and (b) the efficient flow of metabolic intermediates through pathways composed of several consecutive reactions. Throughout the next few chapters, we will encounter many specific examples of both points, but in order to recognize and appreciate their metabolic significance, we must have an understanding of the nature of coupled reactions. Accordingly, we conclude this chapter by focusing on this subject. In so doing, we will briefly reiterate some of the previous material, but we will also develop other aspects not yet mentioned.

Endergonic and exergonic reactions can be coupled in either of two ways, which differ in terms of whether or not there is an actual transfer of energy. Reactions involving an actual transfer of energy occur by one of two mechanisms (Case I and Case II below).

In Case I, where a single reaction is catalyzed by a single enzyme, the energy transfer is effected through the formation of a high-energy intermediate during the course of the conversion of the primary reactant to the primary product. This typifies many ATP-dependent reactions, such as the formation of glutamine and acetyl-SCoA (pp. 351 and 357 of this chapter).

CASE I (ATP dependent). Energy transfer during a single reaction catalyzed by a single enzyme:

$$X + Y + ATP \xrightarrow{E} X—Y + AMP + PP_i$$

or

$$X + Y + ATP \xrightarrow{E} X—Y + ADP + P_i$$

Here the energetically unfavorable process of $X + Y \rightarrow X—Y$ with a $+\Delta G°'$ is coupled to the energetically favorable process of ATP degradation (to AMP and PP_i or ADP and P_i) with a $-\Delta G°'$ via a single reaction in two steps:

Step 1: $$X + ATP \xrightarrow{E_1} [X—AMP] + PP_i$$

Step 2: $$[X—AMP] + Y \xrightarrow{E_1} X—Y + AMP$$

(where X—AMP is a high-energy reaction intermediate)

Net: $$X + Y + ATP \xrightarrow{E_1} X—Y + AMP + PP_i$$

EXAMPLES: $$acetate + CoASH + ATP \longrightarrow acetyl\text{-}SCoA + AMP + PP_i$$

$$amino\ acid + transfer\text{-}RNA + ATP \longrightarrow aminoacyl\text{-}tRNA + AMP + PP_i$$

or

Step 1: $$X + ATP \xrightarrow{E_1} [X—P] + ADP$$

Step 2: $$[X—P] + Y \xrightarrow{E_1} X—Y + P_i$$

(where X—P is a high-energy reaction intermediate)

Net: $$X + Y + ATP \xrightarrow{E_1} X—Y + ADP + P_i$$

EXAMPLE: $$glutamate + NH_4^+ + ATP \longrightarrow glutamine + ADP + P_i$$

This type of coupling is also related to the kinase-catalyzed formation of phosphoesters:

EXAMPLE: $$ROH + ATP \xrightarrow{kinase} ROPO_3^{2-} + ADP$$

$$glucose + ATP \xrightarrow{glucokinase} glucose\text{-}6\text{-}phosphate + ADP$$

A variation of this theme (Case II) occurs when the net reaction results from the consecutive occurrence of two (or more) distinct reactions catalyzed by two (or more) separate enzymes. Here the common intermediate is, in fact, the primary product of one reaction and the primary reactant of the second. The biosynthesis of sucrose is representative of this form of coupling.

CASE II (ATP dependent). Energy transfer accompanying two (or more) separate reactions catalyzed by separate enzymes:

$$X + Y + ATP \xrightarrow{E_1} \xrightarrow{E_2} X-Y + ADP + P_i$$

Here the energetically unfavorable process of $X + Y \rightarrow X-Y$ with a $+\Delta G^{\circ\prime}$ is coupled to the energetically favorable process of ATP degradation (to AMP and PP_i or ADP and P_i) with a $-\Delta G^{\circ\prime}$ via separate reactions:

Reaction 1: $\quad\quad\quad\quad\quad X + ATP \xrightarrow{E_1} X-P + ADP$

Reaction 2: $\quad\quad\quad\quad\quad X-P + Y \xrightarrow{E_2} X-Y + P_i$

Net: $\quad\quad\quad\quad\quad\quad X + Y + ATP \xrightarrow{E_1} \xrightarrow{E_2} X-Y + ADP + P_i$

EXAMPLE (with three enzymes):

$$glucose + ATP \xrightarrow{E_1} glucose\text{-}6\text{-}phosphate + ADP$$

$$glucose\text{-}6\text{-}phosphate \xrightarrow{E_2} glucose\text{-}1\text{-}phosphate$$

$$glucose\text{-}1\text{-}phosphate + fructose \xrightarrow{E_3} sucrose + P_i$$

Net: $glucose + fructose + ATP \xrightarrow{E_1} \xrightarrow{E_2} \xrightarrow{E_3} sucrose + ADP + P_i$

In reactions where an actual energy transfer does not occur (Case III), the coupling is explained purely in terms of equilibrium considerations. Here one of the characteristic differences from the previous two types is that ATP is not involved. However, as in Case II, we are dealing with consecutive reactions catalyzed by different enzymes. As indicated, the unfavorable equilibrium of one reaction can be displaced if the product of that reactant then serves as a substrate (reactant) with a strong tendency to be converted to another material in a subsequent reaction.

CASE III (ATP independent). No actual transfer of energy in two (or more) separate reactions catalyzed by separate enzymes:

$$X \xrightarrow{E_1} Y \xrightarrow{E_2} Z$$

Here an energetically unfavorable process of $X \rightarrow Y$ with a $+\Delta G^{\circ\prime}$ is driven to completion by being coupled to the energetically favorable process of $Y \rightarrow Z$ with a $-\Delta G^{\circ\prime}$:

Reaction 1: $\quad X \xrightleftharpoons{E_1} Y \quad\quad K_{eq_1} << 1 \quad \Delta G_1^{\circ\prime} = +$

Reaction 2: $\quad Y \xrightleftharpoons{E_2} Z \quad\quad K_{eq_2} >>> 1 \quad \Delta G_2^{\circ\prime} = -$

Net: $\quad\quad X \xrightarrow{E_1} \xrightarrow{E_2} Z \quad\quad K_{eq_{net}} > 1 \quad \Delta G_{net}^{\circ\prime} = -$

EXAMPLE: $\quad NAD^+ + malate \xrightarrow{E_1} oxaloacetate + NADH(H^+) \quad \Delta G^{\circ\prime} = +6.7$ kcal

$\quad\quad oxaloacetate + acetyl\text{-}SCoA \xrightarrow{E_2} citrate + HSCoA \quad\quad\quad \Delta G^{\circ\prime} = -9$ kcal

Net: $NAD^+ + malate + acetyl\text{-}SCoA \xrightarrow{E_1} \xrightarrow{E_2}$

$$citrate + HSCoA + NADH(H^+) \quad \Delta G_{net}^{\circ\prime} = -2.3 \text{ kcal}$$

LITERATURE

KLOTZ, I. M. *Energy Changes in Biochemical Reactions.* New York: Academic Press, 1967. A small monograph (available in paperback) on the fundamental concepts of thermodynamics as they apply to biological systems. Written by a chemist for the biologist.

LEHNINGER, A. L. *Bioenergetics.* 2nd edition. New York: W. A. Benjamin, 1971. A superb introductory book (available in paperback) emphasizing the energy relationships in living cells at the metabolic level. Allosteric regulation and recent advances in the molecular organization of membranes, mitochondria, and chloroplasts are also included.

MONTGOMERY, R., and C. A. SWENSON. *Quantitative Problems in the Biochemical Sciences.* 2nd edition. San Francisco: W. H. Freeman and Company, 1976. Chapter 10 summarizes basic principles of biochemical energetics and applies them in problem solving.

MORRIS, J. G. *A. Biologist's Physical Chemistry.* Reading: Addison-Wesley, 1968. Chapters 7, 8, and 9

contain a lucid discussion of the basic principles of classical thermodynamics and their application to biochemical systems.

VAN HOLDE, K. E. *Physical Biochemistry.* Englewood Cliffs, New Jersey: Prentice-Hall, 1971. A more sophisticated though not overpowering treatment of thermodynamics applied to biochemical systems. Available in paperback.

WILLIAMS, V. R., W. L. MATTICE, and H. B. WILLIAMS. *Basic Physical Chemistry for the Life Sciences.* 3rd edition. San Francisco: W. H. Freeman, 1978. Chapters 2 and 3 contain thorough presentations of the principles of thermodynamics and the free-energy concept as they apply to biochemical and biological systems.

See the literature listing at the end of Chap. 13 for a reference to the hypothesis of Atkinson for metabolic regulation via the adenylate energy charge.

EXERCISES

11–1. The standard free-energy change for the hydrolysis of glucose-1-phosphate at pH 7 and 37 °C has been measured as $-5,000$ calories/mole. Calculate the equilibrium constant for this reaction.

$$\text{glucose-1-phosphate} + H_2O \rightarrow \text{glucose} + P_i$$

11–2. On the basis of material presented in this chapter and the principle of the relationship between structure and function, would you predict the $\Delta G^{\circ\prime}$ for the reaction below to be approximately (a) $+3,600$ calories/mole, (b) $-3,600$ calories/mole, (c) $+6,800$ calories/mole, (d) $-6,800$ calories/mole, or (e) $+1,875$ calories/mole?

$$\text{asparagine} + H_2O \rightarrow \text{aspartate} + NH_3$$

11–3. The standard free-energy change for the hydrolysis of glucose-6-phosphate at pH 7 and 25 °C has been measured as $-3,300$ calories/mole. Given this and the information in Exercise 11–1, calculate the $\Delta G^{\circ\prime}$ for the following reaction at pH 7 and 37 °C.

$$\text{glucose-1-phosphate} \rightarrow \text{glucose-6-phosphate}$$

11–4. Which of the reactions given below would be likely to be coupled to the formation of ATP from ADP

	$\Delta G^{\circ\prime}$ (cal)	K_{eq}
(a) phosphoenolpyruvate + H₂O → pyruvate + Pᵢ	–	2.5×10^{10}
(b) 3-phosphoglycerate → 2-phosphoglycerate	–	1.8×10^{-1}
(c) fructose-6-phosphate + H₂O → fructose + Pᵢ	$-3,200$	–
(d) succinyl-SCoA + H₂O → succinate + HSCoA	$-11,000$	–

and Pᵢ? (Assume a pH of 7 and temperature of 37 °C apply to both $\Delta G^{\circ\prime}$ and K_{eq}.)

11–5. The overall standard free-energy change for the reaction

$$\text{pyruvate} + \text{ATP} + CO_2 \rightarrow \text{oxaloacetate} + P_i + \text{ADP}$$

is 1,100 calories/mole. From this information, calculate the $\Delta G^{\circ\prime}$ for the reaction

$$\text{pyruvate} + CO_2 \rightarrow \text{oxaloacetate}$$

(Assume a pH of 7 and temperature of 37 °C for all $\Delta G^{\circ\prime}$ values.) Classify the coupled reaction involving ATP as Case I, Case II, or Case III.

11–6. Dihydroxyacetone phosphate (DHAP) is one of the principal intermediates produced during the degradation of hexoses such as glucose. Under anaerobic conditions, certain bacteria can produce glycerol from glucose due to the action of two enzymes that catalyze first the reduction of dihydroxyacetone phosphate to glycerol-1-phosphate, and then the hydrolysis of glycerol phosphate to yield free glycerol and inorganic phosphate.

	$\Delta G^{\circ\prime}$ (cal)
dihydroxyacetone phosphate + 2Hs → glycerol-1-phosphate	$+8,750$
glycerol-1-phosphate → glycerol + Pᵢ	$-2,400$

Calculate whether the overall sequence from DHAP to glycerol is endergonic or exergonic. Is the overall sequence an example of coupling according to Case I, Case II, or Case III?

11–7. In the cell, the enzyme (a dehydrogenase) that catalyzes the reduction of DHAP to glycerol-1-phosphate utilizes NADH as the source of reducing power. The NADH is in turn oxidized to NAD^+. The complete reaction is

DHAP + NADH + H$^+$ → glycerol-1-phosphate + NAD$^+$

Given that the $\Delta G°'$ for the oxidation of NADH to NAD$^+$ is $-14,760$ calories and the information in Exercise 11–6, calculate the $\Delta G°'$ for the formation of glycerol phosphate according to this reaction. Compare this value to that given in the preceding exercise and explain the difference. What effect does the NADH-dependent reaction of DHAP have on the energetics of the overall conversion of DHAP to glycerol as it occurs in the cell?

11–8. The steady-state concentration of ATP in a red blood cell has been estimated to be approximately 13 times greater than that of ADP. In addition, the concentration of inorganic phosphate has been estimated to be approximately eight times greater than that of ADP. Given this information calculate the value of the free energy change ($\Delta G'$) that would apply to ATP hydrolysis in the red blood cell.

11–9. A buffered (pH 7) solution containing phosphoenolpyruvate (30 millimoles), glucose (20 millimoles), and small amounts of adenosine diphosphate, pyruvate kinase, and hexokinase was incubated at 37 °C. Given the information below, (a) explain what will occur during the incubation process, particularly in the early stages; (b) write the net reaction that would occur; and (c) predict whether or not the net reaction would proceed to completion (i.e., there would be little or none of the original substrates remaining).

$$\text{phosphoenolpyruvate + ADP} \xrightarrow{\text{pyruvate kinase}} \text{pyruvate + ATP}$$

$$\text{glucose + ATP} \xrightarrow{\text{hexokinase}} \text{glucose-6-phosphate + ADP}$$

11–10. Given the thermodynamic information in this chapter, calculate the theoretical yield of ATP from the oxidation of NADH(H$^+$) during the coupled process of oxidative phosphorylation.

11–11. Propose an explanation for the difference between the $\Delta G°$ values for the hydrolysis of glucose-1-phosphate ($-5,000$ calories/mole at pH 7 and 25 °C) and glucose-6-phosphate ($-3,300$ calories/mole, also at pH 7 and 25 °C).

11–12. The $\Delta G°'$ (at pH 7) for the hydrolysis of sucrose to glucose and fructose is $-7,000$ calories/mole.

$$\text{sucrose + H}_2\text{O} \rightarrow \text{glucose + fructose}$$

Given this information, the information in Exercises 11–1 and 11–3, and the answer to Exercise 11–3, calculate the $\Delta G°'$ for the formation of sucrose according to the example given for Case II coupling on p. 363. Then do the same calculation in a simpler way.

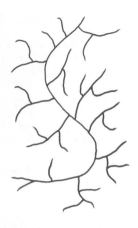

CHAPTER 12

CARBOHYDRATE METABOLISM

PRINCIPLES OF METABOLISM

Metabolism can be defined as the *sum of all the chemical reactions that occur in a living organism.* As the definition suggests, this is a vast subject. The number of reactions alone is staggering, with different organisms, depending on their complexity, characterized by several hundred to several thousand. Collectively, these reactions are responsible for sustaining the very viability of the organism. Although each reaction can be considered as a separate entity of individual importance, the *metabolic expression of the whole organism is due to the integration of each individual reaction into a dynamic reaction circuitry of intricate design, controlled by a host of sensitive regulatory checks and balances.* In general, it can be said that the maintenance and control of this design sustain normal metabolism, whereas its disruption contributes to abnormal metabolism.

The subject of metabolism is not only massive but it is complex as well. Yet great strides have been made in unraveling many of the details (of both a general and specific nature) concerning the metabolic dynamics of the living state. This success is due to many factors, including the development of biochemical methodologies (Chap. 2) and the scientific genius and talent of many researchers. However, the real key to the growth of our knowledge has been provided by the nature of the subject itself, for despite the tremendous metabolic diversity represented by all living forms, there also is a very definite theme of *metabolic unity*. This should not be so surprising, since we have already emphasized in previous chapters that all cells contain the same classes of biomolecules—proteins, nucleic acids, lipids and carbohydrates. In fact, we have already encountered an illustration of this in describing the events of nucleic acid and protein biosynthesis. At this point we are saying that, on the whole, the general metabolism of all classes of biomolecules is in principle basically the same from one organism to another. Indeed, the metabolism of many substances is identical in all organisms.

Metabolism has different aspects. To begin with the distinction is made that all of the reactions comprising the total metabolism of any cell can be broadly segregated into two types—*degradative* and *synthetic*. As mentioned on occasion in previous chapters, the degradative reactions are collectively referred to as *catabolism* and the synthetic reactions as *anabolism*. However, the distinction of degradation vs. synthesis is not the only manner in which we can differentiate the reactions occurring in a living cell. Additional comparisons in terms of whether the reactions involve oxidation or reduction, and of the energetics of the reaction sequences and the nature of the starting materials and end products are listed below.

Catabolism		*Anabolism*
degradative	(1)	synthetic
oxidative in nature	(2)	reductive in nature
energy yielding	(3)	energy requiring
a variety of starting materials with well-defined end products	(4)	well-defined starting materials with a variety of end products

Although catabolic and anabolic reactions are distinct in these respects, they are very much related to each other. Moreover, the relationships are in terms of the very characteristics that constitute their differences.

a. *On the level of oxidation vs. reduction:* Although catabolism is oxidative in nature and anabolism is reductive in nature, in each case most of the oxidative or reductive steps utilize the nicotinamide adenine dinucleotides. More specifically, catabolism utilizes the oxidized forms (NAD^+ and $NADP^+$) and produces the reduced forms (NADH and NADPH), whereas anabolism requires the reduced forms and produces the oxidized forms. The one variation in this pattern is that anabolic reactions use primarily NADPH, producing $NADP^+$. Nevertheless, the general participation of nicotinamide adenine dinucleotides in both processes is a distinct common denominator.

b. *On the level of energetics.* Catabolism is exergonic (energy-yielding), with a net requirement for ADP and a net production of ATP. The ATP then serves as the source of energy for the endergonic reactions (energy-requiring) of anabolism and ADP (and AMP) is produced.

c. *On the level of starting materials, end products, and intermediary metabolites:* The end products and intermediary metabolites that are generated in catabolism generally serve as the starting materials in anabolism. The reverse is also true.

Thus, though the two types of metabolic activity obviously possess different characteristics, we can conclude that they are *integrated, complementary*

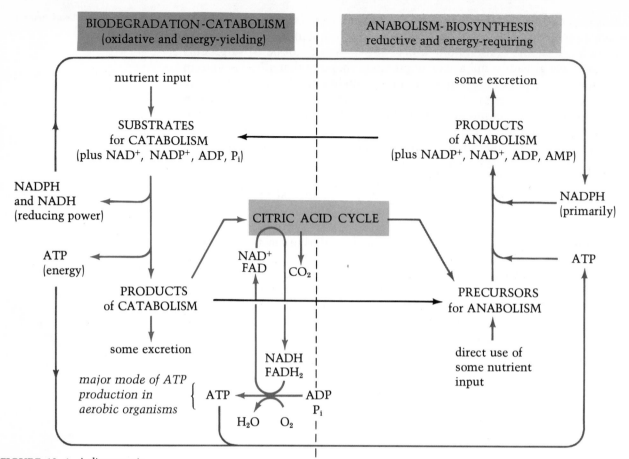

BIODEGRADATION · CATABOLISM
(oxidative and energy-yielding)

ANABOLISM · BIOSYNTHESIS
reductive and energy-requiring

nutrient input

some excretion

SUBSTRATES
for CATABOLISM
(plus NAD⁺, NADP⁺, ADP, P$_i$)

PRODUCTS
of ANABOLISM
(plus NADP⁺, NAD⁺, ADP, AMP)

NADPH
and NADH
(reducing power)

CITRIC ACID CYCLE

NADPH
(primarily)

ATP
(energy)

NAD⁺
FAD

CO_2

ATP

PRODUCTS
of CATABOLISM

PRECURSORS
for ANABOLISM

some excretion

NADH
$FADH_2$

direct use of
some nutrient
input

*major mode of ATP
production in
aerobic organisms*

ATP ← ADP
P$_i$

H_2O O_2

FIGURE 12–1 A diagramatic summary of complementary relationships that provide for an integrated network of catabolism and anabolism. Attention is focused on the nutrient source of carbon because it is the major bioelement of the organic biomolecules. In photosynthetic cells and a few classes of bacteria the required form is CO_2. In all cases (to which this diagram applies) a reduced, organic form of carbon, such as carbohydrates, is required. The central participation of the citric acid cycle and the participation of O_2 in the reoxidation of NADH associated with ATP production will be developed as we proceed. Organisms that require O_2 for this purpose are referred to as *aerobic*. This includes all higher animals, most microorganisms, and nonphotosynthetic plant cells.

processes in that what is produced by catabolism is required by anabolism and vice versa (see Fig. 12–1). This integration of metabolism provides for an optimal level of metabolic efficiency in nature and will serve to unify our analysis throughout the next several chapters.

Our first consideration will be in the area of carbohydrate metabolism. To treat this fully would require much more attention than the scope of this book allows. Rather, we will limit attention to three major reaction pathways, in the following order: (1) *glycolysis* (carbohydrate breakdown), (2) *glycogenesis* (carbohydrate formation), and (3) the *hexose-monophosphate shunt* (see below).

Depending on the organism and/or its growth conditions, the pathway of glycolysis fulfils many functions. In many microbes growing under *anaerobic* (oxygen-absent) *conditions*, it serves as the main energy-yielding catabolic route for carbohydrate substrates, resulting in the production of specific metabolic end products such as ethanol, lactic acid (lactate), and glycerol. This type of process is often referred to as a *fermentation*. Most of the microbial fermentative processes are, however, variations or offshoots of the glycolytic pathway. Fermentation is a much more general term, implying only that the initial substrate is degraded anaerobically.

The most common usage of the term glycolysis is in reference to the anaerobic degradation of carbohydrate substances to yield *lactate* as a metabolic end product. In animals, this accounts for such important processes as the supplying of energy for muscle contraction in skeletal muscle when oxygen is limiting. However, in a broader sense, the reactions of glycolysis have a greater significance, for they comprise the initial phase of carbohydrate degradation that is then linked to the important aerobic phase, which is composed primarily of the set of reactions called the *citric acid cycle*. In this case, glycolysis does not result in the terminal production of lactate, but rather stops one step short, at the level of *pyruvate*, which is the immediate precursor of lactate (see below). Under aerobic conditions the combined action of glycolysis and the citric acid cycle results in the complete oxidation of the reduced carbons in a hexose to CO_2, with an accompanying release of large amounts of potentially useful metabolic energy, largely in the form of reduced nicotinamide adenine dinucleotides (NADH). The NADH is then reoxidized via the oxygen-dependent process of *oxidative phosphorylation*, resulting in the production of ATP. This aerobic phase of metabolism is commonly termed *respiration*, and is the main process whereby oxygen-dependent cells produce most of their needed metabolic energy as ATP.

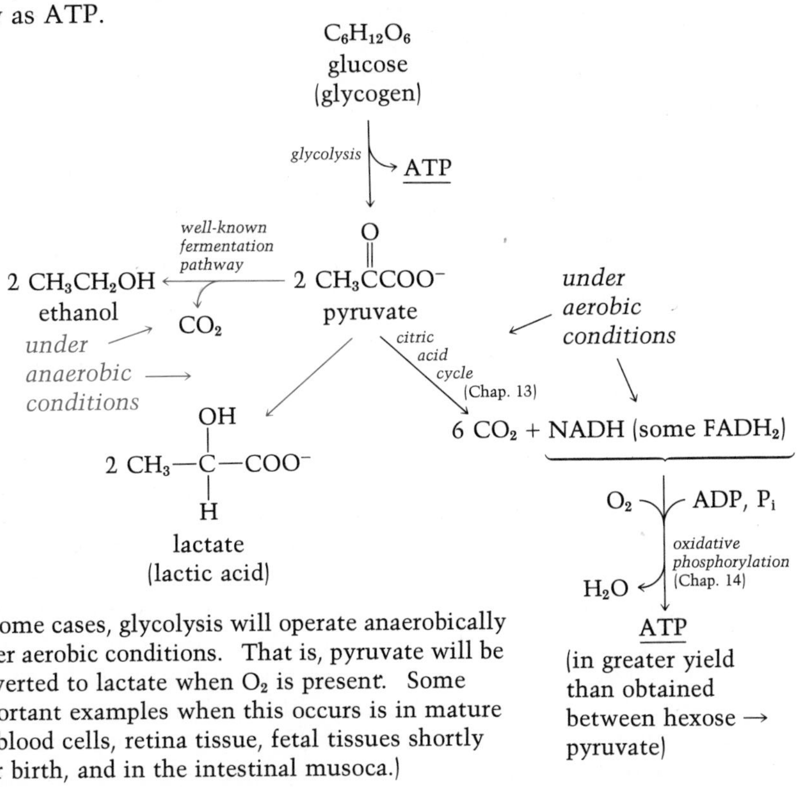

$C_6H_{12}O_6$
glucose
(glycogen)

glycolysis → ATP

well-known fermentation pathway

2 CH_3CH_2OH ← 2 $CH_3\overset{O}{\overset{\|}{C}}COO^-$ *under aerobic conditions*

ethanol CO_2 pyruvate

under anaerobic conditions →

citric acid cycle (Chap. 13)

$$2\ CH_3-\underset{\underset{H}{|}}{\overset{\overset{OH}{|}}{C}}-COO^-$$

6 CO_2 + NADH (some $FADH_2$)

lactate
(lactic acid)

O_2 ⤵ ADP, P_i

oxidative phosphorylation (Chap. 14)

H_2O ⤴

ATP

(In some cases, glycolysis will operate anaerobically under aerobic conditions. That is, pyruvate will be converted to lactate when O_2 is present. Some important examples when this occurs is in mature red blood cells, retina tissue, fetal tissues shortly after birth, and in the intestinal musoca.)

(in greater yield than obtained between hexose → pyruvate)

Although most of this chapter deals with the glycolytic pathway (the citric acid cycle and oxidative phosphorylation will be considered in the next two chapters), we will also analyze the events of an *alternate pathway* of carbohydrate metabolism, the *hexose-monophosphate shunt*. Like glycolysis, the shunt pathway occurs in all types of cells, but generally to a lesser degree. Most interesting, however, is the fact that in many cells the glycolytic pathway and the shunt pathway do not exclude each other. That is, both pathways exist in the same cell. This is not to be interpreted as a measure of metabolic repetition or inefficiency, for, as we shall see, the shunt pathway does have the potential to serve specific metabolic functions that glycolysis cannot.

The consideration of *glycogenesis* will not only complement the coverage of glycolysis but will also provide us with the opportunity to become acquainted with the remarkable network of regulatory features that insure a proper balance or imbalance of degradative (ATP and NAD(P)H yielding) and synthetic (ATP and NAD(P)H requiring) events. The elucidation of this self-controlling capability of living cells is one of the most important developments of modern biochemistry and should stimulate your interest and also aid in your understanding of the subject of metabolism.

Basically, the control mechanisms are of two types, both of which involve the *regulation of enzymes.* One type of popular terminology is to refer to them as (1) the *fine control* mechanisms, which regulate the *activities* of enzymes after they are synthesized, and (2) the *coarse control* mechanisms, which regulate the *synthesis* of enzymes. The fine control mode (with a fast response time) is explained in terms of the allosteric properties, competitive inhibition, cofactor participation, and dissociation and reassociation of oligomeric proteins (see p. 188 in Chap. 6). The coarse control mode (with a slower response time) is a consequence of gene regulation, explained by the theory of induction and repression (see Chap. 8).

This rather lengthy introduction has dealt with the general principles of metabolism, and of carbohydrate metabolism in particular. Do not be concerned if you feel somewhat confused. This state is a natural characteristic of an initial exposure to the subject of metabolism. The many principles will become more coherent chapter by chapter. In other words, there is no such thing as an instant appreciation of metabolism. On the contrary, it develops gradually. (Rote memorization, albeit certainly necessary to develop the basic language, is not the sole answer.)

EXPERIMENTAL HISTORY OF GLYCOLYSIS

The glycolytic pathway was the first metabolic sequence to be elucidated. A historical survey of the experimental events that led to its "discovery" is given below. The time period involved, the late nineteenth century through the early 1940s, covers the early growth of biochemistry as a science and illustrates the slow progress compared to contemporary standards. You should keep in mind that biochemistry was in its infancy, and that specific knowledge concerning the chemical events of the living state was quite limited. Equally significant was the primitive state of biochemical methodology. Delicate techniques for cell fractionation had not been developed; radioisotopes were not available for use as metabolic tracers; and refined chromatographic and electrophoretic techniques had not yet been developed. In this context, the accomplishments of the many participants were nothing short of exceptional.

Chronological Developments in the Study of Glycolysis

1. In 1897 the Buchner brothers accidentally discovered that a cell-free extract of yeast was capable of fermenting sugar to alcohol. This represented the first observation that chemical events associated with life forms were not dependent on intact cells. At the time this must have been a rather startling finding. However, the Buchners did not pursue it further, and the significance of their discovery, namely, that the biochemistry of living cells could be studied without using whole cells, was not recognized for several years.

2. In the early part of the twentieth century (1905–1910), A. Harden and W. J. Young were the first to exploit the discovery of the Buchners. In a systematic study of alcohol production from glucose with cell-free yeast extracts under anaerobic conditions, Harden and Young made four important observations: (a) the process was absolutely dependent on inorganic phosphate; (b) a hexose diphosphate, which they subsequently identified as fructose-1,6-diphosphate, accumulated; (c) the extract could be separated by dialysis into a heat-sensitive component and a heat-stable component, neither of which was individually capable of fermenting glucose; and (d) the production of alcohol could be restored by merely combining the heat-sensitive and heat-stable fractions.

3. Glucose-6-phosphate was isolated (1914) and identified (1931) by A. Harden and R. Robinson.

4. Fructose-6-phosphate was isolated and identified by C. Neuberg in 1932.

5. Glucose-1-phosphate was isolated and identified by C. F. Cori, S. P. Colowick, and G. T. Cori in 1936. (For this and other studies that established the pattern of glycogen degradation through glucose-1-phosphate, Carl and Gerty Cori, who were husband and wife, were jointly awarded the Nobel Prize in 1947. While this has nothing to do with biochemistry, it is interesting to note that the only other husband and wife to receive the Nobel Prize were Pierre and Marie Curie in 1903 for their investigations of radioactive phenomena. Anything in a name?)

6. In 1928 O. Warburg and W. Christian established a linkage between oxidation-reduction and phosphorylation. In 1935 these same workers also identified the nature of the heat-stable component observed by Harden and Young as consisting of adenine nucleotides and the oxidized form of nicotinamide adenine dinucleotide (NAD^+).

7. For about 15 years (1925–1940) Meyerhof, Embden, Parnas, and Warburg studied the interrelationships among the various intermediates and isolated and characterized many of the enzymes involved. One of the most significant discoveries was the isolation and characterization of the enzyme aldolase by Meyerhof in 1936.

To immortalize their contributions in discovering the complete reaction sequence, the glycolytic pathway is sometimes referred to as the *Embden-Meyerhof-Parnas* (EMP) pathway.

The humanistic and professional spirit of traditional name association is quite appropriate, but the omission of Warburg's name is unjust. His contributions in this area were of prime value. In fact, Warburg has been acclaimed as this century's most talented biochemist because of his pioneering discoveries and theories, which played a large part in the development and progress of modern biochemical science. He died at the age of 87 in 1970.

GLYCOLYSIS—AN OVERALL VIEW

Introduction

The entire reaction sequence of glycolysis is shown in Fig. 12–2. If your immediate reaction to the many details of the diagram is one of bewilderment, consider this a normal response. Thousands of other students of biochemistry have reacted similarly. For the moment, it is not the details of glycolysis that are important. Rather, the purpose of the diagram is to provide you with an overall perspective by illustrating that the degradation of the initial substrates (glucose or glycogen are shown) proceeds through a sequence of defined intermediates via several consecutive, enzyme-catalyzed reactions and results in the formation of specific products—lactate or alcohol and CO_2. In the next several pages, we will examine many facets of this pathway with the intent of establishing its metabolic design. When we have finished, the maze of structures and arrows will be better understood. One of the most rational approaches to deciphering a metabolic pathway of this type is to begin with an analysis of *net effects*.

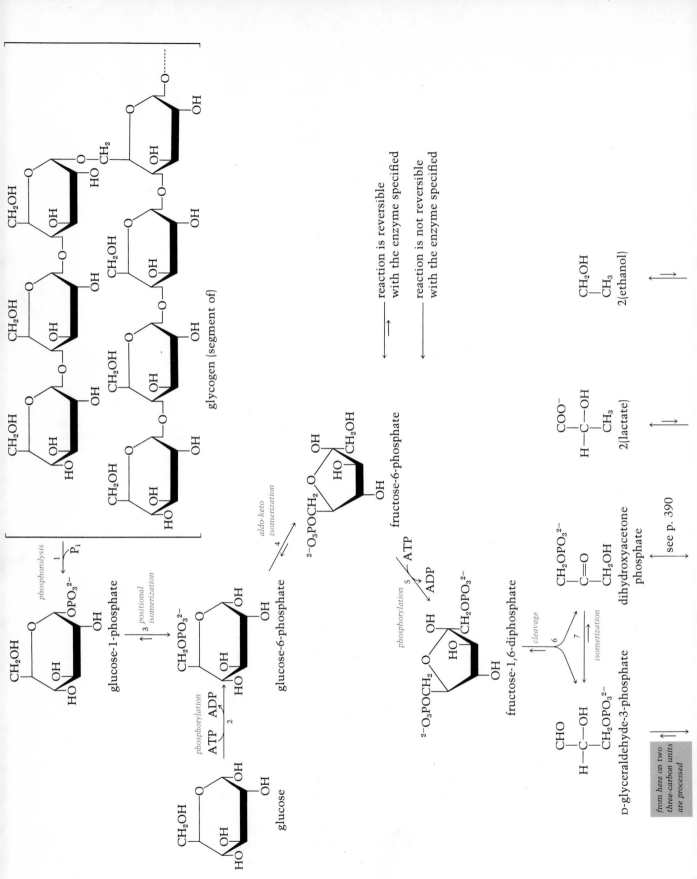

glycogen (segment of)

CH₂OH

phosphorolysis 1 Pᵢ

glucose-1-phosphate

3 positional isomerization

CH₂OPO₃²⁻

glucose-6-phosphate

aldo-keto isomerization 4

²⁻O₃POCH₂ O OH CH₂OH

fructose-6-phosphate

CH₂OH

glucose

phosphorylation ATP ADP 2

phosphorylation 5 ATP ADP

²⁻O₃POCH₂ O OH CH₂OPO₃²⁻

fructose-1,6-diphosphate

cleavage 6

CH₂OPO₃²⁻ C=O CH₂OH

dihydroxyacetone phosphate

isomerization 7

CHO H—C—OH CH₂OPO₃²⁻

D-glyceraldehyde-3-phosphate

from here on two three-carbon units are processed

reaction is reversible with the enzyme specified

reaction is not reversible with the enzyme specified

COO⁻ H—C—OH CH₃ 2(lactate)

CH₂OH CH₃ 2(ethanol)

see p. 390

372

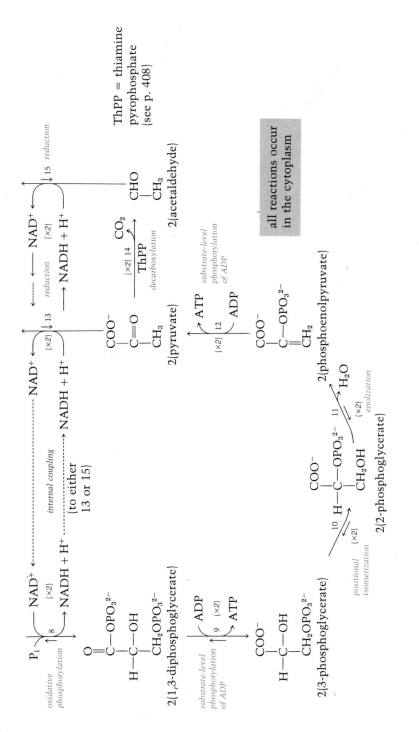

ThPP = thiamine
pyrophosphate
(see p. 408)

all reactions occur
in the cytoplasm

Enzymes:
1. glycogen phosphorylase
2. glucokinase and hexokinase
3. phosphoglucomutase
4. phosphoglucoisomerase
5. phosphofructokinase
6. aldolase
7. triose phosphate isomerase
8. phosphoglyceraldehyde dehydrogenase

9. phosphoglyceric acid kinase
10. phosphoglyceromutase
11. enolase
12. pyruvate kinase
13. lactate dehydrogenase (lactate formation)
14. pyruvate decarboxylase (alcohol formation)
15. alcohol dehydrogenase (alcohol formation)

FIGURE 12–2   A scheme for glycolysis and alcoholic fermentation.

373

Net Chemistry

Under anaerobic conditions, the reactions of glycolysis account for the formation of lactate from glucose or glycogen in some cells, and for the production of alcohol from glucose in others. The same pathway also accounts for a few other conversions in fermenting microbes. Regardless, as Fig. 12–2 clearly illustrates, the majority of the reactions are *identical, with the only difference being the fate of pyruvate.*

The net reactions of glycolysis are given below for each metabolic route. Although the equations are limiting in that they do not indicate the participation of all intermediates, they do clearly specify the essential reactants and the characteristic end products. For each process a family of equations is given, with each successive equation intended to be more descriptive than the previous ones.

A reaction summary of glycolysis

Lactate from free glucose:

$$\text{glucose} \xrightarrow[\text{enzymes}]{11} 2 \text{ lactate}$$
$$1C_6 \qquad\qquad 2C_3$$

$$\text{glucose} + 2\text{ATP} + 4\text{ADP} + 2P_i \longrightarrow 2 \text{ lactate} + 2\text{ADP} + 4\text{ATP}$$
$$\text{glucose} + 2\text{ATP} + 4\text{ADP} + 2P_i + 2\text{NAD}^+ \longrightarrow 2 \text{ lactate} + 2\text{ADP} + 4\text{ATP} + 2\text{NAD}^+$$

Lactate from glycogen glucose:

$$(\text{glucose})_n \xrightarrow[\text{enzymes}]{12} 2 \text{ lactate} + (\text{glucose})_{n-1}$$

$$(\text{glucose})_n + 1\text{ATP} + 4\text{ADP} + 3P_i \longrightarrow 2 \text{ lactate} + (\text{glucose})_{n-1} + 1\text{ADP} + 4\text{ATP}$$
$$(\text{glucose})_n + 1\text{ATP} + 4\text{ADP} + 3P_i + 2\text{NAD}^+ \longrightarrow 2 \text{ lactate} + (\text{glucose})_{n-1} + 1\text{ADP} + 4\text{ATP} + 2\text{NAD}^+$$

Ethanol from free glucose (alcohol fermentation):

$$\text{glucose} \xrightarrow[\text{enzymes}]{12} 2 \text{ ethanol} + 2CO_2$$
$$1C_6 \qquad\qquad 2C_2 \qquad 2C_1$$

$$\text{glucose} + 2\text{ATP} + 4\text{ADP} + 2P_i \longrightarrow 2 \text{ ethanol} + 2CO_2 + 2\text{ADP} + 4\text{ATP}$$
$$\text{glucose} + 2\text{ATP} + 4\text{ADP} + 2P_i + 2\text{NAD}^+ \longrightarrow 2 \text{ ethanol} + 2CO_2 + 2\text{ADP} + 4\text{ATP} + 2\text{NAD}^+$$

The first equation merely defines the overall transformation in terms of the carbon skeletons of the reactants and products. The second equation includes the participation of inorganic phosphate (P_i) and the adenine nucleotides, and by so doing, provides a net analysis of the bioenergetics of glycolysis. The appearance of ATP on both sides of the equation indicates that *ATP is both required and produced* in the overall process. Note, however, that in each case the *amount of ATP produced exceeds the amount required.* Hence, the overall process is *energy-yielding* (see the next section). The third equation, which is the most descriptive, includes the participation of the nicotinamide nucleotides. As with ATP, the involvement of NAD^+ is twofold—it is both *required and produced.* In this case, however, note that there is neither a net requirement nor a net gain.

Before we probe deeper into the participation of ATP, ADP, P_i, NAD^+, you should realize that the last equation in each set is wholly consistent with Harden and Young's early observations. Inorganic phosphate is absolutely required as an essential reactant; the nondialyzable, heat-labile fraction contains the enzymes; and the heat-stable, dialyzable fraction contains, among other things, ATP, ADP, and NAD$^+$. To become familiar with the specific chemical events of glycolysis, it is recommended that you verify the validity of any one of the last equations in each set. To do this you need recall only one point, namely, that a sequence of consecutive reactions such as A $\rightarrow$ B $\rightarrow$ C can be dissected into pattern A $\rightarrow$ B and B $\rightarrow$ C and then added to yield the overall effect of A $\rightarrow$ C.

Overall View of the Energetics of Glycolysis

We have just pointed out that, although glycolysis is both energy-requiring and energy-yielding, the net energy balance favors ATP production, with 3 ATPs being formed from the catabolism of one glucose unit from glycogen, and 2 ATPs being formed from one unit of free glucose. The approach in making this tally is really quite simple. Reference to Fig. 12–2 reveals that there are two ATP-requiring reactions, the formation of glucose-6-phosphate and fructose-1,6-diphosphate; and two ATP-generating reactions, the formation of 3-phosphoglycerate and pyruvate. However, the latter reactions contribute a total of 4 ATPs since each hexose unit is cleaved to two three-carbon units. Overall then there is a requirement of 2 ATPs and a production of 4 ATPs, giving a net gain of 2 ATPs. With glycogen as the initial substrate, ATP is required only in the formation of fructose diphosphate, and thus a net gain of 3 ATPs is realized. This type of analysis is relatively straightforward and there should be no difficulty in understanding its significance, namely, that glycolysis, resulting in partial degradation of carbohydrate substrates, provides a *net gain* of useful metabolic energy as ATP.

Although the preceding analysis is valid—and granted it is significant—it is rather superficial because there is more to the energetics of glycolysis than merely counting ATPs. To appreciate that this is so, let us now concentrate on the energetics of each reaction rather than the net effect. Our approach will be purely thermodynamic. To be more specific, it will be based on a comparison of the known free energy changes ($\Delta G^{\circ\prime}$) of each reaction (see Fig. 12–3). Before using these data, it should be noted that the free-energy changes given in Fig. 12–3 are only for the reactions involving the glycolytic intermediates and do *not* consider any coupling to ATP hydrolysis or ADP phosphorylation. For example, the value of $+3.3$ kcal corresponds to GLU $+$ P$_i$ $\rightarrow$ G6P $+$ H$_2$O and not GLU $+$ ATP $\rightarrow$ G6P $+$ ADP, and the value of -13.1 kcal corresponds to PEP $+$ H$_2$O $\rightarrow$ PYR $+$ P$_i$ and not PEP $+$ ADP $\rightarrow$ PYR $+$ ATP.

An obvious question at this point is, what value do these data have in understanding glycolysis? Well, given the free energy change of each reaction, we can construct an *energy profile diagram* of the complete pathway that should depict the relative differences in free energy among all of the primary metabolic intermediates. Such a diagram is shown in Fig. 12–4. A close study reveals that the tactic in constructing the diagram is really quite simple. Each intermediate, which is involved in at least two consecutive reactions, has been positioned on a free energy scale in terms of its free-energy content relative to those of both its immediate precursor and its immediate product. When this is done, note that there emerges a pattern to the composite energetics of glycolysis in that we can clearly distinguish *two different phases*, one *endergonic* and the other *exergonic*.

Using free glucose as a starting point, we note that the endergonic phase,

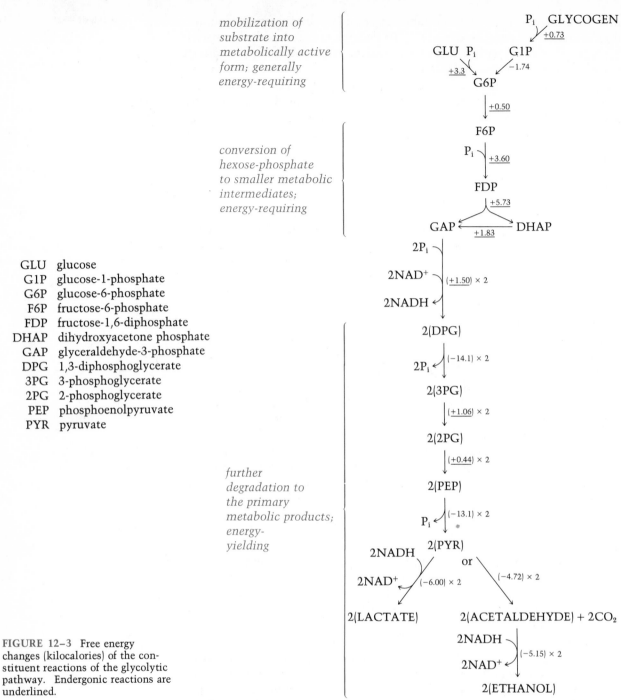

GLU   glucose
G1P   glucose-1-phosphate
G6P   glucose-6-phosphate
F6P   fructose-6-phosphate
FDP   fructose-1,6-diphosphate
DHAP   dihydroxyacetone phosphate
GAP   glyceraldehyde-3-phosphate
DPG   1,3-diphosphoglycerate
3PG   3-phosphoglycerate
2PG   2-phosphoglycerate
PEP   phosphoenolpyruvate
PYR   pyruvate

FIGURE 12–3 Free energy changes (kilocalories) of the constituent reactions of the glycolytic pathway. Endergonic reactions are underlined.

which consists of the initial part of the glycolytic pathway, is comprised of three *energy barriers*. In this context, the rationale for an ATP requirement is now strikingly apparent. Not only does ATP act as a donor of phosphate, but in addition the energy release accompanying the use of ATP is sufficient to overcome the thermodynamically unfavorable formation of G6P and FDP. In other words, the endergonic conversions of the hexoses to their metabolically active form, the phosphate esters, is coupled to the exergonic hydrolysis of ATP.

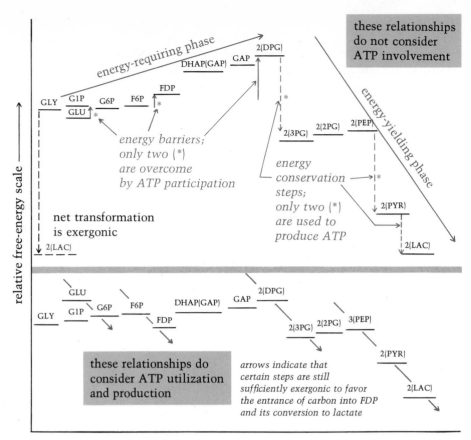

FIGURE 12–4 A skeleton view of the thermodynamic (free energy) relationships among the metabolic intermediates of the glycolytic pathway. Horizontal lines indicate relative energy states of each intermediate. The upper part considers the pathway devoid of ATP involvement. The lower part considers the effect of ATP involvement in the two ATP-requiring and two ATP-yielding reactions. In the latter case, note that three of the reactions, including the last two, are still sufficiently exergonic to promote the flow of intermediates toward lactate formation.

With glycogen as the initial substrate, the profile is different, due to the small differences among the relative energy levels of glycogen, G1P, and G6P. In this case then, the mobilization of a glucose unit to the metabolically active phosphate ester (G1P) is not thermodynamically unfavorable, and thus the need for coupling to ATP hydrolysis does not exist. A direct phosphorylation with inorganic phosphate is sufficient.

Still to be considered, however, is the final barrier from the level of FDP to 2(DPG). It is interesting to note that, though the magnitude of this barrier is rather considerable—being roughly equivalent to the first two combined—it is not coupled to ATP hydrolysis. Nevertheless, during glycolysis this barrier does not contribute to a metabolic blockage, and as a result, FDP does not accumulate. The key to understanding why this accumulation does not occur is provided by the thermodynamics of the very next reaction, which consists of the extremely favorable (exergonic) conversion of DPG to 3PG. (Recall that 1,3-diphosphoglycerate, an acyl phosphate, is an unstable, high-energy compound; see p. 353. Thus, since the reactions are consecutive, the endergonic sequence of FDP through DPG is driven to completion by being *coupled* to the thermodynamically favorable removal of DPG (and hence FDP) as it is formed. Notice that the coupling in this case occurs without an actual transfer of energy (Case III, p. 363).

The conversion of 2(DPG) to 2(3PG) is especially important, because it is not only coupled to the degradation of FDP but also to the phosphorylation of 2 ADPs to yield 2 ATPs. Thus, one exergonic reaction (DPG → 3PG) is coupled to two endergonic reactions (FDP → DPG and ADP → ATP). The cell profits then in two ways: the flow of carbon through FDP is favored and ATP is formed.

The PEP → PYR conversion is also of great metabolic significance. Here again, a high-energy phosphorylated compound PEP (an enoyl phosphate) is used as a donor of phosphate for ATP formation.

It is of interest to note that, although the last reaction of glycolysis (PYR → LAC) has the capacity to be coupled to ATP formation, this does not occur and the energy is not conserved. You might argue that, in view of the thermodynamic design already discussed, this nonconservation site is inconsistent with metabolic efficiency. However, just the opposite may be true. One suggestion is that the appearance of this type of reaction, occurring independently of an energy-requiring process, and positioned at the end of a multistep sequence, may represent added insurance that preceding intermediates will be efficiently converted to the final product. The fact that other metabolic pathways display a similar pattern supports such a suggestion. In so doing, the cell pays a price—energy is not conserved as ATP. Another important metabolic role of the PYR → LAC conversion is described in the next section.

This involvement of ATP in glycolysis (starting with glucose) can be summarized as follows (the −45.4 value is obtained from the data in Fig. 12–3):

$$\text{glucose} + 4P_i \xrightarrow[\text{NADH}]{\text{NAD}^+} 2 \text{ lactate} + 4P_i \qquad \Delta G^{\circ\prime} = -45.4 \text{ kcal} \qquad \text{(available)}$$

$$2\text{ATP} \xrightarrow{\hspace{2cm}} 2\text{ADP} + 2P_i \qquad \Delta G^{\circ\prime} = 2(-8) = -16 \text{ kcal} \qquad \text{(expended)}$$

$$4\text{ADP} + 4P_i \xrightarrow{\hspace{2cm}} 4\text{ATP} \qquad \Delta G^{\circ\prime} = 4(+8) = +32 \text{ kcal} \qquad \text{(conserved)}$$

$$\text{glucose} + 2\text{ADP} + 2P_i \xrightarrow[\text{NADH}]{\text{NAD}^+} 2 \text{ lactate} + 2\text{ATP} \qquad \Delta G^{\circ\prime} = -29.4 \text{ kcal} \qquad \text{(not conserved)}$$

Thus, it would appear that the overall efficiency of utilizing energy available from the conversion of glucose to lactate for ATP production is about 50%. However, under cellular conditions (rather than standard-state conditions) the efficiency is probably higher. Without the ATP expenditure the efficiency would certainly be greater, but then ATP production would be impeded by a significant energy barrier. The evolution of the glycolytic pathway thus represents a compromise of giving some to get more and get it without a serious impediment.

NAD⁺ in Glycolysis

A second characteristic of glycolysis that reflects the remarkable design of metabolism is evident in the participation of nicotinamide adenine dinucleotide. The point becomes rather clear by inspecting Fig. 12–2 and observing that glycolysis, operating anaerobically (be it for lactate or ethanol formation), consists of *two NAD-dependent dehydrogenases*. One *requires NAD⁺ and produces NADH*, whereas the other *requires NADH and produces NAD⁺*. In other words, glyceraldehyde phosphate dehydrogenase and lactate dehydrogenase (or alcohol dehydrogenase) are complementary enzymes. This means that the supply of NAD⁺ is under constant replenishment. This *internal coupling* is of great metabolic importance because NAD⁺ is as essential a substrate of glycolysis as is glucose. Without a regeneration of NAD⁺ from NADH there would be no glycolysis.

SOME INDIVIDUAL REACTIONS OF GLYCOLYSIS

A considerable part of biochemical research—both past and present—is characterized by the detailed investigation of individual reactions. This type of research involves the isolation and purification of the enzyme, followed by extensive studies of its structure and catalytic properties. (Easier said than done.) Nevertheless, all the enzymes of glycolysis have been so studied, and accordingly our current knowledge about this pathway is extensive and detailed. Despite the scope of this knowledge, it is not our purpose to treat the subject in its entirety. This is more appropriate to advanced courses. Alternatively, we will confine our attention to a few select reactions, with a focus on principles such as enzyme specificity, coenzyme participation, and metabolic regulation.

Glycogen Phosphorylase

The first reaction in the catabolism of glycogen is the sequential removal of glucose residues; it is catalyzed by the widely distributed enzyme, *glycogen phosphorylase*. The chemistry of the catalysis is relatively straightforward (see below). In the presence of the enzyme, inorganic phosphate attacks an $\alpha(1 \rightarrow 4)$ glycosidic bond at the *nonreducing end* of the polysaccharide chain, resulting in the formation of G1P and the shortening of the chain by one hexose unit. This proceeds along the chain until an $\alpha(1 \rightarrow 6)$ branch is approached to within approximately three residues. At this point another enzyme (*oligosaccharo transferase*) transfers an intact trigluco fragment to a neighboring branch stem, forming an $\alpha(1 \rightarrow 4)$ bond. A third enzyme, an $\alpha(1 \rightarrow 6)$ *glucosidase*, then removes the glucose unit involved in the $\alpha(1 \rightarrow 6)$ linkage and releases it as free glucose, which would be metabolized further after its ATP-dependent conversion to G6P. Then glycogen phosphorylase action would resume until another branch point is approached.

These events are shown on p. 380.

The result of all this is that glycogen glucose is mobilized as glucose-1-phosphate (primarily). The type of reaction catalyzed by glycogen phosphorylase is nothing more than a *phosphorolysis* of $\alpha(1 \rightarrow 4)$ glucosidic bonds and we will not pursue the mechanism further. There are, however, some fascinating catalytic properties of this enzyme.

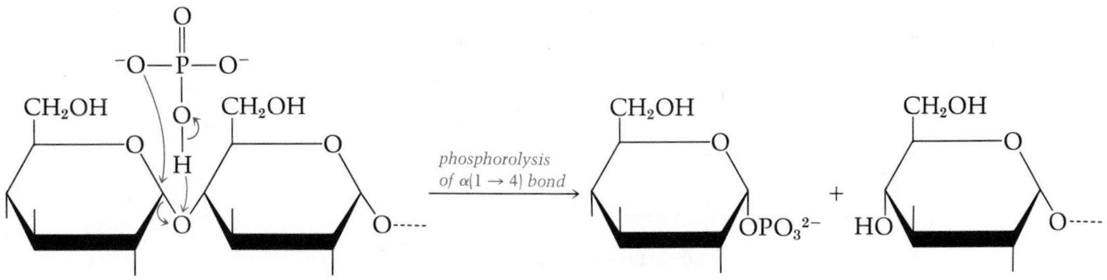

A tip-off to the general nature of both the structure and the catalytic properties of phosphorylase is given in the Michaelis-Menten plot of Fig. 12–5(a). Observe the *sigmoid* nature of the curve, and recall the generalization given in Chap. 6 (p. 190) as to its significance, namely that this kinetic pattern typifies *allosteric* properties of the enzyme. By way of review: according to the theory of allosterism, the data suggest that the enzyme is an oligomeric protein and contains more than one active site and may also contain one or more regulatory sites. The former would account

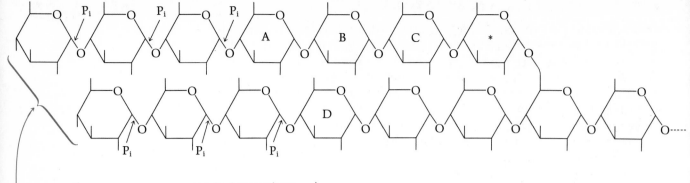

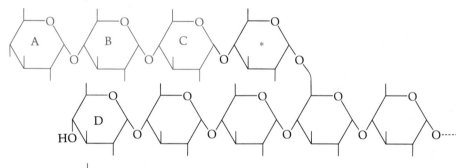

glycogen

proceeding from nonreducing ends there is successive cleavage of α(1 → 4) bonds stopping about 3–4 residues short of an α(1 → 6) branch site

glycogen phosphorylase

units of glucose-1-phosphate

transfer of a small fragment from the shortened branch to the main α(1 → 4) backbone

oligosaccharo transferase

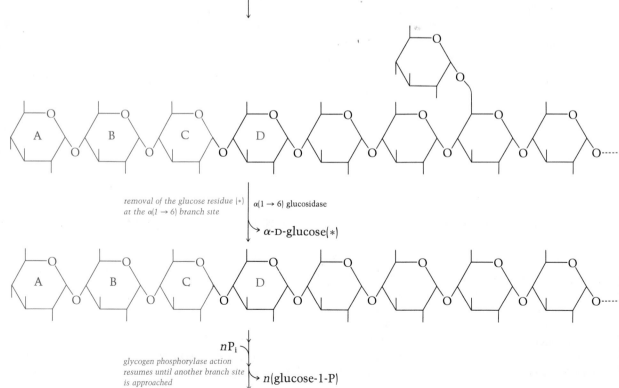

removal of the glucose residue (*) at the α(1 → 6) branch site

α(1 → 6) glucosidase

α-D-glucose(*)

glycogen phosphorylase action resumes until another branch site is approached

$n\mathrm{P_i}$

n(glucose-1-P)

for homotropic effects (as observed for phosphorylase with inorganic phosphate, Fig. 12–5(a)), while the latter would explain heterotropic effects (as observed for phosphorylase with inorganic phosphate in the presence of AMP, with the AMP acting as an activator; Fig. 12–5(b)).

So what, you ask? Well, if the catalytic activity of phosphorylase is so characterized, it is quite probable that the cellular formation of glucose-1-phosphate is regulated. To be more specific, we can argue that the level of phosphorylase activity will be high when the cellular levels of P_i (a substrate that binds cooperatively) and AMP (an activator) are high. When they are low, the rate of formation of glucose-1-phosphate is considerably diminished. Hence, these two variables can act as *metabolic signals that modulate carbohydrate metabolism in general, and the action of phosphorylase in particular.* We are not yet in a position to explain how these signals—varying levels of P_i and AMP—are generated in the cell. At this time, suffice it to say that they will change as the overall metabolism of the cell changes. For now we are interested only in establishing that the phosphorylase step *is a kinetic control point.* However, the AMP effect is just an introduction to and only a small part of glycogen phosphorylase regulation, the totality of which is rather elaborate, to say the least.

Our current understanding of glycogen phosphorylase had its beginning in 1943 when the Coris reported that the enzyme actually exists in two different forms, called phosphorylase *a* and phosphorylase *b*. They are different in both structure and function. This duality of glycogen phosphorylase has guided the study of this enzyme since then and is at the core of the regulatory features that apply to its catalytic operation. The most notable of these features is the *interconversion* of the different structural and functional forms of the enzyme; an interconversion that is *catalyzed by another enzyme.* This novel mode of regulation is now known to apply to several other enzymes. The events associated with the interconversion of glycogen phosphorylases *a* and *b* are fairly well but not completely established. It seems that the more we learn the more complex the issue becomes.

Both forms of glycogen phosphorylase (from muscle, unless otherwise stated) are capable of catalyzing the formation of G1P from glycogen, and both are homogeneous oligomeric proteins composed of the same polypeptide chain, but there are distinct differences between the two. Catalytically the *a* form is much more active than the *b* form—in the absence of activating substances such as AMP. In fact, the *b* form is virtually inactive in the absence of AMP. The activity of the *b* form, however, is substantially increased by AMP whereas the *a* form is relatively unaffected by AMP. The structural distinctions associated with these catalytic differences are as follows: the predominant species of the *a* form is tetrameric, and each of the polypeptide chains contains a phosphorylated serine residue that is essential for activity; the predominant species of the *b* form is dimeric, and each of the two polypeptide chains contains a dephosphorylated serine residue (the same serine as in the *a* form). Both forms contain *pyridoxal phosphate* (see p. 531) covalently attached to the epsilon amino group of a particular lysine residue, a feature that also appears to be essential for the activity of both forms, but for reasons that are not yet clear.

The fascinating dimension of the phosphorylase system is that the *a* and *b* forms are enzymatically interconvertible through the action of two separate enzymes. *Glycogen phosphorylase phosphatase* catalyzes the dephosphorylation of the *a* form, which is also accompanied by a change in the

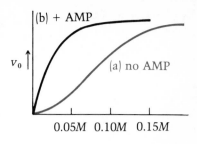

FIGURE 12–5 Michaelis-Menten kinetics of glycogen phosphorylase (*b* form; see text): (a) effect of varying levels of inorganic phosphate concentration in the absence of AMP; (b) as in (a), with $10^{-5}\,M$ AMP present. Constant glycogen concentration (0.25%); temperature of 28 °C. (Data obtained with permission from M. H. Buc and H. Buc, "Allosteric Interactions between AMP and Orthophosphate Sites on Phosphorylase *b* from Rabbit Muscle," in *Regulation of Enzyme Activity and Allosteric Interactions*, E. Kramme and A. Pihl, eds., New York: Academic Press, 1968, p. 118.)

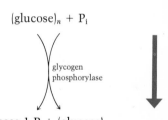

(glucose)$_n$ + P_i

glycogen phosphorylase

glucose-1-P + (glucose)$_{n-1}$

activity of phosphorylase is increased due to

(a) *cooperative binding of P_i at more than one active site (homotropic effect)*

(b) *AMP binding to secondary (regulatory) site effecting activation of enzyme (heterotropic effect); this effect applies only to the b form, however; see text*

state of aggregation (tetramer to dimer). *Glycogen phosphorylase kinase* catalyzes the phosphorylation of the *b* form, which is also accompanied by a change in the state of aggregation (dimer to tetramer). In each case one enzyme alters the covalent structure of another enzyme (the substrate), and this in turn triggers an alteration in the oligomeric composition of the substrate enzyme.

$^{2-}O_3POCH_2$ $CH_2OPO_3^{2-}$

$^{2-}O_3POCH_2$ $CH_2OPO_3^{2-}$

phosphorylase *a*
homogeneous tetramer

$4H_2O$ →(glycogen phosphorylase phosphatase)→ $4P_i$

$4ADP$ ←(glycogen phosphorylase kinase)← $4ATP$

$2\ HOH_2C$————CH_2OH

phosphorylase *b*
dimer

with each polypeptide consisting of a phosphorylated serine residue (active form; no activation by AMP)

with nonphosphorylated serines (less active than *a* form but shows extensive activation by AMP)

(the sphere and square represent the same monomeric polypeptide chain but in different conformations)

Certainly interesting, but the enzymatic interconversion of active and inactive enzymes would mean little unless the enzymes that catalyze the interconversion are themselves under some type of control. The following description reveals that such control does indeed exist. To begin with, the kinase also exists in two forms, of different activity. The highly active form is also a phosphorylated protein, and the other relatively inactive form is a dephosphorylated protein. These two kinase forms are *also* enzymatically interconverted by the action of two other enzymes, *glycogen phosphorylase kinase phosphatase* (active → inactive) *and glycogen phosphorylase kinase kinase* (inactive → active). In turn, the kinase kinase enzyme also exists in active and inactive forms with the allosteric binding of *cyclic AMP* promoting the activation. Finally, the intracellular level of cAMP is itself under control by certain hormones, such as *adrenalin*, which do not enter the cell! The adrenalin (outside)-cAMP (inside) relationship is accounted for by the fact that the enzyme responsible for cAMP formation, *adenyl cyclase*, is *present in the cell membrane*. Adrenalin binds to a receptor site on the external side of the membrane, stimulating the activity of adenyl cylase to produce cAMP, which in turn stimulates the formation of the active form of kinase kinase, which in turn catalyzes the formation of the active kinase, which in turn catalyzes the formation of the active *a* form of glycogen phosphorylase. This elaborate cascade of events is diagramed below. An indication is also given that the conversion of cAMP to AMP by cyclic-AMP phosphodiesterase would remove the activating signal and the deactivating influences would take over. A drop in adrenalin production, or a direct inhibition of adenyl cyclase by some other substance, would produce the same effect.

Despite the advanced state of knowledge in this area, work still needs to be done. In mammals especially, where both physiological variables and different cell types are numerous, there are many unanswered questions. For example, depending on the source of the enzyme, the mechanism of gly-

3',5'-cyclic AMP

adrenalin
(epinephrine)

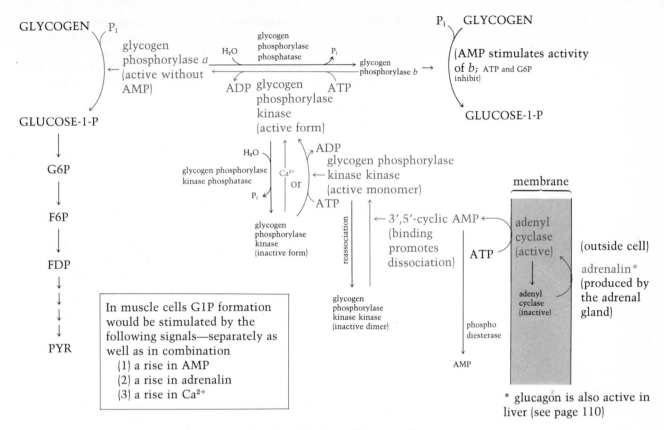

a summary of glycogen phosphorylase (muscle) control

As indicated, the presence of adrenalin in the blood stream would initiate a series of activations that would ultimately give rise to an increased rate of glycogen catabolism; then, as the level of hormone in plasma decreases, the initiating event of the activation cascade decreases and the reverse reactions *would become prominent*

cogen phosphorylase activation is quite diverse. Muscle phosphorylase is activated by adrenalin. Liver phosphorylase, however, is activated by *glucagon* (a polypeptide hormone produced by the pancreas—see p. 110) as well as by adrenalin. Glucagon has little effect on muscle phosphorylase. The effect of glucagon is identical to that of adrenalin—it activates adenyl cyclase in the cell membrane of liver cells. On the other hand, brain phosphorylase does not seem to be affected by either glucagon or adrenalin.

In any event the control of carbohydrate catabolism at the level of the glycogen phosphorylase system is a classic illustration of the *secondary messenger theory*, proposed by Sutherland to explain how certain hormonal substances in very minute quantities could modulate intracellular processes without ever penetrating the cell membrane. Their influence is mediated by an event that involves the participation of a membrane component(s). In many cases cAMP is the secondary messenger, as described here. The messenger theory has revolutionized modern endocrinology, but it does not account for the action of all hormones. Other hormones, such as the steroids, actually enter the cell and exert a direct effect on some process.

Final Notes

1. It may seem that the cell does not really accomplish anything productive because at least 6 ATPs would be consumed to form one molecule of active phosphorylase *a*, but the catabolism of one G1P to lactate produces a net gain of only 3 ATPs. The explanation is, of course, that each molecule of phosphorylase *a* can participate many times as a catalyst, and yield many molecules of G1P, to provide a large amount of ATP in response to the demand that triggered the phosphorylase activation. A little ATP is expended to yield a much greater return.

2. The preceding description of cellular control on glycogen phosphorylase activity focused on the activation of the phosphorylase system. Evidence indicates that the stimulation of the activation cascade may be complemented by the depression of the inactivation of the phosphorylase system. In other words, interconvertible active and inactive forms of the two phosphatases appear to exist, with the inactive forms being stimulated under conditions that stimulate the active form of the kinase.

3. Finally, if the control on the phosphorylase system by either (a) the cAMP-mediated hormonal effect or (b) the increased cellular level of AMP is not enough to impress you, consider that a third mode of stimulation is also proposed. In it Ca^{2+} appears to activate the kinase directly (see diagram). Thus, in muscle cells where an influx of Ca^{2+} stimulates the contractile process, the Ca^{2+} could also contribute to phosphorylase regulation. In other words, in muscle cells at least, there may be *three* separate signals, all of which result in the same effect—an increased rate of glycogen catabolism to produce needed ATP on demand.

Hexokinase

Glucose is by far the most abundant monosaccharide in nature, being utilized by all types of organisms as the primary source of both carbon and energy. In serving this role, glucose participates in many different enzyme-catayzed reactions, but without question the most important is its initial conversion to a phosphoester by an ATP-dependent *kinase*. Recall our earlier mention of the phosphorylated form as the metabolically active form (p. 296). In other words, unless glucose is first converted to a phosphate ester, it will not be metabolized. The importance of the hexokinase enzyme is that simple. Remember also the difference in the permeability of the cell membrane with respect to the free sugars and the sugar phosphates.

As the name implies, hexokinases display a broad range of substrate specificity, although most show a high affinity ($K_m \approx 10^{-5}\ M$) for glucose. In some organisms more than one hexokinasse is present. Mammals, for example, contain at least three hexokinases (isoenzymes), and in some tissues all three are present in the same cell. Although their presence appears to be characteristic of the tissue, the physiological significance is not clear. As further evidence of the biochemical diversity in our biosphere, note that

when the level of G6P in the cell is high, its binding to the enzyme results in a decrease in catalytic activity and thus a reduced rate of glucose mobilization; a high level of ADP has the reverse effect

some organisms also possess a *glucokinase*. Operationally, glucokinase fulfills the same role as a hexokinase, but is characterized by a much more restricted specificity, strongly favoring glucose as a substrate. In mammals it has been proposed that glucokinase serves as a metabolic safety valve in that it operates only when blood glucose levels are high. The basis of this suggestion is that the K_m value of glucokinase is about 1,000 times greater than that for hexokinase, implying that glucokinase will become saturated only in the presence of high levels of glucose. By contributing to the prevention of excessively high glucose levels in blood, the glucokinase helps circumvent many physiological disorders.

A dimeric structure typifies hexokinases from most sources. In general, they are completely inhibited by *p*-chloromercurobenzoate (p. 178), indicating that free sulfhydryl groups are essential for activity. (Optimum catalytic activity is generally also dependent on the presence of a divalent metal cation M^{2+}, such as Mg^{2+}.) They are also typified by regulatory characteristics, with a high concentration of glucose-6-phosphate having a feedback inhibitory effect on many hexokinases. This is not surprising, since metabolic control at this critical point would be an effective way of controlling carbohydrate metabolism, particularly glycolysis. Thus, when the cellular level of glucose-6-phosphate builds up due to the lack of a need for extensive carbohydrate degradation (that is, when the organism is in a resting state with a low energy demand), the rate of mobilization of glucose for catabolism is diminished. A high concentration of the other product, ADP, has the reverse effect; it activates the enzyme. This might seem to be contradictory, but the opposite effects are wholly complementary. Later, we will explain why. The binding of the ATP substate also displays negative cooperativity. All interesting features, yet they comprise only a part of the control existing in the glycolytic pathway.

Phosphofructokinase

The formation of the hexose diphosphate represents the *committed step* of glycolysis. A variety of other reactions involve G1P, G6P, and F6P as substrates, but when F6P is converted to FDP it has only two subsequent metabolic fates: (1) cleavage by aldolase to prime the rest of glycolysis or (2) conversion back to F6P (see p. 396, this chapter). Overall the enzyme is typical of other ATP-dependent kinases. It requires Mg^{2+} for activity (K^+ also increases activity) and does not catalyze the reverse reaction. It differs of course in its high specificity for D-fructose-6-phosphate.

*ADP and AMP
stimulate the reaction
by activating the
enzyme*

$$\text{ATP} + \text{D-fructose-6-phosphate} \xrightarrow[\text{(Mg}^{2+}, \text{ K}^+)]{\text{phosphofructokinase}} \text{D-fructose-1,6-diphosphate} + \text{ADP}$$

*ATP and NADH block
the reaction by
inhibiting the enzyme*

the committed step of the glycolytic pathway

Phosphofructokinase is also distinguished by a host of regulatory characteristics. This makes perfect sense because, being the committed step, the formation of FDP would be an optimum control on the flow of carbon through glycolysis. Indeed, the F6P-FDP interconversion is labeled as a *major control site* in glycolysis. The activity of phosphofructokinase (FDP formation) is stimulated by high levels of ADP and AMP and is depressed by high levels of ATP, NADH, and citrate. Later in the chapter we will examine the metabolic logic of these effects together with complementary regulatory effects on the reverse reaction of FDP → F6P (see p. 397).

Aldolase

Aldolase catalyzes the only degradative step involving a C—C bond of the glycolytic pathway; this results in the cleavage of one 6-carbon unit (a ketohexose) into two 3-carbon units (a ketotriose and an aldotriose). In conjunction with the *trioseisomerase*, the overall effect is specifically the conversion of one unit of FDP to two units of glyceraldehyde-3-phosphate (G3P). Obviously, the isomerization is a key step in glycolysis, since it provides for the complete catabolism of the whole hexose unit rather than just half of it.

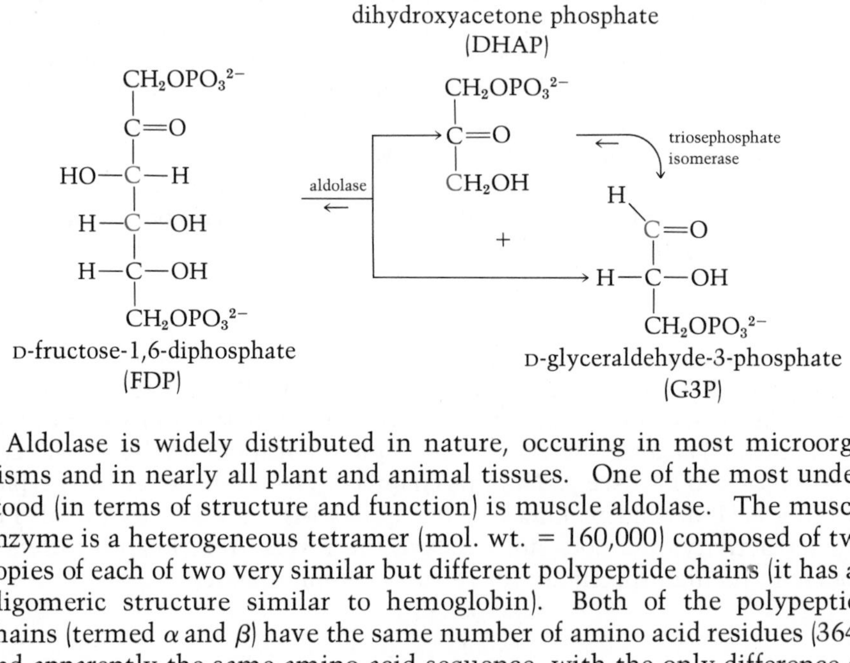

The name aldolase was coined in reference to the ability of the enzyme to also catalyze the reverse reaction (aldehyde + ketone → larger ketone), which in the language of organic chemistry is termed an *aldol condensation*.

Aldolase is widely distributed in nature, occuring in most microorganisms and in nearly all plant and animal tissues. One of the most understood (in terms of structure and function) is muscle aldolase. The muscle enzyme is a heterogeneous tetramer (mol. wt. = 160,000) composed of two copies of each of two very similar but different polypeptide chains (it has an oligomeric structure similar to hemoglobin). Both of the polypeptide chains (termed α and β) have the same number of amino acid residues (364), and apparently the same amino acid sequence, with the only difference so far detected being the fourth position from the C-terminus, and it is a small variation—aspartic acid appears in the α chain, asparagine in the β chain.

Isoenzyme forms of aldolase also occur. These are species composed of different proportions of the two different subunits: α_4, $\alpha_3\beta_1$, $\alpha_2\beta_2$, $\alpha_1\beta_3$, and β_4. The physiological significance of the isoenzyme species of aldolase is not clear. The relative amounts of each species do vary with age and usually from tissue to tissue. The predominant form appears to be $\alpha_2\beta_2$. The enzyme from microorganisms is a homogeneous dimer (mol. wt. = 80,000), which would obviously seem to indicate that the active form in lower organisms is only one-half the size of that found in more advanced life forms.

The mechanism of action is another example of enzymes that enter into a covalent association with the substrate. With aldolase, the active form of the enzyme-substrate complex is proposed to be a *Schiff base* $\left(\diagup C{=}N{-}\right)$ formation involving the carbonyl grouping $\left(\diagup C{=}O\right)$ of the substrate and the epsilon amino group ($H_2N{-}$) of a particular lysine side chain (residue 221). The presence of a single disulfide bond in each chain appears to be especially crucial in confirming a readily accessible surface position of the lysine-containing active site in each chain. As indicated in the diagram below the preferred binding pattern of the hexose substrate to the enzyme is

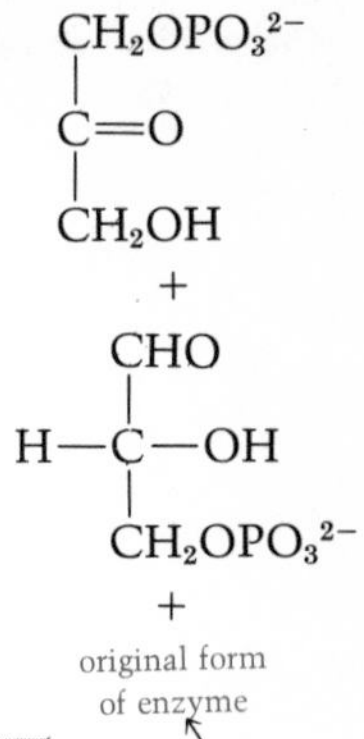

$$CH_2OPO_3^{2-}$$
$$|$$
$$C{=}O$$
$$|$$
$$CH_2OH$$
$$+$$
$$CHO$$
$$|$$
$$H{-}C{-}OH$$
$$|$$
$$CH_2OPO_3^{2-}$$
$$+$$

original form
of enzyme

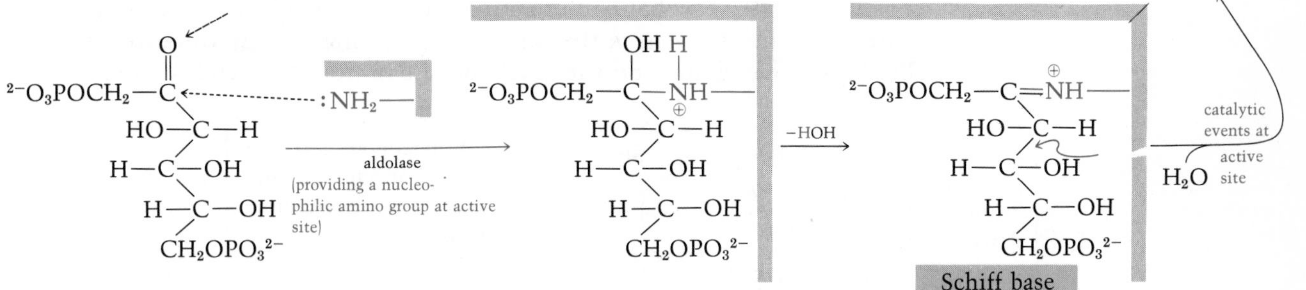

Schiff base
intermediate

catalytic
events at
active site

one with a strong preference for sugars with a *trans* orientation of the C-3 and C-4 hydroxyls. The chemical events of bond cleavage (not shown in the diagram) have been associated with the participation of a second lysine residue, a histidine residue, and a cysteine residue.

Glyceraldehyde-3-Phosphate Dehydrogenase

The conversion of glyceraldehyde-3-phosphate to 1,3-diphosphoglycerate (1,3DPG) represents the *distinct oxidative step* of glycolysis and is one of the two phosphate-requiring steps. The most significant aspect of the reaction is that a high-energy acyl phosphate (diphosphoglycerate) is formed as a product *without* the direct expenditure of energy. That is, *no ATP is required*. The reaction is, however, related to ATP because the conversion of diphosphoglycerate to 3-phosphoglycerate in the next reaction is coupled to *ATP formation*. The next two reactions have the same significance, with the high-energy phosphoenolpyruvate also being formed without ATP. The PEP → pyruvate conversion is then also coupled to *ATP formation*. These two ATP-yielding sites, each involving single enzyme-catalyzed process, are labeled as *substrate-level phosphorylations*, to distinguish them from the ATP production by the more elaborate process of oxidative phosphorylation.

Glyceraldehyde phosphate dehydrogenase is known to be an oligomeric molecule consisting of four identical chains. The protomer of both pig and lobster muscle has been sequenced completely and shown to contain 333

The Schiff base mechanism of muscle aldolase is proposed to apply in general to all of the plant and animal aldolases, but not to aldolases from microorganisms. The mechanism for the latter is proposed to involve a noncovalent enzyme-substrate complex dependent on the obligatory participation of a divalent metal ion. The point of mentioning this is to focus again on the concept that in different cells the same reaction may occur either (a) by an identical or very similar process *or* (b) by a very different process. Thus, an attitude of caution must prevail before drawing comparative conclusions regarding the biochemical similarity of the same processes in different organisms.

An important side reaction in red blood cells that involves 3-phosphoglycerate is discussed later on p. 391.

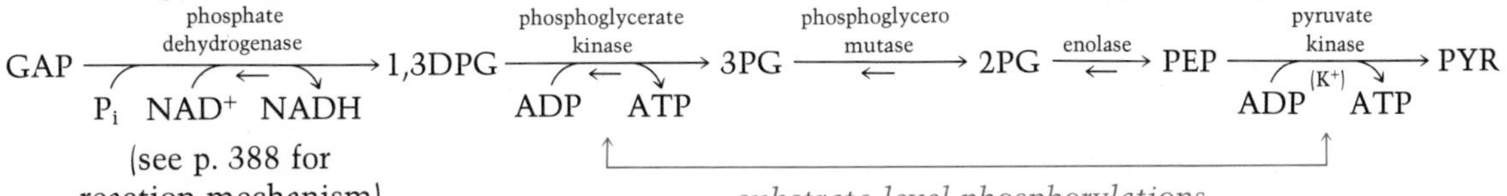

GAP —[glyceraldehyde phosphate dehydrogenase]→ 1,3DPG —[phosphoglycerate kinase]→ 3PG —[phosphoglycero mutase]→ 2PG —[enolase]→ PEP —[pyruvate kinase (K⁺)]→ PYR

P_i NAD^+ NADH

(see p. 388 for reaction mechanism)

ADP   ATP   ADP   ATP

substrate-level phosphorylations of glycolysis

amino acid residues (mol. wt. = 37,000) in each case. Moreover, of the total 333 residues, 240 are identical in each preparation. The extent of homology in amino acid sequence is surprising, because the pig and lobster are widely separated on a phylogenetic scale. This suggests, then, that a great majority of amino acid residues are essential for catalytic activity.

Regardless of specific structural differences in the enzyme from various sources, all preparations studied possess several sulfhydryl groups that are absolutely required for activity. Although the number of SH groups per chain is known to vary, it appears that at least one is part of the active site of each protomer. The active form of the enzyme is also known to contain *bound NAD* in each monomer. Some of the general features for the proposed multistep mechanism of action are diagramed below. An important part of the mechanism is the formation of a high-energy thioester linkage between the enzyme and the aldehyde substrate. This yields an activated

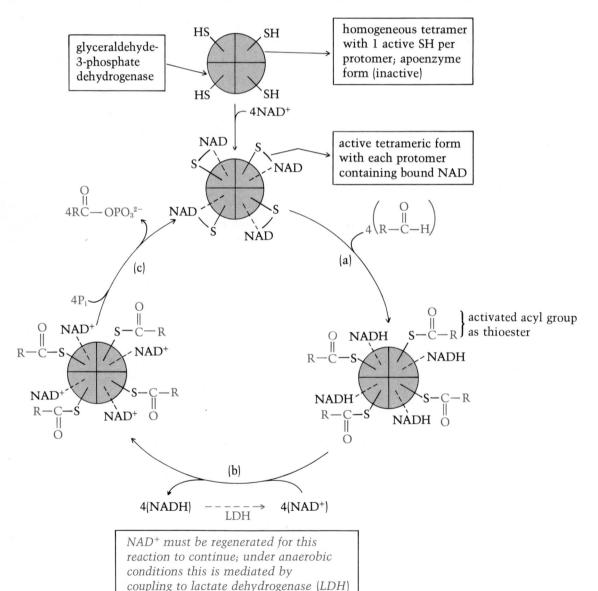

acyl group that is highly reactive towards inorganic phosphate, leading to the formation of the high-energy acylphosphate.

NAD-dependent dehydrogenases are probably the most abundant type of enzymes, and this being our first specific encounter of one, it is a good opportunity to look more closely at these enzymes. One interesting aspect of NAD-dependent dehydrogenases concerns the stereochemical specificity in the way they bind to NAD^+ or NADH. That there could even be a basis for specificity of NAD-binding is made evident by inspection of the photographs and diagrams below. As indicated, the preferred conformation of NAD^+ or NADH in solution is a folded one, stabilized by the hydrophobic interaction of the adenine and nicotinamide ring structures. (This is the same sort of interaction that applies to the stacked bases in the DNA double helix.) However, because NAD can fold in either of two ways, there are two similar conformations of probably equal energy. Yet, they are different in precise structural features, such as which *side* of the nicotinamide grouping is exposed. As the diagram indicates, most dehydrogenases discriminate between these two conformations because of the asymmetry of the binding and catalytic residues on the enzyme surface. GAP-dehydrogenase is known to bind with the conformation that exposes the B face, whereas lactate dehydrogenease and alcohol dehydrogenase prefer the conformation that exposes the A face. This is just one more particular example of the general principle of enzyme specificity.

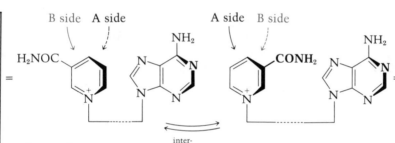

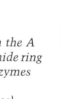

in this conformation the B side of the nicotinamide ring is exposed; some enzymes prefer this (GAP-dehydrogenase)

inter-convertible by bond rotations

in this conformation the A side of the nicotinamide ring is exposed; some enzymes prefer this (lactate dehydrogenase)

very few enzymes will bind with either conformation

Lactate Dehydrogenase

The last reaction of glycolysis has been mentioned earlier relative to its operational significance in terms of internal coupling to regenerate NAD^+ for GAP-dehydrogenase. Thus under anaerobic conditions the glycolytic pathway is self-sufficient. This is not, however, the only metabolic function of lactate dehydrogenase (LDH). Another operational characteristic of importance is that the enzyme also catalyzes the reverse reaction, that is, the formation of pyruvate from lactate. In animals the significance of this is that the lactate generated in the anaerobic metabolism of certain tissues, such as skeletal muscle, can be transported to other more aerobic tissues, such as the liver, where it is reconverted to pyruvate. The pyruvate can then be metabolized further via the citric acid cycle or be reconverted to carbohydrate material as free glucose or stored as glycogen. The latter process,

389

LDH isoenzyme patterns on electrophoresis (pH 8.8)

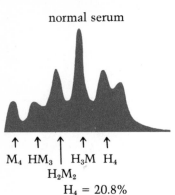

normal serum

↑ ↑ ↑ ↑ ↑
M₄ HM₃ | H₃M H₄
$\quad$ H₂M₂

$H_4 = 20.8\%$
$H_3M = 36.4\%$
$H_2M_2 = 25.2\%$
$HM_3 = 12.0\%$
$M_4 = 5.6\%$

serum after coronary embolism

↑ ↑↑ ↑ ↑
M₄ HM₃ | H₃M H₄
$\quad$ H₂M₂

$H_4 = 37.4\%$
$H_3M = 41.2\%$
$H_2M_2 = 8.6\%$
$HM_3 = 12.8\%$
$M_4 = 0$

involving still other reversible steps of the glycolytic pathway, will be discussed shortly.

Like many of the glycolytic enzymes, LDH is an oligomer, and, specifically, a tetramer. It was the first enzyme to be discovered to exist in multiple *hybrid* molecular forms (isoenzymes; see p. 146) each of which is a tetramer consisting of different proportions of two polypeptide chains. The chains are generally referred to as the H and M chains, relating to their preponderance in either heart or muscle tissue. The five possible combinations are H_4, HM_3, H_2M_2, H_3M, M_4. In humans, LDH is widely distributed in various tissues and body fluids. Furthermore, for reasons which are not yet clear, each source apparently possesses different but characteristically normal levels of each hybrid form. Nevertheless, the phenomenon has been exploited as a sensitive clinical tool, since both the serum level of LDH activity and the isoenzyme pattern are modified in various disease states. The clinical value results from the fact that the changes are often characteristic of a particular disease and of the particular stage of that disease (see margin). Clinical applications of this sort for various other enzymes have been developed, and contribute significantly to the art of modern medical diagnosis of diseases. (For further information on this subject, see Appendix I.)

Glycerol Phosphate Dehydrogenase

We have not mentioned it previously, but there is a third NAD-dependent dehydrogenase associated with the glycolytic pathway, namely, *glycerol phosphate dehydrogenase*. It catalyzes (reversibly) the interconversion of dihydroxyacetone phosphate and glycerol-3-phosphate. As indicated below, it is characterized by absolute stereospecificity, forming and using only the L isomer of glycerol phosphate. This specificity of action also extends to NAD⁺ (NADH), with the preferred conformation being the one with the B side of nicotinamide exposed.

The metabolic significance of this reaction is that it in part accounts for a *link between carbohydrate and lipid metabolism*. (Other aspects of the integration of lipid and carbohydrate metabolism are examined in Chap. 13; see p. 438). Glycerol phosphate can be used directly in the biosynthesis of acylglycerols and phosphoacylglycerols (see p. 518, Chap. 16). Alternatively, when these glycerol-containing lipids are themselves degraded, the glycerol can be degraded further after entering the glycolytic pathway via its conversion to DHAP. (In some microorganisms growth conditions can be manipu-

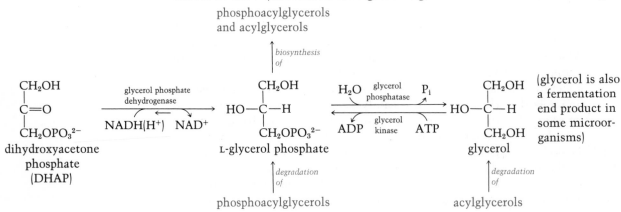

phosphoacylglycerols
and acylglycerols

biosynthesis of

CH₂OH
|
C=O
|
CH₂OPO₃²⁻
dihydroxyacetone
phosphate
(DHAP)

glycerol phosphate dehydrogenase

NADH(H⁺) NAD⁺

CH₂OH
|
HO—C—H
|
CH₂OPO₃²⁻
L-glycerol phosphate

H₂O glycerol phosphatase Pᵢ

ADP glycerol kinase ATP

CH₂OH
|
HO—C—H
|
CH₂OH
glycerol

(glycerol is also a fermentation end product in some microorganisms)

degradation of

phosphoacylglycerols

degradation of

acylglycerols

390

lated to achieve the accumulation of glycerol as a fermentation product.) Another possible role of glycerol phosphate dehydrogenase is described in Chap. 14; see p. 474.

Diphosphoglyceromutase (Biosynthesis of Hemoglobin Regulator)

The mature red blood cell is one place where *aerobic glycolysis* occurs, that is, *lactate is produced even though O_2 is present.* Thus, glycolysis provides the major source of ATP production for these cells. This makes perfect physiological sense since the mature red blood cell is incapable of producing ATP via the mitochondrial process of oxidative phosphorylation—red blood cells have no mitochondria.

In addition to ATP production, the pathway of glycolysis also provides red blood cells with a supply of *2,3-diphosphoglycerate*, which regulates the binding of O_2 to hemoglobin (see p. 197 in Chap. 6). The regulator is formed from 1,3-diphosphoglycerate by the enzyme, *diphosphoglyceromutase.* Another enzyme in red blood cells, diphosphoglycerophosphatase, converts the regulator back to 3-phosphoglycerate for further catabolism but obviously only after the regulator performs its function.

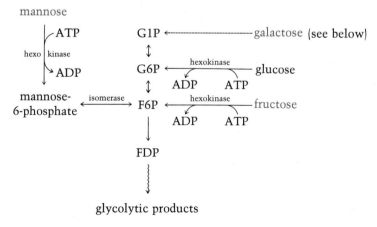

CATABOLISM OF OTHER SUGARS

The operation of the glycolytic pathway is not solely confined to the degradation of glucose or glycogen. On the contrary, the same overall sequence is responsible for the catabolism of most other simple sugars as well. All that is required is a port of entry by conversion of the sugar into one of three phosphorylated intermediates—G1P, G6P, or F6P. This leads to the formation of fructose-1,6-diphosphate. The simplest example is the hexokinase-catalyzed conversion of *fructose* to fructose-6-phosphate. *Mannose* can enter at the same level via a two-step process. The first step involves a kinase-catalyzed conversion to mannose-6-phosphate; this is followed by an isomerase-catalyzed conversion to F6P. The formations of F6P and M6P are identical to the glucose + ATP → glucose-6-phosphate + ADP conversion, and in fact all are catalyzed by the same hexokinase.

In view of the entrance mechanism for mannose, you might expect *galactose* to be processed in a similar fashion, but such is not the case. A completely different process operates here, with galactose being converted to

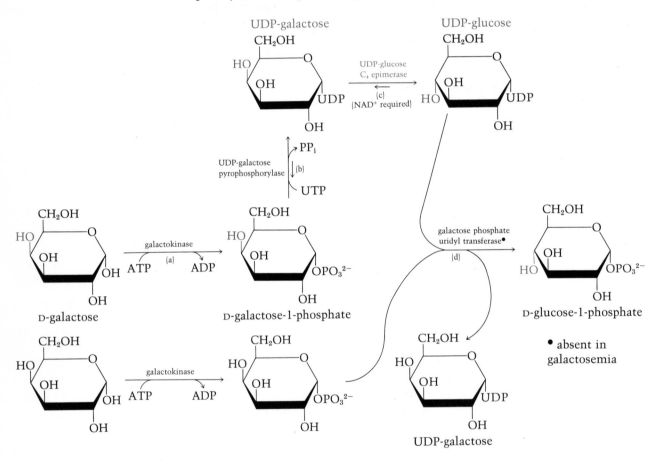

(a)   galactose + ATP ⟶ galactose-1-P + ADP

(b)   galactose-1-P + UTP ⟶ UDP-galactose + PP$_i$

(c)   UDP-galactose ⟶ UDP-glucose

(d) ● UDP-glucose + galactose-1-P ⟶ UDP-galactose + glucose-1-P (not in galactosemia)

Net:*   2 galactose + 2ATP + UTP $\xrightarrow{\text{4Es}}$ UDP-galactose + glucose-1-P + PP$_i$ + 2ADP

Overall effect: 1 galactose ⟶ 1 glucose

 * galactokinase reaction taken twice

glucose through the combined action of four different enzymes (see p. 392). First, galactose is converted to galactose-1-phosphate (step (a)). This involves a kinase with a high specificity for galactose (*galactokinase*) and the phosphorylation of C^1 (rather than C^6, which is typical for most other sugar kinases). Gal-1-P and UTP then act as substrates for the formation of UDP-galactose (step (b)). In the key reaction of this process (step (c)), catalyzed by *UDP-galactose-C^4-epimerase*, the C^4 position of the galactose moiety is epimerized to yield UDP-glucose. In other words, this is the step where the conversion of galactose to glucose actually occurs. In step (d), the glucose unit is released as glucose-1-phosphate in a transfer reaction between UDP-glucose and a second unit of galactose-1-phosphate. The overall process is summarized by the net reaction.

The UDP sequence that is shown is the main process for the normal metabolism of galactose. It is also a pathway exemplifying one of the most classical cases of a metabolic hereditary disorder—*galactosemia*. The disease is characterized by a deficiency of *galactose phosphate uridyl transferase* (d), creating a blockage in the metabolism of galactose. The immediate result is an abnormally high cellular level of galactose-1-phosphate. The situation is quite serious in infants, where the major natural source of dietary carbohydrate is milk lactose, which of course contains galactose (p. 304, Chap. 9). Severe cases of infant galactosemia are characterized by cataract formation, cirrhosis of the liver and spleen, and in some cases mental retardation. Generally, however, all of these clinical manifestations can be avoided by simply eliminating galactose and all galactose-containing substances from the diet. As the infant ages, less dietary control is required, since the adult galactosemic is capable of metabolizing galactose-1-phosphate by alternate routes.

GLYCOGENESIS

In the preceding pages we have seen how the reactions of glycolysis effect the degradation of carbohydrates. With this analysis now complete, the next obvious question to consider concerns the biosynthesis of carbohydrates (termed *glycogenesis*) from smaller carbon fragments such as pyruvate and lactate. (*Note:* For the moment we are focusing attention on the formation of carbohydrates in nonphotosynthetic organisms. The photosynthetic assimilation of CO_2 into carbohydrate material will be discussed separately in Chap. 15.)

Since lactate dehydrogenase is capable of reversibly catalyzing a conversion of pyruvate and lactate, the question is reduced to determining how pyruvate is converted to glucose or glycogen. In addressing ourselves to this question, the immediate tendency would be to suggest that glycogenesis proceeds by a complete reversal of glycolysis. However, this would be incorrect. In that case, then, a logical counter-suggestion would be that a completely different pathway is involved. This is also incorrect. What then, you ask, is the mechanism of glycogenesis? Well, a little thought would lead you to conclude that, if it is neither by a complete reversal of glycolysis nor by an entirely different pathway, then it probably involves a blend of both. In this case you would be correct. Although valid, the conclusion generates a logical uncertainty; namely, to what extent is the glycolytic pathway reversible? That is, which of the reactions are reversible and which are irreversible?

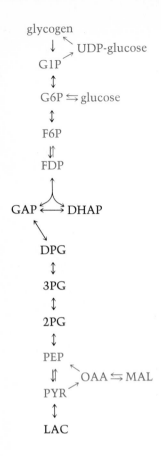

The facts are as follows. The biosynthesis of the metabolically active forms of glucose (G1P and G6P) from smaller carbon units proceeds by a direct reversal of the same reactions of glycolysis with only two exceptions. The two irreversible reactions are those catalyzed by pyruvate kinase and phosphofructokinase. If we consider the biosynthesis of free glucose and free glycogen, then two other reactions are to be included as well, namely, those catalyzed by hexokinase and glycogen phosphorylase. In other words, the *four sites of enzyme irreversibility in the pathway of glycolysis* are $PYR \rightarrow PEP$; $FDP \rightarrow F6P$; $G6P \rightarrow GLU$; and $G1P \rightarrow$ glycogen.

These sites of enzyme irreversibility are logically circumvented. Each conversion involves *enzymes different than those that operate in glycolysis*. The most complicated bypass is for $PYR \rightarrow PEP$, because it occurs in different ways in different cells. In some bacteria, for example, the reaction is catalyzed by a single, ATP-dependent enzyme (*phosphoenolpyruvate synthetase*), and results in the direct formation of PEP from PYR (route 1 on the facing page). In other bacteria and also in the higher plant and animal organisms the conversion is indirect and involves the participation of two enzymes, namely, *pyruvate carboxylase* and *phosphoenolpyruvate carboxykinase* (route 2). The carboxylase catalyzes the ATP-dependent carboxylation of pyruvate (C_1 plus C_3) to yield oxaloacetate (C_4). This reaction requires the participation of *biotin* as an essential coenzyme, as do several other CO_2-fixation reactions. (The role played by biotin in this type of reaction will be examined further in Chap. 16; see p. 509). The carboxykinase then catalyzes a decarboxylation of oxaloacetate, which is accompanied by phosphorylation to yield phosphoenolpyruvate. Depending on the cell type, the carboxykinase shows a distinct preference for utilizing either GTP (guanosine triphosphate) or ITP (inosine triphosphate) as the phosphate donor, rather than ATP.

In the eukaryotic cells of higher organisms the pyruvate carboxylase seems to be localized exclusively in the mitochondrion, regardless of cell type. This creates no problem since pyruvate can pass through the mitochondrial membrane. Depending on the source, however, phosphoenolpyruvate carboxykinase has been found either in the mitochondrion or in the cytoplasm. When it occurs in the mitochondrion, this means that PEP is formed within the mitochondria (route 2). Although the subsequent utilization of PEP in glycogenesis must occur in the cytoplasm (that's where the enzymes are), there is no problem because the phosphoenolpyruvate, like pyruvate, can readily cross the mitochondrial membrane.

When the carboxykinase occurs in the cytoplasm, however, a problem does exist, because oxaloacetate (OAA), the substrate of the carboxykinase, is generated in the mitochondrion and it does not readily cross the mitochondrial membrane. Thus, if the oxaloacetate is formed within the mitochondrion, the question is, how can it be converted to PEP when the necessary enzyme is located outside the mitochondrion? Well, it is proposed that this apparent short circuit is circumvented by the participation of two additional enzymes, *malate dehydrogenase* and *malate dehydrogenase*. No, this is not an error, the enzymes have the same name and they do catalyze the same reaction, namely, a reversible oxidation-reduction reaction between malate (an alpha hydroxy dicarboxylic acid) and oxaloacetate (an alpha keto dicarboxylic acid). Both enzymes are also NAD-dependent. The enzymes differ, however, in terms of their cellular localization in that one occurs *inside* the mitochondrion, and the other occurs *outside* the mito-

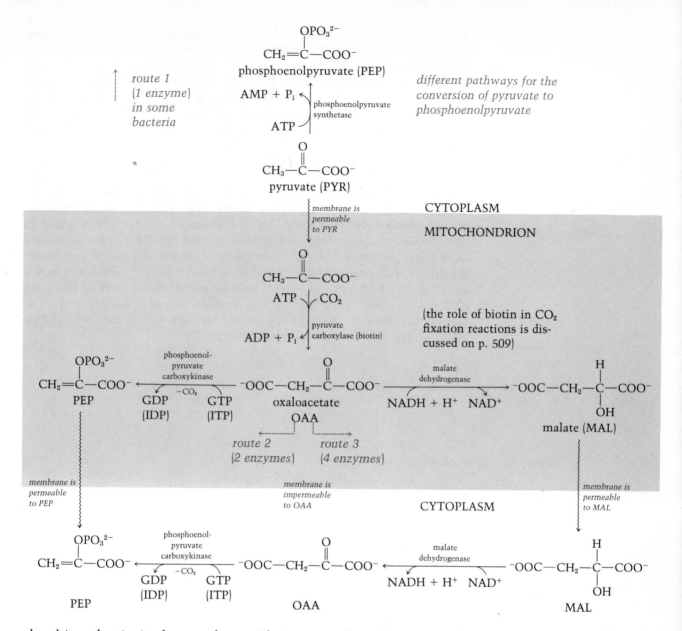

chondrion, that is, in the cytoplasm. Their proposed participation in the PYR → PEP conversion is shown in the diagram as route 3. In this case the link between what occurs in the mitochondrion and in the cytoplasm is provided by the movement of malate across the mitochondrial membrane. When glycolysis occurs under aerobic conditions, this *malate shuttle* can perform another important function by operating in reverse, that is, malate (cytoplasm) → malate (mitochondrion) (see p. 471).

The considerable attention we have given to the PYR → PEP conversion is not to imply any special preeminence for it. Certainly it is an important and key step, but so are several other reactions. The point is that it serves as an excellent example of both the biochemical diversity that exists in nature and the *principle of compartmentalization*. Despite the diversity, however, note that in each route the same net effect is realized. Pyruvate is converted to phosphoenolpyruvate in a process that requires the expenditure of what can be considered as the equivalent of 2 ATP molecules: ATP $\rightarrow$ ADP $\rightarrow$ AMP in bacteria; ATP $\rightarrow$ ADP and GTP(ITP) $\rightarrow$ GDP(IDP) in eukaryotic cells.

After its formation from PYR by either of these routes, PEP is then converted to FDP by a direct reversal of glycolysis. Once FDP is formed, the three remaining irreversible sites of glycolysis are bypassed by processes involving still other enzymes. The first bypass utilizes *fructose diphosphatase* to convert FDP to F6P + P_i (see below). The F6P is then converted to G6P by the same isomerase that functions in glycolysis. *Glucophosphatase* then converts G6P to free glucose and P. Alternatively, the G6P can be converted to G1P by the same mutase that functions in glycolysis. Finally, the hexose moiety of G1P is incorporated into glycogen via a two-enzyme sequence. First, G1P is converted to UDP-glucose by the action of a specific *pyrophosphorylase*, and then the glucose moiety of UDP-glucose is incorporated into glycogen via *glycogen synthetase*. (Recall our earlier mention of the role of NuDP-sugars in polysaccharide biosynthesis.) Thus, we see that although the pathways of glycolysis and glycogenesis involve the *same chemical intermediates* (the two exceptions are oxaloacetate and UDP-glucose) *certain steps do require different enzymes* for the catabolic and anabolic conversions.

$$PYR \rightarrow \rightarrow PEP \qquad \text{(see p. 394)}$$

then by reversal of glycolysis: $PEP \rightarrow 2PG \rightarrow 3PG \rightarrow DPG \rightarrow GAP \rightarrow DHAP \rightarrow FDP$

then: $FDP \xrightarrow[\text{H}_2\text{O}]{\text{diphosphatase}} F6P + P_i$

then by reversal of glycolysis: $F6P \rightarrow G6P$

then: $G6P \xrightarrow[\text{H}_2\text{O}]{\text{glucophosphatase}} \text{glucose} + P_i$

or by reversal of glycolysis: $F6P \rightarrow G6P \rightarrow G1P$

then: $G1P \xrightarrow[\text{UTP} \quad \text{PP}_i]{\substack{\text{UDP-glucose} \\ \text{pyrophosphorylase}}} \text{UDP-glucose} \xrightarrow[(\text{glucose})_n \quad \text{UDP}]{\substack{\text{glycogen} \\ \text{synthetase}}} (\text{glucose})_{n+1}$

OVERALL VIEW OF THE REGULATION OF GLYCOLYSIS AND GLYCOGENESIS

Earlier we described the regulatory characteristics of some of the enzymes that function in glycolysis—glycogen phosphorylase, hexokinase, and phosphofructokinase. Let us now examine in the same context two of the separate enzymes that function in glycogenesis—fructose diphosphatase and glycogen synthetase. This will focus our attention on two steps, $F6P \rightleftharpoons FDP$ and glycogen $\rightleftharpoons$ G1P, and provide a clear illustration of precisely what is meant by the operation of regulatory checks and balances.

Purified preparations of the diphosphatase are strongly inhibited by AMP and ADP. What does this mean? Well, remember that phosphofructokinase is activated by AMP and ADP. Thus, here is a situation where the same effectors (AMP and ADP) have completely *opposite but complemen-*

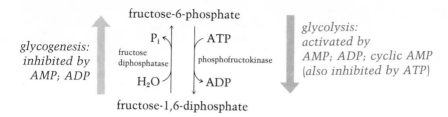

tary regulatory effects on two different enzymes that participate in different directions. The significance of this is not difficult to appreciate.
When a cell is using a lot of energy, that is, when ATP is being rapidly consumed and there is an accompanying rise in the level of ADP and AMP, the
latter serves as a metabolic signal *both to stimulate* the degradation of carbohydrate in order to provide more ATP and *to depress* the biosynthesis of
carbohydrate. Then, on return to a less energy-demanding state, the ATP
level would rise and there would also be a reduction in the levels of ADP and
AMP. Since the activity of purified preparations of phosphofructokinase is
inhibited by high levels of ATP, the rise in cellular ATP would also be a
signal that reduces the rate of glycolysis and further ATP formation.

The principle of complementary control at the F6P ⇌ FDP site also applies to the formation and degradation of glycogen. In fact, *glycogen synthetase* appears to be under the same type of hormonal control as glycogen phosphorylase. In other words, the adrenalin-cAMP activation of
glycogen phosphorylase is accompanied by an inhibition of glycogen synthetase. Yes, the same sort of elaborate relationship that we described for
glycogen phosphorylase, which involved different active forms interconverted by enzymes, which are themselves controlled by the level of
cAMP, also applies to glycogen synthetase. The two synthetase forms
have been labeled *glycogen synthetase D* and *glycogen synthetase I*.
The D form, a phosphorylated oligomer comparable to phosphorylase *b*,
is less active, but is strongly activated by glucose-6-phosphate (D meaning
dependent on G6P). The activity of the I form, which does not require (is
independent of) G6P, is less than that of the G6P-activated D form. cAMP
stimulates the I → D conversion. Thus, in response to an energy demand,
cAMP would stimulate the production of active phosphorylase and a form of
the synthetase that is active only when G6P levels are high, which will not
be the case when glycolysis is operating at an increased rate to satisfy ATP
demand. So in the absence of high levels of G6P the D form of the synthetase is inactive and glycogenesis is depressed.

At this point a diagram is worth a thousand words. Figure 12–6 summarizes all of the regulatory features that we have considered to this point, including one not previously mentioned, namely, the activation of the
PEP → PYR step in glycolysis. This diagram should impress upon you that
certain enzymes of the main pathways of carbohydrate metabolism are most
definitely programed to satisfy the changing needs of an organism in the
most efficient and economical way possible. This principle applies to all of
metabolism, and other examples will be examined in later chapters. In fact,
we will extend our analysis of the regulation of carbohydrate metabolism in
the next chapter to include the effects exerted by NAD^+, NADH, and other
effectors.

Hormonal control of carbohydrate
metabolism is also effected by *insulin*. (This is just one of actions
of insulin.) The insulin control is
twofold. It activates the membrane transport of glucose and also
activates the storage of glucose as
glycogen. Both effects occur subsequent to the binding of insulin
to its receptor protein (IRP) in the
membrane. What happens after
the insulin-binding is not yet
clear, but some evidence favors
either a reduction of intracellular
levels of cyclic-AMP or an increase
in cyclic-GMP levels, or both.

Don't be concerned if you are experiencing some difficulty in
appreciating this material. The
difficulty is not, however, due so
much to the complexity of the
concept, as it is to the fact that we
have yet to examine in detail how
the major metabolic pathways are
integrated with each other. As
your knowledge of metabolism
grows, be assured that your understanding of these metabolic regulatory mechanisms will also grow.

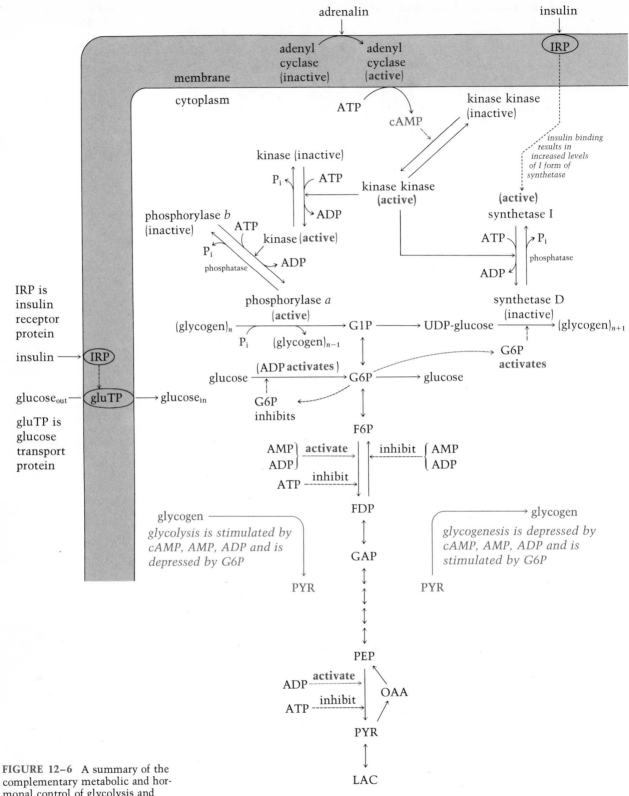

FIGURE 12–6 A summary of the complementary metabolic and hormonal control of glycolysis and glycogenesis. (A partial description. For more information see p. 440 in Chap. 13.)

CARBOHYDRATE CATABOLISM AND MUSCLE CONTRACTION

The phenomenon of muscle contraction has long interested many workers. Originally it was thought to be a purely physical process caused by the mechanical shortening of muscle fibers. This idea was laid to rest in 1954 when A. F. Huxley and H. E. Huxley (not related) independently proposed a molecular mechanism for contraction involving the relative *sliding movement* of protein filaments within muscle tissue. Since then the sliding filament model has been supported and refined by extensive research and is now accepted as the nucleus of our current theory of the contractile process. To understand the proposed mechanism, and ultimately the function of the ATP generated by glycolysis and other energy-yielding processes, it is necessary to first examine some of the details of the ultrastructure of muscle tissue.

As shown in Fig. 12–7, the cells of muscle tissue are bundles of longitudinal fibers called *myofibrils*. As revealed by electron microscopy, each myofibril is composed of a linear arrangement of distinctive repeating units called *sarcomeres*. The structural order of each sacromere is due to a parallel arrangement of several protein *myofilaments*, of which there are two main types. One is an aggregate of several molecules of the protein *myosin*. For obvious reasons, it is called the *myosin filament*. The other is an aggregate of the protein *actin*, and is called the *actin filament*. X-Ray diffraction and high-resolution electron microscopy have shown the myosin filaments to be very electron dense and characterized by short projections regularly spaced all along the surface of the filament. The projections are believed to arise from protruding globular regions of individual myosin molecules, which are primarily fibrous proteins. In the filament itself, the myosin molecules are aligned with each other via interactions among the extended fibrous regions of individual molecules.

By comparison, the actin filament is smaller in diameter (less electron dense) and contains no surface protrusions. It does contain several binding sites along its length that interact with the myosin filament. In the context of these structural characteristics of each myofilament, the electron micrographs of muscle tissue are interpreted as shown in the drawings of Fig. 12–7. In each sarcomere, myosin filaments are located only in a region called the *A band*, whereas the actin filaments traverse the region called the *I band* and protrude longitudinally into the A band. The protrusion contributes to an overlapping of actin and myosin filaments in the A band. See the legend of Fig. 12–7 for additional features.

The sliding filament model proposes that muscle changes length as the overlapping arrays of actin and myosin filaments slide past each other in each sarcomere. When the muscle is contracted, the actin filament is drawn further *into* the A band (see p. 401). In molecular terms Huxley proposed that the key to the movement is the *shifting of binding sites in an active actomyosin complex formed by the contact of the myosin protrusions with the smooth actin filaments.* One suggestion is that the globular myosin projections undergo a conformational change, causing a shift to a new binding site on the actin filaments; this would be analogous to a ratchet-type displacement. The effect would be the relative longitudinal movement (or sliding) of the two filaments, with the myosin filaments providing the so-called *power-stroke* for the displacement. The explanation is consistent with experimental observations that neither the myosin filament

Myosin and actin are the two major proteins in muscle tissue. However, there are two others, *troponin* and *tropomyosin*. The process of muscle contraction actually involves the interaction of all four proteins under the control of Ca^{2+}. The detail of these interactions is not given here. Excellent sources for this information are cited at the end of the chapter (one by Ebashi and another by Murray and Weber).

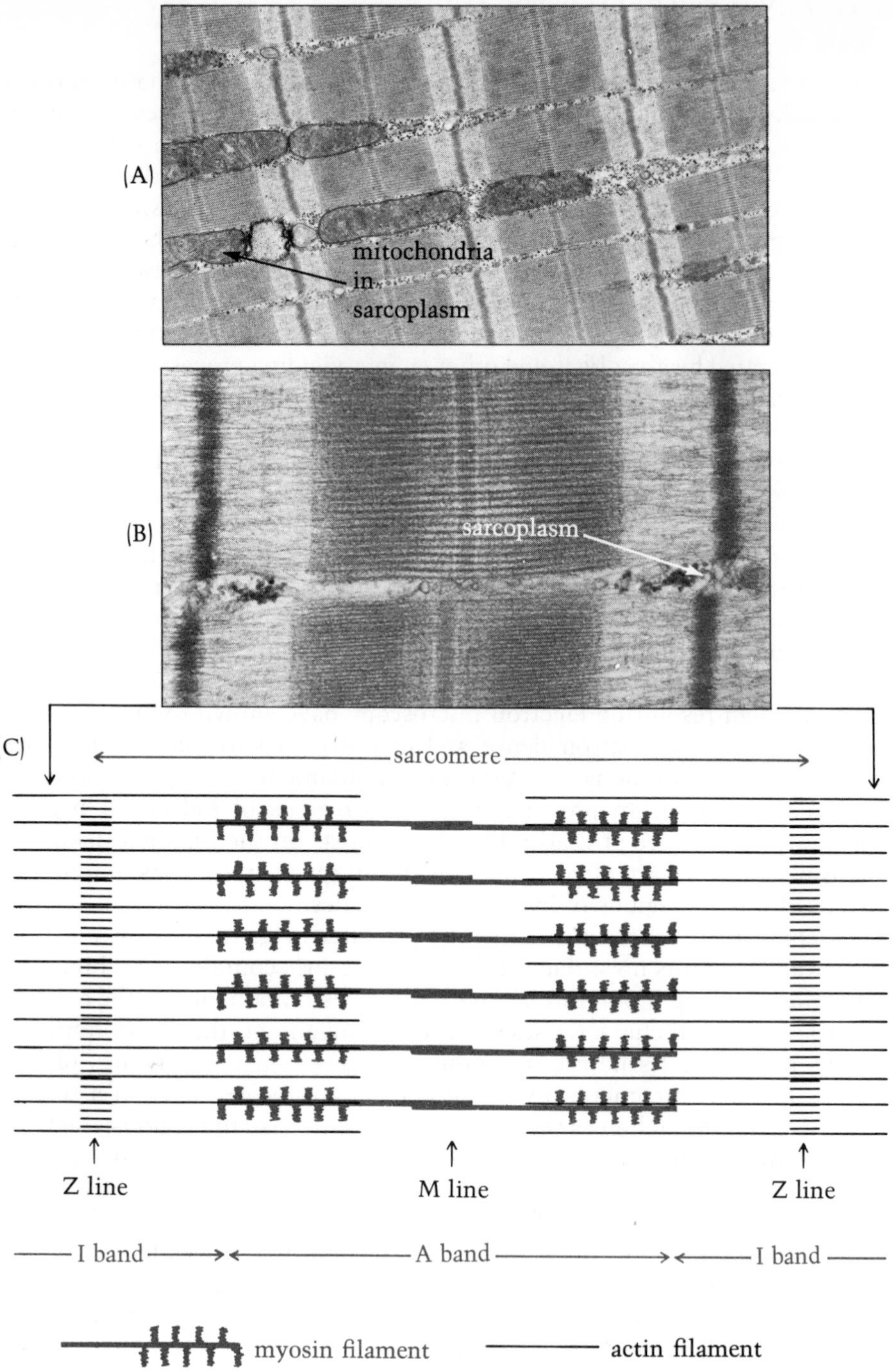

FIGURE 12–7 Ultrastructure of muscle. (A) Electron micrograph of myofibrils in a fiber of papillary muscle tissue of cat heart. The myofibrils run horizontally and are separated by the sarcoplasm, which contains mitochondria and glycogen granules. The latter appear as black dots. Magnification: 25,000×. (B) Electron micrograph of sarcomere unit in two adjacent myofibrils. Magnification: 77,800×. (C) Diagramatic representation of sarcomere unit depicting the parallel arrangement of myosin and actin filaments. Actin filaments occupy the light I band and penetrate some distance into the A band, where they interact with projections of myosin filaments. The A line corresponds to the joining of actin filaments from two adjacent sarcomeres. The M line corresponds to the joining of myosin filaments within the A band. (Reproduced with permission from D. W. Fawcett, *An Atlas of Fine Structure*, Philadelphia: W. B. Saunders Company, 1966. Photographs in (A) and (B) generously supplied by D. W. Fawcett.)

nor the actin filament undergoes a change in axial length. Only their relative positions are changed. It is the repeating sarcomere unit that changes in length.

Whatever the precise molecular details of the sliding phenomenon (see the *Scientific American* article cited at the end of the chapter), the process is definitely known to be ATP-dependent, with the ATP being supplied almost exclusively by the breakdown of carbohydrate. In skeletal muscle where

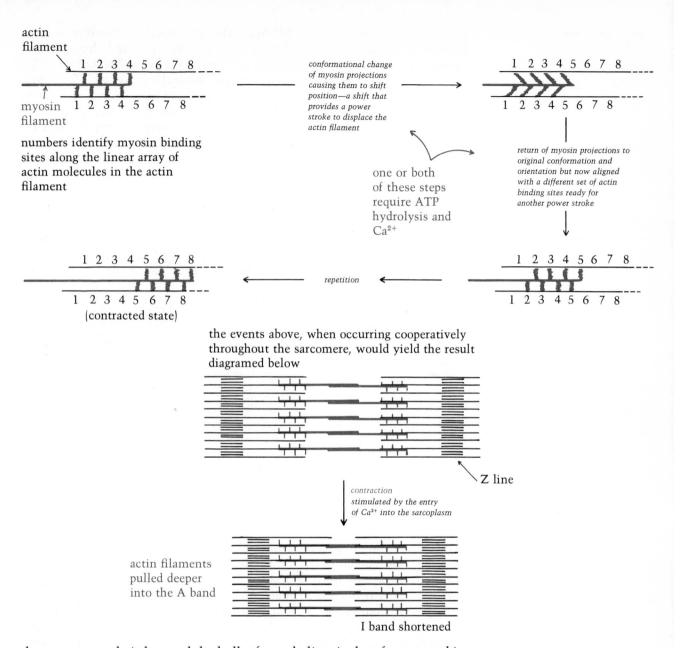

actin filament

myosin filament

numbers identify myosin binding sites along the linear array of actin molecules in the actin filament

conformational change of myosin projections causing them to shift position—a shift that provides a power stroke to displace the actin filament

one or both of these steps require ATP hydrolysis and Ca^{2+}

return of myosin projections to original conformation and orientation but now aligned with a different set of actin binding sites ready for another power stroke

repetition

(contracted state)

the events above, when occurring cooperatively throughout the sarcomere, would yield the result diagramed below

Z line

contraction stimulated by the entry of Ca^{2+} into the sarcoplasm

actin filaments pulled deeper into the A band

I band shortened

the oxygen supply is low and the bulk of metabolism is therefore anaerobic, the primary pathway is glycolysis, with glycogen reserves acting as the main carbon substrate and lactate being the final product. In other muscle tissues, such as cardiac muscle, that are capable of a considerably greater level of aerobic metabolism, the combined action of glycolysis through pyruvate formation and the citric acid cycle is more prevalent. In this instance, the bulk of the needed ATP will be supplied by mitochondrial oxidative phosphorylation. Consistent with this operational distinction between aerobic muscle tissue and anaerobic muscle tissue is the fact that the former type of cells generally contains a much larger proportion of mitochondria than the latter. Approximately 40% of the dry weight of heart represents mitochondria.

Although the immediate source of energy is ATP, the manner in which it

Note: Remember that the enzymes of the citric acid cycle and the molecular apparatus for oxidative phosphorylation are localized in the mitochondria.

is delivered is twofold (see below). The first method involves a *direct transfer* from the energy-producing sites of glycolysis and the combined operations of the citric acid cycle and oxidative phosphorylation. This occurs extensively in periods of prolonged muscular activity. The second method involves the formation of ATP coupled to the hydrolysis of *creatine phosphate, a high-energy phosphoguanidine compound* (see p. 356 and below), which is the molecular form in which energy is stored when the muscle is at rest. The source of energy for the formation of creatine phosphate is, of course, the ATP produced in the degradation of carbohydrate. On demand (a condition known to be signaled by the elevation of the concentration of Ca^{2+} in the sarcoplasm), the hydrolysis of creatine phosphate is sufficiently exergonic to mediate the phosphorylation of ADP. This mode of ATP delivery would typify brief periods of activity and the initial phase of a prolonged stress when stored energy is utilized. As the muscle returns to rest (controlled by a drop in the level of Ca^{2+}), ATP generated by carbohydrate metabolism is then stored again as creatine phosphate.

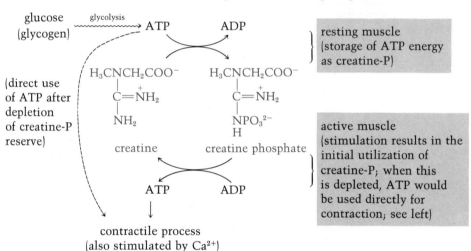

In the contractile process proper, there appear to be two ATP-requiring systems. One involves the conversion of actin from an inactive form (G-actin) to an active form (F-actin). F-actin is the form that binds with myosin. This is not the primary ATP-dependent system, because F-actin normally predominates. The second and more important ATP-dependent system is the actomyosin complex itself, in which it has been shown that the hydrolysis of ATP provides energy for the displacement of the myosin projections to new binding sites on actin, causing the movement of the two filaments past each other.

HEXOSE-MONOPHOSPHATE SHUNT

Although most aerobic organisms utilize the combined reactions of glycolysis and the citric acid cycle as the main catabolic route for the complete oxidation of carbohydrates to CO_2, many possess alternate pathways. This is particularly common in the bacterial world. It is not our intent to investigate all such pathways. Rather, we will consider only one, the *hexose-monophosphate shunt* (HMS). The HMS sequence is widely distributed in animals, plants, and bacteria. Hence, it is the most common alternate pathway of carbohydrate metabolism. However, you will quickly discover that it merits our attention for much more significant reasons. An indication of its importance is provided by extensive observations that many cells possess enzymes of the HMS sequence in addition to those of glycolysis and the citric acid cycle, all of which operate simultaneously but usually to varying degrees. The metabolic logic of this can be appreciated only after an examination of the chemical events that characterize the shunt pathway.

The eight reactions of the hexose-monophosphate shunt are given below (see Fig. 12–8 for structural formulas). Although it is difficult to deduce the metabolic effect of this family of reactions when they are presented in this manner, close inspection reveals that the HMS reactions can be divided into two distinct phases, one *oxidative* and the other *nonoxidative* (Fig. 12–8). The oxidative phase (reactions 1 through 3) is characterized by the conversion of a hexose phosphate to a pentose phosphate and CO_2 via two $NADP^+$-dependent reactions. The nonoxidative phase (reactions 4 through 8) is exclusively characterized by pentose isomerization and transfers of two-carbon and three-carbon units between ketoses and aldoses.

oxidative (1 + 3)

(1) $NADP^+$ + glucose-6-phosphate → 6-phosphogluconolactone + NADPH + H^+
(2) 6-phosphogluconolactone + H_2O → 6-phosphogluconate
(3) $NADP^+$ + 6-phosphogluconate → ribulose-5-phosphate + CO_2 + NADPH + H^+

nonoxidative

(4) ribulose-5-phosphate → ribose-5-phosphate
(5) ribulose-5-phosphate → xylulose-5-phosphate
(6) ribose-5-phosphate + xylulose-5-phosphate →
glyceraldehyde-3-phosphate + sedoheptulose-7-phosphate
(7) sedoheptulose-7-phosphate + glyceraldehyde-3-phosphate →
fructose-6-phosphate + erythrose-4-phosphate
(8) xylulose-5-phosphate + erythrose-4-phosphate →
fructose-6-phosphate + glyceraldehyde-3-phosphate

To understand the potential overall chemical effect, it is necessary only to consider the above reactions occurring in conjunction with other cellular reactions. In this instance, the most appropriate candidates are some of the enzymes of glycolysis (hexose isomerase, triose phosphate isomerase, aldolase) and glycogenesis (diphosphatase). There is nothing invalid in this approach, because the specificity of an enzyme is typified by its action on a certain substrate rather than by its participation in a particular set of reactions in which the substrate is metabolized. Furthermore, in this instance all of the pathways in question occur in the same subcellular compartment, namely, the cytoplasm. In other words, an enzyme is not confined to functioning in only one specific pathway. When this is done, it is possible to assemble the reactions in the fashion illustrated in Fig. 12–8. (Note that the enzymes of glycolysis and glycogenesis would function in the conversion of the two units of glyceraldehyde-3-phosphate to glucose-6-phosphate.) Because of the divergence after ribulose-5-phosphate formation, it is advantageous to start with a nonunit stoichiometry. The convenience of starting with three units of G6P will emerge as you study the various interactions.

Although the diagram appears to be a complicated maze, it does neatly illustrate the metabolic function of the HMS sequence. However, as shown to the right, the latter is revealed most directly through an analysis of the net chemical effect that considers each reaction to interact as diagrammed. Our conclusion as to the metabolic function of the HMS pathway is now quite direct: Given the illustrated theoretical operation of the shunt

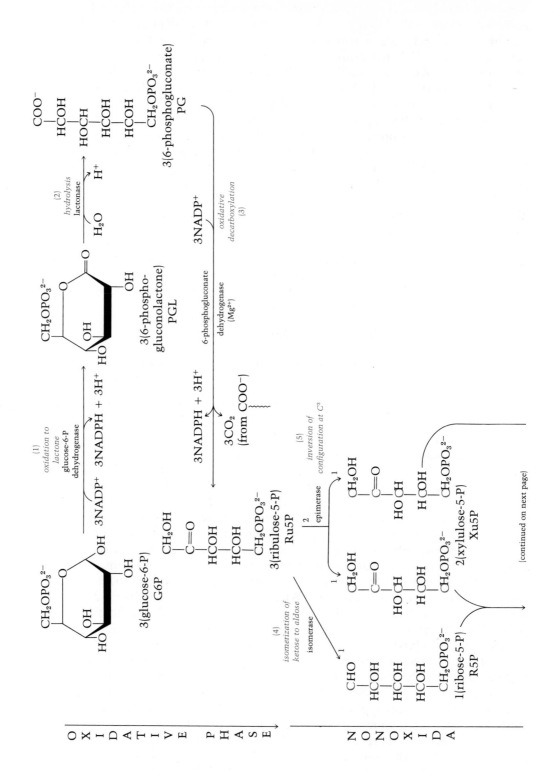

(continued on next page)

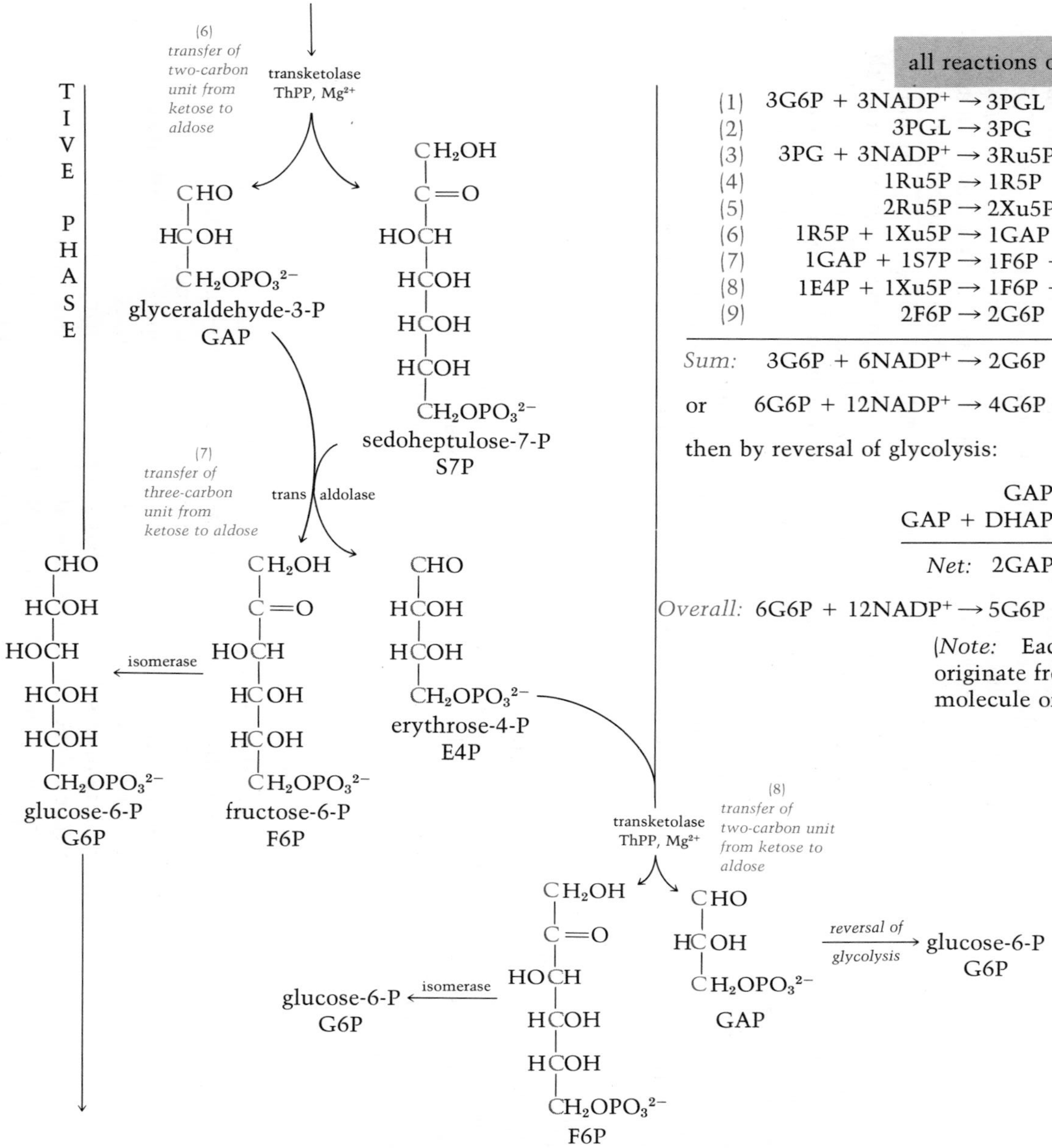

all reactions occur in cytoplasm

(1) $3G6P + 3NADP^+ \rightarrow 3PGL + 3NADPH + 3H^+$
(2) $3PGL \rightarrow 3PG$
(3) $3PG + 3NADP^+ \rightarrow 3Ru5P + 3CO_2 + 3NADPH + 3H^+$
(4) $1Ru5P \rightarrow 1R5P$
(5) $2Ru5P \rightarrow 2Xu5P$
(6) $1R5P + 1Xu5P \rightarrow 1GAP + 1S7P$
(7) $1GAP + 1S7P \rightarrow 1F6P + 1E4P$
(8) $1E4P + 1Xu5P \rightarrow 1F6P + 1GAP$
(9) $2F6P \rightarrow 2G6P$

Sum: $3G6P + 6NADP^+ \rightarrow 2G6P + 3CO_2 + 6NADPH + 6H^+ + GAP$

or $6G6P + 12NADP^+ \rightarrow 4G6P + 6CO_2 + 12NADPH + 12H^+ + 2GAP$

then by reversal of glycolysis:

$GAP \longrightarrow DHAP$

$GAP + DHAP \rightarrow\rightarrow G6P$

Net: $2GAP \rightarrow\rightarrow G6P$

Overall: $6G6P + 12NADP^+ \rightarrow 5G6P + 6CO_2 + 12NADPH + 12H^+$

(*Note:* Each CO_2 unit will originate from a different molecule of glucose-6-P)

FIGURE 12–8 The hexosemonophosphate shunt pathway.

pathway for six units of G6P, the *net chemical effect is seen to be the same as the combined operation of glycolysis and the citric acid cycle, namely, the equivalent of one hexose unit is oxidatively degraded to* $6CO_2$. Also note that the pathway is equally effective in conserving chemical energy, since the 12 units of NADPH may be reoxidized by NAD^+ to yield $NADP^+$ and 12 units of NADH, with the NADH then priming the mitochondrial process of oxidative phosphorylation to yield several ATP molecules. Thus, the hexose-monophosphate shunt is indeed an effective alternate pathway for glucose catabolism.

$$\text{NADPH} + \text{NAD}^+ \underset{\textit{catalyzed by a specific enzyme}}{\overset{\textit{a transhydrogenation}}{\rightleftharpoons}} \text{NADP}^+ + \text{NADH}$$

Despite the theoretical validity of these observations, they probably do not represent the true *in vivo* functions of the HMS sequence. Most evidence suggests that the *more significant* metabolic functions of the shunt reactions are, first, to serve as a major cellular *source of reduced NADP* needed for the reductive reactions of anabolism, and second, to *supply ribose* needed for the biosynthesis of nucleotides and nucleic acids. In addition, in plants and microbes, erythrose-4-phosphate is a biosynthetic precursor of the aromatic amino acids phenylalanine, tyrosine, and tryptophan (see p. 538). All of these possibilities are consistent with the cellular localization of all the enzymes that would be involved. That is, the bulk of biosynthetic reactions (NADPH dependent) occur in the cytoplasm, which, as we stated before, is also where the enzymes of the shunt pathway are localized.

In humans, the shunt pathway is not found in all tissues. In fact, it is relatively unimportant in skeletal muscle where the main reaction sequence is glycolysis. It is more significant in aerobic tissues characterized by considerable biosynthetic activity, such as the liver. In a few cases, such as corneal tissue, it has been implicated as the major metabolic pathway. The marginal note describes the occurrence and significance of the HMS pathway in red blood cells. In no case, however, has a strict quantitative analysis of the relative contributions of the HMS reactions and the combined pathways of glycolysis and citric acid cycle been reported. In plants, a variation of the shunt pathway functions in the photosynthetic fixation of atmospheric CO_2 and its subsequent conversion into carbohydrate (see Chap. 15).

In red blood cells (erythrocytes) the production of NADPH via the shunt pathway is necessary to maintain high levels of reduced glutathione (GSH) in these cells. The GSH in turn is required to sustain the structural integrity of the erythrocyte by preventing the oxidation of —SH-containing proteins and the oxidation of membrane lipids. Severe anemia can result when these events occur. Unfortunately, individuals afflicted with the genetic metabolic disorder, called glucose-6-phosphate dehydrogenase syndrome, provide clinical evidence for these relationships. The problem is a *deficiency of* normal *dehydrogenase* activity, which obviously means the production of less than normal levels of NADPH.

Transaldolase and Transketolase

Before concluding our analysis of the HMS pathway, let us briefly examine the two distinctive enzymes of the nonoxidative phase, namely, *transaldolase* and *transketolase*. Aside from their obvious importance in the HMS sequence proper, the same type of enzymes participate in the carbon cycle of photosynthesis. Even without these justifications, the reactions would merit attention solely on the basis of their interesting chemistry. Both enzymes catalyze the same general type of reaction, which involves the *transfer of a small carbon fragment from a donor ketose to an acceptor aldose to yield an aldose of shorter chain length and a ketose of longer chain length.* Transketolase catalyzes the transfers of a two-carbon *glycoalde-*

hyde unit ($HOCH_2-\overset{\overset{\displaystyle O}{\|}}{C}-$), whereas transaldolase catalyzes the transfer of a

three-carbon *dihydroxyacetone* unit ($HOCH_2-\overset{\overset{\displaystyle O}{\|}}{C}-\overset{\overset{\displaystyle H}{|}}{\underset{\underset{\displaystyle OH}{|}}{C}}-$). Despite the simi-

larity, the details of each each reaction are vastly different.

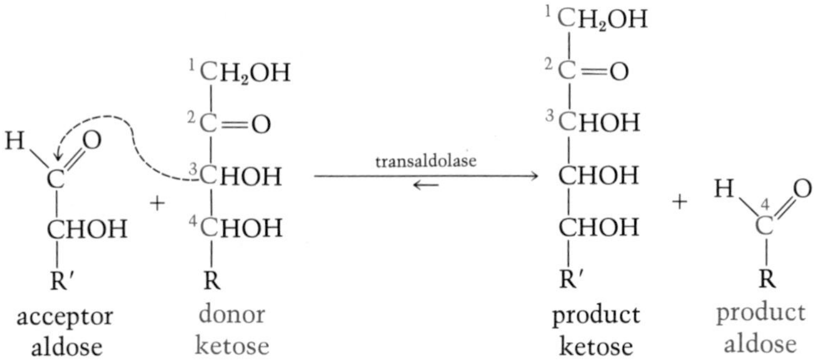

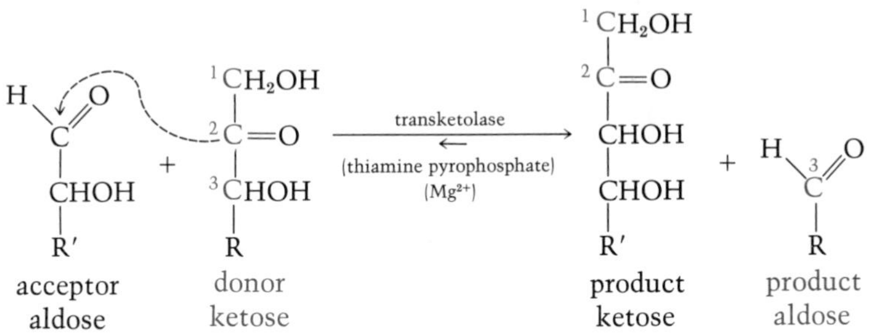

The proposed mechanism of action of transaldolase is very much like that of the aldolase enzyme of glycolysis and glycogenesis. Recall that the main characteristic of this mechanism is the formation of a Schiff-base enzyme-substrate intermediate (see p. 387). We will not discuss the transaldolase enzyme any further.

Transketolase activity involves the obligatory participation of a *metal ion cofactor* (Mg^{2+} is optimal), and more importantly, *thiamine pyrophosphate* (ThPP), which is the metabolically active *coenzyme* form of thiamine (Vitamin B$_1$). Because of a mechanistic relationship of the transketolase reaction to other ThPP-dependent reactions, let us consider the latter topic as a unit. In so doing we will not only explain the transketolase reaction but also describe events associated with the conversion of pyruvate to acetaldehyde (the next to last reaction of alcohol fermentation) and the conversion of pyruvate to acetate (a very crucial reaction that links glycolysis to the citric acid cycle).

Back in Chap. 6 (see Table 6–2, p. 164), thiamine pyrophosphate was

See pp. 408 and 409 for structural formula and space-filling model of thiamine pyrophosphate.

(See diagram, next page.)

described as a coenzyme that participates in *acyl group* ($R-\overset{\overset{\displaystyle O}{\|}}{C}-$) *transfer* reactions. More specifically, the ThPP coenzyme participates in two types of reactions: (1) the decarboxylation of alpha-keto acids and (2) the forma-

407

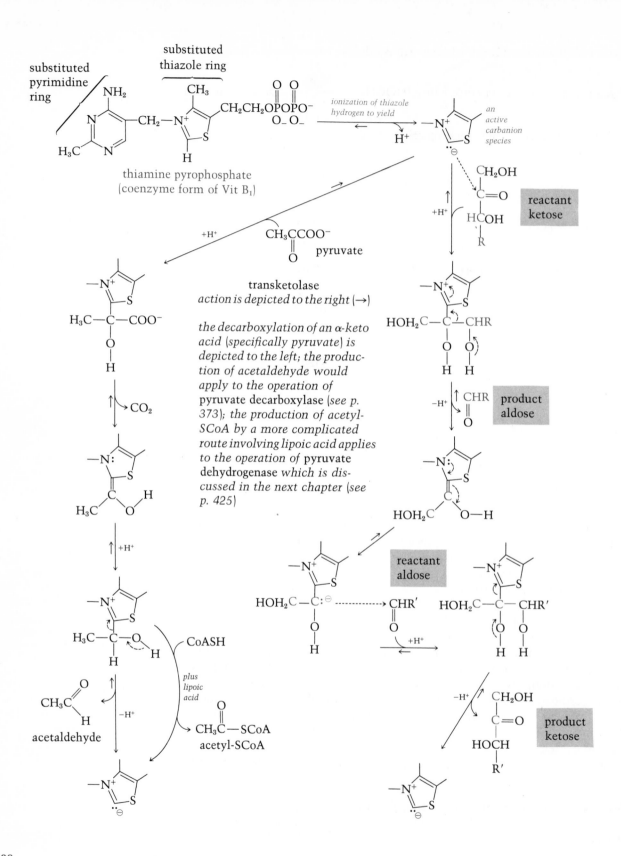

substituted
pyrimidine
ring

substituted
thiazole ring

thiamine pyrophosphate
(coenzyme form of Vit B₁)

*ionization of thiazole
hydrogen to yield*

*an
active
carbanion
species*

reactant
ketose

pyruvate

transketolase
action is depicted to the right (→)

*the decarboxylation of an α-keto
acid (specifically pyruvate) is
depicted to the left; the production of acetaldehyde would
apply to the operation of
pyruvate decarboxylase (see p.
373); the production of acetyl-
SCoA by a more complicated
route involving lipoic acid applies
to the operation of pyruvate
dehydrogenase which is discussed in the next chapter (see
p. 425)*

product
aldose

reactant
aldose

CoASH

*plus
lipoic
acid*

acetaldehyde

acetyl-SCoA

product
ketose

tion of alpha-hydroxy carbonyl linkages, that is, $-\overset{\overset{\displaystyle O}{\|}}{C}-\overset{\overset{\displaystyle OH}{|}}{\underset{|}{C^\alpha}}-$. The transke-

tolase reaction is represented by type 2.

Certain features apply to each type of reaction (see the diagram on p. 408). The coenzyme is bound to the enzyme through several noncovalent interactions, which may include a bridged metal ion. This association with the enzyme enhances the reactivity of the coenzyme, which involves the ionization of C^2 in the thiazole moiety of ThPP and yields a highly reactive *carbanion* species. The carbonyl carbon in the substrate of the reaction then forms a covalent adduct by condensing at the site of the negatively charged carbon. The subsequent fate of this ternary complex of enzyme-(Mg^{2+})ThPP-substrate is then governed by the specificity of the catalytic individuality of the enzyme, as shown in the three paths of the diagram.

The transketolase enzyme directs an electron rearrangement that results in the cleavage of a carbon-carbon bond in the skeleton of the original ketose to yield the aldose product, with the two-carbon fragment corresponding to

the glycoaldehyde $(HOCH_2\overset{\overset{\displaystyle O}{\|}}{C}-)$ terminus of the original ketose still attached. After an electron rearrangement to yield another active carbanion, a new carbon-carbon bond is formed with the carbonyl carbon of the aldose substrate. A final electron rearrangement would release the product ketose and the original state of the enzyme (coenzyme).

The reactions catalyzed by pyruvate decarboxylase (see p. 373) and pyruvate dehydrogenase (see p. 425) represent alpha-keto acid decarboxylation. The events associated with acetaldehyde formation have been diagramed. Those associated with acetyl-SCoA production will be examined in the next chapter.

thiamine pyrophosphate

LITERATURE

AEBI, H. E. "Inborn Errors of Metabolism." *Ann. Rev. Biochem.*, **36,** 271–306 (1967). A review article with special attention given to galactosemia and glucose-6-phosphate dehydrogenase deficiency.

AXELROD, B. "Glycolysis" and "Other Pathways of Carbohydrate Metabolism." In *Metabolic Pathways*, 3rd edition, D. Greenberg, ed., Vol. 1. New York: Academic Press, 1967. Two review articles giving a thorough analysis of reaction pathways.

BALDWIN, E. *Dynamic Aspects of Biochemistry.* 5th edition. London and New York: Cambridge University Press, 1967. For 20 years, which have produced five editions and printings in several foreign languages, this uniquely organized textbook has drawn praise from teacher and student alike. The major emphasis of the book is metabolism, specifically the metabolism of mammalian systems. Chapters 17 and 18 are devoted to the anaerobic metabolism of carbohydrates (glycolysis) as it applies to alcoholic fermentation and muscle contraction.

CLARKE, M., and J. A. SPUDICH. "Nonmuscle Contrac-

tile Proteins: The Role of Actin and Myosin in Cell Motility and Shape Determination." In *Ann. Rev. Biochem.*, **46,** 797–822 (1977). Contractile proteins are present in most types of nonmuscle cells. This article reviews their occurrence, structure, and function.

EBASHI, E. "Regulatory Mechanism of Muscle Contraction with Special Reference to the Ca-Toponin-Tropomyosin System." In *Essays in Biochemistry*, P. N. Campbell and F. Dickens, eds., Vol. 10. New York: Academic Press, 1974. A more detailed review than the Murray and Weber source cited below.

FISCHER, E. H., A. POCKER, and J. C. SAARI. "The Structure, Function, and Control of Glycogen Phosphorylase." In *Essays in Biochemistry*, P. N. Campbell and F. Dickens, eds., Vol. 6, pp. 23–68. New York: Academic Press, 1970. A superb review article on this allosteric enzyme.

HERS, H. G. "The Control of Glycogen Metabolism in the Liver." In *Ann. Rev. Biochem.*, **45,** 167–189 (1976). A thorough review article.

GINSBERG, V. "Sugar Nucleotides and the Synthesis of

Carbohydrates." In *Advan. Enzymol.*, Vol. 26. New York: John Wiley, 1964. A review article dealing with nucleoside diphosphate sugars.

GOLDBERG, N. D. "Cyclic Nucleotides and Cell Function." *Hospital Practice*, **9**, 127–142 (May, 1974). Good discussion of the central regulatory role of the kinase kinase protein, which is proposed to participate in several other processes besides glycogen phosphorylase activation.

HALES, C. N. "Some Actions of Hormones in the Regulation of Glucose Metabolism." In *Essays in Biochemistry*, P. N. Campbell and G. D. Greville, eds., Vol. 3, pp. 73–104. New York: Academic Press, 1967. An excellent review article on an important subject treated only briefly in this chapter.

HORECKER, B. L. "Transaldolase and Transketolase." In *Comprehensive Biochemistry*, M. Florkin and E. H. Stotz, eds., Vol. 15. Amsterdam-New York: Elsevier. A review article on the mechanism of action of these two important enzymes in carbohydrate metabolism.

HOYLE, G. "How is Muscle Turned On and Off?" *Sci-*

entific American, **222,** 84–93 (1970). An article describing the role of calcium ion in the contraction and relaxation of muscle tissue.

HUXLEY, H. E. "The Mechanism of Muscular Contraction." *Science*, **164**, 1356–1365 (1969). An authoritative review describing the possible mechanism of action at the molecular level for the sliding of actin and myosin filaments during contraction.

MURRAY, J. M., and A. WEBER. "The Cooperative Action of Muscle Proteins." *Scientific American*, **230**, 58–71 (February, 1974). A good introductory discussion of current ideas regarding the molecular events of muscle contraction—the interactions of myosin, actin, troponin, and tropomyosin, controlled by calcium ions.

ROACH, P. J. "Functional Significance of Enzyme Cascade Systems." *Trends Biochem. Sci.*, **2**, 84–87 (1977). A short review article.

SEGAL, H. L. "Enzymatic Interconversion of Active and Inactive Forms of Enzymes." *Science*, 180, 25–32 (1973). A good review article on this subject.

EXERCISES

12–1. What is meant by the statement that the reduced forms of nicotinamide adenine dinucleotides have complementary roles in catabolism and anabolism?

12–2. What is the biological significance of catabolic reactions that result in the formation of ATP?

12–3. If radioactive glucose, labeled in positions 3 and 4 with ^{14}C, were incubated with a cell-free liver homogenate under anaerobic conditions, what positions in the lactate produced would be labeled with ^{14}C?

12–4. Verify that $\Delta G^{\circ\prime} = -45.4$ kcal/mole for

$$\text{glucose} + 4P_i + 2NAD^+ \xrightarrow{\text{glycolysis}}$$
$$2\text{lactate} + 4P_i + 2NAD^+$$

12–5. If pure (nondenatured) samples of enzymes 1, 3, 4, 5, 6, 7, 8, 9, 10, 11, 12, and 13 (see code on p. 373) were incubated at 37 °C in a buffered system (pH 7) containing glycogen, ATP, Mg^{2+}, and NAD^+, very little lactate would be produced. Explain.

12–6. The metabolism of sucrose first involves the action of sucrose phosphorylase to yield glucose-1-phosphate and fructose. Assuming that both glucose-1-phosphate and fructose are further metabolized to lactate, (a) how many ATPs would be required, and (b) how many ATPs would be produced?

12–7. Under a growth condition resulting in the production of ethyl alcohol via the glycolytic pathway in fermenting bacteria, what internal coupling mechanism involving NAD would parallel that in muscle cells actively degrading carbohydrate under the same condition? What type of condition is in effect?

12–8. If 0.001 mole of 3-phosphoglyceraldehyde were incubated with 0.005 mole of P_i, 0.0001 mole of NAD^+,

and 3-phosphoglyceraldehyde dehydrogenase, a reaction would occur that would reach equilibrium rather quickly. This equilibrium would be characterized by the presence of a considerable amount of the 3-phosphoglyceraldehyde originally added. Predict what would happen to this equilibrium mixture if 0.005 mole of pyruvate and lactate dehydrogenase were added.

12–9. Write a complete balanced equation that would best describe the net catabolism of mannose to pyruvate.

12–10. Arrange the following proteins in the order in which they participate (beginning with the action of adrenalin) in the conversion of glycogen to glucose-1-phosphate: kinase (active and inactive forms), phosphorylase *b*, adenyl cyclase (active and inactive forms), phosphorylase *a*, and kinase kinase (active and inactive forms).

12–11. Explain the following statement: phosphofructokinase catalyzes the committed step of glycolysis.

12–12. Explain the meaning of the following term: complementary metabolic regulation of the interconversion of two substances.

12–13. Write a complete net reaction for the enzymatic conversion of two lactate units into a glucose unit of glycogen. Assume that the sequence will occur in a eukaryotic cell and that phosphoenolypyruvate carboxykinase is located in the cytoplasm.

12–14. Does the concept of internal coupling (as described for NAD in glycolysis on p. 378) apply to the pathway of glycogenesis? Examine the pathway you worked our for Exercise 12–13.

12–15. The active form of kinase kinase increases the rate of glycolysis but decreases the rate of glycogenesis. To some students this is baffling—how can the active form of the same enzyme activate one process but inhibit another? What is your explanation?

12–16. Is the net requirement of metabolic energy as ATP for the production of glucose-1-phosphate from lactate greater than, less than, or the same as the net production of ATP from the catabolism of glucose-1-phosphate to lactate? (In solving this problem, remember that all of the purine and pyrimidine nucleotides are thermodynamically equivalent. In addition assume that the reactions are those that would occur in eukaryotic cells. Have you done 12–13 yet?)

12–17. In thermodynamic terms, what would be the significance of the reactions between 1,3-diphosphoglycerate and fructose-6-phosphate during glycogenesis?

12–18. If ribose-5-phosphate, uniformly labeled with radioactive carbon, were incubated in a suitably buffered solution containing xylulose-5-phosphate (no ^{14}C), thiamine pyrophosphate, Mg^{2+}, and transketolase, what two new carbohydrates would be produced, and what would be the labeling pattern of carbon 14 in each one?

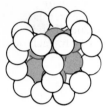

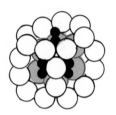

CHAPTER 13

THE CITRIC ACID CYCLE

In the preceding chapter, we referred to the citric acid cycle as the set of aerobic reactions responsible for the complete, energy-yielding, oxidative degradation of the pyruvate produced in glycolysis to carbon dioxide. In that context the citric acid cycle can be considered as a pathway common to carbohydrate metabolism. However, to label the citric acid cycle as fulfilling only this metabolic role would be incorrect. On the contrary, its metabolic significance is much broader in scope. Indeed, in nearly all organisms it serves as (a) *the central pathway integrating the metabolic flow of carbon among all of the main classes of biomolecules* and, in conjunction with the process of oxidative phosphorylation, *as* (b) *the major source of metabolic energy in the form of ATP.*

The diagram below attempts to summarize these principles by illustrating that metabolic intermediates of the citric acid cycle can be both *diverted into biosynthetic reactions* and *formed from degradative reactions* of carbohydrates, lipids, proteins, and nucleic acids. The production of energy occurs when the cycle operates in the degradation of any of these materials, but primarily the hexoses, fatty acids, and amino acids. This and subsequent chapters will provide the details necessary to understand these functions.

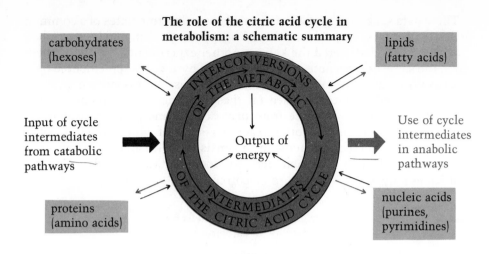

The role of the citric acid cycle in metabolism: a schematic summary

carbohydrates (hexoses)

lipids (fatty acids)

Input of cycle intermediates from catabolic pathways

INTERCONVERSIONS OF THE METABOLIC INTERMEDIATES OF THE CITRIC ACID CYCLE

Output of energy

Use of cycle intermediates in anabolic pathways

proteins (amino acids)

nucleic acids (purines, pyrimidines)

REACTIONS OF THE CITRIC ACID CYCLE

Experimental History

In the development of any scientific discipline, there are what can be called key discoveries. In the biochemical area one such discovery was made in the late 1930s by Hans Krebs, a German-born English biochemist. Krebs was experimenting on the utilization of simple organic acids by pigeon breast muscle, which is characterized by a high level of aerobic metabolism. The experimental design was straightforward. The rate of oxygen consumption (respiration) by muscle homogenates was determined when the muscle was incubated in the presence of glycogen, glucose, or pyruvate, which served as a primary source of carbon and energy, and then compared to that of the same system supplemented with small amounts of various organic acids. In summary form, the important findings were:

a. Regardless of the main carbon source, the additional presence of certain dicarboxylic acids (such as fumaric acid, succinic acid, malic acid, oxaloacetic acid, and α-ketoglutaric acid) and certain tricarboxylic acids (such as citric acid, *cis*-aconitic acid, and isocitric acid) *stimulated* respiration, that is, they increased oxygen consumption and carbon dioxide production. ⋇

b. *In each case*, if malonic acid was also added, the stimulation was inhibited and succinate would accumulate.

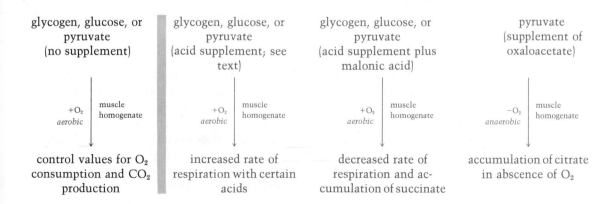

glycogen, glucose, or pyruvate (no supplement)	glycogen, glucose, or pyruvate (acid supplement; see text)	glycogen, glucose, or pyruvate (acid supplement plus malonic acid)	pyruvate (supplement of oxaloacetate)
$+O_2$ *aerobic* / muscle homogenate	$+O_2$ *aerobic* / muscle homogenate	$+O_2$ *aerobic* / muscle homogenate	$-O_2$ *anaerobic* / muscle homogenate
control values for O_2 consumption and CO_2 production	increased rate of respiration with certain acids	decreased rate of respiration and accumulation of succinate	accumulation of citrate in abscence of O_2

413

These data suggested that all of the acids were intermediates of a common pathway that was capable of degrading pyruvate completely to carbon dioxide. Krebs discovered the key in another experiment wherein he incubated the same homogenate in the presence of pyruvate, supplemented only with oxaloacetate, but under *anaerobic* conditions. In this system he found that *citrate* accumulated. Based on these findings, Krebs proposed that when the formation of citrate from oxaloacetate and pyruvate was interpreted in combination with other reactions (some of which were known at that time), the metabolic scheme most consistent with all data was a *cyclic sequence* (see below). Without lengthy documentation, suffice it to say that Krebs's hypothesis proved to be correct.

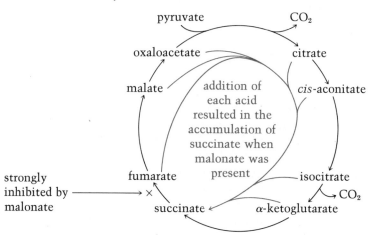

Original proposal of Krebs for cyclic flow of carbon

The discovery by Krebs (Nobel Prize, 1953) was to have a monumental significance to classical biochemistry comparable to the impact of the Watson and Crick proposal for DNA structure on modern biology. The first recognized metabolic function of the *citric acid cycle* (also referred to as the *TCA cycle* or the *Krebs cycle*) was that in conjunction with glycolysis under aerobic conditions it accounted for the complete degradation of hexoses to CO_2. However, it was quickly recognized as a pathway that was the *nucleus of all of metabolism*. Moreover, this importance is not confined to any one select group of cells. On the contrary, the citric acid cycle is probably the most widespread metabolic sequence in our biosphere, being found in all animals, most plants, and many microorganisms.

Overall View

The modern version of the citric acid cycle (with structures of the intermediates) is shown in Fig. 13–1. Note that the flow of carbon differs from the initial proposal of Krebs in two important respects: (1) in what can be considered the *priming step* of the cycle, *acetyl-SCoA* is the compound that reacts directly with oxaloacetate; and (2) the conversion of α-ketoglutarate to succinate involves the intermediate formation of *succinyl-SCoA*. Both modifications were made during the late 1940s and early 1950s, accompanying the isolation and characterization of coenzyme A.

FIGURE 13–1 The citric acid cycle. Each enzyme is coded by a number, with the traditional names given in the center of the diagram. Whenever appropriate, the participation of coenzymes in a reaction is indicated symbolically. It is proposed that *cis*-aconitate is not produced as a free acid but as an enzyme-bound intermediate that is formed prior to isocitrate. In other words, there are two distinct stages to the aconitase reaction. The mechanics of ATP formation coupled to the utilization of the reducing power generated by the cycle are discussed in Chap. 14. General principles of this coupling are discussed in this chapter.

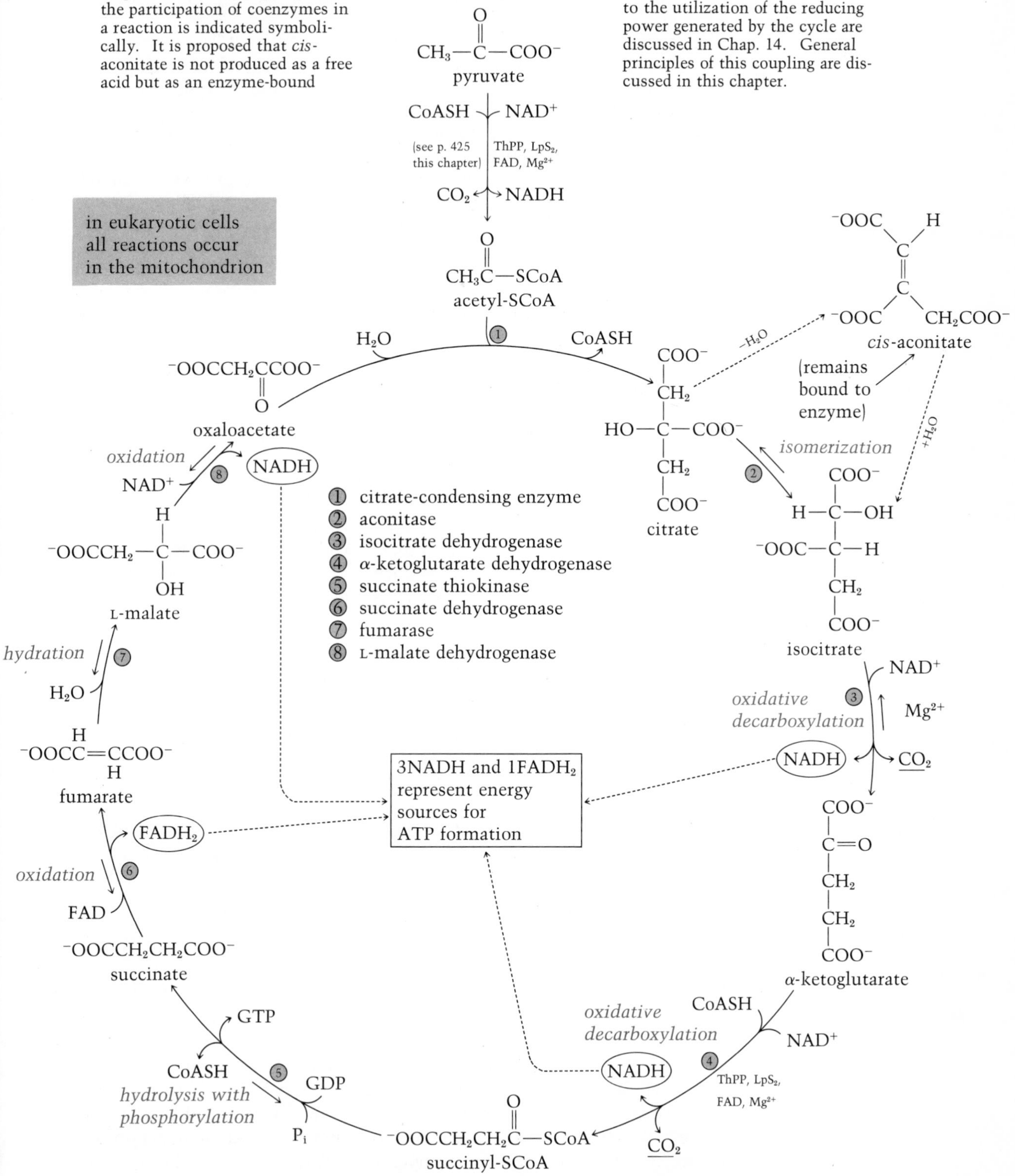

$$CH_3-C-COO^-$$
pyruvate

CoASH NAD$^+$

(see p. 425 this chapter) ThPP, LpS$_2$, FAD, Mg^{2+}

CO_2 NADH

in eukaryotic cells all reactions occur in the mitochondrion

$$CH_3C-SCoA$$
acetyl-SCoA

H$_2$O   CoASH

$^-$OOCCH$_2$CCOO$^-$
oxaloacetate

$^-$OOC, H
C
C
$^-$OOC CH$_2$COO$^-$
cis-aconitate

(remains bound to enzyme)

$-H_2O$ $+H_2O$

COO$^-$
CH$_2$
HO—C—COO$^-$
CH$_2$
COO$^-$
citrate

oxidation
NAD$^+$ ⑧ NADH

H
$^-$OOCCH$_2$—C—COO$^-$
OH
L-malate

① citrate-condensing enzyme
② aconitase
③ isocitrate dehydrogenase
④ α-ketoglutarate dehydrogenase
⑤ succinate thiokinase
⑥ succinate dehydrogenase
⑦ fumarase
⑧ L-malate dehydrogenase

isomerization ②

COO$^-$
H—C—OH
$^-$OOC—C—H
CH$_2$
COO$^-$
isocitrate

hydration ⑦
H$_2$O

H
$^-$OOCC=CCOO$^-$
H
fumarate

FADH$_2$

oxidation ⑥
FAD

$^-$OOCCH$_2$CH$_2$COO$^-$
succinate

3NADH and 1FADH$_2$ represent energy sources for ATP formation

oxidative decarboxylation ③
NAD$^+$
Mg^{2+}
NADH $\underline{CO_2}$

COO$^-$
C=O
CH$_2$
CH$_2$
COO$^-$
α-ketoglutarate

GTP

CoASH GDP
hydrolysis with phosphorylation ⑤
P$_i$

$^-$OOCCH$_2$CH$_2$C—SCoA
succinyl-SCoA

oxidative decarboxylation CoASH
NADH ④ NAD$^+$
ThPP, LpS$_2$, FAD, Mg^{2+}
$\underline{CO_2}$

415

Beginning with the formation of citrate, the characteristic chemistry of the eight reactions of the cycle can be descriptively summarized as follows:

1. The formation of a carbon-carbon bond between a thioester-activated acetyl group (C_2 unit) and a C_4 α-keto dicarboxylic acid to yield a C_6 tricarboxylic acid.
2. A positional isomerization of the C_6 tricarboxylic acid.
3. The first of two successive NAD^+-dependent oxidative decarboxylations, resulting in the conversion of the C_6 isomeric tricarboxylic acid to a C_5 α-keto dicarboxylic acid.
4. The second NAD^+-dependent oxidative decarboxylation, resulting in the formation of a thioester of a C_4 dicarboxylic acid.
5. The hydrolysis of the C_4 thioester to yield a free C_4 dicarboxylic acid in its fully saturated form, coupled to the phosphorylation of GDP.
6. An FAD-dependent dehydrogenation to yield a C_4 unsaturated dicarboxylic acid.
7. The hydration of the unsaturated acid to yield a C_4 α-hydroxy dicarboxylic acid.
8. An NAD^+-dependent dehydrogenation to form the original C_4 α-keto dicarboxylic acid.

Symbolic representations of each reaction are shown below, along with a listing of the standard free-energy changes.

Transformations of the citric acid cycle $\Delta G^{\circ\prime}$ (kcal)

$$
\begin{aligned}
&(1)\ C_2\text{—SCoA(acetyl-SCoA)} + C_4(OAA) + H_2O \xrightarrow{E_1} C_6(CIT) + CoASH && -9.08 \\
&(2)\ C_6(CIT) \xrightarrow{E_2} C_6(ISOCIT) && +1.59 \\
&(3)\ C_6(ISOCIT) + NAD^+ \xrightarrow{E_3} C_5(\alpha KG) + CO_2 + NADH + H^+ && -1.70 \\
&(4)\ C_5(\alpha KG) + NAD^+ + CoASH \xrightarrow{E_4} C_4(SUC\text{-}SCoA) + CO_2 + NADH + H^+ && -9.32 \\
&(5)\ C_4(SUC\text{-}SCoA) + GDP + P_i \xrightarrow{E_5} C_4(SUC) + GTP + CoASH && -2.12 \\
&(6)\ C_4(SUC) + FAD \xrightarrow{E_6} C_4(FUM) + FADH_2 && 0 \\
&(7)\ C_4(FUM) + H_2O \xrightarrow{E_7} C_4(MAL) && -0.88 \\
&(8)\ C_4(MAL) + NAD^+ \xrightarrow{E_8} C_4(OAA) + NADH + H^+ && +6.69
\end{aligned}
$$

$$
Net:\ (1 + 8)\ CH_3\overset{\overset{\text{O}}{\|}}{C}SCoA + 3NAD^+ + FAD + 2H_2O + GDP + P_i \xrightarrow[\text{enzymes}]{8} 2CO_2 + 3NADH + 3H^+ + FADH_2 + GTP + CoASH \qquad -14.82
$$

The overall chemical effect is clearly summarized by the net equation, which states that the operation of the cycle through one complete turn results in

a. The complete oxidation of the acetyl unit to two units of CO_2.
b. The production of three units of reduced NAD and one unit of reduced FAD.
c. The production of one unit of GTP.

Examination of the $\Delta G^{\circ\prime}$ values shows that the composite pathway is exergonic and thus thermodynamically favorable. Yet not every individual reaction is exergonic. Note, however, as was the case with glycolysis, that there is a thermodynamic design to the cycle consistent with the maintenance of an efficient flow of all intermediates. For example, the malate $\rightarrow$ oxaloacetate conversion is strongly endergonic, and by itself would serve as a natural thermodynamic barrier that would favor an accumulation of malate and thus impede the operation of the cycle. A closer analysis, however, reveals that the probability of this occurrence is minimized considerably—indeed, it is eliminated—since the malate → oxaloacetate conversion is *followed in the cycle* by two highly exergonic reactions, namely, the formation of citrate (reaction (1)) and the formation of succinyl-SCoA (reaction (4)). Reaction (1) is quite significant since it is directly coupled to the malate → oxaloacetate reaction. Hence, in the presence of acetyl-SCoA, reactions (1) and (4) will contribute to the efficient removal of oxaloacetate, preventing both a sluggish flow and a gradual buildup of other intermediates. ✻

The citric acid cycle bears another resemblance to our previous discussion of glycolysis on the basis of internal coupling. In the citric acid cycle, the internal coupling is on two levels. The first and most obvious one is due to the unique position of oxaloacetate. In reaction (1) it is a reactant, whereas in reaction (8) it is a product. Of course, this is the key feature that confers a cyclic pattern on the overall reaction sequence. The second is the coupling between reactions (4) and (5) involving coenzyme A. In reaction (4) CoASH is an essential reactant, whereas in reaction (5) it is produced. The significance of these factors should be recognized; namely, the cycle is autocatalytic, and thus self-sufficient as long as catalytic quantities of all intermediates are present.

Of course, the continued operation of the cycle is also dependent on the availability of catalytic levels of NAD^+, FAD, GDP, and P_i. A close inspection of the reaction sequence reveals that the cycle itself is not autocatalytic with regard to any of these substances. Rather, the need is satisfied by the regeneration of these materials in other processes. NAD^+ and FAD are regenerated almost solely by the electron transport chain, with oxygen as the terminal electron and hydrogen acceptor. *It is this participation of oxygen that renders the citric acid cycle an aerobic pathway* (see below). ✻ As we will see later, the cellular localization of the citric acid cycle enzymes and the components of the electron transport chain is consistent with this metabolic link—all are compartmentalized in the mitochondrion.

It is difficult to pinpoint specific processes for the regeneration of GDP. Two likely candidates are given below. One involves the exchange of phosphate between GTP and a nucleoside diphosphate, an exchange catalyzed by a *nucleoside diphosphate kinase*, an enzyme widely distributed in nature. The second involves a GTP-dependent decarboxylation of oxaloacetate to phosphoenolpyruvate that is catalyzed by *phosphoenolpyruvate carboxykinase.* The probability that one or both of these enzymes functions in this capacity is supported by the fact that both are found in the mitochondria, and are thus compartmentalized with the enzymes of the citric acid cycle.

Regeneration of NAD^+ and FAD (see Chap. 14):

$$NADH + H^+ + \tfrac{1}{2}O_2 \xrightarrow[\text{transport}]{\text{electron}} NAD^+ + H_2O + \text{energy}$$

$$FADH_2 + \tfrac{1}{2}O_2 \xrightarrow[\text{transport}]{\text{electron}} FAD + H_2O + \text{energy}$$

$$\longrightarrow \text{\textit{for ATP formation}}$$

Regeneration of GDP:

$$GTP + ADP \xrightarrow[\text{diphosphate kinase}]{\text{nucleoside}} GDP + ATP$$

$$\text{oxaloacetate} + GTP \xrightarrow[\text{carboxykinase}]{\text{PEP}} \text{phosphoenolpyruvate} + GDP + CO_2$$

The contribution of each will vary from organism to organism and will also be controlled by growth conditions. (You will recall that PEP carboxykinase has been proposed to be an enzyme that operates in the reversal of glycolysis, too; see p. 394, Chap. 12, and p. 437, this chapter.)

INDIVIDUAL REACTIONS OF THE CITRIC ACID CYCLE

In this section our attention turns to the individual reactions of the cycle, some of which are very remarkable. As previously explained in describing the enzymes of the glycolytic and hexose monophosphate shunt pathways, the rationale for this approach is twofold. Although the details unique to each reaction will serve as a basis for gaining a deeper insight into the citric acid cycle as a reaction pathway, at the same time they represent general principles of enzyme action. Once again it is worth noting that such an analysis is quite possible since each enzyme has been isolated and studied in pure form. The items we will emphasize are (a) enzyme specificity, (b) coenzyme participation, (c) regulatory properties, and (d) the existence and operation of *multienzyme complexes*.

Citrate Synthetase

The reaction catalyzed (irreversibly) by *citrate synthetase* (sometimes called the *citrate-condensing enzyme* or *citrogenase*) is unique to the cycle, because it represents the only reaction involving the formation of a C—C bond. It may seem unusual for this type of reaction to be part of a pathway responsible for the degradation of acetyl-SCoA. However, when viewed in the context in which one reaction is integrated into the complete sequence, the logic and efficiency of this chemistry become clear.

Another noteworthy point is that, although the reaction involves the condensation of two carbon compounds, a chemical event that requires an input of energy, the overall reaction is *not* ATP dependent. Even then the overall reaction is still very energy-yielding ($\Delta G^{\circ\prime} = -9$ kcal). The reason for this becomes clear if we recall that acetyl-SCoA is a high-energy compound (p. 357). This then is our first encounter with a reaction system that involves a thioester as substrate. In addition to the fact that the thioester (acetyl-SCoA) is more reactive than the free acid (acetate), its hydrolysis during the course of the reaction supplies chemical energy, some of which is utilized for the formation of a carbon-carbon bond in the accompanying production of citrate.

As have all of the enzymes of the citric acid cycle, citrate synthetase has been detected in all aerobic organisms so far examined. In a few cases crystalline preparations have been obtained. The enzyme from pig heart has a molecular weight of approximately 90,000, as determined by ultracentrifugation, and is believed to be oligomeric. Catalytically, the enzyme is rather specific for acetyl-SCoA and oxaloacetate, with fluoroacetyl-SCoA (see note in margin on p. 419) and fluorooxaloacetate being the only known alternative substrates.

Recent kinetic studies suggest that the enzyme is a key regulatory site of aerobic metabolism. In particular, ATP is a strong competitive inhibitor of acetyl-SCoA. Other inhibitors are NADH, succinyl-SCoA, and palmityl-SCoA (the thioester of a C_{16} fatty acid), which are believed to function allo-

Margin figure and caption

$$\overset{O}{\overset{\|}{}}$$
$$^-OOCCH_2\overset{O}{\overset{\|}{C}}COO^-$$
*
oxaloacetate

H_2O → $CH_3\overset{O}{\overset{\|}{C}}$—SCoA

condensing enzyme

CoASH →

$$CH_2COO^-$$
$$HO—C^*—COO^-$$
$$CH_2COO^-$$
citrate

The active site of this enzyme probably promotes the abstraction of H^+ from the methyl group of acetyl-SCoA to yield the carbanion $:\overset{\ominus}{C}H_2COSCoA$, which then attacks the carbonyl carbon (*) of oxaloacetate.

sterically. The significance of these kinetic observations relative to intracellular metabolic control will be explored later, on p. 440.

Individual Reactions of
the Citric Acid Cycle

Aconitase

The conversion of citrate to isocitrate is merely a positional isomerization, but there are special features to this reaction. Most noteworthy is that the reaction is stereospecific in its action, yielding only one product, namely, *threo*-D_s-isocitrate. As shown below the *threo*-D_s designation specifies one of four isomeric isocitrates.

$$
\begin{array}{cccc}
\text{COO}^- & \text{COO}^- & \text{COO}^- & \text{COO}^- \\
\text{H}-\text{C}-\text{OH} & \text{H}-\text{C}-\text{OH} & \text{HO}-\text{C}-\text{H} & \text{HO}-\text{C}-\text{H} \\
{}^-\text{OOC}-\text{C}-\text{H} & \text{H}-\text{C}-\text{COO}^- & \text{H}-\text{C}-\text{COO}^- & {}^-\text{OOC}-\text{C}-\text{H} \\
\text{CH}_2\text{COO}^- & \text{CH}_2\text{COO}^- & \text{CH}_2\text{COO}^- & \text{CH}_2\text{COO}^-
\end{array}
$$

threo-D_s-isocitrate (only product) *erythro*-D_s-isocitrate *threo*-L_s-isocitrate *erythro*-L_s-isocitrate

(none of these three isomers is produced)

Note: The *threo* and *erythro* designations refer to the spatial orientations of the —OH and —COO⁻ groups relative to each other, using the structures of the sugars threose and erythrose as frames of reference. Threo implies a *trans* orientation and erythro a *cis* orientation. The D_s notation refers to the specific configuration of the —OH carbon, using the structure of D-serine as a frame of reference.

Radioisotope studies clearly indicate that the mode of action of aconitase has yet a further distinction. If radioactive acetyl-SCoA (containing a uniformly labeled (^{14}C) acetyl group) and nonlabeled oxaloacetate are incubated in the presence of citrate synthetase, the citrate that is produced will consist of two ^{14}C atoms and four ^{12}C atoms. Now, when this material is then incubated in the presence of aconitase, it is observed that in the *threo*-D_s-isocitrate that is formed the —OH is not attached to a radioactive carbon despite the fact that citrate is a symmetrical molecule containing two identical —CH$_2$COO⁻ groupings. (The presence of ^{14}C in one renders the —CH$_2$COO⁻ groups only physically different, not chemically different. Furthermore, the presence of the radioactive carbon atoms does not interfere with the catalytic action of the enzyme.) Due to the symmetry of ci-

$$
\begin{array}{cc}
\text{CHO} & \text{CHO} \\
\text{H}-\text{C}-\text{OH} & \text{HO}-\text{C}-\text{H} \\
\text{H}-\text{C}-\text{OH} & \text{H}-\text{C}-\text{OH} \\
\text{CH}_2\text{OH} & \text{CH}_2\text{OH} \\
\text{erythrose} & \text{threose}
\end{array}
$$

both are threo-D_s-isocitrate

$$
\underset{\substack{\text{O}\\ \| }}{^-\text{OOCCH}_2\text{CCOO}^-}
$$

$$
\underset{\substack{\text{O}\\ \| }}{^{14}\text{CH}_3{}^{14}\text{C}-\text{SCoA}}
$$

condensing enzyme → H$_2$O CoASH →

$$
\begin{array}{c}
\text{COO}^- \\
\text{CH}_2 \\
{}^-\text{OOC}-\text{C}-\text{OH} \\
{}^{14}\text{CH}_2 \\
{}^{14}\text{COO}^-
\end{array}
\xrightarrow{\text{aconitase}}
\begin{array}{c}
\text{COO}^- \\
\text{H}-\text{C}-\text{OH} \\
{}^-\text{OOC}-\text{C}-\text{H} \\
{}^{14}\text{CH}_2 \\
{}^{14}\text{COO}^-
\end{array}
+
\begin{array}{c}
\text{COO}^- \\
\text{CH}_2 \\
\text{H}-\text{C}-\text{COO}^- \\
\text{HO}-{}^{14}\text{C}-\text{H} \\
{}^{14}\text{COO}^-
\end{array}
$$

but only this is produced

trate, you might expect a 50-50 mixture of two radioactive compounds. Because only one is produced (how this is determined is shown in the margin on p. 421), we conclude that the citrate to isocitrate conversion occurs *asymmetrically.* In other words, by virtue of its asymmetric active site with specific binding and catalytic residues (see p. 180 in Chap. 6 for a review of this principle), the aconitase enzyme has the ability to distinguish between identical chemical groups.

The precise mechanism of aconitase action is still unknown, but the enzme appears to contain bound Fe^{2+} that is essential for activity. It is also proposed that the overall reaction proceeds through two distinct stages:

One very potent inhibitor of aconitase is *fluorocitrate*, which is formed from the condensation of fluoroacetyl-SCoA and oxaloacetate by the action of citrate synthetase. Fluoroacetate is found in certain locoweeds and is used as a coyote poison, with the poisonous effect due to its conversion to the inhibitor, fluorocitrate.

first, the dehydration to yield *cis*-aconitate, which remains bound to the enzyme, and second, the hydration of *cis*-aconitate to yield *threo*-D_s-isocitrate (see Fig. 13–1). There are no known activators or inhibitors (see margin) of common natural occurrence, suggesting that aconitase is not a regulatory enzyme.

Isocitrate Dehydrogenase

The oxidative decarboxylation of isocitrate to α-ketoglutarate is also interesting. One of the more intriguing features is that eukaryotic cells *contain two different types of isocitrate dehydrogenase.* One is NAD^+-dependent, the other $NADP^+$-dependent. Moreover, the former is a particulate enzyme (localized in the mitochondrion) and the latter is a soluble enzyme (localized in the cytoplasm). Although there are some similarities in their action—for example, each has an absolute requirement for Mn^{2+} or Mg^{2+} for optimal activity, and each exhibits the same substrate specificity (see below)—they are nevertheless two distinct enzymes and are proposed to have two different functions even though they catalyze the same reaction. The difference is that they are located in different compartments. In the mitochondrion it is the NAD^+-dependent enzyme that operates in the citric acid cycle, whereas the $NADP^+$-dependent enzyme is thought to function primarily in generating α-ketoglutarate in the cytoplasm for use in anabolic reactions (see p. 434).

Although precise physicochemical data concerning the structure of NAD^+-isocitrate dehydrogenase are not available, the molecule does appear to be very large (mol. wt. $\approx 10^5$) and the native conformation is proposed to be oligomeric. Our knowledge of its catalytic properties is more extensive and refined. In this regard, there are two important characteristics. First, the enzyme displays a high degree of stereospecificity. In fact, the specificity of action may be absolute, since the only effective substrate appears to be threo-D_s-isocitrate. All other stereoisomers are inactive. It is noteworthy that this specificity is completely consistent with the aconitase enzyme, which produces only the threo-D_s isomer in the previous step. Second, the enzyme (particularly the NAD^+-dependent form) is typified by second-site (allosteric) kinetics effecting both the stimulation and the inhibition of activity. ADP and NAD^+ are *activators.* On the other hand, ATP and NADH are *inhibitors.* The effect of ATP is also believed to be allosteric, but the mode of action of NADH is believed to be competitive inhibition with NAD^+. In any event, mitochondrial isocitrate dehydrogenase is proposed to the *major metabolic control point* of the cycle. Further discussion of this point is deferred until later in the chapter (p. 439).

α-Ketoglutarate Dehydrogenase

The oxidative decarboxylation of α-ketoglutarate to succinyl-SCoA is similar to the previous reaction only in that both require an NAD^+-dependent dehydrogenase. Furthermore, not only are the two enzymes different proteins, but the reaction catalyzed by α-ketoglutarate dehydrogenase is much more complex. The tip-off to each of these points is given in Fig. 13–1, which indicates that the activity of α-ketoglutarate dehydrogenase not only *requires NAD$^+$* as a coenzyme and *a metal ion* as a cofactor, but *also requires coenzyme A, thiamine pyrophosphate* (ThPP), *flavin adenine dinu-*

threo-D_s-isocitrate (the ^{14}C-labeling pattern shown would result by using ^{14}C-labeled acetyl-SCoA as previously discussed)

α-ketoglutarate

$$\overset{O}{\underset{\alpha\text{-ketoglutarate}}{^-OO^{14}C^{14}CH_2CH_2\overset{\|}{C}COO^-}} \xrightarrow[\substack{NAD^+}]{\substack{CoASH \\ \xrightarrow{\hspace{1cm}} \\ \alpha\text{-ketoglutarate} \\ \text{dehydrogenase} \\ \text{ThPP, FAD, LpS}_2 \\ Mg^{2+}}} NADH \quad \overset{O}{\underset{\text{succinyl-SCoA}}{^-OO^{14}C^{14}CH_2CH_2\overset{\|}{C}{-}SCoA}} + \overset{*}{CO_2}$$

(see note in margin)

$$\overset{O}{\underset{\text{pyruvate}}{CH_3\overset{\|}{C}COO^-}} \xrightarrow[\substack{NAD^+}]{\substack{CoASH \\ \xrightarrow{\hspace{1cm}} \\ \text{pyruvate} \\ \text{dehydrogenase} \\ \text{ThPP, FAD, LpS}_2 \\ Mg^{2+}}} NADH \quad \overset{O}{\underset{\text{acetyl-SCoA}}{CH_3\overset{\|}{C}{-}SCoA}} + CO_2$$

cleotide (FAD), and *lipoic acid* (LpS$_2$). The latter four substances also function as obligatory coenzymes. The general biochemistry of coenzyme A and thiamine pyrophosphate have been discussed in earlier chapters; hence their role as coenzymes is not being encountered for the first time. Flavin adenine dinucleotide and lipoic acid will be considered shortly.

An obvious question is why one enzyme requires such a large number of different coenzmes. Well, the fact is that α-ketoglutarate dehydrogenase is not a single enzyme, but rather a *multienzyme complex* composed of three different enzymes, with each individual enzyme participating in part of the overall reaction. The remarkable property of this system is that, though each enzyme is functionally and structurally distinct, they do not exist separately from each other in the native state. Rather, they exist as an *ordered aggregate*, that is, as a massive *polyprotein complex*.

We will defer further analysis to a later discussion of the *pyruvate dehydrogenase* system, which is likewise a multienzyme complex very similar in structure and mode of action to α-ketoglutarate dehydrogenase. This similarity is not surprising, since both enzyme systems catalyze the same general type of reaction—the oxidative decarboxylation of an α-keto acid to an acyl thioester. The reason for deferring analysis is simple—we know more about pyruvate dehydrogenase.

Proof of the stereospecific catalysis of aconitase is that the CO$_2$ released in this step originates from oxaloacetate and not one of the acetyl-SCoA carbons that remain in succinyl-SCoA. If the cycle were to begin with ^{14}C-labeled oxaloacetate and unlabeled acetyl-SCoA, the CO$_2$ released in this reaction would be ^{14}CO$_2$, as would the carbon dioxide released in the previous reaction catalyzed by isocitrate dehydrogenase.

Succinyl Thiokinase

The immediate fate of the succinyl-SCoA generated from α-ketoglutarate is hydrolysis to the free acid, with the energy released being conserved through the coupled formation of a nucleoside triphosphate. Remember that reactions of this type are generally referred to as *substrate level phosphorylations* to distinguish them from the production of ATP coupled to electron transport. Also remember that other examples of substrate level phosphorylation are represented by the conversions of 1,3-diphosphoglycerate to 3-phosphoglycerate and phosphoenolpyruvate to pyruvate in glycolysis (see p. 387, Chap. 12).

The unique feature of succinyl thiokinase is that it is rather specific for *guanine nucleotides*. Nevertheless, due to the known facility of phosphate transfer between different nucleoside triphosphates and diphosphates (see nucleoside diphosphate kinase, pp. 355 and 561), the formation of GTP is metabolically equivalent to the production of ATP. Thus, the operational significance of this reaction is that it contributes directly to the production of useful metabolic energy.

This particular site of GTP formation is not to be interpreted, however, as the primary mode of energy production by the citric acid cycle. It plays only a small part. As stated repeatedly, the bulk of energy (ATP) production

$$\overset{O}{\underset{\text{succinyl-SCoA}}{^-OOCCH_2CH_2\overset{\|}{C}{-}SCoA}}$$

$$P_i, GDP \searrow \Big\updownarrow$$

$$GTP \swarrow \Big\downarrow CoASH$$

$$\underset{\text{succinate}}{^-OOCCH_2CH_2COO^-}$$

421

via the citric acid cycle occurs *in conjunction with* the oxygen-dependent reoxidation of NADH and FADH₂ in the process of oxidative phosphorylation.

Succinate Dehydrogenase

The conversion of succinate to fumarate is the third of four dehydrogenations in the cycle. However, succinate dehydrogenase is unique in that the coenzyme hydrogen acceptor is *flavin adenine dinucleotide* (FAD), whereas the other three dehydrogenases are NAD⁺ dependent. Since the FAD is firmly bound to the protein portion, succinate dehydrogenase is frequently referred to as a *flavoprotein* (fp). There are various other flavoproteins.

FAD represents the metabolically active form of *riboflavin* (Vitamin B₂). (Some flavoproteins contain another flavin coenzyme, FMN, flavin mononucleotide.) As shown below, the active portion of FAD (or FMN) is localized in the fused *isoalloxazine* ring system. The unsaturated species repre-

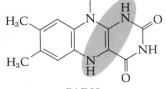

riboflavin (Vitamin B₂)

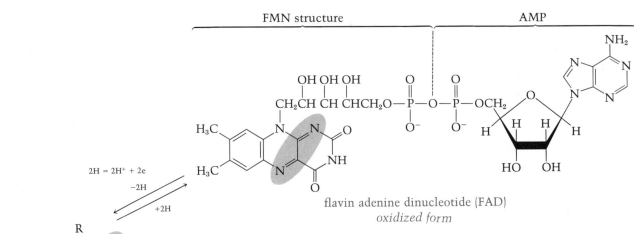

flavin adenine dinucleotide (FAD)
oxidized form

FMN structure   AMP

2H = 2H⁺ + 2e
−2H
+2H

FADH₂
reduced form

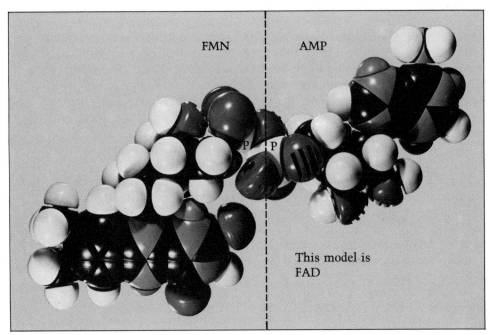

FMN   AMP

P   P

This model is
FAD

sents the oxidized form of FAD, which, upon acceptance of two hydrogen atoms ($2H^+$ and 2 electrons), is converted to the reduced form, $FADH_2$.

On a structural level, succinate dehydrogenase is distinguished by (a) a large molecular weight ($\approx 2 \times 10^5$), suggesting that it is oligomeric; (b) the presence of four to eight tightly bound iron atoms per molecule (their function is unknown); and (c) a covalent attachment of the FAD coenzyme to the protein portion. The enzyme is very difficult to isolate in soluble form since it is tightly embedded in the mitochondrial membrane. Moreover, its integration with all other membrane proteins, both structural and functional, is thought to be highly ordered (see p. 460).

The catalytic action of the enzyme is characterized by absolute stereochemical specificity of hydrogen elimination, with the dehydrogenation being exclusively *trans*. If this specificity did not exist, a 50-50 mixture of two isomeric unsaturated dicarboxylic acids, maleate (*cis*) and fumarate (*trans*), would be formed. Only fumarate is formed, however. The ability of succinate dehydrogenase to discriminate between structurally equivalent hydrogen pairs is another example of asymmetric catalysis of a symmetrical substrate.

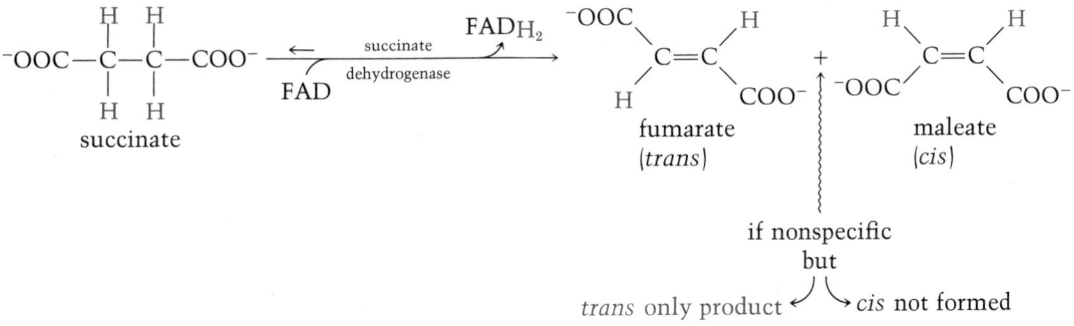

The high level of substrate specificity is also typified by the previously discussed potent competitive inhibition of malonate (p. 176). Although recent studies have shown that the activity of the dehydrogenase is subject to positive and negative allosteric effects, it is uncertain whether the enzyme is a main metabolic control point as isocitrate dehydrogenase is.

Fumarase

The hydration of fumarate to malate, catalyzed by fumarase, is still another example of enzyme specificity. In fact, the enzyme displays absolute specificity, with fumarate as the only active substrate and L-malate as the only product. The isomer, D-malate, is not formed. Extremely pure preparations of fumarase have been obtained, thus permitting detailed physicochemical studies. The enzyme isolated from pig heart appears to be a homogeneous tetramer, with each protomer having a molecular weight of 48,500 (mol. wt. of tetramer $\approx 194,000$). The native enzyme contains no disulfide bonds, but does possess 12 free sulfhydryl groups. Although the —SH groups are not part of the active site, their presence is absolutely required for activity.

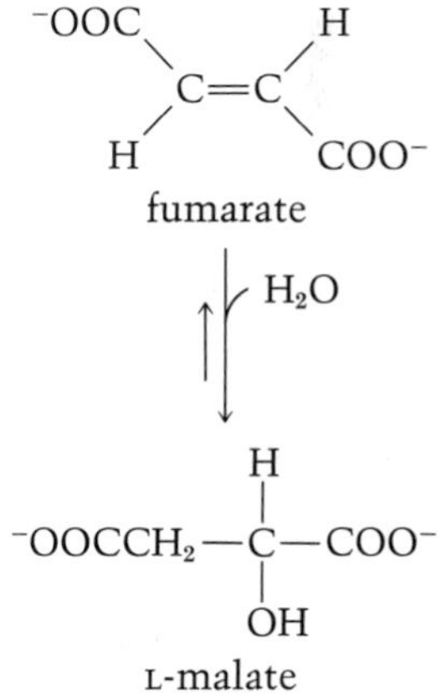

Malate Dehydrogenase

The NAD^+-dependent conversion of L-malate (an α-hydroxy acid) to oxaloacetate (an α-keto acid) is the closing reaction and the fourth oxidative

$$^-OOCCH_2\!-\!\underset{\underset{\displaystyle OH}{|}}{\overset{\overset{\displaystyle H}{|}}{C}}\!-\!COO^-$$

L-malate

$$\big\uparrow \quad \begin{array}{l} \text{NAD}^+ \\ \\ \text{NADH + H}^+ \end{array}$$

$$^-OOCCH_2\underset{\underset{\displaystyle O}{\|}}{C}COO^-$$

oxaloacetate

dehydrogenation of the cycle. Typical of all the cycle enzymes, malate dehydrogenase is highly specific. In fact, absolute specificity is once again proposed.

The biochemistry of L-malate dehydrogenase is similar to that of two other enzymes we have already discussed. First, there is a similarity to lactate dehydrogenase in that hybrid (isoenzyme) forms are known to exist. However, the biological significance of malate dehydrogenase isoenzymes is not clear. The enzyme is also similar to isocitrate dehydrogenase in that it exists in two structurally and catalytically distinct forms in the cells of higher organisms. One form is located in the mitochondrion and the other in the cytoplasm. However, unlike the two isocitrate dehydrogenases that require either NAD$^+$ or NADP$^+$, both malate dehydrogenases are NAD$^+$ dependent.

Of course, the mitochondrial (particulate) enzyme is the one that operates in the citric acid cycle. The participation of the cytoplasmic (soluble) enzyme is similar to but more complex than that of the NADP$^+$-dependent isocitrate dehydrogenase. You may recall that we have already described (see p. 394 in Chap. 12) how the operation of an extramitochondrial malate dehydrogenase can in part account for the biosynthesis of carbohydrate from pyruvate via oxaloacetate. However, some of the same events operating in reverse may be a more important sequence since this would account for the entry into the mitochondrion of NADH that is originally formed in the cytoplasm. The latter process is discussed further in the next chapter (see p. 470).

Postscript

Having completed our analysis of individual reactions of the citric acid cycle, we should now recapitulate our purpose for doing so. That objective can be stated quite simply: we wished to illustrate principles of biocatalysis that apply to enzymes in general and to the operation of the citric acid cycle in particular. Relative to the former, we analyzed the principles of coenzyme participation, substrate specificity, modulation of kinetic activity, and protein structure, with each enzyme of the cycle being uniquely characterized in each respect. The important thing to realize is that, as a consequence of these descriptions, the second part of our objective was fulfilled. In other words, the biochemical essence of the citric acid cycle is derived from the individual properties of its constitutive enzymes. This, of course, is true of any metabolic pathway and, indeed, of the whole metabolism of any organism.

ROLE OF THE CITRIC ACID CYCLE IN METABOLISM

Our analysis of the net chemical effect and the individual reactions of the citric acid cycle provides a base of understanding so that we may now consider in more detail how the cycle functions in metabolism. The material in this section is divided into two parts, providing an analysis of the role of the cycle in both catabolism and anabolism. In each case the material is subclassified into areas dealing with the role of the cycle in the metabolism of the four main classes of biomolecules. It should be emphasized again, however, that the cycle is not a pathway common to either phase of metabolism or to the metabolism of any one specific class of molecules, but rather a central pathway that integrates and unifies the whole of metabolism. It is unfortunate that we must fragment this subject in order to discuss it.

PART A: ROLE OF THE CITRIC ACID CYCLE IN CATABOLISM

Carbohydrate Catabolism

Under normal conditions, most aerobic organisms utilize carbohydrate materials as their major source of carbon and energy. Thus, though the citric acid cycle is not to be considered unique to the catabolism of any one class of compounds, in most organisms it is of great importance in the catabolism of carbohydrates. Recall from the previous chapter, however, that the citric acid cycle itself is not solely responsible for the degradation of carbohydrate, but operates in concert with glycolysis. The latter effects the conversion of hexose phosphates to pyruvate, which under aerobic conditions is then completely degraded to carbon dioxide via the citric acid cycle.

A question that may occur to you is, how can pyruvate be degraded by the citric acid cycle if it isn't one of the metabolic intermediates of the cycle? Although the question is very basic, it is a good one to raise, because it points to the fact that, in order for two or more separate pathways to be integrated, there must exist a metabolic *link*. In this instance the link is provided by the enzyme *pyruvate dehydrogenase*, which catalyzes the oxidative decarboxylation of pyruvate to acetyl-SCoA. ⌊The latter can then condense with oxaloacetate and the cycle is primed, with one complete turn thus resulting in the complete degradation of the C_2 acetyl unit to $2CO_2$. ⌉

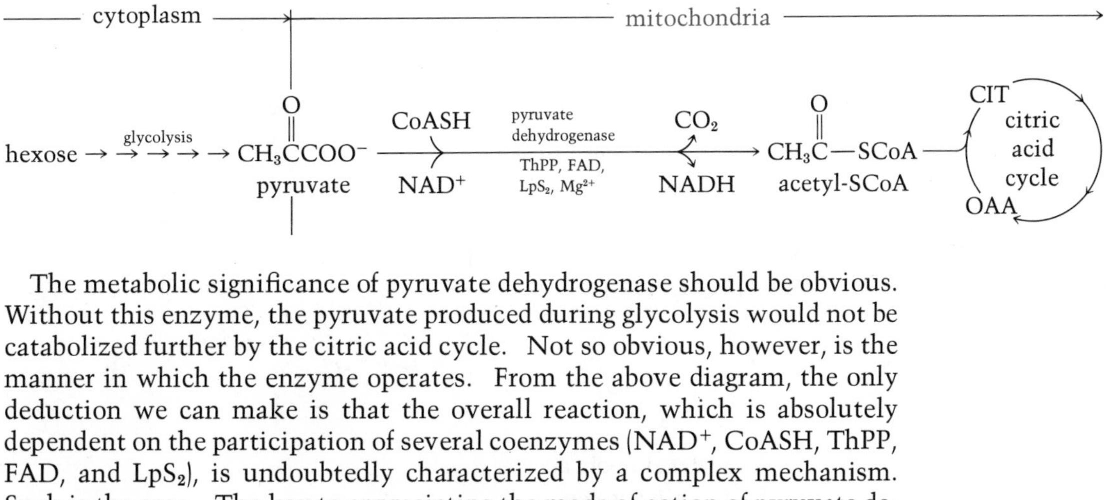

$$\text{hexose} \rightarrow \rightarrow \rightarrow \rightarrow \underset{\substack{\text{pyruvate}}}{\overset{\overset{\displaystyle O}{\parallel}}{CH_3CCOO^-}} \xrightarrow[\substack{\text{ThPP, FAD,}\\ \text{LpS}_2,\text{ Mg}^{2+}}]{\overset{\text{CoASH}}{\underset{\text{NAD}^+}{\text{(pyruvate dehydrogenase)}}}} \underset{\substack{\text{NADH}}}{\overset{CO_2}{\longleftarrow}} \underset{\substack{\text{acetyl-SCoA}}}{\overset{\overset{\displaystyle O}{\parallel}}{CH_3C-SCoA}} \rightarrow \text{citric acid cycle}$$

———— cytoplasm ————|———————— mitochondria ————————→

(CIT / OAA, citric acid cycle)

The metabolic significance of pyruvate dehydrogenase should be obvious. Without this enzyme, the pyruvate produced during glycolysis would not be catabolized further by the citric acid cycle. Not so obvious, however, is the manner in which the enzyme operates. From the above diagram, the only deduction we can make is that the overall reaction, which is absolutely dependent on the participation of several coenzymes (NAD$^+$, CoASH, ThPP, FAD, and LpS$_2$), is undoubtedly characterized by a complex mechanism. Such is the case. The key to appreciating the mode of action of pyruvate dehydrogenase is in its structure. A hint as to the nature of its structure is provided by the term *pyruvate dehydrogenase complex*, which is a more correct way to refer to this enzyme system, since it is known to be an aggregate of more than one enzyme.

Specifically, the pyruvate dehydrogenase complex (PDC) is composed of three different enzymes: (a) *pyruvate dehydrogenase*, sometimes called *pyruvate decarboxylase*; (b) *dihydrolipoyl transacetylase*; and (c) *dihydroli-*

poyl dehydrogenase. Furthermore, the complex contains different amounts of each. The PDC isolated from *E. coli,* which is structurally representative of other preparations including those from mammals, contains 12 dimeric molecules (24 polypeptide chains) of pyruvate dehydrogenase; 24 molecules (24 polypeptide chains) of transacetylase; and 6 dimeric molecules (12 polypeptide chains) of dihydrolipoyl dehydrogenase. The entire complex is obviously quite large. Its particle weight is approximately 4×10^6.

Before examining how each part of this enzyme system participates in the overall reaction, we must first consider *lipoic acid.* This is a low molecular weight, sulfur-containing, carboxylic acid that functions as a coenzyme in two types of reactions. Both functions are based on the presence of two sulfur atoms that exist as either two sulfhydryl (—SH) groups or a single disulfide (—S—S—) linkage. *Hydrogen transfer* is one type of reaction involving the reversible interconversion of oxidized and reduced forms of lipoic acid, a relationship that is very characteristic of sulfur-containing compounds of this type. The second type involves the acylation of oxidized lipoic acid in the presence of a suitable donor, and the subsequent deacylation of the acyl-lipoic acid species in the presence of a suitable acceptor. The net result is described as an *acyl group transfer reaction.* Of course, in any one reaction it is the nature of the enzyme that specifies which way lipoic acid participates. Whatever type of enzyme is involved, the lipoic acid is most always *covalently attached* to it through an amide linkage involving its carboxyl group and the epsilon amino group of a lysine side chain. Now we can proceed to examine the operation of the entire pyruvate dehydrogenase complex.

lipoic acid

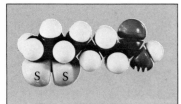

from a hydrogen donor

$$\underset{\substack{\text{oxidized lipoic acid} \\ \text{can function solely} \\ \text{as a hydrogen} \\ \text{acceptor}}}{\text{CH}_2\text{CH}_2\text{CHCH}_2\text{CH}_2\text{CH}_2\text{CH}_2\text{COO}^-} \underset{-2\text{H}}{\overset{+2\text{H}}{\rightleftharpoons}} \underset{\substack{\text{reduced lipoic acid} \\ \text{can function as a} \\ \text{hydrogen donor}}}{\text{CH}_2\text{CH}_2\text{CHCH}_2\text{CH}_2\text{CH}_2\text{CH}_2\text{COO}^-}$$

to a hydrogen acceptor

or

as an acyl group and hydrogen acceptor

acylated lipoic acid acting as acyl donor to HY

Each component enzyme is proposed to act in concert, catalyzing a certain part of the overall reaction (see below). *Pyruvate dehydrogenase* (E_{PDH}), which is dependent on ThPP, catalyzes the decarboxylation of the α-keto acid, with the C_2 unit transferred to ThPP (reaction A). *Transacetylase* (E_{TA}), consisting of bound lipoic acid, then catalyzes the transfer of the C_2 unit to lipoic acid in reaction B. The acetylated transacetylase is then attacked by CoASH (which corresponds to HY in the preceding reactions of lipoic acid) to produce the reduced form of the transacetylase-lipoate enzyme and acetyl-SCoA (reaction C). Thus, reactions A, B, and C account for the complete conversion of substrates (pyruvate and CoASH) to products $(CO_2 +$ acetyl-SCoA). (Refer back to p. 408 in the previous chapter.) The last two steps (D and E) are required to *regenerate* all catalytic components of the complex in their original form. The oxidized form of the transacetylase is regenerated by *dihydrolipoyl dehydrogenase* (E_{LDH}), an FAD-containing flavoprotein. Finally, the oxidized form of the flavoprotein is regenerated through the participation of NAD^+ as a hydrogen acceptor. The net reaction (sum of reactions A through E) is the oxidative (NAD^+-dependent) decarboxylation of pyruvate to the metabolically active, high-energy compound acetyl-SCoA, without the expenditure of metabolic energy as ATP.

Rx A:
$$CH_3\overset{O}{\overset{\|}{C}}COO^- + E_{PDH}\!\frown\!ThPP \xrightarrow{Mg^{2+}} E_{PDH}\!\frown\!ThPP\overset{OH}{-}\overset{|}{C}HCH_3 + CO_2$$

with label "noncovalent linkage" over $E_{PDH}\frown ThPP$

Rx B: $E_{PDH}\!\frown\!ThPP-\overset{OH}{\overset{|}{C}}HCH_3 + E_{TA}-\underset{H}{\overset{O}{\overset{\|}{N}}}-\overset{}{C}(CH_2)_4CH\underset{S-S}{\overset{CH_2\ \ CH_2}{\diagdown\diagup}} \longrightarrow E_{PDH}\!\frown\!ThPP + E_{TA}-\underset{H}{\overset{O}{\overset{\|}{N}}}-C(CH_2)_4CH\underset{S-CCH_3}{\overset{CH_2\ O\ CH_2SH}{}}$

covalent link to a lysine residue

now have an activated acyl group as a thioester

Rx C: $E_{TA}-\underset{H}{\overset{O}{\overset{\|}{N}}}-C(CH_2)_4CH\underset{S-CCH_3}{\overset{CH_2\ O\ CH_2SH}{}} + CoASH \longrightarrow E_{TA}-\underset{H}{\overset{O}{\overset{\|}{N}}}-C(CH_2)_4CH\underset{SH\quad SH}{\overset{CH_2\quad CH_2}{}} + CH_3\overset{O}{\overset{\|}{C}}-SCoA$

Rx D: (regeneration steps) $E_{TA}-\underset{H}{\overset{O}{\overset{\|}{N}}}-C(CH_2)_4CH\underset{SH\quad SH}{\overset{CH_2\quad CH_2}{}} + E_{LDH}\!\frown\!FAD \longrightarrow E_{TA}-\underset{H}{\overset{O}{\overset{\|}{N}}}-C(CH_2)_4CH\underset{S-S}{\overset{CH_2\quad CH_2}{}} + E_{LDH}\!\frown\!FADH_2$

Rx E: $E_{LDH}\!\frown\!FADH_2 + NAD^+ \longrightarrow E_{LDH}\!\frown\!FAD + NADH + H^+$

Net **Rx:**
$$CH_3\overset{O}{\overset{\|}{C}}COO^- + NAD^+ + CoASH \xrightarrow[\substack{(ThPP,\ LpS_2,\\ FAD,\ Mg^{2+})}]{3\ enzymes} CH_3\overset{O}{\overset{\|}{C}}-SCoA + CO_2 + NADH + H^+$$

Despite the inherent complexity of this reaction, the most remarkable properties of the PD complex are its ultrastructural order and its capacity for *self-assembly* from its constituent subunits. All of the component parts are known to be arranged in a certain spatial and geometric pattern that is necessary for optimal activity. The structure (see p. 428) consists of a core of the 24 units of transacetylase, proposed to be arranged as eight trimers to comprise a cube, with a dimeric molecule of pyruvate dehydrogenase bound to each edge of the core and a dimeric molecule of dihydrolipoyl dehydrogenase bound to each face of the core.

The PD complex, which itself is localized in the mitochondrial membrane of higher organisms and thus is appropriately compartmentalized to serve as a link with the citric acid cycle, is but one of several multienzyme complexes known to exist in nature. Mention has already been made of the

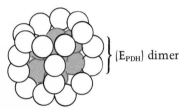

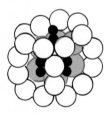
α-ketoglutarate dehydrogenase complex, which is a close relative of the PD complex on both a structural and functional level (p. 420). Additional examples, such as the *fatty acid synthetase complex*, which participates in the biosynthesis of fatty acids, and the *electron transport chain*, will be encountered in subsequent chapters.

Although the precise operational advantage of a multienzyme aggregate remains debatable, it is reasonable to suggest that it would be an extremely efficient metabolic system, since all of the required participants are localized in one position, thus eliminating the need for all to diffuse to a common site. In fact, this may be the prime asset. As shown below the communication among the three enzymes could use the lipoic acid coenzyme as a swinging arm to accept the two-carbon unit from thiamine pyrophosphate, transfer it to CoASH, and then be oxidized back to the active disulfide by the FAD-dependent dehydrogenase.

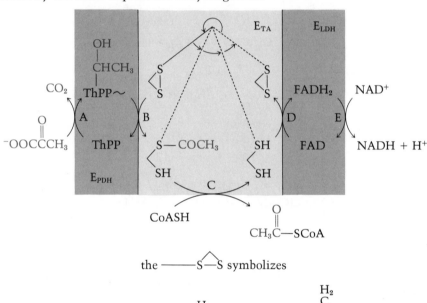

the ——S⌢S symbolizes

the lysyl-lipoic acid extended arm of E_{TA}
free to move from one site to another

Whatever the advantage, recent studies have shown that multienzyme systems such as this are subject to strict *regulatory* controls. The PD complex, for example, is strongly inhibited by ATP and strongly activated by ADP. There are other inhibitors and activators. In a later section, this regulation of the PD complex will be shown to be integrated with other regulatory enzymes in carbohydrate metabolism (see p. 439). The ATP inhibition is another example of an active ⇌ inactive conversion produced by covalent modification. In this case the inactivation results from a phosphorylation by ATP that is catalyzed by *pyruvate dehydrogenase kinase*. This effect is complemented by an activating process of dephosphorylation catalyzed by *pyruvate dehydrogenase phosphatase*, which in turn is activated by Ca^{2+}. Both of these enzymes also appear to be bound to the intact aggregate and in response to the ATP and Ca^{2+} signals regulate the activity of pyruvate dehydrogenase and hence regulate the activities of the entire aggregate.

428

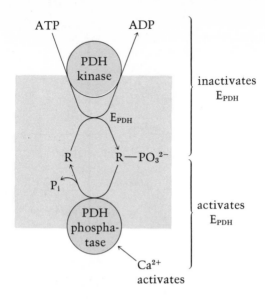

*R in E_{PDH}
represents a specific amino acid
residue susceptible to
phosphorylation*

Our discussion of the metabolic fitness of the pyruvate → acetyl-SCoA + CO_2 conversion should not close without taking note that the overall reaction is quite thermodynamically favorable ($\Delta G^{\circ\prime}$ is approximately $-8,000$ calories/mole), and that it is an irreversible reaction. Both characteristics contribute to a unidirectional flow of carbon from pyruvate to acetyl-SCoA. Moreover, when one considers the pyruvate as originating from phosphoenolpyruvate, with the acetyl-SCoA then entering the citric acid cycle, the net thermodynamics is even more one-sided. That is, all three of the reactions are exergonic (see below), and thus strongly favor the flow of carbon from glycolysis into the citric acid cycle.

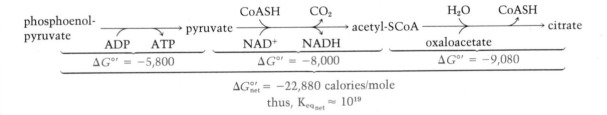

$$\Delta G^{\circ\prime}_{net} = -22,880 \text{ calories/mole}$$
$$\text{thus, } K_{eq_{net}} \approx 10^{19}$$

In the event that our discussion of the details of the PD complex may have detracted from its metabolic significance, let us summarize briefly. This complex is the major metabolic link between glycolysis and the citric acid cycle, which together effect the complete degradation of a hexose unit to $6CO_2$. The overall chemistry, involving a total of 19 distinct reactions, is summarized in the equations on p. 430 (top).

The representation of the NAD^+ and FAD acting as hydrogen acceptors is most relevant, because the *oxygen-dependent reoxidation of the reduced species coupled to the phosphorylation of ADP constitutes the distinguishing feature of aerobic catabolism.* The details of oxidative phosphorylation, as well as the value of n, will be treated in the next chapter. For now it is important only to realize that the bulk of the reducing power produced (largely as NADH) in the degradation of carbohydrates, and hence the bulk of the ATP production, is provided by the enzymes of the citric acid cycle and of the PD complex.

**A summary of the major pathway
for carbohydrate catabolism**

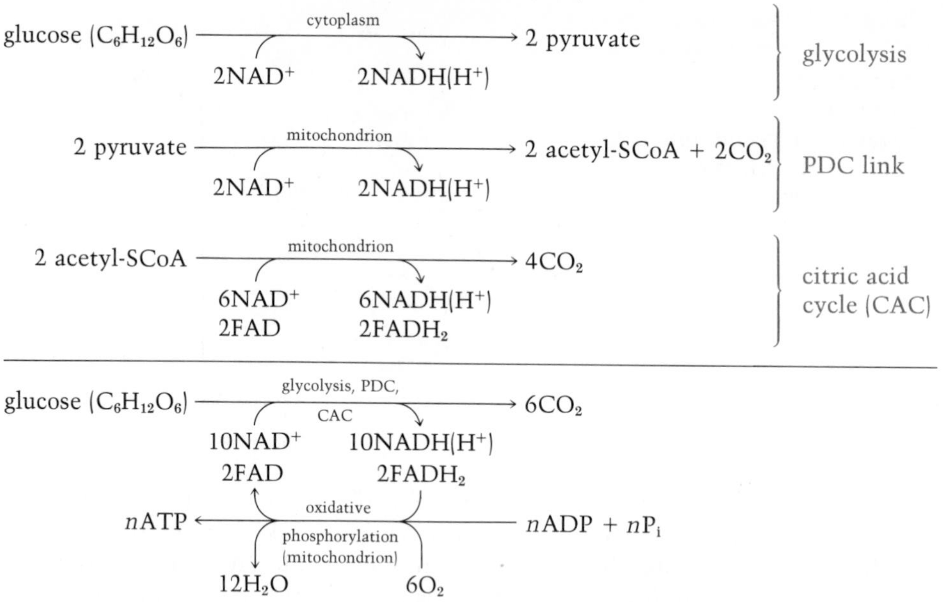

$$\text{glucose } (C_6H_{12}O_6) \xrightarrow[\underset{2NAD^+ \quad 2NADH(H^+)}{}]{\text{cytoplasm}} 2 \text{ pyruvate} \quad \left.\right\} \text{glycolysis}$$

$$2 \text{ pyruvate} \xrightarrow[\underset{2NAD^+ \quad 2NADH(H^+)}{}]{\text{mitochondrion}} 2 \text{ acetyl-SCoA} + 2CO_2 \left.\right\} \text{PDC link}$$

$$2 \text{ acetyl-SCoA} \xrightarrow[\underset{6NAD^+ \quad 6NADH(H^+) \atop 2FAD \quad 2FADH_2}{}]{\text{mitochondrion}} 4CO_2 \quad \left.\right\} \begin{array}{l}\text{citric acid} \\ \text{cycle (CAC)}\end{array}$$

$$\text{glucose } (C_6H_{12}O_6) \xrightarrow[\underset{10NAD^+ \quad 10NADH(H^+) \atop 2FAD \quad 2FADH_2}{}]{\substack{\text{glycolysis, PDC,} \\ \text{CAC}}} 6CO_2$$

$$nATP \xleftarrow[\underset{12H_2O \qquad 6O_2}{\substack{\text{phosphorylation} \\ \text{(mitochondrion)}}}]{\text{oxidative}} nADP + nP_i$$

Lipid (Fatty Acid) Catabolism

Since we have not yet explored the metabolic pathways common to the catabolism of lipids, our treatment of this subject will be necessarily limited. Yet it will not be incomprehensible, because the pertinent facts are few in number and rather easy to understand. In our previous discussion of lipid structure (Chap. 10) it was pointed out that the principal constituents of neutral and compound lipids were the fatty acids. Well, when these lipids are degraded, the respective hydrolysis products can be catabolized further. In the case of the fatty acids, further catabolism proceeds (in virtually all organisms) primarily by the process of β *oxidation* (Chap. 16). The net result of this process is the successive removal of C_2 units, as acetyl-SCoA, from the fatty acid molecule.

$$\underset{\text{fatty acid}}{CH_3(CH_2)_nCOO^-} \xrightarrow[\underset{\text{CoASH}}{}]{\substack{\text{enzymes of} \\ \beta\text{-oxidation}}} \underset{\text{acetyl-SCoA}}{\left(\frac{n+2}{2}\right)^* \overset{\overset{O}{\|}}{CH_3C}-SCoA} \xrightarrow[\text{cycle}]{\substack{\text{enzymes of} \\ \text{citric acid}}} (n+2)CO_2 + \text{energy} \quad \underset{\text{is even}}{(^* \text{ when } n}$$

This then is what links lipid catabolism to the citric acid cycle: the acetyl-SCoA originating from fatty acids can condense with oxaloacetate to yield citrate. Potentially, the combined effect of β oxidation and the citric acid cycle would be the complete degradation of a fatty acid to CO_2. Of course, the process would be energy yielding, due again to the formation of reducing power as NADH and its subsequent reoxidation in the electron transport chain.

Protein (Amino Acid) Catabolism

Whereas the metabolic link between the citric acid cycle and the degradation of carbohydrates and lipids is mediated primarily through just one intermediate, acetyl-SCoA, the input of carbon metabolites from the degradation of amino acids occurs at several sites. Two modes of entry exist. One type involves the entrance of the complete carbon skeleton of the amino acid, whereas the second is typified by the entrance of only part of the carbon skeleton. Since the former is more easily understood, let us consider it first.

Amino acid metabolism is a very extensive subject with each individual amino acid characterized by unique pathways of catabolism and anabolism (Chap. 17). Nevertheless, there are certain chemical transformations common to all amino acids. One such reaction is that of *transamination*. The pertinent chemistry of this reaction is just what the term implies, that is, an amino group is transferred. To be more specific, the *amino group of a donor amino acid is transferred to an acceptor α-keto acid*, resulting in the conversion of the original amino acid into an α-keto acid (below). Enzymes that catalyze this conversion are called *transaminases* (see p. 165) and they require the participation of pyridoxal phosphate as an essential coenzyme (pp. 164 and 531). They are indispensable for normal metabolism.

$$
\underset{\text{amino acid A}}{R_A\!-\!\overset{\overset{\displaystyle NH_3^+}{|}}{C}H\!-\!COO^-} + \underset{\alpha\text{-keto acid B}}{R_B\!-\!\overset{\overset{\displaystyle O}{\|}}{C}\!-\!COO^-} \underset{\substack{\text{(pyridoxal}\\\text{phosphate)}}}{\overset{\text{transaminase}}{\rightleftharpoons}} \underset{\alpha\text{-keto acid A}}{R_A\!-\!\overset{\overset{\displaystyle O}{\|}}{C}\!-\!COO^-} + \underset{\text{amino acid B}}{R_B\!-\!\overset{\overset{\displaystyle NH_3^+}{|}}{C}H\!-\!COO^-}
$$

While all amino acids can participate in transamination reactions, those involving glutamate or aspartate are most important. One particular reason for this that is relevant to our current discussion is illustrated by the reactions given below. You will note that when either *glutamate* or *aspartate* is involved, the product keto acids are *α-ketoglutarate* and *oxaloacetate*, respectively, both of which are intermediates of the citric acid cycle. Consequently, each can enter the cycle for further catabolism. Note, however, that when the cycle is primed at either point, its continued operation will depend on the availability of sufficient acetyl-SCoA to form citrate. The latter can be supplied by the degradation of carbohydrates, fatty acids, or other amino acids.

$$
\underset{\text{aspartate}}{{}^-OOCCH_2\overset{\overset{\displaystyle NH_3^+}{|}}{C}HCOO^-} + R\!-\!\overset{\overset{\displaystyle O}{\|}}{C}\!-\!COO^- \underset{\substack{\text{(pyridoxal}\\\text{phosphate)}}}{\overset{\text{transaminase}}{\rightleftharpoons}} \underset{\text{oxaloacetate}}{{}^-OOCCH_2\overset{\overset{\displaystyle O}{\|}}{C}COO^-} + R\!-\!\overset{\overset{\displaystyle NH_3^+}{|}}{C}H\!-\!COO^-
$$

$$
\underset{\text{glutamate}}{{}^-OOCCH_2CH_2\overset{\overset{\displaystyle NH_3^+}{|}}{C}HCOO^-} + R\!-\!\overset{\overset{\displaystyle O}{\|}}{C}\!-\!COO^- \underset{\substack{\text{(pyridoxal}\\\text{phosphate)}}}{\overset{\text{transaminase}}{\rightleftharpoons}} \underset{\alpha\text{-ketoglutarate}}{{}^-OOCCH_2CH_2\overset{\overset{\displaystyle O}{\|}}{C}COO^-} + R\!-\!\overset{\overset{\displaystyle NH_3^+}{|}}{C}H\!-\!COO^-
$$

The fact that many amino acids can furnish acetyl-SCoA also accounts for the integration of their degradation with the citric acid cycle. Among those amino acids that have *part* of their carbon skeleton converted to acetyl-SCoA are alanine, serine, cysteine, lysine, tryptophan, phenylalanine, and tyrosine. The first three enter via pyruvate and the others via multistep catabolic pathways that yield acetyl-SCoA as one product. Although the details of each conversion are not important at this point, some can be found in Chap. 17.

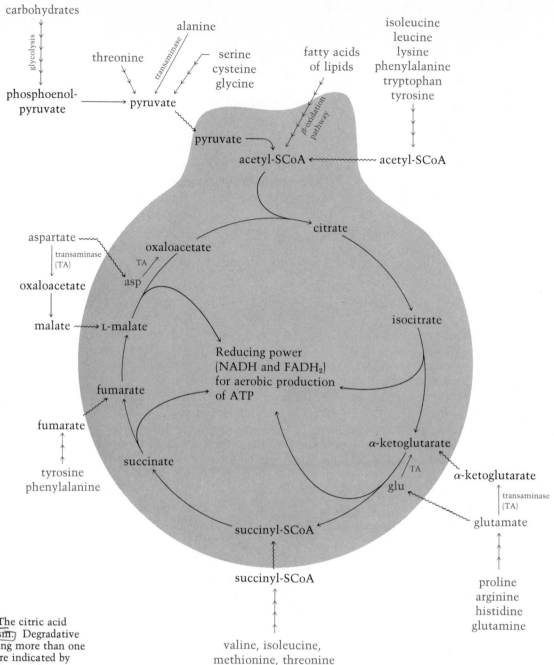

carbohydrates

isoleucine
leucine
lysine
phenylalanine
tryptophan
tyrosine

glycolysis

alanine

threonine

serine
cysteine
glycine

fatty acids
of lipids

phosphoenol-
pyruvate

pyruvate

pyruvate

acetyl-SCoA

acetyl-SCoA

β-oxidation
pathway

aspartate

transaminase
(TA)

oxaloacetate

asp

TA

citrate

oxaloacetate

malate

L-malate

isocitrate

fumarate

Reducing power
(NADH and FADH$_2$)
for aerobic production
of ATP

fumarate

succinate

α-ketoglutarate

tyrosine
phenylalanine

α-ketoglutarate

glu

TA

glutamate

transaminase
(TA)

succinyl-SCoA

succinyl-SCoA

proline
arginine
histidine
glutamine

valine, isoleucine,
methionine, threonine

FIGURE 13–2 The citric acid cycle in catabolism. Degradative pathways involving more than one enzymatic step are indicated by multiple arrows (⇥⇥⇥). The shaded area corresponds to intra-mitochondrial processes. Wavy lines represent movement across mitochondrial membrane. Transaminase (TA) activity is associated with both mitochondrion and cytoplasm.

Rather than continue with a written description of how the carbon skeletons of other amino acids can enter the cycle at other sites, I direct your attention to Fig. 13–2, which diagramatically summarizes the main catabolic relationships of the citric acid cycle. Careful inspection of the diagram, coupled with our discussion of the role of the cycle in catabolism, should give you an appreciation of why the cycle is considered to be a central pathway of catabolism in general, rather than a pathway specific to the degradation of any one compound or class of compounds.

In the preceding section we have described how various metabolites originating from different sources can be shunted into the cycle for further degradation, resulting in the production of ATP. By the same token, the cycle intermediates can be bled off at various points for use as precursors in the biosynthesis of different materials. The unique aspect of this function of the cycle is that the removal of intermediates must occur simultaneously with the *continued* catabolic operation of the cycle for the purpose of supplying the ATP that is also needed for anabolism. In other words, the cycle must simultaneously fulfill two roles.

Such is not the case when the cycle is linked to catabolic reactions, because here the production of energy is, in fact, a natural consequence of the exergonic and oxidative operation of the cycle and its coupling to oxidative phosphorylation. The linkage to anabolic reactions requires, then, that for energy production there always be sufficient catalytic levels of all intermediates, even as they are being bled off into other pathways, to maintain a certain flow of carbon through the reactions of the cycle. The material that follows is addressed to these anabolic relationships. (Figure 13–3 (p. 436) summarizes the material.)

Protein (Amino Acid) Anabolism

Virtually all organisms are capable of synthesizing amino acids, although there are differences in the number of acids synthesized by any one type of organism. For example, most plants and bacteria can synthesize all of the 20 amino acids required for proteins; but animals (including humans) are capable of synthesizing only certain ones, relying on a dietary intake for the rest. Whatever the organism, however, some of the intermediates of the citric acid cycle serve as important precursors for amino acids. Two particularly important compounds are the α-keto acids, oxaloacetate (OAA), and α-ketoglutarate (αKG). A third precursor is pyruvate, also an α-keto acid and of course closely linked to the citric acid cycle. The biosynthetic relationship of each amino acid to the citric acid cycle is diagramatically summarized in Fig. 13–3. The reaction details of some of these conversions are given in Chap. 17.

Since the mitochondrial membrane is impermeable to the movement of OAA and αKG (more so to OAA), these intermediates are bled off in the form of their precursors—MAL and ISOCIT, respectively—which can be transported across the membrane. The action of dehydrogenases in the cytoplasm would be to produce OAA and αKG in the cytoplasm. OAA and αKG could then be acted on by cytoplasmic transaminases to yield ASP (aspartate) and GLU (glutamate), respectively; see the diagram at the top of p. 434. (In some cells transaminase enzymes may be present in the mitochondrion, thus permitting a more direct route of ASP and GLU formation.)

The removal of intermediates for amino acid biosynthesis represents a serious problem to the continued operation of the cycle. Why? Well, if an intermediate is diverted for use in synthesis, the cycle is interrupted and OAA production would fall off. Since OAA is needed to condense with acetyl-SCoA, a process that keeps carbon coming into the cycle, any reduction in OAA production would reduce the activity of the whole cycle and any process related to it. In fact, since the cycle is at the center of all metabolism, the entire metabolic activity of the cell would become sluggish.

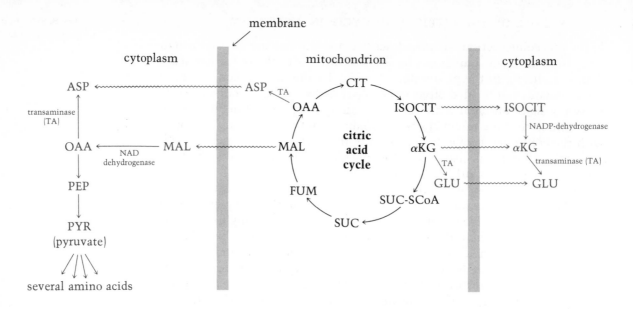

To counteract this possibility, a rather remarkable set of safety valves, called *anaplerotic reactions*, has evolved. The basic function of these reactions is to maintain an adequate supply of a crucial intermediate that participates in a central pathway such as the citric acid cycle when that intermediate is bled off into other metabolic pathways. Relative to our current discussion, the most important reaction of this type—particularly in mammals—is the conversion of pyruvate to oxaloacetate catalyzed by *pyruvate carboxylase*. Recall that this enzyme, which is also localized in the mitochondrion, has already been discussed as the catalyst for the first step in glycogenesis (p. 394). Thus we see how one enzyme can have different metabolic roles. In this case, the anaplerotic function of pyruvate carboxylase is believed to be more important.

$$\underset{\text{pyruvate}}{CH_3\overset{\overset{O}{\|}}{C}COO^-} + CO_2 \underset{\underset{ATP}{\overset{\text{carboxylase (biotin)}}{\longrightarrow}}}{\overset{\overset{ADP + P_i}{\overset{\text{pyruvate}}{\nearrow}}}{\longrightarrow}} \underset{\text{oxaloacetate (OAA)}}{{}^-OOCCH_2\overset{\overset{O}{\|}}{C}COO^-}$$

Being such a crucial enzyme, you might suspect, in view of our previous discussions, that the catalytic activity of pyruvate carboxylase would be under strict intracellular control. Indeed, this has been confirmed through recent kinetic studies on the purified enzyme. These have shown that acetyl-SCoA is a strong allosteric activator. The activation by acetyl-SCoA fits a pattern of metabolic logic. When a cycle intermediate is diverted (for example, when ISOCIT or αKG is used for GLU synthesis), the failure to regenerate OAA would result in an accumulation of acetyl-SCoA because it requires OAA to form citrate. This buildup thus serves as a metabolic signal that acts as an allosteric activator of the anaplerotic enzyme. More OAA is produced, insuring the formation of citrate. Hence, the cycle can continue to supply αKG for glutamate biosynthesis and also permit some of the αKG to flow through the cycle so that the entire sequence is not totally interrupted.

The same principle and the same anaplerotic reaction would operate when any intermediate is diverted from the cycle.

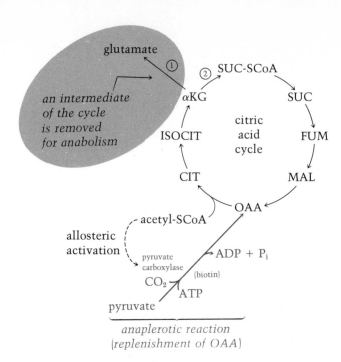

The occurrence of ① would *compete* with the continuation of normal cycle operation ②, resulting in a reduction of OAA and an accumulation of acetyl-SCoA; the rise of acetyl-SCoA concentration would signal the activation (allosterically) of pyruvate carboxylase to replenish OAA

Lipid Anabolism

In Chap. 16 we will discover that the biosynthesis of most lipids originates primarily with acetyl-SCoA. In fact, acetyl-SCoA is the metabolic source of *all* the carbon atoms in the synthesis of fatty acids, carotenoids, and steroids. The anabolic utilization of acetyl-SCoA does not, however, create an operational strain on the normal functioning of the citric acid cycle as does the removal of one of the internal intermediates.

The anabolic utilization of acetyl-SCoA does create one problem; most of the biosynthetic reactions, including those for lipids, occur in the cytoplasm, whereas the bulk of acetyl-SCoA production occurs within the mitochondria. While some acetyl-SCoA may exit from the mitochondrion

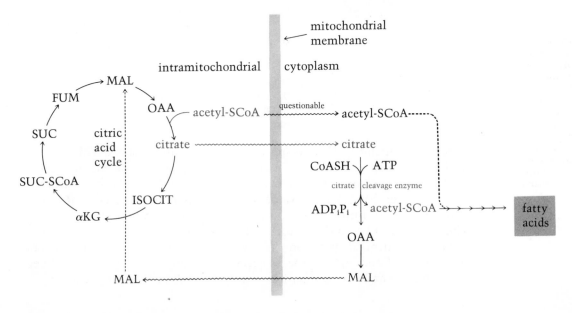

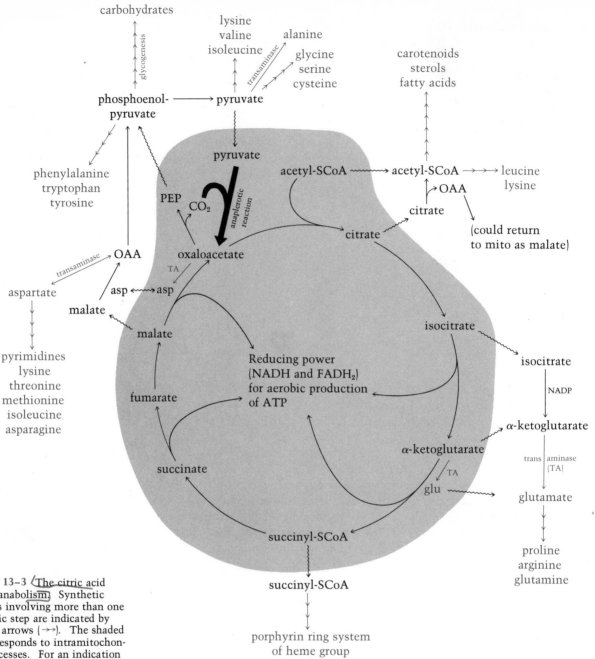

FIGURE 13–3 (The citric acid cycle in anabolism.) Synthetic pathways involving more than one enzymatic step are indicated by multiple arrows (→→). The shaded area corresponds to intramitochondrial processes. For an indication of which amino acids are *not* synthesized in higher animals, see p. 555. The heavy arrow indicates a major anaplerotic reaction. Wavy lines represent movement across mitochondrial membrane.

directly, the major exit route appears to be an indirect one. This route is mediated by a special enzyme found in the cytoplasm. The enzyme, called *ATP:citrate lyase* or *citrate cleavage enzyme*, catalyzes the breakdown of citrate to acetyl-SCoA and oxaloacetate. Thus it is proposed that citrate moves across the mitochondrial membrane and is cleaved in the cytoplasm. The return of OAA to the mitochondrion as MAL would sustain the cycle and allow continued operation. It is worth repeating that continued operation is needed not only to process acetyl-SCoA but also to supply ATP necessary for the anabolic utilization of acetyl-SCoA in the cytoplasm.

Carbohydrate Anabolism

The relationship of the cycle to the biosynthesis of carbohydrates is primarily a result of the operation of one enzyme, *phosphoenolpyruvate carboxykinase.* The enzyme, which is localized in mitochondria, catalyzes the GTP-dependent conversion of oxaloacetate to phosphoenolpyruvate. This enzyme is not new to us. It was previously discussed (p. 394) as being one of two enzymes responsible for effecting the conversion of pyruvate to phosphoenolpyruvate, which is then utilized in the biosynthesis of carbohydrate by the reversal of glycolysis. Our current discussion does not change any of that. But now we are seeing exactly how the enzyme is integrated into the whole of metabolism. The difference from our previous discussion is that, in this integrated framework, it is clear that the oxaloacetate, which is used as the precursor for phosphoenolpyruvate, is not necessarily formed only from pyruvate. Rather, the oxaloacetate can originate from pyruvate *or any* metabolite that can be converted to one of the intermediates of the citric acid cycle, each of which can then be converted to oxaloacetate.

When the carboxykinase is located in the cytoplasm, the link to carbohydrate biosynthesis would involve the exit of malate from the mitochondrion to the cytoplasm. In either case the anaplerotic role of pyruvate carboxylase would be necessary to sustain the level of OAA required for uninterrupted metabolism through the cycle.

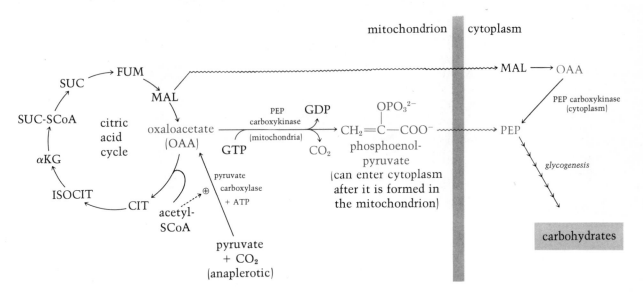

Other Anabolic Relationships

The involvement of the cycle intermediates in biosynthetic pathways is not confined to carbohydrates, lipids, and amino acids. Other critical relationships exist in the biosynthesis of purine and pyrimidine nucleotides and heme groupings. As indicated in Fig. 13–3, nucleotide biosynthesis utilizes aspartate (oxaloacetate) and glutamine (α-ketoglutarate), and heme biosynthesis requires succinyl-SCoA. We will inspect the details of both processes in Chap. 17.

THE CITRIC ACID CYCLE IN METABOLISM—A SUMMARY

At this point, if you feel overwhelmed and perhaps a little confused, it is quite understandable. The chapter is entitled "Citric Acid Cycle," but yet we have digressed into many other areas. You can take solace in one respect, however; our digressions could have been even more extensive.

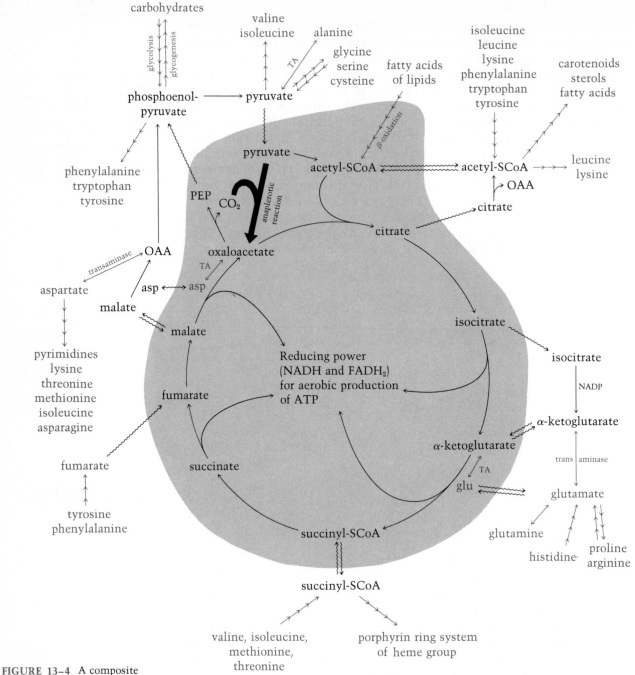

FIGURE 13-4 A composite summary of the catabolic and anabolic functions of the citric acid cycle. See legends of Figs. 13-2 and 13-3 for explanations. (A further summary of relationships to the amino acids can be found on p. 528, Chap. 17.)

Regardless, any uncertainty in your mind as to what the metabolic role of the cycle is can now be removed if you think back over the last several pages and realize that all of our separate digressions were analyzed in terms of the participation of the citric acid cycle. Thus, we have not fragmented and separated metabolism; rather, we have *unified and integrated it.* Therein lies the most important function of the cycle. The whole of cellular metabolism is interconnected, and hence interrelated and interdependent.

Figure 13–4 illustrates this principle. The diagram, of course, is nothing more than a combination of Figs. 13–2 and 13–3. A direct experimental proof of this principle would be to supply an organism with tracer levels of radioactive glucose (^{14}C) and then determine the occurrence of the label in other compounds. If you understand the role of the citric acid cycle, you should be able to predict what the outcome would be and recognize that the same would be true if the original labeled material were glutamate, aspartate, pyruvate, or acetate, and so on.

METABOLIC CONTROL OF CARBOHYDRATE METABOLISM—A SUMMARY

Throughout this and the previous chapter, we have described the catalytic properties of individual enzymes of glycolysis, glycogenesis, and the citric acid cycle. Included were many references to naturally occurring materials that serve as modulators of the activity of certain enzymes by acting as inhibitors or activators. Having done this, we are now prepared to piece together all these separate characteristics and analyze how their collective participation serves as a balanced and integrated network of metabolic control. Although our analysis will be in the particular area of carbohydrate metabolism, the general principle applies to all other areas of metabolism.

In order to do this, it will be useful to briefly recapitulate that most metabolic shifts stem from *alterations in the energy demands* of an organism. Hence, for obvious reasons of efficiency, it is very desirable that a cell be capable of adjusting its energy production to its energy utilization (the principle of supply and demand). As we will discover, this balance is mediated by regulation of energy (ATP) production in particular. Finally, we should recognize that since most aerobic organisms rely on the catabolism of carbohydrate material as the major source of useful metabolic energy, the target pathways of this regulation are glycolysis, glycogenesis, and the citric acid cycle.

The theme of this control is diagramatically summarized in Fig. 13–5, which symbolically illustrates the flow of carbon in the main pathways of carbohydrate metabolism in mammalian systems. (The same general principle and many of the same specific details apply to other types of organisms as well.) For reasons that will be clear shortly, one part of the diagram focuses on the activating and inhibiting influence of ATP, ADP, AMP, NADH, and NAD$^+$. The second part of the diagram highlights the effects of other materials not included in this group. For the sake of simplicity not all reactions are shown. Furthermore, those that are shown are given in an incomplete form; ATP, for example, is not shown as a substrate, and NADH is not shown as a product, and so on. Regulatory enzymes are identified by colored arrows to the side of which is indicated the substances that act as *inhibitors* (−) or *activators* (+). Allosterism (secondary-site effects) explains the action of many of these substances. To analyze the pattern of control requires a clear understanding of the basic principles of bioenergetics so often referred to in this and previous chapters. As an aid in review, the significant principles are summarized on p. 441 (top).

The basic strategy of metabolic regulation involves two major control signals: (a) the cellular ratio of ATP/ADP(AMP), and (b) the cellular ratio of NADH/NAD$^+$. The logic of the strategy is apparent if we recognize that

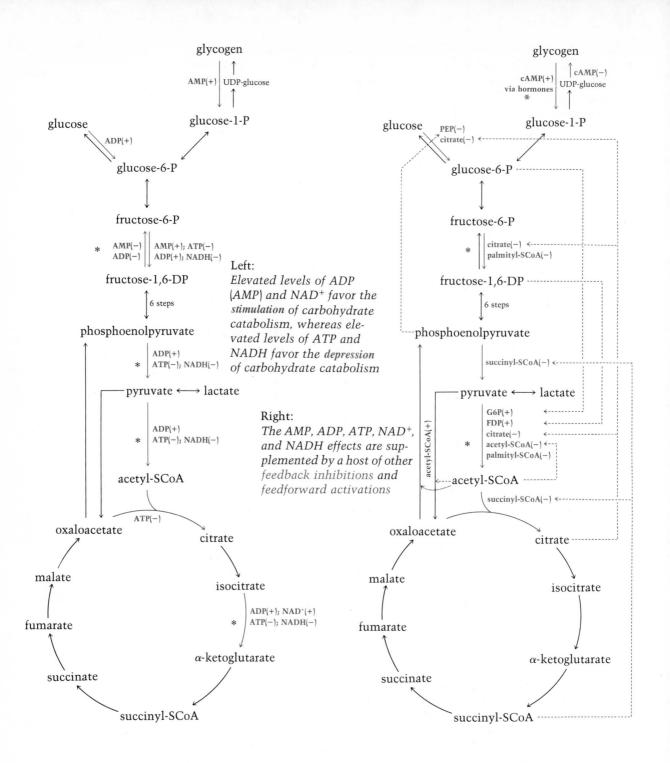

FIGURE 13–5 A summary of regulatory patterns found in the carbohydrate metabolism of mammals. Regulatory enzymes are symbolized by colored arrows. *Left:* control sites involving ATP, ADP, AMP, NADH, and NAD⁺. (+) represents activation; (−) represents inhibition. *Right:* control sites involving hormonal regulation via cyclic AMP and feedback inhibition and feedforward stimulation (dashed lines for both) of metabolic intermediates. (*Note:* Although the text discussion deals primarily with metabolism under aerobic conditions, which would then include the operation of the citric acid cycle, portions of these diagrams would also apply to anaerobic conditions, in which catabolism would terminate with lactate production.) Compartmentalization is not indicated. Asterisks identify major regulatory sites.

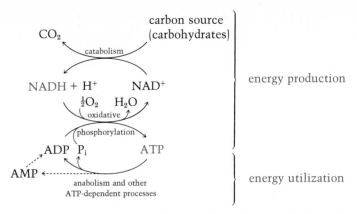

each ratio *reflects the energy needs*—that is, the metabolic state—of the organism. When the cell is very active, and hence the demand for and utilization of energy is high, both ratios will *decrease*. More specifically, the levels of ADP(AMP) and NAD^+ will increase and the levels of ATP and NADH will decrease. By close study of Fig. 13–5, you will observe that the increases in the cellular levels of ADP, AMP, and NAD^+ are all stimulatory signals that would increase the rate of both glycolysis and the citric acid cycle reactions. An increase in the rate of catabolism is further insured by the inhibitory effect of AMP and ADP on the diphosphatase involved in the FDP → F6P step of glycogenesis. The net result of these effects will be an increased rate of production of NADH, which will then enter oxidative phosphorylation and effect the necessary increased production of ATP. In other words, the "turning on" of a few key enzymes of carbohydrate catabolism and the complementary "turning off" of a key anabolic enzyme results in the "turning on" of ATP production. It is important to note that the term "turning on" does not mean that these reactions were not occurring prior to activation. The point is that after activation the catabolism of carbohydrates and ATP production occur at a greater rate in order to keep pace with the demand.

On shifting to a condition requiring less ATP utilization, as, for example, when contracted muscle returns to a relaxed or resting state, the control on metabolism operates in the same fashion, but in reverse. Since the need for energy is decreased, the production of ATP will be in excess of what is needed. Because ATP formation is coupled to NADH oxidation, a decrease in the conversion of ATP to ADP(AMP) will also be accompanied by cellular levels of NADH in excess of requirements. Consequently, the ratios of ATP/ADP (AMP) and $NADH/NAD^+$ will *increase*. In other words, (a) levels of ATP begin to increase while levels of ADP and AMP decrease, and (b) levels of NADH begin to increase while the level of NAD^+ decreases. Inspection of Fig. 13–5 reveals that these changes serve as metabolic signals effecting the *reduction* of the rate of carbohydrate catabolism. In other words, the "turning off" of a few key enzymes results in the "turning off" of ATP production.

The controls exerted by ATP, ADP, AMP, NAD^+, and NADH are reinforced by several other substances whose cellular levels rise and fall with changing states of metabolism. For example, when carbohydrate catabolism is stimulated (by hormones), the increased levels of glucose-6-P and fructose-1,6-diphosphate are proposed to have a feedforward stimulatory effect on their own further catabolism because they activate the conversion of

Cells in a resting metabolic state	⇌	Cells in a very active metabolic state
(energy utilization is less than that of cells in a very active metabolic state)		(energy utilization is greater than that of cells in a resting metabolic state)

high ATP low ADP + AMP high $\left(\frac{ATP}{ADP+AMP}\right)$	⇌	low ATP high ADP + AMP low $\left(\frac{ATP}{ADP+AMP}\right)$

high NADH low NAD^+ high $\left(\frac{NADH}{NAD^+}\right)$	⇌	low NADH high NAD^+ low $\left(\frac{NADH}{NAD^+}\right)$

Note: the analysis here is a simplified application of the *energy charge* and the *energy charge control,* which were mentioned in Chap. 7, p. 212. When energy is needed, ATP is produced at a faster rate. When the need lessens, the rate of ATP production is decreased.

pyruvate to acetyl-SCoA. As a further example, we note that in response to a reduced ATP demand, which would result in a slowing down of the citric acid cycle, the rise in the concentrations of CAC intermediates such as citrate and succinyl-SCoA would serve as feedback inhibition signals. The decreased rate of CAC activity would also reduce the consumption of acetyl-SCoA. This is entirely consistent with the observations that the two enzymes that function in the PYR to PEP conversion for carbohydrate anabolism are activated by acetyl-SCoA. The inhibitory effects of palmityl-SCoA on carbohydrate catabolism contribute to a balance of lipid and carbohydrate metabolism. In general, this means that when lipid is available in a mobilized form (thio ester) for catabolism, glycolysis is reduced (normally).

The entire network of regulatory checks and balances is obviously very elaborate. It may also leave you somewhat bewildered. For one thing it is difficult to indicate a sequence for all of these effects. However, if we assume that *in vivo* the initial influences are due to changing levels of ATP, ADP, AMP, NAD^+, and NADH, then the following generalization may be helpful, even though it is an oversimplification. The effects of ATP and NADH in diminishing the rate of carbohydrate catabolism result in altered levels of other substances such as citrate, acetyl-SCoA, and phosphoenolpyruvate. This alteration would have bonus effects of contributing further to reducing the rate of catabolism as well as favoring the rate of carbohydrate anabolism. In other words, one set of regulatory signals generates a second set that operates in a complementary fashion. A similar analysis would apply to the ADP and AMP effects in stimulating catabolism. This stimulation generates higher levels of glucose-6-phosphate and fructose-1,6-diphosphate and provides secondary signals that further stimulate the rate of catabolism.

However complicated the preceding analysis may appear (patience and close study will render it less complicated), one thing should be obvious: *living organisms are indeed programed for metabolic self-control.* In other words, a living organism is not a conglomerate of chemical processes that operate randomly, but an efficiently coordinated symphony of integrated reactions subject to a remarkable set of regulatory checks and balances.

While these last few pages climax much of the material given in preceding chapters, one point—often overlooked—merits recapitulation and emphasis. *All of the events of metabolism, including both the specificity of the chemical transformations and the control exerted on a select few, are occurring at an enzymatic level.* That is to say, the very essence of cellular metabolism and its characteristic self-control are mediated through proteins. Each has its own specific structure, and consequently a specific catalytic function, and, in some cases, specific regulatory properties. Remember, this fine control at the level of enzyme activity is also augmented by a coarse control exerted at the level of enzyme biosynthesis via induction and repression (see p. 271 in Chap. 8).

The regulatory characteristics of enzymes are established by *in vitro* kinetic assays on purified enzymes. The biochemist then argues, of course, that the behavior of enzymes *in vitro* also applies to the natural *in vivo* environment. However, the interpretation of *in vivo* significance is not necessarily a certainty. Some *in vitro* observations may not apply *in vivo*. Even more noteworthy, however, is the possibility that many aspects of metabolic regulation that do operate *in vivo* are not observed *in vitro*.

LITERATURE

ATKINSON, D. *Cellular Energy Metabolism and Its Regulation.* New York: Academic Press, 1977. A discussion of the interplay between thermodynamic and kinetic factors in maintaining the metabolic stability that underlies life. Atkinson explains his hypothesis of energy charge control.

BALDWIN, E. *Dynamic Aspects of Biochemistry.* 5th edition. London-New York: Cambridge University Press, 1967. Chapters 19 and 20 are devoted to the aerobic metabolism of carbohydrates and the citric acid cycle. See p. 409 for a description of the book.

GINSBURG, A., and E. R. STADTMAN. "Multienzyme Systems." *Ann. Rev. Biochem.,* Vol. 39 (1970). A current review article on the biochemistry of multienzyme complexes including α-keto acid dehydrogenase complexes.

KORNBERG, H. L. "Anaplerotic Sequences and Their Role in Metabolism." In *Essays in Biochemistry,* P. N. Campbell and G. D. Greville, eds., Vol. 2, pp. 1–32. New York: Academic Press, 1966. A good review article on this important phenomenon.

KREBS, H. A. "The History of the Tricarboxylic Acid Cycle." *Perspectives in Biol. & Med.,* **13,** 154 (1970).

LOWENSTEIN, J. M., ed. *Citric Acid Cycle.* Volume 13 of *Methods in Enzymology,* S. P. Colowick and N. O. Kaplan, eds. New York: Academic Press, 1969. A reference source for procedures in the isolation and assay of enzymes of the citric acid cycle including summaries of catalytic properties.

LOWENSTEIN, J. M. "The Tricarboxylic Acid Cycle." In *Metabolic Pathways,* 3rd edition, D. Greenberg, ed. New York: Academic Press, 1967. A review article giving a thorough analysis of this reaction pathway.

MAHLER, H. R., and E. H. CORDES. *Biological Chemistry.* 2nd edition. New York: Harper & Row, 1971. Chapter 14 of this textbook contains an excellent discussion of the individual enzymes of the cycle, the stereospecificity of the cycle, and the integration of the cycle with other areas of metabolism.

EXERCISES

13–1. Write all of the reactions in the citric acid cycle that comprise the oxidative transformations of the pathway.

13–2. Relative to the other reactions of the citric acid cycle, what is unique about the transformations catalyzed by (a) succinyl thiokinase and (b) α-ketoglutarate dehydrogenase?

13–3. If a molecule of uniformly labeled (^{14}C) pyruvate were oxidatively degraded to acetyl-SCoA and the latter then entered the citric acid cycle, which of the label patterns below would correspond to the α-ketoglutarate that would be produced in the first turn of the cycle?

$$^-OO^{14}C^{14}CH_2CH_2\overset{\overset{\displaystyle O}{\|}}{C}COO^-$$

or

$$^-OOCCH_2CH_2{}^{14}\overset{\overset{\displaystyle O}{\|}}{C}{}^{14}COO^-$$

13–4. To continue 13–3: what percent of the radioactivity in the original molecule of labeled acetyl-SCoA would be produced as $^{14}CO_2$ after three complete turns of the cycle? (*Note:* assume that the acetyl-SCoA molecules that would function in the second and third condensations with oxaloacetate are not labeled. *Hint:* the enzymes converting succinate to L-malate do not differentiate between identical carboxyl groupings in succinate of fumarate.)

13–5. The citric acid cycle is frequently described as the major pathway of aerobic metabolism, which means that it is an oxygen-dependent, degradative process. Yet none of the reactions of the cycle directly involves oxygen as a reactant. Why then is the pathway oxygen dependent (aerobic) rather than oxygen independent (anaerobic)?

13–6. Which of the following equations would best describe the net aerobic catabolism of one molecule of pyruvate to α-ketoglutarate?

(a) $PYR + OAA + 2NAD^+ + HSCoA \rightarrow$
$$\alpha KG + 2CO_2 + 2NADH + 2H^+$$

(b) $PYR + 2NAD^+ + HSCoA \rightarrow$
$$\alpha KG + 2CO_2 + 2NADH + 2H^+$$

(c) $PYR + OAA + 2NAD^+ \rightarrow$
$$\alpha KG + 2CO_2 + 2NADH + 2H^+$$

(d) $PYR + OAA + O_2 \rightarrow \alpha KG + 2CO_2 + 2H_2O$

(e) $PYR + 2acetyl\text{-}SCoA + O_2 \rightarrow$
$$\alpha KG + 2CO_2 + 2H_2O$$

(f) $PYR + OAA + \frac{1}{2}O_2 \rightarrow \alpha KG + 2CO_2 + H_2O$

13–7. If the early experiments of Krebs had been performed with purified mitochondria rather than homogenates of pigeon breast muscle, he would have experienced some problems in his study of aerobic metabolism when glycogen or glucose was used as the carbon source. Explain.

13–8. When fluoroacetyl-SCoA condenses with oxaloacetate, fluorocitrate is produced (see p. 419). If the fluorocitrate was acted on by aconitase according to the citrate $\rightarrow$ isocitrate conversion, would the F and OH in fluoroisocitrate be on the same carbon or on different carbons?

13–9. If the catalytic residues in citrate synthetase promoted the formation of the carbanion species of oxaloacetate shown below, what product (show its structure) would likely be formed on subsequent condensation with acetyl-SCoA?

$$^-OOC-CH-\overset{\overset{\displaystyle O}{\|}}{C}-COO^-$$
$$\underset{\ominus}{}$$

13–10. If radioactive CO_2 could be detected when acetyl-SCoA (^{14}C labeled) and oxaloacetate (no ^{14}C label) were incubated in a suitably buffered solution containing NAD^+ and pure samples of citrate synthetase, aconitase, and isocitrate dehydrogenase, it would be an observation contradictory to our understanding of the stereospecificity that applies to this section of the citric acid cycle. Explain.

13–11. If uniformly labeled tyrosine (^{14}C) were incubated with a cell-free liver extract, some of the tyrosine molecules would be metabolized in the following manner:

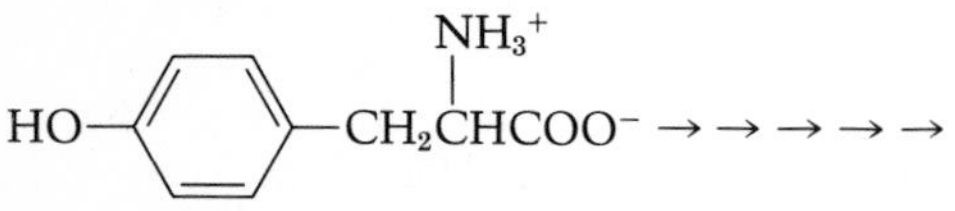

$$HO-\langle\text{ring}\rangle-CH_2\underset{\overset{|}{NH_3^+}}{CH}COO^- \rightarrow \rightarrow \rightarrow \rightarrow \rightarrow$$

(all carbons are ^{14}C)

$$^-OO^{14}C^{14}C\underset{\overset{|}{H}}{\overset{\overset{H}{|}}{=}}{}^{14}C^{14}COO^-$$

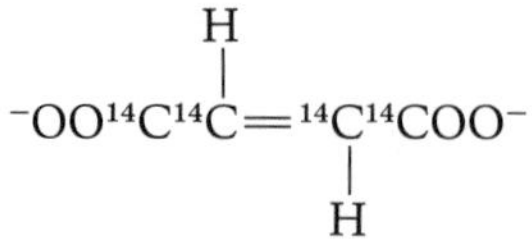

Show all of the necessary reactions that would account for the subsequent appearance in fructose-1,6-diphosphate of the carbons arising from fumarate. Indicate the distribution of radioactive carbons—and nonradioactive carbons, if there will be any—in the hexose.

13–12. If the citric acid cycle were being primed with intermediates originating from aspartate (oxaloacetate) and glutamate (α-ketoglutarate), what compound would have to be generated by some other sources in order for the enzymes of the cycle to continually and efficiently process these intermediates? Explain.

13–13. Write a net equation for the conversion of a molecule of isocitrate to oxaloacetate via the enzymes of the citric acid cycle. Write the equation for the same conversion when it is linked to the mitochondrial electron transport chain.

13–14. If a sample of a freshly prepared muscle homogenate were added to a buffered solution containing pyruvate and oxaloacetate, the amount of CO_2 produced would be greater than if pyruvate were added alone. Propose a sequence of reactions that would explain the stimulation of oxaloacetate without the participation of extramitochondrial NADH.

13–15. If purified mitochondria were separately incubated under each of the conditions listed below, predict which of the conditions would result in (1) the least and (2) the greatest amount of oxygen uptake.
(a) in the presence of succinate and malonate combined
(b) in the presence of succinate alone, followed after two minutes by the addition of malonate
(c) in the presence of fumarate and malonate combined
(d) in the presence of fumarate alone, followed after two minutes by the addition of malonate

13–16. How would you compare the ratios of both $NADH/NAD^+$ and ATP/ADP in heart muscle during periods of sleep and handball playing?

13–17. Summarize your understanding of the central role of the citric acid cycle in metabolism.

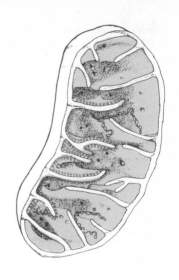

CHAPTER 14

OXIDATIVE PHOSPHORYLATION

In order to exist all living organisms must have a supply of energy from the surrounding environment. Photosynthetic organisms depend on an input of radiant energy. With the exception of a few microbes that can utilize inorganic substances as energy sources, nonphotosynthetic organisms depend on an input of *reduced organic compounds.* Although carbohydrate is the primary exogenous energy source, lipid and protein are also utilized.

Given a source of energy, each type of organism is capable of providing itself with a supply of metabolically useful chemical energy in the form of ATP. Although there are specific differences in the precise details of ATP production in each type of organism, the basic design is the same: *ATP is formed from ADP and P_i in a process coupled to the energy-yielding transfer of electrons from a reduced donor substance of high energy to an oxidized acceptor substance of lesser energy.* This type of chemistry is common to all organisms and is a major unifying biochemical principle of our biosphere.

(Refer to the diagram at the top of p. 446.)

This chapter deals with the formation of ATP in nonphotosynthetic organisms, with a focus on aerobic organisms where molecular oxygen is the terminal electron acceptor. The entire process is called *oxidative phosphorylation.* The prerequisite for oxidative phosphorylation in (most) nonphotosynthetic cells is the production of high-energy electron donors such as

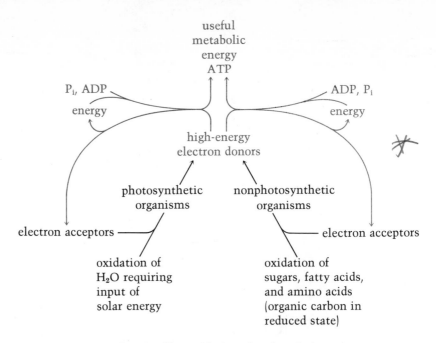

color signifies oxidative phosphorylation

NADH and FADH$_2$ during catabolism—a process to which we have referred on several occasions. The design of oxidative phosphorylation also applies to photosynthesizing cells, but it begins with the production of high-energy electron donors other than NADH, a process brought about by the harnessing of radiant energy. Because of the dependence on an input of radiant energy, the latter process is termed *photophosphorylation*. We will consider it in the next chapter.

ELECTRON TRANSFER SYSTEMS—SOME BASIC PRINCIPLES

Oxidative phosphorylation is a term that collectively refers to two cellular processes: (1) *the exergonic oxidation of reduced species such as NADH and FADH$_2$ via a concerted sequence of electron transport reactions,* and (2) *the endergonic phosphorylation of ADP to yield ATP.* Both processes are membrane localized, with the intracellular site in higher organisms specifically being the inner mitochondrial membrane. Although each is distinct, they are *energetically coupled processes,* with the energy released in electron transport being used for ATP formation. In addition, the two processes are also *spatially coupled,* meaning that the molecules involved in each event are located in the same compartment in a close-packed and highly ordered arrangement. Having already focused on the energetics of ADP + P$_i$ → ATP in Chap. 11, we will begin this discussion of electron transport by analyzing the exergonic phase of oxidative electron transport.

The process of biological electron transport consists of a *set* of several chemical systems, each of which is a distinct component capable of undergoing a reversible oxidation-reduction (*redox*) reaction. All components act in sequence to mediate the *transport of electrons from an initial high-energy donor in the reduced state to a terminal low-energy acceptor in the oxidized state.* Because of this molecular architecture, the process is frequently termed an *electron transport chain*. When oxygen is the terminal acceptor, as it is in all aerobes, the process is termed the *respiratory chain*. A coherent discussion and a meaningful understanding of electron transport demand as a prerequisite a brief digression into certain general physico-chemical principles that underlie the very essence of electron transfer systems. Depending on your own previous training, you will find that such material may or may not be largely a review. Whatever the case, it is extremely important that these principles be understood.

Reduction Potentials

One of the most basic concepts learned in any general chemistry course is that all oxidation-reduction reactions involve the reversible transfer of electrons, with electrons being lost from the component undergoing oxidation and gained by the component undergoing reduction. The theme of this type of process is summarized in the following equations, particularly in the representation of the two *half-reactions*, which together comprise the complete system. Half-reactions are frequently called *couples*, signifying that they are the oxidized and reduced "pair" for a given substance. Two principles are implicit in this type of reaction: (1) the reacting participants differ in terms of their affinity for electrons, and (2) an event of oxidation must be accompanied by an event of reduction. Our major concern below will be addressed to the first principle. Finally, we should note that, like any other chemical transformation, an oxidation-reduction reaction can be described in terms of its overall energetics and whether or not it can occur spontaneously. The point we wish to establish in this section is that both factors will be *controlled by the relative differences in the tendency of the reduced reactant of one couple to lose electrons and be oxidized, and of the oxidized reactant of the other couple to gain electrons and be reduced.*

oxidized A + reduced B → reduced A + oxidized B
from left to right [→]:
A undergoes reduction (electrons gained are donated by B)
B undergoes oxidation (electrons lost are donated to A)

composed of two half-reactions [couples]:
reduced B ⇌ oxidized B + ne
oxidized A + ne ⇌ reduced A

To develop our understanding of these concepts, let us begin by considering the system illustrated in Fig. 14–1. The diagram is a conventional representation of a simple *electrochemical cell*, which by definition is a chemical system capable of generating a flow of current, that is, electricity. The assembly consists of four parts: (a) two containers consisting of a slab of pure metal in contact with a unit molar solution of its ions, for example, Fe with $Fe^{2+}SO_4^{2-}$ and Cu with $Cu^{2+}SO_4^{2-}$; (b) a salt bridge acting as an essential internal circuit connecting the two systems without permitting mixing; and (c) an external circuit consisting of a conducting wire. After assembly and before the circuit key is closed, the bulb will be off, indicating the absence of any current flow. In chemical terms, this means that neither chemical couple is undergoing reaction and no equations are appropriate. When the key is closed, however, the bulb will light, signifying a spontaneous generation of current flow through the external circuit. Putting it an-

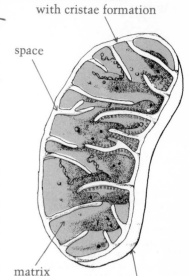

inner membrane
with cristae formation

space

matrix

outer membrane

Some functional features

Outer membrane:
enzymes such as
amino acyl-SCoA synthetase
monoamine oxidase
nucleoside diphospho kinase

Space:
adenylate kinase

Inner membrane:
oxidative phosphorylation
succinate dehydrogenase
carnitine-acyl transferase
ATP transport system

Matrix:
pyruvate dehydrogenase
citric acid cycle enzymes except
SDH
β-oxidation enzymes
phosphoenolpyruvate carboxykinase.

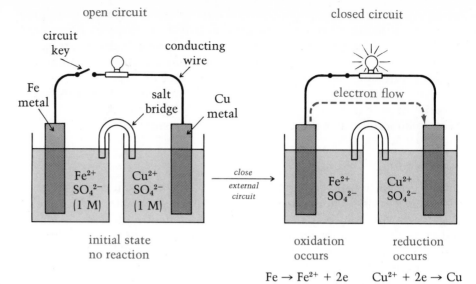

open circuit   closed circuit

FIGURE 14–1 An electrochemical cell. *Left:* circuit open; no reaction. *Right:* circuit closed; electrons will spontaneously flow from iron couple to copper couple. Fe^{2+} would increase in left compartment and Cu^{2+} would decrease in right compartment.

$$Fe \rightarrow Fe^{2+} + 2e \qquad Cu^{2+} + 2e \rightarrow Cu$$

$$Net: \quad Fe + Cu^{2+} \rightarrow Fe^{2+} + Cu$$

other way, the system is *generating useful energy*. It is a simple matter to demonstrate that accompanying this physical change—in fact the very reason for the current flow—is a net oxidation-reduction reaction involving each compartment, with the iron couple undergoing oxidation and the copper couple undergoing reduction. Thus, the concentration of Cu^{2+} in the right compartment will decrease, whereas the concentration of Fe^{2+} in the left compartment will increase. Reflecting on the spontaneity of the process, we can conclude that on a relative basis the copper couple (Cu^{2+}, Cu) has a greater tendency to accept electrons and undergo reduction than does the iron couple (Fe^{2+}, Fe).

In physicochemical terms, the relative capacity of any couple to undergo reduction is expressed in terms of a *standard reduction potential*, symbolized as $\mathcal{E}°$ (with volts as units). The value of each potential is measured against the standard hydrogen couple ($2H^+$, H_2), a universally accepted frame of reference. Under strict standard conditions of unit pressure and unit molar concentrations (pH = 0), the value of reduction potential for the hydrogen couple at 25 °C is arbitrarily set at 0.00 volt. Under conditions more appropriate for biochemical usage, namely, pH = 7, the adjusted potential ($\mathcal{E}°'$) is −0.42 volt.

$$2H^+(1\ M;\ pH = 0) + 2e \rightleftarrows H_2(1\ atm) \qquad \mathcal{E}° = 0.00\ V$$
$$2H^+(10^{-7}\ M;\ pH = 7) + 2e \rightleftarrows H_2(1\ atm) \qquad \mathcal{E}°' = -0.42\ V$$

The measurement of $\mathcal{E}°$ for other couples is quite direct, utilizing a setup similar to that shown in Fig. 14–1, with one compartment being the $2H^+$, H_2 couple and the other the couple to be measured. Both are at standard conditions. The external circuit contains a voltage-measuring device such as a voltmeter or a potentiometer. The observed voltage will be a measure of the capacity for current flow in the complete system. In other words, the voltage will be a positive indicator of the net potential difference between the reduction potential of the two couples. Since the reduction potential of the hydrogen system is arbitrarily set at 0, the observed net voltage will correspond to the reduction potential of the other couple.

$$\begin{matrix} \text{net potential of} \\ \text{complete system} \end{matrix} = \begin{matrix} \text{positive difference between} \\ \text{reduction potentials of each couple} \end{matrix}$$

that is

$$\mathscr{E}^{\circ}_{\substack{net}} = \mathscr{E}^{\circ}_{\substack{couple \\ undergoing \\ reduction}} - \mathscr{E}^{\circ}_{\substack{couple \\ undergoing \\ oxidation}}$$

By convention, a positive (+) sign accompanies the reduction potential of any couple that in fact has a greater tendency to undergo reduction relative to the hydrogen system; a negative (−) sign accompanies the reduction potential of any couple that has a lesser tendency to undergo reduction relative to the hydrogen system and in fact undergoes an oxidation instead. If the couple is pH dependent, appropriate calculations are made to adjust the $\mathscr{E}^{\circ}$ value to a condition applicable to pH ($\mathscr{E}^{\circ\prime}$). The reduction potentials of certain couples, including some of biological importance, are listed in Table 14–1.

Since all reduction potentials are determined in the same manner, that is, against the same reference system, it follows that the relative tendency of any two couples to participate in an electron transfer reaction can be readily predicted. As implied in the above material, the guideline that is used is as follows: *given any two couples, under appropriate conditions the system with the more positive reduction potential will spontaneously tend to gain electrons and undergo reduction.* Consider, for example, the system of Fig. 14–1, consisting of an Fe^{2+}, Fe couple with $\mathscr{E}^{\circ} = -0.44$ volt, and a Cu^{2+}, Cu couple with $\mathscr{E}^{\circ} = +0.34$ volt.

Reduction half-reactions	Reduction potentials
$Fe^{2+} + 2e \rightleftarrows Fe$	$\mathscr{E}^{\circ} = -0.44$ V
$Cu^{2+} + 2e \rightleftarrows Cu$	$\mathscr{E}^{\circ} = +0.34$ V

TABLE 14–1 **A listing of some standard reduction potentials in volts (for $\mathscr{E}^{\circ\prime}$: pH 7; T of 20–30 °C).**

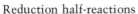

Half-reaction couple	$\mathscr{E}^{\circ}$	$\mathscr{E}^{\circ\prime}$
$\frac{1}{2}O_2 + 2H^+ + 2e \rightleftarrows H_2O$	−	+0.82
$Fe^{3+} + 1e \rightleftarrows Fe^{2+}$	+0.77	−
$Cu^+ + 1e \rightleftarrows Cu$	+0.52	−
$Cu^{2+} + 2e \rightleftarrows Cu$	+0.34	−
$2H^+ + 2e \rightleftarrows H_2$	0.00	−0.42
$Fe^{2+} + 2e \rightleftarrows Fe$	−0.44	−
Cytochrome $a_3 \cdot Fe^{3+} + 1e \rightleftarrows$ cytochrome $a_3 \cdot Fe^{2+}$	−	$+0.3 \rightarrow 0.5^a$
Cytochrome $f \cdot Fe^{3+} + 1e \rightleftarrows$ cytochrome $f \cdot Fe^{2+}$	−	+0.37
Cytochrome $a \cdot Fe^{3+} + 1e \rightleftarrows$ cytochrome $a \cdot Fe^{2+}$	−	+0.29
Cytochrome $c \cdot Fe^{3+} + 1e \rightleftarrows$ cytochrome $c \cdot Fe^{2+}$	−	+0.25
Cytochrome $c_1 \cdot Fe^{3+} + 1e \rightleftarrows$ cytochrome $c_1 \cdot Fe^{2+}$	−	+0.22
Coenzyme Q + 2H$^+$ + 2e $\rightleftarrows$ coenzyme QH$_2$ (in ethanol)	−	+0.10
Cytochrome $b \cdot Fe^{3+} + 1e \rightleftarrows$ cytochrome $b \cdot Fe^{2+}$	−	+0.04
$FAD + 2H^+ + 2e \rightleftarrows FADH_2$	−	-0.12^b
$NADP^+ + 2H^+ + 2e \rightleftarrows NADPH[H^+]$	−	−0.32
$NAD^+ + 2H^+ + 2e \rightleftarrows NADH[H^+]$	−	−0.32
Oxidized ferredoxin + 1e $\rightleftarrows$ reduced ferredoxin	−	−0.43

[a] Very doubtful value; *in vivo* cyt a_3 is complexed with cyt a; the $\mathscr{E}^{\circ\prime}$ of the a,a_3 complex called *cytochrome oxidase*, is approximately +0.29.
[b] *In vivo* FAD exists in a firmly bound state to its protein. The potential for FAD in each of these individual flavoproteins may be more or less negative than −0.12, due to the uniqueness of the association between FAD and the protein. The value of −0.12 given here is a representative average.

A comparison of $\mathscr{E}^\circ$ values predicts that Cu^{2+} will be reduced (its $\mathscr{E}^\circ$ is more positive), that Fe will be oxidized (its $\mathscr{E}^\circ$ is more negative), and that

$$\mathscr{E}^\circ_{net} = \mathscr{E}^\circ_{Cu^{2+},Cu} - \mathscr{E}^\circ_{Fe^{2+},Fe} = +0.78 \text{ volt}$$

Note that this prediction is consistent with our earlier description of what indeed occurs within an iron-copper electrochemical cell.

Finally, let us consider an example of biological importance, namely, a reaction involving the NAD and oxygen couples. Applying the same principle reveals that the oxidative transfer of electrons from NADH to molecular oxygen will be the favored process.

Reduction half-reactions	Reduction potentials (pH 7)
$NAD^+ + 2H^+ + 2e \rightleftharpoons NADH + H^+$	$\mathscr{E}^{\circ\prime} = -0.32$ V
$\frac{1}{2}O_2 + 2H^+ + 2e \rightleftharpoons H_2O$	$\mathscr{E}^{\circ\prime} = +0.82$ V

A comparison of $\mathscr{E}^\circ$ values predicts that O_2 will be reduced and NADH will be oxidized, and that

$$\mathscr{E}^{\circ\prime}_{net} = \mathscr{E}^{\circ\prime}_{O_2,H_2O} - \mathscr{E}^{\circ\prime}_{NAD^+,NADH} = +1.14 \text{ V}$$

Energetics of Electron Transfer   stop

In our description of electron transfer systems, reference was made to the fact that the difference in reduction potentials is linked to the spontaneity of such processes. By recalling the principles of Chap. 11, you should recognize that the overall chemical energetics of electron transfer reactions, embodied by $\Delta G^{\circ\prime}$, ought to be related to the net potential difference ($\mathscr{E}^{\circ\prime}_{net}$). Neglecting any detailed development of this, suffice it to say that such is the case. The exact form of the mathematical relationship has already been given in Chap. 11, but should be briefly reviewed. It is

$$\Delta G^{\circ\prime} = -n\mathscr{F}\mathscr{E}^{\circ\prime}_{net}$$

where n is the number of electrons transferred in the overall process, $\mathscr{F}$ is a proportionality constant equal to 23,060 calories per volt, and of course $\mathscr{E}^{\circ\prime}_{net}$ is the net potential difference between the two couples. The point to note is that, whenever the potential difference has a positive value, the $\Delta G^{\circ\prime}$ will be negative, signifying that the process not only is capable of occurring spontaneously but also is exergonic. Applying this equation to the reduction of NADH via molecular O_2 reveals, for example, that the potential output of energy is $-52,580$ calories, a sizeable energy yield indeed. Shortly you will discover that this relationship, when applied in this context, will make the basis of the coupling of ATP formation to biological electron transport much clearer.

$$NADH + H^+ + \tfrac{1}{2}O_2 \rightleftharpoons NAD^+ + H_2O \qquad \mathscr{E}^{\circ\prime} = +1.14 \text{ volts}$$

When $n = 2$ electrons

$$\Delta G^{\circ\prime} = -(2)(23,060)(1.14) = -52,580 \text{ calories}$$

RESPIRATORY CHAIN OF ELECTRON TRANSPORT

Role of Oxygen in Metabolism

Our current understanding of how molecular oxygen participates in biological oxidations can be summarized by the following reactions. In cases (1) and (2), the oxygen is incorporated directly into the substrate. Depending

Molecular oxygen in biological oxidations

$$(1) \quad SH_2 + O_2 \xrightarrow[]{\text{hydroxylases}} SHOH + H_2O$$

SH_2
reduced
substrate

$DH_2 \quad D$

$$(2) \quad SH_2 + O_2 \xrightarrow{\text{oxygenases}} S(OH)_2$$

SHOH, $S(OH)_2$, and S
are oxidized
products

$$(3a) \quad SH_2 + \tfrac{1}{2}O_2 \xrightarrow{\text{oxidases}} S + H_2O$$

$$(3b) \quad SH_2 + O_2 \xrightarrow{\text{oxidases}} S + H_2O_2$$

on the chemistry involved, the enzymes that catalyze these reactions are
called *hydroxylases* (reaction (1)) and *oxygenases* (reaction (2)). Hydroxylases, sometimes called monooxygenases, catalyze the insertion of just one
of the atoms of molecular oxygen into the substrate (SH_2) as part of a hydroxyl group (—OH). The hydroxylase-catalyzed reactions have a strict requirement for the participation of a second reduced substrate (DH_2), which
itself also undergoes an oxidation, specifically, a dehydrogenation. Although the complete mechanism of a hydroxylase-catalyzed reaction varies
from one hydroxylase to another, in most instances the hydrogen donor
(DH_2) is usually NADH or NADPH. One very important hydroxylase reaction in mammalian systems is the conversion of phenylalanine to tyrosine
(see margin). Some other examples will be encountered in the discussion
of amino acid metabolism in Chap. 17.

Oxygenases, sometimes called dioxygenases, catalyze the insertion of
both atoms of molecular oxygen into the substrate. As indicated in general
equation (2) above, the product may be a dihydroxy derivative of the substrate. In other instances stable dihydroxy derivatives are not produced.
Rather the oxygen atoms are incorporated as part of a carbonyl ($\diagdown$C=O) or
carboxyl (—COO⁻) grouping. An example of the latter reaction type, illustrating the oxidation of homogentisic acid to maleylacetoacetate, is given in
the margin. This reaction is an important step in the catabolism of the aromatic amino acids, phenylalanine and tyrosine (see also Chap. 17, p. 537).

In the third type of biological oxidation involving oxygen (reaction (3)),
the oxygen molecule is not actually incorporated into the substrate.
Rather, the oxygen molecule functions as a hydrogen acceptor. Because
the hydrogen atom is composed of a proton (H^+) and an electron, the role of
oxygen in this type of oxidation is better described as that of an *electron acceptor*. Enzymes catalyzing this type of reaction are called oxidases. As indicated above there are two different types of oxidases, one that produces
H_2O and one that produces H_2O_2. Reaction (3a) is representative of the role
of oxygen in biological electron transport, and so let us pursue it further.
(Two types of peroxide-producing oxidases of importance are described in
Chap. 17; see p. 534.)

The generalized form of the oxidase reaction given above (reaction (3a)) is
somewhat deceptive regarding the role of oxygen in respiration, because it
suggests that the oxidation of a reduced substrate proceeds via a direct dehydrogenation, with oxygen serving as an immediate electron acceptor.
Although there are examples of this type of reaction in cells (the ones just
given are but a few), the participation of oxygen in electron transport is of a
completely different nature, with more than one catalyst involved. The
pioneering work in this area was done independently by H. Wieland,
T. Thunberg, O. Warburg, and D. Keilin in the 1920s and 1930s.
Their ideas resulted in the formulation of the theory of the so-called *respi-*

phenylalanine

$DH_2 \searrow O_2$
phenylalanine
hydroxylase
$D \nwarrow H_2O$

tyrosine

phenylalanine

↓

tyrosine

↓

homogentisic acid

O_2
homogentisic acid
oxygenase

maleylacetoacetate

451

ratory chain or *electron transport chain*, which states that the oxygen-dependent dehydrogenations involve *intermediate electron carriers* that intervene in the flow of electrons between the initial electron donor SH_2 and the terminal electron acceptor O_2. Electron transport would involve the successive interaction of carriers capable of undergoing a reversible conversion between reduced and oxidized states. Each intermediate carrier would first participate in its oxidized state as an acceptor of electrons and then be converted to its reduced state. In the reduced state the carrier as a donor would then transfer electrons to the next carrier in its oxidized state and in so doing would be reconverted back to the original oxidized state. The final carrier would transfer electrons to oxygen, the terminal acceptor, which would be reduced to water.

Transport chain involving intermediate electron carriers

For example,

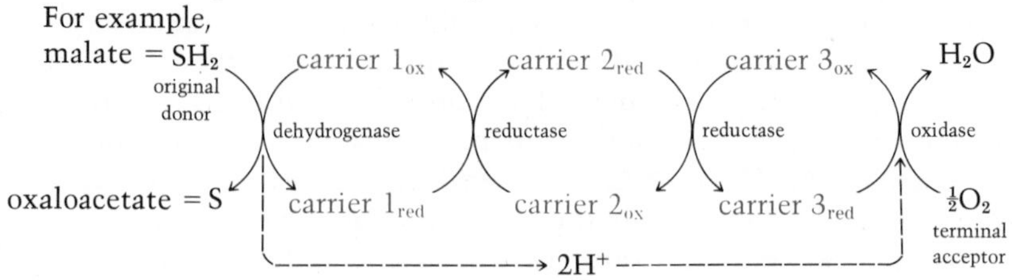

$$\text{Net:} \quad SH_2 + \tfrac{1}{2}O_2 \xrightarrow{C_1, C_2, C_3(...C_n)} S + H_2O$$

The recognition of this pattern of electron flow from donor carrier to acceptor carrier through a successive series of oxidation-reduction reactions (each an electrochemical cell) was a key discovery in biochemistry, especially since the pattern applies to all types of cells. After nearly 50 years of study, many of the details of operation have been discovered but some gaps in our knowledge still remain. We will now consider some of these details.

Electron Carriers

In higher organisms the established intermediate carriers of the respiratory chain are *nicotinamide adenine dinucelotides* (NAD), *flavin nucleotides* (FAD and FMN), *coenzyme Q (CoQ)*, a family of *cytochromes* (designated as *b, c_1, c, a, a_3*) and *nonheme iron proteins* (NHI). It is possible that Vitamins K and E also function as carriers. It is also possible that other carriers exist that have not yet been detected. Extensive evidence suggests the sequence of operation shown below. Nonheme iron proteins are not indicated in this representation because they are proposed to intervene at several positions rather than just one (see p. 456).

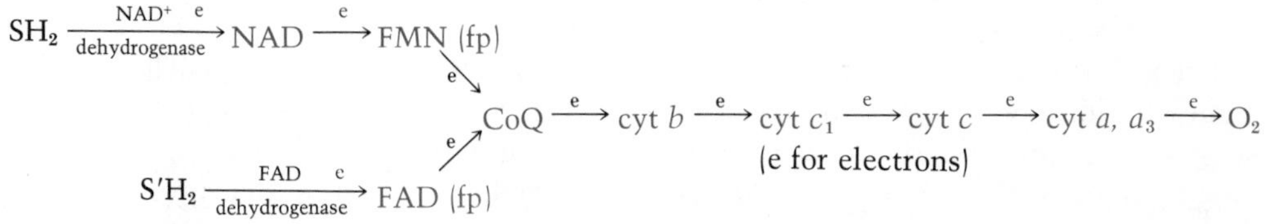

The process begins with a transfer of electrons from a reduced substrate (SH_2 or $S'H_2$) to either NAD or FAD, depending on whether the particular dehydrogenase involved is NAD^+-dependent or FAD-dependent. When the process is initiated at the level of NAD, note that the next carrier is FMN.

Whatever the initial substrate, the electrons from the flavin carrier are then transferred to coenzyme Q, and from there they go through a specific order of the cytochrome carriers, and ultimately are delivered to oxygen, the terminal acceptor. To develop an appreciation of how this shuttling system works, it is obviously necessary to analyze the chemical nature of the participating intermediates.

NAD, FAD, and FMN. The structure and oxidation-reduction chemistry of NAD and FAD (FMN) coenzymes were discussed previously (p. 359 for NAD; p. 422 for FAD-FMN). Now we focus on the fact that the reduced state of each coenzyme (NADH, $FADH_2$, and $FMNH_2$) represents the initial high-energy state of electrons (high-energy relative to oxygen) that prime the respiratory chain. Ordinarily NAD is a free coenzyme that binds to a dehydrogenase enzyme only during the course of the reaction catalyzed by the dehydrogenase. On the other hand, FAD (or FMN) is generally firmly bound to its dehydrogenase and the complete protein is called a *flavoprotein* (fp). Because of the individuality of each flavoprotein, the energy state of the $FADH_2$ is variable (refer back to the footnote *b* in Table 14–1).

Although unrelated to their biological function, remember (p. 57 in Chap. 2) that the reduced and oxidized forms of NAD have distinct absorption spectra permitting the measurement of NADH in the presence of NAD^+. The same type of distinction also applies to $FADH_2$ and FAD and also the cytochromes. These spectral distinctions have benefited the laboratory study of respiring mitochondria (see p. 459 later in this chapter).

Although in our previous discussions of NAD and FAD we referred to each as a hydrogen acceptor coenzyme, the term electron carrier is wholly equivalent. As a hydrogen acceptor, each coenzyme gains the equivalent of two electrons. (*Remember:* Two hydrogen atoms are equivalent to two protons plus two electrons.) With NADH only one proton is incorporated, and the other is gained by the medium, whereas with $FADH_2$ both protons are accepted.

$$SH_2 + NAD^+ \rightarrow S + NADH + H^+$$
$$S'H_2 + FAD \rightarrow S' + FADH_2$$
$$\overrightarrow{}$$
$$-2 \text{ Hs } [2H^+, 2e]$$

In referring to the reduced state of NAD throughout this chapter, the NADH and NADH + H^+ designations are used interchangeably. Dropping the proton is merely a matter of convenience.

Quinones. The coenzyme Q designation refers to a family of *quinone* compounds (Q for quinone) sometimes called *ubiquinones* because of their ubiquitous occurrence in nature. Coenzyme Q molecules differ from source to source in the length of the hydrocarbon side chain, with the value of *n* ranging from 6 (in microbes) to 10 (in mammals). The structure of coenzyme Q is quite consistent with its proposed function as an electron carrier, since the quinone grouping is capable of undergoing a reversible conversion to the hydroquinone grouping. In support of its participation in the respiratory chain is the fact that the bulk of cellular coenzyme Q is found in the mitochondrial membrane (the subcellular site of electron transport), from which it can be readily extracted by chloroform and other lipid solvents. Whether or not the *in vivo* activity of CoQ is dependent on its being complexed to a protein within the membrane is unknown.

Certain bacteria utilize *Vitamin K naphthoquinones* (see p. 340) as intermediate carriers acting between FAD and coenzyme Q, and possibly elsewhere in the sequence. It remains unresolved whether Vitamin K-type compounds play a corresponding role in electron transport in higher organisms. Vitamin E may also participate.

$$H_3CO \overset{O}{\underset{O}{\bigcirc}} CH_3 \quad (CH_2CH=\overset{CH_3}{\underset{}{C}}CH_2)_n H \xrightarrow[-2Hs]{+2Hs} H_3CO \overset{OH}{\underset{OH}{\bigcirc}} CH_3 \quad (CH_2CH=\overset{CH_3}{\underset{}{C}}CH_2)_n H$$

oxidized coenzyme Q (CoQ)
(quinone form)

reduced coenzyme Q (CoQH₂)
(hydroquinone form)

Cytochromes. First detected in the late nineteenth century in various mammalian tissues, *cytochromes* ("cellular pigments") are now known to be present in all types of organisms. They are localized primarily in membranes and in eukaryotic cells, specifically in the mitochondrial membrane. All cytochromes are *hemoproteins,* and are possibly evolutionary relatives of myoglobin and the monomeric unit of hemoglobin. Unlike the latter two proteins, however, the function of the cytochromes is not to bind with and transport oxygen, but to participate in electron transfer reactions with the *iron atom of the heme grouping undergoing a reversible oxidation-reduction.* Recall that the oxygen-binding ability of myoglobin and hemoglobin requires that the heme contain iron in the reduced Fe^{2+} form. With the oxidized Fe^{3+} form, myoglobin and hemoglobin are inactive.

$$Fe^{2+} \underset{+1\ electron}{\overset{-1\ electron}{\rightleftarrows}} Fe^{3+}$$

(reduced state)   (oxidized state)

the most simplified statement of the biochemical role of the cytochromes

Approximately 25–30 different cytochromes are known to exist throughout nature. All types are classified on the basis of distinctive light-absorbing properties, with those of similar spectroscopic behavior segregated into groups designated by lower case letters (*a, b, c,* etc.) Within each group, individual cytochromes with unique spectral properties are then designated by numerical subscripts (b, b_1, b_2, b_3, etc.). The structural factors accounting for the multiplicity of cytochromes are (a) variations in the side-chain substituents of the tetrapyrrole moiety of the heme group (see p. 456), (b) variations in the polypeptide component, and (c) variations in the way the polypeptide component is bound to the heme unit. The first distinguishes one cytochrome class from another, that is, cytochrome *a* from cytochrome *b* from cytochrome *c*; the latter two distinguish among members of the same class, that is, cytochrome *b* from cytochrome b_1, from cytochrome b_2, and so on.

TABLE 14–2 **Spectrophotometric characteristics of some cytochromes. At the left is a typical absorption spectrum, with the usual ranges of the α, β and γ Soret bands in the spectrum of a reduced cytochrome identified. (Molecular weights are also listed).**

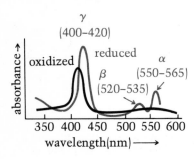

Cytochrome (source)	Molecular weight	Soret bands		
		α	β	γ
b	25,000	563	532	429
c (mitochondria of animals, plants, yeasts, and fungi)	12,500	550	521	415
c_1	37,000	554	524	418
a	≈ 600,000 for	600	absent	439
a_3	a, a_3 aggregate	604	absent	443
b_1 (E. coli)	500,000	558	528	425
b_2 (yeast)	170,000	557	528	424
b_5 (microsomes)	25,000	556	526	423
f (chloroplasts)	?	555	525	423

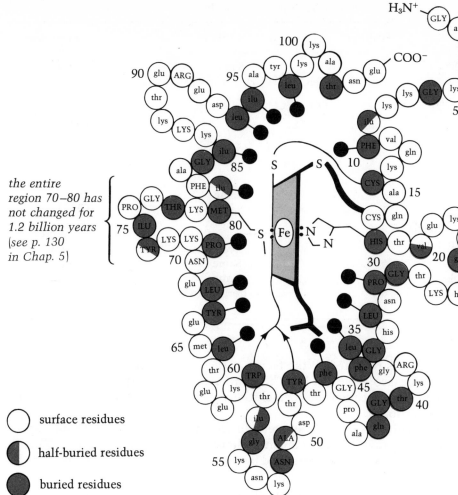

the entire
region 70–80 has
not changed for
1.2 billion years
(see p. 130
in Chap. 5)

surface residues

half-buried residues

buried residues

FIGURE 14–2 Heme-packing diagram of the cytochrome *c* molecule. *Heavy circles* indicate side chains that are buried on the interior of the molecule, and attached *black dots* mark residues whose side chains pack against the heme and interact with it. *Light circles* indicate side chains on the outside of the molecule, and *dark half-circles* show groups that are half-buried at the surface. *Arrows* from tryptophan 59 and tyrosine 48 to the buried propionic acid group represent hydrogen bonds. Residues designated by *capital letters* are totally invariant among the proteins of 38 different species. (Reprinted with permission from · R. E. Dickerson, et al., Ferricytochrome *c*: General Features of the Horse and Bonito Proteins at 2.8 Å Resolution, *J. Biol. Chem.*, **246**, 1511–1535 (1971).)

In the case of cytochrome *c* (see Fig. 14–2) the heme grouping is covalently attached to the polypeptide backbone through two cysteine residues. However, in most cytochromes the binding is exclusively noncovalent. The many other heme-polypeptide interactions in cytochrome *c* are unique in detail to cytochrome *c*, but the same general types of interactions apply to all other cytochromes.

The data of Table 14–2 summarize the general and specific absorption properties of some cytochromes. The data refer to three distinctive maxima in the visible portion of the absorption spectrum for the cytochromes in the reduced state. The precise values of these maxima (called the α, β, and γ *Soret bands*) are used to differentiate among the individual cytochromes. Since the spectrum of the oxidized form of each cytochrome usually lacks these three maxima, it is also possible to distinguish between the oxidized and reduced forms and even determine the relative concentration of the two when they are both present.

The spectrophotometric individuality of each cytochrome is only part of the functional distinction among them. In the context of their participation as intermediate carriers in the electron transport chain, a more significant distinction is that the *variations in cytochrome structure are responsible for variations in their ability to participate in redox reactions*, a distinc-

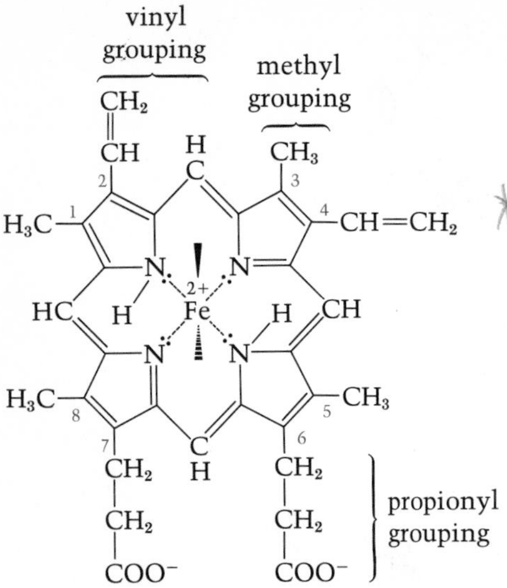

vinyl grouping
methyl grouping
propionyl grouping

heme of all *b* cytochromes and also in hemoglobin, myoglobin, and catalase (wedges represent the fifth and sixth coordination bonds of the heme iron to the polypeptide)

Comparison of heme at left to other hemes in

Position	a cytochromes	c cytochromes
1	Same	Same
2 (in *a*)	$-CHCH_2CH(CH_2)_3CH(CH_2)_3CH(CH_3)_2$ with OH, CH_3, CH_3	
2 (in *c*)		$-CHCH_3$ / $S-protein$ (covalent attachment; see Fig. 14-2)
3	Same	Same
4	Same	$-CHCH_3$ / $S-protein$
5	hydrogen (−H)	Same
6	Same	Same
7	Same	Same
8	$-C=O$ (formyl group) with H	Same

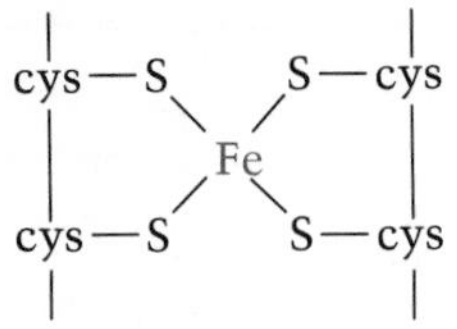

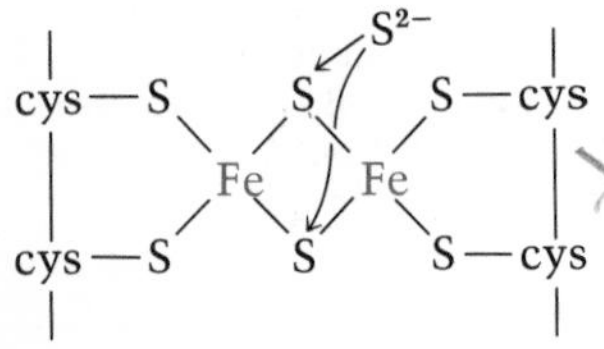

these are just two of the various Fe-S aggregates that occur in nonheme iron proteins

tion that is reflected by the fact that each cytochrome has a characteristic $\mathscr{E}^{\circ\prime}$ value (see Table 14–1). The significance of the respective $\mathscr{E}^{\circ\prime}$ values for the proposed sequence of cytochromes in the respiratory chain will be discussed shortly.

Cytochromes *a* and a_3 are very difficult to isolate as separate substances, suggesting that they exist and function as a tightly knit aggregate. The a,a_3 complex is further distinguished by the presence of *copper* (Cu), which is believed to be bound to the a_3 component and is essential for activity. The unique functional feature of the a,a_3 complex is that it is the only cytochrome system that will react directly with molecular oxygen. Accordingly, the a,a_3 complex is frequently called *cytochrome oxidase.*

Nonheme iron proteins. In many mitochondria most of the iron present is not associated with the heme group of cytochromes but complexed to other proteins called *nonheme iron proteins.* Since the iron is bound through sulfur atoms of cysteine residues or directly to S^{2-} (see margin), these proteins are also called *iron-sulfur proteins.* These proteins are definitely involved in electron transport and apparently in several regions of the transport chain. However, there is still uncertainty about the exact locations and the exact manner in which the iron-sulfur cluster undergoes reversible oxidation-reduction. An iron-sulfur protein of immense importance in photosynthesis is *ferredoxin* (see p. 488 in Chap. 15).

Design for Respiratory Chain

Having described the general nature of the respiratory chain and its major participants, let us now consider the process in greater detail. A more detailed version of what is proposed to occur is diagramed in Fig. 14–3. An understanding of why the NAD → flavin → CoQ → cytochrome → O_2 sequence has evolved is made evident by examining the standard reduction

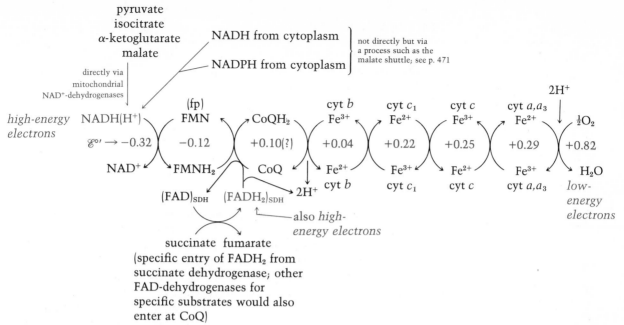

not directly but via a process such as the malate shuttle; see p. 471

FIGURE 14–3 A diagramatic representation of the respiratory chain. ($\mathscr{E}^{\circ\prime}$ values of each half-reaction couple are reduction potentials in volts.)

potentials. Note that with the exception of the CoQ couple, the $\mathscr{E}^{\circ\prime}$ becomes progressively more positive from NAD to oxygen. In other words, the carriers are arranged in order of an *increasing tendency to undergo reduction.* On an energy scale, this means that each carrier in its reduced state is of higher energy than the reduced form of the next carrier in the shuttle. In other words, *electrons proceed to a lower energy level.* Hence, if the chain is primed with NADH, the transfer of electrons to FMN to yield NAD$^+$ and FMNH$_2$ is an *energetically favorable* process with a positive $\mathscr{E}^{\circ\prime}_{net}$,

$$NADH + H^+ \rightarrow NAD^+ + 2H^+ + 2e \qquad \mathscr{E}^{\circ\prime}_{ox} = -(-0.32)$$
$$FMN + 2H^+ + 2e \rightarrow FMNH_2 \qquad \mathscr{E}^{\circ\prime}_{red} = -0.12$$

$$\overline{NADH + H^+ + FMN \rightarrow NAD^+ + FMNH_2 \qquad \mathscr{E}^{\circ\prime}_{net} = +0.20}$$

$$\Delta G^{\circ\prime}_{net} = -n\,\mathscr{F}\mathscr{E}^{\circ\prime}_{net} = -(2)(23,060)(0.20) = -9,220 \text{ calories}$$

and thus a negative $\Delta G^{\circ\prime}_{net}$. The same is true of the FMNH$_2$ and oxidized CoQ involved in the next step. Since this type of pattern is *continuous*, the thermodynamic design is clearly evident: the electron carriers are arranged so that each transfer has the capacity to proceed *spontaneously* and *exergonically*. (*Note:* The nonfit of the coenzyme Q couple should be considered in the context of the conditions under which the $\mathscr{E}^{\circ\prime}$ has been calculated, namely, in 95% ethanol, a condition not exactly representative of the natural environment of a biological membrane. It has been suggested that the true ability of CoQ to function as an electron carrier *in vivo* is more closely approximated by a value of 0.0 volt for $\mathscr{E}^{\circ\prime}$).

The overall chemical and thermodynamic design of the respiratory chain is summarized below, with each event of electron transfer represented as a complete oxidation-reduction reaction. Indeed, this is what happens *in vivo*. Although two electrons are transferred in the overall process, it is not known whether the cytochrome sequence in particular involves one cytochrome molecule functioning twice or two separate molecules transfer-

Respiratory chain reactions	$\Delta\mathscr{E}^{\circ\prime}$ (V)	$\Delta G^{\circ\prime}$ (cal)	
$NADH(H^+) + FAD \rightarrow NAD^+ + FADH_2$	+0.20	−9,220	←
$FADH_2 + CoQ(ox) \rightarrow FAD + CoQH_2(red)$	+0.22	−10,150	←
$CoQH_2 + (cyt\ b\cdot Fe^{3+})_2 \rightarrow CoQ + (cyt\ b\cdot Fe^{2+})_2 + 2H^+$	−0.06	+2,770	increments of energy available
$(cyt\ b\cdot Fe^{2+})_2 + (cyt\ c_1\cdot Fe^{3+})_2 \rightarrow (cyt\ b\cdot Fe^{3+})_2 + (cyt\ c_1\cdot Fe^{2+})_2$	+0.18	−8,300	←
$(cyt\ c_1\cdot Fe^{2+})_2 + (cyt\ c\cdot Fe^{3+})_2 \rightarrow (cyt\ c_1\cdot Fe^{3+})_2 + (cyt\ c\cdot Fe^{2+})_2$	+0.03	−1,380	←
$(cyt\ c\cdot Fe^{2+})_2 + (cyt\ a,\ a_3\cdot Fe^{3+})_2 \rightarrow (cyt\ c\cdot Fe^{3+})_2 + (cyt\ a,\ a_3\cdot Fe^{2+})_2$	+0.04	−1,840	←
$(cyt\ a,\ a_3\cdot Fe^{2+})_2 + \frac{1}{2}O_2 + 2H^+ \rightarrow (cyt\ a,\ a_3\cdot Fe^{3+})_2 + H_2O$	+0.53	−24,440	←

Net: $NADH(H^+) + \frac{1}{2}O_2 \rightarrow NAD^+ + H_2O$ +1.14   −52,580

or $FADH_2 + \frac{1}{2}O_2 \rightarrow FAD + H_2O$ +0.94   −43,360

ring one electron each. The (cyt)$_2$ designation is employed here only for bookkeeping purposes, to account for the *net* transfer of two electrons.

In comparing the net reaction with the multistep pathway, two features are evident. First, the intervention of the intermediate carriers results in the *production of energy in increments*. The second feature is that the entire sequence is *autocatalytic*, not only with respect to the intermediate steps of the chain but also relative to the regeneration of NAD$^+$ and FAD. This insures that the participating NAD$^+$-dependent and FAD-dependent dehydrogenases are continually resupplied with the oxidized form of each coenzyme that is required for the continued operation of these enzymes.

Formulation of the Sequence of Electron Carriers

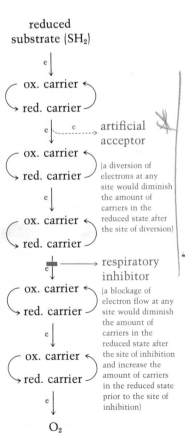

reduced substrate (SH$_2$)

artificial acceptor

(a diversion of electrons at any site would diminish the amount of carriers in the reduced state after the site of diversion)

respiratory inhibitor

(a blockage of electron flow at any site would diminish the amount of carriers in the reduced state after the site of inhibition and increase the amount of carriers in the reduced state prior to the site of inhibition)

O$_2$

Up to now no mention has been made of how the sequence of electron carriers was established. Well, part of the rationale was a presumption that each carrier would be most efficiently arranged in order of an increasing ability to accept electrons and undergo reduction. Hence, knowledge of the $\mathscr{E}^{\circ\prime}$ values was a useful guideline. A key discovery was that cytochrome oxidase (the cyt a,a_3 couple) was the *last* carrier acting directly with oxygen.

Two other tactics were particularly useful. One involved the use of *artificial* (nonbiological) electron *acceptors* with known $\mathscr{E}^{\circ\prime}$ values to divert electrons from the chain at points where specific respiratory carriers were located. The discovery and characterization of *respiratory inhibitors* that act at specific sites along the chain (see p. 462) provided the basis of a second approach. For example, *amytal* and other *barbiturates* are potent inhibitors of electron transfer in the flavin → CoQ (cyt b) region. The antibiotic *antimycin A* specifically inhibits electron transfer at the cyt $b \rightarrow c_1$ site. *Cyanides* (CN$^-$), *azides* (N$_3^-$), and *carbon monoxide* (CO) are potent inhibitors of cytochrome oxidase.

In each instance, it was possible by the use of sensitive spectrophotometric techniques to measure the amount of the various carriers that existed in the reduced state, and to compare these with the levels of reduced carriers in the control system when neither artificial acceptor nor respiratory inhibitor was present. Both the occurrence of divergent electron flow in the presence of artificial acceptors and the inhibition of electron flow in the presence of respiratory inhibitors would be characterized by decreased levels of the carriers that exist in their reduced state *after* the site of divergence or inhibition. Only the inhibition of electron flow by a respiratory inhibitor would result in increased levels of the reduced state of the carriers that exist prior

to the site of inhibition. This increase would not occur in the presence of
an artificial acceptor, since in this case electrons would not build up prior to
the site of action by the acceptor, but would instead be removed by the acceptor. For example, in the presence of antimycin A one would observe
that the amounts of cytochromes c_1, c, a, and a_3 in the reduced state would
decrease. It is possible to detect and measure these changes with refined
spectroscopic techniques because—as was pointed out in the previous discussions of the carriers—each of the carriers displays a different absorption
spectrum in its oxidized and reduced states. For example, reduced cytochrome c shows a strong absorption maximum at approximately 550 nm
whereas oxidized cytochrome c shows less absorption at the same wavelength (the alpha Soret band); see Table 14–2. Thus a measure of the decrease in absorption at 550 nm would indicate the decrease in the amount of
reduced cytochrome c and the increase in the amount of oxidized cytochrome c.

Further proof of the NAD $\rightarrow$ flavin $\rightarrow$ CoQ $\rightarrow b \rightarrow c_1 \rightarrow c \rightarrow a, a_3 \rightarrow O_2$ sequence came in the 1960s with the discovery that intact mitochondria could
be very carefully disrupted to yield submitochondrial fragments still capable
of complete electron transport but incapable of phosphorylation. For obvious reasons this fragment was termed the *elementary particle* of the respiratory chain. A further use of this approach led to further disruption of the
elementary particle into four separate *respiratory complexes* (I, II, III, IV),
each of which was capable of catalyzing a portion of the complete respiratory
chain.

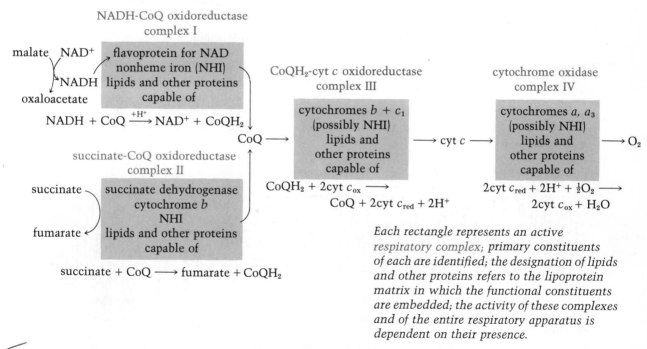

*Each rectangle represents an active
respiratory complex; primary constituents
of each are identified; the designation of lipids
and other proteins refers to the lipoprotein
matrix in which the functional constituents
are embedded; the activity of these complexes
and of the entire respiratory apparatus is
dependent on their presence.*

Complex I (NADH-CoQ oxidoreductase) catalyzed the transfer of electrons from NADH to CoQ; complex II (succinate-CoQ oxidoreductase) catalyzed the transfer of electrons from succinate to CoQ; complex III
(CoQH$_2$-cytochrome c oxidoreductase) catalyzed the transfer of electrons
from CoQH$_2$ to cyt c; and finally, complex IV (cytochrome oxidase) catalyzed the transfer of electrons from cyt c to oxygen. Notice that the cata-

lytic properties of each complex are perfectly consistent with the proposed sequence. To avoid confusion, it should be noted that measurable activity of each complex presupposes the addition of both reactants. In other words, the complexes themselves do not contain appreciable levels of NAD, CoQ, succinate, and cytochrome c. The fact that CoQ and cytochrome c must be added to demonstrate activity is the result of their being extracted during the fractionation procedure and indicates that these two substances are less tightly packaged in the mitochondrial membrane than are the other components. NAD and succinate are soluble substances to begin with.

The isolation and characterization of these respiratory complexes also represented the first decisive proof that the components of the respiratory chain (and probably those materials responsible for the coupling to ATP formation) were highly integrated into some type of super-ordered molecular assembly within the mitochondrial membrane. Further support was provided by successful attempts at partial reconstitution, which was done by simply mixing together any two appropriate complexes to yield secondary units capable of catalyzing the sum of the reactions typical of each complex, as indicated below. We will examine some of these ultrastructural features of the mitochondrial membrane later (see p. 467).

$$\text{NADH(H}^+) + (\text{cyt } c\text{-Fe}^{3+})_2 \xrightarrow{\text{I + III}} \text{NAD}^+ + (\text{cyt } c\text{-Fe}^{2+})_2 + 2\text{H}^+$$

$$\text{succinate} + (\text{cyt } c\text{-Fe}^{3+})_2 \xrightarrow{\text{II + III}} \text{fumarate} + (\text{cyt } c\text{-Fe}^{2+})_2 + 2\text{H}^+$$

$$(\text{cyt } c\text{-Fe}^{2+})_2 + \tfrac{1}{2}\text{O}_2 + 2\text{H}^+ \xrightarrow{\text{III + IV}} (\text{cyt } c\text{-Fe}^{3+})_2 + \text{H}_2\text{O}$$

Topology of Electron Carriers within the Membrane

Although it is not yet possible to describe how all the electron carriers and accessory substances are arranged in the inner mitochondrial membrane, considerable progress is being made in this regard. For example, there is extensive evidence that the topology exhibits distinct sidedness, with some of the carrier proteins located on one side of the membrane and some located

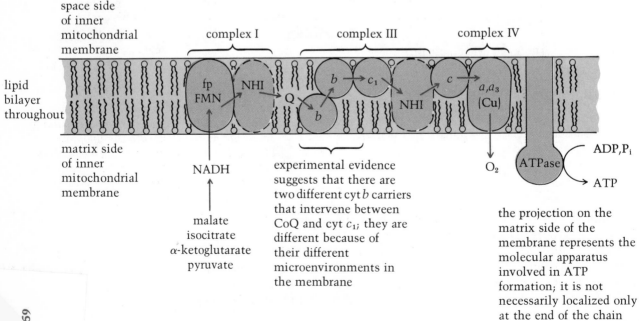

on the other side. In addition, some carriers span the entire width of the membrane. The diagram on p. 460 summarizes the current state of knowledge, including a consideration of the knob-like particles that protrude from the matrix side of the membrane, which are known to be involved in the phosphorylation of ADP. This asymmetric arrangement of the carriers may be a requirement for energizing the mitochondrion prior to ATP formation.

STOP

COUPLING OF THE RESPIRATORY CHAIN TO ATP FORMATION

Phosphorylation Sites

So far we have examined the oxidative, exergonic phase of oxidative phosphorylation. Now we are prepared to consider the endergonic phase of phosphorylation of ADP to yield ATP. Although it is a point to which we have referred on several occasions, we should again note that these *two distinct processes are indeed coupled through an energy transfer from the process of electron transport to the process of phosphorylation.*

The evidence for this coupling is threefold. The first and obvious proof is that the mitochondrial oxidation of NADH or $FADH_2$ is in fact accompanied by the simultaneous formation of ATP. A second proof is the fact that mitochondrial ATP formation can be inhibited without any interference of electron transport by the addition of chemical agents called *uncouplers*. Then, when the uncoupler is removed or deactivated, there is a restoration of ATP formation linked to the respiratory chain. The most classical and very potent uncoupler is *2,4-dinitrophenol* (DNP). A newly discovered class of uncouplers are called *ionophores*, some of which are antibiotics (see p. 466). The third and perhaps most convincing proof is that the extent of mitochondrial ATP formation is significantly reduced in the presence of respiratory chain (that is, electron transport) inhibitors such as antimycin A and the others mentioned above. Each of these points is considered in more detail in subsequent sections.

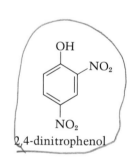

2,4-dinitrophenol

In accounting for ATP formation, the basic factor that must be taken into consideration is the amount of energy required for the formation of ATP form ADP and P_i. If we use the approximate value of $-8,000$ calories for the $\Delta G^{\circ\prime}$ of ATP hydrolysis (given in Chap. 11), this means that about 8,000 calories must be made available for ATP production.

$$ADP + P_i \rightarrow ATP + HOH \qquad \Delta G^{\circ\prime} = +8,000 \text{ calories}$$

Now, by inspecting the steps in the respiratory chain in terms of the thermodynamic summary given earlier (see p. 458), we conclude that four sites could meet this requirement: the NAD-flavin transfer ($\Delta G^{\circ\prime} = -9,220$ calories); the flavin-CoQ transfer ($\Delta G^{\circ\prime} = -10,150$ calories); the cytochrome b-cytochrome c_1 transfer ($\Delta G^{\circ\prime} = -8,300$ calories); and the cytochrome a,a_3-oxygen transfer ($\Delta G^{\circ\prime} = -24,400$ calories). Thus, if the energy at each of these regions were transferred to ATP formation, this would then mean that the transfer of two electrons through the complete respiratory chain from NADH to oxygen would be sufficient to account for $4ADP + 4P_i \rightarrow 4ATP$, with the formation of one ATP coupled to each of the aforementioned exergonic sites. Similarly, if the chain were primed at the level of $FADH_2$ (via the oxidation of succinate with succinate dehydrogenase, for example), complete oxidation of $FADH_2$ through oxygen should account for 3ATP + 3P_i → 3ATP. One less ATP would be formed in this case, since the entrance of electrons bypasses the NAD-flavin region.

whole cells

↓ *rupture*

homogenate

↓ *fractionate*

purified
mitochondria

O_2 ⤙ P_i

NAD^+ ⤙ ADP
(buffer; 37 °C)

malate ⤙

↓

measure amount of
P_i and O_2 consumed
and express as
P/O ratio; in this
case P/O would be
3

The term phosphorylation site
does not necessarily have the literal meaning of a specific step
where ATP is formed directly.
The more general meaning is that
electron flow through these
regions (1, 2, 3) is somehow accompanied by the production of
ATP.

Although the basic principle of this idealized analysis is quite valid, it is not in agreement with experimental studies. Before inspecting some data, let us briefly examine the experimental system that would be used in the study of oxidative phosphorylation. A typical experimental design is given in the margin. Aside from the usual control of pH, temperature, and ionic strength, the basic requirements are that (a) purified mitochondria be used; (b) the reduced substrate added be oxidized by a mitochondrial dehydrogenase; (c) NAD^+ be added if the substrate is to be oxidized to an NAD^+-dependent dehydrogenase; (d) known amounts of ADP and P_i be added; and (e) an accurate measure of the amount of inorganic phosphate consumed or ATP produced, and the amount of oxygen or reduced substrate consumed, be available. Results are conventionally expressed as a *P/O ratio*, which corresponds to the number of *moles of P_i consumed per gram-atom of oxygen consumed*. Alternative interpretations would be the number of moles of ATP produced per gram-atom of oxygen consumed, or the number of moles of ATP produced per mole of substrate consumed.

Studies of this type typically yield data such as those summarized in Table 14–3. Although integral values for the P/O ratio are obtained, implying that specific sites for ATP formation do exist, note that only three sites are implicated when the chain is primed with NADH—one site fewer than was suggested by our idealized analysis. Likewise, two rather than three sites are implicated when the chain is primed with $FADH_2$. Although there have been scattered reports of higher P/O values, the model for oxidative phosphorylation most widely agreed upon is consistent with the data of Table 14–3, that is, it has a *maximum of three phosphorylation sites*.

In view of our earlier conclusion that four potential phosphorylation sites exist, the next logical question is, which are the three actually coupled to ATP formation? Well, we have already implicated one site—the NAD-flavin region. The basis for this conclusion can be seen in Table 14–3, which indicates that, as predicted, the flavin-dependent oxidation of a reduced substrate is indeed accompanied by a P/O value one unit smaller than that observed for NAD^+-dependent oxidation of a reduced substrate. Accordingly, then, we can assign the NAD-flavin region as phosphorylation site 1.

The remaining two sites have been established by measurement of P/O ratios in the presence of respiratory inhibitors with different substrates. For example, the incubation of mitochondria with NAD^+-dependent substrates

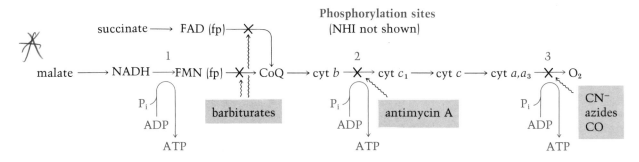

X site of action of respiratory inhibitors

TABLE 14-3 **Oxidative phosphorylation with intact mitochondria.**

Substrate	Dehydrogenase	Inhibitor	P/O ratio
pyruvate	NAD	–	≈ 3
isocitrate	NAD	–	≈ 3
malate	NAD	–	≈ 3
succinate	FAD (flavin)	–	≈ 2
isocitrate	NAD	antimycin	≈ 1
isocitrate	NAD	CN^-	≈ 2
succinate	FAD (flavin)	antimycin	≈ 0
succinate	FAD (flavin)	CN^-	≈ 1

in the presence of antimycin A results in a P/O value of approximately 1. Since the specific site of antimycin A inhibition is the cyt $b \rightarrow$ cyt c_1 transfer, a P/O value of 1 suggests that the flavin $\rightarrow$ CoQ region is *not* a phosphorylation site. If it were, we would predict a P/O of 2, since the NAD-flavin region alone would account for a P/O of 1. Corroborating this conclusion is the observation that very little P_i is consumed when mitochondria are incubated in the presence of antimycin A and succinate. In addition to ruling out the flavin $\rightarrow$ CoQ system, the antimycin effect suggests, then, that sites 2 and 3 are associated with the regions cyt $b \rightarrow$ cyt c_1 and cyt $a, a_3 \rightarrow$ oxygen, respectively. The former is, of course, implicated directly by the antimycin effect, since it is the specific site of antimycin inhibition in the transfer of electrons within the respiratory chain. The cyt $a, a_3 \rightarrow$ oxygen transfer is similarly confirmed by the addition of respiratory inhibitors of cytochrome oxidase, such as cyanide ion, which yield P/O values of approximately 2 when the respiratory chain is primed with NADH and approximately 1 when the chain is primed with $FADH_2$.

Mechanism of Coupling

The coupling of electron transport to phosphorylation of ADP and the maximum yield of 3ATPs produced per pair of electrons transported are firmly established facts of biochemistry. However, it is not yet established how the coupling occurs, that is, how the energy of transport is made available for ATP formation. Since the same type of process is responsible for ATP production (a) in chloroplasts during photosynthesis and (b) in the cell membranes of prokaryotes, this question represents a major unsolved aspect of biochemistry. The biggest obstacle in deciphering the nature of the coupling is the compartmentalization of the entire molecular apparatus in the mitochondrial membrane, wherein each molecular component is likely to be engaged in an extensive network of interactions with other components in the membrane. Despite the inherent difficulties of studying such a complicated system, work progresses—albeit slowly—towards an eventual solution. Whatever the explanation, it must answer two questions: (1) how does electron transport energize the mitochondrion, and (2) how is the energy delivered to the reaction ADP + $P_i \rightarrow$ ATP?

Whether it occurs in mitochondria, in chloroplasts, or in the cell membrane of prokaryotes, there are three major hypotheses proposed to explain the coupling of phosphorylation to electron transport. One is termed a *chemical hypothesis;* the second a *chemiosmotic hypothesis;* and the third a *conformational hypothesis.*

Chemical coupling. The chemical hypothesis is the oldest suggestion, dating back about 20 years. It proposes that the energy is conserved and subsequently transferred via the participation of specific *intermediate compounds,* which have the ability to undergo reversible reactions between low energy and high energy states. One formulation based on this hypothesis is summarized below, where I represents the *unknown coupling intermediate.* The reduced and oxidized carriers of course refer to those participants at one of the phosphorylation sites along the respiratory chain. In simplified form, the model has three steps. First, accompanying the transfer of electrons, the coupling intermediate I forms a *high energy,* carrier-I *complex* with either one of the carriers, with the driving force being the energy released in the course of electron transfer. The specification that I binds with (carrier 1)$_{ox}$ is purely arbitrary. In the second step, the carrier ~I complex undergoes an exchange reaction with P$_i$ producing the free form of the carrier and a *high energy, phosphorylated form of the coupling intermediate* I~P. Finally, the I~P species transfers the activated phosphate to ADP to yield free I and ATP. As indicated in step (3), this reaction would be catalyzed by *ATPase,* the enzyme known to be present in the inner mitochondrial membrane that can reversibly catalyze the interconversion of ATP and ADP. The net effect, of course, is the transfer of energy from the electron transfer between two carriers to the endergonic formation of ATP.

The chemical identity of the coupling intermediate is in dispute. Some workers have proposed that it is a low molecular weight substance such as a quinone, and others have suggested that it is a protein.

the mitochondrion is energized ⟶ (1) $(\text{carrier } 1)_{red} + (\text{carrier } 2)_{ox} + I \rightleftharpoons (\text{carrier } 1)_{ox} \sim I + (\text{carrier } 2)_{red}$

activated phosphate is generated ⟶ (2) $(\text{carrier } 1)_{ox} \sim I + P_i \rightleftharpoons (\text{carrier } 1)_{ox} + I \sim P$

easy transfer to ADP ⟶ (3) $I \sim P + ADP \xrightleftharpoons{\text{ATPase}} I + ATP$

Net: $(\text{carrier } 1)_{red} + (\text{carrier } 2)_{ox} + ADP + P_i \xrightleftharpoons{I} (\text{carrier } 1)_{ox} + (\text{carrier } 2)_{red} + ATP$

The search for I~P has failed, suggesting that the chemical mechanism does not operate. However, it may be that the I~P is turned over so rapidly—as a bound intermediate, for example—that it cannot be isolated in the free state.

The chemical model can account for, in various ways, the action of an uncoupling agent and other inhibitors. For example, the uncoupling effect of substances such as dinitrophenol (which results in abolition of ADP phosphorylation with retention of electron transport) can be explained by proposing that DNP could react with the carrier ~I complex *prior* to the formation of I~P and ATP. Since the electron carrier is regenerated, the operation of the respiratory chain would not be impaired. On the other hand, the inhibition by the antibiotic oligomycin of the combined process of electron transport and phosphorylation can be explained by proposing that the oligomycin-sensitive step is either the formation of I~P or the subsequent exchange reaction of I~P with ADP. Either site of action, but particularly the former, would prevent the regeneration of the free electron carrier in levels sufficient for normal electron transport. Of course, both possibilities would short-circuit ATP formation as well.

Chemiosmotic coupling. In 1961 P. Mitchell proposed something completely different. A key basis for his suggestion was the observation that hydrogen ions (H$^+$) are released from the mitochondrion when they are in an active state of respiration. By assuming that the mitochondrial membrane is impermeable to the free movement of H$^+$, it was argued that the exit of H$^+$

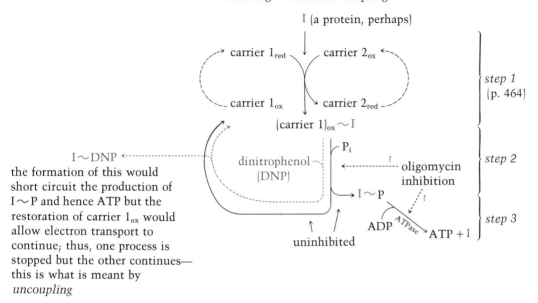

*hypothetical sequence of
events at phosphorylation site
according to chemical coupling*

I (a protein, perhaps)

carrier 1_{red} carrier 2_{ox}

carrier 1_{ox} carrier 2_{red}

$\}$ *step 1*
(p. 464)

(carrier 1)$_{ox}$ ~ I

$I \sim DNP$

the formation of this would
short circuit the production of
$I \sim P$ and hence ATP but the
restoration of carrier 1_{ox} would
allow electron transport to
continue; thus, one process is
stopped but the other continues—
this is what is meant by
uncoupling

dinitrophenol
(DNP)

P_i

oligomycin
inhibition

$\}$ *step 2*

$I \sim P$

ADP $\xrightarrow{ATPase}$ ATP + I

$\}$ *step 3*

uninhibited

from inside the mitochondrion could be an energy-driven process, with the
energy expenditure being provided via the events of electron transport. The
external displacement of H^+ would then result in a concentration difference
(*gradient*) of H^+ across the mitochondrial membrane, with a higher $[H^+]$ on
the outside and a lower $[H^+]$ on the inside. In other words, the mitochondrion is energized by the generation of a concentration gradient of H^+—a
gradient with the potential to do "useful work." In this case the "work" is
to drive the formation of ATP.

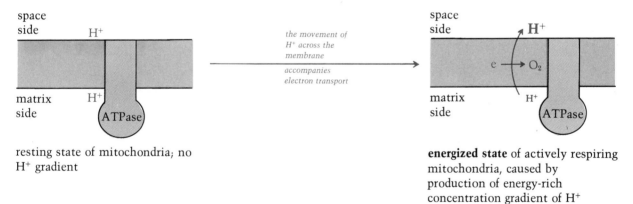

space
side H^+

matrix
side H^+ ATPase

resting state of mitochondria; no
H^+ gradient

*the movement of
H^+ across the
membrane
accompanies
electron transport*

space
side H^+

e $\longrightarrow$ O_2

matrix
side H^+ ATPase

energized state of actively respiring
mitochondria, caused by
production of energy-rich
concentration gradient of H^+

*the movement of H^+ out of the matrix region may involve an
energy-requiring transport system driven by electron flow or it may
be due to the asymmetric orientation of the carriers, some of
which acquire $2H^+$ on the matrix side while others release
$2H^+$ on the space side*

One way (A in diagram below) this type of energy-rich gradient could be
coupled to ATP formation is that it might tend to *pull* the phosphorylation
reaction. As shown this displacement could be accomplished through the

465

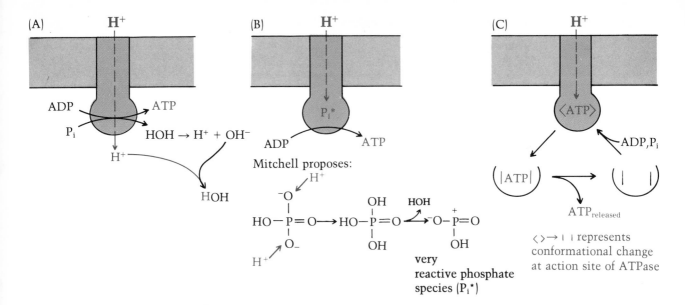

(A)   H⁺

ADP → ATP

P_i

$HOH \rightarrow H^+ + OH^-$

H⁺

HOH

(B)   H⁺

P_i^*

ADP → ATP

Mitchell proposes:

H⁺

$$HO-\overset{\displaystyle{}^-O}{\underset{\displaystyle O_-}{\overset{|}{\underset{|}{P}}}}=O \longrightarrow HO-\overset{\displaystyle OH}{\underset{\displaystyle OH}{\overset{|}{\underset{|}{P}}}}=O \overset{HOH}{\longrightarrow} {}^+O-\overset{\displaystyle }{\underset{\displaystyle OH}{\overset{|}{\underset{|}{P}}}}=O$$

H⁺

very
reactive phosphate
species (P_i^*)

(C)   H⁺

⟨ATP⟩

ADP,P_i

|ATP| → | |

$ATP_{released}$

⟨ ⟩ → | | represents
conformational change
at action site of ATPase

removal of H_2O by the vectorial movement of H⁺ from outside the matrix through "channels" in the ATPase component of the membrane to the low H⁺ area inside the mitochondrion.   Another way (B) that the coupling might work is that the reentry of H⁺ through the ATPase channel may be funneled to the binding site of the ATPase, where it would promote the formation of a highly reactive form of phosphate, which reacts readily with ADP at the active site.   Another argument (C) is that the H⁺ reentry promotes a change in the conformation of the ATPase, resulting in the release of ATP, which was already formed at the ATPase active site.

Albeit an oversimplification, the preceding description covers the major principles of Mitchell's hypothesis.   It has undergone some revision, including suggestions that ions other than H⁺ may be displaced, and also that an electrical potential (a voltage difference) may be established across the membrane.

Some of the strongest experimental evidence in support of the chemiosmotic hypothesis (and against the chemical hypothesis) is found in the action of substances called *ionophores,* molecules that *carry ions* (or otherwise promote the movement of ions) *across membranes.* According to the chemiosmotic theory, the generation of a proton gradient is prerequisite for ATP formation. If the ionophore can move H⁺ ions back across the membrane, the H⁺ gradient is lost and ATP formation is indeed uncoupled from transport.   The antibiotic *valinomycin* is a cyclic peptide that can encapsulate K⁺ ions specifically and carry them across the membrane.   Since valinomycin can uncouple oxidative phosphorylation, this suggests that the establishment of a H⁺ gradient may very well be related to the movement of other ions and other gradients.   The peptide antibiotic *gramicidin* uncouples by embedding itself in the membrane and providing a porous channel through which Na⁺, K⁺, and H⁺ can move.   The uncoupling action of dinitrophenol is also explainable according to the chemiosmotic hypothesis since the ionized form of the phenol could scavenge protons on the outside and carry them back to the matrix.

Conformational coupling.   The most recent hypothesis of the coupling of phosphorylation to electron transport is based on the notion that different energy states are accessible to an entire mitochondrion. More specifically, it is argued that the entire inner mitochondrial membrane undergoes a considerable alteration in the pattern of cristae that permeate the matrix during active periods of electron transport. The idea is that this change in mitochondrial structure represents an energized (high energy) state, which on return to a structure of lower energy somehow releases energy for ATP formation involving ATPase. Electron microscopic evidence depicting different ultrastructures of mitochondria in different states of respiratory activity are consistent with this hypothesis.

Only further research will determine whether one of these ideas, or a variation of one, or a combination of them, is correct. At present each has its proponents and opponents.

Extramitochondrial ATP

Since (a) the complete process of oxidative phosphorylation is *intra*mitochondrial (specifically it is associated with the inner mitochondrial membrane); (b) the bulk of ATP-requiring processes are *extra*mitochondrial; and (c) the mitochondrial membrane is impermeable to a direct diffusion of ATP, an obvious question is, how does the ATP get out of the mitochondria? Well, as you might expect, there is good evidence that the mitochondrial membrane contains a specific transport protein for this function. The interesting feature of this transport system is that it can effect a one-for-one translocation of intramitochondrial ATP and extramitochondrial ADP. In other words, for each ATP molecule transported out of the mitochondrion to the cytoplasm, one ADP molecule is transported from the cytoplasm into the mitochondrion.

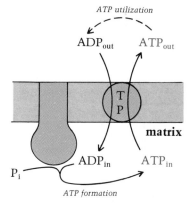

TP = transport protein

ULTRASTRUCTURE OF THE MITOCHONDRIAL MEMBRANE

The isolation of a submitochondrial elementary particle and its constituent respiratory complexes has guided much of the work done to determine the *in vivo* arrangement of all the participating components of oxidative phosphorylation *within* the mitochondrion. Although the complete arrangement of parts is still unknown, progress has been made (refer back to topological diagram on p. 460). High-resolution electron microscopy and the further development of delicate procedures for disrupting mitochondria have revealed in great detail the fascinating organization of these organelles.

The existence of cristae was well known in the 1960s. Closer study revealed that another conspicuous structural feature is the presence on the matrix side of the inner mitochondrial membrane of numerous very small knoblike particles that are attached to the cristae (see Fig. 14–4). As stated previously, it is now quite clear that these particles are essential for the coupling of phosphorylation to electron transport. Among other things they contain the ATPase activity that is proposed to catalyze ATP formation.

An excellent summary and further details of the key experimental developments in this area can be found in the articles by Racker that are cited at the end of the chapter. The information in Fig. 14–5 should also be helpful. Note that submitochondrial fragments that are capable of *both* electron transport and phosphorylation can be isolated, and that the removal of the attached particles destroys their phosphorylating activity. Obviously, this is clear-cut proof that the two processes are coupled, not only in a thermodynamic sense, but also in a literal sense.

467

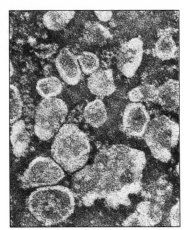

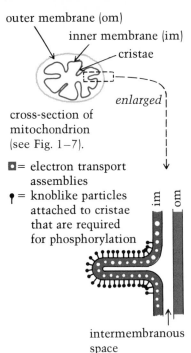

outer membrane (om)

inner membrane (im)

cristae

enlarged

cross-section of
mitochondrion
(see Fig. 1–7).

◻ = electron transport
assemblies
🌱 = knoblike particles
attached to cristae
that are required
for phosphorylation

im om

intermembranous
space

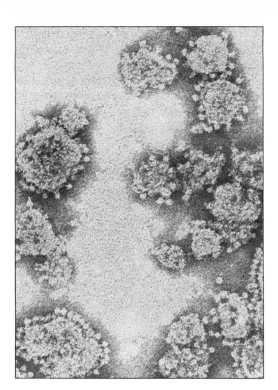

FIGURE 14–4 *Center:* a diagramatic representation of the ultrastructure of a mitochondrial membrane based on high-resolution
electron microscopy. *Right:* an
electron micrograph of submitochondrial particles showing the attachment of numerous spheres and
base stalks. Magnification,
16,000×. *Above:* submitochondrial particles with spheres detached. These particles are represented in the drawing at the
bottom of Fig. 14–5. Magnification, 160,000 ×. (Additional information is provided in Fig. 14–5.)
(Photographs generously supplied
by Dr. E. Racker, Cornell University.)

ATP YIELD FROM CARBOHYDRATE CATABOLISM

Having explored the essential features of oxidative phosphorylation as a
process per se, let us now reexamine its contribution to the main catabolic
pathway of carbohydrates, namely, to the combined operation of glycolysis
and the citric acid cycle. For the sake of consistency, our analysis will
make use of the same type of net equations that were previously used in
Chap. 13 to summarize this phase of metabolism. (Diagram, p. 470.)

Given the accepted quantitative relationships of oxidative phosphorylation concerning the uptake of P_i and O_2, it is now possible to *estimate,* in a
rather straightforward manner, the efficiency of the combined operations of
glycolysis and the citric acid cycle in terms of energy conservation. Since
the oxidation of one $NADH(H^+)$ unit yields 3 ATPs, and of one $FADH_2$ unit
yields 2 ATPs, the yield of the combined route is 34 ATPs. Then consider
(a) that for every molecule of glucose processed through glycolysis there is a
requirement of 2 ATPs and a production of 4 ATPs via substrate-level phosphorylations, and (b) that the complete oxidation of 2 units of acetyl-SCoA
in the citric acid cycle is accompanied by the *production* of 2 GTPs (also by
a substrate-level phosphorylation), which are thermodynamically and metabolically equivalent to 2 ATPs. Hence we can say that the *total net gain* of
high-energy triphosphonucleotides (expressed as ATP) accompanying the
complete oxidation of glucose to $6CO_2$ via glycolysis and the citric acid

	$\Delta G^{\circ\prime}$(calories)	
$38ADP + 38P_i \rightarrow 38ATP$	+304,000	≈44% energy
$glucose + 6O_2 \rightarrow 6CO_2 + 6H_2O$	−686,000	conservation
$glucose + 6O_2 + 38ADP + 38P_i \rightarrow 6CO_2 + 6H_2O + 38ATP$	−382,000	

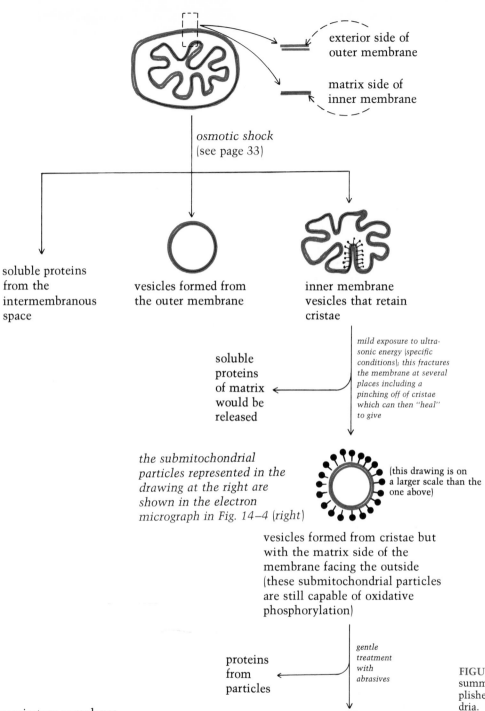

exterior side of outer membrane

matrix side of inner membrane

osmotic shock
(see page 33)

soluble proteins from the intermembranous space

vesicles formed from the outer membrane

inner membrane vesicles that retain cristae

soluble proteins of matrix would be released

mild exposure to ultrasonic energy (specific conditions); this fractures the membrane at several places including a pinching off of cristae which can then "heal" to give

the submitochondrial particles represented in the drawing at the right are shown in the electron micrograph in Fig. 14–4 (right)

(this drawing is on a larger scale than the one above)

vesicles formed from cristae but with the matrix side of the membrane facing the outside (these submitochondrial particles are still capable of oxidative phosphorylation)

proteins from particles

gentle treatment with abrasives

respiratory complexes I, II, III, IV (see page 459) plus solubilized proteins

again, mild exposure to ultrasonic energy

active respiratory vesicle but *non*-phosphorylating (electron transport-yes ATP formation-no)

FIGURE 14–5 A diagramatic summary of what has been accomplished by fractionating mitochondria. Much of the diagram speaks for itself and supplements the material discussed previously. Some investigators are attempting to unravel the secrets of mitochondrial structure and function by isolating individual components from the last two steps, characterizing their separate properties, and then mixing them back together to reconstitute functional submitochondrial particles.

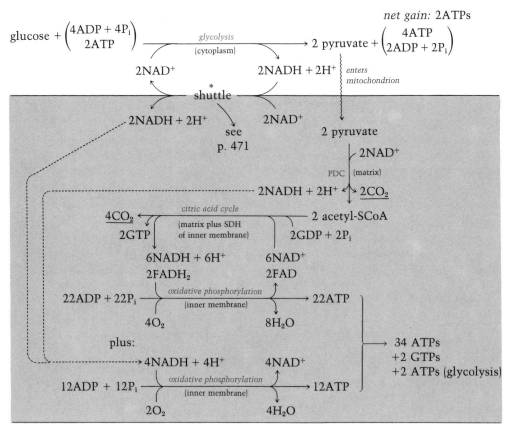

net gain: 2ATPs

$$\text{glucose} + \begin{pmatrix} 4ADP + 4P_i \\ 2ATP \end{pmatrix} \xrightarrow[\text{(cytoplasm)}]{\text{glycolysis}} 2 \text{ pyruvate} + \begin{pmatrix} 4ATP \\ 2ADP + 2P_i \end{pmatrix}$$

$2NAD^+$ $2NADH + 2H^+$ *enters mitochondrion*

*shuttle

$2NADH + 2H^+$ see p. 471 $2NAD^+$ 2 pyruvate

$2NAD^+$

PDC (matrix)

$2NADH + 2H^+$ $2CO_2$

citric acid cycle

$4CO_2$ (matrix plus SDH of inner membrane)   2 acetyl-SCoA

$2GTP$ $2GDP + 2P_i$

$6NADH + 6H^+$ $6NAD^+$
$2FADH_2$ $2FAD$

$22ADP + 22P_i \xrightarrow[\text{(inner membrane)}]{\text{oxidative phosphorylation}} 22ATP$

$4O_2$ $8H_2O$

plus:

$4NADH + 4H^+$ $4NAD^+$

$12ADP + 12P_i \xrightarrow[\text{(inner membrane)}]{\text{oxidative phosphorylation}} 12ATP$

$2O_2$ $4H_2O$

→ 34 ATPs
+2 GTPs
+2 ATPs (glycolysis)

38 ATPs

cycle is 38 (34 − 2 + 4 + 2). Since the formation of 38 ATPs would require 304,000 calories (based again on a $\Delta G^{o'}$ value of $-8,000$ calories for ATP hydrolysis), this means that the cellular efficiency of energy conversion in a metabolically useful form is approximately 44%. Inasmuch as the energy required for *in vivo* ATP formation may be greater than 8,000 calories per mole, the figure of 44% is a rather conservative estimate.

In performing this efficiency analysis there is one item that merits a brief explanation. Note that in the tally made above, the net production of 38 ATPs includes 6 ATPs that would result from the utilization in electron transport of two units of NADH produced during glycolysis. (The other eight NADHs and two FADH$_2$s would be generated in the reactions of the citric acid cycle.) Since in eukaryotic organisms glycolysis occurs in the cytoplasm but oxidative phosphorylation occurs in the mitochondrion, you might ask, how can we legitimately include the NADH generated by glycolysis in the production of ATP? This question is even more pertinent in view of the fact that the mitochondrial membrane is impermeable to a direct transport of NADH. The explanation is that the NADH produced in the cytoplasm can enter the mitochondrion by processes termed *shuttle pathways,* which provide for an indirect movement of molecules across a membrane. In the case of NADH one of the proposed shuttle pathways is shown below. You will note that it depends on the utilization of NADH in the cytoplasm to reduce oxaloacetate to malate, and then the *malate enters* the mitochondrion. The NAD$^+$-dependent oxidation of malate in the mitochondrion completes the *net* transport of NADH. The enzyme catalyzing

470

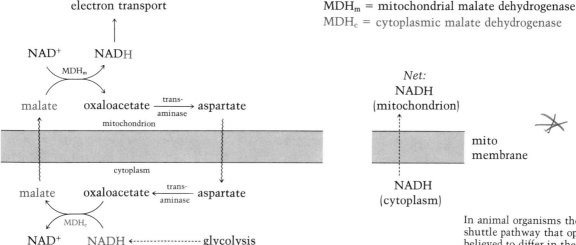

electron transport

NAD^+　　NAD$\underline{H}$

MDH$_m$

malate　　oxaloacetate $\xrightarrow[\text{aminase}]{\text{trans-}}$ aspartate

mitochondrion

cytoplasm

malate　　oxaloacetate $\xleftarrow[\text{aminase}]{\text{trans-}}$ aspartate

MDH$_c$

NAD^+　　NADH $\xleftarrow{}$ glycolysis

reused

MDH$_m$ = mitochondrial malate dehydrogenase
MDH$_c$ = cytoplasmic malate dehydrogenase

Net:
NADH
(mitochondrion)

mito
membrane

NADH
(cytoplasm)

In animal organisms the type of shuttle pathway that operates is believed to differ in the various tissues. Besides the system shown here, there is one that involves glycerol phosphate (see Exercise 14–11. There may also be others.

each reaction would be the malate dehydrogenase, which is contained in each compartment (see p. 424). The oxaloacetate required for the formation of malate in the cytoplasm could be supplied by the transamination of aspartate. Since the transamination can occur in both cytoplasm and mitochondrion, and since the mitochondrial membrane is permeable to aspartate, the shuttle is cyclic, and hence self-sustaining.

OXYGEN TOXICITY AND SUPEROXIDE DISMUTASE

There is considerable evidence that oxygen was not present in the atmosphere of the earth at the time of its formation 4.5–4.8 billion years ago. In fact, it is argued that oxygen became available only about 2 billion years ago as a result of the evolution of the photosynthetic organisms, the first being the blue-green algae. Until then it is believed that the only organisms existing were anaerobic cells using the oxidation of sulfate and nitrate as a means of producing metabolic energy. The gradual accumulation of oxygen accompanying the evolution of photosynthetic organisms was of great evolutionary significance because it provided for the subsequent evolution of aerobic organisms (first appearing about 1.5 billion years ago). By using oxygen as a terminal electron acceptor (oxidizing agent), aerobic cells could extract more energy from reduced food substrates such as glucose because the substrate can be completely oxidized to CO_2.

Remember:

$$\text{glucose} \xrightarrow{\text{anaerobic}} 2 \text{ lactate} + 56 \text{ kilocalories}$$

$$\text{glucose} + 6O_2 \xrightarrow{\text{aerobic}} 6CO_2 + 6H_2O + 686 \text{ kilocalories}$$

Hence, the emerging aerobic life forms had an advantage over anaerobic organisms and the aerobes thrived.

Anaerobic organisms were at a disadvantage also because the presence of oxygen must have killed some (many?). That oxygen is indeed a toxic substance to life forms is known. For example, several obligate anaerobic organisms (soil organisms) are known that thrive only in the absence of oxygen and die in its presence. The toxicity of oxygen to organisms raises two obvious questions: (1) why is it toxic, and (2) why are aerobic organisms able to thrive in an atmosphere containing this toxic substance? These questions can be answered as follows.

The stable ground state of oxygen is not toxic. However, because of its electronic structure, which consists of two unpaired electrons, there are restrictions as to how the oxygen molecule can react as an electron acceptor. For example, in the conversion $O_2 + 4e + 4H^+ \rightarrow 2H_2O$ it is proposed that four univalent (one electron transfer) steps are involved:

$$O_2 + 1e \longrightarrow O_2^- \cdot$$
$$O_2^- \cdot + 1e + 2H^+ \longrightarrow H_2O_2$$
$$H_2O_2 + 1e + H^+ \longrightarrow H_3O_2 \longrightarrow H_2O + HO \cdot$$
$$HO \cdot + 1e + H^+ \longrightarrow H_2O$$
$$\overline{Net: \quad O_2 + 4e + 4H^+ \longrightarrow 2H_2O}$$

If oxygen reacts this way (and it probably does), the production of $O_2^- \cdot$, H_2O_2 (hydrogen peroxide), and $\cdot OH$ as intermediates presents a serious problem to life because they are all potent oxidizing agents. The $O_2^- \cdot$ species is called the *superoxide anion radical* or simply *superoxide*. The $\cdot OH$ species is called the *hydroxy radical* and is the strongest oxidizing agent known. The hydroxy radical can be formed not only by the scheme shown above, but also as follows:

$$H_2O_2 + O_2^- \cdot \rightarrow O_2 + OH^- + \cdot OH$$

 Each species represents a potential threat to living cells because of the damage each can cause to all the classes of biomolecules, notably the proteins and lipids. In other words, the toxicity of oxygen is due to the toxicity of the species ($HO\cdot$, $O_2^- \cdot$, and H_2O_2) that can be formed from it.

In addition to the possible occurrence of the univalent reaction sequence for $O_2 \rightarrow 2H_2O$ in living cells, H_2O_2 and $O_2^- \cdot$ are also known to be produced in several enzyme-catalyzed reactions that utilize O_2 as one of the substrates. Thus the reaction of H_2O_2 and $O_2^- \cdot$ to yield $\cdot OH$ is a real threat to cells. Fortunately self-defense mechanisms serve to scavenge the H_2O_2 and $O_2^- \cdot$ and thus (a) the $\cdot OH$ is unable to form and (b) the danger of H_2O_2 and $O_2^- \cdot$ existing as powerful oxidizing agents in their own right is also removed.

H_2O_2 and $O_2^- \cdot$ can be detoxified (in part) through the intervention of naturally occurring *antioxidants* such as *ascorbic acid, Vitamin E,* and *glutathione.* Recall that this type of function was mentioned earlier for each of these substances (see p. 298, p. 339, and p. 106). H_2O_2 is also enzymatically detoxified by *catalase* and *peroxidase* enzymes:

$$2H_2O_2 \xrightarrow{\text{catalases}} O_2 + 2H_2O$$
$$H_2O_2 + DH_2 \xrightarrow{\text{peroxidases}} D + 2H_2O$$

(where DH_2 is a reduced organic substance, of which there are several, that functions as a hydrogen donor)

A dismutation reaction is one in which a reactant undergoes both oxidation and reduction.

In 1969 I. Fridovich and J. McCord established that the primary mode of detoxifying $O_2^- \cdot$ is also enzymatic, and is a result of the action of *superoxide dismutase:*

$$O_2^- \cdot + O_2^- \cdot + 2H^+ \xrightarrow[\text{dismutase}]{\text{superoxide}} H_2O_2 + O_2$$

Although H_2O_2 is produced, it can be disposed of by antioxidants, catalases, and peroxidases. In view of the fact that $O_2^-\cdot$ (superoxide) is a powerful oxidizing agent it is remarkable that there is an enzyme that can use it as a substrate without being damaged by it. (The same can be said of the catalases and peroxidases.)

Superoxide dismutase has since been *detected in all types of prokaryotic and eukaryotic aerobic cells,* thus supporting its significance as a self-defense mechanism against oxygen toxicity. If this assigned role of dismutase action is correct, what would you predict about its occurrence in obligate anaerobes? Yes, you are right—the enzyme is *not present in obligate anaerobes* and apparently this is the primary reason why O_2 is lethal to such organisms. There are several other recent experimental findings that support this theory of oxygen defense (see the Fridovich article listed at the end of the chapter).

There has been increased interest in superoxide dismutase in recent years because of evidence that granulocytes (a type of white blood cell) liberate large amounts of $O_2^-\cdot$ during a surge of respiratory metabolism accompanying their development into phagocytic cells that ingest and destroy foreign particles, bacteria, and other cells. In addition to participating in the killing of bacteria, $O_2^-\cdot$ may also damage other tissues, contributing to an *inflammation of the tissue.* To combat this inflammation it has been suggested that superoxide dismutase be used (by injection) as an *anti-inflammatory drug.* Some successes for this potential clinical application have already been reported but further research is still necessary.

LITERATURE

Azzone, G. F. "Oxidative Phosphorylation, A History of Unsuccessful Attempts; Is it only an Experimental Problem." *J. Bioenergetics,* **3**, 95 (1972).

Baldwin, E. *Dynamic Aspects of Biochemistry.* 5th edition. London-New York: Cambridge University Press, 1967. Chapters 7 and 8 are devoted to enzymes and electron carriers that function in the oxidation of organic compounds and in the transfer of electrons by the respiratory chain. See p. 409 for a description of the book.

Boyer, P. B., B. Chance, L. Ernster, P. Mitchell, E. Racker, and E. C. Slater. "Oxidative Phosphorylation and Photophosphorylation." In *Ann. Rev. Biochem.,* **46,** 955–1026 (1977). Each author reviews an aspect of energy-coupled phosphorylation in membranes, particularly in mitochondria and chloroplasts. All models of energy transduction are covered.

Dickerson, R. E. "The Structure and History of an Ancient Protein." *Scientific American,* **226**, 58–72 (April, 1972). A concise discussion of the structure of cytochrome *c* and its evolution as a molecule over 1.2 billion years. The amino acid sequence from 38 different species is examined.

Fridovich, I., "The Biology of Oxygen Radicals," *Science,* **201,** 875–880 (1978). A review article on the superoxide radical and superoxide dismutase.

Green, D. E., and H. J. H. Young. "Energy Transduction in Membrane Systems." *Am. Scientist,* **59,** 92 (1971). A review article.

Griffiths, D. E. "Oxidative Phosphorylation." In *Essays in Biochemistry,* P. N. Campbell and G. D. Greville, eds., Vol. 1, pp. 57–90. New York: Academic Press, 1965. An excellent review article summarizing the respiratory chain and the coupled process of ATP formation.

Hinkle, P. C., and R. E. McCarty. "How Cells Make ATP." *Scientific American,* **238**, 104–123 (1978). Excellent article emphasizing the chemiosmotic coupling mechanism and the mode of action of ionophore uncouplers.

Lehninger, A. L. *Bioenergetics.* 2nd edition. New York: W. A. Benjamin, Inc., 1971. Contains an excellent introductory presentation of oxidative phosphorylation.

Mitchell, P. "Protonmotive Chemiosmotic Mechanisms in Oxidative and Photosynthetic Phosphorylation." *Trends Biochem. Sci.,* **3**, N58–N61 (1978). A brief description of the chemiosmotic hypothesis by the original proponent.

Racker, E. "Inner Mitochondrial Membranes: Basic and Applied Aspects." *Hospital Practice,* **9**, 87 (February, 1974), and "The Membrane of the Mitochondrion." *Scientific American* (February, 1968). Excellent introductory review articles of structure and evidence for the chemiosmotic hypothesis of coupling.

————. *A New Look at Mechanisms in Bioenergetics.* New York: Academic Press, 1976. A short book for student reading on mitochondrial ATP production and related processes. Available in paperback.

WILLIAMS, V. R., W. L. MATTICE, and H. B. WILLIAMS.

Basic Physical Chemistry for the Life Sciences. 3rd edition. San Francisco: W. H. Freeman and Company, 1978. Chapter 5 contains an excellent development of principles of oxidation-reduction systems.

EXERCISES

14–1. Calculate the net oxidation-reduction potential and the standard free-energy change for each of the following reactions as written from left to right, and indicate whether or not the reaction would tend to occur spontaneously given the proper conditions. (Assume that pH and temperature are as specified in Table 14–1.)
(a) $NADH + H^+ + CoQ \rightarrow NAD^+ + CoQH_2$
(b) $CoQH_2 + 2(cyt\ b\ (oxidized) \rightarrow$
$\qquad CoQ + 2(cyt\ b\ (reduced)) + 2H^+$
(c) 2 (reduced ferredoxin) $+ NADP^+ + 2H^+ \rightarrow$
$\qquad 2(oxidized\ ferredoxin) + NADPH + H^+$

14–2. The standard reduction potential (pH 7) of the lactate-pyruvate couple is -0.19 volt, which means then that the reaction

$\qquad$ pyruvate $+ 2H^+ + 2e \rightarrow$ lactate $\qquad \mathscr{E}^{\circ\prime}_{red} = -0.19\ V$

is endergonic. Yet, in our analysis of the glycolytic pathway in Chap. 12, we have seen that the reduction of pyruvate to lactate, catalyzed by lactate dehydrogenase, is strongly exergonic. Explain.

$\qquad$ pyruvate $+ NADH + H^+ \xrightarrow[\text{dehydrogenase}]{\text{lactate}}$ lactate $+ NAD^+$

14–3. In physicochemical terms, summarize your understanding of the nature of the respiratory electron transport chain.

14–4. If a respiratory inhibitor functioned at the site of electron transfer between cytochromes c_1 and c, what approximate value for the P/O ratio would be obtained on incubation (aerobically) of intact mitochondria in the presence of malate, ADP, P_i, NAD^+, the inhibitor, and a buffer?

14–5. The incubation of mitochondria in the presence of succinate and malonate would result in less oxygen consumption than in the presence of succinate alone, but the P/O ratios would not be significantly different. Explain.

14–6. If respiratory complexes II and IV were incubated together under aerobic conditions in the presence of added succinate, coenzyme Q, and cytochrome c, what overall oxidation-reduction reaction would occur?

14–7. In the system described in the preceding problem, would you expect to detect the formation of any significant amount of reduced cytochrome c_1 during the reaction? Explain.

14–8. What would be the net yield of ATP from the complete aerobic catabolism of glucose to $6CO_2$ according to the hexose-monophosphate shunt (HMS), and how does the number compare with the ATP yield from glycolysis linked to the citric acid cycle?

$$6(glucose\text{-}6\text{-}P) \xrightarrow{\text{HMS}} 5(glucose\text{-}6\text{-}P) + 6CO_2$$

$$\textit{Net: } 1(glucose\text{-}6\text{-}P) \rightarrow 6CO_2$$

In this case, although reducing power is generated as NADPH and in the cytoplasm, appropriate mechanisms exist in most cells to allow for the transfer of reducing power from NADPH to NADH in the cytoplasm and then the shuttling of the cytoplasmic NADH into the mitochondrion. (Remember, the calculation is to be made on the basis of the catabolism of free glucose.)

14–9. If purified mitochondria are incubated in a buffered solution containing an oxidizable substrate X, cyt c, ADP, and P_i, product Y is produced. In addition, P_i is consumed and O_2 is produced. If representative data for a 30-minute incubation period are 31.3 micromoles of P_i consumed and 15.2 micro-gram-atoms of oxygen produced, what can you conclude about the biochemistry of the X $\rightarrow$ Y conversion?

14–10. When purified mitochondria are isolated and then assayed for oxidative phosphorylation activity, the buffered reaction system (containing the oxidizable substrate, ADP, and P_i) is usually supplemented with a small amount of pure cytochrome c obtained from almost any source. Why do you suppose the cytochrome c addition is necessary, and why is it that it does not have to be obtained from the same source as the mitochondria?

14–11. Some eukaryotic cells contain an NAD-dependent glycerol phosphate dehydrogenase in the cytoplasm as well as an FAD-dependent glycerol phosphate dehydrogenase in the mitochondrion. Both enzymes catalyze the interconversion of dihydroxyacetone phosphate and L-glycerol phosphate, metabolites that can both move across the mitochondrial membrane. What metabolic significance might this compartmentalization of enzymes have?

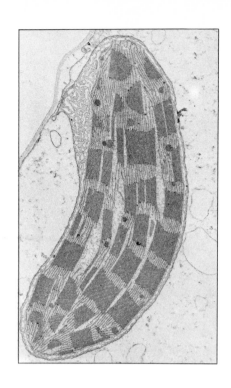

CHAPTER 15

PHOTOSYNTHESIS

Photosynthesis is the most important biological process that occurs on the face of the earth. There are two important reasons for this. First, the major source of carbohydrate $(CH_2O)_n$ which nonphotosynthetic organisms must have as a source of carbon and energy is available in nature *only* through the fixation of atmospheric CO_2 by photosynthesizing cells. Second, the oxygen produced in this reaction is indispensable to the existence of all aerobic organisms. Photosynthesis is the only chemical source of oxygen on this planet.

$$6CO_2 + 6H_2O + \textit{solar energy} \xrightarrow{\textit{photosynthesis}} (CH_2O)_6 + 6O_2$$

The vital importance of photosynthesis to aerobic organisms is mutual, however, since the metabolism of aerobic organisms is equally critical to photosynthetic organisms. Actually the two processes are entirely complementary, inasmuch as the primary carbon products of photosynthesis serve as the primary carbon reactants of aerobic organisms, which in turn yield the primary carbon reactants of photosynthesis. This crucial relationship is generally termed the *carbon cycle*. We emphasize the vital importance of photosynthesis precisely because it begins the carbon cycle, and hence, without it, life as we know it would soon cease to exist.

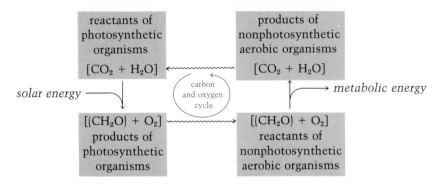

In chemical terms the net effect of photosynthesis is quite simple. Carbon in its *most oxidized, low-energy form,* carbon dioxide, is assimilated by the photosynthetic organism into a *more reduced, higher-energy form,* carbohydrate material. Thus, the process is *endergonic* (requires energy) and *reductive* (requires reducing power). Energy is initially supplied by light from solar radiation, and reducing power is initially contributed by water. The location in the cell where the chemistry of this process takes place is the *chloroplast,* a specialized organelle unique to photosynthetic organisms.

The chemical events occur in two distinct phases, termed the *light reaction* and the *dark reaction.* As the name implies, the light reaction is dependent on an input of radiant energy. The complete process, which represents some of the most remarkable chemistry yet known to man, involves many substances, the most important being *chlorophyll.* As indicated in the margin the light reaction involves three events: the oxidative cleavage of water, the formation of NADPH, and the formation of ATP.

The dark reaction involves the enzymatic assimilation and conversion of CO_2 into carbohydrate. The NADPH and ATP formed in the light reaction are utilized as the metabolic sources of reducing power and energy, respectively.

Our objective in this chapter will be to explore the current state of knowledge concerning each of these processes. To establish a proper foundation for such discussions, we will first develop some basic principles dealing with the nature of light and then examine the ultrastructure of the chloroplast.

results of light reaction:

(directly dependent on radiant energy)

(1) $H_2O \rightarrow (2H^+ + 2e) + \frac{1}{2}O_2$

(2) $NADP^+ + 2H^+ + 2e \rightarrow$

$NADPH + H^+$

(3) $ADP + P_i \rightarrow ATP$

results of dark reaction:

(indirectly dependent on radiant energy)

$CO_2 \xrightarrow[\text{ATP}]{\text{NADPH}} (CH_2O)$

ELECTROMAGNETIC RADIATION

The portion of solar radiation reaching earth that is referred to as visible light (sometimes termed *white light*), comprises a specific region of the complete electromagnetic spectrum (see p. 55, Chap. 2), from about 4,000 Å to 7,000 Å. According to the principles of modern physics, our concept of any type of electromagnetic radiation is that the radiation is propagated through space as a wave with a certain velocity c, amplitude A, and *wavelength* λ. By definition, the wavelength is the distance required for the wave to propagate one complete cycle (see Fig. 15–1). Whereas the velocity of any wave is constant ($c = 186,000$ mi/sec or 3×10^{10} cm/sec), the amplitude and wavelength are variable. Indeed, it is the wavelength that distinguishes one type of radiation from another. This distinction is also expressed in terms of the *frequency* ν, the number of cycles of a moving wave

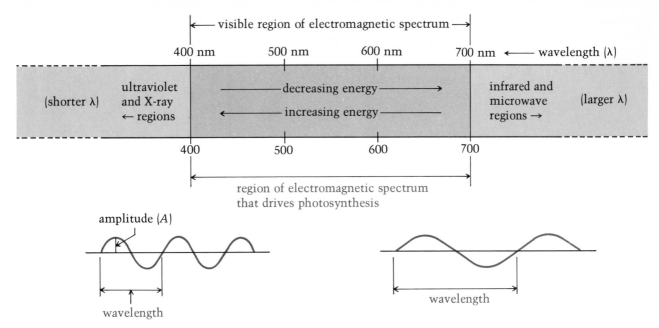

Both rays have same velocity ($c = 3 \times 10^{10}$ cm/sec).
Both rays have same amplitude.
Each ray has different wavelength and frequency.
Each ray has different energy content, with
energy of short λ > energy of long λ

FIGURE 15–1 An analysis of electromagnetic radiation, particularly the visible region.

that pass a given point per unit of time. The mathematical expression for the frequency is

$$\nu = \frac{c}{\lambda}$$

The Quantum Nature of Light

One of the most revolutionary developments in the early days of modern physics was initiated in 1900 by Max Planck's proposal that all types of electromagnetic radiation have a *dual nature;* that is, in addition to having a *wave character,* radiation is also *particulate* in the sense that the energy E of a wave is emitted in discrete packages called *quanta.* Planck further proposed that the energy value E of 1 quantum is directly proportional to the frequency of the radiation.

$$E \propto \nu \qquad E\ (1 \text{ quantum}) = h\nu \qquad E\ (n \text{ quanta}) = nh\nu$$

The proportionality constant h, which for obvious reasons is called Planck's constant, is equal to 6.625×10^{-27} erg/sec. The main statement of Planck's hypothesis is quite understandable. It states that, since the energy content of an electromagnetic wave is dependent only on wavelength λ, the energy content of the entire electromagnetic spectrum is discontinuous, meaning that the energy is constantly changing from values of high energy (low wavelength) to lower energy (higher wavelength). Initially greeted with skepticism, Planck's hypothesis was shortly confirmed by Einstein, who postulated even further that, when matter and electromagnetic radiation interact in a process termed a *photochemical event,* the radiation is absorbed, resulting in the excitation (activation) of molecules to higher energy levels.

excited (high-
energy) states of
M; can be
accompanied by
abstraction of
electron

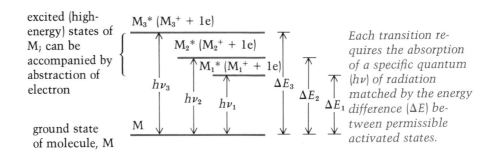

ground state
of molecule, M

*Each transition re-
quires the absorption
of a specific quantum
($h\nu$) of radiation
matched by the energy
difference (ΔE) between permissible
activated states.*

To be more specific, the absorption of radiant energy causes an excitation of electrons within the molecule to higher energy levels. For any such transition, Einstein also suggested that the absorption of 1 quantum of radiant energy with the structure of molecules (see p. 55, Chap. 2). As we will shortly ergy. It was later proposed that each electronic transition between allowable energy levels can occur only when the molecule is exposed to radiation of exactly the same energy (quantum) content as the energy difference between the ground state and the excited state. This, of course, is the basis for spectroscopic investigations that correlate the absorption of radiant energy with the structure of molecules (see p. 55, Chap. 2). As we will shortly discover, the same type of phenomenon occurs in photosynthesis. The absorption of radiant energy by the special photosensitive chlorophyll molecules presumably causes a displacement of electrons from these chlorophyll molecules to special electron carriers.

CHLOROPLASTS—CELLULAR SITE OF PHOTOSYNTHESIS

General Comments

The *complete process* of photosynthesis takes place in the *chloroplasts*, highly specialized, subcellular organelles that distinguish photosynthetic cells from nonphotosynthetic cells. The first experimental demonstration that isolated chloroplasts were capable of supporting the complete process of photosynthesis was achieved by D. Arnon and coworkers in 1954. However, first credit for the successful use of cell-free extracts in photosynthetic studies is to be given to R. Hill, who demonstrated the photoevolution of oxygen with isolated chloroplasts 15 years earlier.

In addition to all animal cells and most bacterial cells, certain plant cells are also nonphotosynthetic and contain no chloroplasts. These include those cells responsible for plant functions other than the harnessing of solar energy and the fixation of CO_2, cells such as those found in plant roots, for example. In plants, the process of photosynthesis occurs primarily in the *green leaf cells* with each cell containing several chloroplasts. On the other hand, since the photosynthetic algae are unicellular organisms, each alga will contain chloroplasts. In contrast to the more advanced algae and green plants, the photosynthetic bacteria (likewise unicellular organisms) do not contain chloroplasts. In these cells the process of photosynthesis occurs in *chromatophores,* which can be thought of as primitive chloroplasts, similar in function but smaller in size and less complex in structure than chloroplasts.

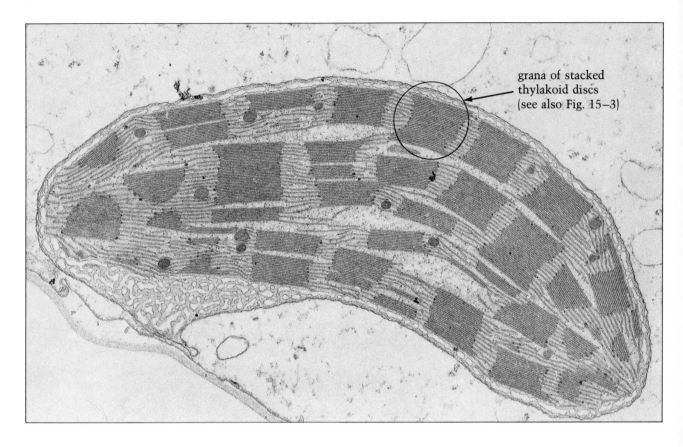

grana of stacked
thylakoid discs
(see also Fig. 15–3)

Chloroplast Ultrastructure

A cross-sectional view of a fully developed chloroplast as seen in electron microscopy is shown in Fig. 15–2. Similar to a mitochondrion, the chloroplast is surrounded by a double (outer and inner) membrane system. (This and other biochemical similarities between chloroplasts and mitochondria have prompted the suggestion that both organelles evolved from the same origin, probably a primitive symbiotic bacterium. The other similarities include the presence of DNA and ribosomes within each organelle and the capacity of each organelle to synthesize protein. Refer to an earlier discussion of mitochondria in Chap. 1, p. 24.) The size, shape and number of chloroplasts vary widely among photosynthesizing cells. In higher plants they are generally cylindrical in shape, ranging anywhere from 5–10 μm in length and 0.5–2 μm in diameter. A comparison of these dimensions with those of mitochondria reveals that the chloroplasts are much larger. Indeed, among the defined organelles of higher cells, only the nucleus is larger.

The most conspicuous feature to be noted in the electron micrograph in Fig. 15–2 is that the chloroplast, like the mitochondrion, possesses a characteristic and highly organized fine level of ultrastructure, evidenced by the ordered array of several electron dense bodies throughout its whole interior. These are interpreted as corresponding to flattened *membranous bodies,* called *lamellae* or, more recently, *thylakoid discs.* As with the cristae of mitochondria the thylakoid discs result from folded protrusions of the inner chloroplast membrane into the core of the chloroplast. In cross-sectional

FIGURE 15–2 The ultrastructure of a fully developed, intact chloroplast as seen in cross-section under the electron microscope. Note the many interconnected grana consisting of stacked thylakoid discs. Samples obtained from mesophyll cells of maize leaf. Magnification: 32,000 ×. (Photograph generously supplied by L. K. Shumway, Department of Botany and Program in Genetics, Washington State University, Pullman, Washington.)

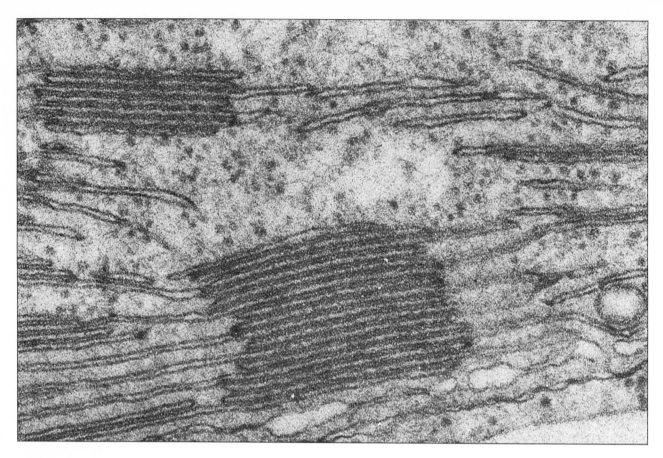

FIGURE 15-3 Electron micrograph showing grana and thylakoid discs at high magnification. The less electron dense background represents the soluble region of the chloroplast, called the lumen. The chloroplast membrane can be seen at the lower right. Magnification: 124,000×. (Photograph generously supplied by L. K. Shumway, Department of Botany and Program in Genetics, Washington State University, Pullman, Washington.)

view, they appear as in Figs. 15-2 and 15-3, with the latter showing the cross section at higher magnification. From a top or bottom view, each thylakoid disc appears as shown in the idealized drawing of Fig. 15-4. The stacked piles of individual lamellae are called *grana*. As shown in Fig. 15-2, the number of stacked thylakoid discs per granum is quite variable.

There are extensive data suggesting that each thylakoid disc contains all of the necessary photosensitive pigments, electron carriers, and accessory components for the crucial light reaction of photosynthesis. The bulk of the enzymes and coenzymes responsible for the assimilation of CO_2 into organic material are found in the soluble portion (*lumen*) of the chloroplast. Despite the fact that each thylakoid disc, and therefore each granum, contains all the materials required for the light reaction, neither is suggested as the basic unit of photosynthesis. High-resolution electron microscopic studies of isolated and partially fragmented thylakoids suggest that the basic unit (termed a *quantosome*) may be a small, somewhat spherical unit, several of which are embedded in the phospholipoprotein matrix of the lamella membrane (see drawing in Fig. 15-4). What the exact composition and arrangement of the quantosomes may be, and whether or not all quantosomes are identical, are questions yet to be answered. Regardless of these uncertainties, it should be quite obvious to you that the ultrastructure of the chloroplast is another example of the fascinating and, indeed, truly remarkable constructions of nature. Even more fascinating, however, is what goes on inside the chloroplast.

The combined effect of the light-dependent ($nh\nu$ is required) phase in all photosynthetic green plants and several algae can be given by the following equation:

$$2NADP^+ + mADP + mP_i + 2H_2O \xrightarrow[nh\nu]{chloroplasts} 2NADPH + 2H^+ + mATP + O_2$$

It is a *net* equation summarizing the four main events of the light reaction; (a) the **photochemical excitation** of chlorophyll; (b) the oxidative cleavage of water—**photooxidation;** (c) the reduction of $NADP^+$—**photoreduction;** and (d) the formation of ATP—**photophosphorylation.** Our objective will be to describe the sequence of events associated with this overall process.

The equation above is written to show the evolution of one unit of molecular oxygen. To avoid unnecessary confusion, this stoichiometry will be used throughout this chapter. Since the production of 1 O_2 involves the transfer of 4 electrons from H_2O to $NADP^+$, the equation also depicts an idealized $2 + 2$ stoichiometry involving H_2O and $NADP^+$. Still uncertain are (a) the number m of ATP molecules produced per molecule of oxygen evolved, and (b) the number n of light quanta required per molecule of oxygen evolved. For ATP formation, several investigators have suggested a value of 4 for m. Others propose that $m = 2$. For the amount of light energy, there are also two suggested values: $n = 4$ and $n = 8$. According to Einstein's law of photochemical equivalence stating that 1 quantum is required to excite 1 electron, the proposal that $n = 4$ would represent 100% quantum efficiency (that is, 4 quanta for a 4 electron process), whereas $n = 8$ would represent 50% efficiency. The controversy over these different values has existed since 1922 when O. Warburg first reported, and defended vigorously, that $n = 4$. The controversy still continues.

Note: A similar equation could be written for the photosynthetic bacteria. In these organisms, however, the source of reducing power is not H_2O, but generally a special inorganic donor such as H_2 or H_2S, or certain reduced organic acids. This variation in the photosynthetic bacteria will not be discussed further.

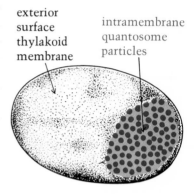

exterior surface thylakoid membrane

intramembrane quantosome particles

FIGURE 15–4 Idealized drawing based on electron microscopic studies of chloroplast thylakoid disc. Sketch depicts a top view of a thylakoid with a portion of surface membrane peeled off to reveal an ordered array of spherical particles. The latter have been proposed to represent the elementary photochemical structural units, termed *quantosomes.*

Primary Photochemical Act

It is now well-established that the *primary photochemical act is the absorption of radiant energy by chlorophyll molecules, resulting in their activation (excitation) to higher energy states.* Thereafter, electrons are transferred from the excited chlorophyll molecules to specialized acceptor mole-

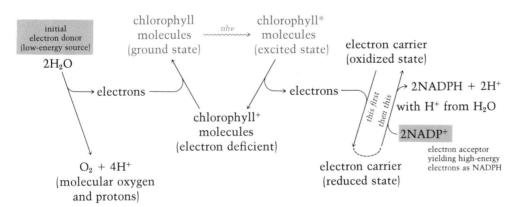

Primary photochemical event

Note: The formation of ATP, coupled to electron transfer reactions, is not indicated in this diagram.

481

Note that the flow of electrons from low-energy H_2O to higher-energy NADPH is just the reverse of the chemistry that occurs in the respiratory chain examined in the previous chapter.

cules, with the latter materials, of course, undergoing reduction (gain in electrons). The electrons are ultimately transferred to NADP⁺. This initial photochemical event is also accompanied by oxidative cleavage of H_2O to yield (a) molecular oxygen, (b) a source of electrons to replenish those transferred from chlorophyll, and (c) a source of protons (H⁺) also required for the reduction of NADP⁺. In other words, as a consequence of the photochemical excitation of chlorophyll, a poor reducing agent (H_2O) is used as an electron source to produce a good reducing agent (NADPH). These statements are summarized in the diagram on p. 481. Some evidence for this participation of chlorophyll as the "action molecule" in the primary act of photosynthesis will be discussed shortly. In view of this critical role, it seems appropriate for us to first examine the chemical nature of chlorophyll.

Chlorophyll

It is a unifying biochemical fact that all photosynthesizing cells contain chlorophyll. Moreover, with the exception of the photosynthetic bacteria and a few algae, it appears that almost all photosynthetic cells contain identical molecular species of chlorophyll. These species of chlorophyll are two in number, with one designated as *chlorophyll a* and the other as *chlorophyll b*. The chloroplasts of all algae and all higher green plants contain chlorophyll *a*. Chlorophyll *b* also occurs in the higher green plants, but only in green algae. Photosynthetic bacteria presumably contain only one type of chlorophyll, which is different from both chlorophylls *a* and *b*, and is called *bacteriochlorophyll*.

The basic structure of all chlorophylls is that of a planar metallotetrapyrrole unit similar to the heme prosthetic group of myoglobin and the cytochromes. The chlorophyll molecule, however, is distinctly different from the heme prosthetic group in three respects: (1) in chlorophyll the metal coordinated to the tetrapyrrole moiety is *magnesium* (as Mg²⁺) rather than iron; (2) chlorophylls are *not* dependent on attachment to a protein grouping for activity; and (3) the chlorophylls contain a characteristic set of pyrrole side-chain groupings, the most notable of which are (a) a large, nonpolar, alcohol grouping (*phytol*) in ester linkage to a propionic acid side-chain residue, and (b) a fused cyclopentanone ring. The nonpolar phytol grouping is particularly noteworthy, since it provides the structural basis for a stabilized integration of chlorophyll molecules into the lipoprotein matrix of the thylakoid membrane. All of these features are depicted below. As indi-

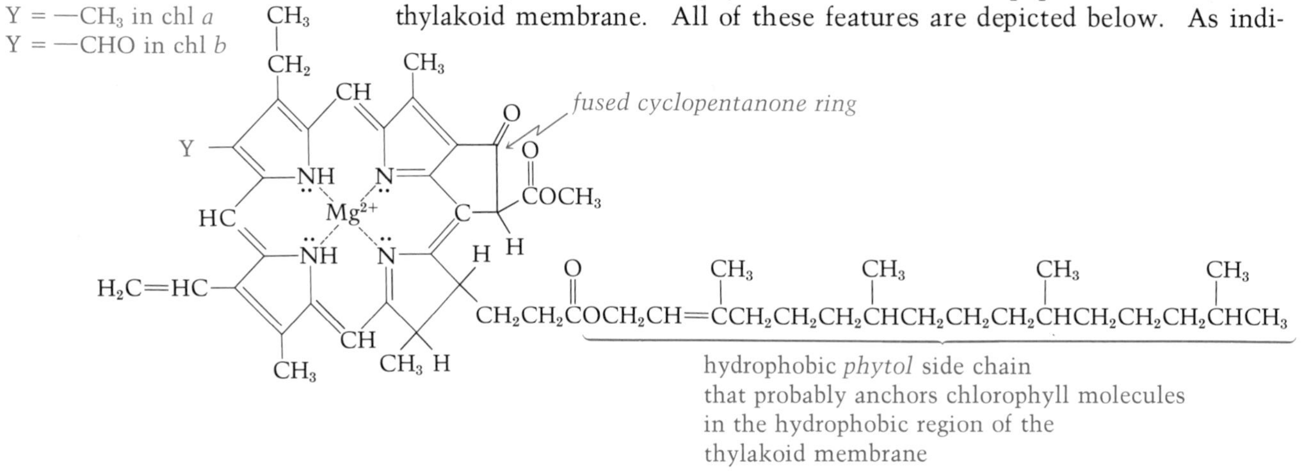

Y = —CH₃ in chl *a*
Y = —CHO in chl *b*

fused cyclopentanone ring

hydrophobic *phytol* side chain
that probably anchors chlorophyll molecules
in the hydrophobic region of the
thylakoid membrane

cated, the structural difference between chlorophyll *a* and chlorophyll *b* is merely a variation in one pyrrole ring substituent.

The distinct physicochemical property of the chlorophyll molecule that is biologically important is, of course, its capacity to absorb light energy from the visible region of the electromagnetic spectrum. In all likelihood, the electrons excited in this process are those in the conjugated double-bond system of the tetrapyrrole unit. In any event, because of the difference in structure between chlorophylls *a* and *b*, each displays a unique absorption (excitation) spectrum (see Fig. 15–5).

Within the confines of the thylakoid membrane there are even differences in absorption among chlorophyll *a* molecules and perhaps also among chlorophyll *b* molecules. These differences arise because of a particular environment in the lipoprotein matrix of the membrane where the chlorophyll molecules reside. Due to a unique location and arrangement, these molecules can participate in unique interactions with other components of the membrane that produce unique light-excitation properties in one or more chlorophyll molecules. For example, whereas most chlorophyll *a* molecules show a maximal excitation at about 665–670 nm, there is good evidence supporting the existence of two specialized molecules that absorb maximally at 683 and 700 nm. Labeling these as specialized chlorophylls means that they occupy a particular location in the quantosome aggregate, where they act as terminal participants in the capture of radiant energy absorbed by other chlorophyll molecules (and possibly other accessory pigments; see below) and then as the immediate donors of electrons to oxidized carriers.

The emphasis on the light absorption characteristics of chlorophyll is quite relevant since there is extensive evidence that the activity of chloroplasts in producing oxygen is maximal at wavelengths in the vicinity of the absorption maxima of chlorophyll (see Fig. 15–5). Observations of this type have been made with chloroplasts from many different sources, and have been interpreted to mean that chlorophyll is the primary photosensitive pigment in the photosynthetic apparatus. Because it is present in larger amounts than *b*, chlorophyll *a* is regarded as the *primary* pigment.

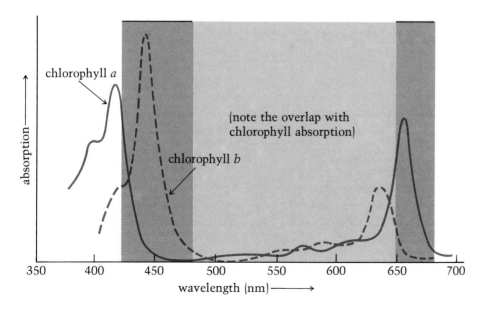

FIGURE 15–5 The absorption characteristics of chlorophyll in the visible region of the electromagnetic spectrum. Chloroplasts exhibit high levels of O_2 production at the light energy values in the two dark grey areas. Smaller but still significant levels of chloroplast activity are observed in the light grey region where some of the accessory pigments absorb.

In addition to chlorophyll, chloroplasts also contain other substances that absorb in the range of the solar spectrum. They are called *accessory pigments* because their absorption of light supplements the action of chlorophyll. These include materials from two classes of compounds, the *carotenoids* (β-carotene is most important in higher plants; see p. 336, Chap. 9) and the *phycobilins,* which are open-chain tetrapyrrole compounds. Although their exact function has not yet been identified, a lot of evidence suggests that by acting as secondary absorbers these accessory pigments assist in the transfer of energy to the specialized chlorophylls within the quantosome reaction center. They may also protect chlorophyll molecules from getting bleached by solar rays.

Emerson Effect and the Dual-photosystem Hypothesis

One of the most significant discoveries regarding the role of the different light-absorbing chlorophyll species in photosynthesis was made in the late 1950s by R. Emerson. He found that the efficiency (see margin) of the photochemical phase was not constant throughout the entire visible spectrum. A distinct reduction occurred when photosynthesizing cells were exposed to monochromatic (single wavelength) light sources at the red end of the spectrum (beyond 680 nm), where the only absorption is due to chlorophyll *a*.

Photosynthetic efficiency is usually measured as the amount of oxygen produced per quantum of radiant energy absorbed.

Since the decrease in efficiency occurs at the red end of the spectrum, the effect is generally referred to as the *"red-drop phenomenon."* Emerson subsequently demonstrated that full efficiency could be restored if the cells were *simultaneously* exposed to another light source with a wavelength of 650 nm, a point of maximum absorption in chlorophyll *b*. In other words, when light absorption by chlorophyll *a* was accompanied by light absorption from chlorophyll *b*, the cells displayed an enhancement (*Emerson effect*) in the photoevolution of oxygen as compared to separate light absorption by either chlorophyll *a* or chlorophyll *b*.

Emerson suggested then that the light-dependent phase of photosynthesis includes *two separate photosystems,* both of which must be activated for maximum efficiency of the light-dependent reaction. One (photosystem I or PSI) contains largely chlorophyll *a* (including the specialized chlorophyll *a* absorbing at 683), the other (photosystem II or PSII) contains both chlorophyll *a* (including the specialized chlorophyll absorbing at 700) and chlorophyll *b*. The hypothesis of two separate photosystems has since been supported by several different lines of evidence and currently governs much of the modern thinking concerning the photochemical apparatus. The present viewpoint is that both systems are separately localized within each quantosome of the thylakoid membrane, and that each system participates in a separate phase of the overall light reaction (see p. 486).

Electron Transport in Photosynthesis

At this point a logical question is, what is the function of each photosystem in the overall light reaction? In order to pursue this subject, it is important that we first recognize the basic type of chemistry that is involved. To analyze this, let us return to the equation given earlier for the net effect of the light reaction. Close examination reveals that the basic chemistry is an

$$2NADP^+ + mADP + mP_i + 2H_2O \xrightarrow[nh\nu]{\text{chloroplasts}} 2NADPH + 2H^+ + mATP + O_2$$

oxidation-reduction reaction involving a transfer of two hydrogen atoms ($2H^+$ and 2 electrons) from H_2O (the reduced donor) to $NADP^+$ (the oxidized acceptor). Recalling our discussion in the previous chapter concerning the difference between the NAD^+-NADH and O_2-H_2O couples, we should also note that the direction of electron transfer involves a transition from a low energy state (H_2O) to a higher energy state (NADPH). It is this unfavorable energy barrier that is overcome by the absorption of light energy, with the accompanying formation of ATP representing a conservation of part of the

$$NADP^+ + 2H^+ + 2e \rightarrow NADPH + H^+ \qquad \mathscr{E}^{\circ\prime}_{red} = -0.32 \text{ V}$$
$$H_2O \rightarrow \tfrac{1}{2}O_2 + 2H^+ + 2e \qquad \mathscr{E}^{\circ\prime}_{ox} = -0.82 \text{ V}$$
$$\overline{NADP^+ + H_2O \rightarrow NADPH + H^+ + \tfrac{1}{2}O_2 \qquad \mathscr{E}^{\circ\prime}_{net} = -1.14 \text{ V}}$$

$\mathscr{E}^{\circ\prime}_{net}$ is negative; therefore, the reaction from
left to right is endergonic; specifically: $\Delta G^{\circ\prime} = +52,580$ calories

Remember that in respiration the reverse of this chemistry applies; with mitochondrial activity $\mathscr{E}^{\circ\prime} = +1.14$ volts and $\Delta G^{\circ\prime} = -52,580$ calories.

absorbed energy. In the pages to follow, our objective will be to elaborate on how these events occur and to explain the suggested participation of photosystems I and II. You will discover shortly that the modes of electron transport and phosphorylation of ADP are *identical in design* to the events of oxidative phosphorylation. In view of our frequent referral to the theme of biochemical unity, you should find this relationship to be a logical one. Our discussion of the whole subject will be somewhat brief, in part because of the restricted limits of this book, but also because the complete details of this process have yet to be determined exactly.

Photoreduction (PSI) and Photooxidation (PSII)

Despite the unknown nature of many phases of the light reaction, there is considerable evidence to suggest that both photosystems fulfill *separate but complementary* roles. These proposed functions are as follows. (We will not review the substance of the experimental proof.) The activation of photosystem I at wavelengths equal to or greater than 680 nm results in a photoreduction ($2NADP^+ + 4H^+ + 4e \rightarrow 2NADPH + 2H^+$), whereas the activation of photosystem II at wavelengths less than 680 nm results in a photooxidation ($2H_2O \rightarrow O_2 + 4H^+ + 4e$). In other words, the complementary relationship between the two photosystems involves reducing power, and may be summarized as follows: *the activation of photosystem II results in a photooxidation step, which generates reducing power required in the photoreduction step, which in turn is mediated by the activation of photosystem I.*

Although the principle of cooperation contained in this statement is basically correct, the implication that the reducing power required for the reduction of $NADP^+$ is supplied directly by the oxidation in photosystem II is partly inaccurate. Actually, a direct linkage between the two systems is proposed to exist only at the level of the protons, with the source of electrons in the photoreduction step of PSI being supplied directly by the photoexcitation of the chlorophyll molecules in photosystem I. What then is the fate of the electrons generated in the photolytic cleavage of H_2O in PSII? According to current interpretations, the electrons generated from the photooxidation of H_2O via PSII are returned to the same photosystem, which becomes electron deficient after undergoing excitation (loss of electrons to a higher energy level) by light absorption. The balance between the two photosystems is restored by the transfer of electrons liberated in the excitation of PSII to the electron-deficient PSI. It is further proposed that this linkage of *electron transport* from PSII to PSI is the exergonic sequence that is *coupled to the phosphorylation* of ADP. The sum and substance of these relationships are represented in the diagram shown below.

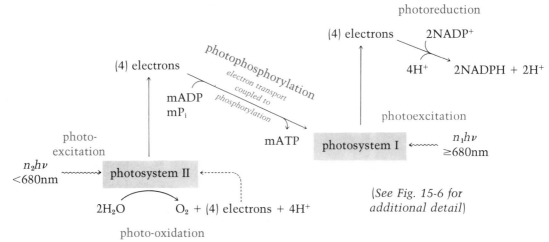

(See Fig. 15-6 for additional detail)

A more detailed summary of the most widely supported proposal for the events occurring in the light reaction is diagramed in Fig. 15–6. The key to understanding the message of this diagram resides in the A, B, and A′, B′ symbolism in the shaded rectangles. These shaded areas correspond to the two active photosystems, each containing the primary photosensitive chlorophyll molecules and other accessory pigments (all of course embedded in a lipoprotein matrix). To put it simply, the materials symbolized as A, B, A′, and B′ are proposed to be highly *specialized electron carriers* unique to each photosystem and required for the processes of photooxidation, photoreduction, and photophosphorylation. Each photosystem is believed to contain a minimum of two such materials, with one acting as an *oxidized electron acceptor* (A_{ox} in PSII and A'_{ox} in PSI) and the other as a *reduced electron donor* (B_{red} in PSII and B'_{red} in PSI).

It is further hypothesized that A_{ox} and A'_{ox} undergo reduction when each receives an electron from a *specialized* chlorophyll *a* molecule in each photosystem that has been excited (Chl_a^*) by the absorption of light energy. On the other hand, B_{red} and B'_{red} are hypothesized to act as immediate electron donors, returning the electron-deficient chlorophyll molecule (Chl_a^+) to its ground state. The resultant formation of the reduced carriers (A_{red}^- and A'^-_{red}) represents the raising of electrons to higher energy levels, from which they are then transferred to still other specialized acceptors. The electrons from A_{red}^- generated in PSII are ultimately transferred "downhill" (that is, to a lower energy level) to PSI, with the accompanying formation of ATP. The electrons from A'^-_{red} generated in PSI are used in the reduction of $NADP^+$. The resultant formation of B_{ox}^+ and B'^+_{ox} represents the formation of low-energy electron acceptors. B_{ox}^+ can act as a suitable electron acceptor to drive the oxidation of H_2O, which is linked to PSII. In PSI, B'^+_{ox} can act as a suitable terminal acceptor in the transport of electrons from the A_{red}^- coming from PSII.

The reference to a specialized chlorophyll *a* concerns one particular molecule of chlorophyll *a* that is in a particular position in the quantosome; this mole-

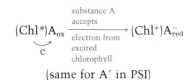

$$(Chl^*)A_{ox} \xrightarrow[\substack{\text{electron from} \\ \text{excited} \\ \text{chlorophyll}}]{\substack{\text{substance A} \\ \text{accepts}}} (Chl^+)A_{red}^-$$

(same for A′ in PSI)

$$B_{red}(Chl^+) \xrightarrow[\substack{\text{electron} \\ \text{to deficient} \\ \text{chlorophyll}}]{\substack{\text{substance B} \\ \text{donates}}} B_{ox}^+ \, (Chl)$$

(same for B′ in PSI)

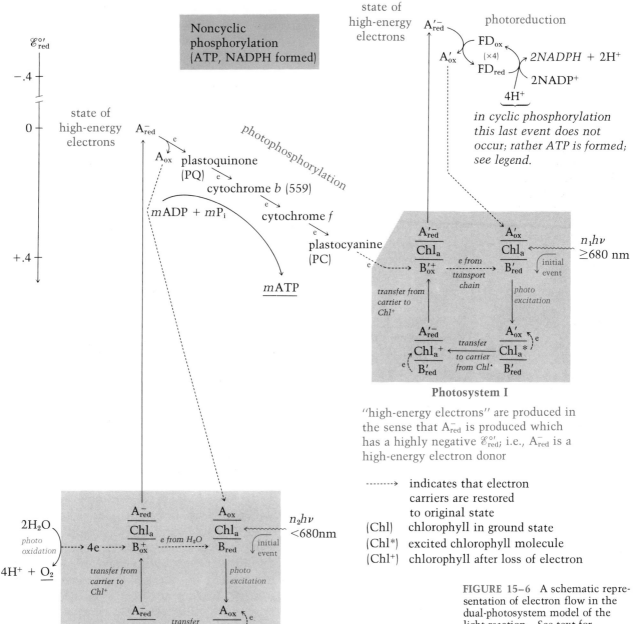

Noncyclic phosphorylation (ATP, NADPH formed)

Photosystem I

"high-energy electrons" are produced in the sense that A^-_{red} is produced which has a highly negative $\mathscr{E}^{o'}_{red}$; i.e., A^-_{red} is a high-energy electron donor

in cyclic phosphorylation this last event does not occur; rather ATP is formed; see legend.

- - - - - → indicates that electron carriers are restored to original state
(Chl)   chlorophyll in ground state
(Chl*) excited chlorophyll molecule
(Chl$^+$) chlorophyll after loss of electron

Photosystem II

"high-energy electrons" are produced in the sense that A^-_{red} is produced which has a highly negative $\mathscr{E}^{o'}_{red}$; i.e., A^-_{red} is a high-energy electron donor

FIGURE 15–6 A schematic representation of electron flow in the dual-photosystem model of the light reaction. See text for description. The model depicted here is referred to as a *noncyclic phosphorylation* mechanism. This means that the electron flow from one photosystem is associated with ATP production, and the electron flow from the other is associated with NADPH production. This pattern is distinguished from what is termed *cyclic phosphorylation*, wherein only ATP is produced. The normal mode of operation is noncyclic flow, but there is evidence that chloroplasts can shift to a cyclic mechanism under certain conditions. (FD represents ferredoxin.)

cule becomes ionized by virtue of being the recipient of energy from a "bucket-brigade–type" of transfer mechanism involving other molecules of chlorophyll *a*, chlorophyll *b*, and possibly the accessory pigments. In other words, the specialized *a* molecule does not absorb light directly. The energy transfer mechanism is probably the least understood aspect of photosynthesis

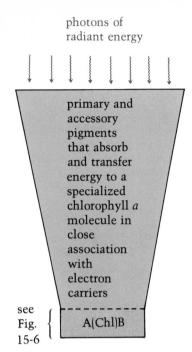

photons of
radiant energy

primary and accessory pigments that absorb and transfer energy to a specialized chlorophyll *a* molecule in close association with electron carriers

see Fig. 15-6 { A(Chl)B

the intention of the diagram is to represent the proposed funneling of energy to the reaction center of each photosystem

$$\mathscr{E}^{o\prime}_{red} \text{ (volts)}$$

$FD_{ox} + 1e \rightarrow FD_{red}$ $\overline{-0.43\ V}$
oxidized   reduced
ferredoxin   ferredoxin

$NADP^+ + 2H^+ + 2e \rightarrow$

$NADPH + H^+$ $-0.32\ V$

and certainly the most remarkable. For one thing it is estimated that it occurs in about 10^{-12} to 10^{-15} seconds. The close association with the electron carriers (A, A′, B, and B′) that permits direct reactions with them is, of course, another unique feature of these specialized molecules.

In recent years, many researchers have claimed that they have identified the electron carriers. At one time, A in PSII was thought to be a *plastoquinone* (PQ) molecule, but the current feeling is that plastoquinone is the first acceptor of electrons from A^-_{red}, whose identity is still unknown. Others have reported evidence that B is a specialized cytochrome carrier whose immediate electron donor in the electron transport chain is *plastocyanine* (PC), a copper-containing protein, which in turn acts as an acceptor of electrons from *cytochrome f*. Another specialized cytochrome, *cyt b* (559) is suggested as likewise operating as a carrier between plastoquinone and cytochrome *f*. The scheme in Fig. 15–6 and its specific features constitute, of course, only a working model of the light reaction. This scheme is widely accepted as representative of what is occurring *in vivo*, because it agrees with most experimental studies made with illuminated chloroplasts and thylakoid preparations. It is, of course, a tremendous oversimplification.

Regardless of the many uncertainties that still exist, such as the nature of A, B, A′, and B′; the location and number of ATP-generating sites; and the possible existence of other electron carriers, there is universal agreement that electron transfer among specific carriers is an integral part of the light reaction. Moreover, in organisms characterized by the operation of two separate photosystems, and presumably this includes virtually all photosynthesizing cells except the photosynthetic bacteria, it should occur to you that, if the scheme in Fig. 15–6 is correct, the electron shuttle system between A^-_{red} and B'^+_{ox} is a crucial step of the light reaction, since it constitutes the link between photooxidation and photoreduction in addition to serving as the driving force for the formation of ATP, that is, photophosphorylation.

To close our discussion of the dual-photosystem model, let us take note of the participation of the substance symbolized as FD in Fig. 15–6, a substance proposed to be the immediate electron acceptor of A'^-_{red} and the immediate electron donor of $NADP^+$ in the photoreduction phase. The FD notation is an abbreviation for *ferredoxin*, an iron-sulfur nonheme protein known to be present in all types of photosynthetic organisms, including the photosynthetic bacteria. In fact, recent studies have shown that the ferredoxins isolated from several different sources are remarkably similar in structure. Regardless of the source, ferredoxin has two distinguishing properties in that (1) it can undergo a reversible oxidation-reduction via electron transfer, and (2) the FD_{ox}-FD_{red} couple has a reduction potential even more negative than that of the $NADP^+$-$NADPH$ couple (see margin and Table 14–1, p. 449). The latter property is one of the strong arguments for the proposed role of FD_{red} as the immediate donor of electrons to $NADP^+$, since the transfer would be energetically favorable.

Although the scheme depicted in Fig. 15–6 is based on an enormous amount of research, it must be emphasized that it is only a working model and an incomplete one at that. No doubt, as work continues, the picture will become more definitive and perhaps even modified.

Introduction

A brief and convenient statement of what happens in the dark reaction is given by the following equation:

$$6CO_2 \xrightarrow[\text{NADPH ATP}]{\substack{\text{chloroplast} \\ \text{enzymes}}} C_6H_{12}O_6$$

The process occurs in the chloroplast, involves many enzyme-catalyzed reactions, requires NADPH and ATP, and results in the formation of carbohydrate material ($C_6H_{12}O_6$) from inorganic CO_2. Although the equation is quite useful in summarizing the overall effect, it can be misleading, since it implies that (a) the carbon atoms of each of six separate molecules of CO_2 become part of the same hexose molecule, and (b) the dark reaction is limited to this one type of metabolic activity. Actually, neither of these implications is correct. Regarding the former, we will shortly discover that the relationship of $6CO_2 \rightarrow C_6H_{12}O_6$ represents only a *net* conversion rather than an actual synthesis of a single hexose molecule from six molecules of CO_2. As for the latter, it should occur to you that the complete enzymatic metabolism of a plant cell must consist of more than a fixation and conversion of CO_2 to carbohydrates. While a significant portion of the assimilated carbon will be stored as sucrose and utilized for the biosynthesis of cellulose, the remainder, by being channeled into central metabolic pathways such as the citric acid cycle, will serve as a chemical source of energy and carbon for the anabolism of other carbohydrates, amino acids, proteins, fatty acids, lipids, purine and pyrimidine nucleotides, nucleic acids, and even the tetrapyrrole moiety for chlorophyll itself. In other words, the original carbon of CO_2 is ultimately incorporated into the entire metabolism of the whole organism.

Experimental History

Investigations into the mechanisms of photosynthetic CO_2 fixation began in earnest only about 35 years ago. The chief pioneering researchers were M. Calvin, A. A. Benson, and J. A. Bassham. Although it had been known since the last century that CO_2 was converted into carbohydrate, there was lively debate as to whether the process was directly dependent on the absorption of light-energy. In this sense, there was considerable significance to Hill's discovery that the light-dependent phase of photosynthesis did not require the presence of CO_2, but instead was characterized by the photooxidation of water to O_2, accompanied by the reduction of $NADP^+$ to NADPH. The nicotinamide adenine dinucleotides were then known to be obligatory coenzymes for dehydrogenase enzymes; this suggested that the reducing power produced as the result of light absorption was made available for the enzymatic conversion of CO_2 to carbohydrate (see margin).

The most fundamental questions regarding the flow of carbon in photosynthesis—indeed, the very questions to which the Calvin group addressed itself—are as follows: (a) What is the identity of the substance that acts as the initial acceptor of CO_2? (b) What is the immediate product that appears after CO_2 fixation? (c) How is the immediate product then converted to simple sugars?

Although it has become customary to refer to the fixation of CO_2 as the dark reaction, this does not mean that the reactions involved do not occur in the presence of light—they do. When illumination ceases, CO_2 fixation will continue to occur for a brief period until the levels of ATP and NADPH become limiting.

489

In attacking these problems, Calvin, Benson, and Bassham studied the flow of carbon in single-celled algae of the genera *Chlorella* and *Scendesmus*. *Chlorella* is particularly useful for photosynthetic studies since it can be easily and reproducibly cultured. The key and indispensable feature of their experiments was the utilization of radioactive CO_2 ($^{14}CO_2$) as the carbon source. Their studies are classic examples of the use of radioactive tracer substances in the elucidation of cellular metabolism (see p. 56, Chap. 2). The overall experimental design was simple but ingenious. $^{14}CO_2$ was injected into illuminated glass tubes through which an algae suspension was flowing. After exposure to $^{14}CO_2$, the suspension was run into hot alcohol. On contact with the alcohol, all enzymatic reactions within the cells were brought to a halt. By adjusting the time between the injection of $^{14}CO_2$ and the final mixing with alcohol, it was possible to limit exposure of the cells to the carbon source to any desired interval, from a few minutes to a fraction of a second. Afterwards, samples of the alcohol solution, which contained dissolved compounds extracted from the cell, were analyzed chromatographically for the appearance of ^{14}C-labeled compounds.

As you might expect, with prolonged exposure times (10 minutes), the extract contained a tremendous assortment of ^{14}C-labeled compounds including many simple carbohydrates (mostly phosphorylated sugars), several amino acids, all of the major nucleotides, and others. Shorter exposure intervals (30 seconds) considerably reduced the number of labeled materials, with virtually all of the ^{14}C being found in a restricted number of phosphorylated carbohydrates. Those identified were trioses (dihydroxyacetone phosphate, glyceraldehyde-3-phosphate, and 3-phosphoglycerate); a tetrose (erythrose-4-phosphate); pentoses (ribose-5-phosphate, ribulose-5-phosphate, ribulose-1,5-diphosphate, and xylulose-5-phosphate); hexoses (fructose-1,6-diphosphate, fructose-6-phosphate, and glucose-6-phosphate); and heptoses (sedoheptulose-7-phosphate, and sedoheptulose-1,7-diphosphate).

When the exposure period was reduced even further, to only a fraction of a second, the extract was found to contain only a single compound with any appreciable ^{14}C content: *3-phosphoglycerate*. Although the initial studies were done with algae, there is now considerable evidence that the initial formation of 3-phosphoglycerate and its subsequent metabolism constitute the

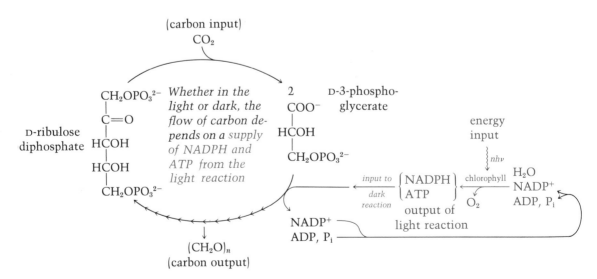

primary metabolic pathway utilized by most green plants as well. Plants utilizing this pathway are called *three-carbon plants*, or simply *C3 plants*, because the immediate product of CO_2 fixation consists of three carbons. Recent studies have uncovered the existence of at least one major alternate pathway in some plants, which we will examine later in the chapter (see p. 495). Photosynthetic bacteria use alternate routes as well.

Although the appearance of 3-phosphoglyerate as the primary product of CO_2 assimilation suggested that the initial acceptor was probably a C_2 compound, Calvin and coworkers subsequently demonstrated that the acceptor molecule was *ribulose-1,5-diphosphate*, a C_5 compound. In other words, the initial flow of carbon was $C_5 + CO_2 \rightarrow 2C_3$. Moreover, the flow of carbon during periods of light and dark indicated that the acceptor $\rightarrow$ product relationship was *cyclic* in nature, with cellular levels of labeled diphosphate decreasing in the dark and levels of phosphoglycerate increasing. The label pattern during dark periods thus suggested that the subsequent metabolism of 3-phosphoglycerate, particularly the regeneration of ribulose-1,5-diphosphate, was limited by factor(s) supplied only during periods of illumination. These factors are now known to be NADPH and ATP. After several years of study, the Calvin group eventually proposed a scheme identifying all the steps in getting from 3-phosphoglycerate to ribulose-1,5-diphosphate. The task was extremely difficult, because the constituent reactions do not occur in a linear sequence but in a complicated, highly branched fashion. For traditional reasons, the sequence is frequently called the *Calvin-Benson-Bassham cycle.*

Primary Path of Carbon in Photosynthesis

The assimilation of CO_2 into carbohydrate material by photosynthetic cells according to the Calvin-Benson-Bassham scheme is shown in Fig. 15–7. At first glance, the composite diagram may strike you as being an incomprehensible maze of structures and arrows. Despite its complexity, don't let it bewilder you, because a distinct metabolic pattern does exist. The first reaction, catalyzed by *ribulose-1,5-diphosphate carboxylase,* is the carboxylation of ribulose-1,5-diphosphate to yield an unstable six-carbon intermediate, which is then cleaved to give two units of 3-phosphoglycerate (reaction (a)). This immediate formation of 3-phosphoglycerate from the CO_2 fixation step is, of course, consistent with the results of the radioactive feeding experiments described earlier. The crucial nature of this reaction is evidenced by the fact that the carboxylase enzyme is known to occur in very large amounts within the chloroplasts, accounting for roughly one-sixth of all the soluble protein in the lumen.

The immediate metabolic fate of 3-phosphoglycerate (in the chloroplast) is a two-step conversion to glyceraldehyde-3-phosphate, involving two enzymes already encountered in glycolysis (phosphoglycerate kinase and triosephosphate dehydrogenase), which function here in reverse. The former requires ATP and the latter NADPH, both of which are supplied for the light reaction. As you can see, the subsequent metabolism of glyceraldehyde-3-phosphate diverges through four distinct *branches* (reactions (d), (e), (g), (j)). Note that the chemistry of these and all subsequent steps involves only the modification and interconversion of various phosphorylated sugars. These reactions include isomerizations, an epimerization, transfers between aldoses and ketoses, dephosphorylations, and a phosphorylation.

Some aspects of the structure and function of the carboxylase enzyme are described later (p. 494).

491

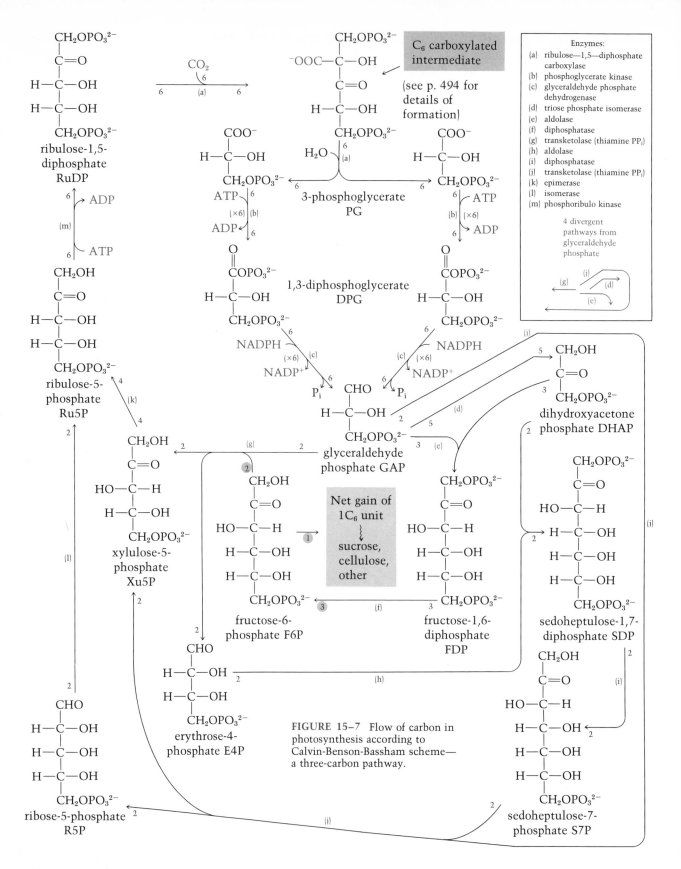

FIGURE 15–7 Flow of carbon in photosynthesis according to Calvin-Benson-Bassham scheme— a three-carbon pathway.

Note also that each type of interconversion was encountered in our earlier discussion of glycolysis and the hexose monophosphate shunt pathway. The characteristic steps of this nonoxidative phase involve the *aldolase* and *transketolase* enzymes.

Although each of the four branches of glyceraldehyde-3-phosphate metabolism is important, the isomerization to (reaction (d)) and condensation with (reaction (e)) dihydroxyacetone phosphate are especially crucial, since they account for the incorporation of the C of the original CO_2 into hexose units. Ultimately, all of these conversions (reactions (c) through (k)) converge on the production of ribulose-5-phosphate. The ribulose-5-phosphate is then phosphorylated in an ATP-dependent kinase-catalyzed reaction to yield the original diphosphate, thus closing the cycle. As in the formation of glyceraldehyde-3-phosphate, the ATP required in this step is supplied from the light reaction.

Since the Calvin-Benson-Bassham cycle is not composed of individual reactions acting in sequence, the metabolic significance of the flow of carbon through the many divergent and convergent steps can best be considered by developing an idealized analysis that utilizes a nonunit stoichiometry. A convenient approach is to begin with the assimilation of six CO_2 units by six units of ribulose diphosphate. Then, by distributing the further metabolism of 12 units of glyceraldehyde-3-phosphate, as indicated in Fig. 15–7, all of the component enzymatic steps of the pathway can be written separately, as shown below. When the net result is tabulated, the overall chemical effect becomes obvious. The *equivalent* of one hexose molecule is produced from six molecules of CO_2, with two molecules of NADPH and three molecules of ATP being required for every CO_2 molecule that is assimilated.

$$6RuDP + 6CO_2 \xrightarrow{(a)} 6[\text{INTERMEDIATE}] \xrightarrow{(a)} 12PG$$
$$12PG + 12ATP \xrightarrow{(b)} 12DPG + 12ADP$$
$$12DPG + 12NADPH(H^+) \xrightarrow{(c)} 12GAP + 12NADP^+ + 12P_i$$
$$5GAP \xrightarrow{(d)} 5DHAP$$
$$3DHAP + 3GAP \xrightarrow{(e)} 3FDP$$
$$3FDP \xrightarrow{(f)} 3F6P + 3P_i$$
$$2F6P + 2GAP \xrightarrow{(g)} 2Xu5P + 2E4P$$
$$2E4P + 2DHAP \xrightarrow{(h)} 2SDP$$
$$2SDP \xrightarrow{(i)} 2S7P + 2P_i$$
$$2S7P + 2GAP \xrightarrow{(j)} 2Xu5P + 2R5P$$
$$4Xu5P \xrightarrow{(k)} 4Ru5P$$
$$2R5P \xrightarrow{(l)} 2Ru5P$$
$$6Ru5P + 6ATP \xrightarrow{(m)} 6RuDP + 6ADP$$

Net: $6CO_2 + 12NADPH(H^+) + 18ATP \xrightarrow[\text{steps}]{13}$

$$\text{fructose-6-phosphate(F6P)} + 12NADP^+ + 18ADP + 17P_i$$

Subsequently:

fructose-6-P $\longrightarrow$ glucose-6-P $\rightarrow\!\rightarrow\!\rightarrow$ sucrose, cellulose, and general metabolism
via glycolysis and citric acid cycle

Ribulose-1,5-diphosphate Carboxylase

The most distinctive enzyme of the Calvin cycle is ribulose-1,5-diphosphate carboxylase (RuDPCase). It is present in large quantities in chloroplasts and is probably the most abundant protein in our biosphere.

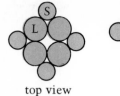

top view
of L_4S_4
layer

side view
of L_4S_4
layer

side view
of L_8S_8
two-layered
molecule
of RuDPCase

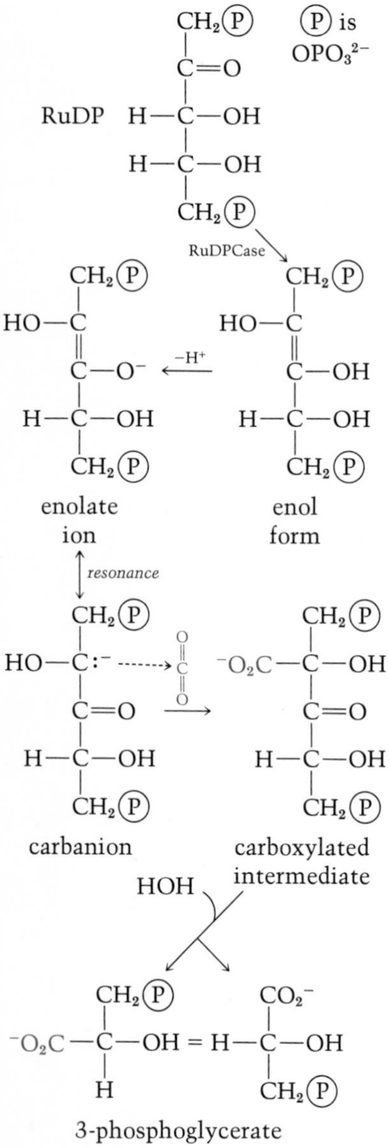

CH_2Ⓟ   Ⓟ is
| OPO_3^{2-}
$C=O$
|
RuDP $H—C—OH$
|
$H—C—OH$
|
CH_2Ⓟ

↓ RuDPCase

CH_2Ⓟ CH_2Ⓟ
| |
$HO—C$ $HO—C$
‖   ←$-H^+$ |
$C—O^-$ $C—OH$
| |
$H—C—OH$ $H—C—OH$
| |
CH_2Ⓟ CH_2Ⓟ

enolate   enol
ion   form

↑ *resonance*

CH_2Ⓟ CH_2Ⓟ
| |
$HO—C:^-$ ······ ⟨C=O O⟩ → $^-O_2C—C—OH$
| |
$C=O$ $C=O$
| |
$H—C—OH$ $H—C—OH$
| |
CH_2Ⓟ CH_2Ⓟ

carbanion   carboxylated
                     intermediate

HOH ╲

CH_2Ⓟ CO_2^-
| |
$^-O_2C—C—OH$ = $H—C—OH$
| |
H CH_2Ⓟ

3-phosphoglycerate

Structure and mechanism of action. The carboxylase molecule (mol. wt. = 560,000) is composed of eight large subunits (L) and eight small subunits (S), that is, its formula is L_8S_8. As implied by the molecular formula, each carboxylase molecule has eight active sites. Recent X-ray diffraction studies show the molecule to be double-layered, with each layer composed of L_4S_4. The two layers appear to be nearly eclipsed (see the idealized drawings in the margin).

The proposed mechanism of action of RuDPCase (see margin), which leads to the formation of a six-carbon intermediate, is based on the fact that the active site of the enzyme promotes the ionization of the enol form of the diphosphate, yielding an enolate ion that can exist as an active carbanion species. After carboxylation, hydrolysis would yield two 3-phosphoglycerate fragments.

Regulation. The action of RuDPCase is inhibited by fructose-1,6-diphosphate (FDP) and activated by fructose-6-phosphate (F6P). These effects are both related to the *activation of the Calvin cycle by light.* As shown in the diagram below, the light activation is mediated by the production of reduced ferredoxin, which in turn activates the diphosphatase, which converts FDP to F6P, the activator of RuDPCase. In other words, when illuminated chloroplasts begin to produce NADPH and ATP, the light also activates the Calvin cycle that uses them. Then, when there is no light, the diphosphatase is rendered inactive, leading to a buildup of FDP, which in turn would deactivate the RuDPCase, thereby diminishing CO_2 fixation. These effects are a remarkable illustration of the fitness and logic of metabolic regulation.

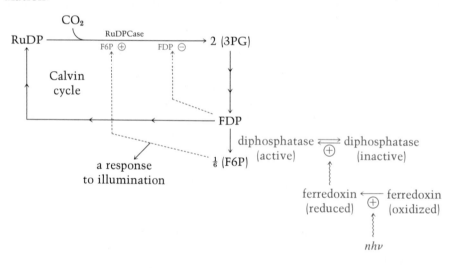

Photorespiration

For reasons not yet understood, most plants are also capable of consuming O_2 in periods of illumination. The process is termed *photorespiration* and differs from mitochondrial respiration, which also occurs in plants. Photorespiration involves O_2 consumption by the chloroplast enzyme *glycolate oxidase,* which converts glycolic acid to glyoxylic acid. The origin of the glycolic acid appears to be ribulose-1,5-diphosphate, with the RuDPCase using O_2 instead of CO_2 to produce a hydroperoxide intermediate. The use of ribulose-1,5-diphosphate for this purpose appears to have no significant

enolate ion of RuDP → (RuDPCase) → hydroperoxide intermediate → H—C—OH + phospho-glycolate + 3PG → (phosphatase) → P_i

glycolic acid (glycolate) → (glycolate oxidase, O_2, H_2O_2) → glyoxylic acid (glyoxylate)

benefit to plants. On the other hand, photorespiration is clearly in competition with the Calvin cycle since it scavenges the diphosphate acceptor of CO_2. Indeed, the occurrence of photorespiration does in fact lower the efficiency of CO_2 fixation. In view of this wasteful competition, it would be highly desirable to discover a way of selectively inhibiting photorespiration, an action that could result in increased crop yields. Obviously, the target enzyme of this control would be glycolate oxidase.

Alternate Pathway of CO_2 Fixation

Although the Calvin-Benson-Bassham sequence is definitely the *major* pathway of CO_2 fixation, it is *not the only* pathway. In 1970 M. Hatch and C. Slack discovered that certain plants are also capable of fixing CO_2 by a different route, as indicated by the rapid appearance of the ^{14}C from $^{14}CO_2$ in four-carbon products such as *malate and aspartate* rather than 3-phosphoglycerate. ^{14}C is eventually detected in phosphoglycerate and phosphorylated sugars, but only after a longer time of incubation. This new route is called the *Hatch-Slack pathway* or the *four-carbon (C4) pathway*.

The explanation of this alternate pathway involves a new CO_2-fixation step, the key feature of which is that the primary acceptor of CO_2 is *phosphoenolpyruvate* (PEP) rather than ribulose-1,5-diphosphate, and that the immediate product is *oxaloacetate* (OAA) rather than 3-phosphoglycerate. Strong evidence in support of this alternate carboxylation step was obtained with the isolation from the chloroplasts of these plants of *phosphoenolpyruvate carboxylase*, the enzyme that catalyzes the reaction. This reaction would, of course, account for the early appearance of both malate and aspartate, both of which could be produced directly from oxaloacetate.

Depending on the type of C4 plant, the immediate fate of oxaloacetate is varied. We will consider just one route in what are termed NADP-ME–type C4 plants. Here the OAA is first reduced to malate by an NADP-dependent malate dehydrogenase and the malate is then decarboxylated by *malic enzyme* (ME) to yield pyruvate and CO_2. The green leaf cells of these plants contain another enzyme, *pyruvate:phosphate dikinase*, capable of catalyzing the conversion of pyruvate to phosphoenolpyruvate, thus *closing the cycle*.

The four-carbon pathway has been demonstrated in about 100 different plants and grasses that have one thing in common—they thrive in hot and arid environments. Crabgrass, Bermuda grass, and many tropical plants such as sugar cane are just a few examples of these plants. It would seem then that the adaptability of plants to such conditions is probably related to the four-carbon route. At first glance the logic of this association appears remote, because the net effect of the four-carbon pathway is just

$$H_2C=C-COO^-$$
$$OPO_3^{2-}$$
phosphoenolpyruvate

CO_2 → (phosphoenolpyruvate carboxylase) → P_i

$$^-OOCCH_2CCOO^-$$
oxaloacetate

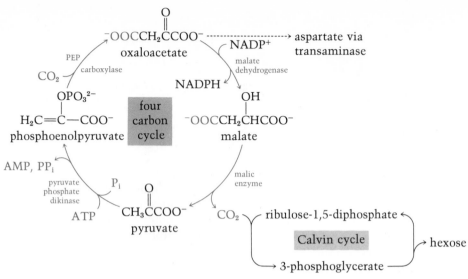

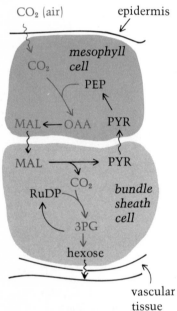

The anatomy of C4 plants is also unique in that they have a double layer of cells between the epidermis and the vascular tissue. *Mesophyll cells* comprise the outer layer and *bundle sheath cells* comprise the inner layer. PEP carboxylase is located in the chloroplasts of the mesophyll cells whereas RuDP carboxylase is located in the chloroplasts of the bundle sheath cells. Thus, the C4 plant anatomy is well-designed for *concentrating the CO₂ level* for RuDP carboxylase.

to fix CO_2 and then release it again to ribulose-1,5-diphosphate, which then enters the Calvin cycle. Why not just fix CO_2 with ribulose-5-phosphate in the first place?

Well, the answer to this riddle of phosphosynthetic metabolism is that the four-carbon pathway increases the efficiency of CO_2 fixation (in the plants capable of it) in regions of maximum solar radiation, high temperatures, and a limited supply of water. In full sunlight, the leaves of C4 plants generally fix CO_2 at about twice the rate of plants that use only the C3 pathway. The explanation proposed for this increased efficiency is rather complicated and we will not go into it in any great depth.

One argument is based on the ability of the pyruvate carboxylase enzyme to react with lower levels of CO_2 than required by the ribulose-1,5-diphosphate carboxylase. By being able to operate in this condition, the pores of the leaf through which the CO_2 enters the plant need only open just enough to maintain these low levels. *The CO_2 would then be concentrated* for use by the RuDPCase. The smaller opening of the surface pores has a bonus effect—*less water vapor leaves the leaf.* In other words, the four-carbon plant can make better use of its water. It can fix CO_2 without being dehydrated when the water supply is low. Another reason for the increased efficiency of CO_2 fixation in C4 plants is that they *do not exhibit photorespiration.*

Because of these special attributes that four-carbon plants have, researchers are attempting to hybridize them with three-carbon plants (Calvin cycle only) in order to increase the growth efficiency of the latter in climates where the three-carbon plants are difficult and expensive to grow—that is, where there is lots of sun and a limited water supply. Perhaps recombinant DNA technology may aid in developing such plants. Successes would have a revolutionary effect on agricultural practices and also contribute significantly to solving the world food problem.

PHOTOSYNTHESIS—A SUMMARY

Because of the complex biochemistry of photosynthesis, let us close our discussion of this subject by briefly reviewing the major events (see diagram below). Photosynthesis consists of two separate but related processes, the

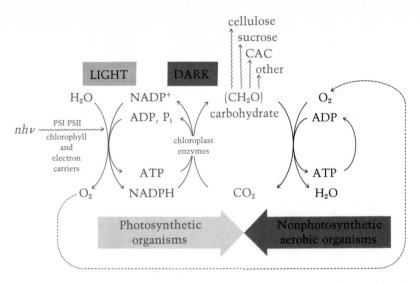

cellulose
↑ sucrose
↑ CAC
↑↑ other

Regarding the origin of life: it is proposed that photosynthetic organisms evolved much earlier than nonphotosynthetic organisms—the emergence of the latter being linked to the O₂ enrichment of the atmosphere by photosynthetic organisms.

so-called light reaction and dark reaction, with the latter being absolutely dependent on the former. In the light reaction, solar light energy is harnessed by chlorophyll-containing photosystems compartmentalized within the lamellae membranes, resulting in the formation of metabolically useful reducing power (as NADPH) and metabolically useful energy (as ATP). The ultimate source of the reducing power is H_2O. In the dark reaction, occurring in the lumen of the chloroplasts, the ATP and NADPH are utilized in the enzymatic assimilation of CO_2 into organic material, with the major acceptor molecule being ribulose-1,5-diphosphate. The subsequent oxidation of the photosynthetically produced carbohydrates by aerobic, nonphotosynthetic organisms completes the major biochemical relationship in our biosphere.

LITERATURE

ARNON, D. I., H. Y. TSUJIMOTO, and B. D. MCSWAIN. "Ferredoxin and Photosynthetic Phosphorylation." *Science*, **214**, 562–566 (1967). Evidence that ferredoxin is associated with ATP formation in photosynthesis.

BJORKMAN, O., and J. BERRY. "High-Efficiency Photosynthesis." *Scientific American*, **229**, 80 (October, 1973). The four-carbon pathway is described.

CHOLLET, R. "The Biochemistry of Photorespiration." *Trends Biochem. Sci.*, **2**, 155–159 (1977). A short review article.

GOVINDJEE, and R. GOVINDJEE. "The Primary Events of Photosynthesis." *Scientific American*, **231**, 68 (1974). The absorption of radiation by chlorophyll and accessory pigments is discussed.

HATCH, M. D. "C₄ Pathway Photosynthesis: Mechanism and Physiological Function." *Trends Biochem. Sci.*, **2**, 199–202 (1977). A short review article.

HILL, R. "The Biochemist's Green Mansion: The Photosynthetic Electron Transport Chain in Plants," In *Essays in Biochemistry*, P. N. Campbell and G. D. Greville, eds., Vol. 1, pp. 121–152. New York: Academic Press, 1965. A review article with historical perspective of the light reaction by one of the pioneering researchers in the field.

KAMEN, M. D. *Primary Processes in Photosynthesis.* New York: Academic Press, 1963. A short paperback offering a thorough introductory treatment of the light reaction of photosynthesis.

LEVINE, R. P. "The Mechanism of Photosynthesis." *Scientific American*, **221**, 58–70 (1969). A good summary of the important features of the light reaction of photosynthesis, some of which are not treated in this chapter.

MACHLIS, L., ed. *Annual Review of Plant Physiology.* Palo Alto: Annual Reviews, Inc. An annual publication containing review articles of current developments concerning various aspects of plant physiology and biochemistry.

RABINOWITCH, E. I., and GOVINDJEE. "The Role of Chlorophyll in Photosynthesis." *Scientific American*, **213**, 74–83 (1965). A description of experiments that resulted in the suggestion of two separate photosystems.

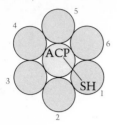

1 = acyl transacylase
2 = β-ketoacyl synthetase
3 = malonyl transacylase
4 = β-ketoacyl reductase
5 = β-hydroxyacyl dehydratase
6 = α,β-enoyl reductase
ACP = acyl carrier protein

CHAPTER 16

LIPID METABOLISM

Lipid metabolism is an extensive subject. There are a large number of structurally distinct lipids, and each could be described in terms of its unique anabolic and catabolic pathways. A consideration of all these is not our objective. We will explore only certain facets, with particular emphasis given to major pathways of *fatty acid metabolism.* In view of the fact that fatty acids are the chief structural components of the simple and compound lipids, the emphasis is a logical one. Moreover, when we speak of an organism utilizing and storing lipid material as a source of carbon and energy, the metabolic facts involve the catabolism (utilization) and anabolism (storage) of fatty acids. In other words, a large segment of lipid metabolism is synonymous with the metabolism of fatty acids.

Although the emphasis will be on fatty acids, we will also investigate other aspects of lipid metabolism, such as (a) the *biosynthesis of prostaglandins,* (b) the *biosynthesis of phosphoacylglycerols,* (c) the *biosynthesis of cholesterol* and other derived lipids, and (d) the ability of many organisms to *convert lipids to carbohydrates.* In addition to some practical implications regarding body weight control, the last subject will provide an opportunity for a brief but informative reinvestigation of the principle of metabolic integration, in this instance with a modified form of the citric acid cycle, the *glyoxylate cycle.*

There is more to be gained from this study than knowledge of a new set of pathways. This chapter also provides a clear example of the general principles that differentiate catabolic and anabolic pathways (see p. 367), another opportunity to analyze the logic of metabolic regulation, another example of multienzyme complexes, and other examples of enzyme stereospecificity. In addition, the coenzyme participation of *biotin* in carboxylation reactions will be discussed.

CATABOLISM OF FATTY ACIDS

Experimental History

The first serious attempt to study the catabolism of fatty acids was made at the turn of the century by F. Knoop, a German biochemist, who proposed that *fatty acids were oxidatively degraded by sequential removal of two-carbon units, proceeding from the carboxyl end of the molecule.* His experimental design was ingenious. Not having access to radioisotopes, he provided a chemical label by synthesizing a series of straight-chain carboxylic acids of even and odd chain length, each containing a *phenyl* grouping at the noncarboxyl end of the molecule. These acids were then fed to experimental animals, after which the urine was collected and assayed to determine the fate of the phenyl group label. The use of the phenyl grouping was important, because short-chain phenyl-substituted acids are not metabolized by animals, but excreted as waste products. The results were as follows. With even-numbered acids *phenylacetate* accumulated, whereas with odd-numbered acids *benzoate* accumulated. Both observations fit the proposed pattern outlined above. Knoop termed the process *β-oxidation*, signifying that the oxidative cleavage occurred at the C^β-C^α bond in the original acid and in each subsequent shorter acid. Acetate would be produced in each chain-shortening step.

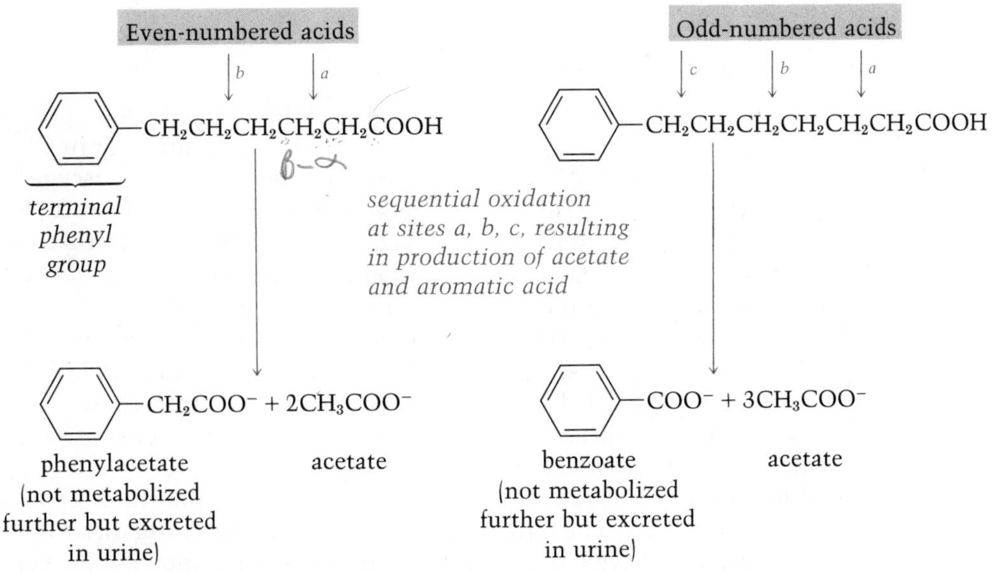

At present, the process of β oxidation is recognized as the primary catabolic pathway of fatty acids in all organisms. While the modern version of β-oxidation incorporates the essential principles of Knoop's original pro-

All of the intermediates and their metabolic sequences have been identified, and all of the enzymes involved have been isolated and studied in purified form.

posal, it was not fully developed until nearly 50 years after his pioneering studies. The key discoveries were that (a) the process occurs *within the mitochondria*, (b) the process is *ATP-dependent*, and (c) the fatty acids as well as all intermediates are processed as *thioesters of coenzyme A* (i.e., CoASH) rather than as free acids.

Cellular Source of Fatty Acids

Obviously, the catabolism of fatty acids presupposes their presence within the cell. However, as pointed out in Chapter 10, free fatty acids do not occur in nature in any significant amounts. They are found for the most part in ester linkage, primarily as acyglycerols and phosphoacylglycerols (see p. 317). Acylglycerols represent the primary endogenous storage form of fatty acids; phosphoacyglycerols are the essential components of biological membranes. The process by which all cells degrade both lipid classes is quite direct, involving the action of two specialized groups of enzymes called *lipases* and *phospholipases*. The only point we wish to make regarding these enzymes is summarized below. Their action is essentially hydrolytic, resulting in the formation of the free fatty acids, which can be reutilized in anabolic pathways or degraded via β-oxidation. Our concern here is with the latter process.

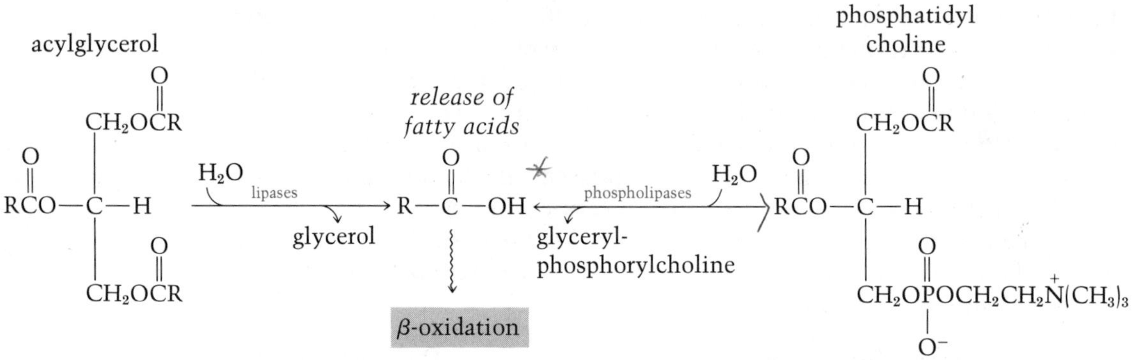

Activation and Mitochondrial Entry of Fatty Acids

Activation. Fatty acids must be converted to thioesters of CoASH prior to their participation in any type of cellular reaction. As pointed out in Chap. 11, the acyl-SCoA metabolites are *high-energy compounds* more susceptible to reaction than the free acids or oxoesters; that is, they are activated. Particular increases in reactivity are observed in (a) reactions involving a condensation at the carbonyl carbon and (b) elimination or addition reactions involving the α and β carbons of the acyl group (see p. 358).

The most widely distributed activation process involves an ATP-dependent *fatty acid thiokinase* (see below). The overall reaction is a classical example of a coupled reaction involving an actual transfer of energy (see p. 362). At least three different thiokinases, varying in substrate specificity, are known to exist in nature. One is highly specific for acetate (C_2), a second for acids of medium chain length (C_4-C_{12}), and the third for long-chain acids $(C_{14}-C_{22})$. The latter two act on both saturated and unsaturated acids. Regardless of type, the activating thiokinases are known to be particulate enzymes localized in cellular membranes. In higher organisms specifically, it is proposed that they are found in the outer mitochondrial membrane.

Entry. The acyl-SCoA species is capable of crossing the outer mitochon-

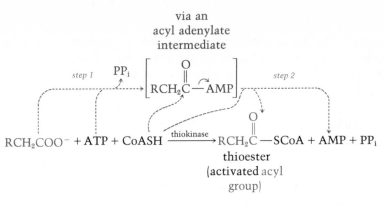

via an
acyl adenylate
intermediate

$$RCH_2COO^- + ATP + CoASH \xrightarrow{\text{thiokinase}} RCH_2\overset{O}{\overset{\|}{C}} -SCoA + AMP + PP_i$$

thioester
(activated acyl
group)

drial membrane, but it does not cross the inner mitochondrial membrane. This impermeability barrier is overcome by the participation of *carnitine*, which serves as an *acyl group carrier*. Two proteins are involved in this operation—*carnitine acyltransferase* (an enzyme) and *carnitine:acylcarnitine translocase* (a transport protein). Both proteins are localized in the inner mitochondrial membrane.

In the intramembranous space the acyltransferase catalyzes the formation of an *acyl-carnitine ester*. The translocase protein then facilitates (in a manner not yet understood) the passage of the acyl-carnitine ester to the other side of the membrane, releasing it into the matrix. Acyltransferase action on the matrix side of the membrane (the enzyme appears to be present on both sides) reforms acyl-SCoA and free carnitine. The return of the free carnitine to the intramembranous space, which completes the transport cycle, may be facilitated by the same translocase protein.

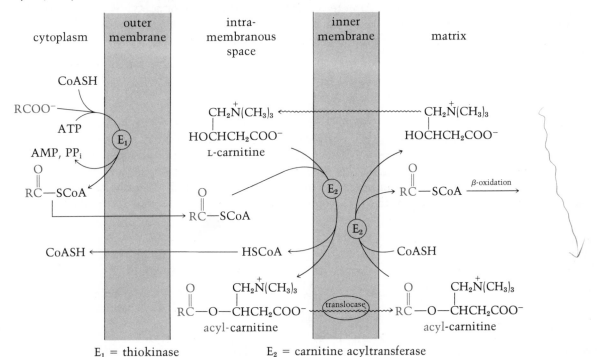

E_1 = thiokinase E_2 = carnitine acyltransferase

excluding thiokinase involvement, the net reaction is

$$R\overset{O}{\overset{\|}{C}} -SCoA_{\text{cytoplasm}} \longrightarrow R\overset{O}{\overset{\|}{C}} -SCoA_{\text{mitochondrion}}$$

Sequence of β-Oxidation

Once inside the mitochondrion, acyl-SCoA compounds are degraded through the action of four enzymes (Fig. 16–1). The chemistry of this set of reactions is straightforward, and follows these steps:

(A) *Elimination of hydrogen* (dehydrogenation) to yield an α,β unsaturated acyl-SCoA.
(B) *Hydration* to yield a β-hydroxyacyl-SCoA.
(C) *Oxidation* (dehydrogenation) to yield a β-ketoacyl-SCoA.
(D) *Thiolytic cleavage* to yield acetyl-SCoA and a second acyl-SCoA now shortened by a two-carbon unit.
(E) Recycling of the shortened acyl-SCoA through steps (A) through (D).

Note that, although the oxidative steps (A) and (C) are catalyzed by dehydrogenases, the first is FAD-dependent and the second is NAD⁺-dependent. Both steps represent sites of conservation of energy that is ultimately utilized in ATP formation (see below). The shortened acyl-SCoA could then go through the same reaction sequence, generating a second acetyl-SCoA unit and another shortened acyl-SCoA, which would be recycled for still another pass. This cyclic pattern of β-oxidation would continue through the formation of the four-carbon β-keto metabolite, *acetoacetyl-SCoA*

$(CH_3-\overset{\overset{\displaystyle O}{\|}}{C}-CH_2-\overset{\overset{\displaystyle O}{\|}}{C}-SCoA)$. Thiolytic cleavage of this material would

yield two units of $CH_3-\overset{\overset{\displaystyle O}{\|}}{C}-SCoA$ and thus complete the process. As indicated in Fig. 16–1, with stearyl-SCoA as a starting material the overall effect would be the complete conversion to nine units of acetyl-SCoA. All of the enzymes have been isolated in pure form. Note the stereospecificities of the enzymes that apply to both product formation and preferred substrate.

Odd-numbered acids. β-Oxidation of a fatty acid having an odd number of carbon atoms would yield several acetyl-SCoA fragments and one fragment

of $CH_3CH_2\overset{\overset{\displaystyle O}{\|}}{C}-SCoA$, propionyl-SCoA. (Right?) The subsequent metabolic fate of the propionyl-SCoA is entrance into the citric acid cycle in the form of succinyl-SCoA according to the following actions, shown in the scheme below: (1) an ATP-dependent carboxylation; (2) an epimerization; and (3) an intramolecular migration. As indicated, the enzyme (described in Chap. 17) in the last step participates in some interesting chemistry.

In animals some acetoacetyl-SCoA is normally hydrolyzed to the free acid, *acetoacetate*, which in turn is reduced to *β-hydroxybutyrate*. Abnormally high blood levels of these two acids cause a lowering of the pH of blood, a condition termed *acidosis*. Diabetes, lipid-rich diets, and starvation are circumstances that can result in acidosis. Persistance of the condition can cause coma and death.

carboxylation

$$CH_3CH_2\overset{\overset{\displaystyle O}{\|}}{\underset{\alpha}{C}}-SCoA \xrightarrow[\underset{CO_2 \quad ATP \quad \underset{P_i}{ADP}}{}]{\underset{\text{(biotin)}}{\overset{\text{propionyl-SCoA}}{\text{carboxylase}}}} CH_3-\overset{\overset{\displaystyle CO_2^-}{|}}{\underset{\underset{H}{|}}{C^\alpha}}-COSCoA$$

propionyl-SCoA

change of configuration

$$\xrightarrow{\overset{\text{methyl}}{\underset{\text{racemase}}{\text{malonyl-SCoA}}}}$$

methyl malonyl-SCoA
(enantiomers)

intramolecular rearrangement

$$CH_3-\overset{\overset{\displaystyle H}{|}}{\underset{\underset{CO_2^-}{|}}{C^\alpha}}-COSCoA \xrightarrow[\text{(B}_{12}\text{ coenzyme)}]{\overset{\text{methyl}}{\underset{\text{mutase}}{\text{malonyl-SCoA}}}} CoASOCCH_2\overset{\alpha}{CH_2}COO^-$$

succinyl-SCoA

Note: This same sequence is of importance to the catabolism of the amino acids, isoleucine, threonine, and methionine, which yield propionyl-SCoA from part of their carbon skeleton; in addition, the catabolism of valine yields methylmalonyl-SCoA as a degradation product; this accounts then for how these amino acids are able to provide part of their carbon skeleton for succinyl-SCoA formation, which can then enter the citric acid cycle for further oxidation; see Chap. 13, Figs. 13–3 and 13–4, and Chap. 17, p. 528.

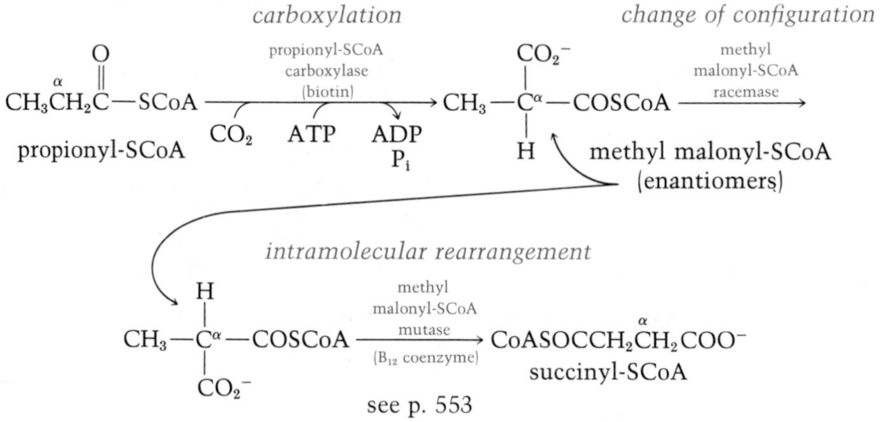

see p. 553

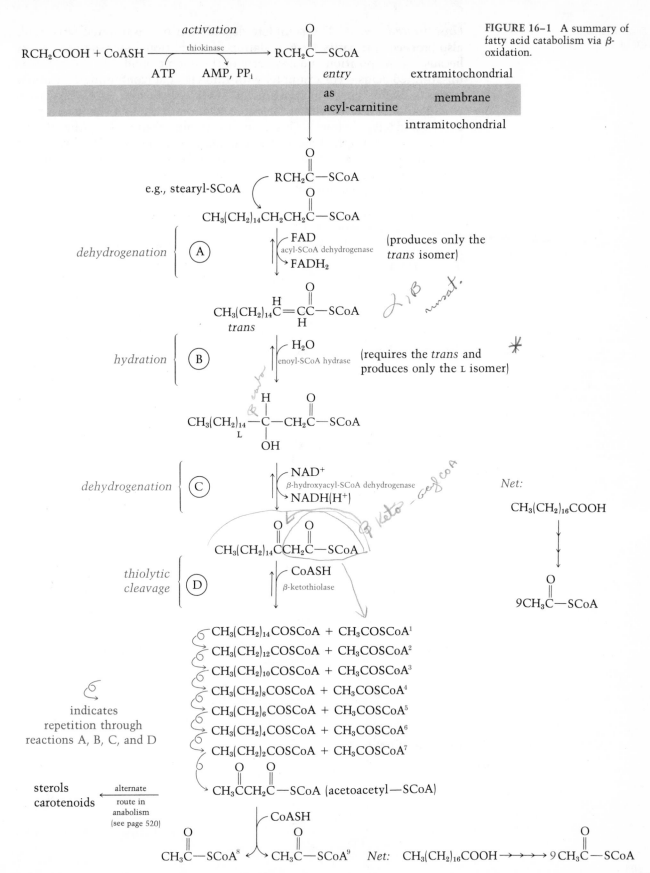

FIGURE 16–1 A summary of fatty acid catabolism via β-oxidation.

activation

$RCH_2COOH + CoASH \xrightarrow[\text{thiokinase}]{} RCH_2\overset{O}{\overset{\|}{C}}-SCoA$

ATP   AMP, PP$_i$

entry   extramitochondrial

as acyl-carnitine   membrane

intramitochondrial

e.g., stearyl-SCoA

$RCH_2\overset{O}{\overset{\|}{C}}-SCoA$

$CH_3(CH_2)_{14}CH_2CH_2\overset{O}{\overset{\|}{C}}-SCoA$

dehydrogenation (A) FAD
acyl-SCoA dehydrogenase
FADH$_2$ (produces only the *trans* isomer)

$CH_3(CH_2)_{14}\overset{H}{\underset{}{C}}=\overset{}{\underset{H}{C}}C-SCoA$

trans

hydration (B) H$_2$O
enoyl-SCoA hydrase   (requires the *trans* and produces only the L isomer)

$CH_3(CH_2)_{14}-\overset{H}{\underset{OH}{\overset{|}{C}}}-CH_2\overset{O}{\overset{\|}{C}}-SCoA$
L

dehydrogenation (C) NAD$^+$
β-hydroxyacyl-SCoA dehydrogenase
NADH(H$^+$)

Net:

$CH_3(CH_2)_{16}COOH$

$CH_3(CH_2)_{14}C\overset{O}{\overset{\|}{C}}CH_2\overset{O}{\overset{\|}{C}}-SCoA$

thiolytic cleavage (D) CoASH
β-ketothiolase

$9CH_3\overset{O}{\overset{\|}{C}}-SCoA$

$\circlearrowleft$ indicates repetition through reactions A, B, C, and D

$CH_3(CH_2)_{14}COSCoA + CH_3COSCoA^1$
$CH_3(CH_2)_{12}COSCoA + CH_3COSCoA^2$
$CH_3(CH_2)_{10}COSCoA + CH_3COSCoA^3$
$CH_3(CH_2)_8COSCoA + CH_3COSCoA^4$
$CH_3(CH_2)_6COSCoA + CH_3COSCoA^5$
$CH_3(CH_2)_4COSCoA + CH_3COSCoA^6$
$CH_3(CH_2)_2COSCoA + CH_3COSCoA^7$

sterols carotenoids

alternate route in anabolism (see page 520)

$CH_3\overset{O}{\overset{\|}{C}}CH_2\overset{O}{\overset{\|}{C}}-SCoA$ (acetoacetyl—SCoA)

CoASH

$CH_3\overset{O}{\overset{\|}{C}}-SCoA^8 \longleftarrow CH_3\overset{O}{\overset{\|}{C}}-SCoA^9$ *Net:* $CH_3(CH_2)_{16}COOH \longrightarrow\longrightarrow 9\,CH_3\overset{O}{\overset{\|}{C}}-SCoA$

503

Unsaturated acids. The complete degradation of unsaturated fatty acids also proceeds primarily by β-oxidation, but additional steps are necessary because of the position and geometry of the double bond(s) in the common unsaturated acids. For example, with linoleic acid (containing 18 carbons with *cis* double bonds at positions 9 and 12) three cycles of β-oxidation would produce a 12-carbon unsaturated chain with *cis* double bonds at positions 3 and 6 (see below). This is an unsuitable substrate for either the dehydrogenase or the hydrase of the β-oxidation sequence. The hydrase in particular requires a *trans* double bond between carbons 2 and 3. As shown, this requirement is satisfied by the participation of an auxiliary *isomerase* enzyme, which catalyzes a double isomerization—both the position and the

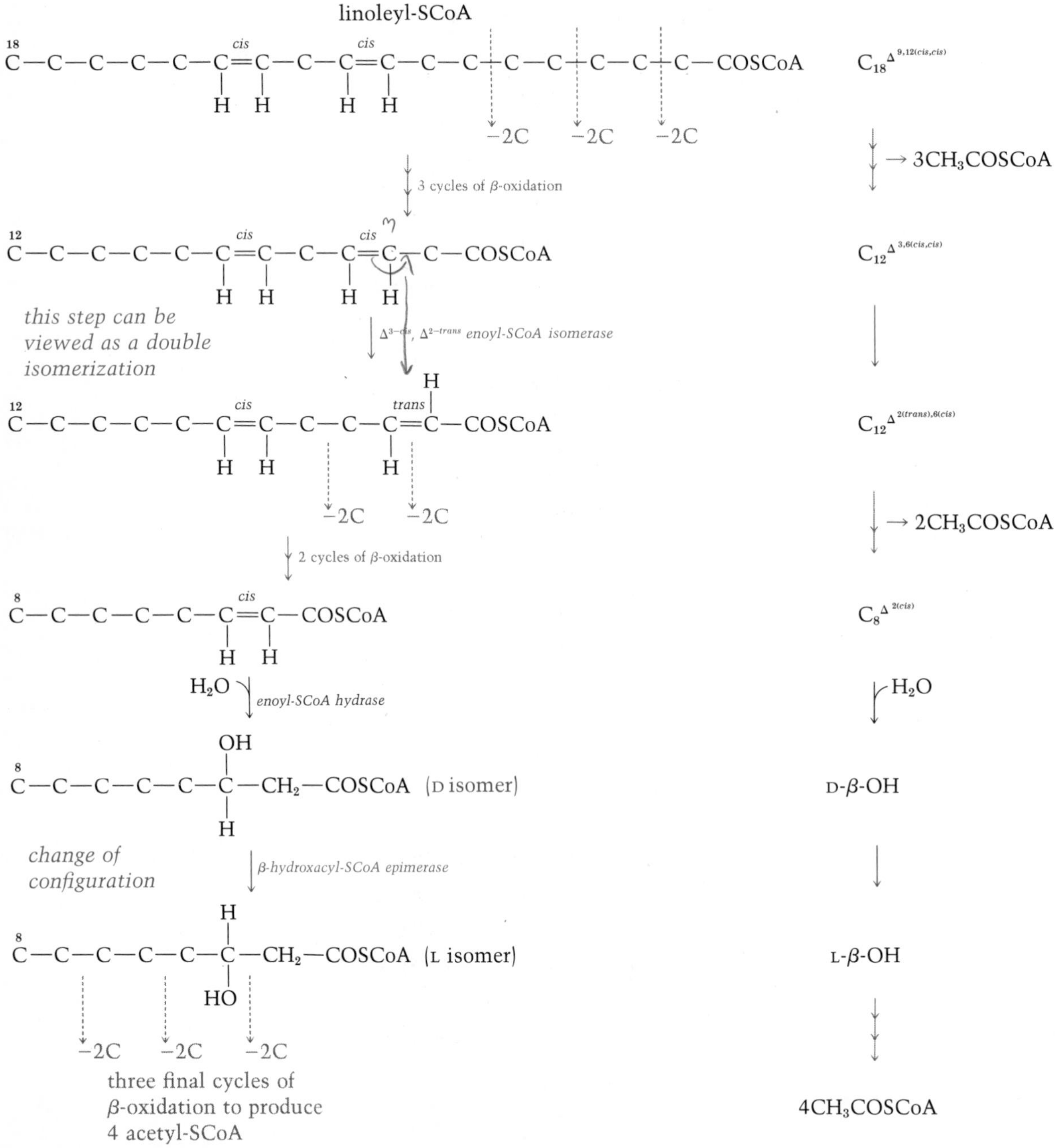

linoleyl-SCoA

this step can be viewed as a double isomerization

change of configuration

three final cycles of β-oxidation to produce 4 acetyl-SCoA

geometry of the double bond are changed.⌋ Once this occurs, β-oxidation can resume. After two additional cycles an unsaturated acyl chain is produced with an α,β position but with a *cis* geometry. This is a suitable substrate for the hydrase enzyme of β-oxidation, but the β-hydroxy product is the D isomer, which is inconsistent with the requirements of the β-oxidation dehydrogenase. This obstacle is overcome by a second auxiliary enzyme, an *epimerase*, which catalyzes a D → L conversion. Thereafter, β-oxidation continues without any additional interruptions.

Energetics of β-oxidation. An ideal analysis of the bioenergetics of fatty acid catabolism requires an assumption that the fate of acetyl-SCoA would be to enter the citric acid cycle where it would be oxidized completely to CO_2. The assumption is not unrealistic. Indeed, such would be the case when the physiological state of the organism and/or dietary factors dictate that lipids rather than carbohydrates be utilized as the primary energy source. Moreover, remember that the enzymes of the citric acid cycle are also localized in the mitochondria. In this context our analysis is straightforward. Stearic acid will be retained as the initial substrate; this is consistent with the information given in Fig. 16–1.

The pertinent equations are as follows. (Note that no distinction is made between the metabolic sources of $FADH_2$ and NADH in the equation for their coupled reoxidation to ATP formation. Regardless of source, they are metabolically equivalent. That is, the P/O ratios are identical—2 for $FADH_2$ and 3 for NADH + H^+.)

A. Balanced equation for activation and β-oxidation:

$$CH_3(CH_2)_{16}COOH + ATP + 8FAD + 8NAD^+ + 8H_2O + 9CoASH \longrightarrow$$

$$\overset{O}{\overset{\|}{9CH_3CSCoA}} + AMP + PP_i + 8FADH_2 + 8NADH + 8H^+$$

B. Balanced equation for citric acid cycle:

$$\overset{O}{\overset{\|}{9CH_3CSCoA}} + 9FAD + 27NAD^+ + 9GDP + 9P_i + 27H_2O \longrightarrow$$

$$18CO_2 + 9CoASH + 9FADH_2 + 27NADH + 27H^+ + 9GTP\ (+9H_2O)$$

C. Balanced equations for oxidative phosphorylation:

$$17FADH_2 + 8.5O_2 + 34ADP + 34P_i \longrightarrow 17FAD + 17H_2O + 34ATP\ (+34H_2O)$$

$$35NADH + 35H^+ + 17.5O_2 + 105ADP + 105P_i \longrightarrow$$
$$35NAD^+ + 35H_2O + 105ATP\ (+105H_2O)$$

A + B + C: Net balanced equation (assuming GDP = ADP and GTP = ATP and AMP + PP = ADP + P_i):

$$CH_3(CH_2)_{16}COOH + 26O_2 + 147ADP + 147P_i \longrightarrow$$
$$18CO_2 + 18H_2O + 147ATP\ (+147H_2O)$$

that is, $$CH_3(CH_2)_{16}COOH + 26O_2 \longrightarrow 18CO_2 + 18H_2O$$

$$147ADP + 147P_i \longrightarrow 147ATP + 147H_2O$$

Note: The specification in reaction B requiring three H_2Os per mole of acetyl-SCoA oxidized by the citric acid cycle includes one H_2O required in the condensation of acetyl-SCoA with oxaloacetate, one H_2O required in the hydrolysis of succinyl-SCoA to succinate, and one H_2O required in the hydration of fumarate to malate.

The H_2O shown in parentheses is that formed during phosphorylation. One of these is borrowed in the net equations to get $18H_2O$.

Using a value of -8 kcal for the $\Delta G^{\circ\prime}$ of ATP hydrolysis as a basis for calculation, we can determine that the net production of 147 units of ATP represents a conservation of energy equivalent to 1,176 kcal (147×8). Since the complete, uncoupled oxidation of stearic acid to CO_2 and H_2O can potentially yield 2,660 kcal, the biological efficiency of energy conservation in fatty acid oxidation is approximately 44%. Once again, however, this figure is a conservative estimate for *in vivo* conditions.

In any case, the fact is that the amount of ATP production accompanying the oxidation of fatty acid carbons is much greater than that accompanying the oxidation of carbohydrate carbons. For example, the oxidation of three glucose molecules (18 carbons → $18CO_2$) would yield $3 \times 36 = 108$ ATP units. We have just seen that stearic acid (also with 18 carbons) yields 147 ATP units. This difference in ATP yield reflects the difference in the state of reduction of fatty acid carbons (mostly —CH_2—) versus carbohydrate carbons (mostly —CHOH). *Fatty acid carbons are in a higher state of reduction and therefore can yield more energy on oxidation.*

Glyoxylate Cycle

The entry of acetyl-SCoA from β-oxidation into the citric acid cycle would seem to provide a basis for the *in vivo* conversion of lipids into other classes of compounds, such as amino acids and hexoses. To determine whether this is so requires that you recognize that this possibility will occur efficiently only if a cycle intermediate, suitable as a biosynthetic precursor, is removed *after* the point of entry of acetyl-SCoA and *before* the cycle is closed with oxaloacetate (OAA) formation. In the case of priming the cycle at the level of acetyl-SCoA, such a situation would exist with α-ketoglutarate. As it was formed, it would be bled off for use as a precursor of glutamate and related amino acids. As long as an anaplerotic reaction was operating to maintain catalytic levels of OAA, the net production of amino acids from the carbon atoms of fatty acids would be quite possible. In fact, such conversions are common to most organisms.

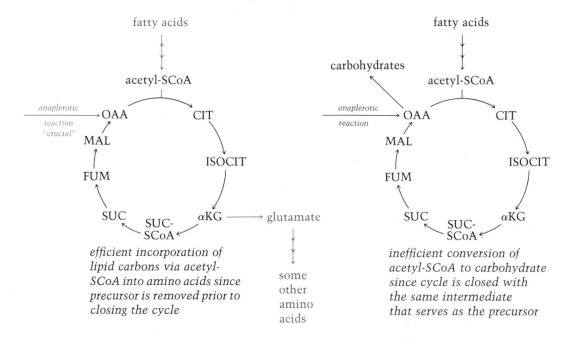

efficient incorporation of lipid carbons via acetyl-SCoA into amino acids since precursor is removed prior to closing the cycle

inefficient conversion of acetyl-SCoA to carbohydrate since cycle is closed with the same intermediate that serves as the precursor

On the other hand, a link of the carbon atoms of acetyl-SCoA to hexoses requires a complete pass through the cycle, to OAA formation. In this case, we have in one complete pass a replenishment of an intermediate used in the initial condensation with acetyl-SCoA. The point is that there would be *no net production of additional* OAA that could be bled off for the biosynthesis of carbohydrates. The OAA that was formed would be required to react with additional acetyl-SCoA. We could argue that this should not make any difference, since the OAA could be bled off, with the anaplerotic reaction then providing the necessary OAA for condensation with acetyl-SCoA. The latter argument presupposes, however, that the cell will select a specific OAA molecule for carbohydrate biosynthesis and another for reaction with acetyl-SCoA. Suffice it to say that living organisms are not that clever. In summary then, a significant net production of carbohydrate from lipid seems to be ruled out. What do we find in nature? Well, in animal organisms this is exactly the case. The fatty acids of lipids are not converted to carbohydrate to any appreciable degree. (Unfortunately, for most of us, the reverse process does occur, and rather efficiently.) There is, however, an efficient conversion of lipid to carbohydrate in many plants and bacteria. Having just determined that this should not occur, we must now explain why it does.

The ability of plants and bacteria to form carbohydrate from fatty acids is due to the operation of two auxiliary enzymes for which there are no known counterparts in animals. The enzymes are *isocitritase* and *malate synthetase*. The reactions they catalyze are shown in Fig. 16–2, which also includes the other enzymes of the citric acid cycle.

There are two key actions in this pathway (commonly called the *glyoxylate cycle*): (a) the shunting of isocitrate away from its usual fate in the citric acid cycle by isocitritase to yield *glyoxylate* and succinate; and (b) the subsequent condensation of the glyoxylate with a second unit of acetyl-SCoA to yield malate, which can then be converted to OAA, thus closing the cycle.

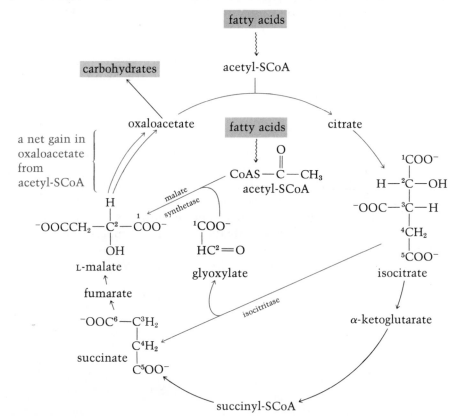

FIGURE 16–2 The glyoxylate cycle (reactions with colored arrows). The activity of malate synthetase and isocitritase results in the formation of OAA in excess of what is required to sustain the formation of citrate. Consequently the excess OAA can be used for carbohydrate biosynthesis. The isocitrate → succinate steps are shown as a reminder of citric acid cycle activity. In fact both cycles can operate simultaneously.

nthesis of Fatty Acids

thetase
ucible
nly in
of
e in-
zyme
of

The important point to realize is that this pathway will be accompanied by a *net increase* in the production of OAA originating from the carbon atoms of acetyl-SCoA. The OAA can then be used in the biosynthesis of carbohydrates and other amino acids as well. Of course, the glyoxylate cycle by itself is also capable of generating reducing power for ATP formation as a result of the NAD^+-dependent conversion of malate $\rightarrow$ OAA and the FAD-dependent conversion of succinate $\rightarrow$ fumarate.

ANABOLISM OF FATTY ACIDS

Relationship to β-Oxidation

It is a well-documented fact that feeding acetate (^{14}C) to any type of organism will result in the production of fatty acids containing the ^{14}C label. In fact, every carbon atom of the newly synthesized acid will be labeled. Given these data, if you had to predict the type of pathway responsible for the biosynthesis of fatty acids from acetyl units, an obvious hypothesis would be that the hydrocarbon chain is assembled by a reversal of β-oxidation. In addition to the fit with the above data, further support for this proposal is provided by the fact that all of the β-oxidation enzymes can catalyze the reverse reactions.

There is, however, one predicament. The biosynthesis of fatty acids from acetyl-SCoA does not occur with purified preparations of mitochondria that contain the β-oxidation enzymes, but rather with soluble (cytoplasmic) extracts. Since it is unlikely that the same complete set of enzymes are localized in two different cellular compartments, it would appear that the biosynthesis of fatty acids in the cytoplasm is accomplished by a different set of enzymes. This is precisely the case. Finally, however, it should be noted that the reversal of β-oxidation within the mitochondria does play a part in fatty acid biosynthesis. In particular, it is believed that the mitochondrial system is the main pathway responsible for the extension of long-chain fatty acids (C_{16} and up) via a direct condensation with acetyl-SCoA and subsequent reduction.

Intracellular Source of Acetyl–SCoA

Although most organisms can utilize externally supplied acetate, under normal conditions the major source (and in many cases the only source) of acetate is intracellular, coming from the reaction pyruvate $\rightarrow$ acetyl-SCoA + CO_2. But, since (a) the decarboxylation of pyruvate occurs in the mitochondria, (b) the enzymes for fatty acid biosynthesis are in the cytoplasm, and (c) the mitochondrial membrane is relatively impermeable to the free diffusion of acetyl-SCoA, we have a problem. How does acetyl-SCoA get across the membrane to participate in anabolism? Well, based on pre-

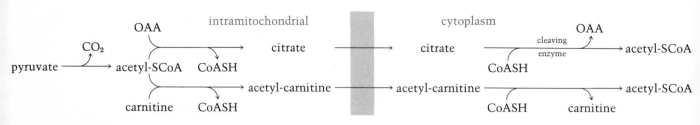

vious material given in this chapter and in Chap. 13, we can propose two possible explanations. One involves the diffusion of citrate from the mitochondria followed by the action of a *citrate-cleaving enzyme* (see p. 435). The second involves the *acyl-carnitine transport* system just described on p. 501.

Soluble Anabolic Pathway

Malonyl-SCoA. The first significant development in understanding fatty acid biosynthesis was the observation by Wakil and coworkers in 1958 that the incorporation of acetyl-SCoA into long-chain acyl groups by soluble extracts was dependent on the presence of ATP, NADPH, and Mn^{2+}, as well as being greatly stimulated by the presence of CO_2. Subsequently, Wakil's group provided the explanation for much of their data by isolating the enzyme *acetyl-SCoA carboxylase*, which catalyzed the ATP-dependent, irreversible carboxylation of acetyl-SCoA to yield *malonyl-SCoA*. It is now well documented that this is the *initial* reaction of fatty acid biosynthesis. In addition to requiring Mn^{2+} for optimal activity, acetyl-SCoA carboxylase, like many other carboxylases, is absolutely dependent on a substance called *biotin.*

Biotin is a water-soluble vitamin of low molecular weight that has now been implicated as being an obligatory coenzyme in approximately 10 different carboxylation reactions. The role of biotin in malonyl-SCoA formation had been implicated by earlier work showing that acetyl-SCoA utilization was strongly inhibited by *avidin,* a protein occurring in egg whites that selectively and firmly binds with biotin.

$$\overset{*}{C}O_2 + CH_3\overset{\displaystyle O}{\overset{\|}{C}}-SCoA + ATP \xrightarrow[Mn^{2+}, \text{ biotin}]{\text{acetyl-SCoA carboxylase}} {}^-OOCCH_2\overset{\displaystyle O}{\overset{\|}{C}}-SCoA + ADP + P_i$$

acetyl-SCoA   malonyl-SCoA

* the active substrate form of CO_2 is bicarbonate, HCO_3^-

Whatever the particular carboxylase, biotin (see margin for structure) is found firmly attached to the enzyme via a covalent linkage to the epsilon amino group of a lysine side chain.) The acetyl-SCoA carboxylase consists of three proteins, only one of which contains bound biotin, and hence this protein component is called the *biotin-carrier protein.* The other two proteins serve as enzymes in the two stages of the carboxylation reaction. Stage (A), in which a *biotin carboxylase* component of the acetyl-SCoA carboxylase is used, involves the energy-dependent carboxylation of the biotin in the biotin-carrier protein. The resultant product is a highly reactive organic form of the one-carbon unit and is termed "activated CO_2." In stage (B), the *carboxyl transferase* component of the acetyl-SCoA carboxylase is used to transfer the CO_2 unit to acetyl-SCoA in a condensation reaction. In other words, the coenzyme function of biotin is to act as a *carrier* (transfer

biotin

covalent attachment to
biotin carrier protein-BCP
(see below)

A $$ATP + CO_2 + \text{biotin}\overset{\displaystyle\downarrow}{-}BCP \xrightarrow[\substack{\text{biotin} \\ \text{carboxylase}}]{Mn^{2+}} ADP + P_i + CO_2-\text{biotin}-BCP$$

B $$CH_3\overset{\displaystyle O}{\overset{\|}{C}}-SCoA + CO_2-\text{biotin}-BCP \xrightarrow[\text{transferase}]{\text{carboxyl}} {}^-O_2CCH_2\overset{\displaystyle O}{\overset{\|}{C}}-SCoA + \text{biotin}-BCP$$

Net: $$ATP + CO_2 + CH_3\overset{\displaystyle O}{\overset{\|}{C}}-SCoA \xrightarrow[\text{carboxylase}]{\text{acetyl-SCoA}} {}^-O_2CCH_2\overset{\displaystyle O}{\overset{\|}{C}}-SCoA + ADP + P_i$$

carboxyl group of biotin is attached to epsilon amino group of lysine in BCP

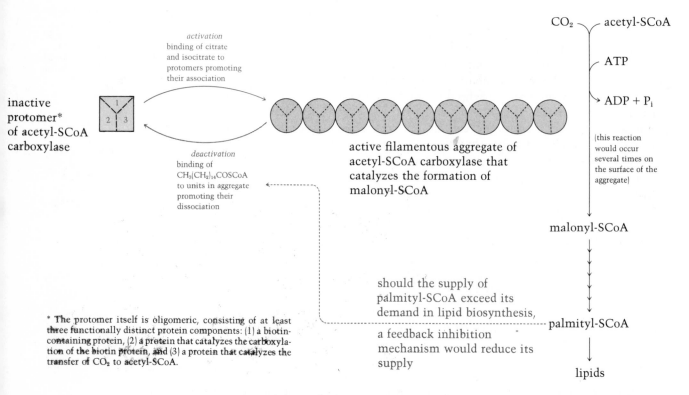

agent) of the one-carbon unit. As indicated, the site of CO_2 attachment to biotin has been shown to be the N atom most distant from the aliphatic side chain.

Acetyl-SCoA carboxylase has been purified from various animal tissues, yeast, and several bacteria. In each source the enzyme exists in two *distinct and isolatable* forms, one inactive and the other active. The inactive form of the avian liver enzyme has a molecular weight of about 400,000, whereas the active form has a molecular weight of about 4,000,000. This difference in size has been clearly established as meaning that the active form is an *aggregate* of the inactive protomer. On the basis of electron microscopic examination (the protein is big enough) the active aggregate has been found to have a filamentous, chainlike appearance, presumably due to an end-to-end alignment of the protomers.

acetyl-SCoA
carboxylase

(1) biotin carrier protein
(2) biotin carboxylase
(3) carboxyl transferase
    (see previous page)

In vivo, it is proposed that the interconversion of these two forms is under strict control, and thus the carboxylase reaction is a crucial regulatory site in lipid metabolism. Known *activators* are citrate and isocitrate. Long-chain acyl-SCoA materials (palmityl-SCoA, for example) are potent feedback *inhibitors.* Presumably, the binding of citrate or isocitrate to the free, inactive protomers causes a change in the conformation of the protomer that triggers its self-association. Conversely, the binding of palmityl-SCoA to the protomeric units in the active aggregate triggers its self-dissociation.

These metabolic signals of activation and inhibition are in perfect harmony, providing a highly efficient control of fatty acid biosynthesis as an integrated pathway within the whole of metabolism. For example, consider the following reasoning regarding the activation by citrate and isocitrate. When the energy (ATP) demands of the organism are great, acetyl-SCoA would best be funneled into the citric acid cycle, since it is the citric acid cycle that provides the bulk of reducing power necessary for ATP formation. As long as the ATP demand remained great, the intracellular levels of citrate, isocitrate, and the other cycle intermediates would not increase despite the continued input of acetyl-SCoA. The concentrations of these materials would remain low and relatively constant because they would be rapidly utilized and reformed in the cycle reactions. When the energy demands of the organism diminish, the need for a rapid oxidation of acetyl-SCoA by the citric acid cycle would also be diminished. This diminished need for energy is signaled by increased ratios of ATP/ADP and NADH/NAD^+. Recall that these signals would result in a negative control (inhibition) of isocitrate dehydrogenase (see p. 439). This, of course, would result in increases in the cellular levels of isocitrate and its immediate precursor, citrate. These increases would in turn act as signals resulting in a positive control (activation) of acetyl-SCoA carboxylase. (In eukaryotic cells these changes would be localized in the mitochondrion and would require the diffusion of citrate and isocitrate out of the mitochondrion into the cytoplasm, where the carboxylase is localized.) Thus, the excess acetyl-SCoA, no longer required to satisfy a large need for metabolic energy, would be diverted into fatty acid biosynthesis for storage.

Acyl carrier protein. The initial formation of malonyl-SCoA is only one of the many features that distinguish the soluble biosynthetic pathway from a reversal of β-oxidation. A second is that none of the acyl-SCoA intermediates is metabolized as a CoASH thioester (except in the carboxylase step), but rather as a special thioester involving a substance called the *acyl carrier protein* (ACP-SH). This is a low molecular weight protein containing an ac-

acyl carrier protein (ACP-SH)

$$\text{ACP-SH} + CH_3\overset{\overset{\displaystyle O}{\|}}{C}-\text{SCoA} \xrightarrow[\text{transacylase}]{\text{acetyl}} \text{ACP-S}-\overset{\overset{\displaystyle O}{\|}}{C}CH_3 + \text{CoASH}$$

$$\text{ACP-SH} + {}^-OOCCH_2\overset{\overset{\displaystyle O}{\|}}{C}-\text{SCoA} \xrightarrow[\text{transacylase}]{\text{malonyl}} \text{ACP-S}-\overset{\overset{\displaystyle O}{\|}}{C}CH_2COO^- + \text{CoASH}$$

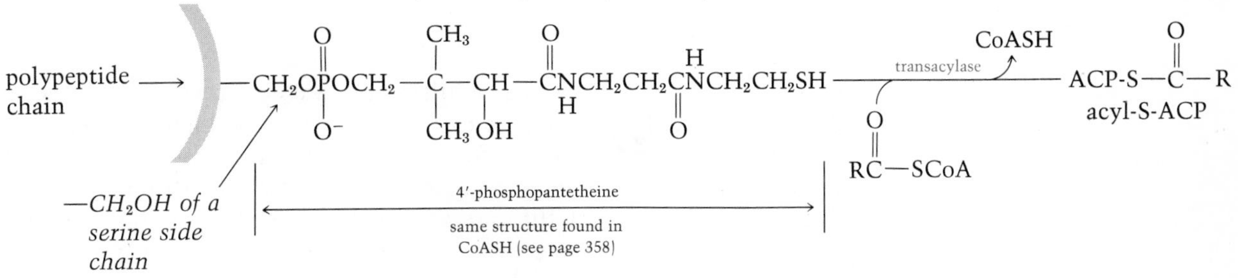

tive sulfhydryl group that reacts with free acyl-SCoA to yield acyl-S-ACP derivatives. The distinguishing feature of the ACP-SH is that it contains a functional sulfhydryl (—SH) grouping contributed not by a cysteine residue, but by a grouping called *4'-phosphopantetheine,* which is covalently attached to the polypeptide chain via a serine residue. The interesting aspect of this is that 4'-phosphopantetheine, you will recall, is also the same grouping found in the structure of CoASH (p. 358). The best way to describe this reaction is as a *transacylation:* an acyl group is transferred from a CoASH thioester to another active —SH group to form a second thioester. The transacylations involving acetyl-SCoA and malonyl-SCoA are known to be catalyzed by separate enzymes—*acetyl transacylase* and *malonyl transacylase,* respectively.

Reaction sequence. Since all of the subsequent reactions involve acyl-S-ACP derivatives, it is now appropriate to direct our attention to the metabolic sequence of the anabolic pathway. In so doing, other distinguishing characteristics will emerge. The complete pathway is illustrated in Fig. 16–3.

The immediate fate of malonyl-S-ACP is condensation with acetyl-S-ACP, which is catalyzed by *β-ketoacyl-S-ACP synthetase,* and results in the production of acetoacetyl-S-ACP. You will note that the chemistry of this reaction, as shown by the use of ^{14}C-labeled substrates in the presence of the purified enzyme, is such that the carbonyl carbon and the alpha (α) carbon of the malonyl group are in the same position in the β-ketoacyl compound. Of prime significance in this reaction is the formation of CO_2 and ACP-SH. Since both materials are required in two previous steps, the pathway is rendered *autocatalytic.* The regeneration of CO_2 is most important, since it is essential for the formation of malonyl-SCoA, which in turn is critical to the initiation of the whole process. Further evidence for the autocatalytic role of CO_2 has been established by radioisotope studies, which have shown that the CO_2 coming off in the synthetase reaction is the same unit of CO_2 used initially in the acetyl-SCoA carboxylase reaction. In other words, the carbon atom of CO_2 does not become part of the carbon skeleton of fatty acids. It merely keeps the pump primed, so to speak.

see p. 513

$$^-OO^{14}C—^{14}CH_2—^{14}C—S-ACP \xrightarrow[(+H^+)]{\substack{\beta\text{-ketoacyl-S-ACP} \\ \text{synthetase}}} CH_3—C—^{14}CH_2—^{14}C—S—ACP + {}^{14}CO_2$$

acetoacetyl-S-ACP (reused)

ACP-SH (reused)

$$CH_3—C—S-ACP$$

The fate of acetoacetyl-S-ACP is ultimate conversion to the fully saturated butyryl-S-ACP via three enzymes acting in sequence: first, a *β-ketoacyl-S-ACP dehydrogenase;* then, a *β-hydroxyacyl-S-ACP dehydratase;* and finally, an *α,β-enoyl-S-ACP dehydrogenase.* Although we are now dealing with different thioesters, you should recognize that the reaction sequence of the acyl moieties is exactly the reverse of β-oxidation. The differences are the enzymes and coenzymes involved and their cellular localization. The two dehydrogenases are particularly noteworthy because they are NADPH dependent. (Recall the requirement for NADPH in Wakil's early work with soluble extracts.) Their counterparts in the reversal of the β-oxidation pathway, on the other hand, would be NADH and $FADH_2$ dependent. Presumably, the source of the required NADPH is the hexose

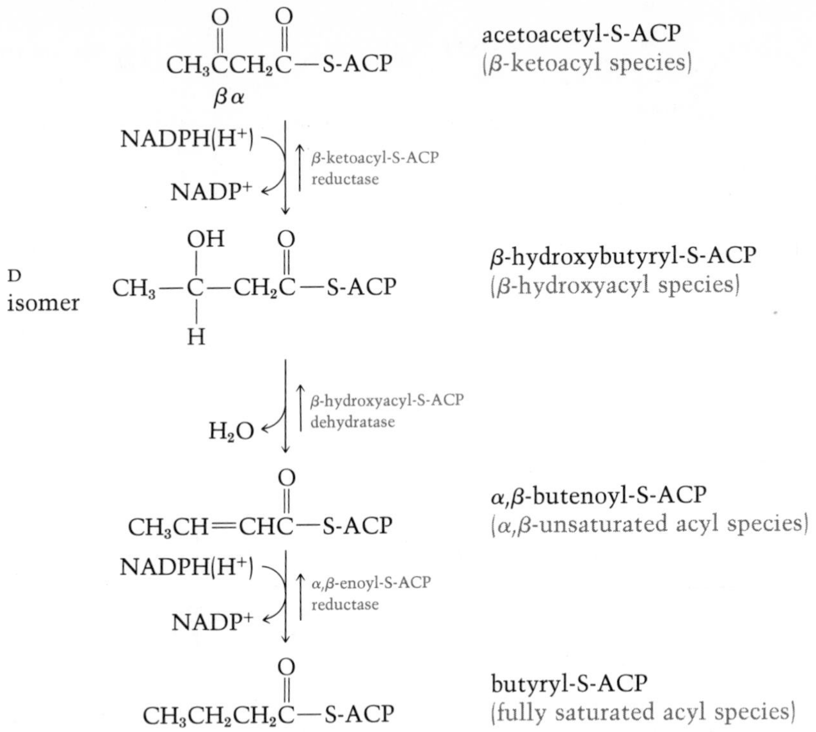

$$\underset{\beta\ \ \alpha}{CH_3\overset{O}{\overset{\|}{C}}CH_2\overset{O}{\overset{\|}{C}}}-S\text{-}ACP$$

acetoacetyl-S-ACP
(β-ketoacyl species)

NADPH(H$^+$) β-ketoacyl-S-ACP reductase

NADP$^+$

D isomer

$$CH_3-\overset{OH}{\underset{H}{\overset{\|}{C}}}-CH_2\overset{O}{\overset{\|}{C}}-S\text{-}ACP$$

β-hydroxybutyryl-S-ACP
(β-hydroxyacyl species)

H$_2$O β-hydroxyacyl-S-ACP dehydratase

$$CH_3CH{=}CH\overset{O}{\overset{\|}{C}}-S\text{-}ACP$$

α,β-butenoyl-S-ACP
(α,β-unsaturated acyl species)

NADPH(H$^+$) α,β-enoyl-S-ACP reductase

NADP$^+$

$$CH_3CH_2CH_2\overset{O}{\overset{\|}{C}}-S\text{-}ACP$$

butyryl-S-ACP
(fully saturated acyl species)

monophosphate shunt, the enzymes of which are also found in the cytoplasm. The complete pathway is diagramed in Fig. 16–3. (Note also that the D isomer of the β-hydroxy species is involved, rather than the L, as in β-oxidation.)

With one pass through this cycle completed, the butyryl-S-ACP would then be recycled by condensation with a second unit of malonyl-S-ACP. As shown, this pattern would presumably continue until palmityl-S-ACP was formed (seven cycles).

Due to the nature of the β-ketoacyl synthetase reaction, it should be clear that the acyl chain is elongated by the successive addition of a two-carbon unit from the terminal methyl carbon found near the carbonyl end. Recall that the reverse pattern is typical of β-oxidation. The important point to

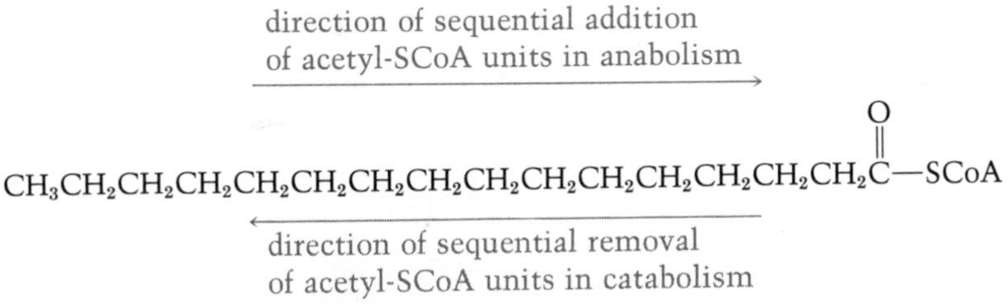

direction of sequential addition
of acetyl-SCoA units in anabolism

$$CH_3CH_2CH_2CH_2CH_2CH_2CH_2CH_2CH_2CH_2CH_2CH_2CH_2CH_2CH_2\overset{O}{\overset{\|}{C}}-SCoA$$

direction of sequential removal
of acetyl-SCoA units in catabolism

note is that, although all carbon atoms ultimately originate from eight units of acetyl-SCoA, there is a direct entry of only one, with the other seven entering via malonyl-SCoA. The final step would be the hydrolytic cleavage of palmity-S-ACP to yield ACP-SH and free palmitate. The latter would then be activated by a thiokinase to yield palmityl-SCoA.

In the cytoplasm the palmityl-SCoA could be utilized directly in the biosynthesis of acylglycerols, phosphoacylglycerols, sphingomyelins, cerebro-

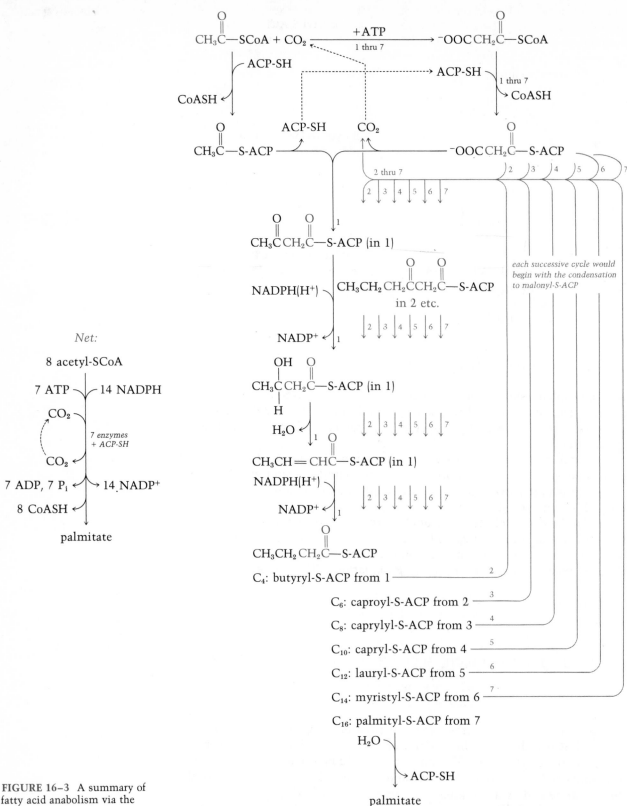

FIGURE 16–3 A summary of
fatty acid anabolism via the
malonyl-SCoA pathway.

sides, and gangliosides. Alternatively, the palmityl-SCoA could *enter the mitochondrion* (via the carnitine transport system) where the enzymes of β-oxidation would participate in the synthesis of longer-chain fatty acids. (The only different step is that the —CH=CH— → —CH₂CH₂— conversion is apparently catalyzed by an NADPH-dependent dehydrogenase rather than the FADH₂-dependent dehydrogenase.) Elongation of palmityl-SCoA may also occur in association with the *endoplasmic reticulum* (also called microsomes). There is evidence that enzymes involved in the production of unsaturated fatty acids are also localized in microsomes (see next section).

Fatty acid synthetase complex. Although all of the enzymes are found in the cytoplasm, they do not exist separately. Instead, *all* (excepting the carboxylase) exist as a polyprotein aggregate, that is, as a *multienzyme complex*. For obvious reasons the aggregate is termed the *fatty acid synthetase system*. The acyl carrier protein serves as the *core* of the aggregate and the six enzymes involved in assembly are on the surface. The phosphopantetheine group of the acyl carrier protein could serve as an extending and pivoting arm that would deliver the acyl group from one enzyme to another. (Remember that the same type of design was encountered with pyruvate dehydrogenase complex (p. 428), with lipoic acid acting as the delivery system.)

Here again is another example, on a molecular level, of the exquisite architectural order found in nature. In support of the fact that this type of arrangement confers a high level (perhaps an optimum level) of metabolic efficiency, and hence a definite metabolic advantage, is that the basic organization of this system has remained unchanged through the billions of years of evolution. All types of cells, from bacteria to mammalian liver tissue, have the same general type of complex and component parts. The system from yeast is the largest, most elaborate, and tightly knit (particle weight is approximately 2,800,000 compared, for example, to a value of about 540,000 for the complex obtained from rat liver).

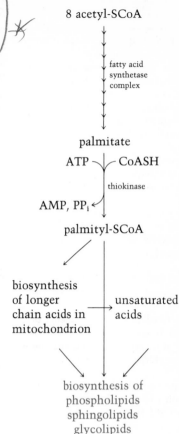

fatty acid synthetase

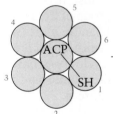
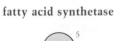

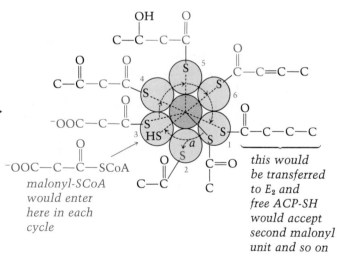

1 = acyl transacylase
2 = β-ketoacyl synthetase
3 = malonyl transacylase
4 = β-ketoacyl reductase
5 = β-hydroxyacyl dehydratase
6 = α,β-enoyl reductase
ACP = acyl carrier protein

515

Anabolism	Catabolism
occurs in cytoplasm[a]	occurs in mitochondrion
all carbons originate from acetyl-SCoA	all carbons converted to acetyl-SCoA
acetyl-SCoA converted to malonyl-SCoA	malonyl-SCoA is not involved
enzymes are part of multi-enzyme complex	enzymes are not part of multi-enzyme complex
NADPH required	NAD$^+$ and FAD required
intermediates are thioesters of acyl carrier protein	intermediates are thioesters of coenzyme A only
first two-carbon unit added is CH$_3$CH$_2$~	first two-carbon unit removed is ~CH$_2$C— ‖ O

[a] Some occurs in mitochondria and some with microsomes.

Review. In the introduction to this chapter, we stated that the metabolism of fatty acids is an excellent example of the contrast between catabolic and anabolic pathways. It should be obvious that this is so. The only distinct similarity is that the pattern of chemical transformations involves the same type of intermediates and begins and ends with a saturated acyl grouping and a β-ketoacyl grouping. The enzymatic machinery that mediates the transformations is completely different. (Refer to the summary in the margin.)

Biosynthesis of Unsaturated Fatty Acids and Prostaglandins

The microsomes of eukaryotic cells contain a multimolecular enzyme system, termed *fatty acid desaturase,* that catalyzes the elimination of hydrogen from long-chain acyl groups to yield unsaturated acids. In effect, the overall process is a miniature electron transport system where molecular oxygen acts as a hydrogen acceptor for two hydrogen donors—the fatty acid and NADPH. Proteins that intervene in the overall process are cytochrome b_5, cytochrome b_5 reductase, and an oxygenase. The result of the action of desaturase system is as follows:

microsomal desaturase system

The location of desaturation along the acyl chain depends on the specific fatty acid involved. For example, linoleic acid (designated as 9,12-C$_{18:2}$, indicating there is a chain length of 18 carbons with two double bonds at carbons 9 and 12) is produced from stearic acid (C$_{18}$) as follows:

$$C_{18} \xrightarrow{-2H} 9\text{-}C_{18:1} \xrightarrow{-2H} 9,12\text{-}C_{18:2}$$

stearyl-SCoA   oleyl-SCoA   linoleyl-SCoA

The desaturase system of mammals is incapable of converting oleic acid to linoleic acid and so the linoleic acid required in the diet is supplied from plant sources. Linoleic acid is essential for mammals because it is a precursor for several prostaglandins, an extremely important group of lipid hormones (see p. 321 in Chap. 10). The linoleic acid to prostaglandin route (see p. 517) begins with its conversion to arachidonic acid, the substrate of *prostaglandin synthetase* (previously referred to as fatty acid cyclo-oxygenase). In a most fascinating reaction, the synthetase converts arachidonic acid to PGG$_2$, the immediate prostaglandin precursor. Refer back to p. 322 for a discussion of subsequent transformations of PGG$_2$ to PGE$_2$, thromboxane, and prostacyclin.

BIOSYNTHESIS OF OTHER LIPIDS

Phosphoacylglycerols

Despite the fact that phosphoacylglycerols occur throughout nature as vital components of biological membranes, very little was known about their biosynthesis until recently. Some of the important relationships that are

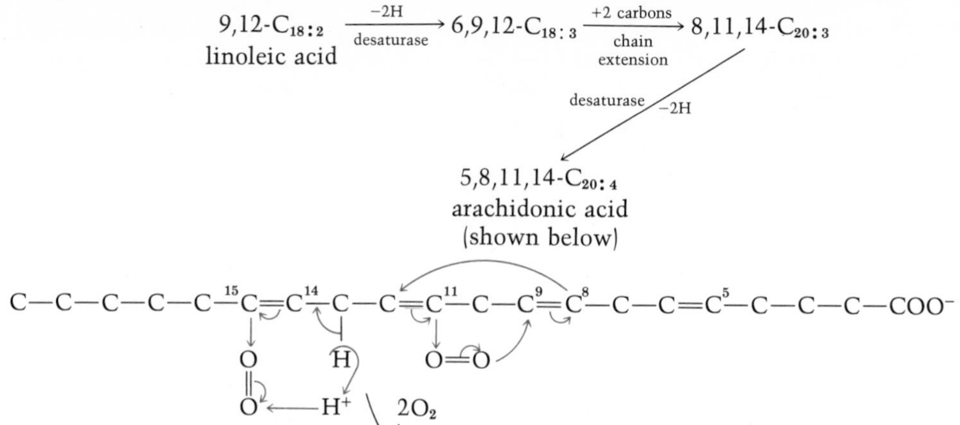

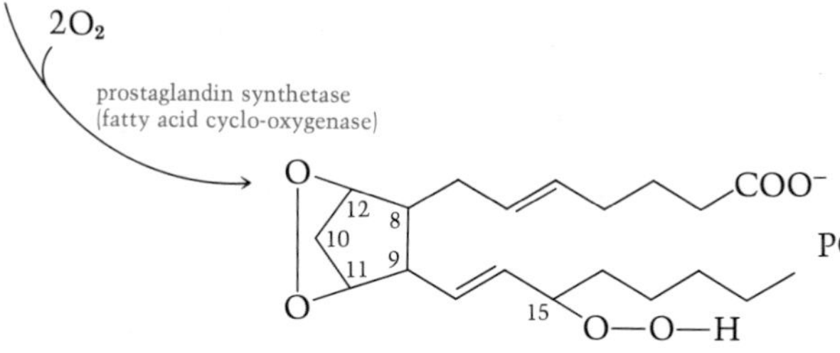

$$9,12\text{-}C_{18:2} \xrightarrow[\text{desaturase}]{-2H} 6,9,12\text{-}C_{18:3} \xrightarrow[\substack{\text{chain}\\\text{extension}}]{+2 \text{ carbons}} 8,11,14\text{-}C_{20:3}$$

linoleic acid

$$\xrightarrow[\text{desaturase}]{-2H}$$

5,8,11,14-C$_{20:4}$
arachidonic acid
(shown below)

C—C—C—C—C^{15}=C^{14}—C—C=C^{11}—C—C=C^8—C—C=C^5—C—C—C—COO$^-$

prostaglandin synthetase
(fatty acid cyclo-oxygenase)

$2O_2$

PGG$_2$ (see p. 322)

the arrows identify the pattern
of chemistry catalyzed by this
enzyme, beginning with the C^{11}
interaction with the first O$_2$
and ending with the C^{15} interaction with the second O$_2$

known are shown in Fig. 16–4. Although they are not specified, there are enzymes that catalyze each reaction.

The key reactions (*) are the formation of *cytidine diphosphate diglyceride* and *cytidine diphosphate choline* (or ethanolamine). It is quite clear then that CTP is an obligatory coenzyme. Moreover, CTP is a quite specific coenzyme in these reactions, since most other nucleotides will not substitute for it. The experimental history behind the discovery of this role of CTP is interesting. In 1955 E. Kennedy and S. Weiss were using cell-free systems in their investigations of the biosynthesis of phosphatidyl choline. One of the components they added to their incubation mixture was ATP. (Previous studies had indicated that the incorporation of choline phosphate into phosphatidyl choline was ATP-dependent.) In the course of their work, it was observed that identically prepared systems gave variable results when incubated under the same conditions, which included adding the same amount of ATP. Eventually the basis for the discrepancy was traced to the purity of the ATP that was added. Specifically, it was discovered that certain commercial sources of ATP that were used in the studies were contaminated with CTP. Noting that the incorporation of choline phosphate into lipids was greater with the CTP-contaminated sources, pure CTP and pure ATP were then utilized. The addition of pure CTP resulted in a considerable stimulation of choline phosphate incorporation, whereas the addition of pure ATP was without effect. Further studies (a) confirmed the specificity of the stimulatory effect of CTP on the utilization of choline phosphate in phosphatidyl choline formation; (b) determined that CTP had a similar effect on the incorporation of ethanolamine phosphate into phosphatidyl ethanolamine; (c) defined the details of how CTP participated in these processes as an essential coenzyme; and (d) determined that the incorporation of phosphatidic acid into other lipids such as phosphatidyl inositol and phosphatidyl glycerol was also stimulated by, and thus dependent on, CTP.

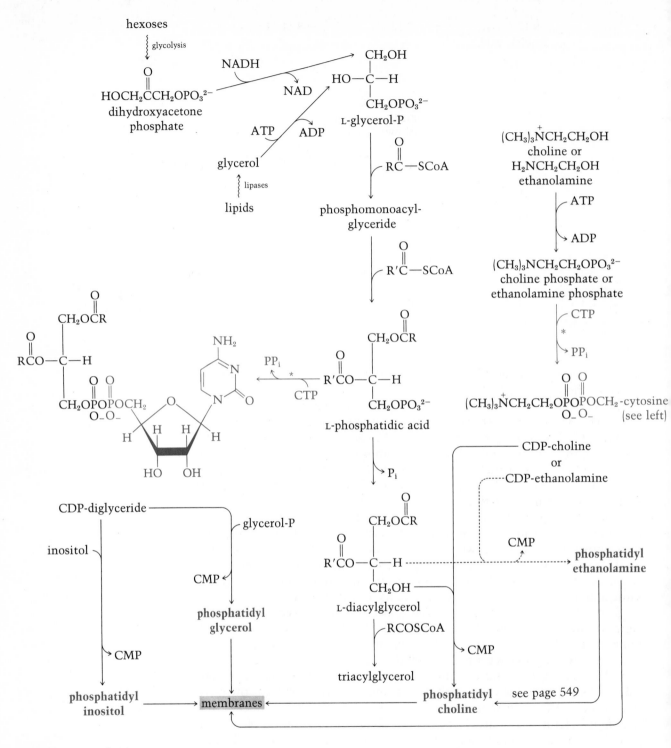

FIGURE 16–4 A brief summary of some anabolic pathways of phosphoacylglycerols. Crucial CTP-dependent reactions are indicated with an asterisk. In the fat cells of adipose tissue the L-glycerol-P → triacylglycerol sequence is a major pathway.

Cholesterol, Steroids, and Carotenoids

Despite the fact that the net chemistry is rather straightforward, the processes of nucleic acid biosynthesis and protein biosynthesis often leave the greatest impression on students of biochemistry. In each process monomeric units are linked together by the formation of identical types of bonding—the phosphodiester bond for nucleic acids and the peptide bond for proteins. Of course, the remarkable feature is that each process is highly programed for the ordered production of a polymer comprised of a specific sequence of monomeric units. From the standpoint of the chemistry involved, however, perhaps the most fascinating anabolic pathway in all of nature is that responsible for the biosynthesis of cholesterol and other related lipids. The primary focus of the following material is on cholesterol biosynthesis.

Before considering the specific sequence of reactions involved, it is helpful to understand their net effect. This is stated rather simply: all of the carbon atoms of cholesterol (a total of 27) originate from acetyl-SCoA. The initial observation of this precursor-product relationship was made in 1950 by studying the utilization of uniformly labeled (^{14}C) acetate by liver cells for the production of cholesterol. Two experimental approaches were employed. One approach involved the feeding of the ^{14}C-acetate to rats whose livers were later removed. The second involved the incubation of fresh liver slices in the presence of the labeled acetate. In both instances the cholesterol isolated from the liver cells was completely labeled with ^{14}C, proving conclusively that the entire carbon skeleton of cholesterol originates from the two carbons of acetate. Subsequent studies determined that the pattern of incorporation was very specific, with 12 specific carbon atoms in cholesterol originating from the carbonyl carbon (c), and the other 15 from the methyl carbon (m) of the acetyl unit.

cholesterol

$CH_3-\overset{\overset{O}{\|}}{C}-SCoA \rightarrow\rightarrow\rightarrow\rightarrow\rightarrow\rightarrow$

carbonyl
carbon (c)

methyl
carbon (m)

The elucidation of the anabolic pathway responsible for this pattern required nearly 10 years. It turned out to be a milestone in the development of biochemistry, because it is a major pathway found in all types of organisms, accounting for the biosynthesis of a host of bimolecules other than cholesterol. The key discovery, made in the late 1950s, was that *mevalonic acid* (3-methyl-3,5-dihydroxyvalerate) was an intermediate of the pathway. The complete sequence of this pathway can be summarized as consisting of three distinct phases: A, the *formation of mevalonate*; B, the *formation of squalene*; and C, the *formation of cholesterol*.

acetyl-SCoA $\xrightarrow[]{\text{phase A}}$ mevalonate $\xrightarrow[]{\text{phase B}}$ squalene $\xrightarrow[]{\text{phase C}}$ cholesterol
  (C$_6$ branched (C$_{30}$ unsaturated
  hydroxy acid)   hydrocarbon)

In phase A, three units of acetyl-SCoA condense to give mevalonate. The process is not direct, but involves (see Fig. 16–5) the intermediate formation of acetoacetyl-SCoA and then *β-hydroxy-β-methylglutaryl-SCoA*, a substituted dicarboxylic acid. The latter is then converted to mevalonate via a

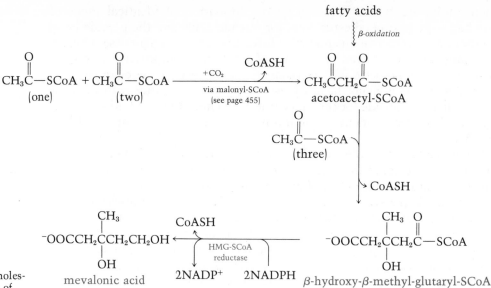

FIGURE 16–5 Phase A of cholesterol biosynthesis; formation of mevalonic acid.

step dependent on two units of NADPH. (The participating enzyme is β-hydroxy-β-methyl-glutaryl-SCoA reductase, or simply HMG-SCoA reductase. This reductase is the key regulatory enzyme of cholesterol biosynthesis; see p. 523 in this chapter.) One NADPH functions in the reduction of the carbonyl grouping to a hydroxyl group, and the second unit participates in the reductive cleavage of the carbon-sulfur bond, which yields CoASH and mevalonate in the free form.

In phase B (Fig. 16–6), mevalonate is first activated by a trio of ATP-dependent kinases to yield *3-phospho-5-pyrophosphomevalonate*. The latter is extremely unstable and undergoes a decarboxylation and a dephosphorylation to yield *isopentenyl pyrophosphate,* which isomerizes further to give *dimethylallyl pyrophosphate*. The latter two compounds are crucial intermediates. In effect, four units of isopentenyl pyrophosphate and two units of dimethylallyl pyrophosphate condense with each other to yield *squalene,* a C_{30} unsaturated hydrocarbon. The condensation is not direct, but rather involves the intermediate formation of two units of *geranyl pyrophosphate* (C_{10}) and two units of *farnesyl pyrophosphate* (C_{15}). One of the farnesyl units is then isomerized to yield *nerolidolpyrophosphate,* which in turn condenses with the second farnesyl unit to finally give *squalene.* Knowledge of this latter process is rather sketchy, with only a requirement of NADPH being well documented.

In most plants and in many microbes, a special branch of this pathway occurs at the point where farnesyl pyrophosphate is formed. It condenses with another unit of isopentenyl pyrophosphate to yield geranyl-geranyl pyrophosphate (C_{20}). A condensation of two units of geranyl-geranyl pyrophosphate then yields a C_{40}, unsaturated, aliphatic carbon skeleton that is the precursor of all *carotenoids* and *xanthophylls.*

The conversion of squalene to cholesterol in phase C (Fig. 16–7) is by far the most complex and least understood part of the whole pathway. First, squalene undergoes a specific oxidation to an epoxide across carbons 2 and 3. Then, by action of an enzyme or enzymes, an extraordinary reaction occurs. The carbon backbone of squalene epoxide undergoes a concerted in-

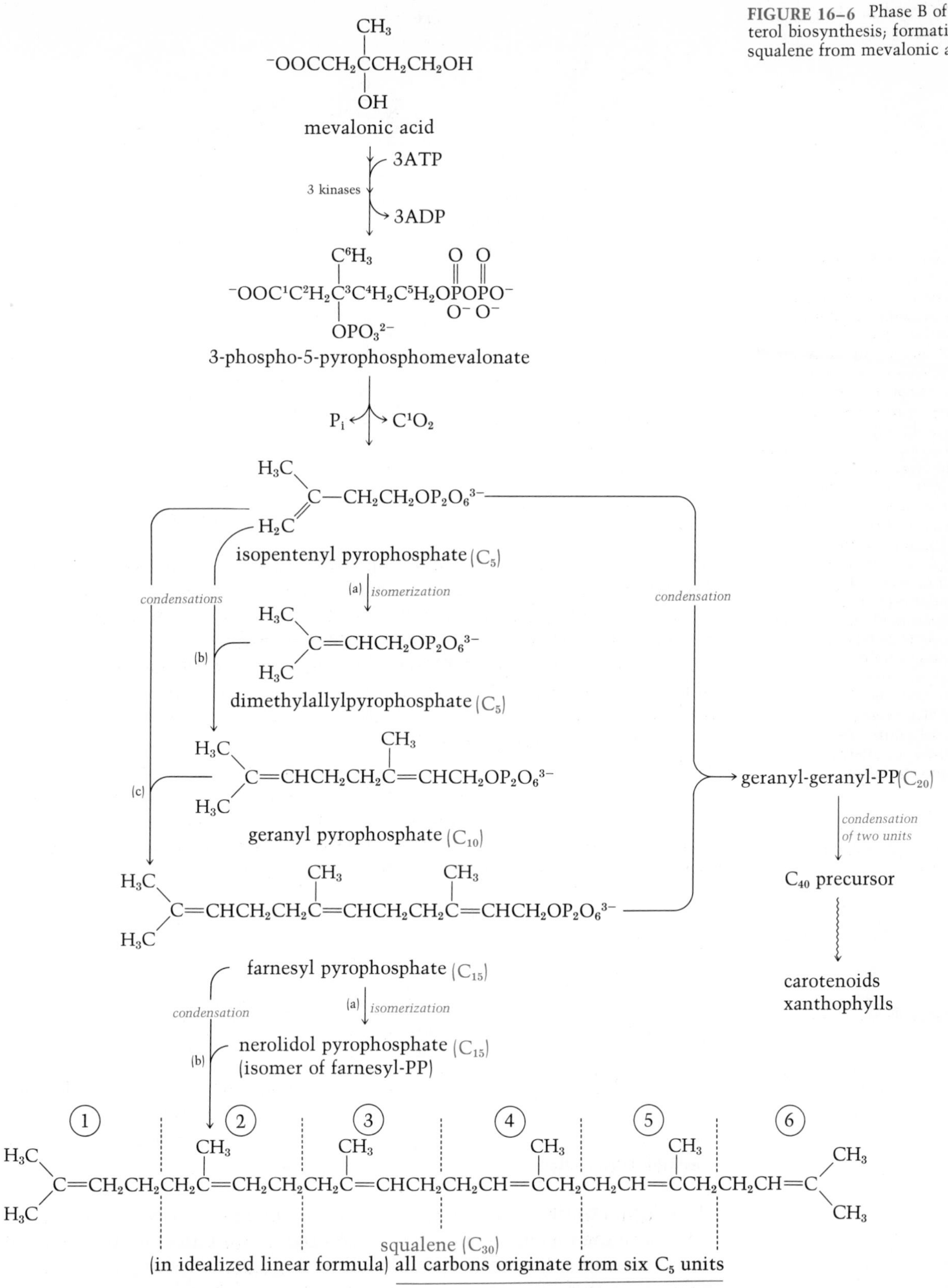

$$CH_3$$
$$^-OOCCH_2\overset{|}{\underset{|}{C}}CH_2CH_2OH$$
$$OH$$

mevalonic acid

3 kinases — 3ATP → 3ADP

$$C^6H_3 \quad\quad O\ \ O$$
$$^-OOC^1C^2H_2\overset{|}{C^3}C^4H_2C^5H_2OPOPO^-$$
$$OPO_3{}^{2-} \quad\quad O^-\ O^-$$

3-phospho-5-pyrophosphomevalonate

P_i ← → C^1O_2

$$H_3C$$
$$C-CH_2CH_2OP_2O_6{}^{3-}$$
$$H_2C$$

isopentenyl pyrophosphate (C_5)

(a) isomerization

(b)

$$H_3C$$
$$C=CHCH_2OP_2O_6{}^{3-}$$
$$H_3C$$

dimethylallylpyrophosphate (C_5)

condensations

(c)

$$H_3C \quad\quad CH_3$$
$$C=CHCH_2CH_2\overset{|}{C}=CHCH_2OP_2O_6{}^{3-}$$
$$H_3C$$

geranyl pyrophosphate (C_{10})

condensation → geranyl-geranyl-PP (C_{20})

condensation of two units

C_{40} precursor

$$H_3C \quad\quad CH_3 \quad\quad CH_3$$
$$C=CHCH_2CH_2\overset{|}{C}=CHCH_2CH_2\overset{|}{C}=CHCH_2OP_2O_6{}^{3-}$$
$$H_3C$$

farnesyl pyrophosphate (C_{15})

(a) isomerization

condensation

(b)

nerolidol pyrophosphate (C_{15})
(isomer of farnesyl-PP)

carotenoids
xanthophylls

① ② ③ ④ ⑤ ⑥

$$H_3C \quad\quad CH_3 \quad\quad CH_3 \quad\quad CH_3 \quad\quad CH_3 \quad\quad CH_3$$
$$C=CH_2CH_2CH_2\overset{|}{C}=CH_2CH_2CH_2\overset{|}{C}=CHCH_2CH_2CH=\overset{|}{C}CH_2CH_2CH=\overset{|}{C}CH_2CH_2CH=C$$
$$H_3C \quad\quad\quad\quad\quad\quad\quad\quad\quad\quad\quad\quad\quad\quad\quad\quad\quad\quad\quad\quad\quad\quad\quad\quad CH_3$$

squalene (C_{30})
(in idealized linear formula) all carbons originate from six C_5 units

FIGURE 16–7 Phase C of cholesterol biosynthesis.

squalene
(*in a more realistic
nonlinear formula*)

squalene epoxide

cholesterol (C$_{27}$)

multistep
$-3CO_2$

lanosterol (C$_{30}$)

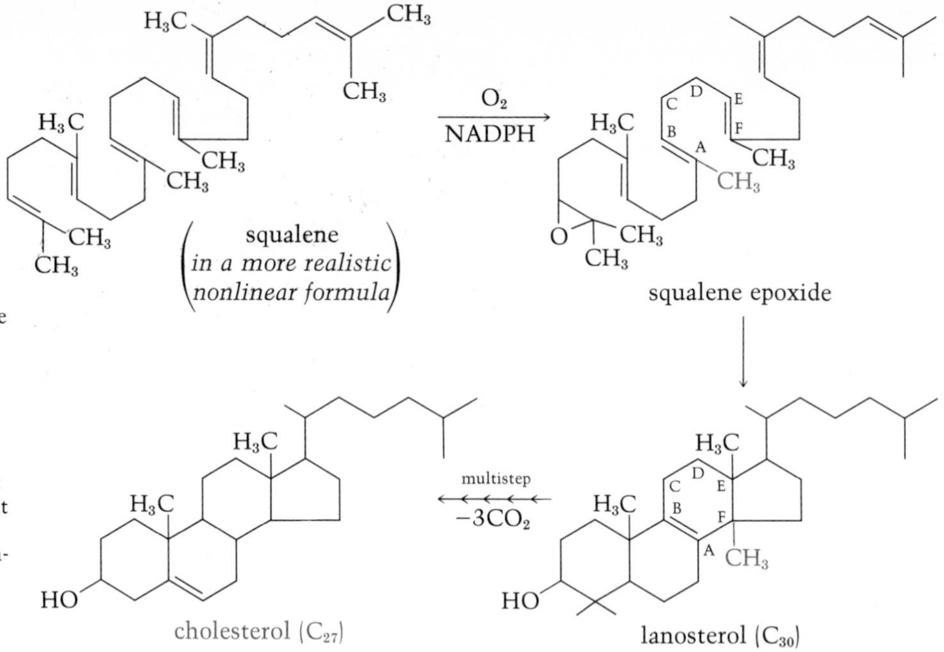

Despite many years of research the identity of several steps of cholesterol biosynthesis from squalene epoxide is presently unknown. Because of the close correlation of high levels of cholesterol in blood to arteriosclerosis and heart disease, there is considerable interest and activity in this field. A recent discovery is that liver homogenates contain a protein essential for the maximum conversion (*in vitro*) of water-insoluble precursors to cholesterol by purified and partially purified enzymes. The protein, termed the *squalene and sterol carrier protein* (SCP for short), is proposed to function by complexing to squalene prior to the participation of enzymes and forming squalene epoxide. The SCP remains complexed throughout all subsequent steps leading to cholesterol. The complexing to SCP probably increases the water solubility of squalene and the sterol precursors. Of greater interest, however, is the possibility that the SCP plays an important part in the regulation of cholesterol biosynthesis as well as in the conversion of cholesterol to other sterols.

tramolecular cyclization to yield four fused rings; this is accompanied by a stereospecific migration of two particular methyl groups. This product is called *lanosterol*. The subsequent conversion of lanosterol to cholesterol may involve a dozen or more additional enzymatic steps.

In mammals, most (about 90%) of the cholesterol synthesis takes place in the liver. Most (about 75%) of this liver-produced cholesterol is utilized to form the so-called *bile acids*, which assist in the digestion of dietary lipid in the intestine by making the lipid droplets smaller and more water-soluble, and hence more susceptible to the hydrolytic action of lipase enzymes. The major bile acid is *cholic acid*. It, like other bile acids, is present in bile in conjugation with glycine or taurine. At the pH of bile, and in the intestine, the active form of these emulsifying and solubilizing substances are the corresponding ionic forms, which are termed *bile salts*.

cholesterol $\rightarrow\rightarrow\rightarrow\rightarrow$

glycine
H_2NCH_2COOH or
$H_2NCH_2CH_2SO_3H$
taurine

cholic acid
$C_{23}H_{36}(OH)_3COOH$

$C_{23}H_{36}(OH)_3\overset{O}{\overset{\|}{C}}-NHCH_2COO^-$
glycocholic acid

$C_{23}H_{36}(OH)_3\overset{O}{\overset{\|}{C}}-NHCH_2CH_2SO_3^-$
taurocholic acid
bile acids (salts)

In addition to this role and its role as a *component of membranes* cholesterol is also the metabolic precursor of all other important steroids, many of which function as hormones (see the following page). Two important intermediates in this area of metabolism are *pregnenolone* and *progesterone*. Pregnenolone is the first steroid derived from cholesterol, and it then serves as the precursor of all other steroids. The subsequent fate of pregnenolone varies from tissue to tissue. The adrenal gland, for example, produces *aldosterone* and *cortisone* (cortisol). The former influences electrolyte and water metabolism, whereas the latter regulates the metabolism of carbohydrates, fatty acids, and proteins. The ovaries and testes elaborate, respectively, the estrogens (*estrone*) and androgens (*testosterone* and *androsterone*), both classes being responsible for the development of sex characteristics (see p. 336). The corpus luteum of the ovaries is the chief site of progesterone production in the female.

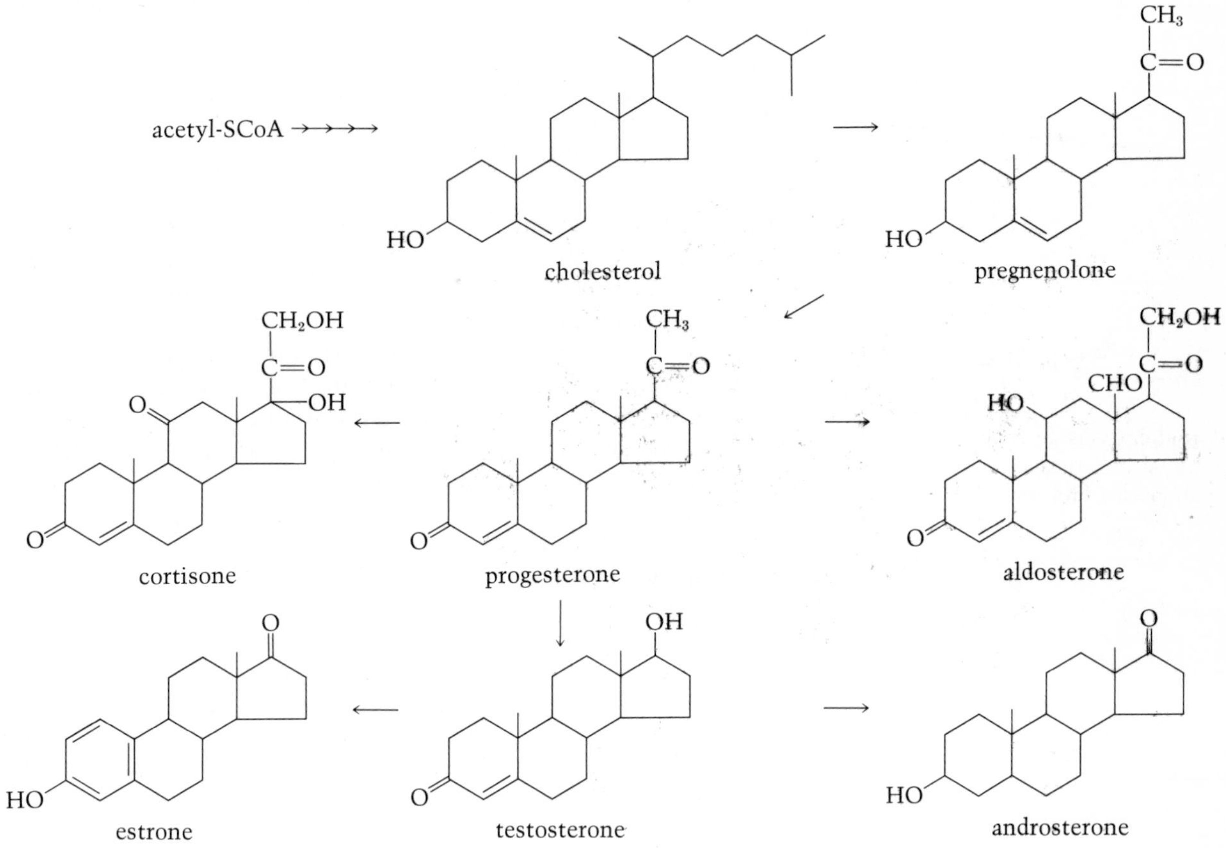

acetyl-SCoA →→→→

cholesterol

pregnenolone

cortisone

progesterone

aldosterone

estrone

testosterone

androsterone

Regulation of Cholesterol Biosynthesis

The normal level of cholesterol in plasma ranges from 150–200 mg/100 ml. A prolonged lipid-rich diet as well as physiological factors (these are not yet clear) can result in abnormally high levels, in the range of 200–300 mg/100 ml. If these levels persist, cholesterol plaques may occur in the aorta and in lesser arteries, a condition known as *arteriosclerosis* (or *atherosclerosis*), which can ultimately contribute to heart failure.

The maintenance of normal levels is accomplished *in vivo* by the regulation of the acetyl-SCoA →→→ cholesterol pathway. The control site is the β-hydroxy-β-methyl-glutaryl-SCCoA → mevalonate reaction and the *regulator is cholesterol* itself. Specifically, the activity of the HMG-SCoA reductase (the control site) is diminished by high concentrations of cholesterol. For several years it was argued that this effect of cholesterol was a typical feedback inhibition process, with the binding of cholesterol to a sec-

ondary site in the reductase causing a transition of the enzyme to a less active conformation. In other words, cholesterol was considered an allosteric
inhibitor of the reductase. However, recent studies suggest that the feedback action of cholesterol inhibits the synthesis of the reductase rather than
the activity of existing reductase. Whatever the mode of action, the effect
is the same: the further production of mevalonate—and ultimately more
cholesterol—is diminished.

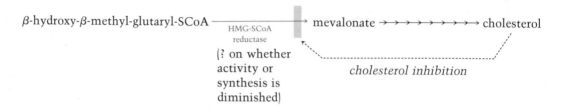

One of the current treatments of high blood cholesterol levels is the administration (orally) of substances that operate by diminishing the body's
capacity to synthesize cholesterol. *Para-chlorophenoxyisobutyric acid* (see
margin) is one such substance, and a very effective one at that. The major
effect of this material is proposed to be inhibition of HMG-CoA reductase.
Research is continuing (a) to develop more drugs for the same purpose that
might be even more effective and not have undesirable secondary effects, (b)
to determine the various factors (besides dietary) that contribute to elevated
levels of cholesterol in the blood, and (c) to analyze the whole question of
plaque formation.

One very promising area of current research in cholesterol biochemistry
deals with its transport and delivery as an ester to other cells. The transport
involves *plasma lipoproteins*. The delivery involves specific receptor molecules in the membranes of target cells that bind to the cholesterol-carrying
lipoproteins. The lipoprotein enters the cell by *endocytosis* and is then degraded by lysosome enzymes to release free cholesterol for use by the cell.
Aberrations in the cell entry process would keep cholesterol in the blood
and obviously contribute to the elevated levels of cholesterol in the blood.

$$Cl-\langle\bigcirc\rangle-O-CH_2CHCOOH$$
$$\underset{CH_3}{|}$$

p-chlorophenoxyisobutyric acid
(in clinical use for depressing
high cholesterol levels)

POSTSCRIPT

If you reflect on both the content of this chapter and our previous analysis of
the metabolic relationships of the citric acid cycle in Chap. 13, you can
appreciate that there is something rather special about the metabolic role of
acetyl-SCoA as an individual metabolite. We can summarize this quite
readily. Acetyl-SCoA is the compound that is the very nucleus of metabolism. It acts both as a metabolic "receiving and shipping department" for
the carbon of all classes of biomolecules, and as a major source of useful
metabolic energy. The general theme of these relationships is diagramatically summarized below. On inspecting the diagram, remember that,
though all processes can and do occur simultaneously, there are (probably in
all organisms) elaborate regulatory checks and balances, as previously
described for the major pathways of carbohydrate metabolism, that modulate the extent to which each process occurs in the presence of the others.

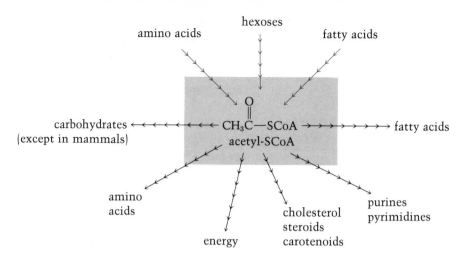

LITERATURE

BREMER, J. "Carnitine and Its Role in Fatty Acid Metabolism." *Trends Biochem. Sci.*, **2,** 207–209 (1977). A short review article.

DAWSON, R. M. C. "The Metabolism of Animal Phospholipids and Their Turnover in Cell Membranes." In *Essays in Biochemistry*, P. N. Campbell and G. D. Greville, eds., Vol. 2, pp. 69–116. New York: Academic Press, 1966. An excellent review article.

DEMPSEY, M. E. "Regulation of Steroid Biosynthesis." *Ann. Rev. Biochem.*, **43,** 967–990 (1974). A very good review article describing current knowledge regarding cholesterol biosynthesis. The squalene carrier protein (SCP) is described.

FISHMAN, P. H., and R. O. BRADY, "Biosynthesis and Function of Gangliosides," *Science*, **194,** 906–915 (1976). A good review article covering (a) the formation of gangliosides, (b) the consequences of impaired formation, and (c) the physiological roles of gangliosides, such as that of mediating the action of cholera toxin.

FRIEDEN, E., and H. LIPNER. *Biochemical Endocrinology of the Vertebrates.* Englewood Cliffs, New Jersey: Prentice-Hall, 1971. A compact introduction to basic biochemical facts on the subject of hormones. Available in paperback.

FULCO, A. J. "Metabolic Alterations of Fatty Acids." *Ann. Rev. Biochem.*, **43,** 215 (1974). A review article focusing on the biosynthesis of unsaturated fatty acids and hydroxyl-containing fatty acids; complemented by the Volpe and Vagelos review cited below.

GATT, S., and Y. BARENHOLZ. "Enzymes of Complex Lipid Metabolism." *Ann. Rev. Biochem.*, **42,** 61–90 (1973). Review article devoted to phosphoacylglycerols and sphingolipids.

GINSBURG, A., and E. R. STADTMAN. "Multienzyme Systems." *Ann. Rev. Biochem.*, **39,** 431–472 (1970). A current review article on the biochemistry of multienzyme complexes, including fatty acid synthetase complexes.

GOLDSTEIN, J. L., and M. S. BROWN. "The Low-density Lipoprotein Pathway and Its Relation to Athero-sclerosis." In *Ann. Rev. Biochem.*, **46,** 897–930 (1977). A good review article describing recent knowledge on the relationship of cholesterol to atherosclerosis.

GREEN, D. E., and D. W. ALLMAN. "Fatty Acid Oxidation" and "Biosynthesis of Fatty Acids." In *Metabolic Pathways*, 3rd edition, D. Greenberg, ed., Vol. 1. New York: Academic Press, 1968. Two review articles giving a thorough analysis of catabolic and anabolic pathways of the fatty acids. Other articles in this volume are devoted to the metabolism of steroids, steroid hormones, carotenoids, and Vitamin A.

GREVILLE, G. D., and P. K. TUBBS. "The Catabolism of Long-Chain Fatty Acids in Mammalian Tissues." In *Essays in Biochemistry*, P. N. Campbell and G. D. Greville, eds., Vol. 4, pp. 155–212. New York: Academic Press, 1968. A review article.

LOWENSTEIN, J. M., ed. "Lipids." In *Methods in Enzymology*, S. P. Colowick and N. O. Kaplan, eds., Vol. 14. New York: Academic Press, 1969. A reference source for procedures in the isolation and assay of enzymes associated with lipid metabolism including summaries of catalytic properties.

MAJERUS, P. W., and P. R. VAGELOS. "Fatty Acid Biosynthesis and the Acyl Carrier Protein." In *Advan. Lipid Res.*, R. Paoletti and D. Kritchevsky, eds., Vol. 5. New York: Academic Press, 1967. A review article.

VOLPE, J. J., and P. R. VAGELOS. "Saturated Fatty Acid Biosynthesis and its Regulation." *Ann. Rev. Biochem.*, **42,** 21–60 (1973). A good review article; complemented by the Fulco review cited above.

WAKIL, S. J., ed. *Lipid Metabolism.* New York: Academic Press, 1970. A comprehensive treatise containing articles that investigate in detail the biochemistry and enzymology of the metabolism of various lipids in animals, plants, and microorganisms. A chapter on the prostaglandins is included.

WOOD, H. G., and R. E. BARDEN. "Biotin Enzymes." In *Ann. Rev. Biochem.*, **46,** 385–414 (1977). A review article.

EXERCISES

16–1. If only the first of eight molecules of acetyl-SCoA used in the biosynthesis of palmitic acid were labeled with ^{14}C, which of the following would represent the location of the label in palmitic acid?

$$^{-}OO^{14}C^{14}CH_2CH_2CH_2CH_2CH_2CH_2CH_2\text{-}CH_2CH_2CH_2CH_2CH_2CH_2CH_2CH_3$$

or

$$^{-}OOCCH_2CH_2CH_2CH_2CH_2CH_2CH_2CH_2\text{-}CH_2CH_2CH_2CH_2CH_2{}^{14}CH_2{}^{14}CH_3$$

16–2. Determine the *net yield* of ATP molecules from the complete aerobic oxidation of one molecule of myristyl-SCoA to CO_2. Also determine the number of ATPs *required* in this process.

16–3. If the palmitic acid produced in Exercise 16–1 were degraded to acetyl-SCoA via β-oxidation, would the labeled acetyl-SCoA unit be produced from the thiolytic cleavage of β-ketomyristyl-SCoA or the thiolytic cleavage of acetoacetyl-SCoA?

16–4. Outline (as shown on p. 504) the complete catabolism of oleic acid, linolenic acid, and arachidonic acid to acetyl-SCoA. (See also Table 10–1.)

16–5. Refer to Eq. A on p. 505. Why does the conversion of stearic acid to nine units of acetyl-SCoA require nine rather than eight units of coenzyme A?

16–6. Why is the ATP yield per six carbons of fatty acid greater than the ATP yield per six carbons of a hexose? (Consider, of course, that the catabolism in each case would proceed all the way to CO_2.)

16–7. The metabolism of aspartate is linked to the citric acid cycle by the following reversible transamination reaction.

$$\text{aspartate} + \alpha\text{-keto acid} \xrightarrow[\text{aminase}]{\text{trans-}} \text{oxaloacetate} + \text{amino acid}$$

Using this information and referring back to Figure 13–4, suggest a reaction sequence that will account for the fact that some of the carbons of aspartate are eventually converted to a metabolite, which can then be used in the biosynthesis of *either* fatty acids or carbohydrates.

16–8. In the situation posed in the preceding exercise, what is the only possible explanation for the conversion of carbons in aspartate to carbons in fatty acids according to the following reaction sequence?

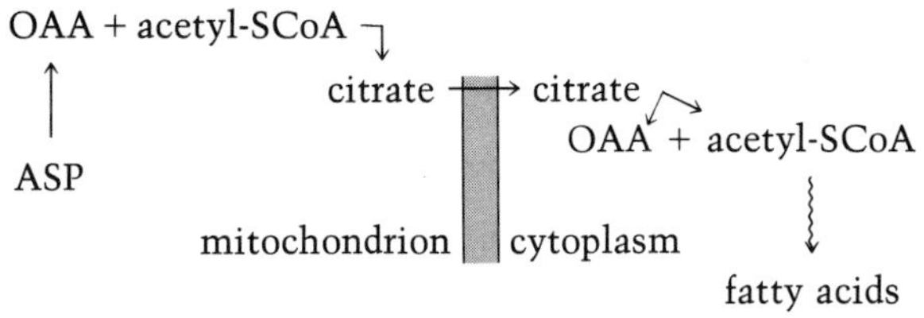

16–9. How do the following reactions differ?

$$RCH_2CH_2\overset{\overset{O}{\|}}{C}CH_2\overset{\overset{O}{\|}}{C}\text{—SCoA} \xrightarrow[\text{mitochondrion}]{\text{in}} RCH_2CH_2\overset{\overset{OH}{|}}{C}HCH_2\overset{\overset{O}{\|}}{C}\text{—SCoA}$$

$$RCH_2CH_2\overset{\overset{O}{\|}}{C}CH_2\overset{\overset{O}{\|}}{C}\text{—SCoA} \xrightarrow[\text{cytoplasm}]{\text{in}} RCH_2CH_2\overset{\overset{OH}{|}}{C}HCH_2\overset{\overset{O}{\|}}{C}\text{—SCoA}$$

16–10. The active site of malate synthetase probably promotes the formation of an active carbanion species of acetyl-SCoA, which then condenses with glyoxylate to form malate. Illustrate the mechanism.

16–11. Write a net reaction for the operation of the glyoxylate cycle (refer to p. 416 for a format to follow). Then compare the ATP yield of the glyoxylate cycle to that of the citric acid cycle.

16–12. If a molecule of uniformly labeled acetyl-SCoA (^{14}C) condensed with acetoacetyl-SCoA to form β-hydroxy-β-methyl-glutaryl-SCoA, which was then converted to mevalonate, what would be the pattern of the ^{14}C label in mevalonate?

16–13. According to the outline of the anabolic pathway given in this chapter, estimate the amount of energy required to produce one molecule of cholesterol from acetyl-SCoA. Make the estimate in terms of the number of high-energy compounds that are required, expressing each in terms of its ATP equivalent. Assume that an acyl-SCoA compound is thermodynamically equivalent to one ATP, and that one molecule of NADH or NADPH is potentially equivalent to three ATPs. Of course, ATP is ATP.

CHAPTER 17

METABOLISM OF
NITROGEN-CONTAINING
COMPOUNDS: AMINO ACIDS
AND NUCLEOTIDES

The use of amino acids, purine nucleotides, and pyrimidine nucleotides in the assembly of proteins and nucleic acids (Chap. 8) is only one aspect of their metabolism—albeit a very important one. Other considerations, to which this chapter is devoted, include (a) their biosynthesis, (b) their degradation, (c) the flow of nitrogen among the amino acids and from amino acids to purines, pyrimidines, and other nitrogen-containing compounds, (d) the excretion of excess nitrogen by organisms, and (e) various other processes that involve specific amino acid transformations. An introduction to some aspects of the metabolism of these compounds was presented in Chap. 13 in describing the amphibolic role of the citric acid cycle. Now we will focus on the many details of individual reactions.

Our coverage of this area of metabolism will not be exhaustive because it is a very broad subject. However, the selection of material does provide adequate exposure to items (a) through (e). In addition, this chapter includes discussion of (f) three other coenzymes not yet examined in previous chapters—pyridoxal phosphate, tetrahydrofolic acid, and Vitamin B_{12}, (g) some principles of nutrition, and (h) the flow of nitrogen in the biosphere.

PART A: AMINO ACID METABOLISM

UNIFYING PRINCIPLES

Although amino acid metabolism is one of the most massive areas for discussion in general biochemistry courses (for example, inspect Fig. 17–2, which summarizes only part of the metabolism of just three amino acids, phenylalanine, tyrosine, and tryptophan), it is possible to unify the subject to some extent.

First, as previously considered in Chap. 13 (see Fig. 13–4) and summarized in the diagram shown below, the metabolism of amino acids is integrated into the mainstream of metabolism by (a) an *input* to the citric acid cycle of a portion of (or the entire carbon skeleton of) nearly all the common amino acids during their catabolism, and (b) an *output* of carbon from the citric acid cycle in the form of cycle intermediates that serve as precursors of the carbon skeletons of many amino acids during their anabolism. Because of the individuality of the pathways involved, a multitude of reactions would be necessary to define each type of link for all the amino acids. While it is not our intent to do that, a few words are in order concerning the general metabolic significance of these relationships. As for the several catabolic links to the citric acid cycle, these account for the fact stated earlier that amino acids can be effective intracellular sources of both metabolic energy (ATP) and carbon needed for the formation of carbohydrates and lipids. Both factors, accompanied by a depression of the hunger sensation, account for the success of weight control diets that are high in protein and low in carbohydrate and lipid. The anabolic link, of course, accounts for the production of amino acids from carbohydrates and lipids. (And yes, the carbons of one amino acid can be used for the production of another amino acid.)

During catabolism, if the amino acid carbons are shunted largely into carbohydrates, the parent amino acids are termed *glycogenic* (carbohydrate

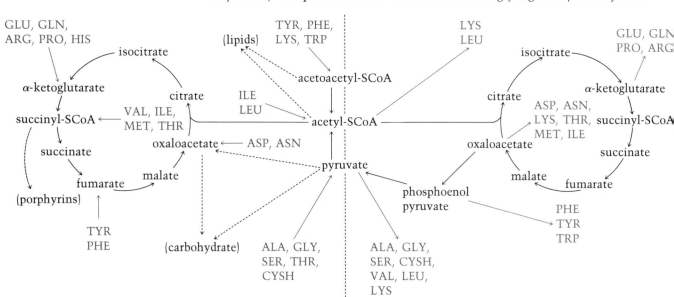

Entrance of carbon into the mainstream of
metabolism from amino acid degradation

Removal of carbon from the mainstream of
metabolism for amino acid formation

forming). If the carbons are more efficiently incorporated into lipid material, the amino acids are termed *ketogenic* (lipid forming). The only requirement for being classified as glycogenic is that the amino acid be converted to *pyruvate* or *oxaloacetate*. This requirement is met by nearly two-thirds of the common amino acids. Those that yield pyruvate directly are glycine, alanine, cysteine, serine, and threonine. Those that yield *oxaloacetate*, or any citric acid cycle intermediate that would eventually be converted to oxaloacetate, are aspartate, asparagine, glutamate, glutamine, arginine, methionine, valine, isoleucine, tyrosine, phenylalanine, histidine, and proline. The potential ketogenic amino acids are those that directly yield *acetyl-SCoA* or *acetoacetyl-SCoA*. Included here are tyrosine, phenylalanine, leucine, isoleucine, lysine, and tryptophan. Note that although several amino acids are only glycogenic, some are both glycogenic and ketogenic. Leucine is the only amino acid that is ketogenic but not glycogenic.

glycogenic	ketogenic
glycine	leucine
alanine	phenylalanine
cysteine	isoleucine
serine	lysine
aspartic acid	tryptophan
asparagine	tyrosine
glutamic acid	
glutamine	
arginine	
methionine	
valine	
isoleucine	
tyrosine	
phenylalanine	
histidine	
proline	

The second element of unity is that, although there are a great number of reactions associated with the metabolism of amino acids as a group, there is considerably less diversity in the *type of reactions* involved. Three reactions of particular importance are *transamination, decarboxylation, and deamination*. Of these, transamination reactions play a major role and hence will be discussed first and in greater depth.

Transamination

Frequently the first chemical event occurring in amino acid degradation and the last step in the synthesis of amino acids is a *transamination*. As the word implies, a transamination reaction is characterized by the transfer of an amino grouping. To be more specific, an *amino grouping is transferred from a donor amino acid to an acceptor α-keto acid to yield the α-keto acid of the donor amino acid and the amino acid of the original α-keto acid acceptor.* The reaction is catalyzed by an enzyme called *transaminase* that requires a metal ion and *pyridoxal phosphate* (see p. 531) for activity. As you might expect, the reaction is readily reversible.

$$\underset{\text{amino acid}}{R-\overset{\overset{\displaystyle NH_3^+}{|}}{C}H-COO^-} + \underset{\text{α-keto acid}}{R'-\overset{\overset{\displaystyle O}{\|}}{C^\alpha}-COO^-} \underset{\substack{\text{(pyridoxal}\\ \text{phosphate)}}}{\overset{\text{transaminase}}{\rightleftarrows}} \underset{\text{α-keto acid}}{R-\overset{\overset{\displaystyle O}{\|}}{C}-COO^-} + \underset{\text{amino acid}}{R'-\overset{\overset{\displaystyle NH_3^+}{|}}{C}H-COO^-}$$

Although several different transaminases occur in nature, they do not generally display an absolute level of substrate specificity. However, many enzymes show a preference for utilizing α-ketoglutarate as the acceptor keto acid, yielding glutamate, or vice versa. Although the reaction as shown from left to right (→) can account for the direct biosynthesis of glutamate from a citric acid cycle intermediate (a very similar transformation and explanation exist for the oxaloacetate ⇌ aspartate set), the metabolic significance of the α-ketoglutarate ⇌ glutamate transamination system is much broader in scope. In amino acid catabolism, for example, it is believed that

$$R-\overset{\overset{\displaystyle NH_3^+}{|}}{C}H-COO^- + {}^-OOCCH_2CH_2-\underset{\text{α-ketoglutarate}}{\overset{\overset{\displaystyle O}{\|}}{C}}-COO^- \underset{\substack{\text{(pyridoxal}\\ \text{phosphate)}}}{\overset{\text{transaminase}}{\rightleftarrows}} R-\overset{\overset{\displaystyle O}{\|}}{C}-COO^- + {}^-OOCCH_2CH_2-\underset{\text{glutamate}}{\overset{\overset{\displaystyle NH_3^+}{|}}{C}H-COO^-}$$

α-ketoglutarate is the major amino group acceptor for most of the amino acid transaminations (path A in the diagram below). In other words, α-ketoglutarate acts as a "collection or receiving compound" for nitrogen. The subsequent metabolic fate of the amino group of glutamate is quite varied.

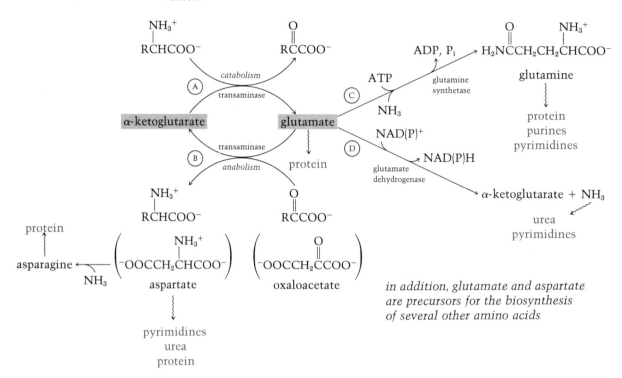

in addition, glutamate and aspartate are precursors for the biosynthesis of several other amino acids

Path (B) is a reutilization of the nitrogen in the biosynthesis of amino acids, also via a transamination reaction. One particular reaction of great importance in this regard occurs when oxaloacetate is the α-keto acid acceptor, yielding aspartate. In addition to its use in polypeptide biosynthesis, aspartate is also a source of both carbon and nitrogen for (a) the biosynthesis of pyrimidines (p. 557), and (b) the biosynthesis of a few other amino acids, and is also (c) a donor of nitrogen for the production of urea, which is the major form of excreted nitrogen in most animal organisms (see p. 535). Path (C), catalyzed by an ATP-dependent *glutamine synthetase,* is an amidation, leading to the formation of glutamine. The glutamine then serves as a nitrogen donor in the biosynthesis of several biomolecules, most notably the purines and pyrimidines (see p. 557). Path (D) results from the action of *glutamate dehydrogenase* (see p. 534), which converts glutamate to α-ketoglutarate and free ammonia. The ammonia can then be used in certain anabolic reactions as an important nitrogen source (glutamine biosynthesis; pyrimidine biosynthesis), or excreted from the organism as ammonia or urea.

Because it does have a central role in amino acid metabolism, and because the reaction mechanism is understood in some detail, let us consider the process of transamination in greater depth. The main feature is the participation of *pyridoxal phosphate* as a coenzyme. Although our immediate concern is its role in transamination reactions, pyridoxal phosphate also functions as a required coenzyme in other types of amino acid transformations, such as decarboxylation and racemization (L-amino acid $\rightleftarrows$ D-amino acid).

Pyridoxal phosphate represents the metabolically active form of a vitamin, specifically, *Vitamin B_6*. As shown below, Vitamin B_6, sometimes called *pyridoxine* or *pyridoxol*, is a substituted pyridine. Oxidation of one hydroxymethyl group (—CH_2OH) to an aldehyde linkage and phosphorylation of the second yields pyridoxal phosphate. The active site of pyridoxal phosphate resides in the aldehyde grouping, which is susceptible to undergoing covalent interaction with free amino groupings (such as that contributed by amino acids, for example) to form a *Schiff base* species with a characteristic imino linkage ($-\overset{\overset{\textstyle H}{|}}{C}=N-$). In fact, it is believed that, prior to reaction with a substrate amino acid, the pyridoxal phosphate is covalently bound to the enzyme by just such a linkage, with the amino grouping being contributed by the ϵ-amino group of a lysine chain in the polypeptide chain. The first stage of a reaction involves formation of the free aldehyde species of pyridoxal phosphate, which then reacts with the amino group of the substrate amino acid resulting in the formation of a new Schiff base species between the coenzyme and the substrate amino acid. The new covalent Schiff base complex probably experiences noncovalent interactions with electrophilic (electron seeking) groupings (E) contributed by the polypeptide.

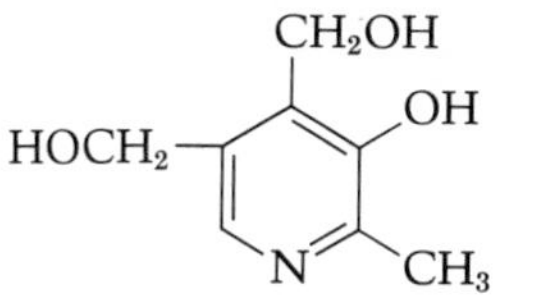

Vitamin B_6
(pyridoxol)

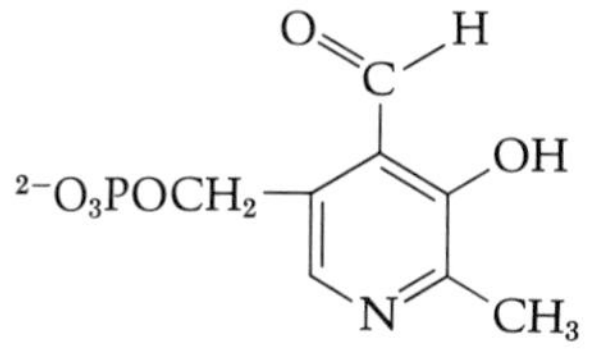

pyridoxal
phosphate

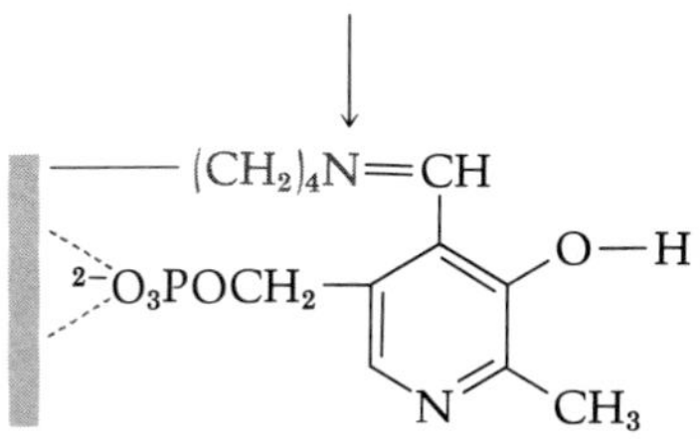

coenzyme covalently attached
to lysine residue of enzyme via
Schiff base linkage

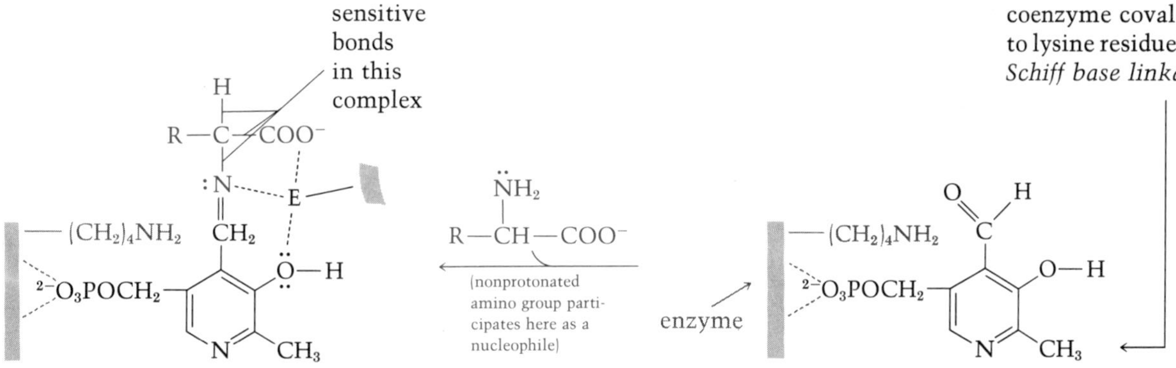

sensitive bonds in this complex

enzyme $\sim$ coenzyme $\sim$ substrate
complex

(dashed lines represent noncovalent
associations to electrophilic (E)
groups in the enzyme)

(nonprotonated amino group participates here as a nucleophile)

enzyme

free aldehyde group
in coenzyme

The subsequent fate of the complex (be it transamination, decarboxylation, or racemization) during the rest of the reaction is controlled by the individuality of the enzyme (that is, the operational specificity of the active site). The events that describe the mode of action of the transaminase enzyme are diagramed below. The proposed mechanism consists of two stages. First, the amino acid (substrate 1) is converted to its α-keto acid by the rearrangement of the imino linkage, which is followed by hydrolysis to yield the α-keto acid (product 1). Note that the amino grouping remains with the enzyme-coenzyme system as *pyridoxamine phosphate*. The second stage is just the reverse of the first. It begins with a Schiff base formation involving the acceptor α-keto acid (substrate 2); then the corresponding amino acid (product 2) is formed. The net result of this cyclic process is the transfer of an amino group from the donor amino acid to the acceptor keto acid.

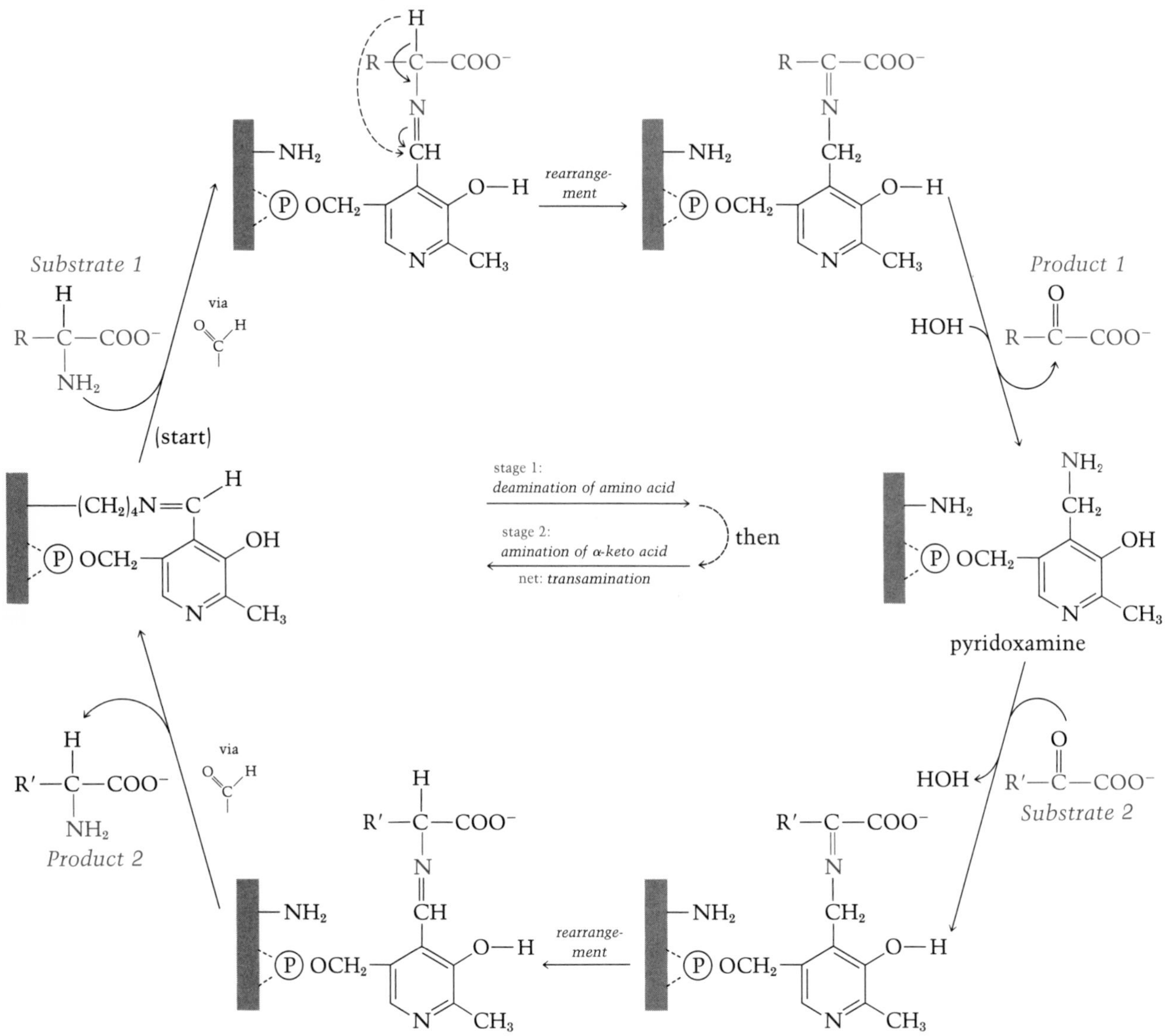

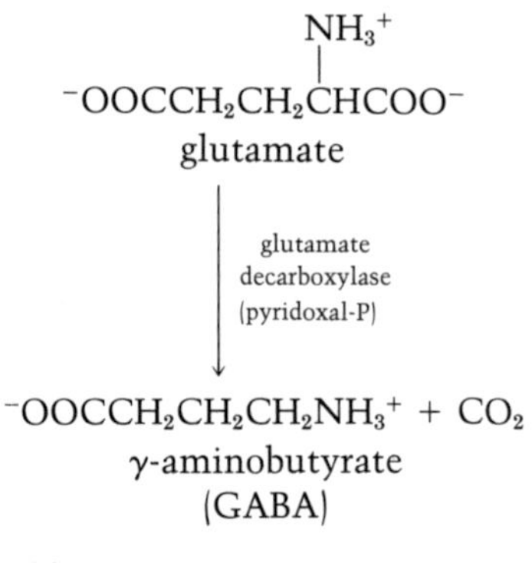

Decarboxylation

NH_3^+

$^-OOCCH_2CH_2CHCOO^-$

glutamate

glutamate
decarboxylase
(pyridoxal-P)

$^-OOCCH_2CH_2CH_2NH_3^+ + CO_2$

γ-aminobutyrate
(GABA)

The removal of the α-carboxyl group of amino acids via pyridoxal phosphate-dependent decarboxylases occurs in all types of organisms. Although acid decarboxylation has a limited occurrence in humans, a few of the reactions that do occur are of considerable importance. Brain cells, for example, contain *glutamic acid decarboxylase*, a highly specific enzyme that produces *γ-amino butyric acid* (GABA) from glutamate. The proposed function of GABA in the brain is to act as a regulator of nerve transmission. Although its precise mechanism of action is unknown, GABA is recognized as a rather potent inhibitor of the transmission of impulses in nerve cells. Glutamate, on the other hand, has an excitatory function.

Thus, the production of glutamate and the subsequent glutamate →
GABA conversion may serve as major control signals in the neurophysi-
ology of the brain.

The decarboxylation of aromatic amino acids such as histidine, tryptophan, 5-hydroxytryptophan, and 3,4-dihydroxyphenylalanine are also of importance. (The latter two amino acids are normal metabolic intermediates
produced during the metabolism of tryptophan and phenylalanine or tyrosine, respectively.) The product of histidine decarboxylation is *histamine.*
Excessive production of histamine occurs during hypersensitive allergic
reactions, and its symptoms are unfortunately familiar to many individuals.
The decarboxylation product of 5-hydroxytryptophan is *5-hydroxy-*
tryptamine, commonly called *serotonin.* The bulk of serotonin production is confined to the brain, intestine, and platelet cells of blood. Although its mode of action is not completely clear, serotonin has long been
implicated as an agent that plays an important role in the regulation of the
nervous system. Tryptophan itself is also decarboxylated in brain cells to
yield *tryptamine,* another neurologically active amine. In plants, tryptamine is a biosynthetic precursor of *indole acetic acid,* a very potent hormone that promotes plant growth. *3,4-Dihydroxyphenylethylamine* (also
called *dopamine*), another amine with neurological activity, is produced
from the decarboxylation of 3,4-dihydroxyphenylalanine (DOPA). In a later
section we will consider some proposals regarding the physiological significance of serotonin, tryptamine, and dopamine production.

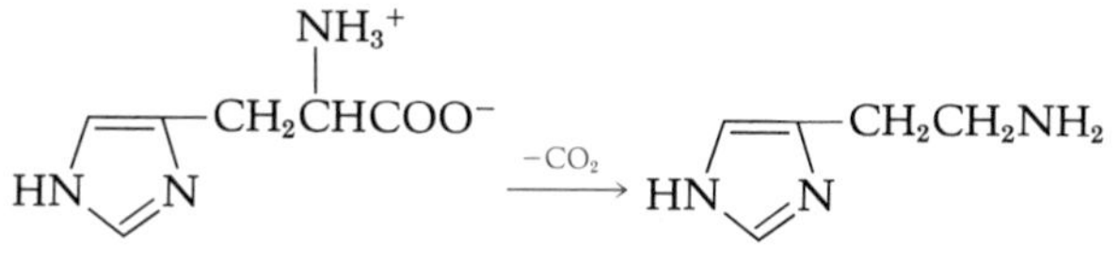

histidine → histamine

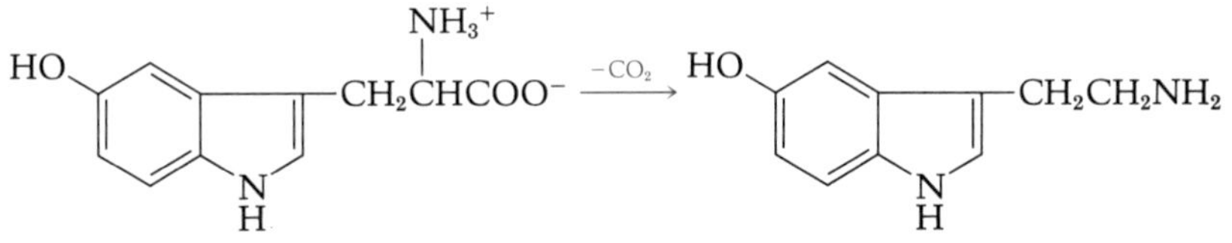

5-hydroxytryptophan → 5-hydroxytryptamine (serotonin)

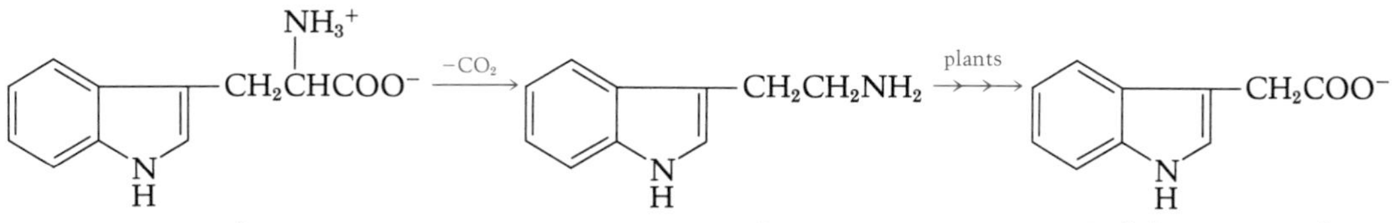

tryptophan → tryptamine → indole acetic acid

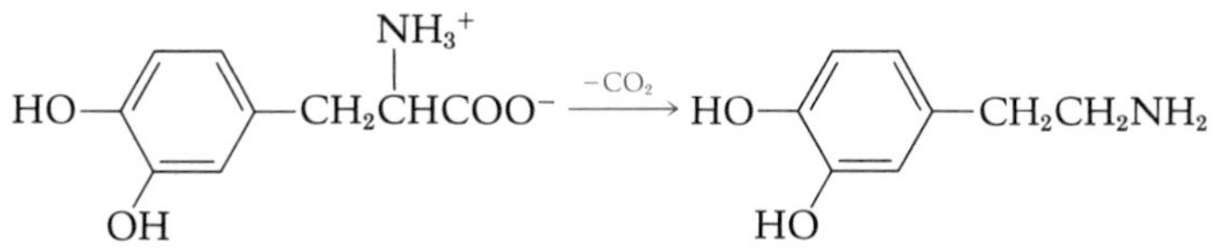

3,4-dihydroxyphenylalanine (DOPA) → 3,4-dihydroxyphenylethylamine (dopamine)

Oxidative Deamination

A secondary route for the conversion of L-amino acids to the corresponding α-keto acids (remember that the primary route is transamination) is *oxidative deamination*, a process catalyzed by either an NAD$^+$-dependent dehydrogenase or a flavin-dependent oxidase. The most important and widespread dehydrogenase is *glutamate dehydrogenase*.

$$\text{L-glutamate} + \text{NAD}^+ \xrightleftharpoons[\text{dehydrogenase}]{\text{glutamate}} \alpha\text{-ketoglutarate} + \text{NH}_3 + \text{NADH(H}^+)$$
$$\text{or} \qquad\qquad\qquad\qquad\qquad\qquad\qquad\qquad \text{or}$$
$$\text{(NADP}^+) \qquad\qquad\qquad\qquad\qquad\qquad\qquad\quad \text{(NADPH)}$$

Although the reaction is readily reversible, its primary role seems to be in deamination. Although the liberated ammonia could be reused in other reactions, such as the amidation of glutamate and asparate to glutamine and asparagine, most is excreted as ammonia or urea (see the next section). An interesting characteristic of this enzyme is that, although glutamate is the favored substrate, the enzyme can utilize either NAD$^+$ or NADP$^+$ as the coenzyme. Hence, in addition to its role in nitrogen metabolism, the enzyme can also serve as an intracellular source of NADPH needed for anabolism.

Two different types of flavin-dependent amino acid oxidases exist. One utilizes FAD as the electron acceptor and the other uses FMN. The FAD oxidase acts on D-amino acids, whereas the FMN-dependent oxidase acts on L-amino acids. The biological significance of these reactions is not clear, since the production of reducing power does not appear to have any metabolic value. Rather, the reduced flavoprotein reacts directly and irreversibly with oxygen, producing hydrogen peroxide, a substance with a potent toxicity. A safety valve does exist, however, in the form of a special enzyme, called *catalase*, which catalyzes the decomposition of hydrogen peroxide to water and oxygen.

FMN (flavin mononucleotide) is a second coenzyme form of riboflavin with a similar but simpler structure than FAD. FMN is composed of riboflavin, ribitol, and one phosphate grouping. FAD has an additional AMP grouping (see page 422, Chapter 13).

FMN

$$\text{L-amino acid} + \text{E—FMN} \xrightarrow{\text{amino acid}}^{\text{oxidase}} \alpha\text{-keto acid} + \text{NH}_3 + \text{E—FMNH}_2$$

$$\text{D-amino acid} + \text{E—FAD} \xrightarrow{\text{oxidase}} \alpha\text{-keto acid} + \text{NH}_3 + \text{E—FADH}_2$$

$$\text{E—FMNH}_2 + \text{O}_2 \longrightarrow \text{E—FMN} + \text{H}_2\text{O}_2$$
$$\text{(E—FADH}_2) \qquad\qquad \text{(E—FAD)}$$

$$\text{H}_2\text{O}_2 \xrightarrow{\text{catalase}} \text{H}_2\text{O} + \tfrac{1}{2}\text{O}_2$$

An enzyme widespread in the animal kingdom and important to normal brain metabolism is *monoamine oxidase*. This enzyme, which is related in action, but not in structure, to the amino acid oxidases, catalyzes the oxidative deamination of primary amines to the corresponding aldehydes. The significance of this reaction is discussed further on p. 542.

$$\text{RCH}_2\text{NH}_2 + \text{O}_2 \xrightarrow[\text{oxidase}]{\text{monoamine}} \text{RCHO} + \text{NH}_3 + \text{H}_2\text{O}_2$$
$$\text{(amine)} \qquad\qquad\qquad\qquad \text{(aldehyde)}$$

Excess or unused nitrogen resulting from amino acid degradation is excreted from an organism in the form of ammonia, urea, and uric acid. As was pointed out above, a major source of ammonia is deamination of glutamate by glutamate dehydrogenase. Urea is formed by a special group of enzymes whose combined operation constitutes the *urea cycle,* which will be examined here. Uric acid is not a direct end product of amino acid metabolism, but rather originates from the degradation of purines (p. 566). The inclusion of uric acid as an excretory form of amino acid nitrogen is based on the fact that the nitrogen in purines originates from amino acids. Although the major end product varies from one type of organism to another, certain generalizations are possible. For example, most bacteria, plants, and fish excrete ammonia, whereas birds and most invertebrates excrete uric acid. In most mammals, including humans, the primary product is urea, due to the opertion of the enzymes of the urea cycle. In mammals these enzymes are localized in the liver. From the liver, the urea is passed into the circulating blood, which is eventually dialyzed in the kidneys, resulting in loss of the low molecular weight urea to the urine.

The urea cycle itself is composed of only five reactions, each catalyzed by a different enzyme. As indicated in the following net reaction, (a) the immediate source of urea nitrogen is twofold, namely, ammonia and the amino group of aspartate; (b) the urea carbon originates from CO_2, which thus can be thought of as the nitrogen acceptor; and (c) the process is significantly endergonic, requiring 3 ATPs per molecule of urea produced.

$$CO_2 + NH_3 + H_2O + \text{aspartate} \xrightleftharpoons[\substack{(N) \quad 3ATP}]{\substack{\text{5 enzymes} \quad AMP, PP_i \\ 2ADP, 2P_i}} H_2N-\overset{\overset{O}{\|}}{C}-NH_2 + \text{fumarate}$$
$$\text{urea}$$

As we have found on other occasions, however, a net reaction relates nothing at all about the sequence of events that intervene between reactants and products. These are diagramed in Fig. 17–1. An analysis of how this metabolic sequence results in urea formation best begins by considering the formation of *carbamyl phosphate* from CO_2, NH_3, and ATP as the first reaction. The enzyme involved is *carbamyl phosphate synthetase* and, as you can see, two molecules of ATP are required. Once formed, carbamyl phosphate condenses with *ornithine* to form *citrulline.* Both of these substances are basic alpha-amino acids found in most organisms, but not known to occur in any of the cellular proteins. In the second ATP-dependent reaction, citrulline condenses with aspartate (the second N source) to yield *argininosuccinate,* which in turn is converted to arginine and fumarate. The key and unique step in the urea cycle is the hydrolysis of arginine by the enzyme *arginase.* Not only is it the step wherein urea is formed, but the product, ornithine, also serves to close the cycle. Non-urea-producing organisms lack this enzyme. The other enzymes are widely distributed in nature and participate in pathways other than the production of urea. Carbamyl phosphate synthetase, for example, catalyzes the first step in the biosynthesis of the pyrimidine nucleotides (see p. 558).

Although ammonia and aspartate are the immediate nitrogen sources, note that both can be traced back to glutamate. Glutamate dehydrogenase

For humans the average daily excretion of urea is approximately 30 grams, an amount representing about 80–90% of the total urine nitrogen. This is roughly 20 times the combined amounts of ammonia and uric acid, and 150 times the amount of amino acids, excreted in the same period. An obvious question is, what happens to all this urea? Well, eventually it is used as a nitrogen source by plants and bacteria, which possess the enzyme *urease* that converts urea to ammonia and carbon dioxide. Why isn't the urea reused by mammals? Simply because mammals do not produce urease.

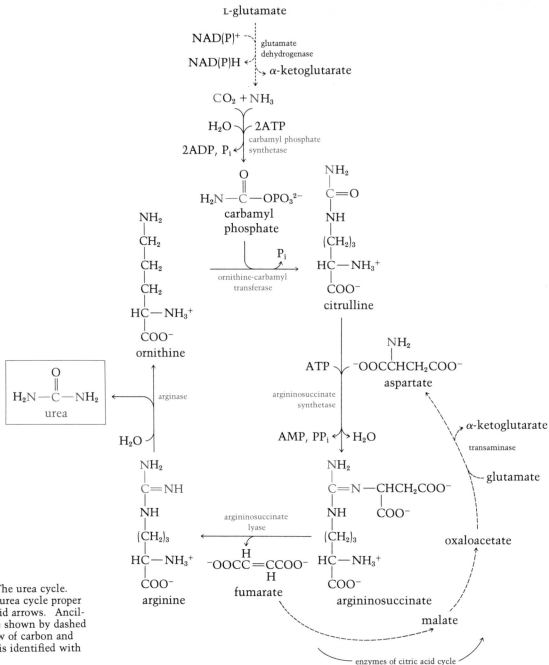

FIGURE 17–1 The urea cycle. Reactions of the urea cycle proper are shown by solid arrows. Ancillary reactions are shown by dashed arrows. The flow of carbon and nitrogen to urea is identified with color.

accounts for the formation of ammonia. Aspartate can acquire its amino nitrogen from glutamate by a simple transamination involving oxaloacetate as the acceptor keto acid. Since oxaloacetate can be formed from fumarate by enzymes of the citric acid cycle, none of the intermediates of the urea cycle is wasted.

METABOLISM OF SPECIFIC AMINO ACIDS

Having examined the major types of metabolic transformations common to most amino acids, we will now consider some of the specific reactions describing the metabolism of some individual amino acids.

One aspect of amino acid metabolism that students of biochemistry find interesting is the myriad of reactions associated with the metabolism of the three aromatic acids, phenylalanine, tyrosine, and tryptophan. A quick glance at Figs. 17–2 and 17–3, which summarize the major metabolic relationships (a few minor ones also exist) of these amino acids, indicates why. Though the composite diagram of the two figures may initially appear as an incomprehensible maze of reactions, it is not that at all. Briefly, Fig. 17–2 summarizes the anabolism of the three amino acids and the major catabolic pathway of tyrosine and phenylalanine, and Fig. 17–3 summarizes the principal catabolic pathways of tryptophan and additional catabolic pathways of tyrosine and phenylalanine, each of which has some special biological significance.

The best place to begin is with the precursor-product relationship of phenylalanine → tyrosine. This is the only way in which tyrosine is produced, and obviously, then, a very significant step. The participating enzyme is *phenylalanine hydroxylase.*

As you might expect, the enzyme is specific in its action. Hydroxylation occurs only at the para position of the phenyl ring, which yields parahydroxyphenylalanine (that is, tyrosine). If it were not specific, hydroxylation would also occur at any of the other four positions in the phenyl ring, yielding both orthohydroxyphenylalanine and metahydroxyphenylalanine. This strict specificity for one of several possible hydroxylation sites in the substrate is characteristic of hydrolyases in general. Recall that hydroxylases are also typified by a requirement for a cooxidizable substrate, which acts as a hydrogen donor (see Chap. 14, p. 451). In the case of phenylalanine hydroxylase this role is fulfilled by a substance called *tetrahydrobiopterin.* During the course of the reaction it is oxidized to dihydrobiopterin.

> The structures of both biopterin species are shown below. Chemically, the biopterin structure would be described as that of a *substituted pteridine,* the parent heterobicyclic compound (see margin). *Tetrahydrofolic acid,* another pteridine-derived coenzyme of immense importance in the metabolism of amino acids, purine, and pyrimidines, will be discussed later.

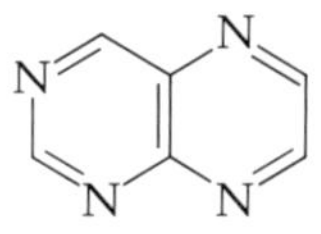

pteridine

Although the immediate cooxidizable substrate for phenylalanine hydroxylase is tetrahydrobiopterin, the reducing power ultimately originates from NADPH. As shown in the diagram, this is accomplished when the re-

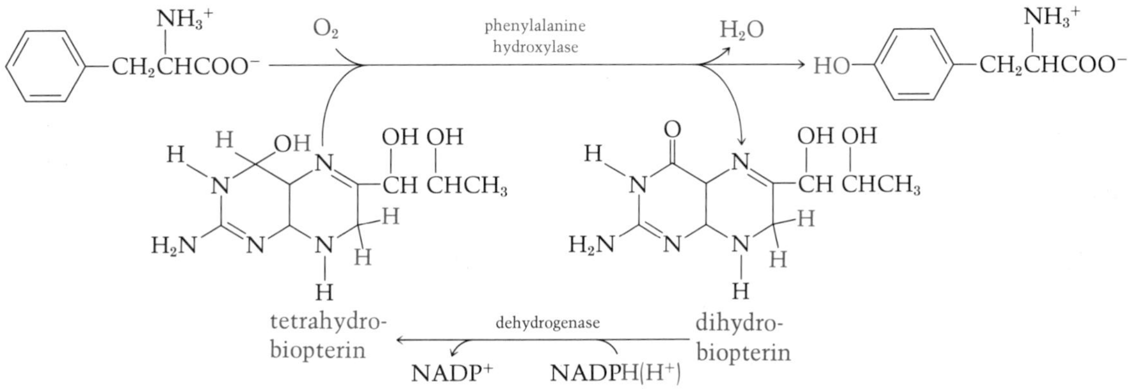

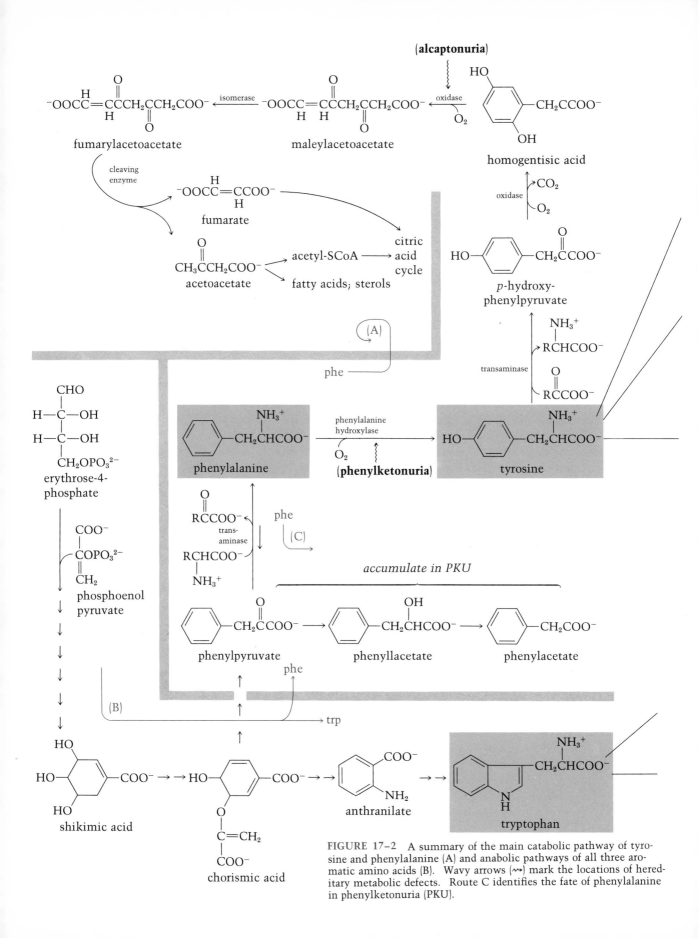

FIGURE 17–2 A summary of the main catabolic pathway of tyrosine and phenylalanine (A) and anabolic pathways of all three aromatic amino acids (B). Wavy arrows (⤳) mark the locations of hereditary metabolic defects. Route C identifies the fate of phenylalanine in phenylketonuria (PKU).

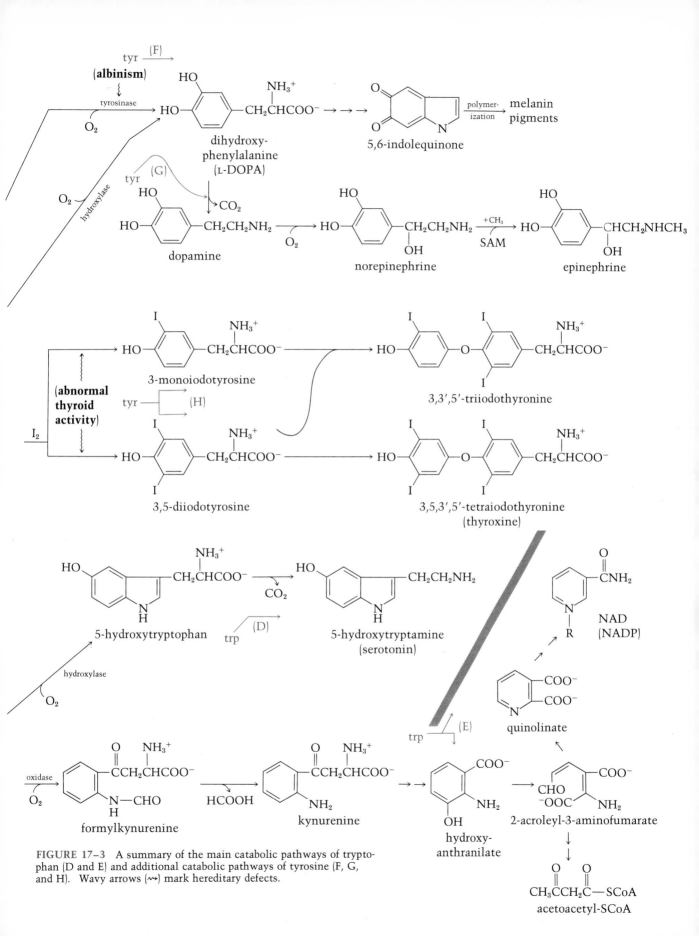

FIGURE 17–3 A summary of the main catabolic pathways of tryptophan (D and E) and additional catabolic pathways of tyrosine (F, G, and H). Wavy arrows (⤳) mark hereditary defects.

ducing power in NADPH is utilized to form tetrahydrobiopterin from dihydrobiopterin. Note that the NADPH:dihydrobiopterin reaction is catalyzed by a specific dehydrogenase and not by phenylalanine hydroxylase.

Phenylalanine hydroxylase occurs in all types of organisms, including humans. Accordingly, tyrosine is *not* classified as an *essential amino acid* that must be supplied in the daily diet of humans. On the other hand, phenylalanine is required, since the enzymes for the biosynthesis of phenylalanine (Fig. 17–2) are not present in humans. In fact, the dietary requirement of phenylalanine applies to most mammals, with the necessary anabolic enzymes being found primarily in plants and bacteria.

The complete amino acid dietary requirements of humans are summarized later in this chapter (see p. 555).

Unfortunately, because of a hereditary genetic defect, not all humans possess the hydroxylase enzyme, or they produce it in levels too low to support normal metabolism. Consequently, a metabolic blockage is created, resulting in elevated levels of phenylalanine in the body tissues and fluids. This situation in turn results in an increase in the levels of phenylpyruvate, phenyllactate, and phenylacetate (route C in Fig. 17–3). Under normal conditions the production of these acids is very small. The complication results from the fact that these substances are toxic, with the damaging effects being primarily directed against developing brain cells. Severe cases of this condition, called *phenylketonuria* (PKU), are characterized by an extensive degree of mental retardation and an early death. Since the brain damage is particularly rapid and irreversible in infants, an early diagnosis is mandatory in order for treatment to be effective. Fortunately both are possible. The diagnosis, which is now generally a routine procedure done shortly after birth, is based on a simple diaper test, which detects the presence of high levels of phenylalanine in the baby's urine. The treatment for PKU infants is to limit the intake of phenylalanine by the use of specially formulated diets with a low phenylalanine content. In adult life there is less need for dietary control since the toxic products apparently are excreted with greater efficiency.

Recent studies provide evidence that the biochemical basis of brain damage associated with phenylketonuria may be an inhibition, specifically in brain cells, by phenylpyruvate, of pyruvate dehydrogenase, the multienzyme complex that catalyzes the conversion of pyruvate to acetyl-SCoA (see Chap. 13, p. 425). Since this is the primary source of acetyl-SCoA formation in brain cells, this means that significantly reduced amounts of acetyl-SCoA would be available for entering the citric acid cycle for energy production. In addition, less acetyl-SCoA would be available for the biosynthesis of fatty acids and cholesterol, both of which are important constituents of nerve tissue in the brain. In short, the inhibition of the pyruvate → acetyl-SCoA conversion would severely affect the normal metabolism, development, and function of the brain.

Phenylketonuria is but one of several hereditary diseases associated with disorders of amino acid metabolism. In fact, there are four different diseases associated with the metabolism of phenylalanine and tyrosine alone (see below). Furthermore a number of metabolic diseases exist that are associated with the metabolism of compounds other than amino acids. For example, recall our earlier discussions in Chap. 9 of hereditary metabolic diseases associated with the abnormal degradation of cerebroside and ganglioside lipids. These and several other conditions are described in the reference book by Stanbury et al. (see p. 572).

Rather than proceed with a blow-by-blow account of the other pathways shown in Figs. 17–2 and 17–3, only a brief synopsis of each is given below. Although the mention of a few specifics cannot be avoided, a thorough analysis of all of the details is avoided.

1. Route A (Fig. 17–2) depicts the main catabolic pathway of tyrosine and phenylalanine. The first reaction is a transamination to yield *parahydroxyphenylpyruvate*, the α-keto acid of tyrosine. The keto acid is then oxidized to *homogentisic acid*, which in turn is ultimately degraded to *fumarate* and *acetoacetate*. The fumarate can then enter the citric acid cycle for further degradation to CO_2 or incorporation into carbohydrate. Thus, phenylalanine and tyrosine are glycogenic. The acetoacetate fragment, after conversion to acetoacetyl-SCoA, can be used directly in the biosynthesis of fatty acids or sterols. Hence, phenylalanine and tyrosine are also ketogenic.

Alternatively, the acetoacetyl-SCoA can be converted to acetyl-SCoA for entrance into the citric acid cycle.

2. The degradation and synthesis of an amino acid generally proceed by different pathways. Only the catabolism and anabolism of glutamate, aspartate, and alanine involve the same reactions operating in reverse. In each of these instances, the amino acid is converted to and formed from its corresponding α-keto acid in a transamination reaction: α-ketoglutarate ⇌ glutamate; oxaloacetate ⇌ aspartate; pyruvate ⇌ alanine. In the case of phenylalanine production (route B in Fig. 17–2) the carbons arise from *phosphoenolpyruvate* and *erythrose phosphate,* both of which in turn originate from the catabolism of hexoses. As indicated in Fig. 17–2, the complete biosynthetic sequence involves many steps and several intermediates, with *shikimic acid* and *chorismic acid* being the most characteristic. Note that the last reaction is a transamination step. Note also that part of the same pathway (through the formation of chorismic acid) is utilized in the biosynthesis of tryptophan. This sharing of part of a pathway for the production of two or three amino acids is not uncommon, and takes place with other groups of amino acids. Such sharing also occurs in degradative pathways.

3. Let us begin our discussion of Fig. 17–3 by considering the degradation of tryptophan (routes D and E). Since this is one of the most complex areas of amino acid metabolism, only a skeleton summary is given here. First note that the initial reaction in each pathway is oxidative in nature. In route D, tryptophan is hydroxylated on the benzene ring to yield *5-hydroxytryptophan,* which is then decarboxylated to yield *5-hydroxytryptamine,* generally called *serotonin.* Since serotonin is known to stimulate strongly the constriction of smooth muscle, it has been the subject of much study. In addition, serotonin may play an important role in the physiology of the brain by acting as a regulator of neural transmission. Despite years of study, neither of these actions nor their physiological significance is yet clearly understood.

In the other catabolic pathway (route E), the indole ring is oxidatively cleaved to yield *formylkynurenine,* which in turn is converted to *kynurenine.* The loss of carbon as formic acid (HCOOH) in this step links the catabolism of tryptophan to an important area of metabolism termed *one-carbon metabolism.* We will return to this subject in a later section (p. 547). The further catabolism of kynurenine produces either (a) *acetoacetyl-SCoA,* which can then enter the citric acid cycle or be converted to lipid, or (b) *quinolinate,* which then serves as the biosynthetic precursor of the *nicotinamide* moiety of NAD and NADP. The tryptophan → NAD(NADP) pathway operates under normal conditions, and supplements the dietary requirement of nicotinic acid, which is used directly in the formation of NAD and NADP. It does not, however, replace the dietary requirement.

4. Route F depicts one of the specific roles of tyrosine metabolism that occurs in a specialized group of cells variously called *melanocytes, melanoblasts,* or simply *pigment cells.* These cells are responsible for producing the so-called *melanin* pigments, which are the substances imparting the coloration of skin, eyes, and hair. Melanin pigments appear to be polymeric forms of *indole-5,6-quinone.* As shown, the formation of melanin initially depends on the hydroxylation of tyrosine to yield *3,4-dihydroxy-L-phenylalanine* (L-DOPA). This process is catalyzed by the enzyme *tyro-*

Alkaptonuria is a rare metabolic disease that results from an inability to produce homogentisic acid oxidase, causing an accumulation of homogentisic acid, which is then excreted in the urine. On exposure to alkali and oxygen, the urine turns black, due to the oxidation of homogentisic acid. This condition is not known to be associated with any adverse physiological effects.

Abnormalities in tryptophan metabolism in brain may be associated with symptoms of schizophrenia.

sinase, which, as you might expect, is quite abundant in melanocytes. An inability of the melanocytes to produce tyrosinase is responsible for the condition known as *albinism;* this inability is caused by a defective (and hereditary) gene.

5. Route G summarizes a second special feature of tyrosine metabolism, the formation of physiologically active hormones, termed the *catecholamines,* which include L-*dopamine, norepinephrine* (noradrenalin), and *epinephrine* (adrenalin). Recall that we have already discussed the major metabolic effect of epinephrine, that is, its ability to accelerate the catabolism of carbohydrates by stimulating the production of cyclic AMP, which in turn stimulates the action of glycogen phosphorylases (see p. 383). The production of catecholamines is particularly prominent in the cells of brain tissue, nerve tissue, and the adrenal gland. As in melanin production, the key reaction is the initial formation of L-DOPA, which then undergoes a decarboxylation to yield L-*dopamine.* The hydroxylation of L-dopamine and a subsequent methylation reaction (see a later section, p. 549) yield norepinephrine and epinephrine, respectively. Although the melanin and catecholamine pathways both begin with the conversion of tyrosine to L-DOPA, note that two different enzymes are involved. The hydroxylase catalyzing the transformation in the catecholamine pathway appears to be quite similar to the hydroxylase involved in the conversion of phenylalanine to tyrosine. The fact that the administration of L-DOPA offers relief to individuals suffering from Parkinson's disease suggests that this affliction is related to an inability of certain cells to produce adequate levels of L-DOPA. It further suggests that in these cells tyrosine hydroxylase may be (a) produced in low levels or not at all, or (b) produced in a defective form.

The subsequent fate of dopamine, norepinephrine, and epinephrine is primarily excretion in the urine. However, only small levels of the free amines are found in the urine. For the most part they are excreted as the corresponding acids and, in the case of norepinephrine and epinephrine, also as the methylated derivatives of the corresponding acids. The acids are produced by the action of an NAD^+-dependent dehydrogenase on the corresponding aldehyde, which in turn is produced from the free amine by the action of monoamine oxidase (see top of next page and p. 534).

Schizophrenia may also be associated with low levels of monoamine oxidase activity, which would result in elevated body levels of various catecholamines.

6. The thyroid gland is another site of a specialized and extremely important facet of tyrosine metabolism, the formation of a group of low molecular weight iodinated compounds called the *thyroid hormones* (route H). Although the exact sequence of events is more complex than that shown in Fig. 17–3, the process can be broken down into two stages; (a) *iodination,* to yield 3-monoiodotyrosine and 3,5-diiodotyrosine, and (b) *conjugation* of these species to yield 3,5,3'-triiodotyrosine, and 3,5,3',5'-tetraiodothyronine (*thyroxine*). The latter two materials are the active thyroid hormones. In the thyroid gland, triiodothyronine and thyroxine are present almost exclusively in peptide linkage in a protein called *thyroglobulin.* When the thyroid is stimulated, the thyroglobulin molecule is hydrolyzed by proteolytic enzymes in the gland, and the hormones are secreted into the circulating blood and carried to other tissues. By processes yet largely unknown, the thyroid gland profoundly influences a board spectrum of metabolic and physiological processes. Actually, there are few phases of the general growth and development of humans that are not regulated to some extent by these hormones.

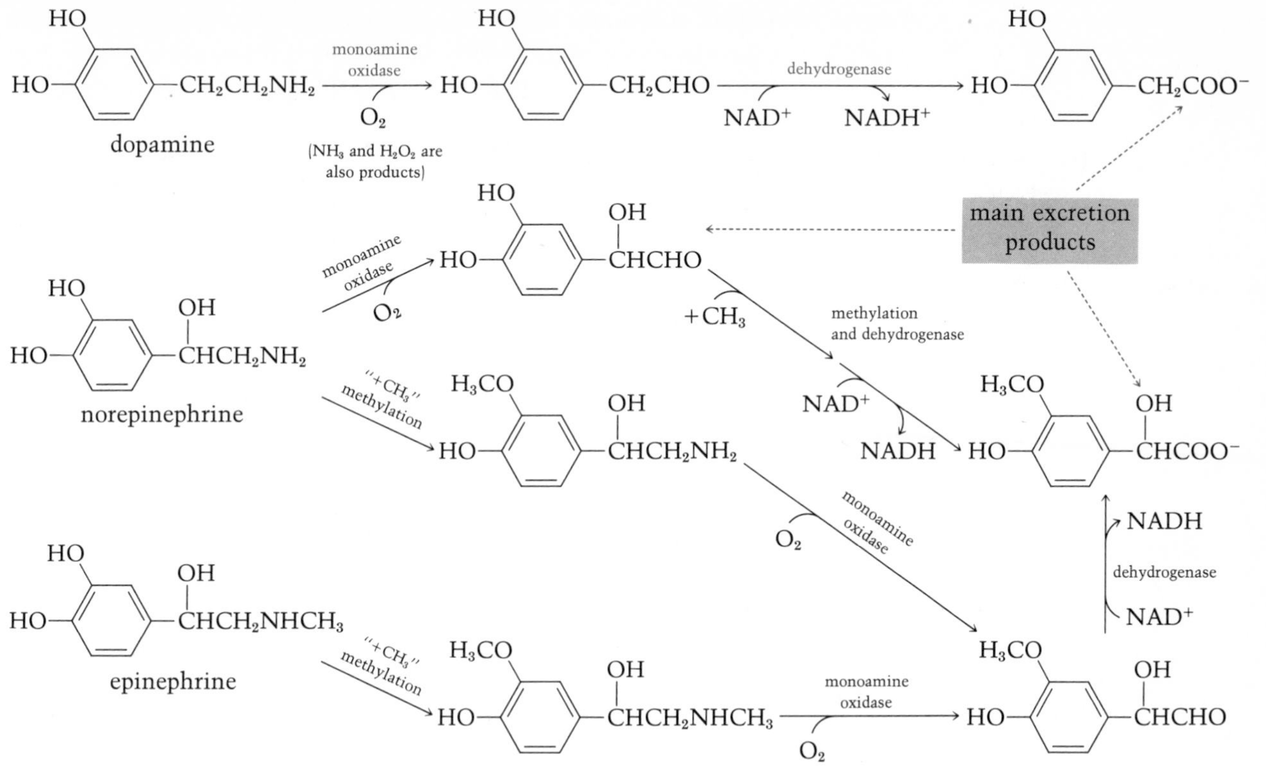

dopamine $\xrightarrow[\substack{O_2 \\ \text{(NH}_3 \text{ and } H_2O_2 \text{ are} \\ \text{also products)}}]{\text{monoamine oxidase}}$ (3,4-dihydroxyphenyl)$-CH_2CHO$ $\xrightarrow[NAD^+ \quad NADH^+]{\text{dehydrogenase}}$ (3,4-dihydroxyphenyl)$-CH_2COO^-$

norepinephrine $-CHCH_2NH_2$ $\xrightarrow[O_2]{\text{monoamine oxidase}}$ $-CHCHO$ $\xrightarrow[+CH_3]{}$ ... $\xrightarrow[NAD^+ \quad NADH]{\text{methylation and dehydrogenase}}$ $-CHCOO^-$ (H_3CO ...)

"$+CH_3$" methylation → $-CHCH_2NH_2$ (H_3CO ...) $\xrightarrow[O_2]{\text{monoamine oxidase}}$

epinephrine $-CHCH_2NHCH_3$ $\xrightarrow[]{\text{"}+CH_3\text{" methylation}}$ $-CHCH_2NHCH_3$ (H_3CO ...) $\xrightarrow[O_2]{\text{monoamine oxidase}}$ $-CHCHO$ (H_3CO ...)

dehydrogenase ($NADH$ / NAD^+)

main excretion products

Glycine and Tetrapyrrole Biosynthesis

The functional importance of the porphyrin (tetrapyrrole) grouping in all types of living cells cannot be questioned. So far we have seen that it is the active structural moiety of hemoglobin, myoglobin, chlorophyll, and all the cytochromes. It is also the structural backbone of Vitamin B_{12} (see p. 553). Accordingly, it is only appropriate that we consider its biosynthetic origin. The subject is well suited to this chapter, since the sole anabolic precursors are *glycine* and *succinyl-SCoA*. A few of the main steps are shown in Fig. 17–4. The process begins with a condensation of succinyl-SCoA and glycine to form *α-amino-β-ketoadipic acid.* This in turn is decarboxylated (the CO_2 lost comes from the glycine carboxyl group) to yield *δ-aminolevulinic acid.* The next step involves an ordered intermolecular condensation of two molecules of aminolevulinic acid to yield *porphobilinogen,* a substituted pyrrole. Four molecules of porphobilinogen are then condensed to form a *linear tetrapyrrole,* which then undergoes a ring closure to produce a *cyclic tetrapyrrole* species. Depending on the specificity of the enzymes that catalyze this latter step, the ring closing can occur either by a direct joining of the two ends of the linear tetrapyrrole or, as shown in Fig. 17–4, be accompanied by a flipping of the pyrrole unit at the nonamino end into a different position before ring closure. The latter thus causes the pyrrole side chains to be arranged in a nonalternating fashion; that is, A-P; P-A; P-A; P-A, rather than P-A; P-A; P-A; P-A. In other words, a specific isomer is produced, called *uroporphyrinogen III.* From this point, a specific combination of reactions involving modification of the side chains and reductions of the —CH_2— pyrrole links will yield the various porphyrins.

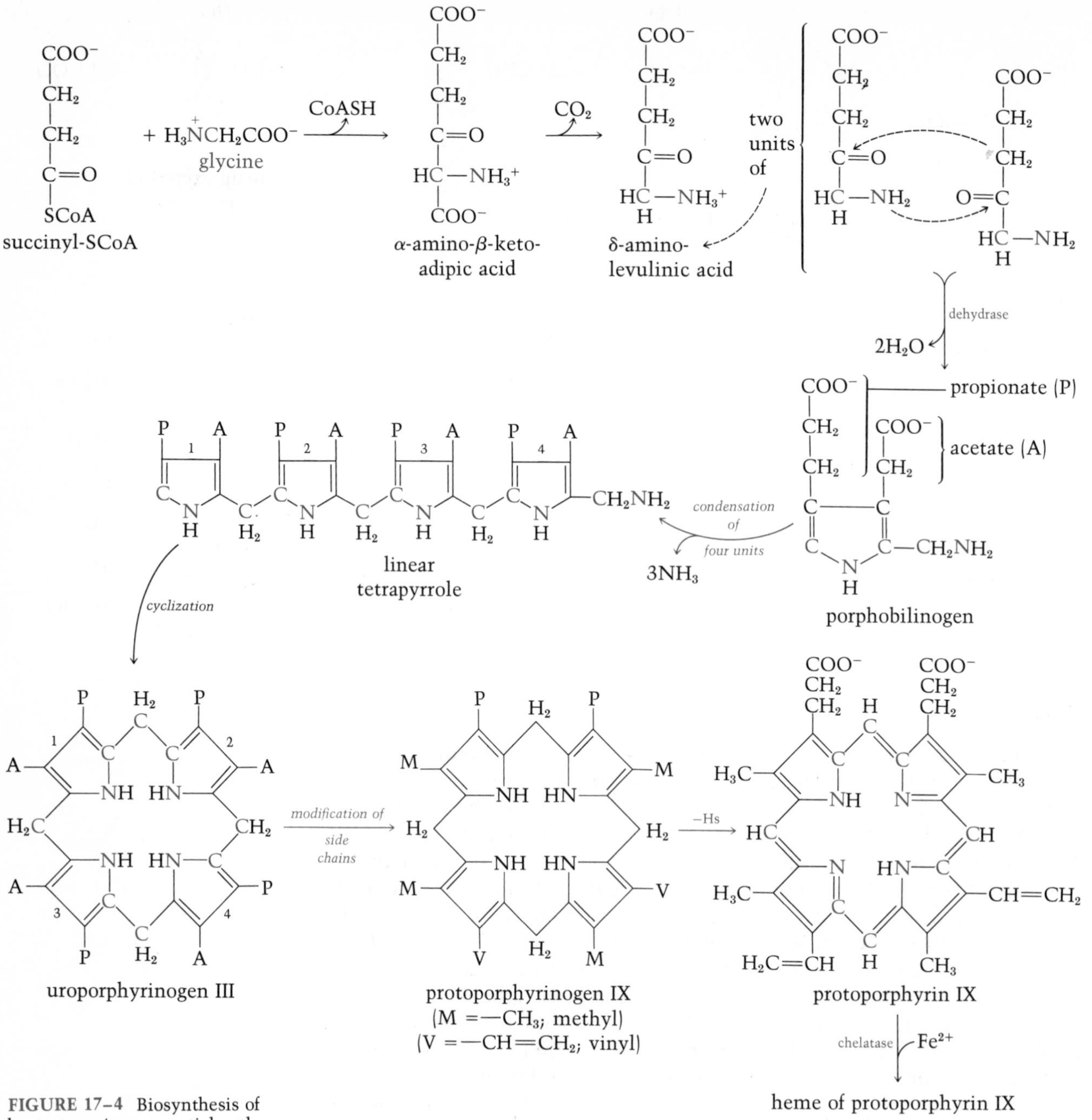

FIGURE 17–4 Biosynthesis of heme groupings; a partial pathway. A free amino group (—NH₂) is involved in condensations.

The structure of *protoporphyrin IX* is shown because it is the most common porphyrin system, being found in hemoglobin, myoglobin, and some cytochromes (see p. 456). The chelation of the porphyrin with the corresponding metal completes the formation of the active metalloporphyrin. Note that all four nitrogen atoms, the four methenyl links (—CH=), and four pyrrole carbons come from glycine. All of the other carbons originate from succinyl-SCoA. Since succinyl-SCoA is an intermediate of the citric acid cycle, the ultimate source of these carbons could be carbohydrate, lipid, or any amino acid that could be degraded to an intermediate of the citric acid cycle.

Cysteine and Methionine—Sulfur Metabolism

Sulfur is one of the six major elements (C, H, O, N, P, S) of which most biomolecules are composed. So far we have encountered sulfur in several substances of major importance to all living organisms. These substances include *glutathione, thiamine pyrophosphate, biotin, lipoic acid, coenzyme A. chondroitin sulfates, sulfolipids, non-heme-iron proteins,* and those proteins in which *cysteine* and *methionine* are present. There are several other sulfur-containing compounds that occur naturally.

The ultimate source of sulfur for most organisms is inorganic sulfate SO_4^{2-}. After entering a cell, sulfate is mobilized for metabolism by conversion first to *adenosine-5'-phosphosulfate* (APS) and then to *3'-phosphoadenosine-5'-phosphosulfate* (PAPS). Each step (see below) requires ATP. The chemical significance of the conversions is that inorganic sulfate in the free unreactive state SO_4^{2-} is converted to a more highly reactive *sulfo-phospho anhydride* species.

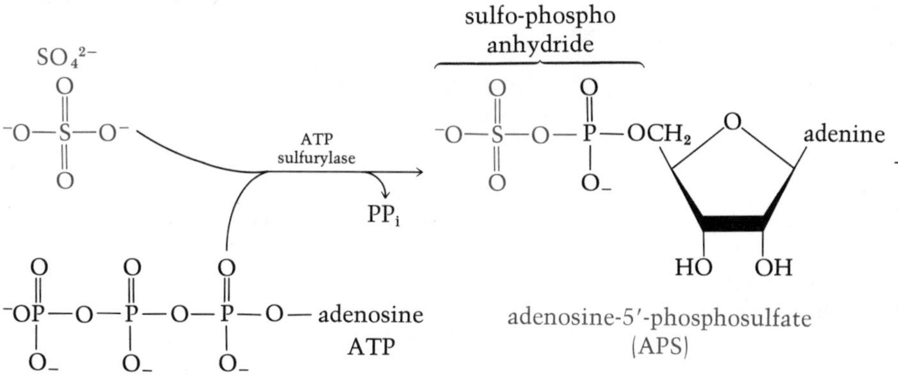

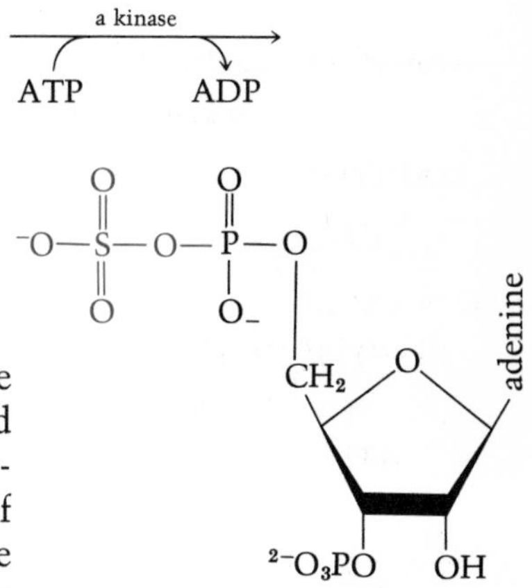

3′-phosphoadenosine-5′-phosphosulfate (PAPS)

In humans and other animals PAPS serves (see margin) as a donor of the sulfate group in the formation of organic sulfate esters such as heparin and the chondroitin sulfates (see p. 309 in Chap. 9). Animal organisms, however, are incapable of utilizing PAPS as a donor of sulfur in the formation of other sulfur-containing compounds. This requires that the sulfur of the sulfate grouping be reduced to sulfide S^{2-}, a conversion that does not occur in animals. Enzymes capable of the SO_4^{2-} to S^{2-} conversion do exist in plant and bacterial organisms. The proposed sequence is shown below. It involves two separate reductase enzymes, each using NADPH as an electron donor. A total of four NADPHs (that is, eight electrons) is required.

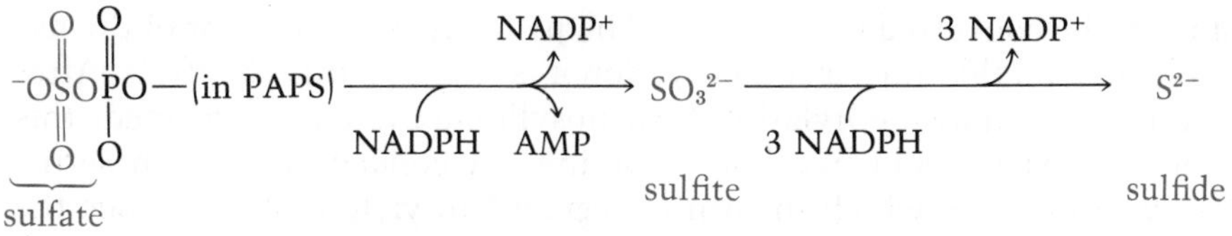

(both processes are enzyme catalyzed)

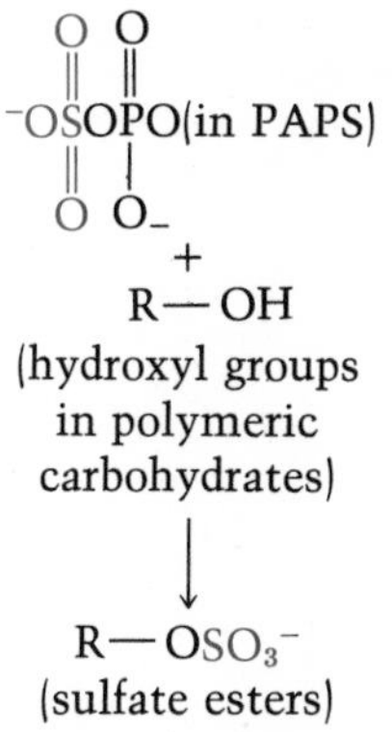

In plants and bacteria the major pathway proposed for the *S^{2-} to cysteine* conversion utilizes the carbon skeleton of serine, which is first activated by conversion to an O-acetyl derivative. Cysteine is then the source of sulfur in the biosynthesis of methionine, with the carbon skeleton of methionine originating from *aspartic acid.* The transfer of sulfur involves the condensation of cysteine with O-succinyl-homoserine (which originates from aspartate and succinyl-SCoA) to yield *cystathionine.* A cleavage to *homocys-*

teine and a transmethylation reaction, using as methyl donor a methylated adduct of tetrahydrofolic acid (CH_3—FH_4; see p. 550 in this chapter), finally *yields methionine.*

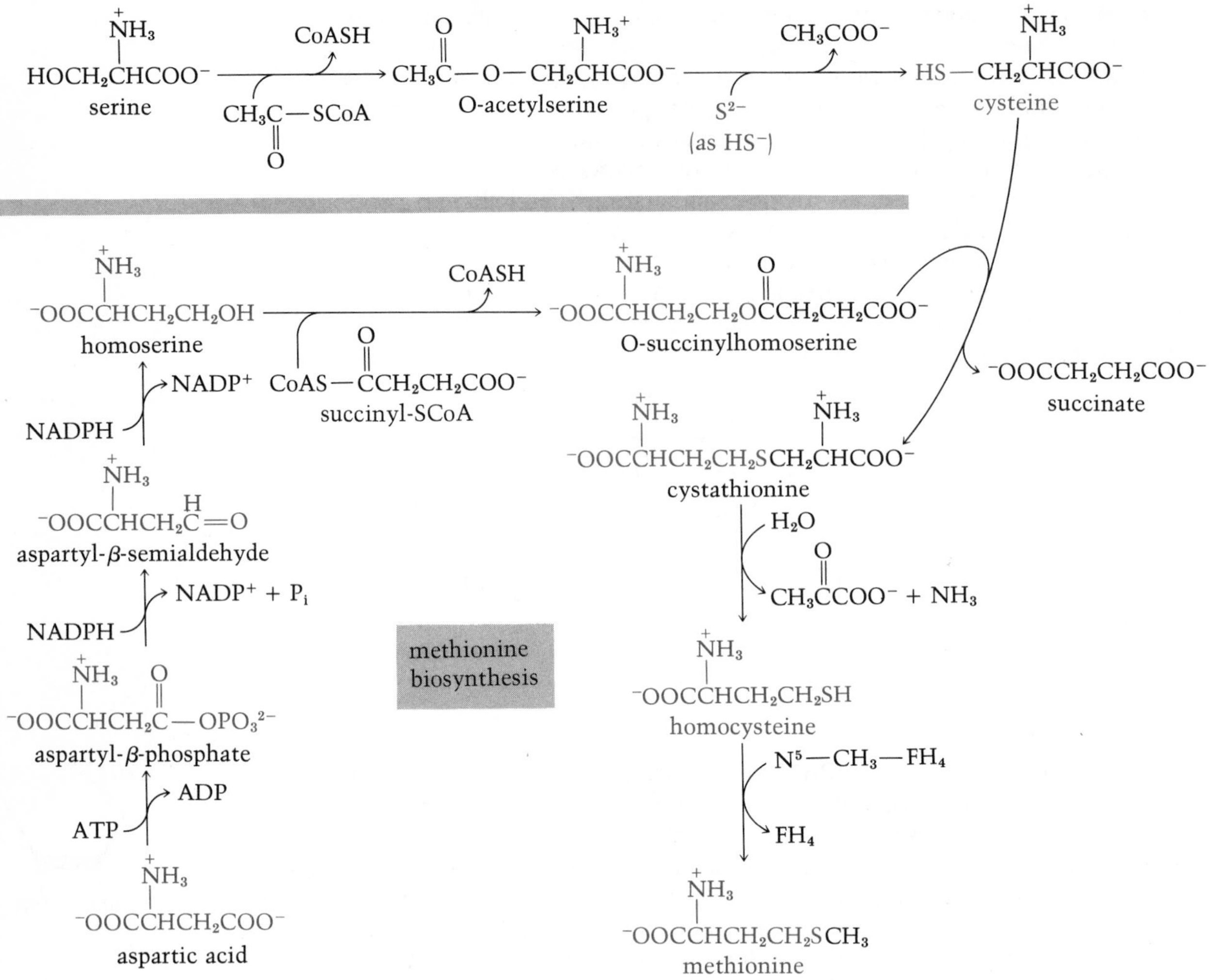

$$\underset{\text{serine}}{HOCH_2\overset{\overset{+}{N}H_3}{\underset{|}{C}}HCOO^-} \xrightarrow[\underset{O}{\underset{\|}{CH_3C}}-SCoA]{} \underset{\text{O-acetylserine}}{\underset{O}{\underset{\|}{CH_3C}}-O-CH_2\overset{\overset{+}{NH_3}}{\underset{|}{C}}HCOO^-} \xrightarrow[\underset{\text{(as HS}^-)}{S^{2-}}]{CH_3COO^-} \underset{\text{cysteine}}{HS-CH_2\overset{\overset{+}{N}H_3}{\underset{|}{C}}HCOO^-}$$

---

methionine biosynthesis

$$\underset{\text{homoserine}}{^-OOC\overset{\overset{+}{N}H_3}{\underset{|}{C}}HCH_2CH_2OH} \xrightarrow[\underset{\text{succinyl-SCoA}}{CoAS-\underset{O}{\overset{O}{\overset{\|}{C}}}CH_2CH_2COO^-}]{CoASH} \underset{\text{O-succinylhomoserine}}{^-OOC\overset{\overset{+}{N}H_3}{\underset{|}{C}}HCH_2CH_2O\overset{O}{\overset{\|}{C}}CH_2CH_2COO^-}$$

$$^-OOCCH_2CH_2COO^- \quad \text{succinate}$$

$$\overset{\text{NADPH}}{\nearrow}\text{NADP}^+$$

$$\underset{\text{aspartyl-}\beta\text{-semialdehyde}}{^-OOC\overset{\overset{+}{N}H_3}{\underset{|}{C}}HCH_2\overset{H}{\underset{}{C}}=O}$$

$$\overset{\text{NADPH}}{\nearrow}\text{NADP}^+ + P_i$$

$$\underset{\text{aspartyl-}\beta\text{-phosphate}}{^-OOC\overset{\overset{+}{N}H_3}{\underset{|}{C}}HCH_2\overset{O}{\overset{\|}{C}}-OPO_3^{2-}}$$

$$\overset{\text{ATP}}{\nearrow}\text{ADP}$$

$$\underset{\text{aspartic acid}}{^-OOC\overset{\overset{+}{N}H_3}{\underset{|}{C}}HCH_2COO^-}$$

$$\underset{\text{cystathionine}}{^-OOC\overset{\overset{+}{N}H_3}{\underset{|}{C}}HCH_2CH_2S\,CH_2\overset{\overset{+}{N}H_3}{\underset{|}{C}}HCOO^-}$$

$$\overset{H_2O}{\searrow}\underset{CH_3\overset{O}{\overset{\|}{C}}COO^- + NH_3}{}$$

$$\underset{\text{homocysteine}}{^-OOC\overset{\overset{+}{N}H_3}{\underset{|}{C}}HCH_2CH_2SH}$$

$$\overset{N^5-CH_3-FH_4}{\searrow}FH_4$$

$$\underset{\text{methionine}}{^-OOC\overset{\overset{+}{N}H_3}{\underset{|}{C}}HCH_2CH_2S\,CH_3}$$

Unable to convert cysteine to methionine, animals depend on a dietary supply of methionine. Although animals are also unable to convert SO_4^{2-} to cysteine, cysteine is not an essential dietary amino acid because methionine can be converted to cysteine. The pathway is an offshoot of the use of S-adenosylmethionine as a methyl donor (see p. 549 this chapter). After the reaction of transmethylation S-adenosylhomocysteine is formed; this can be cleaved to yield free homocysteine. A condensation with serine gives cystathionine, which in turn is degraded to yield cysteine. (See the top of the next page.) Thus methionine serves as the sulfur source for cysteine and serine serves as the carbon source for cysteine.

Coenzyme A from Valine, Aspartic Acid, and Cysteine

The *biosynthesis* of *coenzyme A* is illustrated in the diagram on p. 548. As indicated, individual parts of the structure originate from *ATP, valine, aspartic acid,* and *cysteine.* Whereas plants and bacteria utilize the entire pathway, animal organisms can perform only the conversions starting with

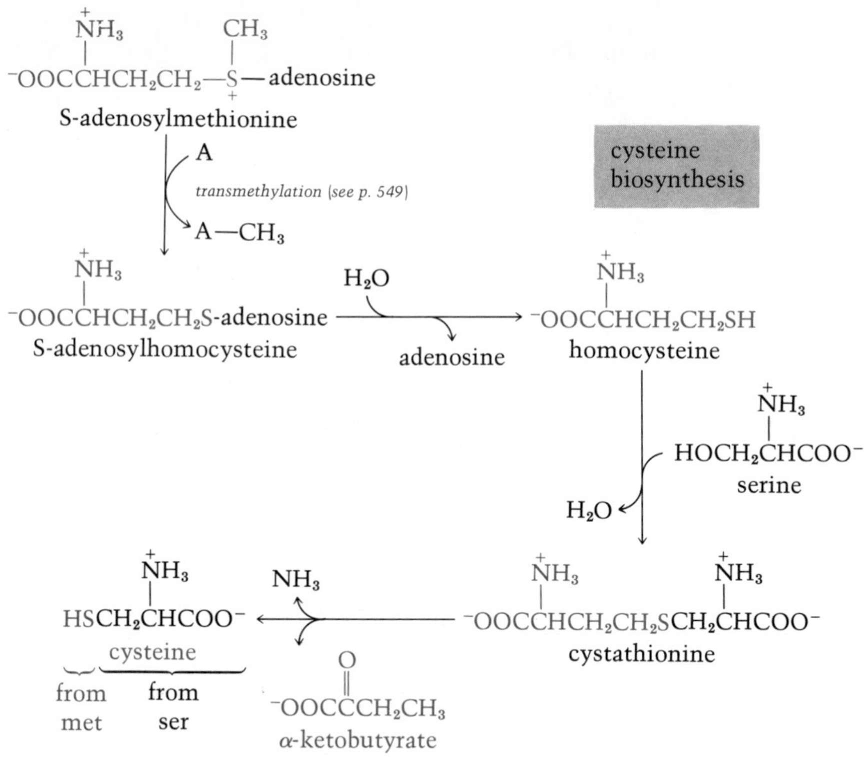

$$\overset{+}{N}H_3 \qquad CH_3$$
$$^-OOCCHCH_2CH_2\!-\!\underset{+}{S}\!-\!adenosine$$

S-adenosylmethionine

A

transmethylation (see p. 549)

A—CH$_3$

cysteine
biosynthesis

$$\overset{+}{N}H_3 \qquad\qquad H_2O \qquad\qquad \overset{+}{N}H_3$$
$$^-OOCCHCH_2CH_2S\text{-adenosine} \longrightarrow\qquad ^-OOCCHCH_2CH_2SH$$

S-adenosylhomocysteine

adenosine

homocysteine

$$\overset{+}{N}H_3$$
$$HOCH_2CHCOO^-$$
serine

H$_2$O

$$\overset{+}{N}H_3 \qquad NH_3 \qquad\qquad \overset{+}{N}H_3 \qquad\qquad \overset{+}{N}H_3$$
$$HSCH_2CHCOO^- \longleftarrow\qquad ^-OOCCHCH_2CH_2SCH_2CHCOO^-$$

cysteine

cystathionine

from met   from ser

$$O$$
$$^-OOCCCH_2CH_3$$
α-ketobutyrate

pantothenic acid. Obviously *pantothenic acid* is then an essential component of a normal balanced diet of animals. It is classified as a *vitamin*.
Note that the phosphopantethenic grouping would also be utilized in the
formation of the acyl carrier protein (ACP—SH; see p. 511 in Chap. 16).

Methionine and One-carbon Metabolism

Apart from carboxylation and decarboxylation reactions, there occur in all
living cells a small but very important group of transformations involving
the transfer of a one-carbon unit between two substrates. Included are reactions involving the transfer of *methyl* (—CH$_3$), *hydroxymethyl* (—CH$_2$OH),
formyl (—CHO), and *formimino* (—CH=NH) groups. The most abundant
type is the methyl group transfer reaction, termed a *transmethylation*. A
general representation of a transmethylation reaction can be given as
follows. Under the influence of a *transmethylase* (sometimes called a
methyltransferase), a complete methyl group is transferred from a donor to
an acceptor molecule.

$$A \quad+\quad D\!-\!CH_3 \xrightarrow{\text{transmethylase}} A\!-\!CH_3 \quad+\quad D$$

acceptor   methyl   methylated   demethylated
substrate   donor   acceptor   donor

The reason for considering transmethylation reactions in this chapter is
that the primary methyl donor is *S-adenosylmethionine*, often symbolized
simply as SAM. This substance is present in all types of cells, where it is
formed directly from methionine and ATP. Chemically, SAM would be
classified as an unstable *sulfonium* compound in which the $-\overset{+}{\underset{|}{S}}\!-\!CH_3$ link-

age is quite labile. Accordingly, S-adenosylmethionine is frequently

547

Coenzyme A biosynthesis

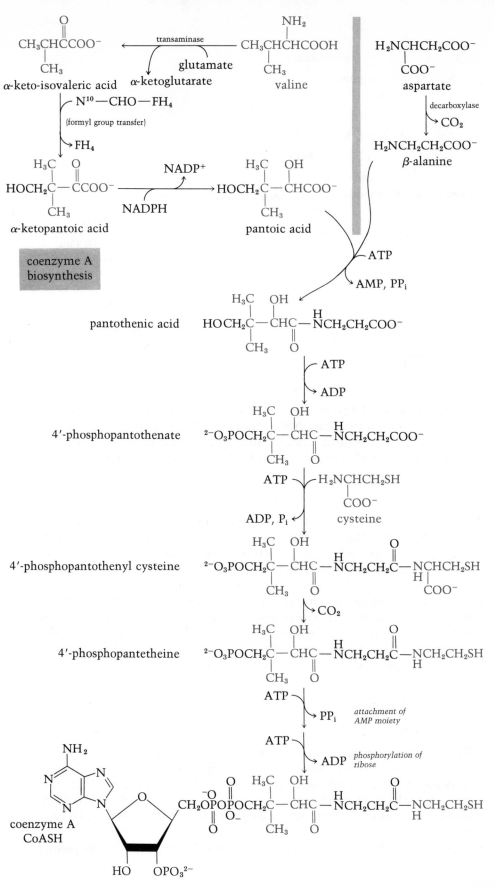

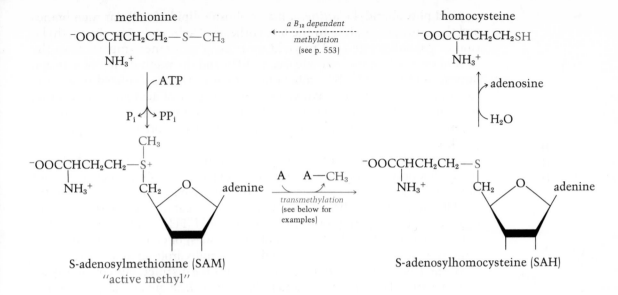

termed an *active methyl species*. Note that S-adenosylmethionine is converted to *S-adenosylhomocysteine* (SAH) during the transmethylation reaction. S-Adenosylhomocysteine is then enzymatically hydrolyzed to homocysteine and adenosine. The homocysteine is then converted to methionine by another type of transmethylation reaction. However, the methyl donor in this reaction is not S-adenosylmethionine. We will discuss this reaction shortly.

A few specific transmethylation reactions of considerable importance are given below. These include the methylation of phosphatidyl ethanolamine

O
‖
CH₂OCR

O
‖
R'CO—C—H

O
‖
CH₂OPOCH₂CH₂NH₃⁺
|
O⁻

phosphatidyl ethanolamine

SAH   SAH   SAH
↗   ↗   ↗
⟶   ⟶   ⟶
SAM   SAM   SAM

*three successive
methylations*

O
‖
CH₂OCR

O
‖
R'CO—C—H

O
‖
CH₂OPOCH₂CH₂N⁺(CH₃)₃
|
O⁻

phosphatidyl choline

arginine + glycine ⟶ H₂N—C—N—CH₂COO⁻

NH₂⁺
‖
H₂N—C—N—CH₂COO⁻
|
H

guanidoacetate

SAH
↗
⟶
SAM

NH₂⁺
‖
H₂N—C—N—CH₂COO⁻
|
CH₃

creatine

H₃N⁺CH₂CH₂CH₂COO⁻

SAH   SAH   SAH
↗   ↗   ↗
⟶   ⟶   ⟶
SAM   SAM   SAM

γ-aminobutyric acid
(from glutamate, see p. 532)

(H₃C)₃N⁺CH₂CH₂CH₂COO⁻

γ-butyrobetaine

O₂
⟶

(H₃C)₃N⁺CH₂CHCH₂COO⁻
|
OH

carnitine

one-carbon–FH₄ adducts

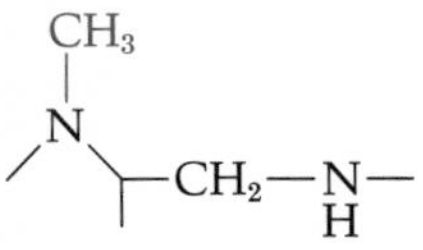

N^5-methyl-FH₄

N^5 — CH₃ – FH₄

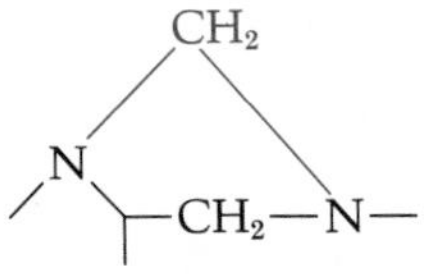

N^5,N^{10}-methylene-FH₄

N^5,N^{10} — CH₂ — FH₄

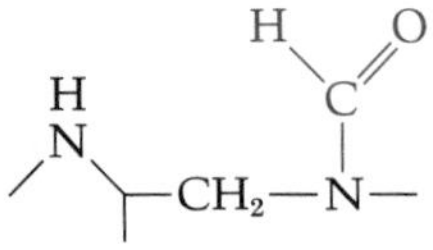

N^{10}-formyl-FH₄

N^{10} — CHO – FH₄

to yield phosphatidyl *choline,* a major phospholipid of cellular membranes and the source of choline in the biosynthesis of sphingolipids; the methylation of guanidoacetic acid to yield *creatine,* a substance important in the flow of energy in muscle tissue (see p. 402); and the methylation of γ-amino butyric acid to yield γ-butyrobetaine, which in turn is oxidized to yield L-*carnitine,* the substance involved in the transport of acyl groupings across biological membranes (see p. 501). Another reaction is the formation of *epinephrine* from norepinephrine, which we discussed previously (item 5 on p. 542).

The role of S-adenosylmethionine as a methyl donor in transmethylation represents only one facet of one-carbon metabolism. Mentioned above but not yet considered are those reactions involving transfer of —CH₂OH, —CHO, and —CH=NH groups. Enzymes that catalyze these reactions all require the participation of *tetrahydrofolic acid* (FH₄) as a coenzyme. The key to understanding the function of FH₄ resides, of course, in its chemical structure (see below). As indicated, the tetrahydrofolic acid molecule is conveniently considered to be composed of three distinct, but covalently attached, parts: (1) a *substituted* and *reduced pteridine moiety,* (2) *paraaminobenzoic acid,* and (3) *glutamic acid.* Again, as is the case with most other coenzymes, tetrahydrofolic acid represents the metabolically active form of a vitamin essential to most mammals, including humans. In this case, the parent vitamin is *folic acid,* which is converted to the tetrahydro state through a dihydro species of two successive NADPH hydrogenations, both catalyzed by dihydrofolic acid reductase. Of the three species, only the tetrahydro form has activity as a coenzyme. The participation of FH₄ as a

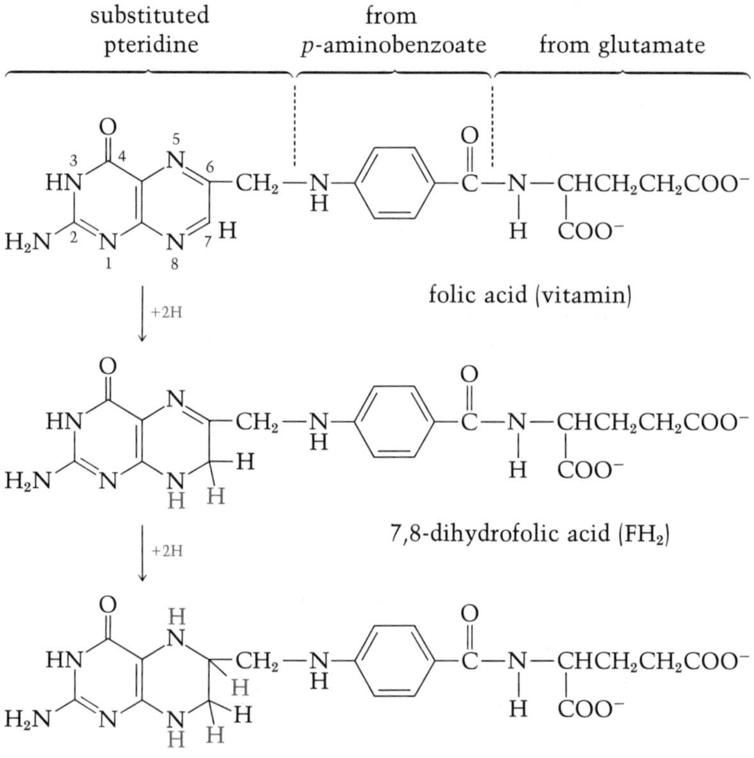

folic acid (vitamin)

7,8-dihydrofolic acid (FH₂)

5,6,7,8-tetrahydrofolic acid (FH₄)
(active coenzyme)

transfer agent of one-carbon compounds is due to its ability to form either single-bonded covalent adducts at the nitrogen positions 5 and 10 or a bridged adduct between positions 5 and 10.

A brief analysis of the role of tetrahydrofolic acid in metabolism is difficult, since it is involved in several reactions and in many different ways. The approach employed here begins by considering the participation of FH_4 in the conversion of serine to glycine (see Fig. 17–5). Although the reaction is characterized by a complex mechanism (the enzyme also requires pyridoxal phosphate as a coenzyme), note that the overall effect is the loss of the β carbon of serine to FH_4 to yield N^5,N^{10}-methylene-FH_4. While it may appear from Fig. 17–5 that the subsequent metabolic fate of N^5,N^{10}—CH_2—FH_4 is quite varied, actually only two main routes are indicated. In one the methylene derivative is reduced to N^5-methyl-FH_4, which then functions as the methyl donor for homocysteine in the biosynthesis of methionine, a process originating with aspartic acid and cysteine (details on p. 546). The overall reaction is rather complex and also requires the participation of the coenzyme form of Vitamin B_{12} (see p. 553, this chapter).

The second route involves the utilization of the one-carbon fragment in the biosynthesis of all purines and of one pyrimidine, thymine (see p. 564 in this chapter). These relationships occur at the level of N^5,N^{10}-methylene-FH_4, N^5,N^{10}-methenyl(—CH=)-FH_4, and N^{10}-formyl-FH_4. As indicated in Fig. 17–5, the latter two species are enzymatically and sequentially formed from the methylene derivative. The supplying of one-carbon fragments for thymine and purine biosynthesis is the most vital aspect of the biological function of FH_4. An indication of how vital this is can be seen in the effect of many sulfa drugs, which act as potent inhibitors in

Although we have considered the serine → glycine conversion as the major starting point, note that the FH_4 intermediates can also be produced from other sources. The basis for this is the reversibility of the reactions involving the interconversions of the FH_4 derivatives. One source is formic acid, a product of tryptophan catabolism. Another is formiminoglutamate, a product of histidine catabolism.

FIGURE 17–5 A partial summary of the role of tetrahydrofolic acid in one-carbon metabolism. All transformations are enzyme catalyzed.

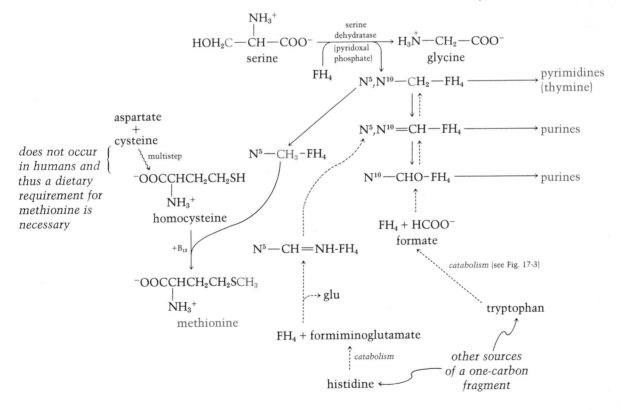

the microbial biosynthesis of folic acid. Presumably, the basis of the inhibition is competition, which prevents the incorporation of the paraaminobenzoic acid moiety into the compound (see p. 177, Chap. 6). The blockage is fatal to the organism. Humans cannot synthesize folic acid and hence it must be supplied from another source. Recommended daily dietary levels are rather low, however, since the bulk of what is required is satisfied by synthesis carried out by the bacteria of the intestinal tract.

Vitamin B_{12} and B_{12}-dependent Reactions

Since its discovery in 1948 (simultaneously by K. Folkers and E. Smith), Vitamin B_{12} has attracted the interest and fascination of many researchers (both chemists and biochemists) and at the same time has been a very unappealing subject to many others. It is difficult to be indifferent. It is a very complex substance in terms of both structure and biological function. Seven years of intensive study were required to establish the structure (for which Dorothy Hodgkin received the Nobel Prize in 1964) and knowledge of its participation in cellular processes is still developing. To date approximately 12 B_{12}-dependent reactions have been documented, most occurring in bacteria. Some of these reactions are described below. The daily intake requirement of B_{12} for humans is not known, but the condition of pernicious anemia brought on by a deficiency of the vitamin can be prevented by an intake of approximately 0.1 μg (0.0000001 gram) per day. This low requirement in part reflects the small number of B_{12}-dependent reactions in humans but is also due to the presence of B_{12} supplied to the body by the intestinal bacteria capable of B_{12} synthesis.

The biosynthesis of B_{12} is a rather routine task for bacteria because they are genetically programed to produce all of the necessary enzymes. In contrast, R. B. Woodward and A. Eschenmoser devoted nearly 11 years to develop a chemical synthesis of B_{12}. Their achievement is probably the most elegant in the annals of organic synthesis and may have practical application. Currently all of the B_{12} that is used for nutritional and therapeutic purposes is harvested from large scale bacterial fermentations.

The B_{12} structure is basically that of a tetrapyrrole compound, albeit a rather elaborate one (see below). Compared to the more common tetrapyrrole grouping in heme the distinctive features of B_{12} structures are (a) there is one less methenyl (—CH=) bridge between pyrroles, so that two pyrroles are linked directly; to distinguish it from the porphyrin ring of heme this system is called a *corrin* ring; (b) the corrin tetrapyrrole system is coordinated to *cobalt*; (c) one of the pyrrole side chains contains a *5,6-dimethylbenzimidazole* grouping that is also coordinated to the fifth position of the central cobalt; and (d) the pyrrole side chains consist of *amide* rather than acid groupings. When B_{12} is isolated from fermentation broths, it is usually recovered with CN^- (cyanide) complexed to the sixth position of cobalt. This is the form that is commonly referred to as Vitamin B_{12}. It is also called *cyanocobalamin* or *cyanocobamide*. One of the *active coenzyme forms* contains a *5'-deoxyadenosyl* grouping coordinated to cobalt at the sixth position rather than cyanide. This B_{12}-coenzyme is called *5'-deoxyadenosylcobamide*. Other forms contain OH^-, H_2O, or a methyl group in place of deoxyadenosine.

Two B_{12}-dependent reactions of importance in mammals are the methylation of homocysteine to yield methionine and the isomerization of methylmalonyl-SCoA to yield succinyl-SCoA. The formation of methionine uses N^5—CH_3-FH_4 as the methyl donor; Vitamin B_{12} is proposed as the initial methyl acceptor that then transfers the —CH_3 group to homocysteine—all under the influence of the protein component of the enzyme. The homocysteine → methionine conversion is a rather important

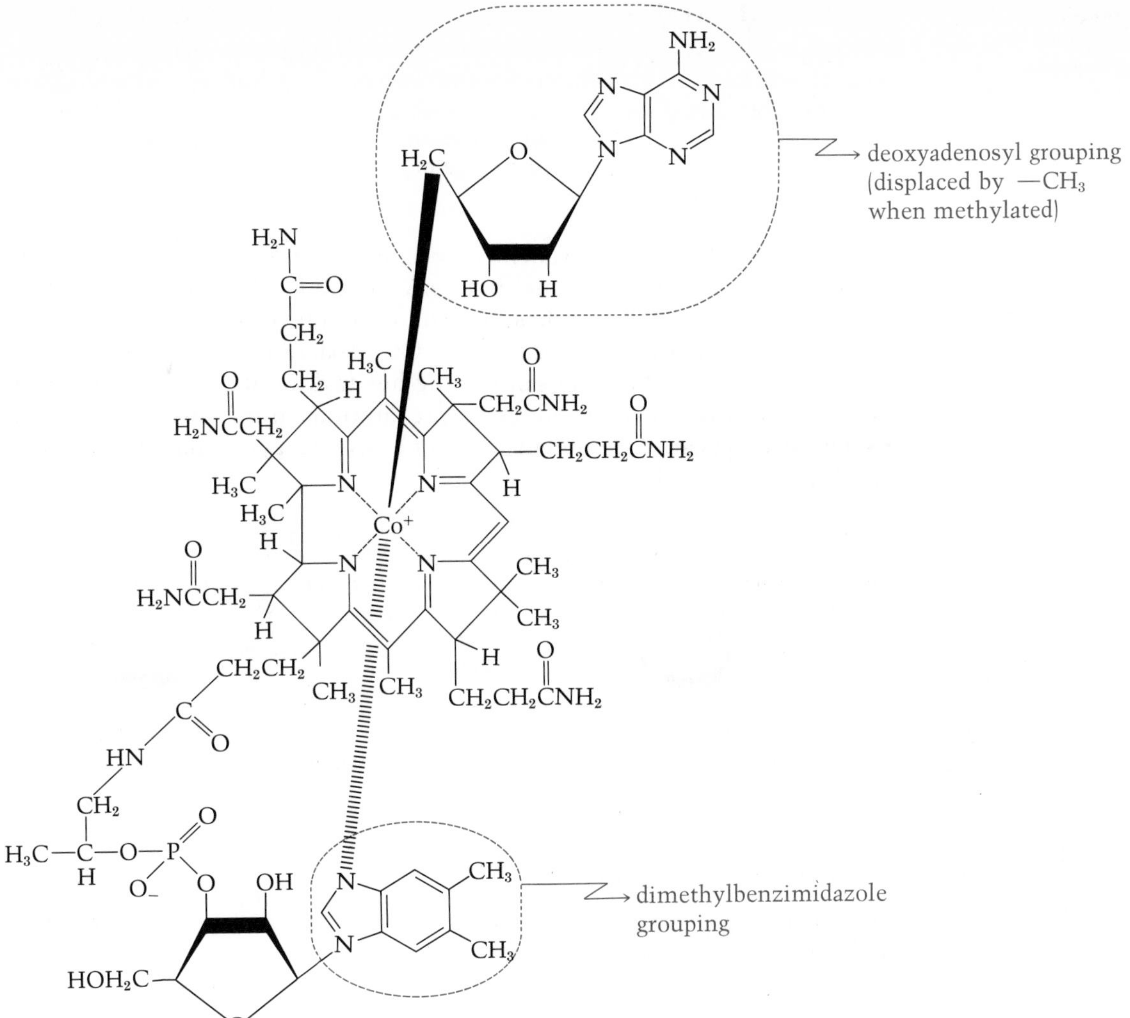

NH_2

H_2C — O

HO — H

deoxyadenosyl grouping
(displaced by —CH$_3$
when methylated)

H_2N
C=O
CH_2
CH_2 H_3C CH_3
H

H_2NCCH_2 O ——— $-CH_2CNMH_2$ O

H_3C ——— $-CH_2CH_2CNH_2$
H

H_3C
H

Co$^+$

H_2NCCH_2 O

H CH_3

CH_3

CH_2CH_2 H O

CH_3 CH_3 $CH_2CH_2CNH_2$

C O

HN

CH_2

$H_3C-C-O-P$ O

H O$^-$ OH

N CH_3

dimethylbenzimidazole
grouping

HOH_2C N CH_3

O

Vitamin B$_{12}$

one since it accounts for the regeneration of methionine after it is utilized as a methyl donor in transmethylation reactions. Obviously, nothing would be gained if homocysteine were converted to methionine with S-adenosyl-methionine functioning as the methyl donor.

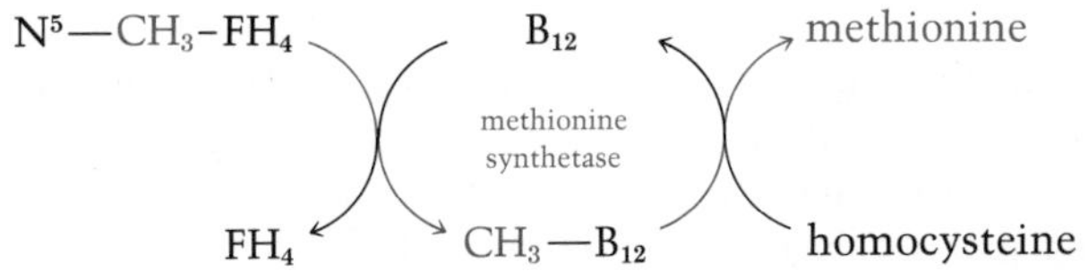

N^5—CH$_3$–FH$_4$ $\qquad$ B$_{12}$ $\qquad$ methionine

methionine
synthetase

FH$_4$ $\qquad$ CH$_3$—B$_{12}$ $\qquad$ homocysteine

Abnormally low levels of either of these reactions have not yet been directly or indirectly linked to the symptoms of pernicious anemia, of other anemias, or of neurological disorders that are associated with a dietary B$_{12}$ deficiency. This is still a relatively unknown area of biochemistry.

The methylmalonyl-SCoA → succinyl-SCoA conversion utilizes the 5′-deoxyadenosyl form of B$_{12}$, which under the guidance of the enzyme participates in what is proposed to be an intramolecular migration. Recall that this reaction is part of a set that accounts for the further metabolism of propionyl-SCoA after its formation from the β-oxidation sequence of fatty acids having an odd number of carbons (see p. 502 in Chap. 16).

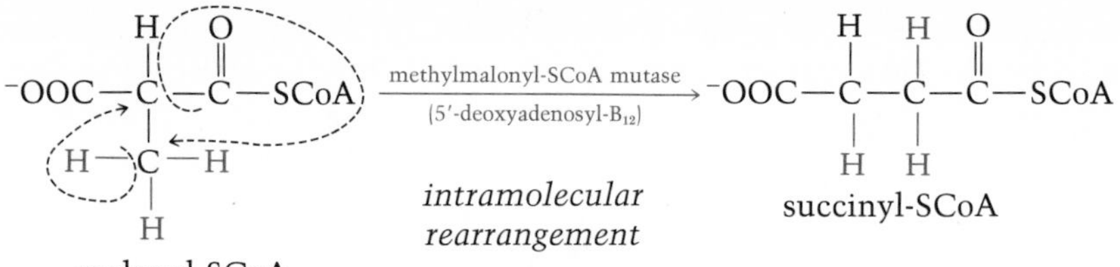

$$\text{malonyl-SCoA} \xrightarrow[\text{(5'-deoxyadenosyl-B}_{12})]{\text{methylmalonyl-SCoA mutase}} \text{succinyl-SCoA}$$

*intramolecular
rearrangement*

In some bacteria 5'-deoxyadenosylcobamide is the coenzyme for ribonucleotide reductase, the enzyme that converts ribonucleotides to deoxyribonucleotides. This reaction has obvious importance to DNA biosynthesis and we will discuss it later (see p. 563). Other bacteria contain a ribonucleotide reductase that does the same thing but it does not require a B_{12} coenzyme. Partially purified reductase preparations from some mammalian tissues also appear to be independent of a B_{12} coenzyme, but the subject is anything but closed.

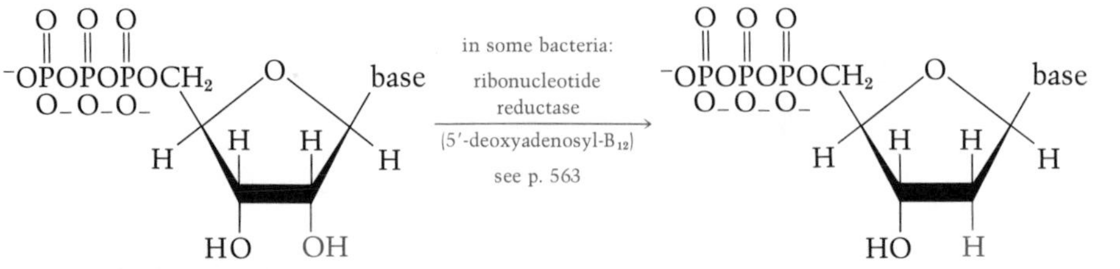

$$\text{5'-triphosphoribonucleotide} \xrightarrow[\substack{\text{reductase} \\ \text{(5'-deoxyadenosyl-B}_{12}) \\ \text{see p. 563}}]{\substack{\text{in some bacteria:} \\ \text{ribonucleotide}}} \text{5'-triphospho-2'-deoxyribonucleotide}$$

NUTRITIONAL CONSIDERATIONS

Obviously every organism requires a supply of the amino acids for normal growth and development. In many cases, this requirement is satisfied by the organism itself due to an enzymatic ability to synthesize each of the amino acids from other sources. Most microbes and the photosynthetic plants are prime examples. In fact, given supplies of carbon, energy, inorganic nitrogen, and varied inorganic nutrients, these organisms can produce not only all of the amino acids but also every other substance necessary for normal growth and development. On the other hand, animal organisms—and humans in particular—are not quite so versatile. Many substances—the exact ones varying from one organism to another—must be supplied in the diet, since the organisms lack the necessary enzymes for their biosynthesis.

Two especially critical aspects of human nutrition concern an intake of the vitamins and several amino acids. As we have noted on several occasions throughout the last few chapters, the vitamins are required for many life processes in the area of general metabolism where they function as coenzymes. At this point it is suggested that you return to our first treatment of coenzymes, where the relationship to vitamins was initially discussed (pp. 163–164 in Chap. 6). In view of your understanding of how coenzymes participate in metabolism, the review can now be made with a greater perspective. Moreover, certain vitamins are involved in specialized physiological processes; Vitamin A in vision, Vitamin D in bone development, Vitamin E in the function of the kidneys and male genital organs, and Vitamin K in blood clotting. The need for amino acids is, of course, obvious. Thousands of different proteins must be synthesized, and, as evidenced by the last section, many amino acids have specialized metabolic functions. The major exogenous supply of amino acids is protein. The amino acids are made available during digestion by the action of the proteolytic enzymes of the gastrointestinal secretions—trypsin, chymotrypsin, pepsin, carboxypeptidase, and aminopeptidase. The free amino acids are then transported across the intestinal wall and carried throughout the whole body in the blood stream.

An obvious question at this point is, which of the 20 common amino acids are required by humans? The answer is provided in Table 17–1 and, as you can see, the list is not small. Also note that the amounts required for an adult are appreciably less than those for an infant. In addition, the number of amino acids required is one fewer for the adult.

The effects of malnutrition are varied, and many are not even understood. A deficiency of each vitamin is usually manifested by a complex set of symptoms. A prolonged deficiency of nutritive protein, that is, protein rich in the essential amino acids, results in a particularly tragic disease called *kwashiorkor.* If untreated in early life, this is most frequently an irreversible and terminal state. Since most of the common food sources contain some of the essential amino acids, the general cause of this condition is *not* a total *lack of all* the essential amino acids in the diet. Although such a situation would certainly cause this condition, kwashiorkor can also result from the absence or subminimal presence of *only some* of the essential acids. To satisfy the need, the body tissues begin to degrade their own protein. The results are impaired development and function of all vital organs. An early death is quite common, particularly in infants feeding from a mother who herself is subsisting on a deficient diet. The hideous problem of malnutrition is widespread, but is especially severe in developing countries. Efforts to increase the production of high-quality nutritive food from the oceans, or in the form of algae or other microbes grown on plentiful resources such as crude petroleum deposits and waste products, have the potential to rid our world society of these intolerable conditions.

TABLE 17–1 **Approximate minimum daily intake requirements[a] of essential amino acids.[b]**

L-*Amino acid*	*Infants* (mg/kg body wt.)	*Male adult* (g)	*Female adult[d]* (g)
Histidine	30	0	0
Tryptophan	20	0.25	0.16
Phenylalanine[c]	90	1.1	0.22
Lysine	100	0.80	0.50
Threonine	90	0.50	0.31
Methionine	45	1.1	0.35
Leucine	150	1.1	0.62
Isoleucine	130	0.70	0.45
Valine	110	0.80	0.65

[a] Recommended levels are generally twice the minimal values.
[b] Small quantities of arginine are required under certain conditions.
[c] Since much of the physiological function of phenylalanine requires its conversion to tyrosine, roughly 75% of the phenylalanine requirement can be covered by tyrosine.
[d] Greater intake of all acids is recommended during pregnancy and lactation.

Biosynthesis of UMP and IMP

The biosynthetic pathways of both pyrimidine and purine nucleotides have been completely elucidated. Much of the early work, started over 20 years ago, was done with mammalian systems and later extended to other types of organisms. Now there is extensive evidence that the same basic steps occur in nearly all organisms. Actually two separate pathways are involved, one accounting for the biosynthesis of pyrimidine nucleotides, and the other for the biosynthesis of purine nucleotides. The end product of the pyrimidine pathway is *uridine-5'-monophosphate* (UMP), which then serves as the precursor of all other pyrimidine ribonucleotides and deoxyribonucleotides. The end product of the purine pathway is *inosine-5'-*

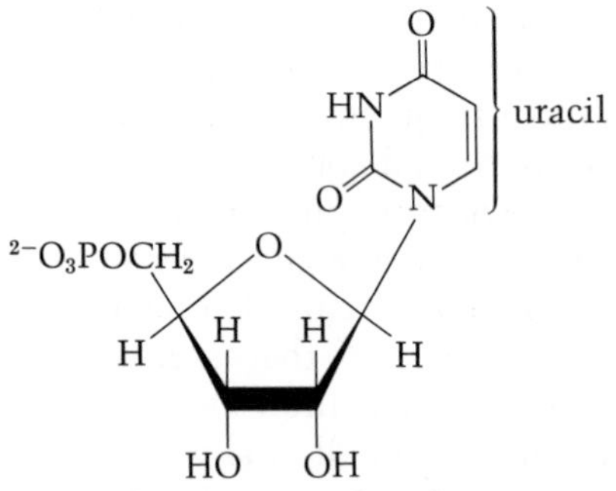

uridine-5'-monophosphate
(UMP)
*precursor of all
pyrimidine
nucleotides*

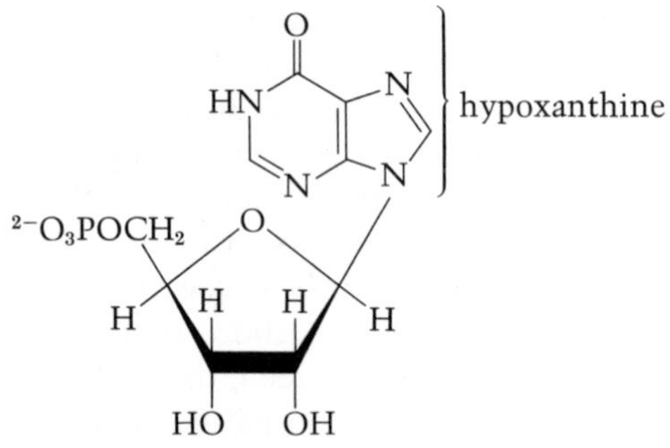

inosine-5'-monophosphate
(IMP)
*precursor of all
purine
nucleotides*

monophosphate (IMP), which is then converted to all other purine ribonucleotides and deoxyribonucleotides.

Although the end product of each pathway is a monophosphate nucleotide, it is the heterocyclic nitrogen base moiety that is actually being assembled—*uracil* in UMP and *hypoxanthine* (6-oxopurine) in IMP. In both processes, the phosphoribose component is contributed by *5-phosphoribosyl-1-pyrophosphate* (PRPP), which in turn originates from ribose-5-phosphate in an ATP-dependent pyrophosphorylation reaction. The ribose phosphate in turn originates from glucose by such pathways as the hexose-monophosphate shunt (see Chap. 12). Consequently, it is logical to examine each pathway in the context of how the pyrimidine and purine ring systems are put together.

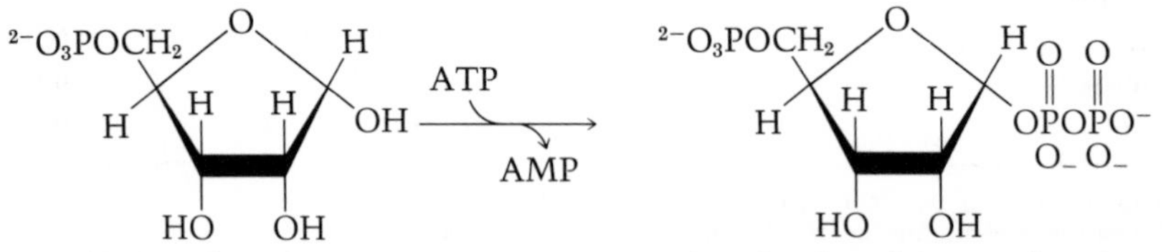

α-ribose-5-phosphate

α-5-phosphoribosyl-1-pyrophosphate
(PRPP)

As we have seen on several occasions now, the end products of anabolic pathways are derived from smaller precursors: pyruvate ⟶⟶ glucose (glycogen); acetyl-SCoA ⟶⟶ fatty acids; acetyl-SCoA ⟶⟶ cholesterol. Moreover, different precursors are often involved, as in the biosynthesis of the tetrapyrrole grouping from glycine and succinate, and the formation of the aromatic amino acids from phosphoenolpyruvate and erythrose-4-phosphate. Two of the clearest examples of both these principles are represented in the production of the uracil and hypoxanthine ring systems. The former is built up from aspartate, CO_2, and NH_3. As shown below, three

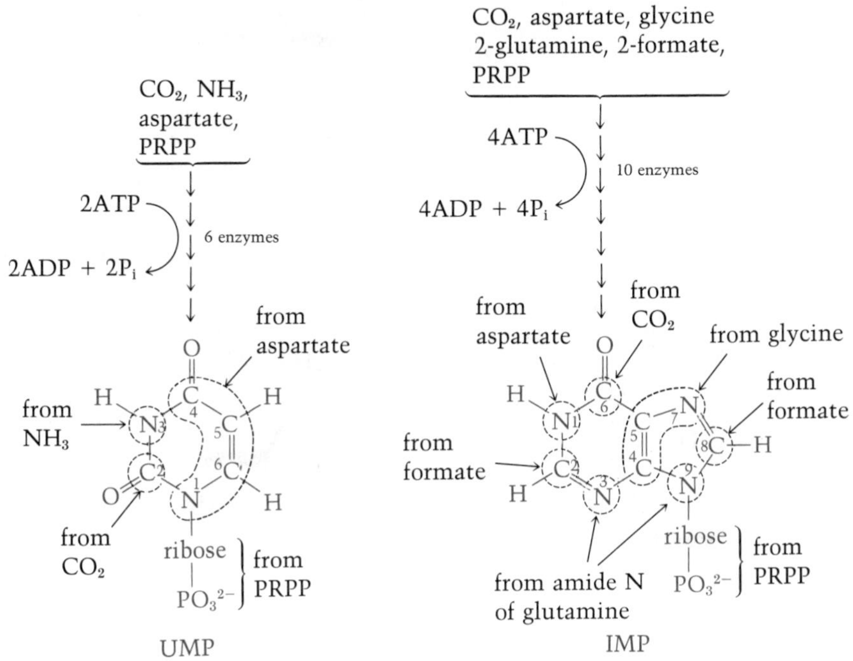

CO₂, NH₃, aspartate, PRPP — 2ATP → 2ADP + 2Pᵢ, 6 enzymes → UMP

from aspartate (C); from NH₃ (N³); from CO₂; ribose–PO₃²⁻ from PRPP

CO₂, aspartate, glycine, 2-glutamine, 2-formate, PRPP — 4ATP → 4ADP + 4Pᵢ, 10 enzymes → IMP

from aspartate; from CO₂; from glycine; from formate; from amide N of glutamine; ribose–PO₃²⁻ from PRPP

carbons (C^4, C^5, C^6) and one nitrogen atom (N^1) originate from aspartate. The second nitrogen atom (N^3) originates from NH_3, and the last carbon atom (C^2) arises from CO_2. The purine ring is even more diverse in origin, requiring glycine, formate, glutamine, aspartate, and CO_2. In this instance, aspartate contributes one nitrogen atom (N^1); the amide grouping from each of two units of glutamine contributes separately to two other nitrogen atoms (N^3 and N^9); the final nitrogen (N^7) and two carbons (C^4 and C^5) originate from an entire glycine molecule; the carbons of two formate molecules contribute to two other positions (C^2 and C^8); and CO_2 contributes the last carbon (C^6). Of course, the enzyme-catalyzed reactions responsible for these relationships and the sequence in which they occur comprise the essence of both pathways. The complete pathways are diagrammed in Figs. 17–6 (UMP) and 17–7 (IMP).

Aside from the fact that in both cases it is the heterocyclic nitrogen base moiety that is being assembled, the two pathways are completely different and the greater complexity of the IMP pathway is obvious. The first specific difference between the two is the point of insertion of the phosphoribose moiety. In the IMP pathway, the α-PRPP is involved in the very first

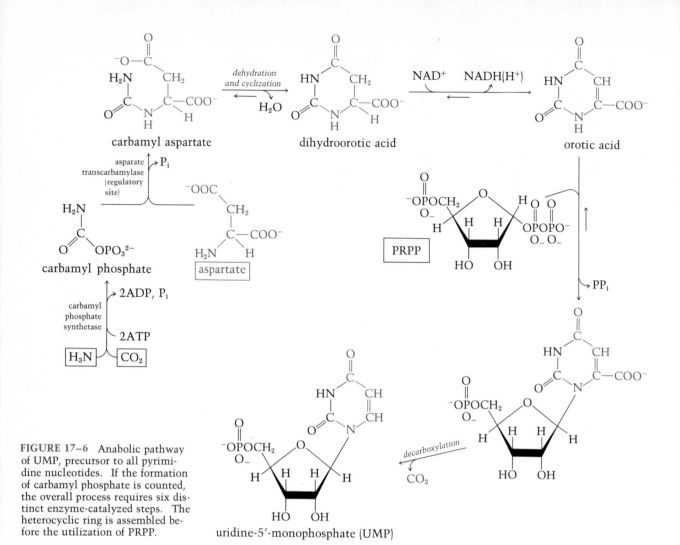

FIGURE 17–6 Anabolic pathway of UMP, precursor to all pyrimidine nucleotides. If the formation of carbamyl phosphate is counted, the overall process requires six distinct enzyme-catalyzed steps. The heterocyclic ring is assembled before the utilization of PRPP.

reaction (an amino group transfer), resulting in the formation of *β-5-phosphoribosyl-1-amine*. In other words, the phosphopentose component is inserted *prior* to the formation of the purine ring. All subsequent reactions involve the progressive build-up of the purine ring from this position. Included among these reactions are two ring closures. Eventually the carbon-nitrogen linkage in *β*-5-phosphoribosyl-1-amine becomes the characteristic $C^{1'}(\beta)\text{-}N^9$ glycosidic bond of the purine nucleotide. In the UMP route, on the other hand, the *α*-PRPP is utilized in the next to last reaction *after* the pyrimidine ring (as orotic acid) is formed. Note here the formation of the $C^{1'}(\beta)\text{-}N^1$ glycosidic linkage common to pyrimidine nucleotides. The *inversion of configuration* at the anomeric carbon in each of the two PRPP-dependent reactions is also noteworthy (remember, the glycosidic linkage in naturally occurring nucleotides is characterized by the *β* configuration) and typifies the specificity of the two enzymes involved.

Another contrasting feature is that the IMP pathway is distinguished by the participation of *tetrahydrofolate* (FH₄) as an obligatory coenzyme in two separate *formylation* reactions. In one case the N^{10}-formyl-FH₄ species is

A special role tetrahydrofolate plays in pyrmidine biosynthesis will be discussed later.

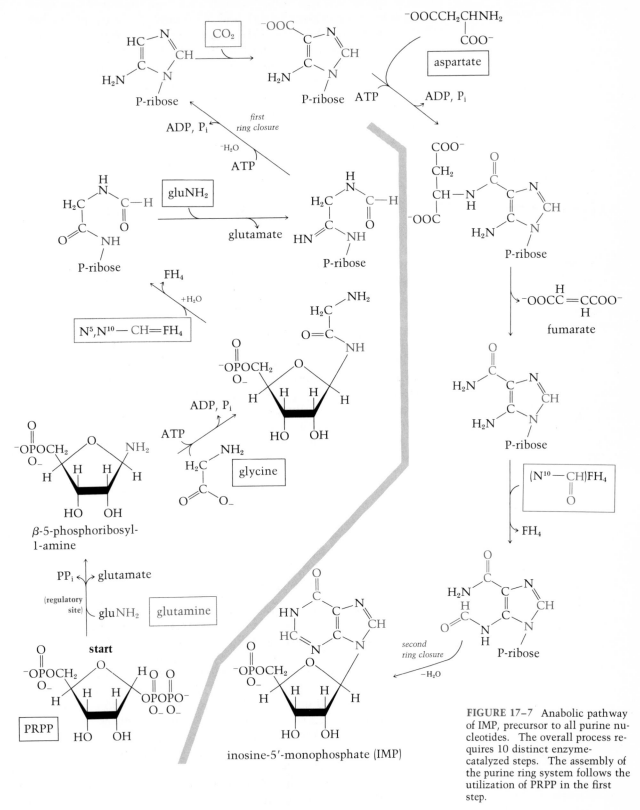

FIGURE 17–7 Anabolic pathway of IMP, precursor to all purine nucleotides. The overall process requires 10 distinct enzyme-catalyzed steps. The assembly of the purine ring system follows the utilization of PRPP in the first step.

inosine-5′-monophosphate (IMP)

β-5-phosphoribosyl-1-amine

used directly, whereas a conversion to the N^5,N^{10}-methenyl-FH_4 form is required in the second reaction. Although we have indicated that formate is the source of both the formyl and methenyl carbons, remember that N^{10}—CHO-FH_4 and N^5,N^{10}=CH—FH_4 can originate from other sources.

A final distinction is observed in that the UMP pathway produces reducing power (as NADH) in the formation of orotic acid from dihydroorotic acid. Reducing power is not generated in the IMP pathway. By considering the potential reoxidation of NADH via oxidative phosphorylation as yielding three ATPs, we can propose the interesting hypothesis that the UMP pathway is self-sufficient in terms of an energy balance, since only two ATPs are required.

The net reactions of UMP and IMP biosynthesis are given below.

pyrimidine pathway (see Fig. 17-6):

$$CO_2 + NH_3 + \text{aspartate} + PRPP + 2ATP + NAD^+ \longrightarrow UMP + 2ADP + 2P_i + PP_i + NADH + H^+ + CO_2$$

purine pathway (see Fig. 17-7):

$$CO_2 + 2\,\text{glutamine} + 2HCOO^- + \text{aspartate} + \text{glycine} + PRPP + 4ATP \xrightarrow{FH_4}$$
$$\text{formate}$$

$$IMP + \text{fumarate} + 2\,\text{glutamate} + PP_i + 4ADP + 4P_i$$

In addition to summarizing Figs. 17–6 and 17–7 and earlier discussions, these equations also draw attention to the formation of other products besides UMP and IMP. In considering the pyrimidine pathway, we have already discussed the possible fate of NADH and its significance. The production of CO_2 (occurring in the last reaction; see Fig. 17–6) is also significant because it renders the pathway autocatalytic. In the purine pathway, the fumarate arising from aspartate is likewise not wasted, since it can be shunted into the citric acid cycle and be catabolized further as a source of carbon and energy. In fact, after its conversion to oxaloacetate, followed then by a transamination, it can be reconverted to aspartate. Many things can happen to the glutamic acid arising from glutamine. However, by the action of glutamine synthetase, it would be reconverted to glutamine for further use in IMP production.

Biosynthesis of UTP, CTP, GTP, and ATP

The biosynthesis of UMP and IMP represents only the first of three equally important phases of nucleotide biosynthesis (see margin). The second phase consists of the conversion of these ribonucleotides to the *triphosphoribonucleotides*: UTP and CTP from UMP, and GTP and ATP from IMP. In addition to having specialized metabolic roles (some of which we have already examined in other chapters, for UTP in carbohydrate metabolism and CTP in phospholipid metabolism) the four triphospho species are also the required substrates in transcription for the biosynthesis of RNA. The third phase involves the formation of the *triphosphodeoxyribonucleotides*. This phase is especially important because the deoxy species are required for DNA biosynthesis. The ribo → deoxy conversions and the formation of the third pyrimidine nucleotide, dTTP, will be discussed later. Our immediate concern is the formation of the triphosphoribonucleotides.

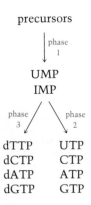

precursors

phase 1

UMP
IMP

phase 3 phase 2

dTTP	UTP
dCTP	CTP
dATP	ATP
dGTP	GTP

It should be noted here that these reactions, particularly those involved with the formation of ATP, are distinct from the production of ATP by phosphorylation coupled to the electron transport process or substrate level phosphorylation reactions. They are distinct in two respects. First, the reactions to be discussed here are catalyzed by soluble enzymes in the cytoplasm rather than being associated with membranous systems such as the mitochondria and the chloroplasts. Second, the primary function of these reactions is to produce triphosphoribonucleotides, including ATP, for purposes other than supplying useful metabolic energy to the cell. Indeed, this would be rather foolish, since the eventual production of UTP, CTP, GTP, and ATP through UMP and IMP would require more energy than would be produced. Their synthesis for use as substrates in nucleic acid biosynthesis is of greatest importance. The formation of useful metabolic energy, remember, is the primary function of electron transport systems coupled to the phosphorylation of ADP.

The conversion of UMP to UTP proceeds via two successive ATP-dependent phosphorylations. Both relations are catalyzed by the same enzyme, *nucleotide kinase,* which displays a broad degree of specificity, acting equally well on any of the mono- or diphosphonucleotides.

$$\boxed{\text{UMP}} \xrightarrow[\substack{\text{ATP} \quad \text{ADP}}]{\substack{\text{nucleotide} \\ \text{kinase}}} \boxed{\text{UDP}} \xrightarrow[\substack{\text{ATP} \quad \text{ADP}}]{\substack{\text{nucleotide} \\ \text{kinase}}} \boxed{\text{UTP}}$$

The formation of CTP then occurs by a direct amination of UTP at position 4 of the pyrimidine ring, with ammonia as the nitrogen source. The enzyme catalyzing this endergonic reaction is likewise ATP-dependent, and also requires GTP for optimum activity, suggesting an allosteric regulation. Indeed, this reaction is but one of several regulatory sites in nucleotide biosynthesis. Other major regulatory sites are discussed in a later section.

$$\boxed{\text{UTP}} \xrightarrow[\substack{\text{NH}_3 \quad \text{H}_2\text{O}}]{\substack{\text{ATP} \quad \text{ADP, P}_i}} \boxed{\text{CTP}}$$

The biosynthesis of the triphospho purine nucleotides, GTP and ATP, involves two different pathways from the same IMP precursor. Both are diagramed below. In each case, the hypoxanthine ring of IMP is first modified to yield the appropriate monophosphonucleotides, AMP and GMP. The latter are then phosphorylated (as described above for UMP) to ultimately yield ATP and GTP. Note that AMP and GMP are likewise formed by amination reactions, but each in a fashion unlike that of the UTP → CTP transformation. The amination to yield AMP utilizes aspartate as a nitrogen source and requires two separate steps. After a necessary oxidation of the hypoxanthine ring to form a carbonyl grouping at position 2 (yielding xanthosine-5'-monophosphate—XMP), the amination step to GMP—as it occurs in higher organisms—utilizes glutamine as the nitrogen donor. Lower organisms, on the other hand, employ ammonia directly. In any case, the effect is the same: XMP → GMP.

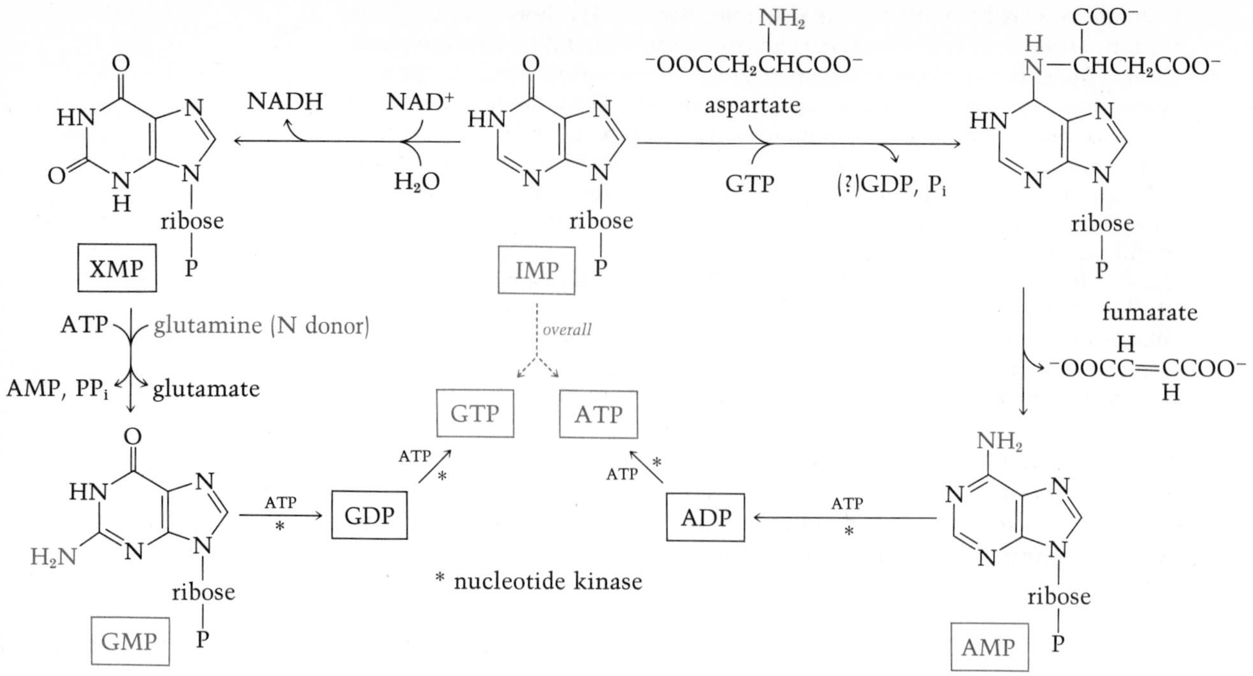

The diagram shows:

XMP (structure) ←— NADH / NAD⁺ with H₂O —— IMP (structure) —— aspartate / GTP / (?)GDP, P_i —→ adenylosuccinate (structure with N—CHCH₂COO⁻ and COO⁻)

ATP → glutamine (N donor); AMP, PP_i ← glutamate

GMP (structure) —ATP*→ GDP —ATP*→ GTP

overall: GTP, ATP (from IMP)

ADP —ATP*← ATP*

fumarate, H—OOCC=CCOO—H

AMP (structure) —ATP*→ ADP

* nucleotide kinase

Biosynthesis of dCTP, dGTP, dATP, and dTTP

DNA biosynthesis depends on a supply of the four triphosphodeoxyribonu-cleotides, dCTP, dGTP, dATP, and dTTP, as substrates. Although it was proposed at one time that the reaction pathways responsible for bio-synthesis of the deoxynucleotide precursors were similar to but independ-ent of the UMP and IMP pathways, it is now quite clear that the major process (if not the only process) for the formation of the deoxynucleotides proceeds directly from the preformed ribonucleotides. In other words, the reductive conversion of β-D-ribose to β-2-D-deoxyribose occurs at the nu-cleotide level rather than at the level of the free sugars. The formation of dCTP, dGTP, and dATP proceeds by a two-stage process originating from the corresponding diphosphoribonucleotides, which in turn, as we have just seen, originate from UMP and IMP. The route to dTTP follows a slightly more complex course.

The key breakthrough in deciphering the basis of deoxyribonucleotide biosynthesis was made by P. Reichard and coworkers in 1964 with the dem-onstration that a soluble protein component obtained from *E. coli*, called *thioredoxin,* could serve as a hydrogen donor in the enzymatic reduction of ribonucleotides to deoxyribonucleotides. Thioredoxin is a small protein (mol. wt. = 11,700) composed of a single polypeptide chain containing one disulfide bond. However, the protein does not function in this, the oxidized state, symbolized below as TR-S₂. First, the disulfide bond is reduced to give reduced thioredoxin (TR-(SH)₂), which contains two thiol groupings. This process requires NADPH and the enzyme *thioredoxin reductase.*

(1) oxidized thioredoxin (disulfide bond) —S—S—   TR-S₂ ——thioredoxin reductase—→ TR-(SH)₂   reduced thioredoxin

NADPH(H⁺)   NADP⁺

Reduced thioredoxin then functions as the hydrogen donor in the reduction of a ribonucleotide to a deoxyribonucleotide. This reduction involves the participation of another enzyme, *ribonucleotide reductase*, which has different characteristics depending on the source. Several bacteria contain a 5'-deoxyadenosylcobalamin-dependent ribonucleotide reductase (2(a)) that acts on any of the four *tri*phosphoribonucleotides. On the other hand, the ribonucleotide reductase (2(b)) in *E. coli* and apparently also in mammalian cells does *not* require 5'-deoxyadenosylcobalamin (it contains nonheme iron) and utilizes *di*phosphoribonucleotides as substrates. In either case both utilize thioredoxin as a hydrogen donor and the effect is the same; the ribosyl moiety is deoxygenated specifically at the 2' position. As for the combined effect of thioredoxin and ribonucleotide reductase, note that the reduced thioredoxin would be oxidized to the disulfide state, which would then be reduced once again to keep the reaction circuit closed.

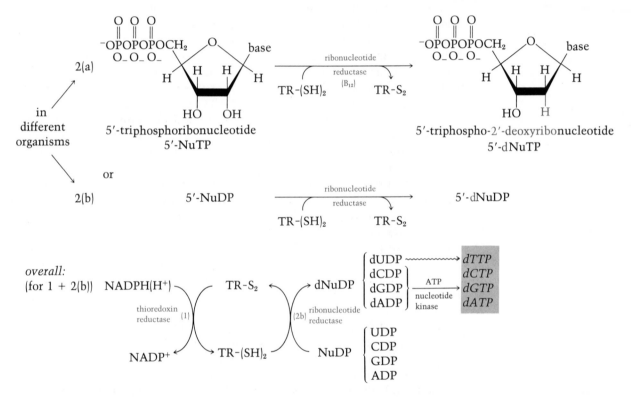

When diphosphonucleotides are utilized, the final step in the biosynthesis of dCTP, dGTP, dATP, and dTTP is a kinase-catalyzed phosphorylation of the corresponding diphospho species, three of which (dCDP, dGDP, and dADP) are produced directly in the thioredoxin reaction. What about dTTP? Indeed, we have not yet indicated how thymine itself is produced, despite the fact that we have already examined the formation of the pyrimidine nucleotides. The reason for delaying discussion of thymine formation until now was that it occurs at the level of a deoxyribonucleotide rather than of a ribonucleotide. First, dUDP is hydrolyzed to dUMP. Then the uracil moiety is methylated to yield dTMP via a process utilizing N^5,N^{10}-methylene-FH$_4$ as the methyl donor. (Recall our mention of this earlier, see p. 551). The stepwise phosphorylation of dTMP to yield dTTP completes the transformation.

Without the participation of the tetrahydrofolate adducts as donors of one-carbon units, the production of ATP, GTP, dATP, and dTTP would be blocked. In turn, this would impede DNA and RNA biosynthesis directly and protein biosynthesis indirectly. Recall our earlier discussions concerning the mechanism of action of the sulfa drugs, which function as effective chemotherapeutic agents in combating infectious organisms by inhibiting the biosynthesis of tetrahydrofolic acid (see p. 176, Chap. 6). At this point, the basis of their potency should be appreciated.

dUMP

dTMP

UDP $\xrightarrow{\text{TR-(SH)}_2}$ dUDP $\longrightarrow$ [dUMP structure, P_i]

$N^5,N^{10}-CH_2-FH_4$ ... FH_2 [dTMP structure, CH_3] $\xrightarrow{\text{(ATP)}}$ dTDP $\xrightarrow{\text{(ATP)}}$ dTTP

(dUMP: HN–C=O, O=C–N, deoxyribose–P)

(dTMP: HN–C=O with CH$_3$, O=C–N, deoxyribose–P)

Regulation of Nucleotide Biosynthesis

The biosynthesis of pyrimidine and purine nucleotides is under strict intracellular control. The basic strategy used is that of *feedback inhibition*, with certain of the nucleotide end products having the ability to reduce the catalytic activity of certain key enzymes, thus reducing the overall rate of their own formation. For example, CTP is a potent inhibitor of aspartate transcarbamylase (ATCase). This inhibition would lead to a decrease in the production of UMP; this in turn would result in a decreased production of UTP, CTP, dCTP, and dTTP. The CTP-ATCase control point is not new to us. Recall that we used it as a model system in developing the theory of allosterism, which explains the regulatory properties of enzymes in terms of secondary binding sites in active and inactive conformational states of proteins (Chap. 6, pp. 190–193).

Since our introduction of this phenomenon, we have encountered several examples of how this tactic is employed in nature to guarantee the efficient production and utilization of energy and other resources within the living cell. The regulation of nucleotide biosynthesis is another example of this principle, but a rather remarkable example at that. In addition to the CTP effect on the control of UMP production, other regulatory signals and sites are known to exist. The major ones are summarized in Fig. 17–8 and include (a) the stimulation by ATP of ATCase, and thus of pyrimidine biosynthesis; (b) the feedback inhibition by adenine and guanine nucleotides—AMP, ADP, ATP, GMP, GDP, and GTP—of the enzyme catalyzing the first reaction in the biosynthesis of IMP; (c) the feedback inhibition by AMP and GMP of the enzymes responsible for converting IMP to each of these purine nucleotides; (d) the stimulation of AMP and GMP formation from IMP by the presence of GTP and ATP, respectively; and (e) the inhibition by dATP of all the ribonucleotide-deoxyribonucleotide conversions. All of these effects appear to be allosteric in nature.

The patterns of regulation are, of course, based on the study of the catalytic properties of purified enzymes *in vitro,* and it might be argued that they do not necessarily reflect the true *in vivo* pattern. However, in this regard it is more likely that the *in vivo* situation is even more complex and elaborate than what is observed *in vitro* and summarized in Fig. 17–8. Whatever the real situation, there is overwhelming evidence that the many reactions responsible for the biosynthesis of nucleotides are highly tuned to the changing needs of the cell, and that there is no waste of metabolic resources. In periods of active nucleic acid biosynthesis, the enzymes of the anabolic pathways are actively churning out the required triphosphonucleotides. In this situation, the nucleotides do not operate as allosteric feedback

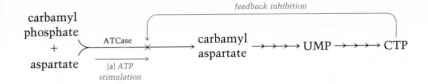

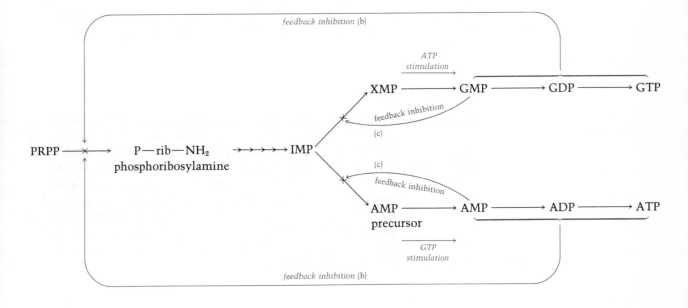

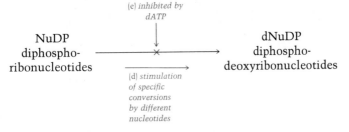

(e) for B$_{12}$-independent reductase; dTTP is a potent
  inhibitor of the B$_{12}$-dependent reductase
(d) applies to both types of reductases

FIGURE 17–8 A diagramatic summary of metabolic regulation in the three major phases of nucleotide biosynthesis.

inhibitors, but instead are preferentially utilized as substrates in the assembly of polynucleotides. When, for one reason or another, the requirement for nucleic acid synthesis is reduced, the substrate nucleotides built up then serve as a group of signals that, figuratively speaking, turn off their own biosynthesis. When the level of DNA and/or RNA biosynthesis picks up again, the inhibitory signals are removed by the utilization of the nucleotides and the anabolic pathways are turned on.

Nucleotide Catabolism

Both DNA and RNA are acted on by various *nucleases* to yield monophosphonucleotides. (In Chap. 7 the specificity and biological roles of nuclease enzymes were discussed.)

$$\text{nucleic acids} \xrightarrow[\substack{\text{of exonucleases} \\ \text{and endonucleases}}]{\text{combined action}} \text{monophosphonucleotides}$$

Although the monophosphonucleotides can be reused for RNA and DNA biosynthesis, our objective here is to consider their further degradation to other products. Owing to the fact that several types of reactions are associated with the degradation of deoxyribonucleotides, we will narrow our treatment even further by specifically examining the catabolism of ribonucleotides.

Purine Catabolism and the Gouty Condition

In most organisms the primary degradative pathways for GMP and AMP involve the same set of separate reactions that converge in the production of *xanthine* (see below). The conversion of guanine from GMP to xanthine is rather direct, whereas the formation of xanthine from adenine of AMP is indirect. After a dephosphorylation of AMP, the adenine moiety of adenosine is then converted to hypoxanthine to yield *inosine*. Inosine is then hydrolyzed to yield *hypoxanthine* as the free base, which is finally converted to xanthine. Although the details are not considered here, xanthine could also originate from the guanine and adenine of the corresponding deoxyribonucleotides, dGMP and dAMP, respectively.

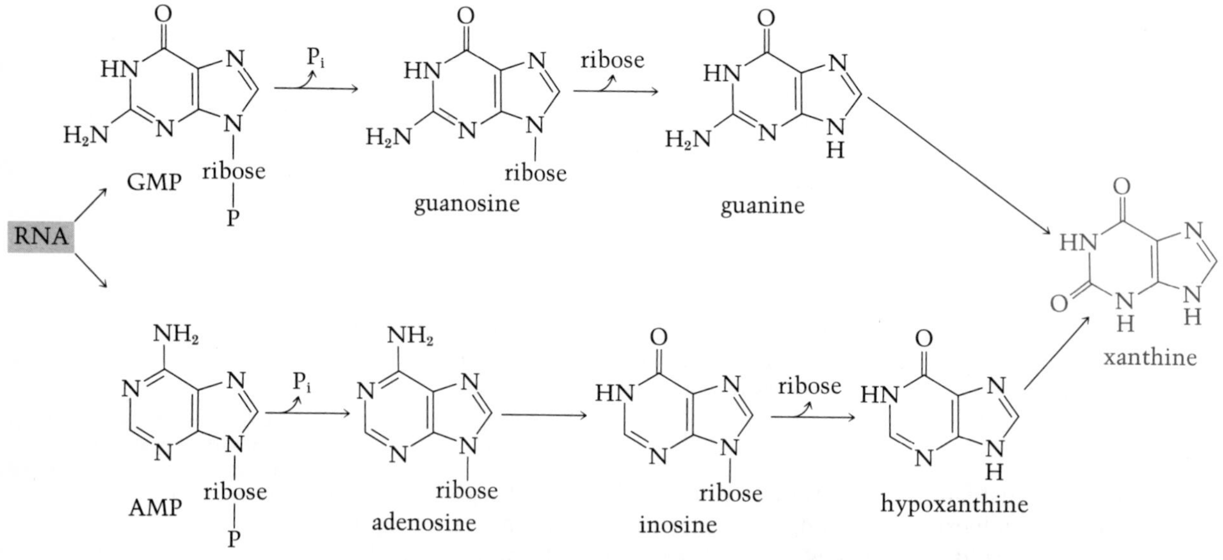

Depending on the organism, the subsequent fate of xanthine is varied (see below). In most primates (including humans) birds, certain reptiles, and the majority of insects, xanthine is converted to *uric acid*, which is excreted as the final end product of purine catabolism. In all other land animals *allantoin*, which is formed by the further oxidation of uric acid, is the final end product. In amphibians and fish allantoin is further degraded to *allantoic acid*. In many microorganisms allantoic acid is converted to *glyoxylate* and *urea*. All of these reactions are catalyzed by specific enzymes and clearly illustrate the themes of biochemical unity and diversity.

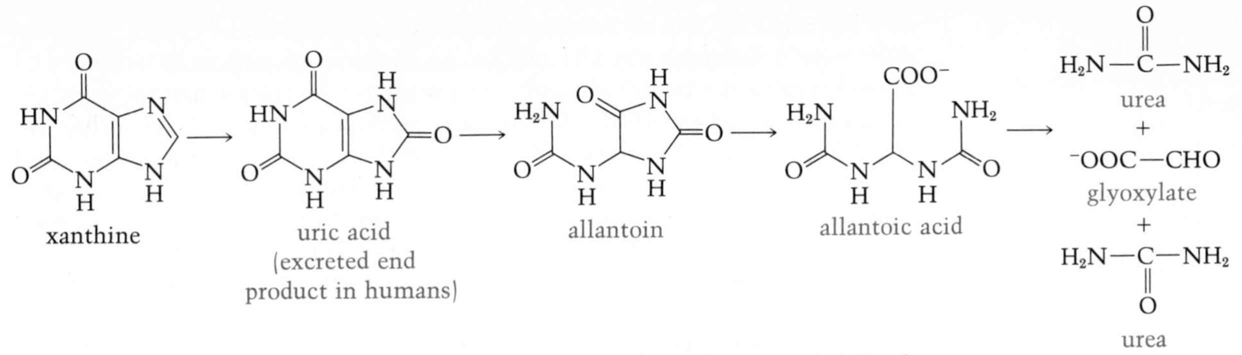

$$\text{xanthine} \longrightarrow \text{uric acid} \longrightarrow \text{allantoin} \longrightarrow \text{allantoic acid} \longrightarrow \text{urea} + \text{glyoxylate} + \text{urea}$$

xanthine

uric acid
(excreted end
product in humans)

allantoin

allantoic acid

urea

glyoxylate

urea

The accumulation and crystallization of uric acid in the synovial fluid around bone joints are responsible for the painful condition known as *gout*. A hereditary disease, gout results from an overproduction of uric acid plus a failure of the kidneys to eliminate uric acid in large amounts. The biochemical basis for the overproduction of uric acid is not yet known. There is some evidence, however, that in some individuals this metabolic abnormality may be due to an inability to depress the biosynthesis of purine nucleotides by the normal feedback inhibition. Recall that the activity of the enzyme catalyzing the initial reaction of the anabolic pathway of purine nucleotides, that is, the formation of β-5-phosphoribosyl-1-amine from α-5-phosphoribosyl-1-pyrophosphate (PRPP) and glutamine, is normally under strong negative control (it is inhibited) by GMP, AMP, and IMP (see p. 565). The malfunction in this normal control point may be a defective enzyme, defective in the sense that it is not capable of binding to and hence not capable of being regulated by the feedback inhibitors.

Another explanation would be that the AMP and IMP cellular levels required for inhibition to occur are not attained despite the fact that purine biosynthesis is occurring at an increased rate. Indeed, this type of malfunction has recently been documented as being the basis of the *Lesch-Nyhan syndrome*, a rare genetic disease related to gout. Among other things, the disease is characterized by excessive anxiety, aggressiveness, mental retardation, and a compulsion toward self-mutilation of the lips, tongue, and fingers. The biochemical defect in Lesch-Nyhan patients is the lack of *hypoxanthine-guanine phosphoribosyltransferase (HPRT)*, an enzyme that normally catalyzes the reuse of the free purine bases, guanine and hypoxanthine, in the biosynthesis of GMP and IMP, respectively. The reactions, both of which utilize PRPP as the phosphoribosyl donor, are as follows:

$$\text{guanine} + \text{PRPP} \xrightarrow[\text{phosphoribosyltransferase}]{\text{hypoxanthine-guanine}} \text{GMP} + \text{PP}_i$$

$$\text{hypoxanthine} + \text{PRPP} \xrightarrow{\text{same enzyme}} \text{IMP} + \text{PP}_i$$

A separate enzyme that specifically catalyzes a similar reaction for adenine to yield AMP is present in both normal individuals and victims of the Lesch-Nyhan syndrome.

$$\text{adenine} + \text{PRPP} \xrightarrow[\text{phosphoribosyltransferase}]{\text{adenine}} \text{AMP} + \text{PP}_i$$

Under normal conditions these *"purine salvage reactions,"* as they are sometimes called, would contribute to (a) elevated cellular concentrations of GMP and IMP and (b) reduced cellular concentrations of PRPP. Both results would diminish the rate of formation of phosphoribosyl-1-amine, and thus diminish the rate of formation of purine nucleotides when they are not required in great quantity. When, as is proposed in Lesch-Nyhan disease, the salvage reactions are not operating, both effects are not achieved, and there is no control of purine nucleotide biosynthesis. More IMP is produced than is required, and rather than accumulate, the excess IMP is converted to uric acid. These points are diagramatically summarized in Fig. 17–9. The above information and the summary in Fig. 17–9 provide an obvious example of the complexity of metabolism and its control. In addition, we see an interesting example of how one specific biochemical defect can have a severe disruptive effect on a whole area of metabolism and

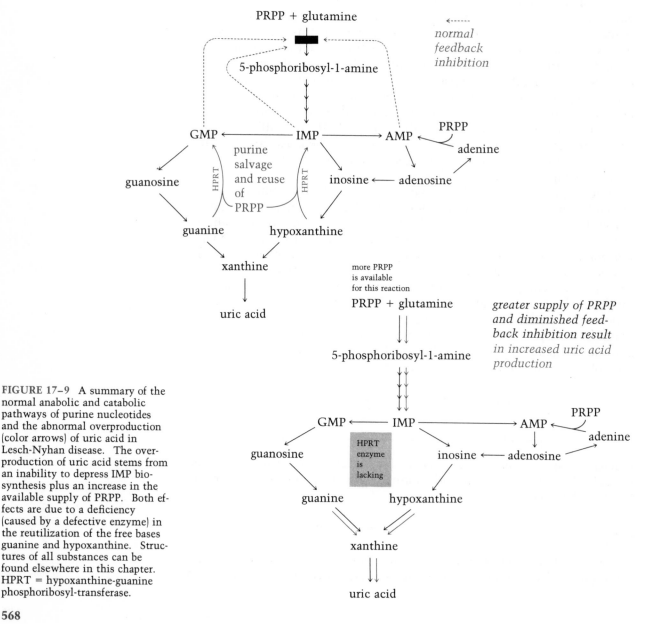

FIGURE 17–9 A summary of the normal anabolic and catabolic pathways of purine nucleotides and the abnormal overproduction (color arrows) of uric acid in Lesch-Nyhan disease. The overproduction of uric acid stems from an inability to depress IMP biosynthesis plus an increase in the available supply of PRPP. Both effects are due to a deficiency (caused by a defective enzyme) in the reutilization of the free bases guanine and hypoxanthine. Structures of all substances can be found elsewhere in this chapter. HPRT = hypoxanthine-guanine phosphoribosyl-transferase.

normal body growth and development. The details associated with the HPRT deficiency and the neurological abnormalities in Lesch-Nyhan disease are not yet known.

Pyrimidine Catabolism

The catabolism of pyrimidine bases originating from pyrimidine-containing nucleotides apparently proceeds through one of several different pathways, depending on the organisms. For example, one of the major pathways, occurring in humans and other animals, involves the reduction of uracil or thymine to yield a fully hydrogenated heterocyclic ring. The degradation of cytosine follows the same route after an initial deamination to yield uracil. Ring cleavage of the product produced by uracil (cytosine) yields *carbamyl-β-alanine,* which is further hydrolyzed to CO_2, NH_3, and *β-*alanine. All of the products would be either excreted as waste products or reused in other areas of metabolism. For example, *β*-alanine could be reused in the biosynthesis of coenzyme A.

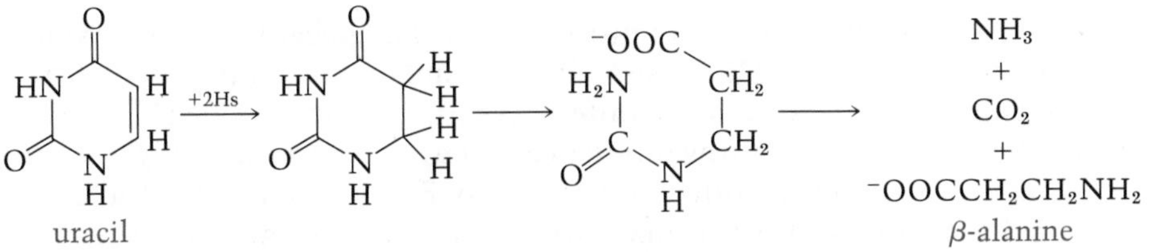

Thymine would yield: CO_2, NH_3, and $^-OOCCHCH_2NH_2$ with CH_3 branch.

PART C: NITROGEN FIXATION AND THE NITROGEN CYCLE

Amino acids (polypeptides) and purines and pyrimidines (nucleic acids) all have one thing in common: they are nitrogen-containing compounds. Where does the nitrogen originate from? The CO_2 in the atmosphere is of course the ultimate source of carbon, with photosynthetic organisms converting the inorganic carbon source into organic materials. Well, the atmosphere is also the major natural source of nitrogen in the form of N_2 gas, with various microorganisms responsible for the assimilation process. This process of direct *nitrogen fixation* is supplemented in our modern agricultural age by the use of fertilizers. In fact it has been estimated that the current input of nitrogen into the biosphere through the use of fertilizers is on the order of 30 million tons per year (and this is growing), which is about the same amount as is estimated to be contributed by the natural process of nitrogen fixation. Although the ramifications of our contributing further to this imbalance are not clear, they certainly merit serious consideration.

The thousands of microbes that are capable of nitrogen fixation include certain anaerobic and aerobic soil bacteria growing independently or in a symbiotic fashion with plants, photosynthetic bacteria, and blue-green algae. The name given to the enzyme system that is responsible for the fixation process is *nitrogenase.* The overall reaction catalyzed by the nitrogenase system is the following reduction:

$$:N\equiv N: \xrightarrow[\substack{\text{6 electrons} \\ \text{6 protons (6H}^+)}]{\text{nitrogenase}} 2\left(:N-H \atop \substack{H \\ | \\ | \\ H}\right)$$

(N_2)

The first breakthrough in understanding this system came in 1960 (some 70 years after nitrogen-fixation was established as a natural occurrence) with the successful isolation of cell-free extracts from the anaerobic bacterium *Clostridium pasteurianum* that were capable of supporting N_2 fixation *in vitro*. Similar successes with other sources, and many studies that characterized various operative features of the cell-free extracts, quickly followed. The isolation from extracts of active nitrogenase protein preparations was achieved for the first time in 1965. It has been carried out several additional times using various sources since then. The body of knowledge resulting from the study of these purified preparations is considerable, but we still do not know precisely how the nitrogenase system operates. The following description covers some of the important facts that have been established.

The gross structural features of nitrogenase so far appear to be rather similar, whatever the source. It consists of two components, both of which are required for activity. Aside from differences in size and amino acid composition, the most distinguishing difference is that one contains molybdenum (Mo) and iron (Fe) and the other contains only Fe. The type of metal complexing that is involved is yet unknown, but the iron in each is of the nonheme variety. Although no details have yet been established, there is no doubt that the Mo and Fe are involved in the mode of action of nitrogenase and little doubt that at least Mo participates directly in the reaction. Other important features concerning the operation of the nitrogenase system that have been established are as follows:

1. ATP and Mg^{2+} are essential for activity. The current hypothesis suggests that binding of at least one ATP to the Fe protein is required for the transfer of one electron to the N_2 that may be bound to the Mo-Fe protein. Thus, at least six ATPs would be required for the complete $N_2 \rightarrow 2NH_3$ conversion.
2. Proteins having a highly negative reduction potential $(\mathscr{E}^{o'}_{red})$—meaning that they are strong reducing agents—are the immediate source of electrons. Two types of proteins have been implicated: *ferredoxins* and *flavodoxins*. A ferredoxin is an Fe-S—containing protein and a flavodoxin is an FMN-containing protein. (Remember that ferredoxins also participate as electron transfer agents in the light reaction of photosynthesis.) Since nitrogen-fixing organisms contain both of these proteins it is argued that they comprise part of an electron carrier sequence that operates prior to the nitrogenase acceptor.
3. The original source of electrons is an intracellular metabolite, the specific identity of which can vary from organism to organism. Furthermore, different materials can be used by the same organisms. Anaerobic organisms can utilize pyruvate, H_2, or formate. A common source in aerobic organisms is the NADPH produced in the dehydrogenation of glucose-6-phosphate, isocitrate, or α-ketoglutarate.

These relationships are outlined below. The events depicted are for the anaerobe, *Cl. pasteurianum*, using pyruvate as a source of electrons. The oxidative decarboxylation of pyruvate involves a thiamine pyrophosphate-dependent dehydrogenase, which may use oxidized ferredoxin as an electron acceptor. The subsequent conversion of acetyl-SCoA to acetate and ATP is

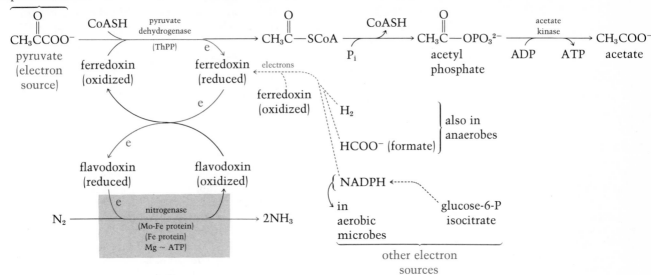

noteworthy because it means that pyruvate oxidation can supply both electrons and some energy for the nitrogenase system. Additional ATP would be supplied from other sources.

The NH_3 product of microbial nitrogen fixation can be utilized directly by plants for the production of all nitrogen-containing organic compounds. Plants are then ingested by animals. Ammonia is regenerated by animal excretions and by the death and decay of both plants and animals. Certain soil bacteria can also utilize NH_3 as a source of electrons (energy source) by oxidizing it to NO_2^- and to NO_3^-. These oxidative processes of *nitrifica-*

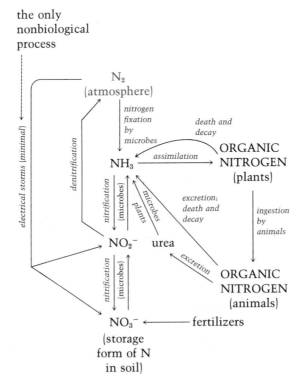

tion are complex and not very well understood. Because NH_3 is a toxic substance the nitrifying bacteria play an important role when they convert it to nontoxic NO_3^-, the storage form of nitrogen in the soil. Many microbes are capable of the reverse of nitrification (that is, the reductive conversion of NO_3^- and NO_2^- ultimately to NH_3), which would resupply the pool of NH_3. Other microbes are also capable of *denitrification* ($NO_2^- \rightarrow N_2$), which would of course return N_2 to the atmosphere. All of these interrelationships comprise the *nitrogen cycle* (see diagram on p. 571, bottom).

LITERATURE

BENKOVIC, S. J., and C. M. TATUM. "Mechanisms of Folate Cofactors." *Trends Biochem. Sci.*, **2**, 161–163 (1977). A short review article.

BLAKELY, R. L., and E. VITROLS. "The Control of Nucleotide Biosynthesis." *Ann. Rev. Biochem.*, **38**, 210–224 (1968). A review article summarizing the many regulatory enzymes associated with the pathways of nucleotide biosynthesis.

COOPER, J. R., F. E. BLOOM, and R. H. ROTH. *Biochemical Basis of Neuropharmacology.* 2nd edition. New York: Oxford University Press, 1974. A short book in neurobiochemistry. Available in paperback.

FRIEDEN, E., and K. H. LIPNER. *Biochemical Endocrinology of the Vertebrates.* Englewood Cliffs, New Jersey: Prentice-Hall, 1971. A compact introduction to basic biochemical facts about hormones. Available in paperback.

GREENBERG, D. M., and W. W. RODWELL. "Carbon Catabolism of Amino Acids" and "Biosynthesis of Amino Acids and Related Compounds." In *Metabolic Pathways*, 3rd edition. D. Greenberg, ed., Vol. 3. New York: Academic Press, 1969. Two review articles giving a thorough analysis of catabolic and anabolic pathways of all the amino acids. Other articles in this volume are devoted to nitrogen metabolism of amino acids, sulfur metabolism, and metabolism of porphyrin compounds.

MEISTER, A. *Biochemistry of the Amino Acids.* 2nd edition. New York: Academic Press, 1965. A treatise in two volumes containing a comprehensive treatment of amino acid metabolism.

MEISTER, A. "On the Enzymology of Amino Acid Transport." *Science,* **180,** 33–39 (1973). A description of the γ-glutamyl cycle for membrane transport of amino acids.

SAFRANY, D. R. "Nitrogen Fixation." *Scientific American,* **231,** 64 (October, 1974).

SOBER, H. A., ed. *Handbook of Biochemistry.* 2nd edition. Cleveland: The Chemical Rubber Company, 1970. A listing is given on pages K-50 through K-52 of the recommended daily dietary allowances of protein and vitamins for humans, as revised in 1968 by the Food and Nutrition Board of the National Academy of Sciences and the National Research Council.

STANBURY, J. O., J. B. WYNGAARDEN, and D. S. FREDRICKSON, eds. *The Metabolic Basis of Inherited Disease.* 2nd edition. New York: McGraw-Hill, 1966. All types of metabolic diseases caused by genetic defects are discussed.

STADTMAN, T. C. "Vitamin B_{12}." *Science,* **171,** 859–867 (1971). A good review article.

STREICHER, S. L., and R. C. VALENTINE. "Comparative Biochemistry of Nitrogen Fixation." *Ann. Rev. Biochem.,* **42,** 279–302 (1973).

UMBARGER, H. E. "Regulation of Amino Acid Metabolism." *Ann. Rev. Biochem.,* **38,** 323–370 (1969). A review article describing metabolite-mediated regulation of enzyme activity in multifunctional pathways in general, and pathways of amino acid metabolism in particular.

EXERCISES

17–1. Complete each of the following reactions:

(a) arginine + α-ketoglutarate $\xrightarrow{\text{transaminase}}$

(b) valine + α-ketoglutarate $\xrightarrow{\text{transaminase}}$

(c) L-lysine + FMN $\xrightarrow[\text{oxidase}]{\text{amino acid}}$

(d) cysteine + α-ketoglutarate $\xrightarrow{\text{transaminase}}$

(e) leucine $\xrightarrow[\text{decarboxylase}]{\text{leucine}}$

(g) pyruvate + NADH + H^+ + NH_3 $\xrightarrow{\text{dehydrogenase}}$

17–2. Is the structure shown on p. 534 for flavin mononucleotide that of the oxidized or reduced form?

17–3. What relationship exists between the reactions given below, known to occur in liver, and the lack of a dietary requirement for arginine in adults?

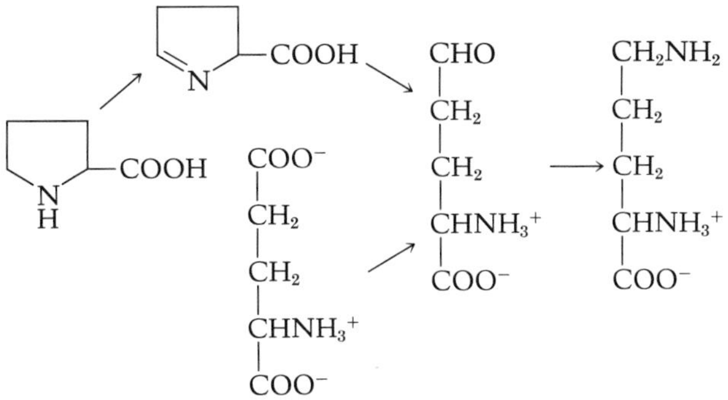

Δ^1-pyrroline-2-carboxylic acid / proline → glutamic-γ-semialdehyde ($CHO-CH_2-CH_2-CHNH_3^+-COO^-$) → glutamate ($COO^- -CH_2-CH_2-CHNH_3^+-COO^-$) → $CH_2NH_2-CH_2-CH_2-CHNH_3^+-COO^-$

17–4. Propose a series of reactions to explain how the carbon atoms of glucose could be converted to γ-aminobutyrate.

17–5. Propose a series of reactions to explain how a carbon atom of tryptophan could eventually be utilized in the biosynthesis of epinephrine. What specific carbon atom of tryptophan is involved?

17–6. In many organisms, the immediate biosynthetic precursor of L-lysine is α,ε-diaminopimelic acid (structure below). What type of enzyme would catalyze this reaction, what coenzyme would be required, and what type of enzyme-substrate complex would be formed?

$$^-OOCCHCH_2CH_2CH_2CHCOO^- \xrightarrow{\text{E}} \text{L-lysine}$$
$$\text{NH}_3^+ \qquad\qquad \text{NH}_3^+$$
α,ε-diaminopimelic acid

17–7. Write a net reaction for the operation of the urea cycle, linking it to ancillary reactions of the citric acid cycle and glutamate dehydrogenase. Does this integration of metabolism have any relationship to item (c) on p. 535?

17–8. On p. 553 it is stated: Obviously nothing would be gained if homocysteine were converted to methionine with S-adenosylmethionine functioning as the methyl donor. Explain.

17–9. Summarize your understanding of the structure and function of those bimolecules that contain a cyclic tetrapyrrole grouping. (*Note:* a thorough response to this question covers a sizeable amount of basic biochemistry.)

17–10. The catabolism of valine involves (in the order given): (1) a transamination to give its alpha keto acid A; (2) oxidative decarboxylation of A in the presence of CoASH to yield a four-carbon thioester B; (3) dehydrogenation of B to yield an unsaturated derivative C; (4) hydration of C to yield a β-hydroxy thioester D; (5) hydrolysis of D to give the free acid E; (6) an NAD$^+$-dependent oxidation of the —CH$_2$OH group in E to yield F, which contains —CHO; (7) conversion of F to methylmalonyl-SCoA; and (8) the last step, catalyzed by methylmalonyl-SCoA mutase. Illustrate the pathway, giving structures of all the intermediates.

17–11. Tryptophan is known to be converted into indole acetic acid, a potent growth hormone in plants, by the action of tryptophan decarboxylase, monoamine oxidase, and an NAD-dependent dehydrogenase, in that order. Reproduce the complete pathway, showing the structures of all intermediate metabolites.

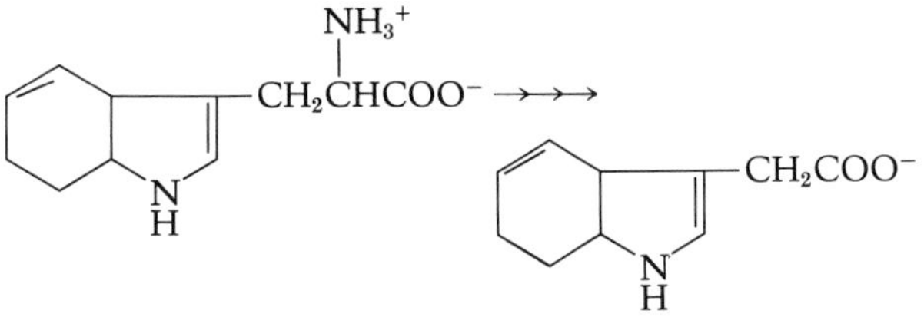

$$\text{tryptophan (indole-}CH_2CHCOO^-, NH_3^+) \longrightarrow\longrightarrow\longrightarrow \text{indole-}CH_2COO^-$$
indole acetate

17–12. Propose a reaction sequence to account for the utilization of carbon from fatty acids in the biosynthesis of porphyrin compounds.

17–13. If patients suffering from alcaptonuria were fed homogentisic acid, what do you predict would happen to this compound after it entered the blood stream?

17–14. Mastocytosis is a physiological disorder resulting from the infiltration of mast cells into skin and such organs as the liver, spleen, and kidney. Clinically, a patient with this condition may possess an enlarged liver and spleen containing unusually high amounts of histamine. It has been suggested that the latter is caused by an excessive production of histamine as a result of the invading mast cells, rather than by a block in the degradation of histamine, which would likewise result in its accumulation. If this is so, what enzyme would you expect to find in large amounts in mast cells?

17–15. How do you account for the fact that certain of the carbon atoms of glucose-6-phosphate ultimately appear as the carbon atoms in the pyrimidine ring that are known to be contributed by aspartate?

17–16. What possible role does the hexose monophosphate shunt play in the conversion of ribonucleotides to deoxyribonucleotides?

17–17. What is your prediction of the relative ability of system A and system B (see below) to form dADP? Explain.

System A	System B
5'-ADP	5'-ADP
thioredoxin (—SH)$_2$	thioredoxin (—SH)$_2$
ribonucleotide reductase	ribonucleotide reductase
buffer	p-chloromercurobenzoate
	buffer

APPENDIX I

(ISO)ENZYMES IN CLINICAL DIAGNOSIS

In deciding on the treatment to be administered to a patient the physician must first make a diagnosis of what is wrong. The diagnosis must be accurate and in many cases it must be quick. In the past 10–15 years the assay of enzyme levels in body fluids and tissues has emerged as a valuable diagnostic tool, supporting other clinical information by (a) identifying the presence of a disease, (b) distinguishing the specific disease from other diseases, and (c) providing an estimate of the severity and duration of the disease.

Assays for the *isoenzyme levels* of specific enzymes in blood serum have been particularly useful in contributing to the correct diagnosis of various diseases. For example, marked increases in the serum levels of the H_4 and H_3M species of *lactate dehydrogenase* (LDH; see p. 390 in Chap. 12) indicate myocardial infarction; a marked increase in the M_4 species of LDH is observed in hepatitis and infectious mononucleosis. The basis for such interpretations is that each body fluid and tissue has a normal and characteristic isoenzyme profile. In the case of heart tissue, the H_4 and H_3M species predominate in heart cells; thus one would expect that when there is heart damage and the cells die and release their enzymes to the blood, the normal serum levels of H_4 and H_3M would be inflated.

The H_4 and H_3M increases are not, however, absolutely indicative of myocardial damage. The main reason is that although isoenzyme profiles reflecting the predominance of one or more forms are relatively distinctive, all five LDH isoenzymes are present in most human tissues. For example, the predominate heart LDH isoenzyme H_4 is also present in red blood cells, and in the kidney, brain, stomach, and pancreas. Thus, a patient with hemolytic anemia would also show elevated H_4 and H_3M serum levels.

A much more specific enzyme criterion for diagnosis of myocardial infarction has been devised in recent years. It involves the isoenzyme species of *creatine phosphokinase* (CPK). Three dimeric isoenzymes of CPK exist: MM, MB, and BB. The MM-CPK species predominates in skeletal muscle, while the BB-CPK species predominates in brain. The only tissue where there is a sizeable amount of MB-CPK is in heart, where it represents about 15% of the total CPK activity. The correlation of elevated MB-CPK serum levels to myocardial damage appears to be a remarkably specific index of this disease. A major limitation, however, is that the level of MB-CPK in serum, after reaching a maximum at six hours after the infarction, remains elevated only for about 12 hours longer. Thus the CPK analysis must be done as soon as possible after a suspected infarction. On the other hand, elevated LDH isoenzymes remain stable for at least 14 days after the infarction. However, because the elevated MB-CPK levels do occur quickly with heart damage, the MB-CPK evaluation is particularly useful in monitoring a patient during surgery to detect an infarction brought on by the treatment—a possibility that cannot be reliably monitored by electrocardiography (ECG).

In the future it is very likely that more diagnoses will involve enzyme criteria. New enzyme markers indicating a particular disease are sure to be discovered and advances are constantly being made in the development of the methodology and instrumentation to perform fast, accurate, and precise quantitative measurements of enzyme activity. Many of the procedures have also been automated.

The following references contain additional information on this subject.

GOODLEY, E. L., ed. *Diagnostic Enzymology.* Philadelphia: Lea and Febiger, 1970.

ROBERTS, R., and B. E. SOBEL. "CPK Isoenzymes in Evaluation of Myocardial Ischemic Injury." *Hospital Practice,* 55–62 (January, 1976).

WILKINSON, J. H., ed. *The Principles and Practice of Diagnostic Enzymology.* Chicago: Year Book Medical Publishers, 1976.

APPENDIX II

A PARTIAL LISTING OF HUMAN GENETIC DISORDERS AND THE MALFUNCTIONAL OR DEFICIENT PROTEIN OR ENZYME*

Disease	Protein or Enzyme
Acatalasia	Catalase (red blood cells)
Albinism	Tyrosinase
Alkaptonuria	Homogentisic acid oxidase
Cystathioninuria	Cystathionase
Fabry's disease	Ganglioside-hydrolyzing enzyme
Galactosemia	Galactose-1-phosphate uridyl transferase
Gaucher's disease	Glucocerebroside-hydrolyzing enzyme
Glycogen storage disease	Different types:
	α-amylase
	debranching enzyme
	glucose-1-phosphatase
	liver phosphorylase
	muscle phosphofructokinase
	muscle phosphorylase
Goiter	Iodotyrosine dehalogenase
Gout and Lesch-Nyhan syndrome	Hypoxanthine-guanine phosphoribosyl transferase
Hemolytic anemias	Different types:
	glucose-6-phosphate dehydrogenase
	glutathione reductase
	phosphoglucoisomerase
	pyruvate kinase
	triose phosphate isomerase
Hemophilia	Clotting protein (antihemophilic factor) in blood

* Detailed descriptions of the clinical symptoms, biochemical characteristics, and treatments of these diseases are available in the following sources, particularly in the book edited by Stanbury et al.

AMPOLA, M. G. "Errors of Amino Acid Metabolism." In *Handbook of Biochemistry and Selected Data for Molecular Biology,* H. A. Sober, ed. Cleveland: The Chemical Rubber Co., 1970. B105–B111.

DAVIE, E. W., and O. D. RATNOFF. "The Proteins of Blood Coagulation." In *The Proteins,* vol. 3, Hans Neurath, ed. New York: Academic Press, 1965.

STANBURY, J., J. WYNGAARDEN, and D. FREDRICKSON, eds. *Metabolic Basis of Inherited Disease.* 2d ed. New York: McGraw-Hill Book Company, 1966.

Disease	Protein or Enzyme
Histidinemia	Histidase
Homocystinuria	Cystathionine synthetase
Hyperammonemia	Ornithine transcarbamylase
Hypophosphatasia	Alkaline phosphatase
Isovalericacidemia	Isovaleryl-SCoA dehydrogenase
Maple syrup urine disease	α-Keto acid decarboxylase
McArdle's syndrome	Muscle phosphorylase
Metachromatic leukodystrophy	Sphingolipid sulfatase
Methemoglobinemia	NADPH-methemoglobin reductase and NADH-methemoglobin reductase
Niemann-Pick disease	Sphingomyelin-hydrolyzing enzyme
Phenylketonuria	Phenylalanine hydroxylase
Pulmonary emphysema	α-Globulin of blood
Sickle cell anemia	Hemoglobin
Tay-Sachs disease	Ganglioside-degrading enzyme
Tyrosinemia	Hydroxyphenylpyruvate oxidase
Von Gierke's disease	Glucose-6-phosphatase
Wilson's disease	Ceruloplasmin (blood protein)

APPENDIX III

TABLE OF LOGARITHMS

N	0	1	2	3	4	5	6	7	8	9
10	0000	0043	0086	0128	0170	0212	0253	0294	0334	0374
11	0414	0453	0492	0531	0569	0607	0645	0682	0719	0755
12	0792	0828	0864	0899	0934	0969	1004	1038	1072	1106
13	1139	1173	1206	1239	1271	1303	1335	1367	1399	1430
14	1461	1492	1523	1553	1584	1614	1644	1673	1703	1732
15	1761	1790	1818	1847	1875	1903	1931	1959	1987	2014
16	2041	2068	2095	2122	2148	2175	2201	2227	2253	2279
17	2304	2330	2355	2380	2405	2430	2455	2480	2504	2529
18	2533	2577	2601	2625	2648	2672	2695	2718	2742	2765
19	2788	2810	2833	2856	2878	2900	2923	2945	2967	2989
20	3010	3032	3054	3075	3096	3118	3139	3160	3181	3201
21	3222	3243	3263	3284	3304	3324	3345	3365	3385	3404
22	3424	3444	3464	3483	3502	3522	3541	3560	3579	3598
23	3617	3636	3655	3674	3692	3711	3729	3747	3766	3784
24	3802	3820	3838	3856	3874	3892	3909	3927	3945	3962
25	3979	3997	4014	4031	4048	4065	4085	4099	4116	4133
26	4150	4166	4183	4200	4216	4232	4249	4265	4281	4298
27	4314	4330	4346	4362	4378	4393	4409	4425	4440	4456
28	4472	4487	4502	4518	4533	4548	4564	4579	4594	4609
29	4624	4639	4654	4669	4683	4698	4713	4728	4742	4757
30	4771	4786	4800	4814	4829	4843	4857	4871	4886	4900
31	4914	4928	4942	4955	4969	4983	4997	5011	5024	5038
32	5051	5065	5079	5092	5105	5119	5132	5145	5159	5172
33	5185	5198	5211	5224	5237	5250	5263	5276	5289	5302
34	5315	5328	5340	5353	5366	5378	5391	5403	5416	5428

N	0	1	2	3	4	5	6	7	8	9
35	5441	5453	5465	5478	5490	5502	5514	5527	5539	5551
36	5563	5575	5587	5599	5611	5623	5635	5647	5658	5670
37	5682	5694	5705	5717	5729	5740	5752	5763	5775	5786
38	5798	5809	5821	5832	5843	5855	5866	5877	5888	5899
39	5911	5922	5933	5944	5955	5966	5977	5988	5999	6010
40	6021	6031	6042	6053	6064	6075	6085	6096	6107	6117
41	6128	6138	6149	6160	6170	6180	6191	6201	6212	6222
42	6232	6243	6253	6263	6274	6284	6294	6304	6314	6325
43	6335	6345	6355	6365	6375	6385	6395	6405	6415	6425
44	6435	6444	6454	6464	6474	6484	6493	6503	6513	6522
45	6532	6542	6551	6561	6571	6580	6590	6599	6609	6618
46	6628	6637	6646	6656	6665	6675	6684	6693	6702	6712
47	6721	6730	6739	6749	6758	6767	6776	6785	6794	6803
48	6812	6821	6830	6839	6848	6857	6866	6875	6884	6893
49	6902	6911	6920	6928	6937	6946	6955	6964	6972	6981
50	6990	6998	7007	7016	7024	7033	7042	7050	7059	7067
51	7076	7084	7093	7101	7110	7118	7126	7135	7143	7152
52	7160	7168	7177	7185	7193	7202	7210	7218	7226	7235
53	7243	7251	7259	7267	7275	7284	7292	7300	7308	7316
54	7324	7332	7340	7348	7356	7364	7372	7380	7388	7396
55	7404	7412	7419	7427	7435	7443	7451	7459	7466	7474
56	7482	7490	7497	7505	7513	7520	7528	7536	7543	7551
57	7559	7566	7574	7582	7589	7597	7604	7612	7619	7627
58	7634	7642	7649	7657	7664	7672	7679	7686	7694	7701
59	7709	7716	7723	7731	7738	7745	7752	7760	7767	7774

N	0	1	2	3	4	5	6	7	8	9	N	0	1	2	3	4	5	6	7	8	9
60	7782	7789	7796	7803	7810	7818	7825	7832	7839	7846	80	9031	9036	9042	9047	9053	9058	9063	9069	9074	9079
61	7853	7860	7868	7875	7882	7889	7896	7903	7910	7917	81	9085	9090	9096	9101	9106	9112	9117	9122	9128	9133
62	7924	7931	7938	7945	7952	7959	7966	7973	7980	7987	82	9138	9143	9149	9154	9159	9165	9170	9175	9180	9186
63	7993	8000	8007	8014	8021	8028	8035	8041	8048	8055	83	9191	9196	9201	9206	9212	9217	9222	9227	9232	9238
64	8062	8069	8075	8082	8089	8096	8102	8109	8116	8122	84	9243	9248	9253	9258	9263	9269	9274	9279	9284	9289
65	8129	8136	8142	8149	8156	8162	8169	8176	8182	8189	85	9294	9299	9304	9309	9315	9320	9325	9330	9335	9340
66	8195	8202	8209	8215	8222	8228	8235	8241	8248	8254	86	9345	9350	9355	9360	9365	9370	9375	9380	9385	9390
67	8261	8267	8274	8280	8287	8293	8299	8306	8312	8319	87	9395	9400	9405	9410	9415	9420	9425	9430	9435	9440
68	8325	8331	8338	8344	8351	8357	8363	8370	8376	8382	88	9445	9450	9455	9460	9465	9469	9474	9479	9484	9489
69	8388	8395	8401	8407	8414	8420	8426	8432	8439	8445	89	9494	9499	9504	9509	9513	9518	9523	9528	9533	9538
70	8451	8457	8463	8470	8476	8482	8488	8494	8500	8506	90	9542	9547	9552	9557	9562	9566	9571	9576	9581	9586
71	8513	8519	8525	8531	8537	8543	8549	8555	8561	8567	91	9590	9595	9600	9605	9609	9614	9619	9624	9628	9633
72	8573	8579	8585	8591	8597	8603	8609	8615	8621	8627	92	9638	9643	9647	9652	9657	9661	9666	9671	9675	9680
73	8633	8639	8645	8651	8657	8663	8669	8675	8681	8686	93	9685	9689	9694	9699	9703	9708	9713	9717	9722	9727
74	8692	8698	8704	8710	8716	8722	8727	8733	8739	8745	94	9731	9736	9741	9745	9750	9754	9759	9763	9768	9773
75	8751	8756	8762	8768	8774	8779	8785	8791	8797	8802	95	9777	9782	9786	9791	9795	9800	9805	9809	9814	9818
76	8808	8814	8820	8825	8831	8837	8842	8848	8854	8859	96	9823	9827	9832	9836	9841	9845	9850	9854	9859	9863
77	8865	8871	8876	8882	8887	8893	8899	8904	8910	8915	97	9868	9872	9877	9881	9886	9890	9894	9899	9903	9908
78	8921	8927	8932	8938	8943	8949	8954	8960	8965	8971	98	9912	9917	9921	9926	9930	9934	9939	9943	9948	9952
79	8976	8982	8987	8993	8998	9004	9009	9015	9020	9025	99	9956	9961	9965	9969	9974	9978	9983	9987	9991	9996
N	0	1	2	3	4	5	6	7	8	9	N	0	1	2	3	4	5	6	7	8	9

INDEX

Testosterone, 335–336
Tetrahydrobiopterin, 537, 540
Tetrahydrofolate, in pyrimidine biosynthesis, 558, 560, 563
Tetrahydrofolic acid, 164, 176, 527, 537
  biosynthesis, inhibition, 563
  as coenzyme, 550–551
  metabolic role, 551–552
3,5,3',5'-Tetraiodothyronine, 539, 542
Tetranitromethane, reaction with tyrosine, 102
Tetrapyrrole:
  biosynthesis, 543–544
  cyclic, 543
  linear, 543, 544
Tetroses, in photosynthesis, 490
Thermodynamics, 345–350
  principles, 346–350
Thermodynamic state function, 347
Thiamine, coenzyme form, 407–408
Thiamine pyrophosphate, 164, 407–408, 409, 426
  and α-ketoglutarate dehydrogenase, 420–421
  sulfur in, 545
Thin-layer chromatography, 54–55
Thioesters, 320
  as high-energy compounds, 357–359
  in β-oxidation, 500
Thioethanolamine, 357
Thiokinases, 162, 320, 357
Thioredoxin, 562–563
Thioredoxin reductase, 562
4-Thiouracil, 205, 207
Threonine:
  catabolism, 502
  chirality, 92
  conversion to pyruvate, 529
  structure, 90
Threose, 419
D-Threose, 290
Thrombin, 187, 340
Thromboxane A, 322
Thunberg, T., 451
Thylakoid discs, 479–480, 481, 482
Thymidine, 208
Thymidylic acid, 209
Thymine:
  biosynthesis, 551
  degradation, 569
  in DNA and RNA, 205, 206, 221, 248–249
  in pyrimidine cleavage, 218
Thymine dimers, 260, 261
Thyroglobulin, 542
Thyroid gland:
  abnormal activity, 539
  tyrosine metabolism, 542
Thyroid hormones, 542
Thyroid-stimulating hormone, 110
Thyrotropin-releasing factor, 110
Thyroxine, 90, 92, 542
Tocopherol, 339
Trailer sequences, in mRNA, 227, 228
Transacetylase (see Dihydrolipoyl transacetylase)
Transacylation, 512
Transaldolases, 162, 406–407
Transaminases, 431, 433, 529–530

mode of action, 531–532
Transamination, 431, 529–532
Transcriptase, 237–238
  reverse, 246
    and cancer, 246
Transcription, 226, 236 (see also Genes, transcription)
Transferases, 163
Transfer-RNA, 226, 227–230
  cloverleaf conformation, 229
  formation, 237
  hyperchromicity, 227–228
  intrachain base-pairing, 228
  molecular weight, 203
  processing, 242, 244
  sequencing, 217
  structure, 227–230, 263
  synthesis, 258
  X-ray diffraction, 229
Transition state theory, 167–170
Transition temperature, 225
Transketolases, 162, 406–409
  in photosynthesis, 492
Translation, 226, 237 (see also Genes, translation)
Transmethylase, 547
Transmethylation, 547–550
Transport proteins, 341–342
Triacylglycerols, 324–325
Trifluoroacetyl chloride, reaction with amino acids, 102
3,5,3'-Triiodothyronine, 539, 542
Trioseisomerase, 386
Triosephosphate dehydrogenase, in photosynthesis, 491
Triosephosphate isomerase, 403
Trioses, in photosynthesis, 490
Tripeptides, 100
  and sickle cell disease, 115
Triphosphoanhydride, 208
Triphosphodeoxyribonucleotides, 554, 560, 562–564
Triphosphonucleotides, 209
  hydrolysis, 356
  as substrates, 250, 258
Triphosphopyridine nucleotide, 359
Triphosphoribonucleotides, 554, 560–561
Triplet coding ratio, 271
TRIS (see Tris(hydroxymethyl)aminomethane)
Tris(hydroxymethyl)aminomethane:
  as buffer, 74, 76
  pK_a, 75
N-Tris(hydroxymethyl)methyl-2-aminoethane sulfonate:
  as buffer, 76
  pK_a, 75
N-Tris(hydroxymethyl)methylglycine, pK_a, 75
Tritium, 57, 248
Tropocollagen, 138, 139
Tropomyosin, 399
Troponin, 399
Trypsin, 162, 186, 187
  in protein analysis, 125, 127
Trypsinogen, 186
Trypsinolysis, 127
Tryptamine, 533
Tryptophan:

and acetyl-SCoA formation, 431, 432, 529
  anabolism, 538
  decarboxylation, 533
  degradation, 539, 541
  destruction by acid, 122
  and lipid formation, 529
  metabolism, 538, 539
    and schizophrenia, 541
  structure, 90
Tuberculostearic acid, 318
Two-state model, 194, 195
Tyrosinase, 539, 541–542
Tyrosine, 183, 248
  and acetyl-SCoA formation, 431, 432, 529
  catabolism, 538, 540–541
  conversion to oxaloacetate, 529
  ionization, 94, 96
  and lipid formation, 529
  metabolism, 538, 539, 541–543
    and schizophrenia, 542
    and thyroid activity, 539, 542
  from phenylalanine, 541, 537
  reaction with phosphomolybdotungstic acid, 102
  reaction with tetranitromethane, 102
  structure, 90
Tyrosine hydroxylase, 542

Ubiquinones, 453
UDP-N-acetyl-D-galactosamine, 301
UDP-N-acetyl-D-glucosamine, 301, 302
UDP-galactose, 393
UDP-galactose-C⁴-epimerase, 393
UDP-glucose, 297, 303
UDP-muramic acid, 301
UDP-sugars, 296–297
Ultracentrifugation, 37, 38–43
  density gradient, 38–39, 42–43
    isopycnic technique, 42–43
    rate zonal technique, 42
  differential sedimentation velocity technique, 39–42
  theory, 41
Ultracentrifuge:
  analytical, 38, 39–42
  preparative, 38
Ultramicrotome, 16
Ultrasonic vibrations, in cell disruption, 32, 33
Ultraviolet light, as mutagen, 260
Uncouplers, 461, 466
Ungar, G., 112–113
Unit-mitochondrion hypothesis, 24
Uracil, 205, 206
  degradation, 569
  production, 556, 557
  in RNA and DNA, 205, 206
Urea:
  from amino acid degradation, 535
  as denaturing agent, 148
  as nitrogen source, 535
  production, 530
  in purine catabolism, 566, 567
Urea cycle, 535–536
Urease, 535
  isolation, 162